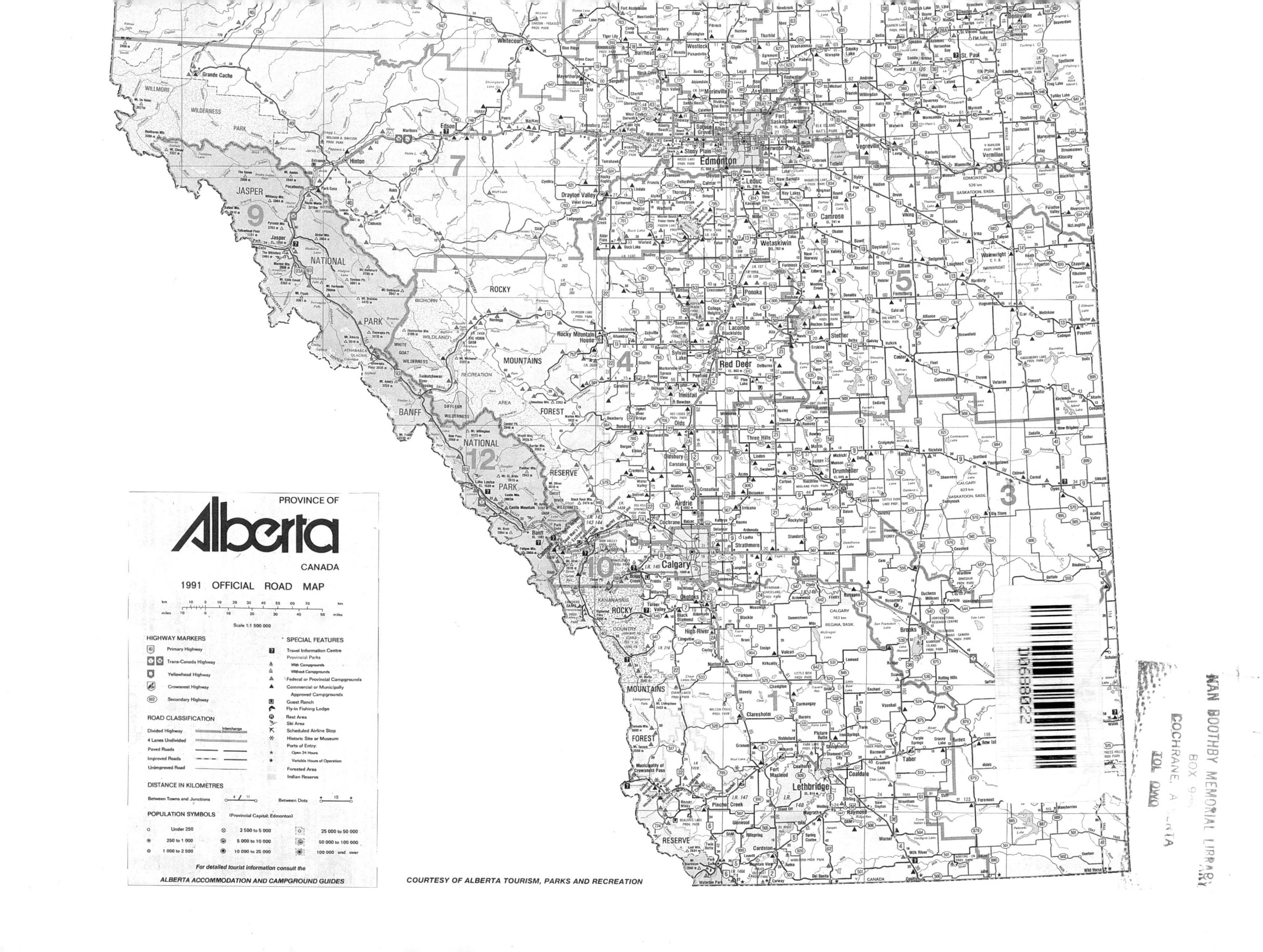
PROVINCE OF
Alberta
CANADA
1991 OFFICIAL ROAD MAP
Scale 1:1 500 000
HIGHWAY MARKERS
Primary Highway
Trans-Canada Highway
Yellowhead Highway
Crowsnest Highway
Secondary Highway
ROAD CLASSIFICATION
Divided Highway
Interchange
4 Lanes Undivided
Paved Roads
Improved Roads
Unimproved Road
DISTANCE IN KILOMETRES
Between Towns and Junctions
Between Dots
POPULATION SYMBOLS
(Provincial Capital: Edmonton)
Under 250
250 to 1 000
1 000 to 2 500
2 500 to 5 000
5 000 to 10 000
10 000 to 25 000
25 000 to 50 000
50 000 to 100 000
100 000 and over
SPECIAL FEATURES
Travel Information Centre
Provincial Parks
With Campgrounds
Without Campgrounds
Federal or Provincial Campgrounds
Commercial or Municipally Approved Campgrounds
Guest Ranch
Fly-in Fishing Lodge
Rest Area
Ski Area
Scheduled Airline Stop
Historic Site or Museum
Ports of Entry:
Open 24 Hours
Variable Hours of Operation
Forested Area
Indian Reserve
For detailed tourist information consult the
ALBERTA ACCOMMODATION AND CAMPGROUND GUIDES
COURTESY OF ALBERTA TOURISM, PARKS AND RECREATION
Edmonton
Calgary
Red Deer
Lethbridge
Jasper
Banff
Hinton
Edson
Whitecourt
Grande Cache
JASPER NATIONAL PARK
BANFF NATIONAL PARK
ROCKY MOUNTAINS FOREST RESERVE
WILLMORE WILDERNESS PARK
KANANASKIS COUNTRY

The Atlas of Breeding Birds of Alberta

Alberta Recreation, Parks and Wildlife Foundation

Distribution of this book, *"The Atlas of Breeding Birds of Alberta"*,
to all libraries in the province
has been undertaken by the
Alberta Recreation, Parks and Wildlife Foundation
with the valued assistance of **TransAlta Utilities**,
Edmonton Power and **The City of Medicine Hat Electrical Utility**.

The Foundation and its partners believe the subject and quality of this book will contribute to our citizens' understanding of the fragile beauty of all wildlife, and its habitat and will help to further the objectives of the Foundation, which are:

- to develop and maintain recreational programs, services and facilities
- to develop and maintain parks
- to manage and conserve fish and wildlife

All concerned acknowledge with gratitude the co-operation and assistance provided by the Honorable Jim Dinning, Minister of Education, and the Honorable Don Sparrow, Minister of Tourism, Parks and Recreation in making this project possible.

With Gratitude To:

Gold Sponsor

Edmonton Power

Owned and Operated by
The City of Edmonton

Silver Sponsor

The Atlas of Breeding Birds of Alberta

Edited by
Glen P. Semenchuk

Published by the Federation of Alberta Naturalists

ISBN 0-9696134-0-7

Published by the Federation of Alberta Naturalists
P.O. Box 1472
Edmonton, Alberta
T5J 2N5

Additional copies of The Atlas of Breeding Birds of Alberta, or the use of FAN's data base for scientific research or other purposes, may be obtained by contacting FAN at the above address.

Printed and bound at Friesen Printers
Electronic Separation and Page Assembly by
Colour Images Graphics Inc.

All photographs courtesy of the Edgar T. Jones collection at the Provincial Museum of Alberta, unless otherwise indicated.

Canadian Cataloguing in Publication Data
Main entry under title:
The Atlas of Breeding Birds of Alberta

Includes bibliographical references and index.
ISBN 0-9696134-0-7

1. Birds - Alberta - Geographical distribution.
2. Birds - Alberta - Breeding - Maps.
3. Birds - Alberta - Nests - Maps.
I. Semenchuk, Glen P. (Glen Peter), 1949-
II. Federation of Alberta Naturalists.
QL685.5.A4A84 1992 598.256'097123 C92-091598-1

Printed in Canada

TABLE OF CONTENTS

LIST OF FIGURES

LIST OF TABLES

FOREWORD

The goal of this project, which began almost seven years ago following a workshop on endangered species at the Provincial Museum, was to provide a systematic assessment of the breeding distributions of the birds of Alberta. Other objectives, related to fund-raising or the publication itself, established as the project proceeded, were always secondary to our goal. There is no doubt in my mind that our goal was achieved, along with many of our objectives.

The maps in this Atlas are based on far more data than any previous distribution maps—such as those found in the Birds of Alberta (Salt and Salt 1976) and the Birds of Canada (Godfrey 1986). Those authors were forced to intuit breeding distributions of birds of the province based, at times, on only a handful of specimen records and informal observations.

Even though our goal was met, the Atlas maps are not complete. Many areas of the province remain to be surveyed and, in a few years, areas will have to be resurveyed for changes. This project has provided the baseline, however, against which all future assessments of the birds of Alberta will be based. Every subsequent ornithological study in Alberta will benefit from the Atlas and, I suspect, find reason to reference its findings.

Atlas staff and volunteers can take pride in the establishment of a province-wide network for collecting Natural History data, compiling, evaluating, and mapping data leading to the Atlas itself, and a willingness to continue acquiring data. The Atlas demonstrated that Albertans are concerned about their natural environment and are willing to volunteer their time to help understand it.

The geography and demography of Alberta do not lend themselves to a project of this sort. Over half of all Albertans live in Edmonton and Calgary. Without a doubt, the maps produced reflect, to a certain extent, the preponderance of birders in these two cities. Other apparent concentrations of birds are around cities like Lethbridge, Medicine Hat, Red Deer, Fort McMurray and Grande Prairie. By contrast, much of the Boreal Forest Natural Region is inaccessible by car. Travel in mountainous regions of Alberta is virtually impossible, except along main routes. Even the Grassland region can be difficult to census in poor weather or because of restricted access on private or crown land. Interpreters of these atlas maps have to consider the contribution of logistics to lower bird densities in remote areas.

The coverage of Alberta, especially the remote regions, while not complete, reflects the strong commitment of Atlassers to the project and the dedication of Atlas staff and Regional Coordinators in surveying as much of the province as was feasible. A final intensive year of surveys, aimed at remote regions, added almost 40% to the accumulated data base.

There is an acknowledgments section elsewhere in this volume. It is extensive and lists the many areas of involvement from corporate sponsor to technical reviewer to casino volunteer. A special comment needs to be made however, regarding the contributions of the Recreation, Parks and Wildlife Foundation. This organization was committed to the Atlas before it was guaranteed to be successful. The continued financial support of the Foundation provided the long-term continuity to the project it needed for success.

The Atlas resulted from the combined vision of the Federation of Alberta Naturalists, The Canadian Wildlife Service, Alberta Fish and Wildlife Division, and the Provincial Museum of Alberta, as well as the hard work of a small group of paid staff and over 1200 birders. That all these parties could see beyond their differences and work towards a common goal is a remarkable achievement. It also means that other cooperative projects which benefit the study and preservation of Alberta's Natural History are likely to be similarly successful. Besides the book you hold in your hand, this level of collaboration is a highlight of the project.

As a scientist, I feel the major contributions of the Atlas are still to come. I look forward to the many research publications that I am sure the Atlas data will generate. These analyses will be essential to a full understanding of the information gained over the last six years. As a birder, I know the Atlas will be a source of information and enjoyment for both serious and casual birdwatchers. All who contributed to the project can be justifiably proud of its success.

W. Bruce McGillivray
Chairman, Management Committee

SPONSORING ORGANIZATIONS

The Provincial Museum of Alberta

The Provincial Museum of Alberta has for 25 years, as part of its mandate, acquired specimens to document the status, distribution, and geographic variation of the flora and fauna of Alberta. Through its exhibits, publications, educational programs, and encouragement of research, it has worked to inform Albertans about their natural history heritage. As one of the founding partners in the Bird Atlas Project, the Provincial Museum of Alberta provided staff to help establish the goals of the project and develop field and technical protocols. The head office of the project for its first five years was at the Provincial Museum, which provided space, resources, and clerical support. Through this institutional connection, the Atlas had a base and an association that helped build credibility in the early days of the project.

The Provincial Museum is proud to be affiliated with the Alberta Bird Atlas Project. This current record of the distribution of Alberta's breeding birds, when contrasted with the historical records from collections data, will provide vital clues to the status and needs of species as they react to environmental change. The Museum has benefitted as well, through increased association with the Federation of Alberta Naturalists and fellow government agencies involved in natural history preservation. We look forward to building on this atlas database and on other joint projects in the future.

The Canadian Wildlife Service

The Canadian Wildlife Service of Environment Canada oversees the protection and management of migratory birds, endangered species, and nationally significant habitats, and the control of international trade in endangered species.

As well, this organization conducts ongoing research into these and other matters of significance to wildlife nationally in Canada, and which are the responsibility of the federal government. The Service collaborates with the Canadian Parks Service, other federal agencies involved in wildlife research and management, and with provincial and territorial governments.

The Canadian Wildlife Service made significant financial contributions to the Alberta Bird Atlas Project. In addition, many staff members (noted in the Acknowledgments Section) volunteered their time atlassing, reviewing drafts, and contributing text as technical advisors.

Alberta Department of Forestry, Lands and Wildlife

The Alberta Department of Forestry, Lands and Wildlife is a founding sponsor of the Alberta Bird Atlas Project, and has supported the program through several of its Divisions. The Fish and Wildlife Division was instrumental in promoting the initial Atlas concept, and in bringing together the coalition of agencies and organizations which have worked so well to complete this program. As the agency responsible for wildlife management in Alberta, the Fish and Wildlife Division has also contributed manpower, funding, and computer hardware for collecting, analyzing, and reporting on the massive database developed by the Atlas Project. The Land Information Services Division provided manpower and funding to develop and refine the management and reporting software necessary to convert field observations into usable maps and reports. The Alberta Forest Service and Fish and Wildlife Division both contributed manpower, funding, and logistic support to assist atlassers in reaching the many remote areas in northern and western Alberta. The Department believes that the Alberta Bird Atlas Project has been very successful in building partnerships between government and volunteer organizations, in providing a valuable and productive channel for naturalists to contribute to wildlife conservation, and in significantly improving the understanding of the distribution of breeding birds in Alberta.

Recreation, Parks and Wildlife Foundation

The Recreation, Parks and Wildlife Foundation was established in 1976. The objective of the Foundation is to provide opportunities for individuals and organizations to donate real or personal property to the Foundation, to be used for:

- the development or maintenance of recreational programs, services, and, to a lesser extent, facilities;
- the development or maintenance of parks;
- the management, conservation, or preservation of fish and wildlife.

The Foundation may make grants to any person or organization for projects within the objectives of the Foundation.

The Foundation believes that the subject and quality of this book will contribute to an understanding of the fragile beauty of all wildlife.

In addition to substantial cash contributions towards field work during the Atlas Project, the Recreation, Parks and Wildlife Foundation, through its *Youth-in-Action* program, is also purchasing the Atlas and distributing it to each school in the province.

ACKNOWLEDGMENTS

The publishing of *The Atlas of Breeding Birds of Alberta* was possible through the combined efforts of organizations, corporations, and over 1,000 individuals who contributed time, expertise, and resources to the Alberta Bird Atlas Project. This complex project involved a major field survey component that relied on 1,000 or more atlassers, compilers, regional coordinators, and office support staff, who gathered, reported, and verified data from Alberta's grasslands, mountains, and forests. Many of our province's fields, sloughs, and woodlots would not have been explored for their rich and varied bird life were it not for the cooperation of Alberta's landowners. The atlas also benefitted from the naturalists, photographers, and biologists who allowed access to their notebooks, records, and expertise, and the agencies that provided access to unpublished reports and libraries.

The continued support of volunteers, organizations, and corporations in Alberta allowed this publication to become a reality. Where recorded, their names are listed. If not recorded, their contributions to the project are still greatly appreciated.

This book is dedicated to all those people who contributed in some fashion to the success of this project, and to those individuals and groups who will continue the work that began with Alberta's first bird atlas.

PATRON

Honourable J. W. Grant MacEwan

SPONSORING ORGANIZATIONS

Alberta Fish and Wildlife Division
Canadian Wildlife Service, Environment Canada
Federation Of Alberta Naturalists
Provincial Museum of Alberta
Recreation Parks and Wildlife Foundation

MAJOR DONORS

Alberta Tourism, Parks and Recreation
Canada Employment and Immigration
Dow Chemical Canada Inc.
Interprovincial Pipe Line Inc.
The James L. Baillie Memorial Fund
TransAlta Utilities Corporation

SUPPORTERS

Brad Arner
Angela Bell
Carl Belyea
Myrna Belyea
William Benford
T. Birch
Chris Bjornson
John Campbell
Hugh Campbell
Graham Chambers
Linda Chorney
Donna Clanfield
Gordon Clements
Doug Collister
D. Craig
D. Crawshaw
Evelyn Dickie
Elizabeth Dowd
Dave Ealey
Ruth Fleming
Dr. Bob Gainer
Rob Gardner
William Glasgow
Cathy Glosky
Angela Gottfred
Delcie Gray
Graeme Greenlee
John Gulley
John Hemstock
W. Hewlett
Edith Hilsop
George Hilsop
Dr. E. Otto Höhn
Slavoj Hontela
A. Jenkins
Derek Johnson
Beverly Kissinger
E. Kissinger
Peter Lancaster
Peter Lee
Lloyd Lohr
Mary Lore
F. Lowe
Ann MacDonald
Ian MacDonald
Janet MacKay
Patricia MacKonka
Robyn Maerz
Bert Martens
Dan McAskill
Ronald McElhaney
Lois McKillop
Dorothy Monner
Betty Nelson
Phil Norris
Esther Ondrack
J. Paget
Daphne Richey
John Rintoul
Thomas Sadler
June Seidel
Richard Sharpe
Darrell Smith
Stephan Spisak
Donna Stefura
H. Stelfox
H. Sutherland
Margaret Sware
Mr. Swisher
Hubert Taube
Phyllis Thomson
John Towers
Bob Turner
Donna Wakeford
Cleve Wershler
Henry West

Alberta Culture and Multiculturalism
Alberta Fish and Game Association
Alberta Forest Service
Alberta Vocational Centre
Alberta Government Telephones
Alberta Oil Sands Technology & Research Authority
Alberta Natural Gas Company Ltd.
Bank of Montreal
Beaverhill Lake Nature Centre
Bushnell Division, Bausch and Lomb Canada Inc.
Canadian Forces Base - Cold Lake
Canadian Forces Base - Suffield
Canadian Petroleum Association
Canadian National Sportsman Shows/World Wildlife Fund
Cardinal River Coals Ltd.
Carthy Foundation/Loram Group of Companies
Chevron Canada Resources
Chieftain International Inc.
Co-op Press Ltd.
Coal Association of Canada
Cornell Laboratory of Ornithology
Daishowa Canada
Department Of Phys. Ed., University of Alberta (G. Hanna)
Ducks Unlimited
Edmonton Bird Club
Edmonton Journal
Edmonton Power
Field & Stream
Fording Coal Ltd.
Friesen Printing
General Systems Research
Harris McConnan Chartered Accountants
Home Oil Company Ltd.

Inglewood Bird Sanctuary
John Janzen Nature Centre (Edmonton Parks and Recreation)
Junior Forest Wardens
Medicine Hat - The Gas City
Mobil Oil Canada
North Canadian Oils Ltd.
Petro Canada Ltd.
Pitney Bowes of Canada Ltd.
Ranger Oil Ltd.
Royal Bank of Canada
Saskatchewan Museum of Natural History
Sceptre Resources
Sweetgrass Consultants
Syncrude Canada Ltd./Northward Developments Ltd.
The Electric Scribe
The City Of Fort McMurray
The City Of Lethbridge
The Wildbird General Store
Trimac Ltd.
Weldwood of Canada Ltd., Hinton Division

MANAGEMENT COMMITTEE
Chairmanship rotated among members of the sponsoring agencies
Steve Brechtel
Pat Clayton
Doug Culbert
Jim Goodwin
Margo Hervieux
Geoff Holroyd
Bruce McGillivray
Chuck Moser
Glen Semenchuk
Phil Stepney
Don Stiles

TECHNICAL COMMITTEE
Peter Boxall
Doug Collister
Loney Dickson
Ross Dickson
Geoff Holroyd
Gerry Lunn
Nash Marani
Bruce McGillivray
Mike Melynk
Robert Storms
Terry Thormin
Eric Tull
Cleve Wershler

NEWSLETTER EDITOR
Loney Dickson

PUBLICATION

EDITOR
Glen Semenchuk

ASSOCIATE EDITOR
Jane E. Livingston

FAN PUBLICATION COMMITTEE
Pat Clayton
Dave Ealey
Derek Johnson

AUTHORS
Edrea Daniel
Leland Ferguson
Ross Hastings
Bruce McGillivray
Glen Semenchuk
Trevor Wiens

TECHNICAL ADVISORS
David Boag
Steve Brechtel
Eldon Bruns
Loney Dickson
Lynne Dickson
Dave Ealey
Rainer Ebel
Gordon Elbrond
Gary Erickson
Cam Finlay
Richard Fyfe
Paul Goossen
Wayne Harris
Otto Höhn
Geoff Holroyd
Bruce McGillivray
Dave Moore
David Prescott
Blair Rippin
Len Shandruk
Al Smith
Daryl Smith
Wayne Smith
Phil Stepney
Ed Telfer
Bruce Turner

DATA MANAGEMENT
Eric Ellehoj
Trevor Wiens

BASE MAP PRODUCTION
Erik Ellehoj
Rick Haag
Trevor Wiens

RESEARCHERS
Karen Aldred
Michael Braeuer
Selena Cole
Michael Cenkner
Karen Dudley
Winona Eads
Tasha Greer
Laura Gretzinger
Craig Machtans

COMPUTER DATA ENTRY
Della Clish
Shirley Trott

COMPUTER ASSISTANTS
Jamie Bater
Gulam Jamal

ILLUSTRATIONS
Ann Stefura

COVER DESIGN
Dick Dekker

MARKETING
Karen Seto

FIELD PROGRAMS

EXECUTIVE DIRECTORS
Albert Brule
Jack Clements

EXECUTIVE ASSISTANTS
Dave Adie
Carolyn Seburn
Petra Stubbs

REMOTE AREA ADVISOR
Dave Ealey

REGIONAL COORDINATORS
Brad Arner
Rick Bonar
Iris Davies
Rob Gardner
Peter Kennedy
Richard Klauke
Lloyd Lohr
Gerry Lunn
Myrna Pearman
Andy Raniseth
Bill Sharp
Bob Storms
Deborah Taylor
Nancy Twynam
Mike Vandyk
Kevin Van Tighem
Terry Thormin

OFFICE STAFF
Yassin Boga
Mattie Brown
Gordon Burns
Leland Ferguson
Bob Gerlock
Jan Hagen
Mike Kelly
Seanna MacKenzie
Erik Soderstrom

FIELD STAFF
Karen Aldred
Michael Braeuer
Selena Cole
Karen Dudley
Boyd Fischer
Tasha Greer
Laura Gretzinger

Michael Harrison
Craig Machtans
Pat Marklevitz
Scott Millar
Robert Storms

EVENT VOLUNTEERS AND OFFICE ASSISTANCE

Gordon Armitage
Louis Beens
Joe Bourke
Natalie Bourke
Ernest Bozman
Len Bradley
Col. A.N. Brown
Sean Brown
Jim Butler
Paola Chiovelli
Olive Clark
Dick Clayton
Beverly Cooper
Susan da Silva
Tony da Silva
Jim Dallow
Jeff Dixon
Joe Dragon
Bruce Duffin
Sue Fast
Dennis Fitzgerald
Joyce Gray
Jan Hagen
Gord Holton
Col. D. Jurkowski
Mike Kelly
Chuck Knight
Frank Mansell
Don McGalddey
Sheila McKay
Roger Merriot
Mable Nevin
Wayne Nordstrom
Arlene Ohlsen
Ken Patraschuk
Florence Phillips
Laura Pretula
Richard Quinlan
Mike Quinn
Claire Radke
Anne Rimmer
June Seidel
Marilyn Sergei
Bryan Shantz
Darrel Smith
Kirby Smith
Ted Smith
Noreen Tate
Phyllis Thomson
Tom Webb
Earl Wilson

ATLASSERS

Tricia Abbott, Charles Adolewski, D. Akitt, Don Albright, Karen Aldred, S.E. Alexander, David Allan, Kim Allan, Barry Allan, James Allen, Daniel Allen, Robina Allen, M. Allen, Jim Allen, Rodney Allison, Helmut Amelang, Tracy Anderson, Bruce Anderson, Jim Anderson, Kara Anderson, Breya Anderson, Lula Anderson, Harry Anema, Ron Antill, Gary Archibald, Ruth Archibald, H. Armbruster, Peggy Armstrong, Marjorie Arndt, Brad Arner, L.M. Arthur, Gordon Ashacker, Chairul Aslam, Jason Attwell, Ruth Attwell,

Marlene Bainard, Bary Bainard, Sabrina Bainard, Steven Bainard, Al Ballantyne, Marilyn Ballantyne, Kevin Ballantyne, Lee Bannister, Reid Barclay, Dennis Baresco, A. Bartsch, Michael Baumgartner, Rosemary Baxter, Elisabeth Beaubien, James Beck, Barbara Beck, Bev Beck, G. Beckman, Guy Belanger, Joseph Belicek, Sharon Belicek, Debra Belmonte, William G. Benford, Craig W. Benkman, Lloyd Bennett, Suzanne Benoit, Bjorn Berg, Sid Berginhajon, Carrie Berry, Ray Bertram, Freda Bertram, Joshua Bertwistle, Matt Besko, Pet V. Bidwell, J.P. Bielech, Myrtle Biggs, Dylan Biggs, Mary Biggs, Tom Biggs, Charles Bird, Garth Bishop, Cheryl Bissell, Ron Bjorge, Herb Blades, Yvonne Blades, Chris Blomme, Dale Blue, Tom Blunden, Tason Blunden, Barry Blunden, Jan Blunden, David Bly, Hans Boerger, Paul Boisvert, Rick Bonar, Norma Bonar, John Bourne, J.D. Bovell, Peter Boxall, Calvin Boze, Wes Bradford, Leonard Bradley, Michael Braeuer, Sheila Braun, Steve Brechtel, Kelly Brett, R. Brewin, Martin Brilling, C. Broatch, Kevin Brooks, Maxine Brown, William M. Brown, Anne Brown, E.C. Brown, D. Brown, Ken Browne, Chris Browne, Eldon Bruns, Barb Bryson, John Bryson, P. Bunford, Gordon Burns, Rod Burns, Mary Burpee, Vivian Busat, Jim Butler, Christy Butt, M. Butt, TedButterwick, David Butterwick,

Wendy Calvert, Hugh Cambpell, Ingrid Campbell, David Campbell, J. Campbell Jr., Dick Cannings, Lu Carbyn, Doris Carleton, Bill Carey, Doreen Carey, Don Carlyle, Robert Carroll, Jan Carroll, W. Cavan, Peter Chapa, Linda Charlton, George Cheeseman, Barb Chisholm, Jeffrey Chorney, R.K. Christen, Ted Christensen, Marilyn Christiansen, Dave Christiansen, Jessie Christiansen, George Christianson, A. Claiter, Kim Clark, Barbara Clark, Lester Clark, Jim Clark, Berta Clarke, Gary Clarke, R.C. Clayton, Jack Clements, John D. Clements, Booke Clibbon, Marie Cliffe, Dann Cline, Norma Cole, Selena Cole, Richard Coleman, Paulin Coleman, Trevor Coleman, Alex Coleman, Barbara Collier, Doug Collister, Pat Connor, Frances Connor, Carl Cooper, Lorna Cordeiro, Judy Cornish, B. Cornish, Mike Courtney, Julian Coward, Gordon Cowie, Eileen Cowtan, Jim Critchley, Willie Critchley, Claire Critchley, Ray Crome, R. Cromie, Patricia Crossley, Lara Crowther, Jean Crump, Susan Crump, Doreen Cuddie, Doug Culbert, Lloyd Cumming, Beth Cumming, Minnie Cunningham, Sean Curry, Chris Curry, Ron Curtis, David Cuthiell,

Mr. Dahlgren, Barbara J. Danielson, Robert M. Danielson, W. Dann, Tate David, Jim Davies, Brian Davies, Iris Davies, Ralph Davies, Al De Groot, John C. De Groot, Luke De Wit, Joan Degeer, Heather Dempsey, Peter Demulder, Tony Derby, Carly Deschuymer, Mark Desormeaux, Dorothy Dickson, Lynne Dickson, Loney Dickson, Ross Dickson, Mary H. Dixon, J. Dixon, Teresa Dolman, Doug Dolman, Trudy Downy, David Doyle, Sue Draxler, Olga Droppo, Larry Dubitz, Peter Duck, Karen Dudley, J. Duffy, Paul Dunning, Donna Dunning, Eva Dzuris,

Dave Ealey, Rainer Ebel, A. Eberl, Lorie Eddy, Angie Edgerton, Janet Edmonds, Brad Edwards, Gerry Ehlert, Heidi Eijel, J. Elliott, David Ellis, Dave Elphinstone, Larry Enders, Don Enright, Norman Erhardt, Ruth Erhardt, Gary Erickson, Cheryl Ewaschuk, Ernie Ewaschuk, Ferna Ewaschuk, Byron Erwin,

Edward Faechner, Dan Farr, Ron Feniuk, Leland Ferguson, Cam Finlay, Joy Finlay, Warren Finlay, Boyd Fischer, G. Fjordbotten, Freeda Fleming, B. Fletcher, Simone Flynn, Erika Foley, J. Folinsbee, Corrie Fordyce, Airlie Fowler, C. France, A. Frank, George Freeman, Bob Frew, Beth Frith, Ken Froggatt, F. Fullerton,

Darren Gabert, Larry Gabert, Ben Gadd, Flo Gagne, Kent Gallie, John Gallimore, Merve Gammel, Coraline Gardner, Rob Gardner, B. Gauthier, D. Geremia, Garry Giffen, Kathleen Giffen, Evelyn Gillis, John Girvan, Erling Glabais, Margaret Glabais, Ken Glover, Greg Goble, Bob Goddard,

Peter Goddard, B. Godsalve, Andrea Goodall, Christine Goodall, Jesse Goodall, Jim Goodall, Phyllis Goodall, Rob Goodall, Shannon Goodall, Jim Goodwin, PaulGoossen, Valorie Goossen, Pilar Gosselin, Angela Gottfred, Jeff Gottfred, Evling Grabais, Margaret Grabais, Tony Graham, K. Graig, Marc Granger, Pat Greenaway, Phil Greenaway, Betsy Greenlee, D. Greene, Graeme Greenlee, Lloyd Greenlee, Patricia Greenlee, Tasha Greer, Andrew Gregg, Jim Gregg, Rene Gregg, Laura Gretzinger, Deirdre E. Griffiths, Doug Grover, Jim Guilbault, Louis M. Guillemette, John Gulley, G. Gunderson, Ellamae Gunn,Sharon Gunn,
Michael Hagel, Joel Hagen, Greg Hale, Ian Halladay, L. Halmrast, Terry Hanak, D. Hansen, R. Hanses, Lorraine Hardin, Wayne C. Harris, Margaret Harris, Michael Harrison, Ferrell Haug, Dorothy Hazlett, R. Hegland, Mary Heily, Hilary Hellum, R.L. Hemsing, Malcolm Hendersen, Phil Henderson, Andrew Hendry, Alice Hendry, Grant Henry, M.C. Herbst, Steve Herman, Kay Hernes, R. Peter Heron, Margot Hervieux, Dave Hervieux, Doreen Higgin, L. Higgins, Boyne Hill, Alan Hingston, Dave Hobson, Louis Hochachka, Wolfgang Hoffman, Ed Hofman, Otto Höhn, C.F. Holloway, Pat Holloway, Robert Holmbers, D. Holowoychuk, Geoff Holroyd, Michael Holroyd, Doris Hopkins, Eric Hopkins, Rolf Hopkinson, Phil Horch, Garry E. Hornbeck, Ed Hotte, M. Hougen, S. Howard, Hazel Hudson, H. Hughes, O. Hughes, Robert B. Hughes, Simon Hum, Dan Hungle, Carl Hunt, Laurie Hunt, S. Hunt, Darrell Hutchinson, Donna M. Hutchinson, Glen Hvenegaard,

Sandra M. Irwin,

Sheila Jahraus, E. Jans, Louise Jarry, Phyllis Jarvin, Bob Jason, Robert V. Jason, J. Jellicue, Marje Jensen, Colleen Jevne, Roger Johnson, Gordon Johnson, Pat Johnson, Ellen Johnson, Van Johnson, Jean Johnson, Will Johnson, Lois Johnson, J. Derek Johnson, Clark Johnston, Tom Johnston, Stuart Johnston, Mary Johnstone, Kate Jones, Larry Jones, Edgar T. Jones, Jeanne W. Jones, Jon Jorgenson, Stefan Jungkind,

Gerry Kadey, Peggy Kadey, Brenda Kane, Glen Kane, J.A. Kariuk, Allexis Karne, Valerie Keer, Jasper Keizer, Grant Kelly, Carol Kelly, Erin Kelly, Peter Kennedy, Sherry Kennedy, Joan Kerr, Valerie Kerr, Lex Khrne, H.P.L. Killiaan, Jessie Kinley, J. Kinnaird, Eileen Kinsman, George Kinsman, Eva Kirtley, Mike Klassen, Richard Klauke, Margot Klauke, Michael Klazek, Ruth Kleinbub, J.A. Klem, Rudolf F. Koes, Kim Krause, Glen Kroeker, Doug Krystofiak, Woody Kuehn, Floyd Kunnas, E. Kuyt,

Tom Lahring, Heinjo Lahring, Lana Laird, Don Laishley, Jane Lancaster, Jill Lancaster, T. Landra, Harv Lane, Jim Lange, Ken Lange, Sandra Lapointe, Ken Larsen, Linda Larsen, Randy Lee, E. Lee, Andre Legris, Donna Legris, K. Lehr, R. Lein, L. Lengyel, Neil Lennie, Ron Levagood, Frank Lind, Don Lind, Jackie Lind, G. Logan, Lloyd Lohr, Richard Loken, S. Lord, Gerry Lubbers, Mike Luchanski, George Luco, A. Luft, Ken Lumbis, David Lumley, Jerry Lund, S. Lundberg, Gerry Lunn, Giselle Luzon, Doris Lysons,

Tom Maccagno, Beth Maccallum, Chel Macdonald,Rory Macdonald, T. Machacek, Craig Machtans, Murray Mackay, J. Mackenzie, Patricia E. Mackonka, Angus Macleod, Bohr Macmullan, Shannon Macneill, F. Madge, Kelly Maguire, E. Mah-Lim, Laura Mahowich, Scott Main, Ray Makowecki, Richard Mann, Avard Mann, Joyce Mannen, Pat Marklevitz, W.P. Marshall, Anastasia Martin-Stilwell, D. Mather, Jennifer Mather, L. Mather, M. Mattern, Bill Mattes, J. Dan McAskill, David McCorquodale, Robin McDonald, Malcolm McDonald, Joan McDonald, Vivian McElhinney, Bruce McGillivray, Bill McGrath, Blaine McGrath, Danny McGrath, Rev. Anne McGrath, Marilyn McIlvena-Sergi, D. McIntyre, Mike McIvor, Gordon McIvor, B. McKay, Gerald McKeating, Del McKinnon, Robert McMeekin, Karen McMeekin, Bohr McMullan, Vern McNally, Irene McNally, Darren McNally, Deborah McNally, Dean McNally, Michael McNaughton, Dale McQuid, Florence McQuid, Raymond McQuid, Peter Mehrer, Hank Melax, V. Merrel, Myrna Meyer-Field, Edith Middleton, Roy Middleton, Scott Millar, Julia Millen, Don Mills, Alex Mills, Blake Mills, Ken Molloy, Carmen Molloy, David A. Moore, Gudrun Moore, Lynn Moore, Richard Moore, John Moore, John Morel, Lee Morin, Therese Morris, J. Mouser, Ross Munro, Sharon Murphy, Andrew Murphy, Shelagh Murphy, Sandy Murray, Dorothy Murray, Anne Murray, Gary Murray, Bob Mutch, Cynthia Mutch, Sandra Myers,

David Nadeau, Noreen Nadin, Paul Nedza, Avis Nedza, Marlene Neiman, Ruth Nelson, Wayne Nelson, Colleen Nelson, Tony Nette, Eldon Neufeld, Garry Newton, Heng-Joo Ng, James Nicholl, Joanna Nicholl, Brian Nicolai, Lennie Niel, Grant Nieman, Bert Niwa, Wayne Nordstrom, Grace Norgard, Hugh Norris,

Micheal O'Shea, Ken Oakes, L. Obbagy, E. Odland, Arleen Ohlsen, Gerald Ohlsen, Mehari Ohlsen, Cliff Olbenberg, Edna Oldenberg, Elson Olorenshaw, Lloyd Olsen, Sharon Olsen, Nola Opp, F. Orcutt, Peter Orlicki, Renata Osterwoldt, Micheal Otto, E. Ottohohn, C. Owen, S. Owen, Mark Oxamitny,

Roger Packham, T. Pahl, Roger Painchaud, Percy Palmer, Oemie Palmer, Jessie Palmer, Clyde Park, Jack L. Park, Norm Parkins, Fiona Parkinson, Shirley Paustian, Mary Payne, Myrna Pearman, Elizabeth Peck, Albert Peck, Bett Peck, B. Pelham, David Penner, Satu Pernanen, Sheila Perry, Ryan Peruniak, Carole Petersen, Shirley Pfefferle, Florence Phillips, Dave Pick, K. Pickett, T.W. Pierce, Marie Pijeau, Harold Pinel, Erhard Pletz, Jamey Podlubny, Dodie Pollard, Marian Porter, Bill Post, Cindy Post, Esther Potter, J. Potter, Marie Prjeau, Stephany Proudfoot, Lorne Proudfoot, Myrna Pruden, M. Prusky, Jim Purdy, Margo Pybus,

Richard Quinlan, Michael Quinn,

Anne Rae, Faye Rae, Lou Rae, Glenda Rajewski, Brad Ramstead, Andy Raniseth, Marilyn Raniseth, H. Rath, Eleanor Reddecliffe, Kerry Rees, Alan Reid,Wendy Reid, Marge Reine, Joni Renovf, R. Reynolds, Doris Rhodes, Pat Rhude, Larry Rhude, Brian Richard, Bill Richards, John Richardson, Daphne N. Richey, Francis Ries, Edward Ries, David Rimmer, John Rintoul, Blair Rippin, Ross Risvold, Jim Robertson, Anna Robertson, Tim Robina, Michelle S. Rodrigue-Poscen, Trevor Roper, K. Rose, Norma Ross, L. Ross, Irma Rowlands, K.L. Russell, Frank Russell,

Thomas Sadler, Elizabeth Saunders,

Liz Savoy, Volker Schelhas, Mark Schiebelbein, Lisa Schmidt, Ken Schmidt, G. Schmitke, Joe Schmutz, Candi Schopfer, Colin Schopfer, Kurt Schopfer, Neil Schopfer, Tim Schowalter, Frances Schultz, Wilf Schurig, Fred Schutz, Miles Schwartz, Darlene Schwartz, Bonnie Schwindt, James Schwindt, Geoff Scott, Faye Scott, L. Scott, Alf Scott, Chuck Scott, Rick Scott, Rod Scout, Carolyn Seburn, David Seburn, Duanne Sept, Marilyn Sergi, Don Shanty, Angela Shanty, W. H. Sharp, H. Shierman, Bill Shipton, Ivan Shukster, Chris Siddle, Therese Siemers, Shirley Sillito, Hilah L. Simmons, Jan Simonson, Gwen Simpson, Andrea Sissons, Bob Skarra, Marilyn Skinner, Carl Skostengaard, Shelley Skrepnek, Ray Skrepnek, Andrew Slater, Donna L. Sloan, Bob Smith, Catherine Smith, Cyndi M. Smith, Darrell Smith, Hugh Smith, Janice Smith, Jean Smith, Jeanne Smith, Joyce Smith, Kirby Smith, Lorna M. Smith, Ralph J. Smith, Roy Smith, Terry Smith, Joan Snyder, Jean Snyder, B. Sproule, Mark Stabb, M. Stalder, John B. Steeves, Ann Stefura, Mark Steinhilber, Harry Stelfox, David Stelfox, Teresa Stelfox, John G. Stelfox, Phil Stepney, R. Stewart, Don Stiles, William V. Stilwell, P. Stockdale, J. Stokes, Katherine Storey Smith, Bob Storms, R. Strangways, Petra Stubbs, Kathy Stubbs, Paul Stubbs, Kelly Sturgess, Michael Sullivan, Joan Susut, Joanne Susut, Fred Sutherland, J.A. Sutherland, Robert Swainger, G. Swanson, Andrea Sweetnam, Gordon Sweetnam,

Charles Tait, B. Tanghe, Phyllis Tarvin, Corinne Tastayre, David Tate, Deborah Taylor, P. Taylor, Alex Taylor, Duncan Taylor, Roger Taylor, Doug Taylor, Ed Telfer, Ken Tenove, Frances Thacker, K. Thiessen, John W. Thompson, Jeff Thompson, John Thompson, B. Thompson, R. Thompson, Natalia Thor, Terry Thormin, Kevin Timoney, Phil Trefrey, Helen Trefrey, Bruce Treichel, Elaine Trembley, Wilbur Tripp, Eric Tull, Marj Tunney, A. Turnbell, K. Turnbull, Howard Turner, Linda Turner, David Twist, Gordon K. Tye,

Martin Urquhart,

A. W. Van Pelt, Kevin Van Tighem, Gail Van Tighem, Jay Vandergaast, D. Vandervelde, Ben Velner, Michael Velner, Kari Vepsalainen, D. Versluys, Suzanne Visser, Anne Vos,

Bruce Wakeford, Donna Wakeford, Jim Walbridge, M. Walentowitz, Chris Walker, D. Walker, Cheryl Wall, Eric Wallace, Dave Walty, Robert Wapple, Terry Waters, Donald Watson, Marian Wauch,Donald A. Weidl, Bill Weimann, Bill Wesbitt, Pat Wescott, Brent Wescott, Erin Wescott, R.L. West, Henry E. West, George Weston, Gretchen Whetham, Fred Whiley, Andy White, Jean Whitney, Ivan Whitson, Linda A. Whittingham, Jay Wieliczko, Pat Wiens, Paul Wilkinson, Kathleen Williams, R. Williams, Barry Wilson, Laurie Wilson, Bill Windsor, Kevin Wingert, Thea Wingert, Jeff Wingert, Rick Wolcott, Gwen Wood, Dylan Wood, Lois Wooding, Ray Woods, Ken Woollard, Dale Worawa, G. Wright, Marian Wright, Jack Wright, Bob Wynes,

Janet Yoneda, Pat Young,

J. Zeller, Rick Zroback, Ken Zurfluh

ATLASSING GROUPS

Beaverhill Bird Observatory Staff and friends
Blockbusters - various areas
Brooks Fish and Wildlife Staff
Clearwater River Expedition
Conklin Lookout Staff
CWS Ottawa Bird Banding Office
Footner Lake Forestry Staff
Keane Tower Summer Staff
Mountain View Prairie Dusters
Whitney Lakes Staff
Winifred Tower Summer Staff

The History of Ornithology in Alberta: An Overview

In his book, *Key to North American Birds*, Elliot Coues (1903) wrote "The history of American Ornithology begins at that time when men first wrote upon American birds; for men write nothing without some reason . . . " Today, we would fault Coues for failing to consider female ornithologists and for failing to recognize the undoubtedly tremendous knowledge of birds, including those of Alberta, passed orally from generation to generation within various aboriginal cultures for 12,000 years (Pielou 1991) before "Ornithology" began. The quote serves as a reminder, however, that the history of a scientific discipline is more than the simple listing of a series of discoveries separated from their context.

Ornithology in Alberta grew as a byproduct of the fur trade, subsequent scientific explorations, and settling of the province. It is remarkable that the harsh conditions of the fur trade or pioneer homesteading in Alberta should provide fertile ground for the growth of a science. Yet despite the primitive conditions in early Alberta, ornithological knowledge kept pace with the rest of the world.

It is known that Indians provided information to Hudson's Bay Company (HBC) employees that was added to ornithological reports of Andrew Graham and Thomas Hutchins (see Houston and Houston 1990, 1991) in the 1770s. As well, the accounts of Indian names for birds (Höhn 1962, 1973) and the uses of birds by natives (e.g., Ross 1861, 1862) imply a depth of ornithological knowledge. We can only speculate on the importance of birds to prehistoric Indian and paleo-Indian cultures. Historic evidence from museum collections confirms that birds were used in adornment and ritual as well as serving as a food source. The lack of sophisticated weapons and optical aids in these cultures indicates that Natives would have required detailed understanding of behavior before they could have captured adult birds or found nests.

The written record of ornithology in Alberta begins with the travels of Anthony Henday to southern Alberta in 1754-1755 on behalf of the HBC. Henday, the first European to set foot in Alberta made natural history observations generally related to food. Birds mentioned include "pheasants" (Sharp-tailed Grouse), magpies, pigeons, ducks, woodpeckers, geese, and swans (Burpee 1907). Not unexpectedly, Henday also commented on Alberta mosquitoes.

Previously, in 1670, the HBC had been granted a trade monopoly and the right to govern all lands that drained into the Bay. By the 1730s, and continuing for decades, employees of the HBC collected specimens and recorded their observations about the flora and fauna they encountered. The three earliest collectors, Ferdinand Jacobs, James Isham, and Andrew Graham, worked out of forts in Ontario and Manitoba that bordered Hudson and James bays (Moose Factory, Albany [James Bay]; Severn, York, and Churchill [Hudson Bay]). Specimen information and other data from these three individuals were included by Pennant (1785) in his seminal *Arctic Zoology*, Vol. 2.

Moses Norton succeeded James Isham as governor at Churchill (Fort Prince of Wales). At that time, Indians from an unknown region west of the fort had been bringing samples of native copper to trade. Norton convinced the HBC to send Samuel Hearne out with native guides to find the source and determine its commercial potential. Over the course of two aborted trips and one successful journey from 1769-1772, Hearne likely passed close to the northern boundary of Alberta on his return from the mouth of the Coppermine River on the Arctic coast. The qualifier is needed because Hearne was not a skilled geographer and made numerous errors (Glover 1958). Nonetheless, his natural history observations (written in 1795, see Hearne 1958) are relevant to the Alberta region and show that Hearne was perhaps the finest observer of nature in North America until the time of Audubon (Glover 1958, see also Houston and Houston 1987).

In 1792, twenty years after Hearne, Alexander Mackenzie, of the rival North West Company worked in the vicinity of Fort Chipewyan before his voyage to the Pacific Ocean. There are scattered references to the natural history of the region in his account published in 1802. David Thompson explored much of northern and central Alberta from 1793-1812 for the North West Company. His records (Hopwood 1971) are full of hunting accounts, notes on bird and mammal behavior and botanical and ethnological observations.

To put these dates in perspective, it is worthwhile to note that the science of ornithology was only beginning in Europe. The first general natural history books on birds were written in this period by two famous French naturalists, Malthurin-Jacques Brisson (1760) and Comte Georges-Louis Leclerc de Buffon (1770). The major American ornithologist of the time was William Bartram (remembered in the generic name of the Upland Sandpiper *Bartramia*), who included a catalogue of the birds of the eastern United States in his *Travels through North and South Carolina* (1791).

The next major set of ornithological observations for Alberta came with the land and sea expeditions of Sir John Franklin. These voyages were outfitted for scientific exploration (not for expansion of the fur trade) with the goal of finding a navigable route to Asia. The first expedition spent the spring of 1820 in and around Fort Chipewyan on Lake Athabasca. Accompanying Franklin was his surgeon-naturalist Dr. (later Sir) John Richardson. Also on the first Franklin expedition was Robert Hood, who painted a number of birds including the type specimen of the North American subspecies of the Black-billed Magpie and four species new to science (Black-backed Woodpecker, Yellow-headed Blackbird, Hoary Redpoll and Evening Grosbeak). On the second Franklin expedition in 1825-1827, Richardson's assistant, Thomas Drummond, worked independently west to Edmonton, Fort Asisinboine, and Jasper. His collections of birds, other animals, and plants were the first from central Alberta. The observations from these expeditions plus those of the earlier Arctic explorers James Ross and William Parry formed the basis for the magnificent series *Fauna Boreali-Americana*. Volume II of this series titled *The Birds*, by William Swainson and Richardson was published in 1837. The specimens collected by Thomas Drummond were also reported in *Flora Boreali-Americana* and in Charles Lucien Bonaparte's *American Ornithology* (1825-1833).

In 1859, the year of the publication of Darwin's *Origin of Species*, a 24-year-old naturalist, Robert Kennicott, was sent into the territories owned by the Hudson's Bay Company by Spencer Fullerton Baird of the Smithsonian Institution. Kennicott's mission was to collect birds, eggs, mammals, plants, and ethnographic artifacts for the research collections of the Smithsonian (Lindsey 1991). In the three years Kennicott was in Canada, and in subsequent years afterwards, over 12,000 specimens were sent to the Smithsonian from various HBC posts.

The fur trade also brought Catholic and Anglican missionaries to the newly established communities. Rev. Emile Petitot, a French missionary travelled extensively in northern Alberta and north to Great Slave Lake in the 1870s. He wrote on the biology, geology, meteorology and inhabitants of the region in several papers including articles in the journal *Canadian Record of Science* (1884) and five subsequent books.

In the United States over the 50-year period 1820-1870, ornithology flourished. The contributions of men such as Wilson, Audubon, Nuttall, Townsend, Baird, Brewer, Cassin, Lawrence, and Gambel, if not well known, are remembered in the common and scientific names of a number of birds.

Despite Henday's initial visit to southern Alberta, the first 100 years of recorded observations of birds in Alberta revolved around the fur trade and focussed on the northern half of the province. The exceptions were the two Franklin expeditions, which incorporated detailed scientific observations into a search for a commercial trade route to Asia. The Palliser expeditions of 1857-1859 had a similar mix of commerce and science. Captain Palliser, along with engineers and scientists, was sent by the British government to assess the quality of lands west of the Canadas (Upper and Lower) and to find a land route to the Pacific completely F(on British soil. Palliser was instructed to find a route, through the mountains south of the known Yellowhead Pass, that could be traversed by horses. Accompanying Palliser was Lieutenant Thomas Blakiston, whose main role was to measure the strength of the earth's magnetic field. Blakiston also collected birds and made natural history observations. After exploring the North and South Kootenay passes, he left the expedition to join another into China. In 1863, he published "On the birds of the interior of British North America". Sadly, his notes and collections left at Woolwich Institute in London have been lost.

Blakiston's ornithological work was rarely cited by subsequent writers on Canadian birds and the various reports prepared by Blakiston, Dr. Hector (geology), and M. Bourgeau (botany) disappeared into government files. Palliser's report concluded that a large proportion of the Canadian prairies was an extension of the Great American Desert and ill-suited for agriculture. A Canadian-sponsored expedition (1857-1858) under H.Y. Hind came to the same conclusion. This did not sit well with a Canadian government that was encouraging settlement in the prairies to expand the Dominion and protect it from American interests. No popular account of the expedition was written until 1963 (Spry 1963).

An opposite opinion on the southern prairies was espoused by botanist and ornithologist John Macoun and Rev. George Grant who accompanied Sandford Fleming in 1872 in an effort to find a route for a rail line that would link British Columbia with the rest of Canada. A rail link had been promised to British Columbia as a condition for its entry into Confederation in 1871. Rev. Grant wrote glowingly about the Canadian prairies and the suitability of the Yellowhead Pass for a rail line in his 1873 book *Ocean to Ocean*. Macoun, the expedition's botanist, filed reports on the flora and fauna encountered. Some of his observations appear in Grant's book.

A separate book could be written on the relationship of politics to science in this period of Canadian history. The scientific explorations were adjuncts to the political aspirations of the fledgling Dominion. Rivalries formed over issues, like the adequacy of the prairies for settlement, influenced appointments for other surveys and government positions, which in turn shaped the growth of ornithology in Alberta.

Four additional explorations provided natural history observations but little substantive ornithological data. W. Milton and W.B. Cheadle (1865), considered Alberta's first tourists, reported on the flora and fauna of the Edmonton area, before exploring the Yellowhead route to the Pacific. Dr. G.M. Dawson (1875), of the British North American Boundary Commission Canada, wrote on the geology and resources of the 49th parallel from Lake of the Woods to the Rocky Mountains. As did Elliot Coues (1878) who was surveying the border from the U.S. side. A.R.C. Selwyn (1874), of the Geological Survey of Canada, described his journey from Manitoba to Rocky Mountain House. George Dawson, of the Geological Survey of Canada, returned toAlberta in 1879 to explore the Peace country to see if it offered advantages for passage of the Canadian Pacific Railway.

John Macoun served as botanist with Selwyn on an 1875 exploration that descended the Peace River from Fort St. John to Lake Athabasca. He then travelled extensively in central Alberta in 1879 and to southeastern Alberta in 1880. His ornithological observations constitute Chapter 21 of his book, *Manitoba and the Great North-west*, published in 1883. It represents the first systematic advance in the ornithology of Alberta since *Fauna Boreali-Americana* (Swainson and Richardson 1832).

In the 1880s, the Geological Survey of Canada sent several surveying parties into northern Alberta. John Macoun's son James, also an ornithologist, accompanied Thomas Fawcett of the Dominion Lands Survey in 1888 on an exploration of the Athabasca region from the mouth of Lesser Slave River to the junction of the Athabasca River with the Clearwater River. Macoun made detailed observations of the birds, which unfortunately were not published for 20 years (Macoun and Macoun 1909). This collection of natural history information as part of the Geological Survey Expeditions continued in the 1890s under the direction of J. W. and J. B. Tyrrell (brothers) and in the early 20th century by Dr. Robert Bell and Charles Camsell. Bell acquired a number of bird specimens from northern Alberta which were sent to the Victoria Memorial Museum in Ottawa.

A comment on collected specimens is in order. John Macoun recounts in his autobiography (1922: 219). "In 1880, while on the prairie I decided to take up the study of birds and adopted the plan I had followed when I started teaching school; I described the birds as they were shot and, when I found the books, I compared my descriptions with them and so learned the rudiments of ornithology . . . This was how matters stood when I was ordered to report at Ottawa to be a resident [to serve as Botanist to the Geological and Natural History Survey of Canada] there in the autumn of 1882.

"The skins of birds, that I brought, were spread out by me on a table in the Long Room, where the draughting group was, and, after a time, were placed in long drawers that were in the old Museum and, in course of time, with Dr. Bell's specimens of birds, collected on Hudson's Bay were destroyed almost wholly by insects, and remainder burned." After this, Macoun took charge of all his collected specimens and built the basis of the National Museum collection inherited by Percy Taverner. I am not aware of the number of specimens collected in Alberta in the 1800s, nor how many are still in existence but it is disturbing that so much data has been lost through poor management of museum collections.

The first Canadian book devoted entirely to birds was published in 1887. Entitled *A Catalogue of Canadian Birds*, it was written by Montague Chamberlain of New Brunswick,who was to be the naturalist on the 1875 Selwyn expedition until Macoun was given the post. Chamberlain begins by quoting American sources who lament the lack of knowledge of Canadian birds. He goes on "I am quite aware that this opinion regarding the narrow limits of our knowledge of Canadian birds is opposed to that held by some of the leading scientific men of the Dominion, who consider that all that can be learned about our fauna is now known to science."[!] Several interesting species accounts are found in Chamberlain. For instance, for the Eskimo Curlew: "This species is common . . . "; the Piping Plover: "This species is reported as breeding abundantly . . . "; and the Passenger Pigeon: "The "Wild Pigeon" . . . is abundant on the Plains." Saskatchewan and Alberta were divided into the northern "prairies" and the southern "fur-countries". Most of the Alberta observations in Montague's book seemed to be from Elliot Coues' work as part of the U.S. Army Northern Boundary Commission expeditions along the 49th parallel in 1873-1874.

A significant but poorly known figure in Albertan and Canadian ornithology is William Spreadborough. Spreadborough signed on with John Macoun in 1889 as a cook but his aptitude of collecting birds and preparing study skins so impressed Macoun that he became a zoological collector on all subsequent Macoun expeditions. Spreadborough was a consummate collector whose work formed the basis for the bird collections of the National Museum of Canada. In Alberta, he collected around Banff in 1891, around Medicine Hat in 1894, the region south of the Canadian Pacific Railroad line in 1895, in 1897-98 in the Rocky Mountains south to the Crowsnest Pass and along the Yellowhead Pass, and in 1903 in the Peace River region. Spreadborough worked throughout Canada and his collections and those of John and James Macoun contributed greatly to the second *Catalogue of Canadian Birds* by John and James M. Macoun (1909; see also Macoun 1900, 1903, 1904) This work is differentiated from Chamberlain's limited catalogue by the more extensive geographic coverage and the wealth of behavioral and ecological observations.

Americans were also interested in the birds of Alberta. As mentioned earlier, several thousand bird specimens collected by HBC employees were sent to the Smithsonian Institution. Other collections headed south as well. In 1892, Miss Elizabeth Taylor made a trip on HBC boats down the Athabasca, Slave and Mackenzie rivers. She collected birds, mammals, fish, insects, and plants. Some of the birds ended up at the U.S. National Museum (Smithsonian). In the summer of 1893, Frank Russell, working for the University of Iowa, collected ethnological artifacts and natural history specimens from Edmonton to Fort Chipewyan. Much of his time was spent collecting birds in the Peace-Athabasca Delta (Preble 1908).

During the summers of 1894-1896, J. Alden Loring collected bird and mammal specimens from central Alberta for the U.S. Biological Survey. Loring worked in the vicinity of Edmonton, Strathcona, St. Albert, west to Jasper and the Yellowhead Pass, and the region up to 200 km north of Jasper. The observations and specimens collected by Loring are detailed in Preble (1908). Preble, accompanied by Ernest Thompson Seton, passed through northeastern Alberta on his way to the Northwest Territories. Preble had visited the area twice before in 1901 and 1903-04.

The Smithsonian Institution sent collectors with the Alpine Club of Canada to Jasper Park in 1911. The Alpine Club, founded in 1906, organized tours for naturalists and wealthy tourists to the Canadian Mountain Parks. The 1911 trip, which included Mt. Robson, was primarily scientific and was accompanied by William Spreadborough. The bird observations were published in Riley (1913).

Shortly after the turn of the century, ornithological work proliferated in Alberta on several fronts. Geological Survey of Canada work continued in northern Alberta and in the national parks, professional ornithologists from eastern Canadian and American museums made regular visits to the province, and Albertans began to contribute through egg collection and publication of local natural history studies.

As part of the Athabasca-Great Slave Lake Expedition of the Geological Survey of Canada in 1914, Francis Harper (1915) provided bird records from north of Lake Athabasca in the Kazan Uplands. An unusual observation was of a Rock Wren at Fort Chipewyan. Elliot Coues (1903) published his *Key to North American Birds*, which included many observations from his work with the U.S. army along the Canadian border in the 1870s.

Professional ornithologists (i.e., those based out of a museum or university but working independently, and not as part of a government survey team) first worked in Alberta around the turn of the century. Frank M. Chapman, well-known Curator of Ornithology from the American Museum of Natural History in New York, travelled in western Canada in 1901 and 1907. Most of his time was spent in Manitoba and Saskatchewan, but he visited eastern Alberta and Banff in 1907. Chapman's (1908) book, *Camps and Cruises of an Ornithologist*, presented the bird life of the prairies in a popular format. Arthur Cleveland Bent, editor and compiler of the *Life Histories of North American Birds*, collected birds in southwestern Saskatchewan in 1906 and wrote two papers for the Auk on the Cypress Hills region (Bent 1907, 1908).

The first documented dinosaur fossils from Alberta were collected in 1874 by G.M. Dawson (1875). The golden age of dinosaur collecting in Alberta began in 1909 when Barnum Brown of the American Museum of Natural History began his annual explorations (Spalding 1988). The Geological Survey of Canada soon hired Charles H. Sternberg to find dinosaurs on behalf of the government.

In 1915 and 1916, George Sternberg of the VictoriaMemorial Museum and Geological Survey of Canada was collecting dinosaur fossils under the direction of his father Charles in the Edmonton and Belly River formations of the badlands of the Red Deer River. He also collected a few birds which prompted Percy Taverner, the Dominion Ornithologist to travel west in 1917. Taverner floated the Red Deer river from just east of the town of Red Deer to Steveville (300 km). Hundreds of specimens were acquired on this trip and are still in the collections of the Canadian Museum of Nature. An interesting observation from this trip is Taverner's delight at encountering Black-billed Magpies. He quotes Frank Farley, "No one knew this bird 10 years ago and for the past few years a month does not pass that some one does not ask about it. I think this about its limit line as I never saw or heard of one farther north than 10 miles from Camrose." (Taverner 1919).

The art of egg collecting and the study of eggs is known as Oology. Thomas Blakiston of the Palliser Expedition was likely the first scientific egg collector in Alberta. Walter Raine (1892) wrote a book entitled *Bird-nesting in north-west Canada*, which included observations on egg collecting in Alberta. Egg collecting for food had undoubtedly been practised in Alberta for centuries however. Up to the turn of this century, the eggs of wild birds were featured on the menus of hotels. The collecting of eggs as a hobby rather than a dietary supplement flourished in the 1890s (Houston and Bechard 1982).

For about 40 years, until 1940, the records of egg collectors appeared in species accounts of ornithologists such as Percy Taverner. Others were published separately in journals such as the Ottawa Naturalist, the Oologist and the Oological Record. Egg collecting was more involved than coin or stamp collecting because of the effort required in finding nests, documenting the find, preparing the eggs and then selling, trading or preserving the specimens. The popularity of the pastime was enormous. The Oologist had as many as 2500 subscribers, which is about 50% of the current membership in the American Ornithologists' Union. Significant Alberta collectors included Archibald Henderson, T.E. Houseman, G. F. Dippie, George Cook, Tom Randall (a warden of Elk Island National Park), Evan Thompson, N.V. Fearnhough, Frank Farley, and Charles Horsbrugh. Although many of the collections of these individuals are in Canadian and American Museums, others including the Henderson collection, are still in private hands.

Probably the best known Alberta Oologist was Archibald Henderson who lived near Belvedere, northwest of Edmonton, and collected primarily in that region. He published 48 papers over a 26-year period (1915-1941) in ornithological and oological journals (see Houston and Bechard 1990 for complete list). That these egg collectors were also naturalists and contributed significantly to Alberta ornithology is evident from the titles of some of their papers. For example, Raine (1904) reported on the discovery of nest and clutch of the Solitary Sandpiper by Evan Thompson. Henderson wrote on "The breeding waders of the Belvedere district, Alberta Canada" (1941); "Bonaparte's Gull nesting in northern Alberta" (1926); "Cycles of abundances and scarcity in certain mammals and birds" (1923a); "The return of the Magpie" (1923b). Other papers include Fearnhough (1940) "Is the Long-billed Curlew doomed? " and Randall (1933) "A list of the breeding birds of the Athabasca district, Alberta".

Of the early Alberta oologists and other naturalists such as Norman Criddle, or C.B. Horsbrugh (see Horsbrugh 1915), Francis La Grange (Frank) Farley deserves special mention. Although born in Ontario in 1870, Farley moved to Red Deer at the age of 22 and settled in Camrose in 1907. Farley maintained detailed notes of the birds of the Camrose region, wrote numerous articles on birds for the local paper and published 36 scientific papers on the birds of Alberta. His major contributions include *Birds of the Battle River region* (1932), "Changes in the status of certain animals and birds during the past 50 years in central Alberta" (1925), and "Summer Birds of the Lac La Biche, Ft. McMurray region" (1922). Frank Farley's book collection is the foundation of the ornithology library in the Provincial Museum of Alberta. His influence on his disciple, Bert Wilk, was so great that the first edition of the Birds of Alberta (Salt and Wilk 1958) was dedicated to him. Farley's name is remembered in a subspecies of Boreal Chickadee (*Parus hudsonius farleyi*) named by Earl Godfrey and in part through the writings of his grand-nephew — Farley Mowat.

The Canadian government undertook a series of detailed studies of the fauna of Banff, Jasper, Elk Island, and Wood Buffalo national parks during the World War II years in Alberta. Although Banff Park was declared in 1887, the others were comparatively recent (Jasper 1907, Wood Buffalo 1922 and Elk Island 1930). J. Dewey Soper reported on the fauna of Elk Island (1940) and Wood Buffalo (1942) parks and on the sum of his Rocky Mountain experiences since 1913 (1947). In his role with the Department of Mines and Resources: Lands, Parks and Forest Branch, National Parks Bureau, Soper studied the faunas of other Alberta regions including the Grande Prairie-Peace River area (1949a), and Nemiskam National Park (established 1914, closed 1947) (1949b). Soper also studied waterfowl, monitoring their declines on the prairies, and discovered the nesting grounds of the Blue Goose on Baffin Island. Clarke and McTaggart-Cowan (1945) published an extensive review of the Birds of Banff National Park including the Ya Ha Tinda ranch.

The National Museum of Canada continued to build collections through research trips around the country. A.L. Rand published a detailed analysis of the birds of southern Alberta (1948) and W.E. Godfrey (1952) provided extensive observations on the Lesser Slave Lake and Peace River areas.

A major source of ornithological data for the province inthe early and middle parts of this century developed from the work of various University of Alberta professors. The most notable was William Rowan, who in 1920 was hired to build a Department of Zoology. Rowan is considered to be Canada's first avian biologist (Ainley 1987). His pioneering research into bird migration brought him world attention. He spawned several generations of graduate students including W. Ray Salt, Lloyd Keith, and Al Oeming. His fieldwork at Beaverhill Lake detailed in Lister (1979), set the stage for its recognition as an ornithological paradise and his acquisitions began the ornithology collection at the University of Alberta. Ironically, Rowan's private collection did not end up at the University of Alberta but was sold to the University of California at Berkeley to help finance his research. Rowan was known as a first class shot who did not hesitate to use his gun. A subspecies of Sandhill Crane (*Grus canadensis rowani*) was named for him. He also named a subspecies of Short-billed Dowitcher (*Limnodromus griseus hendersoni*) after his friend Archibald Henderson. Rowan was instrumental in bringing Otto Höhn to Alberta. Höhn made contributions as an avian physiologist but also provided numerous distributional notes on the birds of Alberta.

Höhn, along with Rowan and Bob Lister, were effective in founding the Edmonton Bird Club in 1949. The first field trip for the club was to the, now long-since drained, McKernan slough within Edmonton city limits (Höhn 1981). Bob Lister kept birders up-to-date through a weekly column on birds for the Edmonton Journal for nearly 20 years beginning in 1962. The Calgary Bird Club was formed in 1955, published regular accounts on bird sightings in the Calgary region, then underwent an expansion of its mandate to become the Calgary Field-Naturalists' Society in 1969 (Myres and Stiles 1981).

The end of the early phase of Alberta ornithology can be marked by the appearance of the *Birds of Alberta* (Salt and Wilk 1958). Although bird distributions change, their documentation will be considered refinements to the work of Salt and Wilk. W. Ray Salt moved to Calgary when he was six years old. His interest in nature led him to collect, mount and preserve birds. By selling his bird skin collection he was able to raise enough money to complete his B.Ed. degree. Despite qualifications as a zoologist, his M.Sc. thesis on the cloacal protuberance in passerines (Salt 1948) garnered him an appointment in 1949 as Assistant Professor of Anatomy at the University of Alberta and through his administrative talents he quickly became the Assistant Dean of Medicine. Even with this full-time job, Salt found time to band, collect, and study birds. His publications span six decades.

The ornithology collection of the University of Alberta grew through purchase of Ray Salt's and E.T. Jones' collections. Specimens were added by David Boag, Bert Wilk, Otto Höhn, and their colleagues and students. David Boag's private collection was donated to the zoologydepartment in the 1970s.

In a massive undertaking, David M. Ealey and Martin K. McNicholl (1991) compiled 7444 records on publications on Alberta birds. The vast majority of these have been published recently from staff and students of the Universities of Alberta, Calgary and Lethbridge, from professional biologists with the Canadian Wildlife Service, Fish and Wildlife Division of Alberta Forestry, Lands and Wildlife, the Provincial Museum of Alberta, amateur naturalists generally affiliated with the Federation of Alberta Naturalists, and by biologists working for consulting firms.

It is not appropriate to assess the impact of contributions from one's friends and colleagues, and I will not do it here. Many ornithologists active in the 60s and 70s are still publishing and will make contributions for years to come. Suffice to say that in a few decades we will be in a better position to assess the contributions of scientists such as Drs. David Boag, Fred Zwickel, Vic Lewin, Susan Hannon, and their students of the University of Alberta; Tim Myres, Ross Lein, and their students from the University of Calgary, Geoff Holroyd, George Scotter, Ernie Kuyt from the Canadian Wildlife Service; and Phil Stepney and subsequent ornithology curators from the Provincial Museum of Alberta.

There are a few trends and societal changes worthy of comment. I think it is fair to say that Professors Salt and Höhn were the last university-based staff in Alberta who regularly commented on bird distributions. In Alberta, as elsewhere, ornithologists shifted their attention to topics such as behavior, song, physiology, breeding ecology, mating systems, reproductive strategies, and conservation. The distribution torch has been passed to firms doing extensive surveys for development purposes and to the Provincial Museum, which has conducted natural history surveys of several remote areas (McGillivray and Hastings 1988, 1990, 1992).

The opening of the Provincial Museum of Alberta (PMA) in 1967 provided a provincial base for bird specimens and resulted in the preservation of numerous private egg collections. In 25 years, the Ornithology collection has grown to be the third largest in Canada. It provides a permanent record of the bird life of this province in the form of study skins, mounts, eggs, nests, skeletons, and soft tissue. Sadly, because of the relatively late start of the Museum, tens of thousands of specimens from Alberta are scattered in other museums in North America, Europe, and in private hands. The second head of the Natural History section of the PMA, David A. E. Spalding wrote several papers about early participants in Alberta natural history (see Spalding 1981 and McNicholl et al. 1981). He was a key figure, himself,in the establishment of the Federation of Alberta Naturalists.

The most notable societal change is the participation of volunteer ornithologists in large data-gathering projects such as this Atlas as well as the Christmas Bird Count and Breeding Bird Surveys. The popularity of birding today isreminiscent of the interest in egg collecting in Alberta earlier this century. Birders, however, have many more organizations, publications and facilities to support their hobby. Undoubtedly, the culmination of volunteer involvement in natural history in Alberta to date is this Atlas. Because of the efforts of thousands of volunteers, we know more about the distribution of breeding birds in this province than at any previous time.

I think it is fitting that when science is beginning to use sophisticated methods to assess and measure human-induced change in the environment, the question being asked about a region is still "What species are there?" and "Why are they there?" — the same questions naturalists attempted to answer 150 years ago in Alberta. Of course, we can add a third question now: What must be done to preserve these species and their habitats in the wild? In 1887, according to Chamberlain, some scientists felt that we knew all that it was possible to know about the birds of Canada. What this Atlas demonstrates is how much more we have to learn even 105 years later. It is gratifying to know that when the birds of Alberta are to be examined in the future, along with the data in museum collections and historical surveys, a significant resource will be the Atlas data gathered by volunteers over the last six years.

This compilation was possible thanks to the bibliographic research undertaken for me by Josef Martha in 1989. I am grateful for the comments of Drs. David Boag, Stuart Houston, Pat McCormack and David Spalding on an earlier draft of the manuscript. They improved the chapter immensely and corrected many errors. All remaining errors and omissions are mine alone.

W. Bruce McGillivray, Ph.D.
Assistant Director
Natural History and Collections Administration
Provincial Museum of Alberta

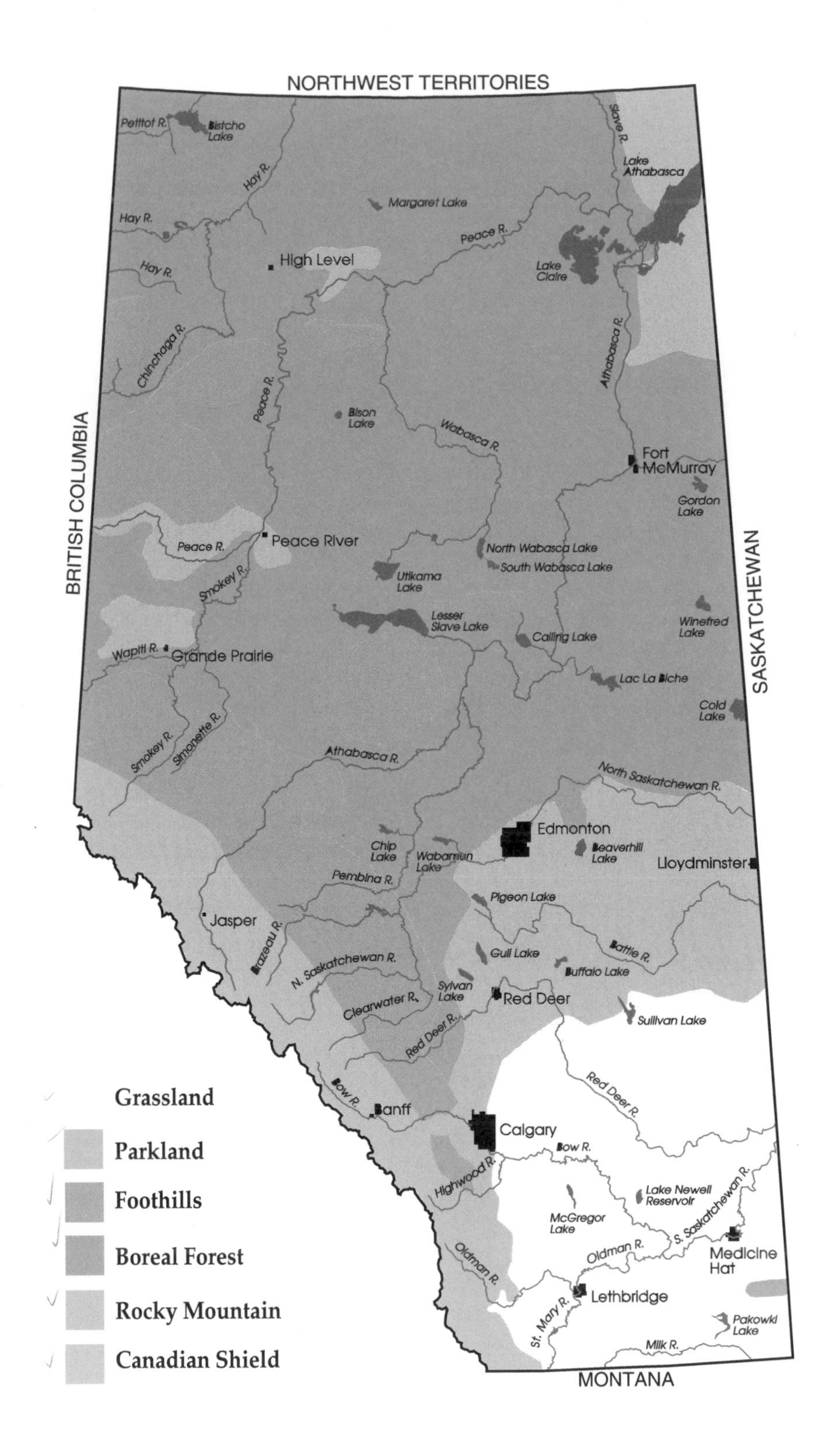

Map of the Natural Regions of Alberta

Biogeography of Alberta

INTRODUCTION

The following subdivisions of the Natural Regions of Alberta have been abstracted from a report by Achuff et al. (1988). Their report was based on previous documents prepared by the Natural Areas Program and Alberta Recreation and Parks. The subdivisions are based on recurring distinctive landscape patterns of vegetation, soils, landforms, and, to a limited extent, by wildlife. Each Natural Region is further divided into Sections based on recurring landscape patterns relative to other parts of the Natural Region. Application of the system divides large areas into progressively smaller units. This approach was chosen because the information base is too incomplete to allow aggregation of small areas into larger units. The result is a classification system which represents the broad diversity of the province, yet is easily recognizable to the untrained observer.

Available vegetation information is concerned primarily with modal (mesic, moderately well-drained, upland) or "climax" types. Mesic, moderately well-drained upland is only one of many landscape types which characterize a Section and little is known of the vegetation of other types. In the Boreal Forest Natural Region, large areas are devoid of "modal" vegetation because of poor or excessive drainage.

GRASSLAND NATURAL REGION

The Grassland Natural Region occupies a broad area in southern Alberta. It extends to the Rocky Mountain Cordillera in the west, and north to the southern boundary of the Parkland in central Alberta. This region is a broad, flat to gently rolling plain with few major hill systems. Where valleys are carved deeply into the bedrock, extensive badlands have developed. Numerous coulees and ravines are associated with these valley systems.

The Grassland Natural Region contains three Sections -Mixed, Northern Fescue and Foothills. These Sections are separated primarily on the basis of climatic factors. The Mixed Grassland is the most extensive grassland region in Alberta, occurring from the U.S.A. border in the south to the fescue grasslands in the north and west. Fescue grasslands of the Foothills Grassland and Northern Fescue Grassland Sections occupy narrow belts along the western and northern margins of the Mixed Grassland. There is also a disjunct occurrence of Foothills Grassland on the plateau of the Cypress Hills in southeastern Alberta.

Climate:

The climate of the Mixed Grassland is the warmest and driest in the province. Because of the warm summer temperatures and high average wind speed, the rate of evaporation is high throughout the summer months. It is a typical continental climate characterized by cold winters, warm summers, and low precipitation. The climate of the Northern Fescue Grassland is transitional between that of the Mixed Grassland and Central Parkland. The climate of the Foothills Grassland differs from that of the Northern Fescue Grassland in several ways. Winter chinooks are more frequent in the foothills, resulting in a milder, warmer climate. There is also a greater amount of snowfall in the late winter and early spring.

Vegetation:

The name "Mixed Grassland" is derived from the fact that both short and mid-grasses are predominant. The most widespread grass species of the uplands are the mid-grasses (Spear Grass, Western Wheat Grass and June Grass) and the short-grass (Blue Grama).

The Northern Fescue Grasslands are dominated by Rough Fescue. The majority of this region has been converted into agricultural cropland, and most of the remaining sites of native vegetation occur either on rugged hummocky moraine or on Solonetzic soils.

The Foothills Grasslands are dominated by Rough and Idaho Fescues and Parry's and Intermediate Oat Grasses. They occur primarily as a narrow band between Mixed Grassland and Foothills Parkland vegetation. These grasslands are characterized by a much greater variety and cover of forbs than exists in the Northern Fescue Grassland.

Wetlands in the Grassland Natural Region are primarily wet meadows and fresh to saline shallow marshes occurring in depressions. Natural lakes are rare in this region and are mostly intermittent.

Permanent streams are scarce but there are many seasonal streams along the coulees and ravines leading into the main river valleys. Seasonal streams carry spring meltwaters and flow briefly during persistent summer rains or violent thunderstorms. Riparian habitats, consisting of Plains cottonwoods and a variety of tall and low shrub communities, are well-developed along a few permanent streams with broad active floodplains. Sagebrush flats and valley grasslands occur along the older parts of the valley floor.

PARKLAND NATURAL REGION

The Parkland Natural Region, with the exception of the Peace River Parkland Section, is transitional between the drier grasslands of the plains and the coniferous forests of the Boreal Forest and Rocky Mountains. It is comprised of three Sections - the Central, Foothills, and Peace RiverParklands. These sections are separated on the basis of geographic location and major floristic differences.

In general, the Parkland Natural Region has more relief than the Grassland Natural Region. The topography ranges from nearly level in large areas of glacial lake deposits and ground moraine, to strongly rolling in areas of deformed bedrock along the foothills.

The Parkland Natural Region is the most densely populated of all the Natural Regions; land-use patterns have altered much of the native vegetation.

Climate:

The climate accommodates both woodland and grassland vegetation. There is a moderate level of precipitation, slightly higher than that of the Grassland Natural Region and slightly lower than that of much of the Boreal Forest Natural Region.

The climate of the Foothills Parkland, particularly in the scuthwest, differs from that of the Central Parkland. The winters are milder because of the influence of chinooks and there is a prevalence of late winter and early spring snowstorms.

The climate of the Peace River Parkland differs form other Parkland Sections in that it is typified by shorter, cooler summers and longer, colder winters. The high latitude results in long hours of daylight from May to August.

Vegetation:

Perhaps the most distinctive feature of the Parkland Natural Region, particularly the Central Section, is the great variety and extent of marsh and meadow vegetation. The deeper more permanent marshes are dominated by bulrushes, cattails, and sedges. Shallower ponds have sedge, Reed Grass, and Slough Grass marshes. Most of these wetlands have a border of tall willow shrubs. Boreal fens and bogs occur along the northern and western boundaries of the region.

Permanent streams, lakes, and wetlands are much more abundant here than in the Grasslands Natural Region. River valleys harbor a great diversity of vegetation, including badlands and white spruce woods, as well as a variety of grasslands, deciduous woodland and shrublands.

In the Central Parkland, the dominant upland vegetation types are Aspen and Balsam Poplar woodlands and Rough Fescue grasslands similar to those of the Northern Fescue Grassland. A number of shrub species dominate the edges of the woodlands. Within the Central Parkland there are extensive grasslands dominated by many Mixed Grassland species.

The Foothills Parkland occupies a very narrow band between the Foothills Grassland and Montane Sections. It is most extensive in the area surrounding Waterton Lakes National Park and in the area from Calgary south to the Porcupine Hills. Foothills Parkland is characterized by extensive willow communities on moist sites in the northern part, and by Aspen woodland with fescue and oat grass grasslands throughout the Section on mesic sites. The grasslands are similar to those described for the Foothills Grassland.

The Peace River Parkland occurs as disjunct islands of grassland within the Boreal Forest. The woodlands reflect the surrounding region whereas the grasslands contain numerous species typical of southern and western grasslands in North America. The grasslands occur mostly on glacial lake deposits and are similar in most respects to the Northern Fescue Grasslands, except Rough Fescue is absent.

FOOTHILLS NATURAL REGION

The Foothills Natural Region is essentially a transition zone between the Rocky Mountains and Boreal Forest Natural Regions. It is divided into two Sections, the Main and Northern Outliers. The Main Foothills includes most of the "geological" Foothills Belt, a band of folded sedimentary rock along the eastern edge of the Rocky Mountains, as well as adjacent uplands on flat-lying sedimentary rock. Northern Outliers consist of isolated flat-topped hills which are erosional remnants in the western part of the Boreal Forest. These outliers occur in the Swan Hills, Mount Watt, Pelican Mountain, and Buffalo Head Hills. They are separated from the Main Foothills by fairly broad areas of Mixed Wood forest.

Climate:

The climate of the Foothills is variable; it is a mixture of Cordilleran and Boreal climates. In general, it can be classed as humid at higher elevations and moist subhumid in lower areas. From east to west and from south to north there are increases in precipitation, although there is presumed to be a major rainshadow immediately adjacent to the Rocky Mountains. From east to west and from north to south there are increase in the average winter temperatures. Winter temperatures in the Foothills tend to be slightly warmer than those of the Boreal Forest.

The Northern Outliers, especially at higher elevations, are cooler and moister with shorter summers, as exhibited by the dominance of coniferous trees. They can be classed as humid at higher elevations and moist subhumid at lower areas. Summer precipitation is higher than in the surrounding regions because of the effect of higher elevation.

Vegetation:

The Foothills vegetation is a mixture of Rocky Mountain and Boreal Forest species and types. The Foothills are extensively forested with Lodgepole Pine. White Spruce and Aspen Poplar occur more frequently along the borders with the Boreal Forest Natural Region. Black Spruce-feather moss forests are predominant on uplands on imperfectly drained sites. Spruce-fir forests occur in northern regions of the Main Foothills and in the Northern Outliers at higher elevations. Poorly drained sites are dominated by Black Spruce peat bogs. Lakes and marshes are relatively uncommon in the Foothills, the most extensive wetlands being the numerous bogs, fens, and swamps found in depressions and along the valley systems.

BOREAL FOREST NATURAL REGION

The Boreal Forest is the largest Natural Region in Alberta. It is comprised of broad lowland plains and discontinuous but locally extensive hill systems. The bedrock is deeply buried by glacial deposits. As in other parts of the province, the land generally slopes down to the northeast, but the most prominent uplands are located in the north. The Boreal Forest Natural Region contains four Sections - Sub-Arctic, Hay River, Peace River Lowlands, and Mixed Wood. These four Sections are separated on the basis of climate, predominant landscape-forming processes, and typical vegetation patterns.

The prevalence of wetlands is a major characteristic of the Boreal Forest Natural Region. The Boreal Forest of northern Alberta comprises some of the wettest terrain in Canada, with the world's largest inland freshwater delta occurring at the confluence of the Peace and Athabasca Rivers. Extensive portions of the lowlands may be almost totally occupied by wetlands. Bogs, fens and swamps are abundant and marshes are locally prevalent. Permafrost wetlands are locally common in the Sub-Arctic Boreal Forest.

Climate:

Length of growing season and total annual precipitation are probably two of the most important factors in determining the vegetation pattern in this region, but there is very little long-term climatic data. The macro-climate of the Boreal Forest can be characterized as cool and subhumid to humid. Upland areas are cooler and wetter than lowland sites.

The climate of the Sub-Arctic Section is boreal cold-temperate continental characterized by short, cool, moist summers and long cold winters. However, due to airmass inversions, the winter temperatures are thought to be higher in these sub-arctic uplands than in the surrounding lowlands. The seasonal differences in climate from the surrounding Hay River and Peace River Lowlands Sections Probably are a result of the higher elevations of the Sub-Arctic Section.

The climate of the Hay River Section is characterized by moist cool summers and long cold winters. The average snow cover lasts for 185 days, one of the longest periods in Alberta. The climate of the Peace River Lowlands Section is characterized by a short growing season, cool summers and long cold winters. The climates of the Peace River Lowlands and lower Hay River Sections are warmer during the summer than the surrounding uplands.

The climate of the Mixed Wood Section is similar to other Boreal Forest and Canadian Shield areas, except that it is characterized by a pronounced peak in the mid-summer precipitation, and there is a tendency for a more moderate climate. The climate is subhumid continental with long cold winters and short cool summers.

Vegetation:

The coniferous forest character increases with elevation, from south to north, and west to east. The dominant forests of the uplands are mixed forests of poplar and spruce on intermediate-textured morainal materials and Jack Pine on coarser textured sand dune and outwash materials.

Larger lakes are common in the Boreal Forest Natural Region and they include some of the largest lakes in Alberta (Lake Claire, Lesser Slave Lake). Bog lakes are abundant.

The Sub-Arctic Section is characterized by extensive permafrost and open Black Spruce bogs. Black Spruce-feather moss woodlands occur on mesic sites, whereas woodlands dominated by Aspen Poplar, White Spruce or pine occur on drier sites.

The Hay River Section is typified by large areas of glacial lake deposits and outwash which are overlain extensively by organic soils. The characteristic forest vegetation which occurs on these deposits is a Black Spruce-feather moss type. Extensive wetlands are typical of the Section, especially in the west, where there are large swamps and fens.

The Peace River Lowlands consist of the massive Peace-Athabasca Delta and its floodplain, which is bordered by a series of older terraces. Large White Spruce trees dominate the forests along the rivers. There are extensive and diverse wetlands comprised largely of fens and marshes.

The Mixed Wood Section can be broadly divided into upland units forested with a mixture of White spruce, Balsam, and Aspen Poplar, and lowlands which are comprised largely of peatland complexes. Dominant species of the bogs and swamps are: Black Spruce, Tamarack, willows, Dwarf Birch, sedges, and brown and peat mosses. Jack Peat is dominant on sandy soils, with the most extensive stands occurring in the eastern portion.

ROCKY MOUNTAIN NATURAL REGION

The Rocky Mountain Natural Region is part of a major uplift, trending along the western part of the province, which forms the Continental Divide. It is separated from the Foothills Natural Region primarily on the basis of the structural geology of the bedrock. There are two major mountain ranges—the Front and main Ranges. They are composed almost entirely of thrust faulted sediments. A distinctive feature is the northwest-southeast trending valleys which are separate the ranges. The southwest slopes of the Front Ranges are typically long and smooth, whereas the northeast faces are characterized by steep escarpments with exposed strata. The Rocky Mountain Natural Region is the most rugged of all the Natural Regions in the province. Elevations generally rise from east to west, from a low of 2,100 m in the Front Ranges to a high of 3,700 m in the Main Ranges. The highest mountains occur in the central area along the Continental Divide, the lowest in the far north and far south.

Many of the larger rivers in Alberta originate in this area. Stream courses are characterized by steep to precipitous gradients with numerous waterfalls and rapids.

The numerous macro-climates and micro-climates, created as a result of the broad elevational and latitudinal extents, have enabled the establishment and maintenance of the most diverse assemblage of vegetation types in Alberta. These range from open dry grasslands, ridge systems, and deciduous woodlands to lush coniferous forests, dry coniferous forests, and alpine tundra vegetation. The three major subdivisions of the Rocky Mountain Natural Region (Montane, Sub-Alpine and Alpine) result primarily from the influence of climatic differences.

Climate:

The climate of the Montane Section is the warmest and driest in the Rocky Mountain Natural Region. The Montane Section is found in the warmer parts of the southern Rocky Mountain Natural Region and, in more northerly locations, it is found in association with major east-west valleys. These valleys channel warmer air (chinook winds) during the winter. The chinooks significantly raise overall temperatures and parts of this Section are intermittently snow free in winter. This winter warming trend is probably a major factor in the distribution of Montane habitats.

The climate of the Sub-Alpine Section is cooler and moister than that of the Montane. It is characterized by greater winter precipitation than other areas of the province. Precipitation in the eastern Front Ranges is markedly lower than that in the western Main Ranges. Subsequently, the Front Ranges support extensive areas of grassland and dry shrub vegetation.

The Alpine Section is the coldest of the three Sections. It has a more severe Cordilleran climate than the Sub-Alpine, with strong winds, long winters and cool summers. There is a wide variety of microclimates created by the various exposures to sun and wind. There does seem to be a general trend towards drier conditions in the southern and eastern ranges and wetter conditions in the northern and western ranges. The dampness of far northern ranges is enhanced because of cooler temperatures and reduced solar radiation.

Vegetation:

The Montane Section occurs in the southern portion of the "geological" Foothills Belt, the Porcupine Hills, the Cypress Hills, and along major valleys within the Sub-Alpine Section. It is characterized by fescue-oat grass grasslands, open woodlands dominated by Douglas Fir and Limber Pine, and more closed forests of Douglas Fir, Lodgepole Pine, White or Engelmann Spruce, and Aspen Poplar. Grasslands are most common on steep south-facing slopes. Limber Pine vegetation occupies the driest portions of the rocky ridge tops. Douglas Fir occurs on sheltered sites within exposed ridge systems, and on south and west-facing sites within the mountain valleys.

The Sub-Alpine Section occurs at altitudes above either the Montane Section or Main Foothills Sections. It is distinguished from the Foothills Section mainly by the prevalence of Engelmann Spruce and Subalpine Fir in the "climax" vegetation. Fire-maintained Lodgepole Pine forests are common at lower altitudes.

The upper Sub-Alpine is cooler and wetter, with greater snowfall, later snowmelt, and a shorter growing season than the lower Sub-Alpine. The upper Sub-Alpine forest is dominated by spruce, Subalpine Fir, Subalpine Larch, and Whitebark Pine. Timberline occurs at the climatic forest-line; the upper elevational limit of continuous tree growth. Dwarfed Engelmann Spruce, Subalpine Fir and Whitebark Pine, and open stands of Alpine Larch are typical species.

In the Alpine Section, Yellow Heather occurs on areas of moderate snow accumulation, whereas wetter sites are dominated by White Mountain Heather. Lush herb and willow areas are found where there is abundant seepage. Sedge communities dominate on sites with deep snow which melts in late summer. Lichens are prevalent throughout the Alpine but dominate in stonefields. Grasses and kobresia are found on more exposed warmer sites.

Deep marshes and open water wetlands are locally extensive, particularly along the major stream valleys where there are numerous backchannels. Wet sedge meadows are local throughout the Rocky Mountains. High elevation seepage areas contain unique organic wetlands which are similar to arctic types.

CANADIAN SHIELD NATURAL REGION

The Canadian Shield extends only peripherally into Alberta. The Canadian Shield Natural Region is comprised of two distinct Sections - the Kazan Upland, which includes most of the exposed Canadian Shield in Alberta north of Lake Athabasca, and the Athabasca Plain, which includes part of the north shore of Lake Athabasca and the Canadian Shield south of Lake Athabasca. In the Kazan Upland Section, the glaciated bedrock plays a major role in the landscape appearance and in determining the biological character. Generally, there is low relief with numerous lakes. Extensive sand plains, sand dune complexes, and kame moraines form a distinctive landscape in the Athabasca Plain Section. Lakes are common in much of the Section.

Climate:

The climate of the Canadian Shield is cool and subhumid. The climate of the Kazan Upland is characterized by cold winters and warm, dry summers. The mean annual precipitation is relatively low. The climate of the Athabasca Plain Section is characterized by a warm moderately dry growing season and long cold winters.]

Vegetation:

As a result of the shallow soils and high proportion of exposed bedrock, the vegetation includes extensive areas of open Jack Pine woodland and rock barrens. The ground cover of lichens and mosses is a notable feature of this Natural Region north of Lake Athabasca in the Kazan Upland. Black Spruce bogs predominate in the depressions.

Literature Cited

Achuff, P., J. Godfrey and C. Wallis. 1988. A systems planning natural history framework and evaluation system for Alberta Recreation and Parks. Vol. 1 - Natural region overviews and natural history themes. 27 pp.

METHODS

Conduct of the Survey

Project Scope:

The initial goal of the Alberta Bird Atlas was to record the breeding status and the relative abundance of the breeding bird species of Alberta within the five-year period between 1987 and 1991. Data collection was conducted between February and July of each year. A sixth year was allotted for data compilation and publication of the results. Early collections in February and March primarily involved owls. Initial surveying focused on urban areas of the province, generally in the Grassland and Parkland natural regions. A greater emphasis was placed in the final two years of the project on rural and remote regions in the Foothills, Rocky Mountain, and Boreal Forest regions.

The Grid System

In order to gather breeding evidence in a systematic manner, the province was divided into 10 regions (Figure 1). Each region was assigned a Regional Coordinator (RC) who was required to provide project materials and expertise to volunteers, and collate results. These RCs assigned atlassers to 10 km x 10 km squares of the Universal Transverse Mercator (UTM or Military) grid system. The UTM system was chosen for the project to allow comparisons to other North American atlas projects. Atlassers were encouraged to consult 1:50,000 National Topographic Series (NTS) maps, as well as wildlife, habitat, and forest cover maps, prior to undertaking data collection in each square.

Each 10 km x 10 km square can be found on the appropriate National Topographic Series map. A complete square designation is composed of a zone, block and square code (Figure 2). The zone and block codes are given in an inset box in the margin of all NTS maps. (The area to the east of 114 degrees West longitude is zone 12U: west of 114 degrees longitude is zone 11U. The two letter block code can be read from the NTS map.) Squares are numbered from 00 in the southwest corner of the block to 99 in the northeast corner.

To correct for the inaccuracies of a two dimensional grid laying over the earth's curved surface, adjustments are made
along zone lines. In Alberta, this is the 114th meridian. On either side of this line and along provincial boundaries, partial squares result. For purposes of the Atlas Project, data for squares smaller than 50 km^2 was added to neighboring full squares for ease of surveying and mapping.

Coverage Goals

In Alberta, there are over 6,000 10 km x 10 km UTM squares, many in remote locations. To have censused all the breeding birds of these squares was beyond the scope of this project. Hence, it was decided that one square out of each block of four (20 km x 20 km block) should be termed a "priority square". Regional Coordinators in southern Alberta were asked to pick priority squares as those having the greatest habitat diversity in the 20 km x 20 km block. In northern Alberta, coverage was greatly hindered by limited access and low atlasser population, hence a minimum of one square for each 100 km x 100 km block (a 1:100 ratio) was assigned priority. However, squares were chosen to ensure sampling from all northern areas and, in most cases, several squares per block were surveyed.

To ensure adequate square coverage, atlassers were encouraged to survey all habitat types in a square and consult expected species lists compiled by Regional Coordinators. A minimum of 20 hours of atlassing effort per priority square was required. However, variations in habitat diversity, topography, and forest type could increase this figure. Atlassers were encouraged to record 75% of expected species, confirming breeding for at least 50% of this number, and placing 35% and 15% in the probable and possible categories.

Non-priority squares were also atlassed, usually on an atlasser availability basis and often when atlassers travelled to and from priority squares. Rare or colonial species were recorded regardless of square location.

To ensure adequate provincial coverage, atlassers were encouraged to visit less populated areas or to organize "square bashing parties" through their local naturalist clubs. These "parties" involved several atlassers banding together to cover a difficult square more completely. However, by 1991, coverage still appeared heavily concentrated in urban areas. Hence, an intensive effort to achieve wider coverage in remote and rural areas was undertaken in the fifth year. Strategic planning sessions identified a minimum number of squares to be atlassed in the remaining field season in order to ensure comprehensive provincial coverage. This strategy encompassed five data collection methods including: the traditional volunteer effort; a campaign that solicited data from other researchers; a Remote Areas Program that assisted volunteers in planning a variety of field trips to northern Alberta; a Blockbusters program whereby student employment programs were utilized to hire atlassers; and a secondment program that borrowed employees from donor agencies.

Data Collection Methods

Atlassers were provided with handbooks, data cards, and rare bird forms. Atlassing "techniques" were standardized through a quarterly newsletter, contact with the Regional Coordinators, and regional atlasser meetings. Atlassers filled out one data card for each assigned square per year. They recorded location, hours and dates of atlassing, species, breeding, and abundance codes. Rare and hard to identify species were indicated on the data card to alert the atlassers to file a rare bird report or to be thorough with respect to these species.

Four categories of breeding evidence and 15 breeding codes are outlined in Table 1. Atlassers were encouraged to record only breeding species, however, the code "x", in the observed column, acted as a safety net for separating migrants from potential breeding species as indicated by Regional Coordinators. Alberta is a very large area with a relatively small population; since this was the first large-scale data collection attempt, observational data was deemed to be of significance. This reflects the limited documentation to date on the distribution of Alberta birds and the need to consider migration routes or future breeding sites as areas of concern.

A special effort to document Alberta's rich and varied owl life was also made. Volunteer owl prowlers were provided with an owling supplement to the Atlasser handbook, as well as an owl tape obtained from the Library of Natural Sounds, Cornell University.

Abundance data was also collected in the first year of the project. An analysis of the first year's data showed that data was not collected consistently, thereby reducing the accuracy. Recording of this data was suspended in subsequent years of the Project.

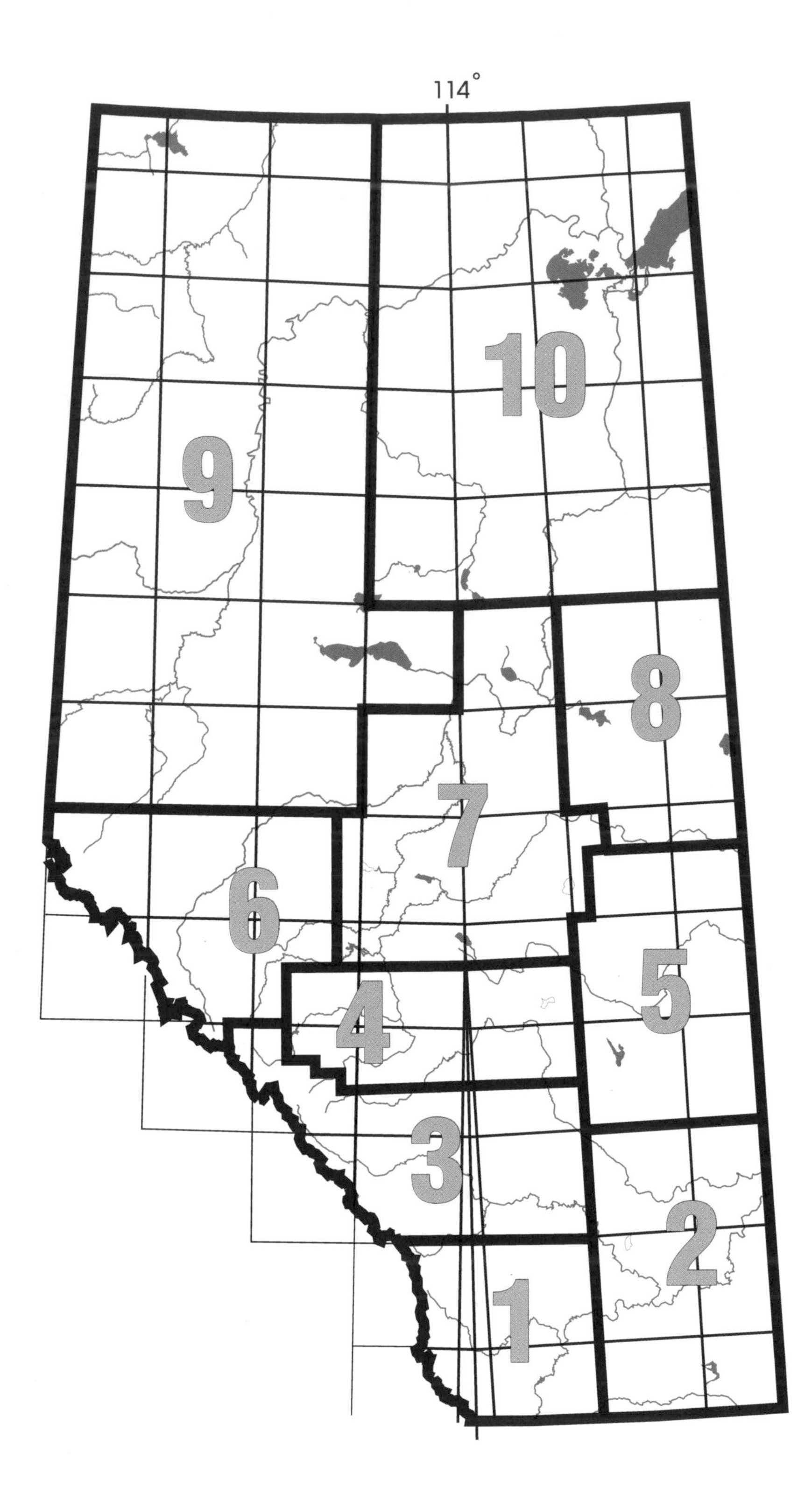

Fig. 1 Atlassing Zones

TABLE 1
Breeding codes used by the Alberta Bird Atlas Project

The objective of the atlas was to obtain the strongest breeding evidence for as many species as possible within each square. There are four levels of evidence coded as follows:

O - species observed in breeding season but no indication of breeding;
PO - possible breeding;
PR - probable breeding;
CO - confirmed breeding.

Within each of these levels, there are categories of evidence denoted by a letter-code representing behavioral and empirical evidence. All of these codes apply to a species seen or heard during its breeding season:

Observed (O)

X - species identified, but no indication of breeding.

Possible (PO)

H - species observed, or breeding calls heard, in suitable nesting HABITAT.

Probable (PR)

P - PAIR observed in suitable nesting habitat.
T - TERRITORY presumed through territorial nesting behavior in the same location on at least two occasions a week or more apart.
C - COURTSHIP behavior between a male and a female.
V - VISITING probable nest-site, but no further evidence obtained.
N - NEST-BUILDING or excavation of nest hole by wrens and woodpeckers.

Confirmed (CO)

NB - NEST-BUILDING or adult carrying nest material; used for all species except wrens and woodpeckers.
DD - DISTRACTION DISPLAY or injury feigning.
UN - USED NEST or eggshells found.
FL - recently fledged young or downy young.
ON - OCCUPIED NEST indicated by adult entering or leaving nest-site or adult seen incubating.
CF - CARRYING FOOD; adult seen carrying food or faecal sac for young.
NE - Nest with EGGS.
NY - Nest with YOUNG.

Data Processing and Verification

Completed data cards were submitted to appropriate Regional Coordinators (RCs), who retained master files and requested rare bird forms as required. Data cards were forwarded to Project headquarters, serialized in a master logbook, and entered onto a Compaq 386-20 computer. While the data input process contained several check systems, a final manual verification of computer files against data cards was conducted before those cards were stored.

Although all sightings were entered on the computer, files were periodically indexed and sorted to provide a set of master files with updated breeding and abundance codes. The database system is comprised of 12 databases and over 60 individual programs for data entry, data editing, report production and map production.

Digital species maps were produced using MAPINFO and dBase IV. MAPINFO was chosen because of its ability to work with Dbase files directly, and its ability to produce maps in a number of digital formats. A program written in Dbase reads the Bird Atlas database files and produces a dBase file which is later read by MAPINFO to produce the map for that individual species. These maps are stored as .CGM (metafiles) files.

The symbology used for the breeding categories was derived by examining previous atlas efforts and the cartography literature. This literature indicated that the visual importance of any symbol should be equivalent to the relative or absolute importance of the object being mapped. In this manner, confirmed sightings are most important and therefore take visual priority. Each symbol is centred in a square with a maximum area of 100 km^2. Squares with areas greater than 50 km^2 and smaller than 100 km^2 are shown with symbols that extend beyond the boundary of the square for the sake of consistency.

The natural regions base map was created by digitizing the natural regions of Alberta map showing biophysical zones, rivers, lakes, etc. This information was stored in GEM Artline, a graphics illustration program, and, in this program, appropriate colors were assigned to the different entities on the map. This information was used as a base for all species maps because of the expected correlation between habitat and species sightings.

The individual species digital maps (in .CGM format) produced by MAPINFO were converted to a .GEM format using HiJaak graphic conversion software. The base map and species map were overlaid using GEM Artline. These final maps were converted to a Postscript format using HiJaak. These postscript files were merged with the text and scanned photo to form a page in the atlas using desktop publishing software.

Records Confirmation

Rare bird documentation forms were requested for all rare species. These forms were sent on to headquarters, but it was not until May of 1991 that a committee was formed to adjudicate these and other records of regionally rare and hard to identify species. The Alberta Bird Atlas Records Confirmation Committee (ABARCC) consisted of a Chairman and several voting members chosen from both northern and southern Alberta—such that members could contribute on the basis of their respective areas of ornithological expertise.

The committee's first task was to review all species maps generated after the final field season. From these, the ABARCC requested documentation forms for over 2,000 sightings. These were reviewed over the period of October 1991 to March 1992.

Five categories of species were recognized by the committee. Ninety-eight species common or widespread to the province (example - the Blue-winged Teal) required no further review by the committee. The remaining 260 species fell into one of the following categories: regionally rare or extralimital to its historical boundary; historically rare to the province; new to the province (species formerly did not appear on the species list or was recorded for the first time breeding); or species considered easy to confuse with another (eg. Common and Forster's Tern).

As a result of time constraints, not all of the sightings falling outside historical ranges were reviewed by the committee. All probable migrant species sightings or new to the province records were reviewed. Other records were reviewed in the following manner. After reviewing historical boundaries of each species taken from Godfrey (1986) and Salt and Salt (1976), the committee drew in new boundaries where logical extensions occurred, incorporating information gained from local sources. In many cases, these new range boundaries were acknowledged by amateur and professional birders but not yet published or documented widely.

All breeding records outside the new boundaries were reviewed by the committee regardless of whether or not documentation forms were received. (If no forms were received from the observer, the committee rejected the records.) Non-breeding observations (**x**) were not reviewed by this committee unless the sighting was of a rare accidental or new species (eg. Northern Cardinal).

The committee did not require "hard evidence" (a photo, carcass, or egg shells) for any record to be acceptable, but rather followed the European style of reviewing reports of written descriptions. Each record required approval of both species identification and breeding code. In the first round of review, a record required complete approval to be accepted. If a minimum of one vote dissented, a second round was entered. In the second round, the record was circulated along with comments received from members in the first round. If the second round did not reach unanimity, review went to a third round unless a majority voted against the record, in which case it was rejected. In the third round, the vote had to have 75% support for acceptance. First time breeding records for Alberta required unanimous acceptance of identification and a breeding status of at least probable or confirmed.

When reviewing species maps, one should keep in mind that all sightings within historical range, though equally subject to identification and breeding status errors, have not been reviewed by the Records Confirmation Committee. All records rejected by the committee, though pulled from the maps, have been retained in a file for future reference

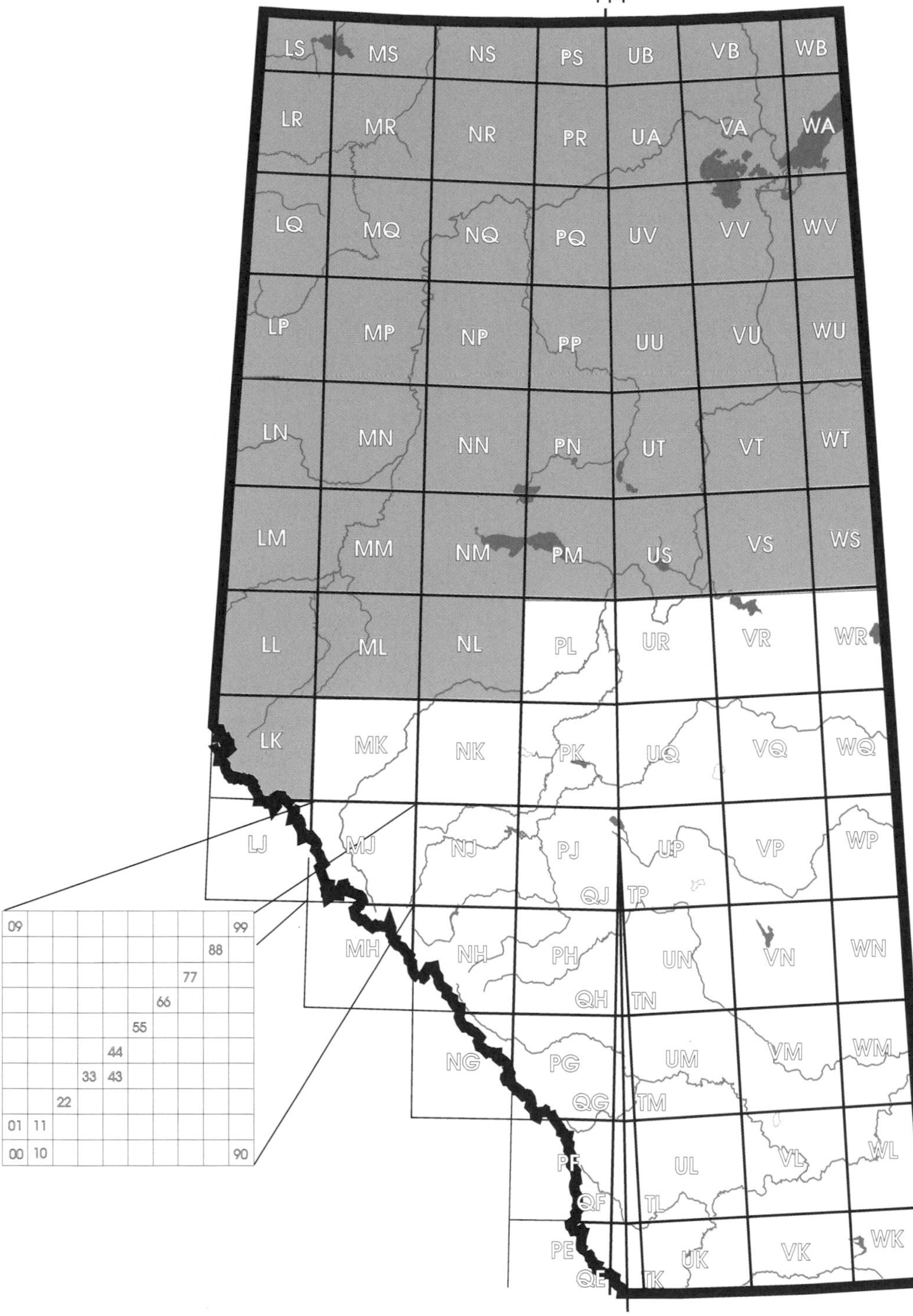

(shaded region denotes remote areas)

Fig. 2 Square Designation

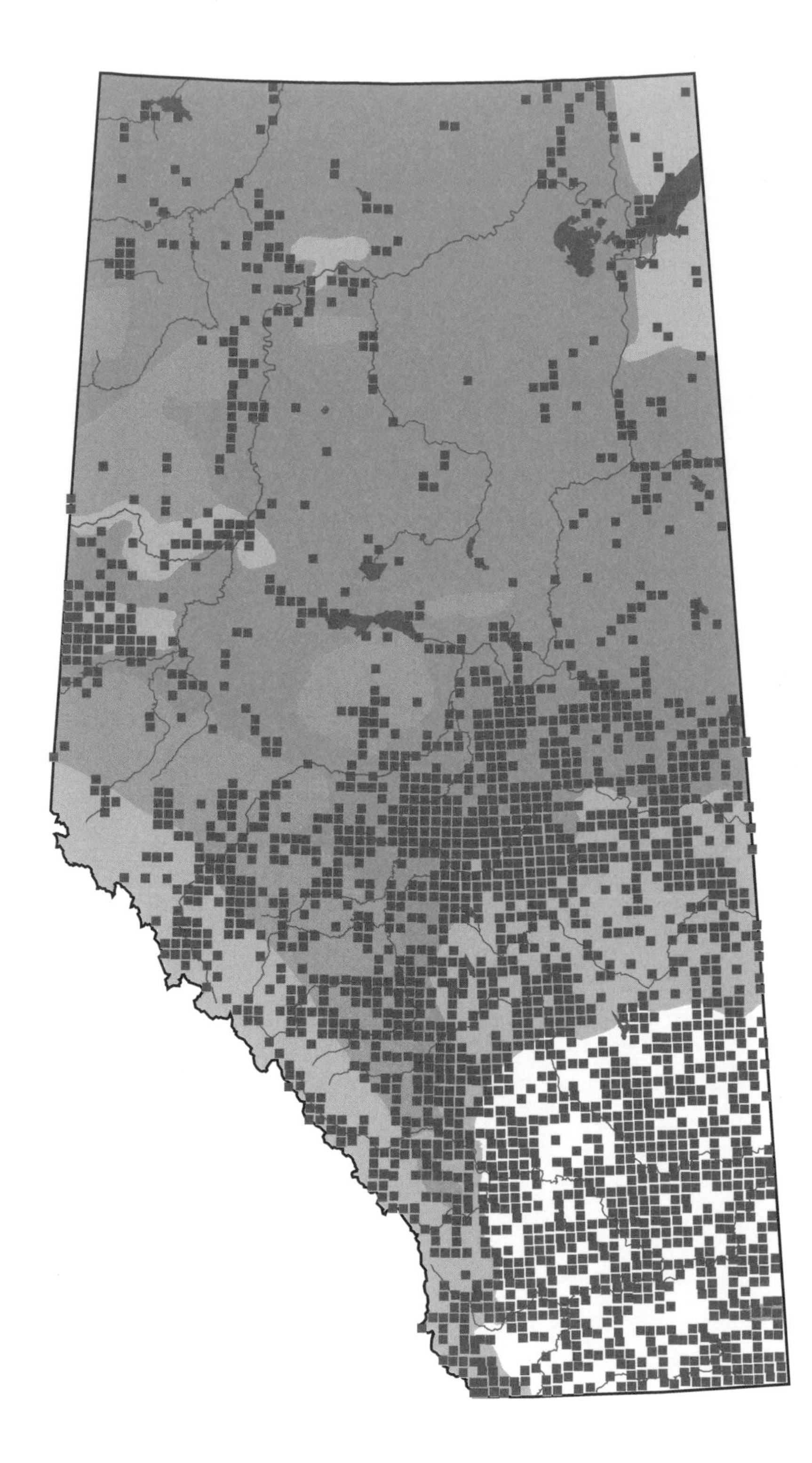

Fig. 3 Survey Coverage Map

Atlas Highlights

The five field seasons of the Alberta Bird Atlas Project provided a new view of birds in Alberta, with some interesting results. There were a total of 122,400 breeding records submitted by 943 atlassers. This represented over 40,000 hours of survey effort.

Some species were known to have bred in Alberta previously, but were not recorded by the Atlas Project.

- The Greater Prairie Chicken was classified as extirpated prior to the inception of the Atlas Project and its absence in the Atlas data confirms this evaluation.
- Lewis' Woodpecker was known to breed in Alberta but was last recorded in the 1940s and was also not recorded in the Atlas Project.
- Godfrey (1986) suggested that the Red-headed Woodpecker may breed in south eastern Alberta in the Cypress Hills, but no records were found to confirm this.
- Nashville Warbler was not recorded as a confirmed breeder during the Project, but confirmed breeding of this species was recorded previously at Link Creek.
- The Gray-cheeked Thrush was not recorded during the Project, but confirmed breeding of this species was recorded previously, in both Jasper and the Caribou Mountains.

Other species not previously confirmed as breeders in the province were confirmed as breeders during the Atlas surveys.

- Alberta's first confirmed breeding record of the Sage Thrasher was documented in 1988 south of Medicine Hat (O'Shea 1991).
- Although confirmed breeding of the House Finch has not yet been obtained in Alberta, this species is currently experiencing a North America-wide range expansion. There are records of this bird in summer in the province. Birders in the area of Crowsnest Pass, south to Cardston and east to Lethbridge are encouraged to confirm all finch sightings in order to document the spread of the House Finch into Alberta.

During the Atlas Project, many species were found to have expanded their range from the historically documented ranges. This information was confirmed through the rigorous work of the Records Confirmation Committee.

- The Black-throated Green Warbler expanded south over the Athabasca River into the Foothills.
- The Calliope Hummingbird was confirmed breeding in the upper Wapiti River. It had previously only been recorded as far north as Jasper.
- The Caspian Tern, historically recorded in the extreme northeastern corner of Wood Buffalo National Park, was found breeding around Vauxhall in southern Alberta.
- The Common Tern, which had not been recorded in the extreme northwest corner of the province, was recorded at Bistcho Lake during the Project.
- American Coots were found breeding in Banff and Jasper National Parks.
- Salt and Salt (1976) stated that the Eastern Phoebe was not found in the Rockies; it was recorded in 3% of the surveyed squares in that region. Furthermore, this bird was found in 18% of surveyed Foothills squares.
- Forster's Tern, which had previously been recorded only as far north as Edmonton, was found in an area that included Lac La Biche and Cold Lake, and a probable sighting was accepted at Winagami Lake near Fairview.
- The Great Crested Flycatcher was found to be establishing itself in east central Alberta, with extralimital records in the Canadian Shield Natural Region.
- The Magnolia Warbler was found south to Red Deer and Sundre; it had been previously recorded only as far south as the North Saskatchewan River and in the mountains to Banff.
- The Pied-billed Grebe was found nesting in the Jasper area.
- Sandhill Cranes were observed to be extending their range south through Rocky Mountain House to the Bottrell area in 1976 by Salt and Salt, as a reclamation of their original range. The Atlas data confirmed this movement.
- The Yellow Rail was found to be further south than previously recorded.

Analysis of the database has provided additional information generated by Atlas surveys. Table 2 lists those squares where the highest number of breeding species were recorded. As expected, these squares were in the central part of the province, and represent the Parkland and Boreal Forest natural regions. Table 3 lists those squares where highest number of species were recorded. This represents both breeding records and observational data recorded during survey efforts. Table 4 is a breakdown of Atlas coverage by natural region. The coverage in the Boreal Forest and Canadian Shield regions reflects the remoteness of these areas. The coverage obtained in these regions required a major logistical effort in the final year of surveys. Table 5 lists the species recorded breeding in the greatest number of squares in the province.

Figure 4 represents the number of species recorded per surveyed square. The first level, 0-25 species, is included to highlight those squares that contained only casual reports or that received inadequate coverage. The second level, 26-50 species, was selected to portray good coverage of squares inthe Grassland Natural Region or incomplete coverage in any other natural region. The level containing from 51-100 species included those squares that received comprehensive coverage in any given natural region. Squares having over 100 species recorded contained a wide diversity of species and habitats. However, direct correlations between the values included on the map and expected numbers should be avoided. A number of important variables lead to these values, such as the experience of the birder, accessibility to all areas of the square, habitat diversity within the square, the remoteness of the square, the time in the breeding season, and daily weather conditions. Despite these points, it does effectively depict extreme values. On the low end, inadequate coverage, except in alpine regions, is undoubtedly the cause. High values do highlight areas which obtained excellent coverage and likely are ecologically significant to avian activities.

Squares With the Highest Number of Breeding Species			
Block	**Square**	**# Species**	**Location**
VR	63	139	Fork Lake
UQ	62	136	Cooking Lake
PK	65	135	Lac Ste. Anne
WR	53	133	Cold Lake
UQ	91	132	Beaverhill Lake
NM	26	129	Winagami Lake Provincial Park
PH	60	122	Water Valley
UQ	84	120	Elk Island National Park
WR	44	120	Cold Lake
VR	37	120	Sir Winston Churchill Provincial Park

Table 2

Squares With the Highest Number of Species Recorded (Including Observations)			
Block	**Square**	**# Species**	**Location**
UQ	91	227	Beaverhill Lake
VL	22	212	Old Man River (north of Taber Lake)
VL	12	199	Taber Provincial Park
WR	53	197	Cold Lake
UQ	62	190	Cooking Lake
WL	24	172	Bullshead (south of Medicine Hat)
UP	51	167	Spotted Lake
VL	21	167	Taber Lake
VT	78	161	Fort McMurray
PK	65	160	Lac Ste. Anne

Table 3

Coverage by Natural Region			
Natural Region	**# of Squares**	**# of Squares Surveyed**	**% of Coverage**
Grassland	907	633	69.79
Parkland	765	426	55.69
Foothills	854	244	28.57
Boreal Forest	3431	687	20.02
Rocky Mountain	483	200	41.41
Canadian Shield	183	16	8.74
TOTALS	6623	2206	33.32

Table 4

Species Breeding in the Greatest Number of Squares	
Species	**# of Squares**
American Robin (*Turdus migratorius*)	1159
Barn Swallow (*Hirundo rustica*)	1090
Red-winged Blackbird (*Dolichonyx oryzivorus*)	1081
Mallard (*Anas platyrhynchos*)	1025
Killdeer (*Charadrius vociferus*)	919
Clay-colored Sparrow (*Spizella pallida*)	913
American Crow (*Corvus brachyrhynchos*)	884
Yellow Warbler (*Dendroica petechia*)	874
Red-tailed Hawk (*Buteo jamaicensis*)	874
Chipping Sparrow (*Spizella passerina*)	868

Table 5

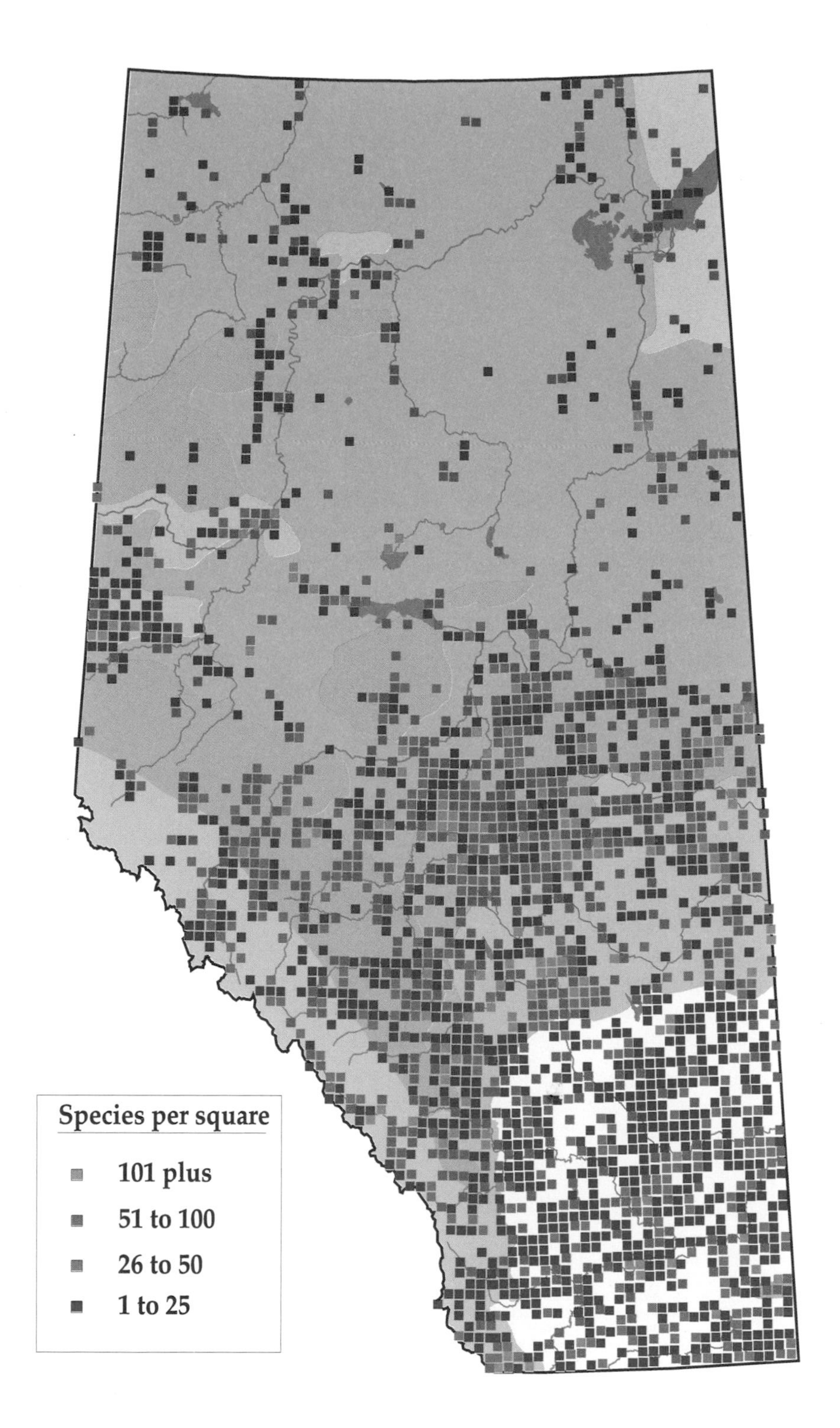

Fig. 4 Number of Species per Surveyed Square

How To Use This Book

As mandated by the Alberta Bird Atlas Project Management Committee, *The Atlas of Breeding Birds of Alberta* is a written account of the surveys conducted during the five year period of the Atlas Project. It is intended to be a popular representation of the database that was created through the cooperative efforts of volunteers, supporting agencies, and government. Targeting a wide range of perspectives ranging from the backyard birder and avid naturalist to the professional biologist and policy maker, this publication is intended to support and stimulate all interests. The main feature of the book, the species account, has three main components: a photograph, text, and a distribution map. Each photograph is an attractive visual representation of the species as it appears in Alberta. The text, which includes the subjects of status, distribution, habitat, nesting, and remarks, provides a background of the species' life history within the province. This consists of a description of where the bird nests locally and within the province, its breeding season and migratory activity, status, and conservation. The distribution map plots the raw data collected during the Atlas Project on a backdrop of the six natural regions of Alberta: The Boreal Forest, Canadian Shield, Parkland, Foothills, Rocky Mountain, and Grassland regions. This publcation is the summary of a five year "snapshot" of avian history in the province, as well as including baseline information against which future efforts can be compared and conclusions more soundly drawn.

The Atlas begins with the Foreword, written by Dr. W. Bruce McGillivray, chairman of the Alberta Bird Atlas Project Management Committee. His experience, direction, and dedication to the Atlas Project, shared by other members of the Management Committee, have been instrumental in the successful completion of the Project. In the Foreword, the original purpose for this effort is stated, as is a background to its execution and an acknowledgment of the financial, professional, and technical contributors and the volunteers.

The second section is dedicated to the supporting agencies of the Atlas Project, The Provincial Museum of Alberta, The Canadian Wildlife Service, the Recreation, Parks and Wildlife Foundation, and the Alberta Department of Forestry, Lands and Wildlife. A description of each agency, its mandate and activities, and its general contribution to the Project is outlined.

The Acknowledgments section is dedicated to the innumerable volunteers, organizations, and corporations which assisted in this cooperative venture. Although this list is intended to be comprehensive, it may be possible that, over the course of six years and a number of different staff and coordinators, an oversight has been made. In such a case, please accept our sincerest apologies and thanks.

Beginning on page nine, Dr. W. Bruce McGillivray chronicles the history of ornithology in Alberta. Initially, ornithology was a work of almost secondary importance to explorers and traders of commerce, opening up the west and foretelling of the incredible impact European man would eventually have on the landscape. Before this time, and largely unavailable to us, was the role of birds in the lives and folklore of aboriginal peoples. True scientific study began with Blakiston in 1857, heralding a new era of bird study by naturalists, geologists, botanists, oologists, and various museum groups. Dr. McGillivray goes on to trace the evolution of Albertan ornithologists, including Rowan, Lister, and Höhn. This period peaked with the landmark provincial publication, *The Birds of Alberta*, (Salt and Wilk 1958). Since that time, professional research has largely been carried out by universities, museums, and government wildlife agencies. The passage ends where this publication begins, with the ongoing change in the roles of volunteer contributors. While providing a link to ornithological study in the past, Dr. McGillivray has succinctly placed in perspective this publication and how a new era is poised to herald new results.

A very important subject that is essential to using and understanding the information given for each species is Biogeography. Displayed in each distribution map, the six natural regions are discussed individually in this section. Within the discussion of each natural region, three topics are covered: the sections of each particular natural region, each region's climate, and its vegetation. In the first topic, the location and general topography of the natural region is introduced, including a brief discussion of each section within the natural region. Next, the general climate of each region is described, as well as how it may be influenced by topography. In the vegetation discussion, the dominant macrovegetations are listed, and moisture conditions are examined. In trying to understand the presence or absence of a species in a particular region, it is important that the observer be able to correlate a particular biogeography, and habitat, to the life history of that particular species.

In describing the scope of the Project and the manner in which it was carried out, the Methods section details the field and office procedures implemented to achieve the final result. The Universal Transverse Mercator (UTM) grid system was used to divide the province into atlassing squares and administrative blocks (Figures 1 and 2). This approach was adopted because of its systematic progression and its standard use by other atlas projects across North America. Coverage goals were established in relation to volunteer activity, priority squares, and remoteness of areas. As a result of the concentration of populations in only a few areas of the province, the strategy was amended in the final field season to emphasize coverage in remote areas. Figure 3 is a Survey Coverage Map that details the squares which obtained at least some measure of atlassing. Standardized data cards (page 340 and 341) were used to promote the recording of all information in a systematic manner, for both comparative and handling purposes. Table 1 details the levels of breeding evidence used for the Project. This relatively standard approach recognizes the fact that a species may be breeding in the area but, because of time and place, the observer was unable to confirm it. Because atlassing hours per square was not consistently noted, such statistics are not analyzed in the publication. The sub-section entitled Data Processing and Verification describes the computer hardware and software used to verify records, create the database, and produce the distribution maps for the Atlas. Records confirmation was an arduous task of investigating the reports of species that were either new to the province, historically rare in the province, extralimital, or easily confused with another species. The Alberta Bird Atlas Records Confirmation Committee (ABARCC) studied the available information and validated results that would be accepted by the ornithological community. Dedicated cooperation was required of atlassers to substantiate these reports. In cases where the Committee was unable to elicit a timely response, the recorded was not tabulated or included in the data set used to generate the distribution map. However, these records remain in the database, flagged for further investigation.

The final subject in the preliminary part of the Atlas is the Atlas Highlights. Selecting from the entire section of Species Accounts, this summary briefly discusses those species for which an apparent change in provincial range has been documented by the Atlas Project, and outlines other interesting results of the survey effort. In the Highlights section, Figure 4 details, on a provincial map, the number of species recorded per square.

Preceding each of the member families is an Order page, which lists the families, common names, and groups which belong to that Order. To better understand how these families and species are related, similarities, and sometimes differences, in habitat use, anatomy, physiology, and behavior are described. Each Order page also includes an attractive line drawing of a member of the Order.

The member species of the appropriate Order, arranged by families, follow the order pages. Each page individually discusses a species, its occurrence, life history, and health in the province. The Species Accounts cover page closely details the structure followed for each account.

Following the Species Accounts are the Appendices. A discussion of migrants includes the 32 most commonly seen and visible throughout the province. A short paragraph on each includes their occurrence in the province, their habitat use here and on the breeding grounds, and their breeding and winter ranges. An appendix of Winter Visitors is handled in the same manner. Figure 5, a map of Canada which names the various districts and major Arctic islands of northern Canada, permits the reader to relate to geographical references in these appendices.

The Birds of Alberta Species List is an updated account, based on the Atlas, of the species which breed in Alberta. It, too, is in taxonomic order and includes any name changes in compliance with the American Ornithologists' Union Checklist.

The example provided of the Field Data Card is an exact reprint of that used by atlassers in the field. Produced before the first field season, it included all species that had recently bred in the province, as well as a few that were rare.

On page 342 there is a line drawing of a sparrow-like bird, equipped with a widely diverse plumage. The extensive markings are included in order to describe all possible parts of a bird.

The enclosed Glossary includes many of the more specialized terms that appear in the Atlas and are of an ornithological nature.

An appendix called Future Records is included to provide direction to those who would like to continue recording data on bird populations in Alberta. It is important that a common approach be used in handling these records in order that they contribute to the maintenance and update of the database.

The Personal Communications appendix includes all individuals who are cited in the Atlas as providing unpublished knowledge of given species to the Atlas.

The extensive Bibliography includes all the references reviewed during the production of the Atlas, many of which were cited. It includes a General Reference guide, as well as a reference guide that is taxonomically ordered.

The Atlas concludes with three indexes of bird names in alphabetical order, one each in English, French, and Latin.

The Species Accounts

The Species Accounts section progresses taxonomically, by family and by species. It includes 270 species that have historically bred and are still considered to breed, or have been confirmed breeding in the province. Each species account begins with the species' common name and latin (or scientific) name. Below this is a color photograph of the species. The majority of photos come from a comprehensive collection of photo submissions from various Alberta photographers, and they were graciously donated for inclusion in the Atlas. For ease of identification in the field, and unless otherwise indicated, males in breeding plumage were selected. Below each species photograph, the American Ornithologists' Union (AOU) species code is given, as is the name of the photographer. Any photographs that do not have a photo credit are courtesy of the Edgar T. Jones collection at the Provincial Museum of Alberta.

The text begins with a section on Status. This section contains a statement on the health of the provincial population and conservation measures undertaken. Because of the number of species in Alberta and in Canada, detailed analysis of every species' status has not been possible. In trying to determine the status of each species, a variety of sources were used, including the Committee on the Status of Endangered Wildlife in Canada (COSEWIC) and the National Audubon Society's Blue List (1986), as well as these local references: Salt and Salt (1976), Holroyd and Van Tighem (1983), Pinel et al. (1991 and in prep.), and the Wildlife Management Branch (1991). Where more accurate information was not available elsewhere, the Wildlife Management Branch Report (1991) was selected as the most recent determination of a species' status in Alberta.

In the Distribution section, the place names for areas that mark an expanded or reduced provincial range are given, based on an historical distribution as provided by Salt and Salt (1976), Godfrey (1986), and Pinel (1991 and in prep.). Where mentioned, northern Alberta refers to the area generally north of Athabasca to the northern border. The area south of Red Deer to the United States border is referred to as southern Alberta. The area between Athabasca and Red Deer is generally referred to as central Alberta. Reference is often made to the percentage of squares in a given natural region for which at least one record was obtained. This percentage is always in respect to the number of squares surveyed in that region. As used in the text, "records" is defined as any sighting in which the species exhibited breeding level activity that was recorded as possible, probable, or confirmed. It does not include observations.

In the field, an understanding of the habitat use of a species by the observer is vital in forecasting its presence. The Habitat section describes the preferred breeding areas for that species in Alberta and, when available, the areas used during migration. Where available, local references are used. If site specific information was not available for the province, then, where applicable to Alberta, generic habitat information from Godfrey (1986) was included.

Essential to any breeding bird atlas is a discussion on nesting activities. The Nesting section includes: nest site and construction, the number and color of eggs, incubation time and by whom, fledging time (first flight), distraction displays, number of broods, parasitism, and any unique nesting features particular to an individual species.

The Remarks section includes all other pertinent points about the species. Physical appearance is described, briefly highlighting differences between sexes, and emphasizing distinguishing features and differences from similar species in Alberta. If used, songs and calls are included for some species, as are the foraging habits and food of the species. The Remarks section includes spring and fall migration dates and records of overwintering of normally migratory species.

The Distribution maps plot the survey data generated by the Atlas Project. They include each of the natural regions, as discussed in the Biogeography section, as well as major lakes and rivers. As described on the inside back dustjacket of this book, there are three symbols on the maps representing breeding level activity: possible, probable, and confirmed. The observation data generated during the surveys is also included, reporting presence of the species during the breeding season. In the case of partial squares along provincial borders, if greater than 50% of the square existed then the square stands alone and is represented as an individual unit. Where less than 50% of the square exists, the data is absorbed and depicted by the immediately adjacent, inward square. Interpretation of the apparent distribution as depicted on the maps should be tempered by incomplete coverage of the province. Generally, remote areas and many of the squares in northern Alberta are not represented. Thus, in these squares, if the species is present in similar habitat elsewhere within the region, it may be present in the square in question, given that the square was not surveyed adequately, or at all.

Order Gaviiformes

Although they are members of the avian community, loons share adaptations which assist in their mostly aquatic lifestyle. Their very short legs and webbed feet (first, second, and third toes) are situated near the very rear of their bodies, permitting flipper-like efficiency underwater as they swim or dive for prey. A strong, pointed bill, short stiff tail, and small pointed wings help create aquadynamic streamlining. Bones with marrow, heavier than the air-filled bones of other birds, allow these birds to sink more easily. Dives normally last less than a minute, although they have been recorded greater than five minutes. However, these aquatic refinements have caused the loons to be very awkward on land, where they can only push themselves, and costly in flight, which is made difficult by high wing loading. Land activites are restricted to copulation and nesting, both at the water's edge. These species only become airborne after a difficult long and pattering take-off. Locally seen singly or in pairs, the sexes share a similar dense and watertight plumage. Normally sitting low in the water, loons share with the grebes the ability to partially submerge themselves when sufficiently wary. Once alarmed, they will dive to elude detection.

Red-throated Loon

Gavia stellata

RTLO photo credit: G. Beyersbergen

Status:
The Red-throated Loon appeared on the North American Blue List of threatened birds from 1973 to 1976. It is a very rare breeder in Alberta, where it is at the periphery of its range.

Distribution:
In North America, this species breeds in arctic and subarctic regions from Ellesmere Island south to about the 60th parallel. The only authenticated breeding record in Alberta is from the Margaret Lake area in the Caribou Mountains (Höhn 1976). The only Atlas data for this species over the five year survey period was one possible breeding record from the Margaret Lake area. There were scattered observations in the central part of the province. This loon is most often observed during fall, when birds have been sighted as far south as Waterton Lakes National Park.

Habitat:
Red-throated Loons frequent freshwater ponds and lakes in both tundra and subarctic forested areas. Those waterbodies chosen for nesting are generally smaller and shallower than those used by the Pacific Loon, with which it shares a similar range. They nest on small shallow waterbodies and feed in marine waters on very large lakes, such as Great Slave Lake (D.L. Dickson pers. comm.).

Nesting:
The nest of this loon is found on the shoreline of a pond or small marshy islet. In dry areas, the nest is a hollowed area lined with small amounts of aquatic vegetation. When nesting in shallow water, birds build up a platform of sedges and other vegetation.
The female lays 1-2 eggs (usually two), glossy olive-buff or, sometimes, greenish or dark brown, with dark spots. At Margaret Lake, one nest contained an egg late in June, another was still empty on July 1. Incubation is 24-29 days, by both sexes, but mainly by the female. Young are tended by both parents, with fledging at 46-50 days. Red-throated Loons raise one brood per season.
This species shows general territorial fidelity. It usually returns to the same teritory to nest, but may not use the same nest site. Pairs probably mate for life, but evidence is very slim, due to lack of banded birds (D.L. Dickson pers. comm.).

Remarks:
A chestnut-red throat patch, gray head with striping on the hind neck, and a plain, uncheckered back characterize the summer plumage of this species. A slender, up-turned bill is also diagnostic. This is the smallest of the loons, and its take-off is quicker and less labored than that of other loons, enabling it to inhabit quite small waterbodies. The Red-throated Loon is the only loon which can take flight directly from land.
Like all loons, the Red-throated Loon is superb in the water, where it feeds on fish and aquatic invertebrates. It catches its prey by diving, then rapidly pursuing prey underwater. Dives of this species are shallower and shorter than other loons. Unique to its family, the Red-throated Loon does not require a food supply in the nesting pond. The birds may make foraging flights of up to several kilometres to sources of prey fish.
The Red-throated Loon is the most gregarious of loons, being sociable in all seasons. A few nesting pairs may gather together on one large lake to feed, and, in winter, birds will congregate on open water in flocks of considerable size.
Calls include a loud pulsing coo, a prolonged wail, a rapid, crow-like guttural bark, and a low moan.
The fall migration through Alberta may occur over an extended period of time (Pinel et al. 1991). There is one overwintering record from the Banff area.

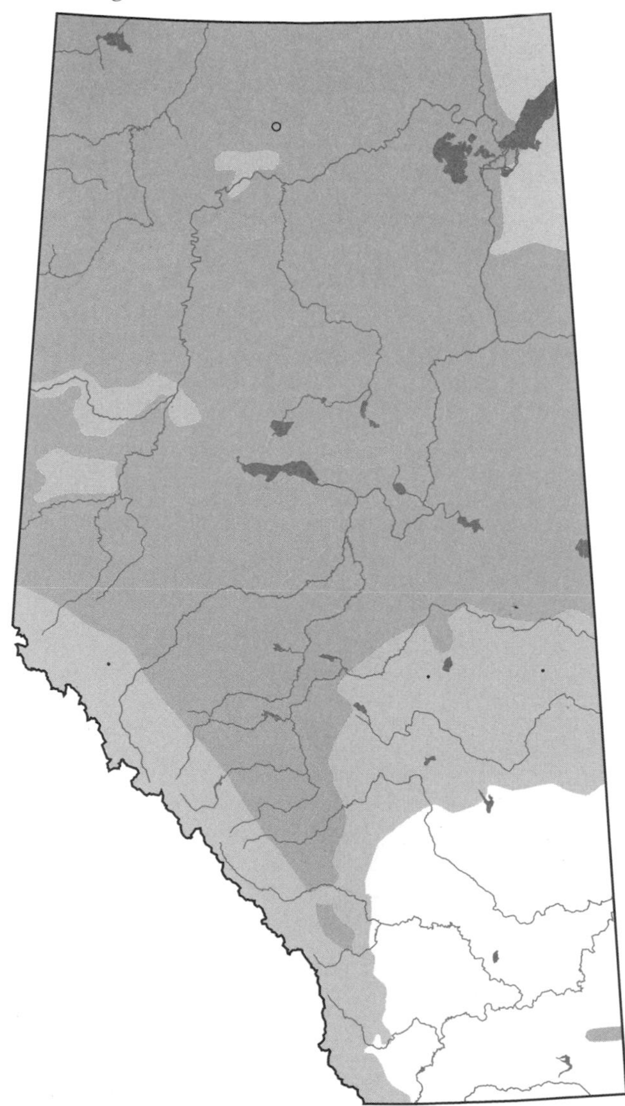

Pacific Loon

Gavia pacifica

PALO

Status:

This former subspecies of the Arctic Loon (*Gavia arctica*) was accorded full species status in 1985 (American Ornithologists' Union 1985). There are no population estimates available for this species for North America but, like other loons, it is relatively vulnerable to oil pollution on its migration and wintering areas along the west coast (Johnsgard 1987). It is a rare breeder in Alberta, where it is at the southern periphery of its range.

Distribution:

There are two documented breeding records for the Pacific Loon in Alberta, both from the Canadian Shield region in the province's northeast corner (Höhn 1972; Kuyt and Goossen 1986). A pair of adults with one downy young were observed in early July at Leland Lakes (Höhn 1972), and one nest that contained two eggs was observed in early June at Four Mile Lake (Kuyt and Goossen 1986). There were no breeding records for this species over the five year survey period of the Atlas Project.

Habitat:

The Pacific Loon breeds on deeper and larger lakes in tundra and Boreal Forest and, more rarely, on smaller waterbodies. While not gregarious in the breeding season, it congregates in large numbers off the west coast during the winter.

Nesting:

These loons are highly territorial and intolerant of conspecifics during the breeding season (Johnsgard 1987). Nesting is on a small island, sometimes in wet grassy areas on the lake shore, very close to the water's edge. The nest may vary from a simple scrape in flat ground to a raised mass of dead vegetation. The nest found at Four Mile Lake was a soggy mat of green vegetation with a base of moss. It was situated on a low flat nest island measuring 9 m by 4.5 m (Kuyt and Goossen 1986). They use the previous year's vegetation to build the platform (D.L. Dickson pers. comm.). The clutch consists of 1-2 eggs, which are greenish olive to brown and have black spots. Both adults incubate the eggs for a 28-30 day period, and both also care for the young. The hatchlings are led from the nest after 1-2 days and are brooded in various locations for about two weeks. The young are know to alternate swimming beside an adult and riding on its back. They fledge at approximately 60 days. The young are fed after fledging, sometimes for as long as 90 days (Johnsgard 1987). The Pacific Loon is single-brooded. Young loons stay in wintering coastal areas until they reach breeding age (two years). Pair bonding is for life.

Remarks:

This sleek handsome bird is smaller than the Common Loon and is easily distinguished from that species by its silvery gray crown, forehead, and back of the neck. The back is extensively spotted with white patterning, including two large areas of white on the wings and two smaller areas on the upper back (Johnsgard 1987). Its black underneck once earned it the name of Black-throated Diver.

Its calls include long, pulsed raven-like croaks, a prolonged mournful cat-like "meow" (particularly in early morning and dusk), and a dog-like yelp given just before it dives.

Like other loons, the Pacific Loon captures its prey through long sustained dives. Most dives average about 45 seconds, and only rarely do they exceed two minutes.. The maximum duration of the dive is approximately five minutes (Johnsgard 1987). It feeds on fish, crustaceans, molluscs, and aquatic plants and insects.

There is no information available on the migration dates of the Pacific Loon.

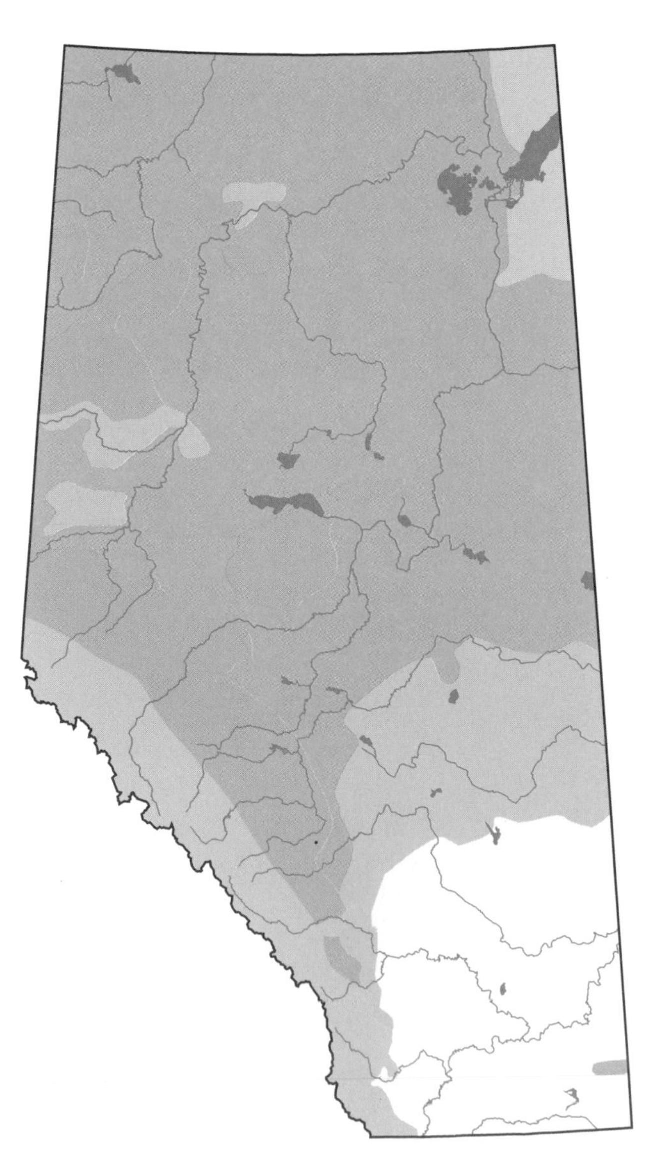

Common Loon

Gavia immer

COLO photo credit: G. Horne

Status:
This species is fairly common in the province, with the exception of the southern portion of the Parkland region and the Grassland region. It has disappeared from much of its southeastern Canadian range, because of loss of nesting habitat to campers and cottages, and its sensitivity to human disturbance (Wildlife Management Branch 1991).
In 1980 and 1981, the Common Loon was on the North American Blue List of threatened species. Though deleted from the list in 1982, it has been maintained as a species of "special concern", primarily because of the detrimental effects of recreational activity—particularly boating—on nesting success and because of the threat of acid rain in eastern provinces.

Distribution:
The Common Loon breeds in suitable habitat throughout Alberta. In the Grassland region, this loon is only very locally distributed, with nesting reported at Lake Newell (near Brooks) and in the Cypress Hills.

Habitat:
Optimum habitat for the Common Loon is a clear, open lake (or large river), where there is minimal shoreline development, little recreational activity, and an abundance of small fish. Loons are solitary nesters and, on smaller lakes, only one pair will be found. On larger lakes, pairs may lay claim to different bays, but always nest some distance from one another.

Nesting:
This species nests on the ground close to deep water, which allows a speedy underwater exit. The birds may seek out an island or a protected spot on a promontory or in a sheltered bay. Occasionally, they will nest on the remains of a muskrat house or on a mound of floating vegetation. Common Loons have a strong attachment to their accustomed territory and will return to the same nest site year after year.
The nest is usually a platform consisting of a heaped mass of aquatic vegetation, though eggs are sometimes laid on bare ground. The female typically lays two eggs, which are olive-green or olive-brown with splashes of black. Incubation is 29 days; both male and female take turns incubating the eggs and are very attentive to the nest. Parents brood newly hatched chicks until the down dries, at which time the chicks take to water. For a week or two, chicks will be seen riding on their parents' backs. After six weeks, the chicks can feed themselves. Fledging is at 70-80 days. Common Loons raise one brood per season.

Remarks:
The Common Loon is probably best known for its lonely echoing calls, synonymous, for many people, with the peacefulness of unspoiled wilderness. It can stay underwater for long periods, up to a minute while feeding, and longer when escaping danger. Like all loons, it can change its specific gravity, allowing it, when alarmed, to lie low in the water or to sink slowly out of sight. Because its legs are placed so far back along the body, the loon is very cumbersome on land and can take flight only from water, where it requires a long splashing run to get aloft.
Common Loons eat mainly fish, along with crustaceans, molluscs, frogs, and some vegetation. One pair with two chicks will consume over 1,000 kg of small, coarse fish over a breeding season.
Common Loons arrive on the breeding grounds early in May as soon as the ice has melted. They are already paired and it is likely that they mate for life. The loons winter at the seacoasts, leaving the province by late October. Individuals have been recorded overwintering at Calgary, Wabamun Lake, and Waterton Lakes.

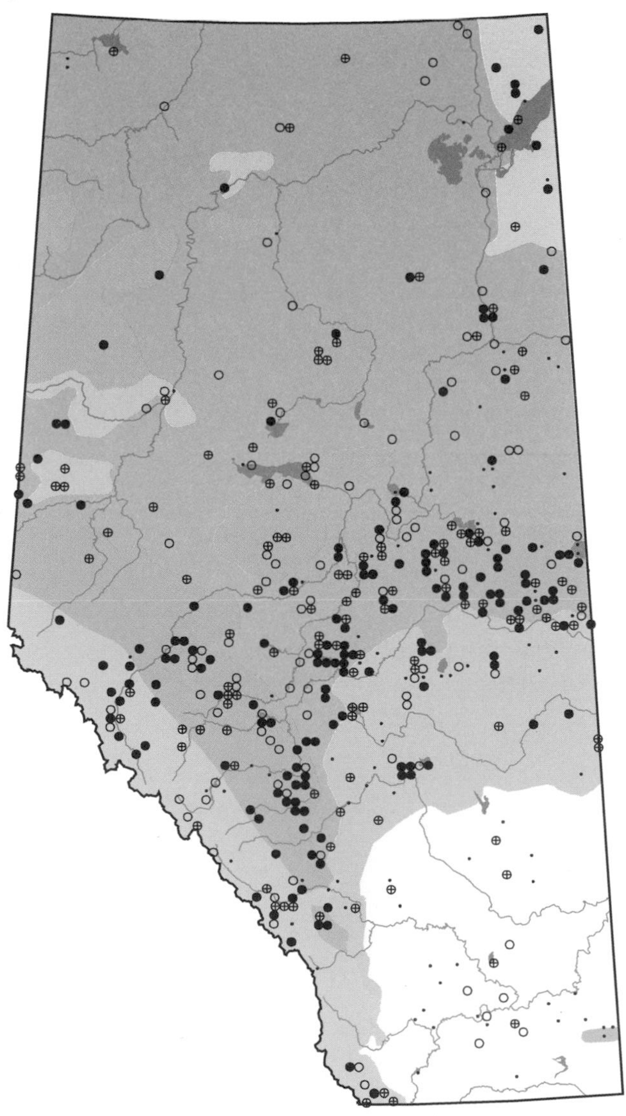

Order Podicipediformes

Despite being smaller in size, this highly aquatic order shares a similar anatomy, behavior, and physiology with the loons. These include short stubby wings, indistinct tails, rear-positioned legs, sharply pointed bills, the ability to partially submerge by expelling air from their plumage, strong diving ability, and limited flight beyond migration requirements. As well, most species are solitary, although they do carry out elaborate courtship displays. Take-offs are long and arduous, requiring the birds to "sprint" along the surface. The most obvious differences from the loons are the grebes' long necks, lateral tufts on the head, and lobed toes, which open during the power stroke and collapse on the recover, or forward, stroke. Sexes are alike. They also have the peculiar habit of swallowing feathers, which may help filter the skeletal parts of swallowed prey items before their expulsion as pellets.

Pied-billed Grebe

Podilymbus podiceps

PBGR photo credit: R. Gehlert

Status:
The Pied-billed Grebe is fairly widespread in Alberta and is the most common grebe in North America. This species' wariness and habitat adaptability have probably been major contributors to its success. However, it faces the same threats as other wetland species in Alberta, namely wetland clearing and drainage for agriculture.

Distribution:
This species breeds in suitable habitat throughout most of Alberta. It is found most commonly in the wetlands of the Parkland and Boreal Forest regions, where it was recorded in 17% and 14% of surveyed squares, respectively. It was also recorded in 8% of surveyed squares in the Foothills region, 7% in the Grassland and 6% in the Rocky Mountain region. Salt and Salt (1976) indicated the species was absent from Jasper National Park, but Atlas records show this grebe is now nesting in the Jasper area.

Habitat:
Preferred habitat of the Pied-billed Grebe is a pond, wetland or prairie slough with shoreline or islands dense with emergent growth. Occasionally, the bird will be found in the backwater of a river, shallow bay of a large lake, a slow-running stream, or large irrigation ditch. It is rarely seen in the open except in the morning and, even then, remains close to cover (Holroyd and Van Tighem 1983).

Nesting:
The Pied-billed Grebe is a solitary nester. The nest is a solid structure of rotting and green marsh vegetation, either floating and attached to stands of emergent vegetation or built-up from the shallow water bottom.
Eggs are bluish white or greenish white, soon nest-stained. Clutch size is 6-8 (Salt and Salt 1976). Incubation is by the female at first, then shared, then again by the female alone during hatching. Parents will leave the nest unattended for long periods, but leave the eggs covered.
After hatching, the young follow their parents and climb up on a parent's back to brood. Adults often dive with chicks aboard. Both sexes feed and care for the young. The fledging period for this species is unknown.

Remarks:
This small, shy grebe is brownish black or grayish black with a black throat patch. Its brown eyes and stubby chicken-like bill with black ring help distinguish it from others in its family. Pied-billed Grebe chicks are striped black and white, in contrast to their parents.
Like all grebes (and similar to the loons) the Pied-billed Grebe's legs are housed largely in the body skin and are located well back along the body. These characteristics make the bird clumsy on land and prevent take-off except from water. However, when on water, this grebe can change its specific gravity and sink without a trace, hence its nicknames, "hell-diver" and "water witch". Similar to other grebes, this species has lobed rather than webbed feet. These lobes flare out on the propulsive stroke of a dive, then fold back on the return stroke. Foraging dives are usually in shallow water and are of short duration.
Pied-billed Grebes eat mainly insects, fish, and crustaceans, but will occasionally feed on amphibians, molluscs, and aquatic plants. Also, chicks are first fed insects which are broken up for easier swallowing.
This species arrives in the province during the last week of April. While solitary and reclusive during the breeding season, in the fall birds congregate in flocks on larger lakes. Fall migration is prolonged, with the last birds lingering into late October and early November. This species has been observed overwintering at Wabamun Lake and Calgary.

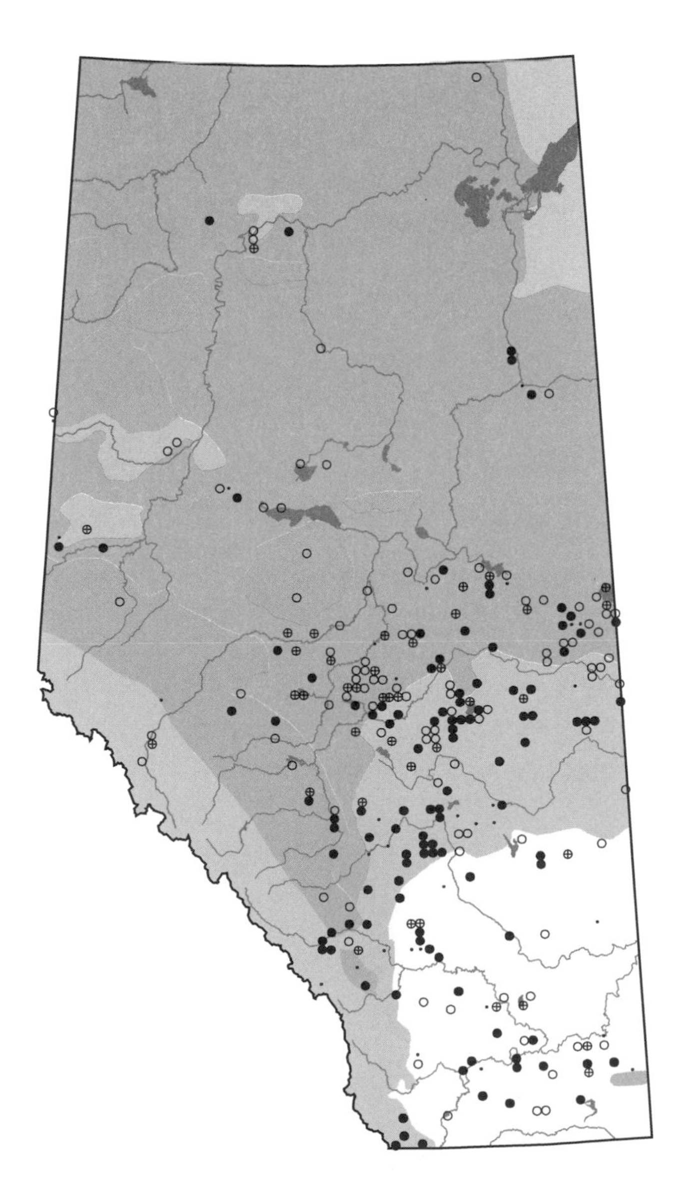

Horned Grebe

Podiceps auritus

HOGR photo credit: T. Webb

Status:
The Canadian populations of the Horned Grebe appear to be generally stable. It is, however, the grebe with perhaps the lowest population densities in Alberta (B. Rippin pers. comm.). In 1986, this species appeared for the first time on the North American Blue List, mainly because numbers of migrant and wintering birds were greatly down in the northeast (Tate 1986).

Distribution:
The Horned Grebe breeds throughout much of Alberta, but generally prefers the wetlands of the Parkland region, where it was recorded in 31% of surveyed squares. It was also recorded in 21% of Boreal Forest squares, though it is much more scarce and local in the northern part of this region (Pinel et al. 1991), and 10% in the Grassland region. The species is found locally in the Foothills (5% of surveyed squares) and the Rocky Mountain (3% of surveyed squares) regions. The only confirmed nesting record for the mountain parks is from the Waterton Lakes National Park area.

Habitat:
This grebe nests in both open and forested areas, preferring those ponds, sloughs, and lakes with extensive marshy vegetation. These water bodies are usually less than five ha in size. If ponds are semi-permanent and dry up, breeding pairs will move and nest again (Riske 1976).

Nesting:
The nest is built by both sexes and is a floating mass of decayed and fresh aquatic vegetation built up in shallow water and usually anchored to reeds.

The female lays 3-7 eggs, usually four (Riske 1976). The eggs are bluish white or greenish white, and are soon discolored. Incubation is 23-24 days, by both sexes, although, once the clutch is complete, the female spends appreciably more time on the nest than the male (Johnsgard 1987). This species is highly protective of the nest, and may rush at intruders, display, or hiss from the nest. Young leave the nest when they hatch and can swim and dive feebly almost immediately. They are tended by both parents. For several days, hatchlings will brood on the parents' backs or under their wings, and will sometimes be carried underwater when their parents dive. Little is known about this species' fledging period. One brood is raised per season.

The Horned Grebe is not colonial and usually only one pair is found on a small pond or several pairs may nest at widely separated points on a larger body of water (Salt and Salt 1976). However, when nesting sites are at a premium, or food is abundant, loose groups of nesting birds may occur on the same pond.

Remarks:
This attractive little grebe is easily identified by its broad, buffy ear tufts or "horns", glossy black crown and neck ruff, and chestnut neck. It is not a very gregarious bird, but is considered to be the least wary member of its family. Like other grebes, it does not move easily on land, and is also a weak flyer, so it rarely leaves the water, except for migration.

This species eats mainly fish, insects, and crustaceans, and usually forages in shallow water, along beds of emergent vegetation. It will dive for food, skim prey from the water's surface, and snatch insects from the air or overhanging plants (Johnsgard 1987). Like other grebes, it eats its own feathers, probably for roughage.

Spring arrival for this species is mid- to late April, with the peak of fall migration occurring in September (Pinel et al. 1991). A few stragglers will linger into mid-November. Horned Grebes have been observed overwintering at Lethbridge, Wabamun Lake, and Waterton Lakes.

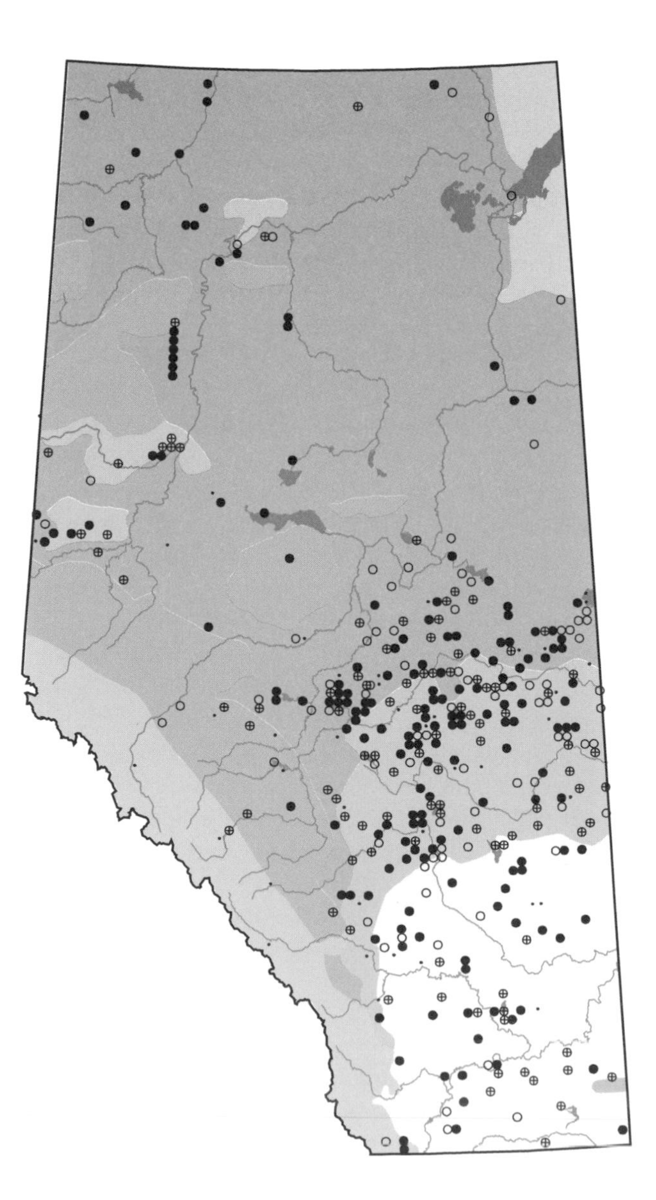

Red-necked Grebe

Podiceps grisegena

RNGR photo credit: T. Webb

Status:

Populations of Red-necked Grebes have suffered substantially in the last 100 years, from pothole drainage, land clearing, environmental contamination and human recreational activities on lakes. These activities continue to occur in Alberta's northern Parkland and southern Boreal Forest regions, where the majority of the province's Red-necked Grebes nest. The species is still reasonably common, and is probably Alberta's most common grebe. There are indications of a slow downward trend in parkland areas (Wildlife Management Branch 1991). Riske (1976) documented a major decline in Pine Lake, Alberta populations over the period 1971-1976, which was attributed to increased lake recreational activity and cottage development. About half of Canada's breeding adults of this species reside in Alberta (De Smet 1982).

Distribution:

Early observers found Red-necked Grebes in most prairie sloughs, as well as lakes of the northern forest country. The species continues to nest throughout Alberta, but is now found most abundantly in the southern Boreal Forest and Parkland regions, where water levels are stable. The species is scarce and nests only very locally in the Grassland, Foothills and Rocky Mountain regions. Low numbers recorded in the northern Boreal Forest region are more a reflection of survey effort in that area. Largest nesting concentrations occur near Edmonton, Calgary, and Lesser Slave Lake.

Habitat:

This species is found on small, shallow lakes greater than 2 ha in size, or in shallow protected areas and bays of larger lakes. Suitable lakes usually have extensive emergent and submergent vegetation. While birds are most often found on permanent water bodies, small semi-permanent lakes and potholes may also attract a nesting pair. This grebe needs about 50-60 m of open water for takeoff.

Nesting:

Nests are often built within or next to emergent vegetation and fastened to reeds in areas protected from wave action.
They are an accumulation of moist, decaying vegetation and they may float or be anchored to the lake bottom or other support with mud, sticks, and reeds. They may be supported by willow stumps, fallen trees, branches, beaver caches, and mudflats. Nests are located to provide birds with easy access to water of swimming depth. This grebe is primarily a solitary nester, but loose aggregations of nests occur on some larger lakes.
Red-necked Grebes lay 2-9 eggs, usually 3-6 (Riske 1976), light blue to chalky white, which are quickly stained brown. Incubation is 22-25 days, by both sexes. Unlike Horned Grebes, this species shows little hostility to intruders and, if disturbed, quietly retires from the scene. Male and female tend the young. Similar to other grebes, hatchlings ride on their parents' backs and may go along on a dive. Fledging is at 8-10 weeks. Red-necked Grebes are single-brooded.

Remarks:

The most common diving bird on most lakes in central Alberta, the Red-necked Grebe is easily recognized by its broad black crown, white cheek patches, and chestnut neck. It eats the same fare as other grebes; aquatic insects, small fish, crustaceans, and the occasional salamander and tadpole.
This large noisy grebe arrives on the breeding grounds in Alberta in the last two weeks of April. Grebes appear to migrate at night and rarely fly once they have reached their nesting destination. In September, Red-necked Grebes flock in considerable numbers on large lakes, often in the company of other grebes and loons. Most birds have left the province by mid-October or early November. Overwintering has been observed at Wabamun Lake.

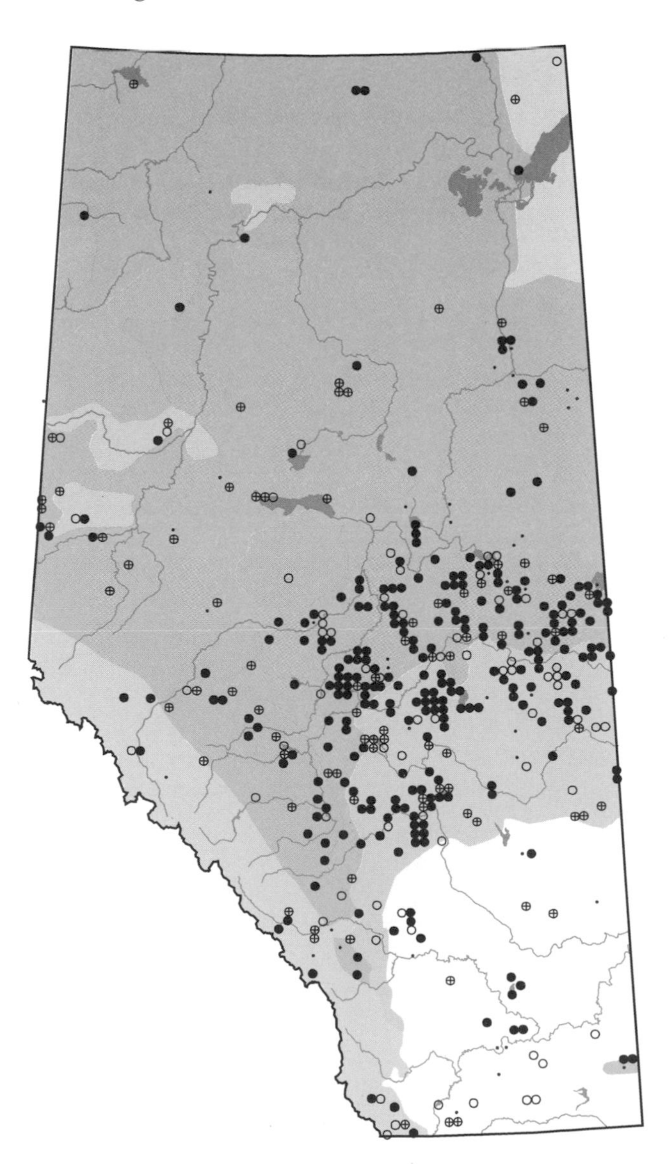

Eared Grebe

Podiceps nigricollis

EAGR photo credit: G. Beyersbergen

Status:

Although the Eared Grebe is fairly common throughout its range, its populations are thought to be in decline (Koonz and Rakowski 1985). This species is not tolerant of water level changes, so is particulary threatened by drainage of marshes and lakes. The population over large areas may fluctuate widely according to the availability of suitable nesting water (B. Rippin pers. comm.). It is also susceptible to herbicide and pesticide use, and human disturbance at the nest site.

There is no accurate estimate of the North American population. However, up to 750,000 Eared Grebes have been observed on Mono Lake, California, a major staging area for coastal wintering birds (Johnsgard 1987). The status of this species in Alberta is not well known, but Atlas records indicate it is relatively common in the Parkland region, where it was recorded in approximately 30% of surveyed squares.

Distribution:

According to Salt and Salt (1976), this species' breeding range coincides roughly with the Grassland and Parkland regions of the province, with evidence of a northward trend in recent years. The northern limit of the range appears to be southeast from Peace River, to Lesser Slave Lake and Lac La Biche, with one northern nesting record from the Fort Vermilion area.

Habitat:

Eared Grebes are found in shallow lakes, potholes and sloughs, usually larger than 4 ha, and with extensive areas of emergent vegetation.

Nesting:

This species is generally colonial, nesting in colonies of various sizes, in small groups or single pairs. Large lakes with rich feeding areas in open water and sheltered areas of emergent growth will support significant colonies. The largest colony in Alberta, at Rich Lake, has up to 10,000 nests in one season.

Nests are soggy mounds of rotting aquatic vegetation, either floating or built up from the bottom, and anchored to stems of emergents or to piles of submerged vegetation. Some nests are completely exposed over large areas, while others are located in sedge beds and are relatively hidden from lateral view.

Clutch size is 1-6 eggs, usually 2-3 (Riske 1976). Eggs are bluish or greenish white and quickly discolored. Incubation is 20-22 days, by both sexes. This species covers its eggs when it leaves the nest. After hatching, a brood may be divided between the parents who then tend to avoid one another. The fledging period of this species is not accurately known. One brood is raised per year.

Remarks:

This very social grebe is distinguished by a crested black head, yellow feathered "ears" and reddish brown sides. In the last century, it was killed by the thousands by plume hunters who sought its thick, satiny breast feathers. This "grebe fur" was used to decorate women's hats, capes, and muffs. The species is well known for its courtship antics which include a "penguin dance". In this display, the male and female face each other, paddle and stand upright with only their tails and feet in the water. They raise their crests, fluff their body and neck feathers, and stretch out their necks. They chitter excitedly, heads turning continuously and very quickly (Palmer 1962).

Insects and their larvae form a major part of the Eared Grebe's diet, along with small fish, tadpoles, frogs, and shrimp.

Eared Grebes arrive in Alberta in mid- to late April (Pinel et al. 1991) and linger as late as November. Overwintering has been reported at Waterton Lakes and Wabamun Lake.

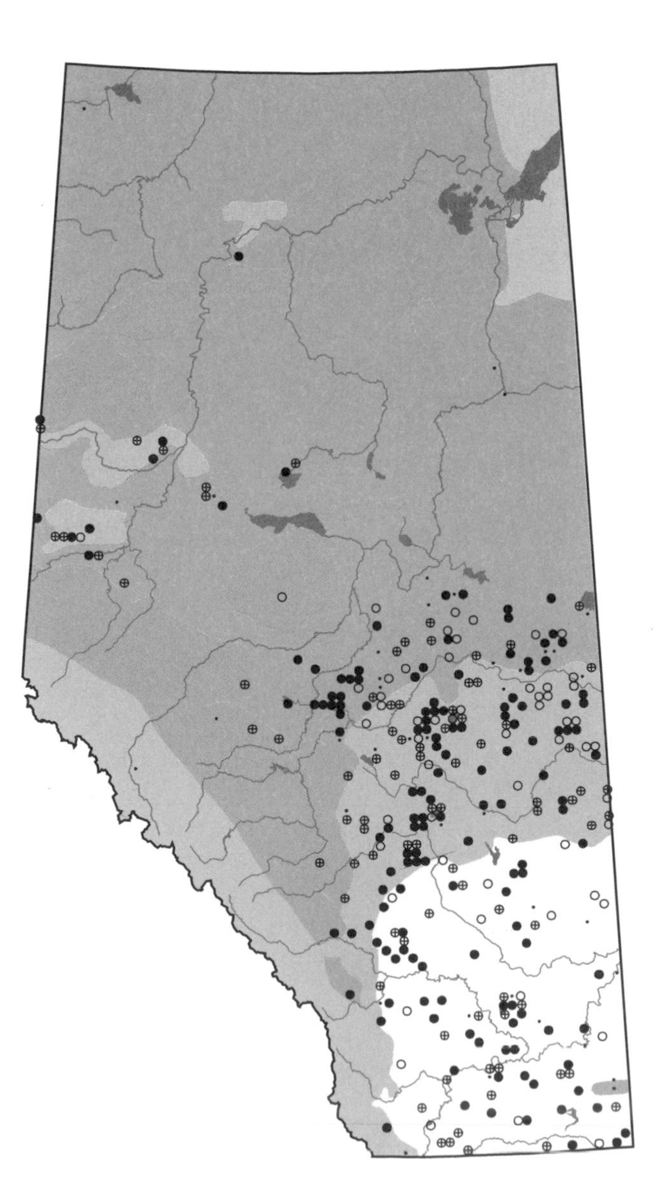

Western Grebe

Aechmophorus occidentalis

WEGR

Status:

The Western Grebe was on the North American Blue List from 1973-1986. However, by 1982 it was believed that the population was stabilizing at a reduced level. In 1986, the Western Grebe was delisted to a species of "special concern" (Tate 1986). In spite of this, Canadian populations are thought to be in decline (Koonz and Rakowski 1985). They were recorded by Atlassers in 8% of surveyed squares in the Boreal Forest and Parkland regions, 6% in the Grasslands, and 2% in the Foothills region.

Distribution:

Western Grebes are presently breeding in medium to large colonies on many of the fish-bearing lakes of Alberta's Boreal Forest and Parkland regions, as well as on larger prairie reservoirs (B. Rippin pers. comm.). Salt and Salt (1976) located the southern boundary of this species' range at Namaka Lake and Frank Lake. However, Atlas records show this species breeding south to the American border. It is possible that some of these southern sightings are Clark's Grebe, which were considered a light phase subspecies of the Western Grebe until 1985.

Habitat:

Western Grebes are colonial nesters, usually found on medium to large lakes. These grebes require access to open water deep enough for diving and with substantial fish populations. For nesting grebes, lakes need to have extensive stands of emergent vegetation in areas protected from wind and wave action.

Nesting:

Much like other grebes, the nest of this species is a mass of decaying vegetation, either floating or built up from the substrate, and anchored to emergent or submerged vegetation. Nests are usually well concealed in reeds. Some colonies have thousands of birds nesting in close proximity.

Clutch sizes vary according to colony. Riske (1976) reported clutches of 1-5 eggs, usually 3-4, at Lac Ste. Anne and Kristensen and Nordstrom (1984) reported clutches of 1-8, most commonly 3-5, at Cold Lake. Eggs are a dull bluish white and nest-stained.

Incubation is 23 days, by both parents. Young leave the nest when their down is dry and will brood on their parents' backs or under a wing. Fledging is at 49-51 days. After adult Western Grebes leave the nest with their young, they spend most of their time in deep water, far from shore (B. Rippin pers. comm.). Western Grebes are single-brooded.

Remarks:

This most gregarious of all the grebes is easily recognized by its sharply contrasting black and white coloration, thin green-yellow bill and long slender neck. Sometimes called the Swan Grebe, this graceful bird was once a major victim of plume hunters.

The diet of this grebe consists mainly of small fish and a few insects. As with other grebes, its gizzard contains a mat of feathers. Unique to the Western and the Clark's grebes, the male extensively feeds the female before nesting.

Western Grebes arrive in the last half of April and in early May. In September, the birds congregate on large lakes and most have departed the province by mid-November. In Alberta, overwintering of the Western Grebe has been recorded at Wabamun Lake, Waterton Lakes, and Lethbridge.

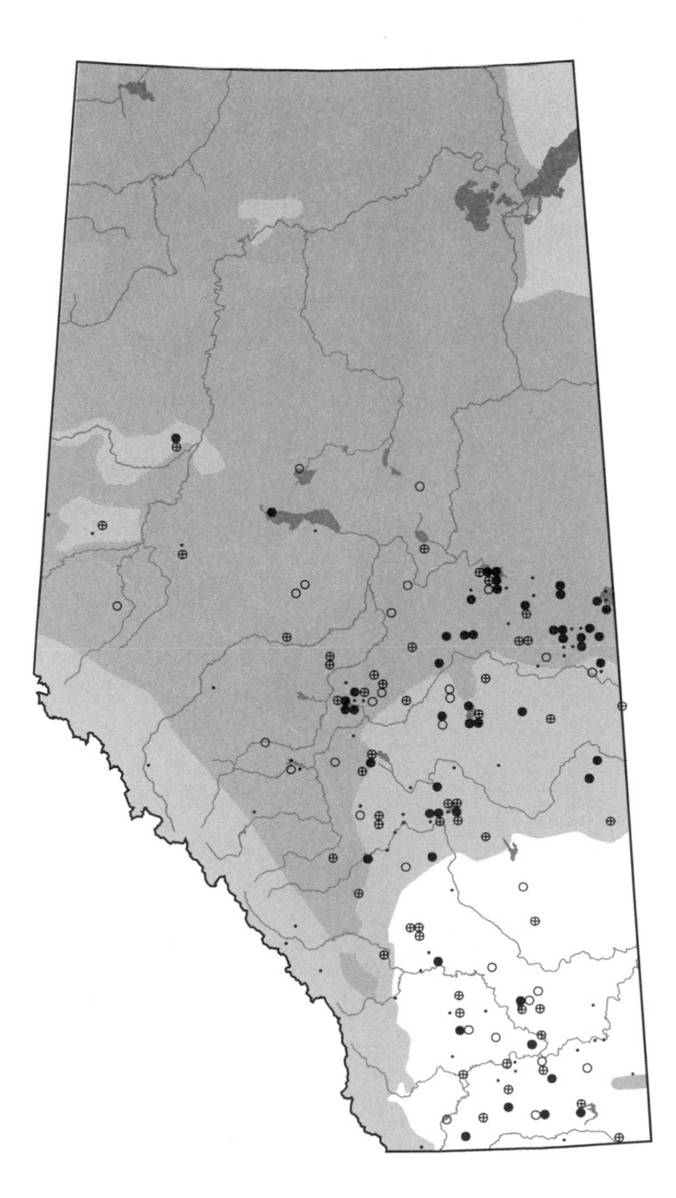

Clark's Grebe

Aechmophorus clarkii

CLGR photo credit: W.A. Paff/VIREO

Status:
This former "light phase" of the Western Grebe was accorded full species status in 1985 (American Ornithologists' Union 1985). The population status in Alberta and Canada is unknown (Wildlife Management Branch 1991).

Distribution:
The North American ranges of Clark's Grebe and the Western Grebe overlap, with the Clark's Grebe occupying the southwestern and western portions of the former Western Grebe range. In Alberta, Clark's Grebe breeds only in the extreme south of the province, with one confirmed nesting record from Crow Indian Lake (Dolman 1988). This grebe had two confirmed records of breeding in the same area in southern Alberta during the five year duration of the Atlas Project. Alberta is on the very periphery of this grebe's range, and its occurrence in the province is presently limited to the Crow Indian Lake reservoir (B. Rippin pers. comm.). One specimen was collected from Buffalo Lake in 1914 (Taverner 1919).

Habitat:
As with the Western Grebe, the Clark's is found on larger lakes having abundant fish populations and lakeside emergent vegetation. Water levels need to be stable and human interference minimal for nesting to be successful.

Nesting:
Clark's Grebes nest in colonies, sometimes with Western Grebes where their ranges overlap. In such mixed colonies, species are partially segregated.
The nest is a mound of wet vegetation, floating or built up from the lake bottom, attached to nearby emergents, and usually well-concealed. Clutches are usually 3-5 eggs, blue-white and soon discolored. Incubation is 23 days by both sexes; young are also tended by both parents. Nesting behavior is generally similar to the Western Grebe, though the Clark's Grebe's advertising call, used to attract mates, is a single "creeet" while that of the Western Grebe is a two-noted "creet creet" (Godfrey 1986). Both species leave and approach the nest underwater, but do not cover the eggs when leaving the nest.
Chick plumage differs significantly between the two species. At 20-50 days of age Clark's Grebe chicks are snowy white, while Western Grebe chicks have a charcoal black back. Clark's Grebe chicks are tended by both parents and will brood on a parent's back or under a wing. Fledging of this species occurs at 63-77 days. Clark's Grebe is single-brooded.

Remarks:
This species is very similar to the Western Grebe in size, shape and color. However, in the Clark's Grebe the black crown does not extend down to reach the eyes or lores as in the Western Grebe. The Clark's Grebe's bill is orange-yellow, while that of the Western Grebe is green-yellow (Godfrey 1986).
Both the Clark's Grebe and the Western Grebe are well known for their elaborate and showy courtship displays. In the "water dance" or "racing" display, two birds will rear up vertically and race side by side, wings partly raised and feet pattering rapidly over the water.
Food species include fish and insects. Clark's Grebe tends to forage in somewhat deeper water than the Western Grebe and use "spring" dives more often to reach its food. In this species, mate feeding, with the male feeding the female, occurs regularly between mated pairs prior to nest initiation (Neuchterlein and Storer 1989).
Spring migration takes place in late April to early May, with birds arriving just as the ice is breaking up on the larger lakes. Most birds leave the province again by mid-November. It is unknown whether this species overwinters in Alberta.

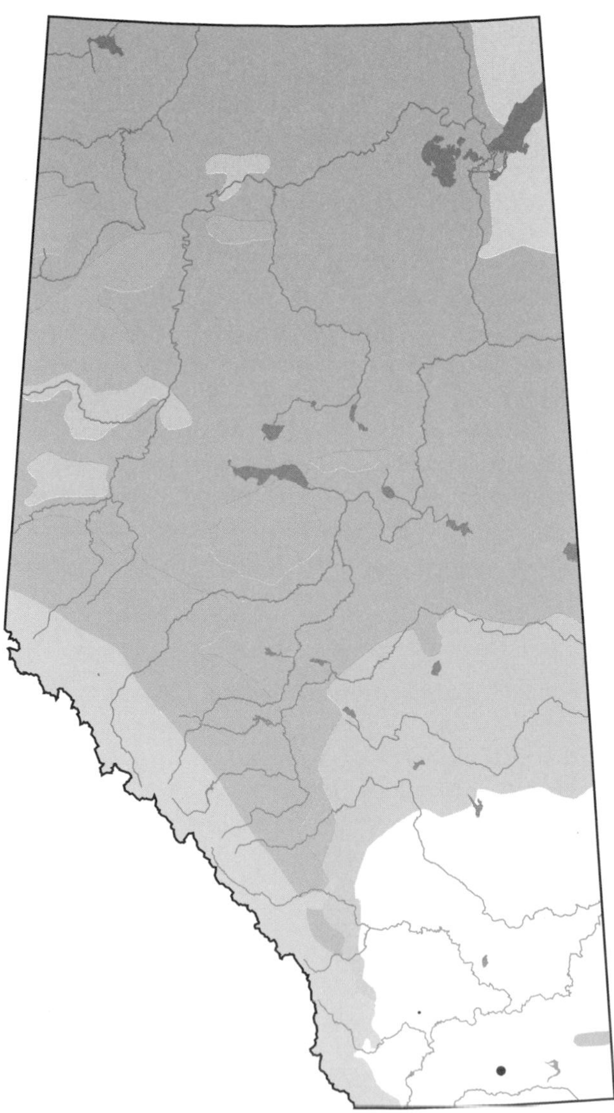

Order Pelecaniformes

(Family Pelecanidae, Family Phalacrocoracidae)

The presence of a throat pouch, whether large like the pelican's or small like the cormorant's, is the common element connecting members of this order. American White Pelicans, among the largest birds in the world, use their pouches and long bills for display, cooling, and eating. A wing span of 2.5 m or more allows for an easy glide during landing and for spiralling upward during soaring before cross-country flying. Before conservation measures were implemented, pelicans were persecuted because of their competition with local fisheries. The Double-crested Cormorant also has webbed toes and, although reduced, a throat pouch, which is orange. The "sea crow" is relatively large, with a significant tail and a long strong beak with a pronounced hook at the end of the upper mandible. While perched, it may be distinguished by its upright posture and general bowling pin shape. In flight, it may be mistaken for a goose, although it is silent, and flies lower, swifter, and only in loose formation. Nesting colonies are in the isolated and inaccessible areas of large lakes, where they are conspicuous by the large accumulation of excrement, and often occur in association with herons, pelicans, and gulls.

American White Pelican

Pelecanus erythrorhynchos

AWPE photo credit: G. Beyersbergen

Status:

Populations of the American White Pelican declined in Canada until the 1970s and the species was declared "threatened" in Canada in 1978 (COSEWIC 1978). Under protective provincial legislation, the Canadian breeding population increased from less than 16,000 breeding pairs in the mid-1970s to 50,000 pairs ten years later (Koonz 1987). About 2,000 of these breeding pairs are in Alberta. This allowed the species to be removed from the COSEWIC list in 1987.

In Alberta, the species continues to be designated as "endangered". As the population is increasing, but the number of active colonies decreasing, comprehensive colony protection is essential. Several key colony islands have been designated as Seasonal Wildlife Sanctuaries, thereby prohibiting all public access on or within approximately 1 km of the nesting island between April 15 and September 15 annually (S. Brechtel pers. comm.).

Distribution:

This species nests in the eastern half of the province. At present, there are least seven active colonies in Alberta, the northernmost being at Slave River Rapids, 3 km south of the 60th parallel. Seventeen lakes contained breeding colonies in at least one year between 1980 and 1985.

Habitat:

Typical habitat for the American White Pelican is a shallow, turbid lake remote from human activity with extensive shallows near shore and good rough fish populations. Birds will occasionally colonize deep, clear lakes and rivers. Nesting is on low flat islands which are generally treeless, protected from wave action and free from mammalian predators.

In northern Alberta, most colonies have been occupied consistently for an extended period. In southern Alberta, traditional colonies are used only sporadically according to such factors as fluctuating water levels and human disturbance.

Nesting:

The American White Pelican is a colonial nester; colonies in Alberta have a few to several hundred pairs. The nest is either a shallow scrape in the ground lined with sticks, rushes or grass, or a heaped up mound of dirt, sticks, reeds, and debris.

The female lays 1-4 white eggs, usually two (Vermeer 1969). Incubation is 29 days by both sexes. Young birds are very vulnerable to any disturbance which may cause adults to leave the colony, exposing chicks to extreme temperatures or predation by gulls. Rarely does more than one chick survive per nest, mainly because of rivalry for food. After three to four weeks, the young congregate in juvenile groups, or "pods", where they remain for several more weeks. Young fly at 10 weeks.

Remarks:

The American White Pelican is among the world's largest birds, weighing 5-8 kg and having a wing span of over 2 m. Its pouch or gular sac can hold 13 litres of fish and water. This Pelican feeds from the surface. It plunges its head into the water, scoops up fish in its gular sac, squeezes out the excess water, and swallows. It never flies with fish in its pouch. The Pelican's main prey are minnows and slow-moving non-game fish.

Pelicans are very sensitive to human disturbance; untimely visits on or near a colony can result in heavy predation or the total abandonment of a nesting island. Human disturbance, habitat loss, and toxic contamination are the main causes of population declines this century.

Birds arrive in southern Alberta in late April or early May, travelling in long lines or in V-formation. Most birds depart for wintering grounds in mid to late September.

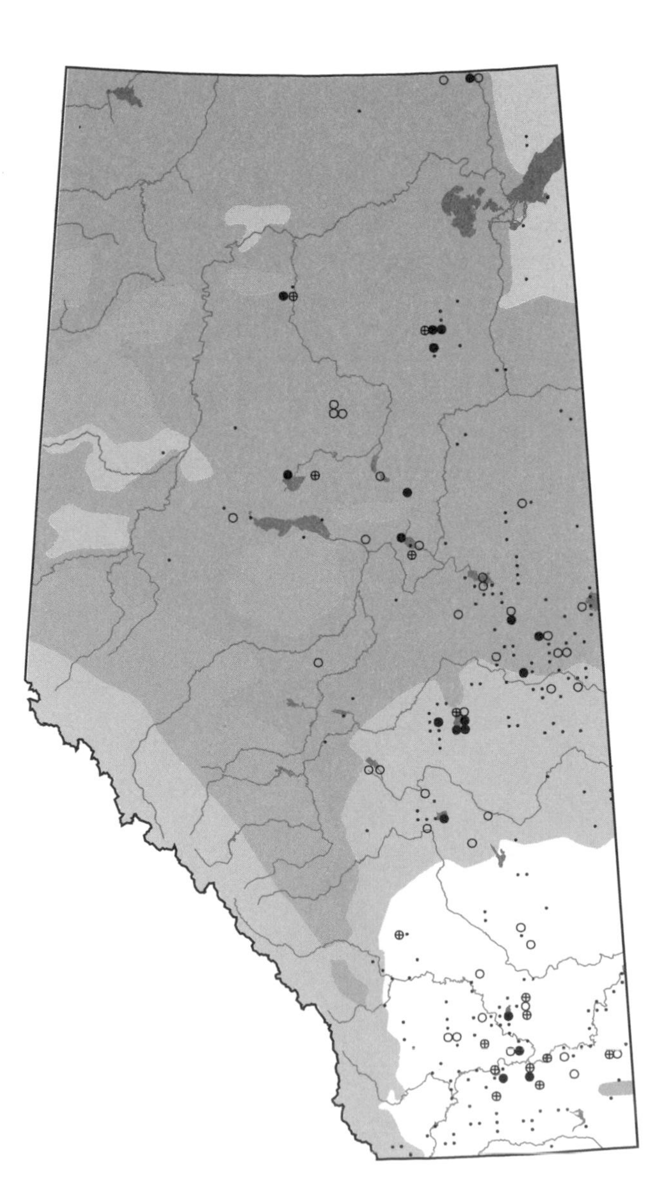

Double-crested Cormorant

Phalacrocorax auritus

DCCO

Status:

At one time, the Double-crested Cormorant had disappeared from large areas of its Canadian range and it was designated as "endangered" in Alberta in 1977. Since then, increasing numbers of colonies and individuals have allowed it to be removed from the endangered list. In Alberta, there were an estimated 2,300 breeding pairs in 27 active colonies in 1984, a 1,200% increase from 1967 (Fish and Wildlife 1984).

Although not protected under the Migratory Birds Convention Act because of the erroneous view that it consumes large amounts of commercial and game fish, it is protected under The Wildlife Act in Alberta.

Distribution:

Cormorants are found in the central and southern parts of the province east of the foothills, reaching the northern and western limit of its continental range. It has been colonizing new areas within its range, generally choosing locations used by other colonial birds. Colonies have been established recently at Prefontaine Lake, Antoine Lake, and Lac La Biche. The largest cormorant colony is at Frog Lake, with 2,000 pairs (1987).

Habitat:

The cormorant typically breeds on larger lakes or reservoirs where there are both good fish populations and undisturbed, low-lying islands for nesting. While cormorants prefer traditional nesting areas, to which they return year after year, they have recently shown a great ability to expand into unoccupied habitat. The tendency to nest on islands leaves rookeries highly susceptible to water level fluctuations, particularly on reservoirs and in periods of drought.

Nesting:

Cormorants usually nest on the ground on low-lying islands in close-packed colonies, often in the company of gulls, pelicans, and herons. Several colonies, established in the past 15 years, have taken to nesting in trees on wooded islands.

Nests vary in size and durability; new nests are usually shallow and flimsy and those occupied and refurbished over several years can be 1 m high and wide. Nest materials include sticks, rushes, reeds, coarse grass, and weeds, and are reinforced with guano.

The number of eggs laid depends on the colony and year; the average is 3-4. Brechtel (1983) reports clutches of 1-6 eggs for six colonies near Brooks, Alberta. Eggs are bluish white with a chalky coating, and are soon nest-stained. Incubation is 28 days by both parents. Newly hatched young are very vulnerable to any human disturbance, which may chase away the adults, leaving young exposed to heat, cold, or predation by gulls. After 3-4 weeks youngsters leave the nest, form bands and wander through the colony. Young enter the water only when fully feathered. Fledging is at 35-42 days. Double-crested Cormorants raise only one brood, but will lay a second clutch if early eggs are lost.

Remarks:

Cormorants are jet-black with a bronze sheen, long sinuous necks, and yellow-orange throat pouches. In spring, adult birds have long fine feathers behind each eye, forming the double crests for which these birds are named.

Being excellent swimmers, they will dive from the surface to catch their food by pursuit. Cormorant wing feathers are not water proof so, after hunting, the birds perch and extend their wings to dry. Their food is mostly coarse fish, complemented by a few crustaceans and amphibians.

This species arrives in southern Alberta in late April and early May. Most birds have departed by early October. Overwintering has been recorded at Wabamun Lake and Calgary.

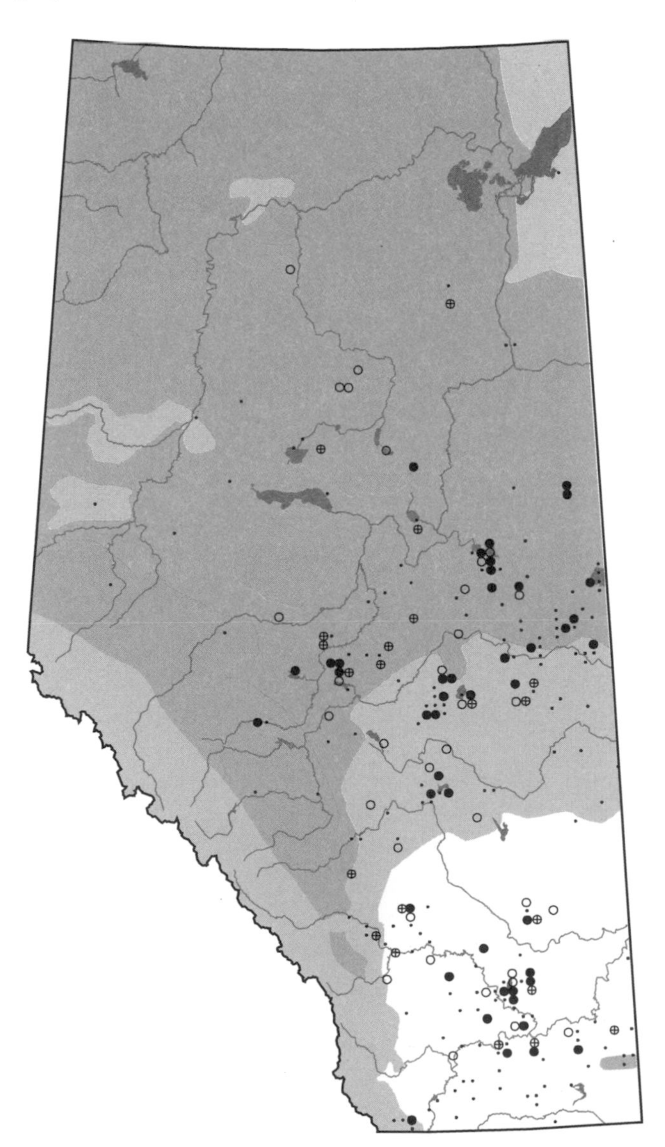

Order Ciconiiformes

(Family Ardeidae, Family Threskiornithidae)

The American Bittern, Great Blue Heron, and Black-crowned Night Heron are the breeding members of this order in the province. Common features of these species include long pointed bills, long necks, large wings, short tails, long legs, and long toes. Their loose plumage includes unique down feathers which continuously fray at the tip to produce a powder used for preening. The sexes have the same appearance. In flight, the legs are trailed behind the tail and the neck is curled back and positioned between the shoulders. In search of fish and other aquatic animals, these waders stalk in marshes or along the shallow fringes of lakes and streams. Their hunting strategy consists of standing inconspicuously or slowly wading forward and darting at their prey with a quick thrust from their long necks.

American Bittern

Botaurus lentiginosus

AMBI photo credit: R. Gehlert

Status:
Though nowhere especially abundant, the American Bittern is widely and generally distributed over nearly all of North America, where suitable habitat exists (Bent 1963c). This is generally true for Alberta, although the Wildlife Management Branch (1991) reported that declines in some areas are suspected. Drainage, consolidation and cultivation of wetlands are a continuing threat to this species.

Distribution:
The Bittern breeds in suitable habitat throughout Alberta with confirmed breeding records only in the Grassland, Parkland, and southern Boreal Forest natural regions. Salt and Salt (1976) reported that this species had been seen in the breeding season at Banff, Jasper, and Hinton, although it had not yet been shown to breed. Holroyd and Van Tighem (1983) reported one nesting record in Banff National Park in 1911. This trend continues; there were no confirmed breeding records over the five year Atlas surveying period in the Rocky Mountain Natural Region.

Habitat:
This species breeds in marshes, swamps, moist meadows, wet alder or willow thickets, and occasionally in drier meadows, but always in areas with a dense growth of emergent vegetation or tall grasses. It prefers secluded bogs and swamps where it can lead its rather solitary existence, foraging alone in shallow, muddy areas.

Nesting:
The bittern nest is on the ground, or raised on a tussock, usually in marshy areas of tall vegetation. It will occasionally be found in drier areas of tall grass. The nest is a matted platform of reeds, rushes, small sticks, and coarse grass, 0.3 m or more in diameter, and built a bit above the water level. The female builds the nest and clears both entrance and exit pathways to it. The female never flies directly to or from the nest, but uses the end of these pathways to land and take off.

Two to six eggs are laid, usually 4-5, buffy olive or buffy brown. The incubation period is 24-28 days; the young leave the nest after 14 days and are tended nearby. The female alone incubates and feeds the young as males may be polygamous, with several females nesting separately within a territory (Harrison 1978).

Remarks:
The bittern is a bird both solitary and secretive. When alarmed, it proves itself to be a master of concealment. It will draw its plumage in tightly, point its bill straight in the air, and remain motionless for long intervals. Its streaked neck and breast and yellowish-brown coloration allow it to blend with its marshy environment.
A carnivorous species, the bittern subsists mainly on frogs, snakes, small fish, and crustaceans. Occasionally, it will venture to open fields to feed on grasshoppers and mice. It hunts in typical heron fashion, walking slowly and deliberately among the reeds, or waiting motionless until some creature comes into thrusting range of its sharp bill.
This species has many local names, most of which refer to the "booming" or "pumping" sound made by the male during the mating season to warn off rival males and attract potential mates. The call is a hollow deep "pump-er-lunk", repeated several times and heard mainly in the evening or dawn (Godfrey 1986).
This species arrives in the province in late April or early May. Most birds leave by early September, with some lingering late into October (Pinel et al. 1991).

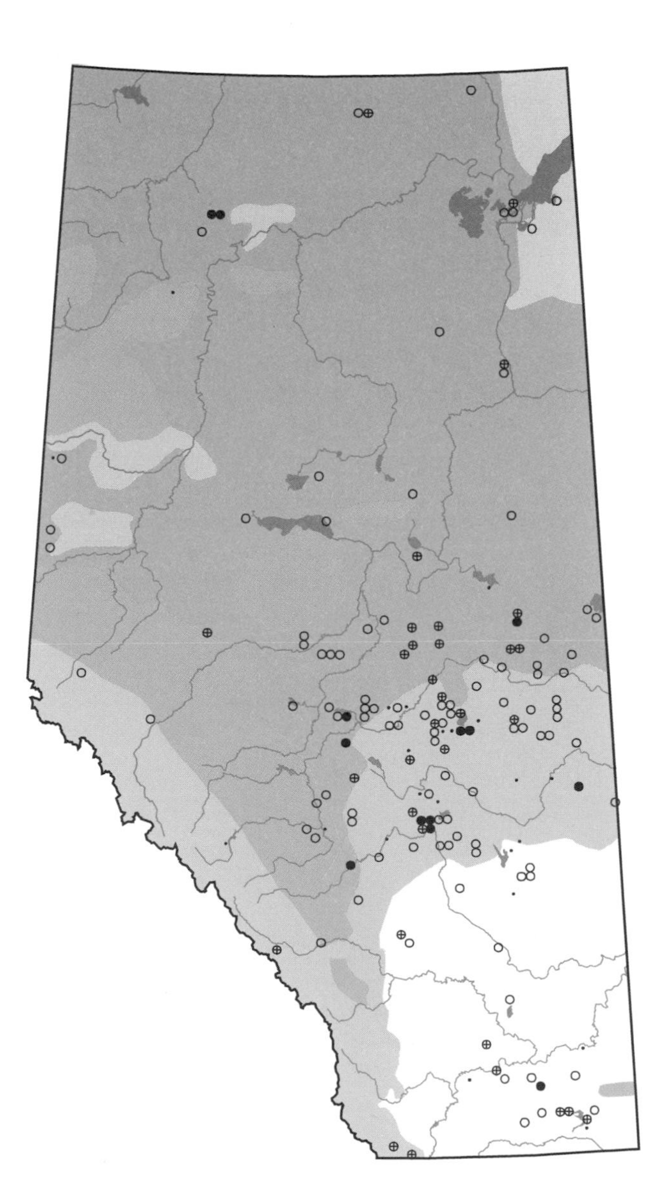

Great Blue Heron

Ardea herodias

GBLH photo credit: G. Horne

Status:

Salt and Salt (1976) reported that the range of this species had not changed over the last century, but the number of breeding colonies had significantly diminished. At present, the number of colonies appears to be expanding, with an estimated 75 active colonies, up from 27 in 1967. According to 1980 data, these colonies included 1500 active breeding pairs (Wildlife Management Branch 1991). However, this estimate may be low. Brechtel (1981) reported that the northeast region (Fish and Wildlife Division boundary) breeding population had been estimated at 615 pairs. Estimates taken in 1991 in this region were between 880 and 950 pairs on at least 43 and up to 50 colonies. Similar increases probably occurred elsewhere in the province (D. Moore pers. comm.).

Populations still remain under pressure in more settled parts of the province, primarily from rural subdivision development, recreational land use demands, and agriculture (Paulsen 1982). The entire Alberta population is dependent on fewer than 100 known nesting sites, so management of these key habitats and protection from human disturbance is essential (Wildlife Management Branch 1991).

Distribution:

The Great Blue Heron nests mainly in central and southern Alberta east of the foothills. The range may now be expanding northward. The northernmost breeding colony in Alberta (and Canada) occurs where the Birch River meets Lake Claire in Wood Buffalo National Park (Kristensen 1981). Other northern colonies are located at Winefred Lake, Goodwin Lake, and Pelican Lake.

Habitat:

This species is found in and about open shallow water at the edges of lakes, streams, rivers, ponds, sloughs, ditches, marshes, and mudflats.

Nesting:

Nesting is usually near water on lake islands, lake shorelines, or near creeks and rivers. Great Blue Herons prefer to nest high in trees, but sometimes nest in bushes and bulrushes, very rarely on the ground. South of 51 degrees latitude (Calgary) the birds nest most often in cottonwoods; north of this latitude they nest in aspen, balsam poplar, and spruce. Nesting is often in a dead tree, the birds' excrement contributing to the tree's demise. These herons are colonial, and return to the same nest sites year after year. Colonies average about 20 breeding pairs. New nests are delicate, flat platforms of interlaced twigs, while older nests, built up over several years, are large bulky structures up to a metre across. Nests have a central cavity which the birds line with twigs, moss, lichens, rushes, or conifer needles.

The Great Blue Heron lays 3-5 pale, greenish-blue eggs. Incubation is by both sexes, 26-27 days, and both parents tend the young. Chicks are brooded for one week; for the next few weeks parents take turns hunting and tending the young. Fledging is at 53 days.

Remarks:

This is the largest and most widely distributed heron in Canada. It is a sociable species that flocks on its feeding grounds as well as nesting in colonies. Stilt-like legs enable it to forage well out into shallows, where it stands motionless waiting for prey to come within range of its long, sharp bill. It also stalks food with the slow, deliberate pace characteristic of herons. Foods include fish, frogs, salamanders, water snakes, large insects, mice, small birds, and plant seeds.

This heron arrives in Alberta in the last half of March or early April. Most birds leave by mid-October, with the occasional few lingering into November in some years. Overwintering has been observed in the Calgary area, Medicine Hat, and Lethbridge.

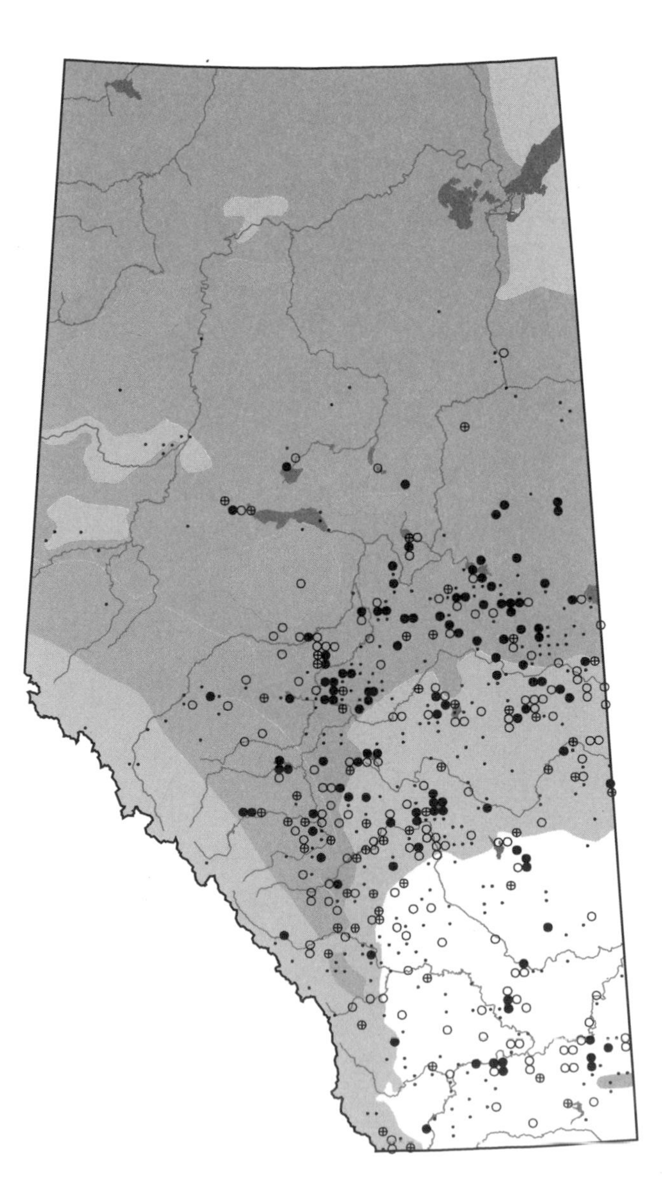

Black-crowned Night-Heron

Nycticorax nycticorax

BCNH photo credit: F. Whiley

Status:
The Black-crowned Night-Heron was first observed in Alberta in 1958 at Strathmore, near Calgary. It has remained in the province as a local breeder, but the breeding population is sparse and there are relatively few colonies. Colonies that exist tend to fluctuate in size, and are sometimes abandoned for no apparent reason. Still, the species has been expanding its range in the southern part of the province. On the North American Blue List from 1972-1981, this marsh-loving bird is particularly susceptible to land clearing, drainage, and human disturbance.

Distribution:
Colonies of this species are found in Alberta's Grassland, Parkland, and southern Boreal Forest natural regions.

Habitat:
These herons tend to colonize relatively large bodies of water with dense emergent vegetation. In the northern part of their Alberta range, the birds are found on natural, marshy lakes and ponds, while farther south they occur in suitable habitat primarily composed of irrigated farmland with man-made impoundments.

Nesting:
The Black-crowned Night-Heron, a colonial species, is a tree nester in most parts of its range. In the central part of the province, it often nests in large willows, 2 m to 3 m above the ground (D. Moore pers. comm.) However, in other areas of Alberta, nesting is usually in dense emergent vegetation, often on islands of emergents wholly or partially surrounded by water. These nesting sites are free of mammalian predators but are susceptible to high predation from gulls.

Nests are crude platforms with little or no cup, often built within centimetres of the water surface. Most are made of cattails, bulrushes, twigs, and other plant materials.
Clutches are of 1-6 eggs; a few found with 7-8 eggs are probably the product of two females. Eggs are a pale, bluish green. Incubation is 21-28 days (Wolford and Boag 1971), by both sexes and the young are also assisted by both parents. Fledging is at six weeks. Towards the end of the summer, the young leave the nest to wander alone or in small pods. Black-crowned Night-Herons are single-brooded.

Remarks:
This heron is a small, stocky wader recognizable by its greenish black crown and long, slender white head plumes. It is most active at night and its resounding "quawk" may be heard at dusk as the bird flies to its feeding grounds. This species feeds at the marshy shallows of lakes, streams, and irrigation ditches. While an expert still-fisherman, this heron is also a very active stalker. It may also alight in deep water, seize its prey and take wing again. The heron's main foods are fish, nestlings, frogs, insects, small rodents, and vegetation. In daylight hours, these herons are likely to be found roosting in nearby trees.
This species arrives in southern Alberta in mid-April to early May. Most birds leave by mid-September, though a few stragglers may remain into October. Overwintering is recorded from the Calgary area.

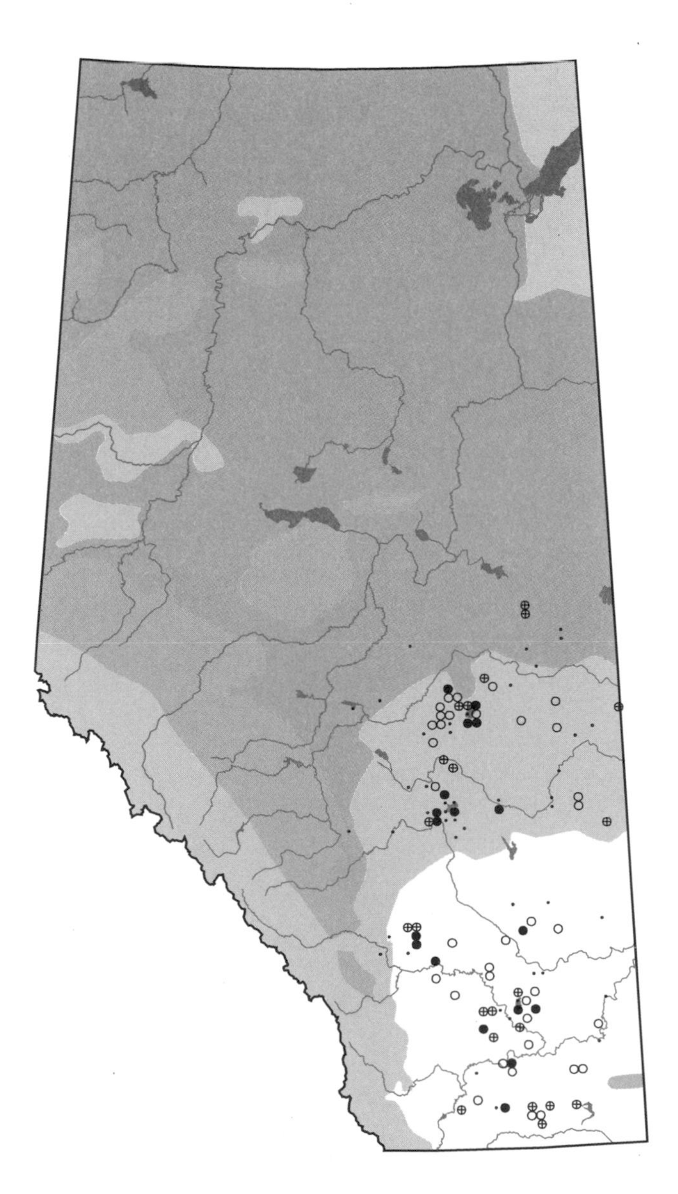

White-faced Ibis

Plegadis chihi

WFIB photo credit: H. Cruickshank/VIREO

Status:
The first authenticated record of this species in Alberta was a sight record of a single bird photographed at Pakowki Lake in 1974 (Salt and Salt 1976). Since that time, the species has regularly been observed at Pakowki Lake and occasionally at other lakes in southern Alberta. Nesting has been confirmed at Pakowki Lake and at one of the Minor Lakes (Goossen et al. in prep.)
In the United States, the White-faced Ibis is considered to be a "sensitive" species in its main breeding area of the Great Basin. It is vulnerable to marsh drainage, human disturbance, and, in its winter range in Central and South America, shooting, as well as pesticide contamination (United States Fish and Wildlife 1985). The species was on the Blue List from 1972 to 1979, but was delisted in 1981 (Tate 1981). Although evidence suggests the species may now be expanding its range to the north, there are no recent population estimates. Over 80 records of this species are known for western Canada from 1984-92 (Goossen et al. in prep.).

Distribution:
The usual range of the White-faced Ibis is from central California, eastern Oregon, southern Idaho, southern South Dakota, and Nebraska south to the coastal region of Texas and Louisiana, also Central and South America to Argentina and Chile.
In Alberta, this species has been observed at approximately a dozen wetlands including Pakowki Lake, Frank Lake, Crow Indian Lake, and Stirling Lake, all in the extreme south of the province (Goossen et al. in prep.).

Habitat:
The White-faced Ibis breeds in marshes of larger lakes and forages on mudflats, in shallow marshes, sloughs, flooded pastures, meadows, and in irrigated fields. A soft muddy substrate and shallow water are usually required at feeding sites. Birds roost at night in dense marsh.

Nesting:
This species nests in close colonies, often with herons or other colonial birds, in the dense emergent vegetation of marshes. Nests, made of coarse pieces of dead emergents and lined with finer vegetation and grasses, are placed on mats of trampled dead stems or attached, floating, to new growth emergents. The nests are usually deeply cupped and may be built up 1 m or more above the water. On rare occasions, this species will nest on the ground in low vegetation. Both male and female construct the nest.
The female lays 2-4 green-blue eggs which are incubated for 21-22 days by both sexes. Upon hatching, the young are closely brooded for five days. Young are tended by both parents and fledge at about 28 days. One brood is raised per year.

Remarks:
The White-faced Ibis is a medium-sized stork-like bird. It is dark chestnut with a purple and green gloss and a long down-curving bill. In the breeding season, its red eye and white facial border are characteristic. It is considered to be the least aggressive of all colonial wading birds, and loses most encounters with other birds.
This species probes for its food along shallow margins of muddy pools, ponds, and marshes, as well as flooded fields and pastures. Its diet is mostly invertebrates, including insects and their larvae, leeches, snails, spiders, earthworms, some vegetation, and the occasional amphibian, crustacean, or small fish. It likes to feed in large groups, and also travels in groups.
In Alberta, the White-faced Ibis has been seen as early as May 6 and as late as September 20 (Goossen et al. in prep.).

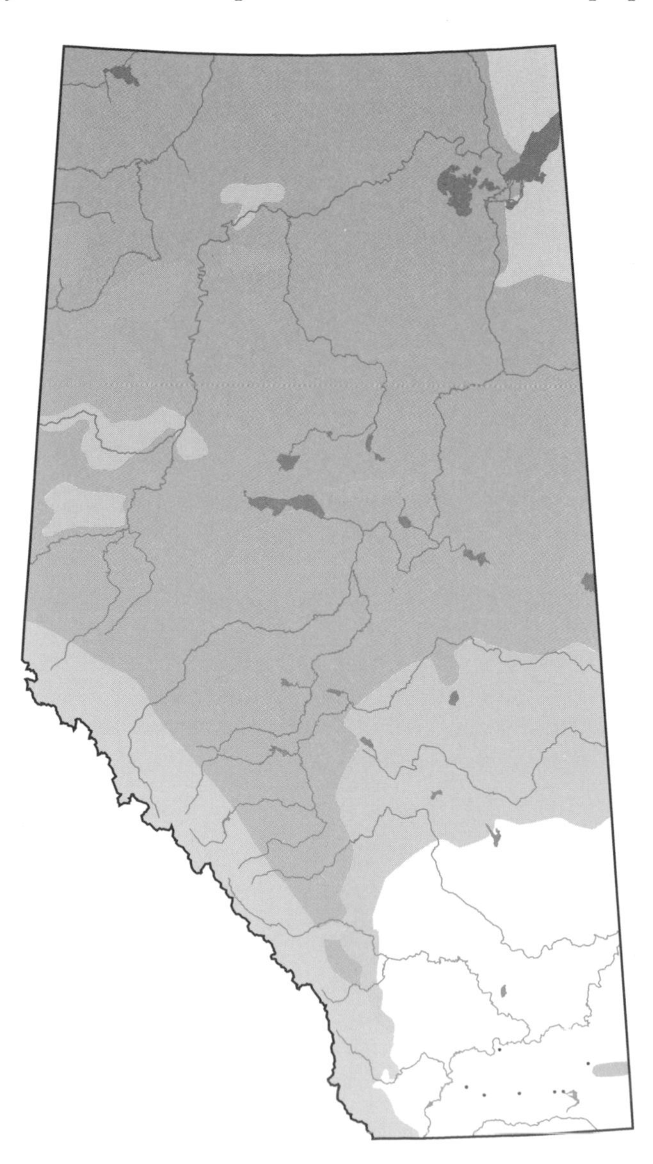

Order Anseriformes

This familiar order of aquatic birds includes swans, geese, and ducks. Although all have webbed feet, feeding strategies include diving, straining water, and upland grazing, which requires a variety of neck, leg, and bill structures. Swans, the largest waterfowl, have short legs and long necks that permit these birds to dabble well below the water's surface in search of roots and weeds. Geese, comprising the next smallest members of the order, have longer legs and shorter, stout bills adapted to grazing on green shoots of grass and weeds in upland regions. Smallest in physical size, the duck family has evolved the widest variety of bills and feeding strategies. Surface-feeding ducks have typically wide duck bills, which are used to dabble within reach below the surfaces of grassy sloughs, marshes, and shallow lake edges. Bristles on the bills' margins assist in straining excess water. However, many species do feed on land. These ducks are noted for their strong take-offs. Diving ducks are heavy-set and have longer and narrower serrated bills, used to capture fish. To assist in diving, the mergansers have relatively short wings, requiring them to sprint a short distance over the water in order to take-off.

Trumpeter Swan

Cygnus buccinator

TPSW

Status:

Trumpeter Swans were once widespread across North America. Excessive hunting and habitat loss depleted the population so that, by the 1930s, only small remnant flocks remained in the northwest part of the range.

Protective legislation and management have been crucial in rebuilding a portion of the population. There are now about 15,000 birds continent-wide, with 475 in Alberta, including 50 breeding pairs (Shandruk 1991). The Alberta flock is growing at a rate of about 7% per year.

The management and expansion of wintering habitat in the United States appears to be the most critical factor in the continued growth of the Alberta population. The Trumpeter Swan was designated as a "rare" species in 1978 by the Committee on the Status of Endangered Wildlife in Canada; it is now classified as "vulnerable" because of its limited and restricted breeding range in Canada (Shandruk and McKelvey 1990). It is considered "threatened" in Alberta.

Distribution:

The majority of Alberta's Trumpeters are found in the Peace River District near Grande Prairie. Small pioneering populations are located in the Edson/Whitecourt area, Elinor Lake, Fawcett Lake, Otter Lake, the Chinchaga River/Clear Hills area, and Waterton Lakes National Park. Cygnets have recently been observed at Bistcho Lake near the N.W.T. border and attempts are being made to establish a breeding population at Elk Island National Park.

Habitat:

In Alberta, this species is found on small to medium-sized shallow, isolated lakes that have well-developed emergent and submergent plant communities.

Nesting:

Trumpeter Swans are highly territorial in the spring, so it is rare to see more than one pair per lake. The nest is a down-lined bulky heap of marsh vegetation, which may be built on an old beaver or muskrat house, a floating sedge mat, or in emergents. The same nest site may be used for several years. Clutch size ranges from 2-9, with an average of 5-6. Eggs are dull white and are often nest-stained. Incubation is about 32 days, by the female alone, although the male remains on guard close by. Both tend the young, which leave the nest to feed after a day or two. Fledging is at about 100 days, and the cygnets stay with their parents until the following spring migration.

This species mates for life, or until one of the pair dies, when the surviving bird mates again.

Remarks:

The Trumpeter Swan is the largest of all North American waterfowl, with a wingspan over 2.5 m and a weight of 12 kg. The bird's long twisted windpipe acts like a resonating chamber to produce this species' loud, bugle-like call, which helps distinguish it from the very similar Tundra Swan.

It obtains stems, leaves, roots, and tubers of aquatic plants by vigorous use of the feet and by searching the shallows with its long neck. Cygnets initially eat high protein food such as insects, crustaceans, or molluscs.

Trumpeter Swans arrive in southern Alberta in mid-April and most of them pass through in two weeks (Pinel et al. 1991). In the fall, most birds have departed by mid-October. Overwintering has been reported during mild winters at Calgary, Carseland, and Waterton Lakes.

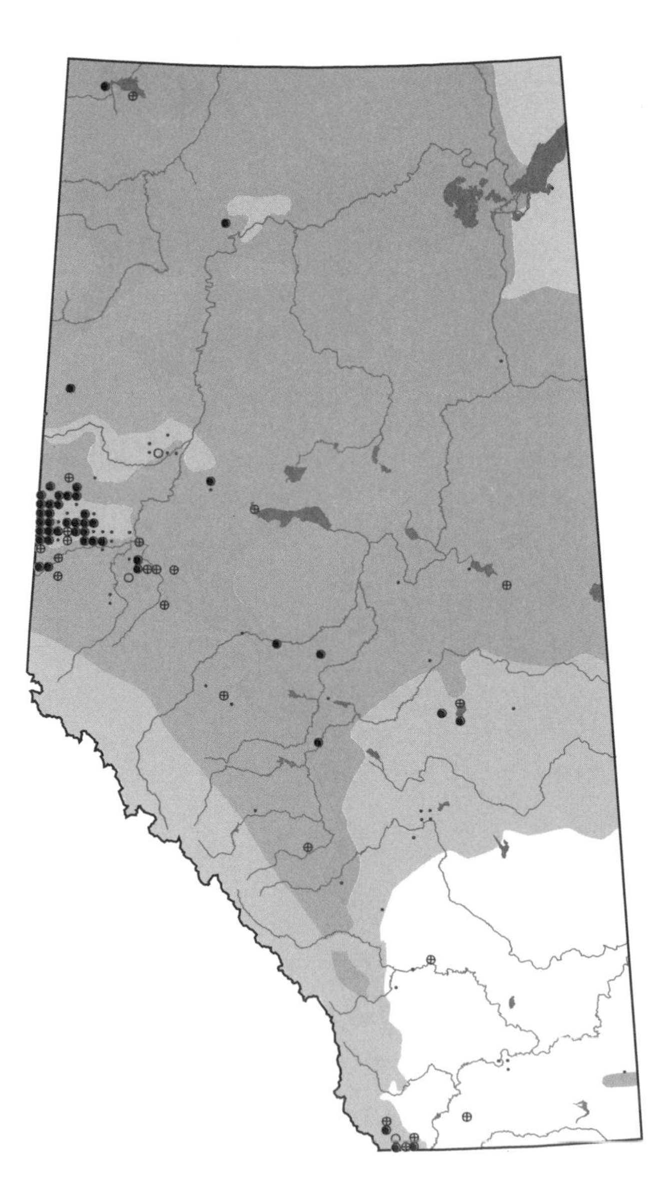

Canada Goose

Branta canadensis

CAGO photo credit: T. Webb

Status:

There are several distinct races of the Canada Goose in Canada, varying in size, color, and geographic location. Some of these pass through Alberta as migrants, but it appears only the "large" Canada Goose (*B. c. moffitti*) breeds here.

During the 1930s, breeding Canada Geese all but disappeared from Alberta. Drought dried up their habitat and birds were subject to heavy predation and hunting pressures. More recently, with management and reintroduction programs, plus regulation of hunting, these geese have once more become abundant. The province now has a healthy spring population of about 50,000 breeding birds.

Distribution:

Canada Geese breed locally throughout Alberta, with concentrations in the central and southern parts of the province. Resident populations have increased their breeding range considerably in recent years.

Habitat:

A highly adaptable bird, the Canada Goose usually congregates and nests near lakes, ponds, sloughs, rivers, and irrigation reservoirs.

Nesting:

Canada Geese generally breed singly but, in sanctuary areas or on islands, they sometimes nest in loose colonies. Nests are close to water, with islands preferred. Most nests are built on the ground, but some have been found in low shrubbery, in open meadows, or on old muskrat and beaver houses. Geese also nest in raptor or heron nests, on cliff edges, and, where provided, on man-made platforms. The female builds the nest, using a variety of vegetation including grasses, reeds, cattails, twigs, sedges, and moss. After she lays a few eggs, she lines the nest with down. Nests may be used by the same pair for several years.

In Alberta, clutches are usually 4-6 eggs, which are creamy white and nest-stained (Vermeer 1970). Incubation varies with subspecies; geese in southern Alberta are reported to have an incubation period of 25-28 days (Vermeer 1970). The female incubates, while the male stands guard nearby. Young leave the nest a day after hatching, but may be brought back at night to brood for a few days. Parents are very attentive to their young. Occasionally, large "gang" broods will form, with several adults in charge of 50-60 goslings. Young fledge at nine weeks.

Canada Geese exhibit strong pair and family bonds. Young fly south with their parents in the fall, and families are together until the return to the breeding grounds the following spring.

Remarks:

Canada's best known goose is easily recognized by its long black neck and black head with white cheek patches. It is a conspicuous bird, and is tolerant of human activity.

Though geese occasionally tip up, they are primarily grazers, and most feeding activity occurs in upland areas. Their diet consists mainly of shoots of grasses and sedges, seeds, berries, and grains, plus small amounts of insects, larvae, crustaceans, and molluscs.

Nesting geese tend to arrive in the province early in March, while the small northern races pass through in peak numbers a month later (Pinel et al. 1991). In late May, nonbreeding birds move through southern Alberta on their way to the northern moulting grounds. Large numbers of birds congregate in late September and early October in staging areas in the Parkland and Grassland natural regions. They usually leave the province by mid- to late November. Overwintering has been recorded at Edmonton, Wabamun Lake, Calgary, Lethbridge, Medicine Hat, Snake's Head, Claresholm, and Waterton Lakes National Park.

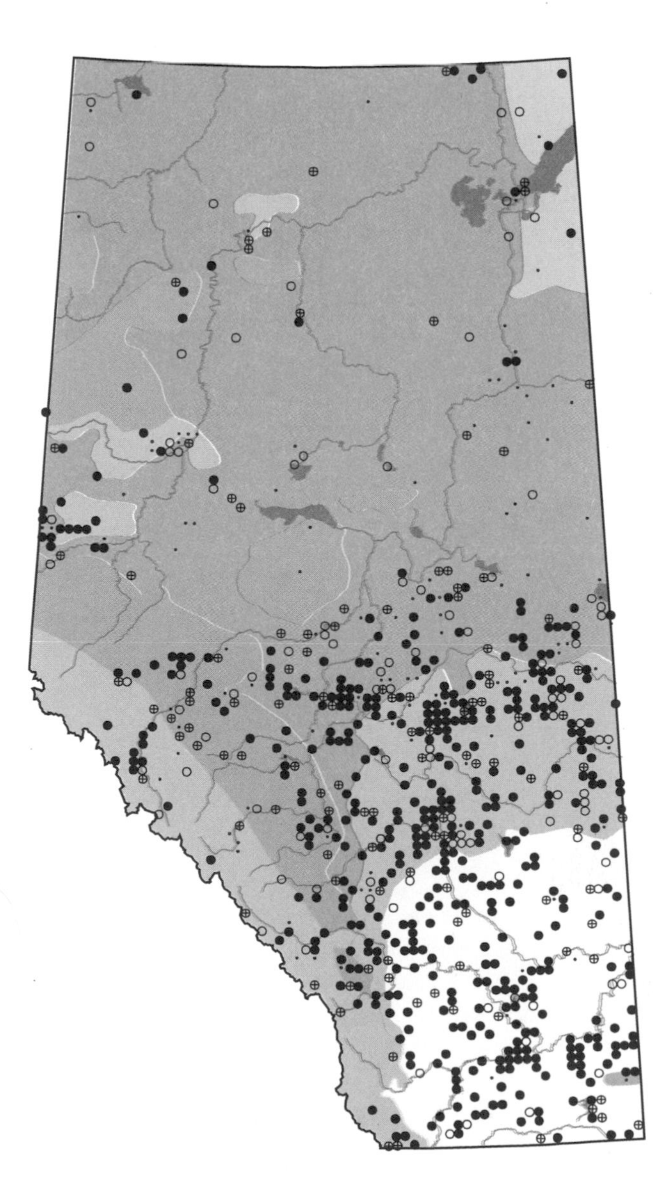

Wood Duck

Aix sponsa

WODU photo credit: L. Bennett

Status:
The Wood Duck was a threatened species in North America at the turn of the century, but benefitted from early conservation legislation when hunting restrictions were put in place as a result of the Migratory Birds Treaty of 1917. The species has good potential to increase its numbers because it usually out-produces other duck species (Bellrose 1976). This is related to a large clutch size, high nesting success, and a strong re-nesting tendency. Continental populations are stable and it is believed that this species is in a relatively secure position at present. It is currently ranked as the third most harvested duck in North America. Presence of this species has increased in Alberta in the last decade, but it is viewed as an uncommon game species with minor importance in the province.

Distribution:
Alberta is at the northern limits of this duck's normal breeding range. The Wood Duck is mainly found in southern Alberta in the Grassland region, but has also been recorded in the Parkland and Rocky Mountain regions.

Habitat:
The Wood Duck breeds near wooded lakes, ponds, lowland sloughs, and streams, especially where large willows and cottonwoods are present. Preferred breeding habitat must contain suitable cover, available water, adequate food sources, and suitable brood-rearing locations. Breeding cover, including trees and shrubs having low branches, should be flooded.

Nesting:
These ducks nest in the hollow trunk or limb of a tree, or in nest boxes. If a natural cavity is not available, they will use the deserted nesting hole of one of the larger woodpeckers or flickers. The presence of trees at least 40 cm in diameter, having cavities with entrances at least 9 cm wide and interiors at least 20 cm in diameter, appear to be minimal nesting requirements (Johnsgard 1975). Apparently, there is a preference for nesting in rows or clusters of large trees and open stands are preferred over dense cover. The use of artificial nest boxes has contributed to the recovery of the species and these are still widely used to boost production. Usually, 13-15 white or creamy-white eggs are deposited in the down-lined nest. Eggs are incubated for 28-30 days, by the female. In response to the call of the female, the young leave the nest, usually the day after hatching, by climbing with their sharp claws to the nest hole and dropping to the ground unaided. The young are tended only by the female and fledge at an average of 60 days. The Wood Duck is the only species of duck to raise two broods per season.

Remarks:
The brightly-colored male, often called our most beautiful duck, has an iridescent green and blue head with a prominent crest, a purple face with two white marks extending up from the white throat, and a darker breast and body. The female has a white eye-ring and a light colored throat and crest; the rest of the body is dull brown mottled with white.
This is Canada's only perching duck and mated pairs can often be seen perching in trees when looking for nest sites. Their food preference is the fruits and nuts of woody plants such as dogwood and elm, as well as the seeds of aquatic plants. They also consume a small amount of aquatic insects.
This species has been introduced into Alberta on several occasions, with the first recorded breeding record near Midnapore in 1956 (Can. Field-Nat. 1957). Alberta does not appear to be within the Wood Duck's regular range and, as a result, it occurs sporadically (Pinel et al. 1991). There are two winter records, one in Calgary and one in Edmonton.

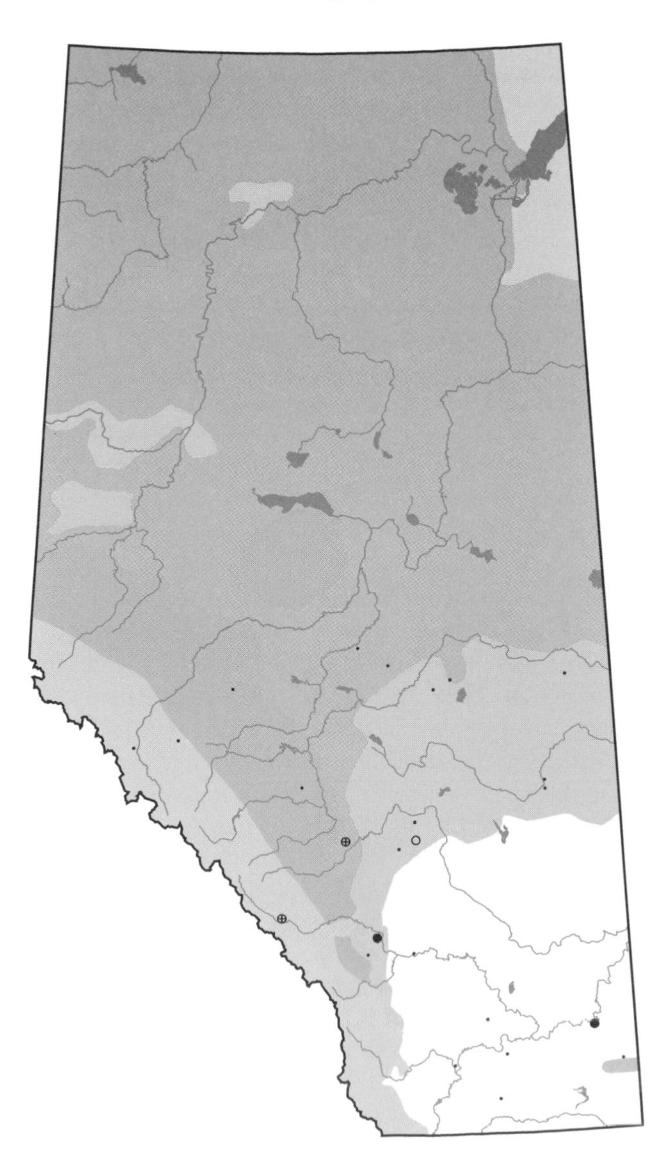

Green-winged Teal

Anas crecca

GWTE photo credit: W. Burgess

Status:
The results of the duck breeding population surveys indicate that the Green-winged Teal population in North America is relatively stable and near its average over the past 35 years. However, the Alberta population has decreased by about 40% in southern Alberta during the last decade (Wildlife Management Branch 1991).

Distribution:
The Green-winged Teal breeds from middle Alaska and the northwestern and middle Canadian mainland, and was found breeding in all the natural regions in the province. It was recorded in 35% of the surveyed squares in the Parkland region, 32% in the Boreal Forest region, 22% in the Foothills region, and 16% in the Grassland region. Although it was previously reported as not a common species in the Rocky Mountain region (Salt and Salt 1976), it was recorded in 12% of surveyed squares in that region. The continent's highest breeding densities of this species occur in the Athabasca Delta, the Slave River parklands, and east of Great Slave Lake. The aspen parkland has the next highest density (Johnsgard 1975).

Habitat:
Green-winged Teals use wooded ponds and streams in parkland and boreal areas. They nest primarily in upland areas with dense grass, weeds, or brush. In mixed prairie habitat in southern Alberta, most nests were on low ground near sloughs or, occasionally, on dry prairie (Keith 1961).

Nesting:
The nest is on the ground, usually near fresh water, but sometimes a fair distance from it. It is well concealed by grass or shrubbery (Godfrey 1986), and is built in a shallow scrape which is lined with soft grasses, weeds, leaves, and down from the breast.
There are typically 8-10 white to pale olive buff eggs which are incubated in 21-23 days by the female. The female, who tends the young alone, adds a considerable amount of down to the nest through incubation and uses it to cover the eggs when she is absent from the nest. The male deserts the female shortly after incubation begins. The young leave the nest shortly after hatching and are led to the nearest water. Females defend their young with remarkable intensity, distracting the intruder while the brood hides (Johnsgard 1975). The young are fledged at approximately 44 days.

Remarks:
This is the smallest duck in North America. The breeding male is grayish with a chestnut head and green face patch, a vertical white bar on the side of the body in front of the wing and a bright green wing patch. The adult female is dusky brown with a white throat and dusky stripe through the eye. It also has a small green wing patch.
The Green-winged Teal is known as one of the fastest flying ducks; documented flying speeds exceed 70 km per hour.
The small bill of the Green-winged Teal limits the size of material it can consume, and plant seeds such as pondweed, sedges and grasses are an important part of its diet. Animal foods, which make up less than 20% of the diet, include mainly the larvae and pupae of aquatic insects, small mollusks, and crustaceans.
The first spring migrants generally arrive in southern Alberta at the end of March and most have left the province by late October. Pinel et al. (1991) reported that the Green-winged Teal overwinters regularly in some areas, and records exist from Calgary, Banff, Canmore, Galahad, Edmonton, and Wabamun Lake.

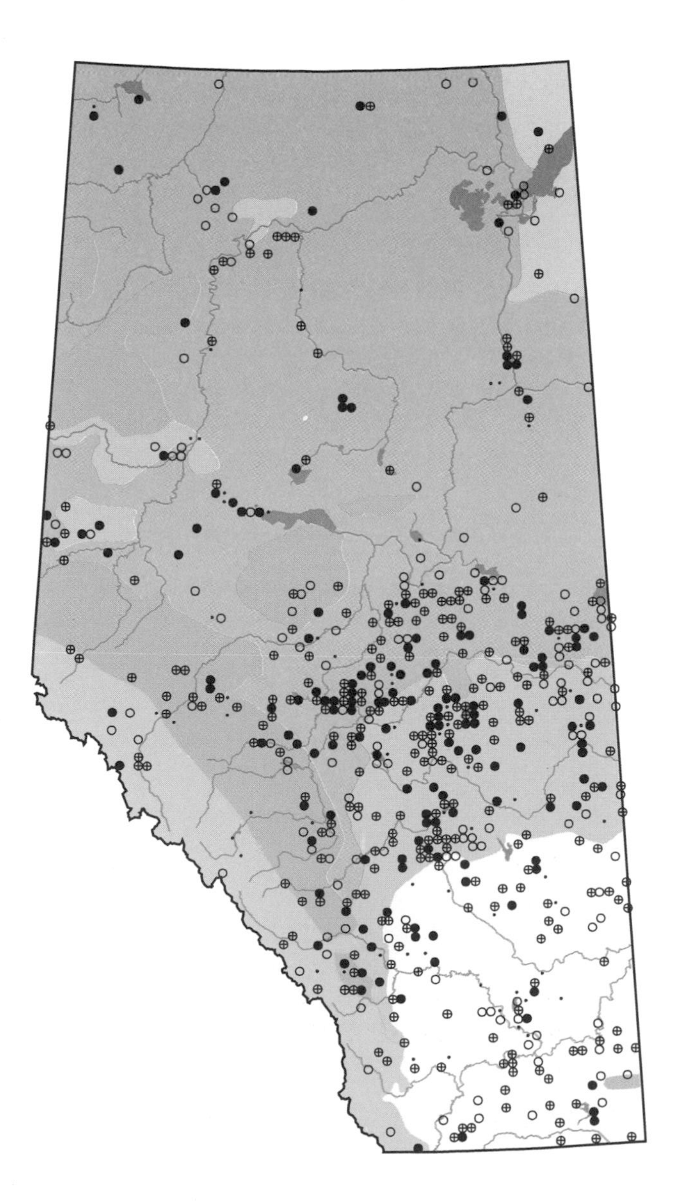

Mallard

Anas platyrhynchos

MALL photo credit: K. Morck

Status:
The Mallard is the most widely distributed and abundant duck within Canada and the United States, with its highest breeding densities in the prairie provinces. During recent years, the population has been declining. Breeding population indices were lower in the 1980s than throughout the late 1960s and 1970s. The 1991 population estimate for North America is 27% below the 1955-1990 average. Mallard numbers have varied most widely in the Grassland region, down 37% in Alberta, as a result of the greater annual variability in weather and habitat conditions. During dry conditions on the prairies, puddle ducks, including Mallards, fly further north to find suitable wetland conditions.

Distribution:
Atlas data indicate that the Mallard breeds throughout the province wherever suitable habitat is available. During the breeding season, it occurs in the greatest abundance on potholes in the Grassland and Parkland regions, with lower densities on the lakes and marshes of the Boreal Forest and in the valleys of the Rocky Mountain and Foothills regions.

Habitat:
The Mallard is a highly adaptable duck, frequenting marshes, ponds, the margins of small and large lakes, islands, quiet waters of rivers, ditches, and flooded land in both treeless and wooded country. It also forages on land, especially grain fields (Godfrey 1986). Mallard numbers in prairie Canada have traditionally varied directly with the number of ponds available for breeding. The presence of shallow-water feeding areas and the availability of suitable nest sites appear to be the only critical limiting factors for breeding.

Nesting:
The nest is usually built on fairly dry ground where there is tall vegetation for cover. Recent studies on the prairies indicate that preferred nesting habitat was grass-brush, followed by right-of-way, odd area, woodland, wetland, cropland, hayland, and grass (Greenwood et al. 1987). Mallards have also been recorded nesting in stumps and trees. The nest is made up of grasses, reeds, and leaves, and is lined with down.
There are normally 8-10 pale greenish-blue eggs which are incubated by the female for 26-29 days. The female leads the young to water soon after hatching, where she tends them for most of the eight week period required for the young to obtain flight.

Remarks:
The male Mallard is unmistakeable in breeding plumage, with his brilliant green head and upper neck. A white collar separates the green head from the chestnut breast. The remaining underparts and sides are light gray. The female has a buff colored head, brown back, and buff breast streaked with darker brown.
The Mallard in its wild form is widely known in the north-temperate zones of the world, and, in captivity, it is the ancestor of most varieties of domestic ducks, several of which still resemble it closely (Godfrey 1986). It also seems to be more inclined to hybridism than any other species, particularly with the American Black Duck.
In water, it feeds on the stems and leaves of aquatic vegetation and the tuberose roots of plants growing in the shallows, which it generally reaches by tipping up. On land, it eats grain, seeds, and green shoots, with barley and wheat being preferred. It also consumes a small amount of animal matter, mostly insects and their larvae.
This duck usually arrives in late March or early April and leaves by late November. It regularly overwinters in some numbers at Edmonton, Wabamun Lake, Forestburg, Stauffer Creek, Dickson Dam, Medicine Hat, Calgary, Waterton Lakes, and Fort McMurray.

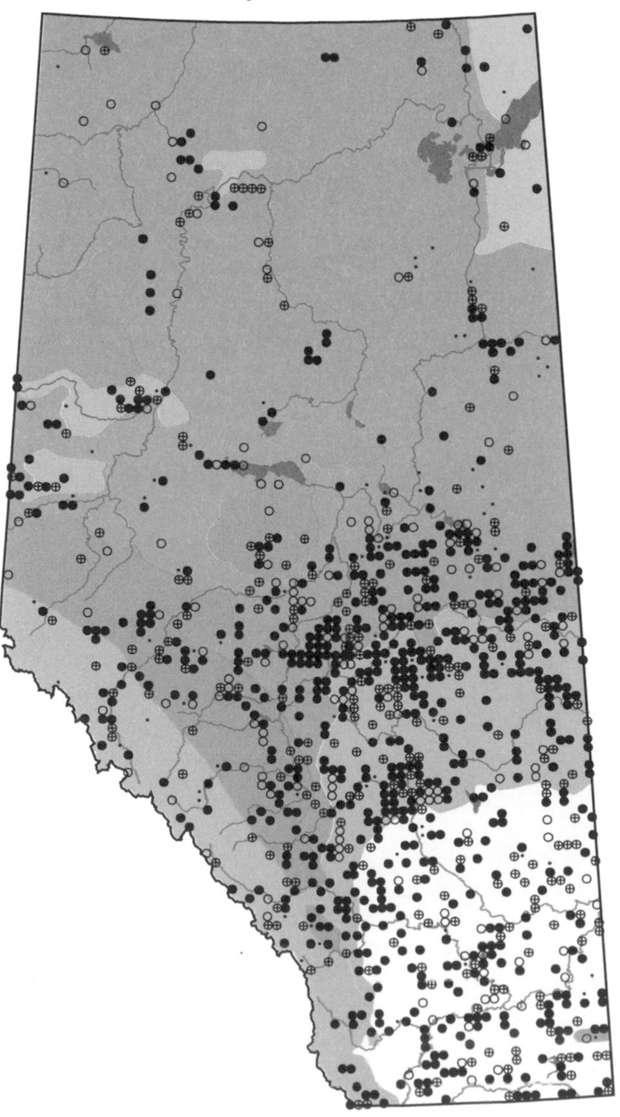

Northern Pintail

Anas acuta

NOPI photo credit: G. Beyersbergen

Status:
The most recent duck breeding population and habitat surveys indicate that pintail numbers across the prairie provinces are at a record low. Poor recruitment is the primary cause of the population decline.
Alberta is a major production area for this important game bird. However, breeding populations have decreased about 87.5% throughout the province (B. Turner pers. comm.). Northern Pintail recovery in the near future is dependent on a return to better water conditions and the successful implementation of habitat management strategies designed to improve the quality of upland nesting habitats.

Distribution:
The Northern Pintail breeds throughout the province where suitable habitat is found. It was recorded in 40% of surveyed squares in the Parkland region, 25% in the Grassland region, and 18% in the Boreal Forest region. It also nests in the Rocky Mountain and Foothills natural regions, where it was recorded in 6% of the squares surveyed.

Habitat:
The Northern Pintail favors open terrain with shallow ponds, sloughs, marshes, and reedy shallow lakes, usually with drier margins. These wetlands often provide an abundance of food and good visibility for avoidance of predators and other disturbances (Cross 1988). In the Boreal Forest region, open sedge and grass meadows are preferred for nesting.

Nesting:
Pintails usually arrive on the breeding grounds earlier in the spring than other ducks. Nesting is started soon after the wetlands become ice-free. They select areas with low or sparse vegetation and build their nests on the ground. Timbered or extensive brushy areas are generally avoided. The nest is a hollow lined with grasses, weeds, and down. It may be concealed in low vegetation, but is often located in open areas. The Pintail builds the nest further from water than do most other duck species, with nests in the Grassland region reported commonly situated 1-2 km from water (Duncan 1986).
There are usually 7-10 pale green to olive buff eggs, which are incubated by the female in 23-26 days. The young follow the mother from the nest a few hours after hatching. Fledging time varies with latitude and is undoubtedly influenced by the length of daylight and the daily time available to forage (Cross 1988). In most cases, young pintails fledge at 42-46 days.

Remarks:
The Northern Pintail is a large wary duck characterized by its slim-bodied and long-necked profile and sharply pointed tail. The color differences between the sexes is clearly displayed in the accompanying photograph.
Pintails prefer to feed in very shallow water at the surface or by up-ending; they rarely submerge entirely (Palmer 1976). In agricultural areas or during migration, the birds fly into grain fields to feed. Their diet is usually about 90% vegetation, made up of seeds of pondweed, sedges, grasses, and grain. Molluscs, crustaceans, and insects form the animal component of the diet. During egg laying, the animal component of the female's diet will increase to 50-60%.
The Pintail usually arrives in Alberta in mid-March. It is one of the few duck species that occurs in spectacular flocks in the spring (Pinel et al. 1991). Most of these ducks leave southern Alberta by the end of September or early October. Pinel et al. (1991) reported overwintering records in the province in nine of 10 years, including sightings at Calgary, Wabamun Lake, Jasper, and Banff.

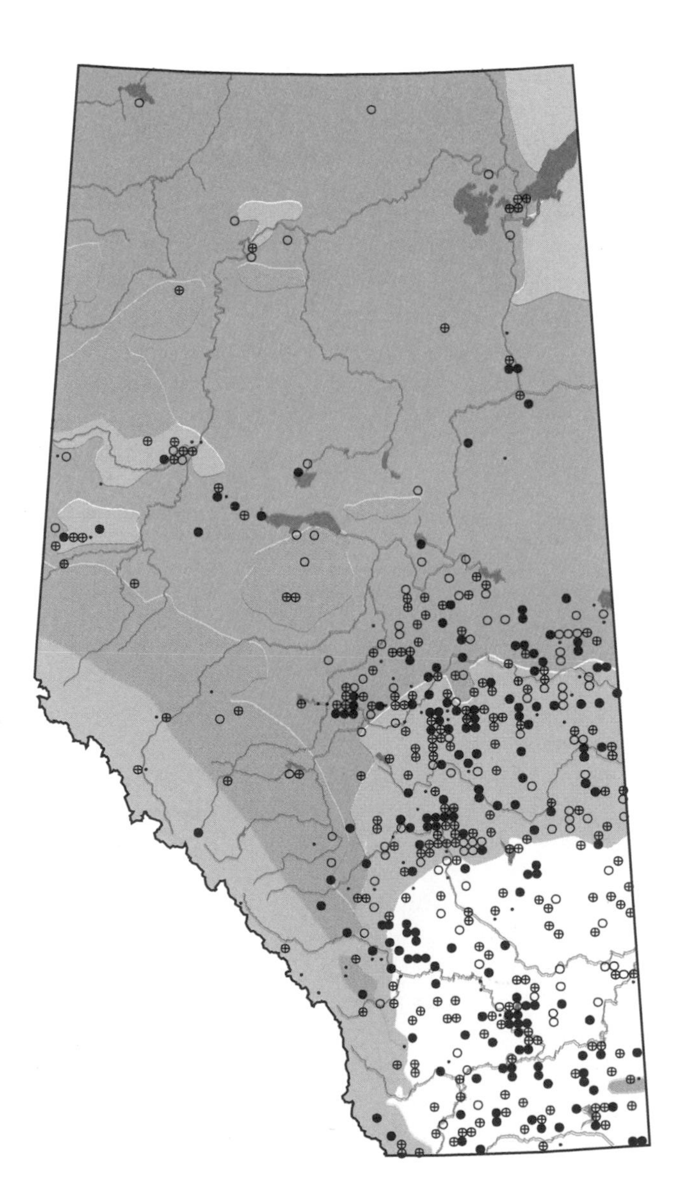

Blue-winged Teal

Anas discors

BWTE photo credit: K. Morck

Status:
The breeding population of the Blue-winged Teal has decreased about 25% throughout Alberta during the past decade, mainly due to serious habitat degradation resulting in nesting habitat losses (Wildlife Management Branch 1991). However, the population estimates for North America in 1991 show only a 10% drop over 1955-1990 averages.

Distribution:
Atlas data confirm that this is the common teal of the Grassland and Parkland regions. They also show that the Blue-winged Teal is more common in the south and central areas of the province than in the north. It was recorded in 54% of surveyed squares in the Parkland region, 31% in the Grassland region, and 38% in the Boreal Forest region. In the Foothills and Rocky Mountain regions, it was recorded in 23% and 15% of surveyed squares, respectively.

Habitat:
This duck prefers marshes, sloughs, ponds, dugouts, and ditches. It is also found along the weedy edges of lakes and slow moving rivers and streams. It is more of a shoreline inhabitant than one of open water (Palmer 1976).

Nesting:
The nest of this teal exhibits a wide diversity of construction, but is usually a basket-like structure composed of dry dead grass and nearby vegetation. It is generally lined with down and placed in a hollow in the ground with the top of the nest flush with the surface. The nest site is normally dry, but is seldom far from water. It is well-concealed where there is lush vegetation, as in meadows, drier sedge habitat, and hay fields in the vicinity of sloughs and ponds. These areas provide good cover quite early and, later, the vegetation often forms a canopy over the nest (Palmer 1976).
There are typically 8-12 dull white to pale olive-white eggs which are incubated 22-24 days by the female alone. The female leads the brood from the nest within 24 hours of hatching and takes them to heavy brooding cover. The female tends the young until they fledge at 40-44 days.

Remarks:
The Blue-winged Teal is one of our smallest ducks. The adult male is identified by the large white crescent in front of the eye on the otherwise grayish head. The forewing is light blue with a white border. The female has a rather uniform brown head, but with a whitish or buffy mark just behind the bill.
This teal is a dabbling duck and spends most of its time foraging on the surface or tipping in shallow water. Approximately two-thirds of its diet consists of plant food, including seeds of grasses and sedges and the seeds, stems, and leaves of pond weeds. The animal part of its diet consists of snails, aquatic insects, and crustaceans.
Blue-wings are reported to have the highest annual mortality of all the dabbling ducks, reaching 65% (Bellrose 1980). Contributing factors to this elevated death rate are that it is a favorite of hunters early in the season, it encounters high risk when migrating over the oceans to South America and lax hunting regulations in Central and South America (Root 1988).
This species usually arrives in southern Alberta in the last two weeks of April and most birds leave by the end of September (Pinel et al. 1991). Overwintering is highly unusual, with records coming from Calgary, Banff, Jasper, and Wabamun Lake.

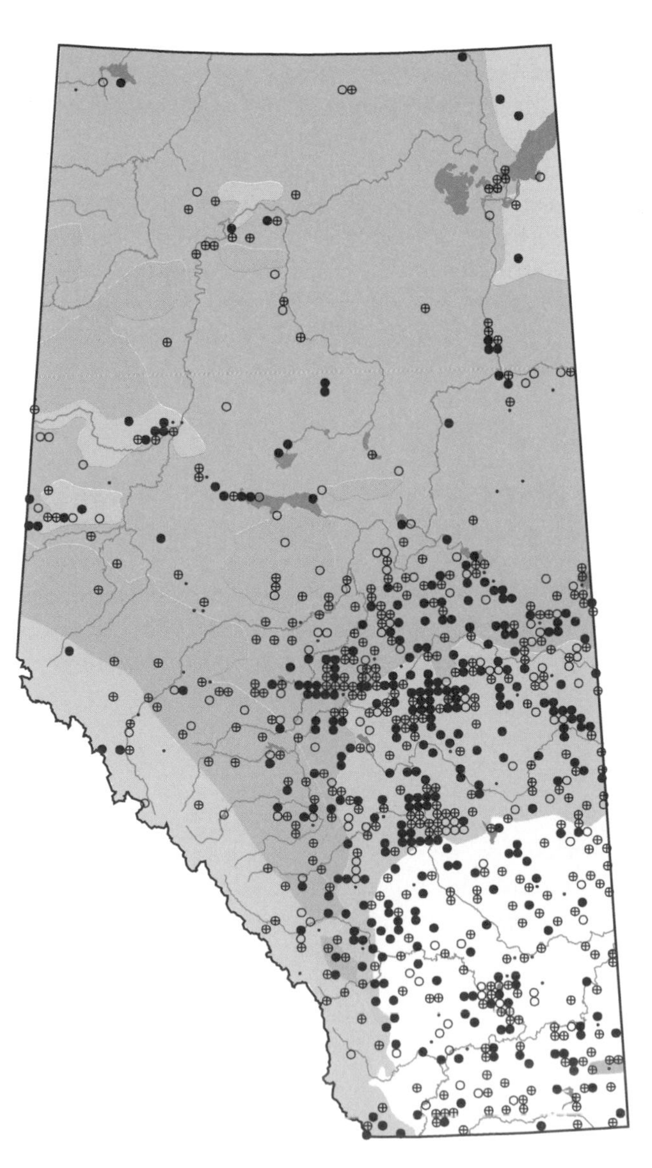

Cinnamon Teal

Anas cyanaptera

CITE photo credit: W. Burgess

Status:
Since about 1950, the Cinnamon Teal has occurred in increasing numbers in Alberta and Salt and Salt (1976) reported it as a regular but uncommon summer resident throughout the southern half of the province. The species continues to be uncommon in Alberta and there are no specific management concerns (Wildlife Management Branch 1991).

Distribution:
In the 1970s, this species continued to expand its range in Alberta. The most northerly records for the province were from Lesser Slave Lake, McClelland Lake, Fort McMurray, and the Grande Prairie area (Pinel et al. 1991). The data gathered by the Atlas Project confirm that the Cinnamon Teal has expanded its range from the Grasslands into the Parkland region, where it was recorded in 16% of surveyed squares, and sporadically north into the Boreal Forest region, where it was recorded in 4% of surveyed squares. Atlas data also confirm the first breeding record for this species in the Rocky Mountain region of Alberta in Waterton Lakes National Park. It was recorded in 4% of squares surveyed in this region.

Habitat:
This teal prefers shallow lake margins, marshes, ponds, ditches, dugouts with emergent vegetation, and muddy shorelines. On larger bodies of water, it is never found very far from shore. Large lakes do not usually attract them, nor are they often seen in wooded areas (Salt and Salt 1976).

Nesting:
There is great variety in sites chosen for the nest by this species. The nest may be in cover by the water's edge, or at some distance. Usually, it is very well concealed, deep in growing herbage, or in marsh or waterside plants on the ground (Harrison 1978). The nests range from neat basket-like structures, woven skilfully of soft grasses and carefully concealed in heavy reeds, to mere depressions in the ground, scantily lined with bits of trash and without any thought of concealment.
The female lays 8-12 whitish creamy buff eggs, which she incubates alone for 22-25 days. The Cinnamon Teal nest is commonly parasitized by the Redhead; these nests often have smaller clutches and desertion is more frequent (Palmer 1976). The male is known to stay in the vicinity of the nest and drakes often defend this territory.
The young are led from the nest, soon after hatching, to beds of reeds or other emergent vegetation, where they find shelter. They are frequently tended by both parents and fledge at approximately seven weeks. One brood is raised per season.

Remarks:
The male in breeding plumage is a very conspicuous cinnamon red, as the photo indicates. It also has a bright blue wing patch. The females are similar to female Blue-winged Teals, but are more heavily marked on the sides of head and chin, with a faint tinge of chestnut on underparts (Salt and Salt 1976).
The Cinnamon Teal usually feeds in shallow water, where its diet consists of about 80% vegetation and 20% animal matter. This includes the seeds and other parts of sedges, pondweed, and grasses, combined with molluscs and insects.
Most Cinnamon Teals arrive in southern Alberta during the last half of April and leave the province in late September or early October. There are no overwintering records for this species in Alberta. There are very few fall observations of Cinnamon Teal, but this may be due to the fact that the males are no longer distinctively marked and closely resemble the Blue-winged Teal.

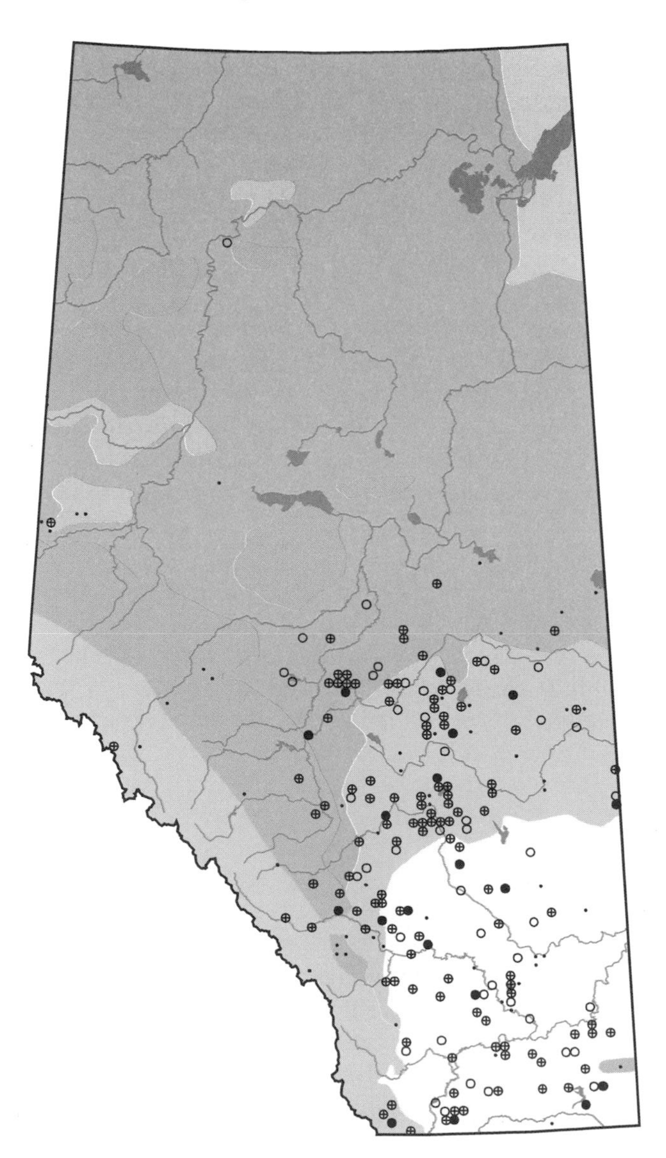

Northern Shoveler

Anas clypeata

NOSV

Status:
The efforts of game departments and private organizations across Canada, the United States, and Mexico to preserve and enhance waterfowl habitat have been beneficial to the Northern Shoveler. These projects have protected and created preferred habitat for this species. Breeding populations in Alberta seem to be stable.

Distribution:
The Northern Shoveler nests in all parts of Alberta. Atlas data indicate that the preferred breeding areas are in the Grassland and Parkland regions of the province. The data also confirm the scarcity of this species in the northern Boreal Forest region and the Rocky Mountain region, where suitable habitat is lacking.

Habitat:
This duck utilizes open marshy areas with shallow waterways, muddy freshwater lakes and sloughs with good cover, overgrown pools, bogs, or slow creeks. It is preferable if these contain abundant aquatic vegetation. These water bodies, which may become stagnant or even dried-out and mostly mud, produce an abundance of food for the shoveler. It can be found in water that is clean and clear, muddy, flowing or stagnant, considerably alkaline, or even heavily polluted with sewage or industrial waste (Palmer 1976).

Nesting:
The shoveler prefers a dry site, sheltered by some grass or growing vegetation, where it nests on the ground normally not far from water. The nest is a hollow that is lined with plant material, down, and some feathers (Harrison 1978). There are usually 8-12 dull, pale olive to pale greenish-gray eggs which are incubated by the female for 22-25 days. Shoveler nests have been reported to be parasitized by the Redhead and the Lesser Scaup. The female shoveler leads her young away from the nest within 24 hours after hatching.
The male stays in the vicinity of the nest and frequently remains with the brood for a short time after hatching, but does not assist with the raising of the young. The young fledge at approximately eight weeks.

Remarks:
The bill of this dabbler is long and narrow at the base and becomes wider at the end, resulting in the duck's local name of "spoonbill". The breeding male has a green head, a dull black throat, a white breast and shoulders, and a blue forewing. The female is buff and brown except for the blue forewing.
The shoveler is essentially a surface feeder, filtering the water to obtain small animal life and seeds. It paddles slowly, moving its head from side to side as water is steadily drawn in at the tip of the bill and expelled at the base. About 60% of the diet is vegetation, mainly grasses, sedges, pondweed, and algae. The animal part of its diet is mainly molluscs, aquatic insects, and larvae.
This species arrives in southern Alberta in late March and early April. It is one of the earliest migrants in the fall; most Northern Shovelers leave southern Alberta by the end of September. They have been recorded overwintering at Wabamun Lake (Pinel et al. 1991).

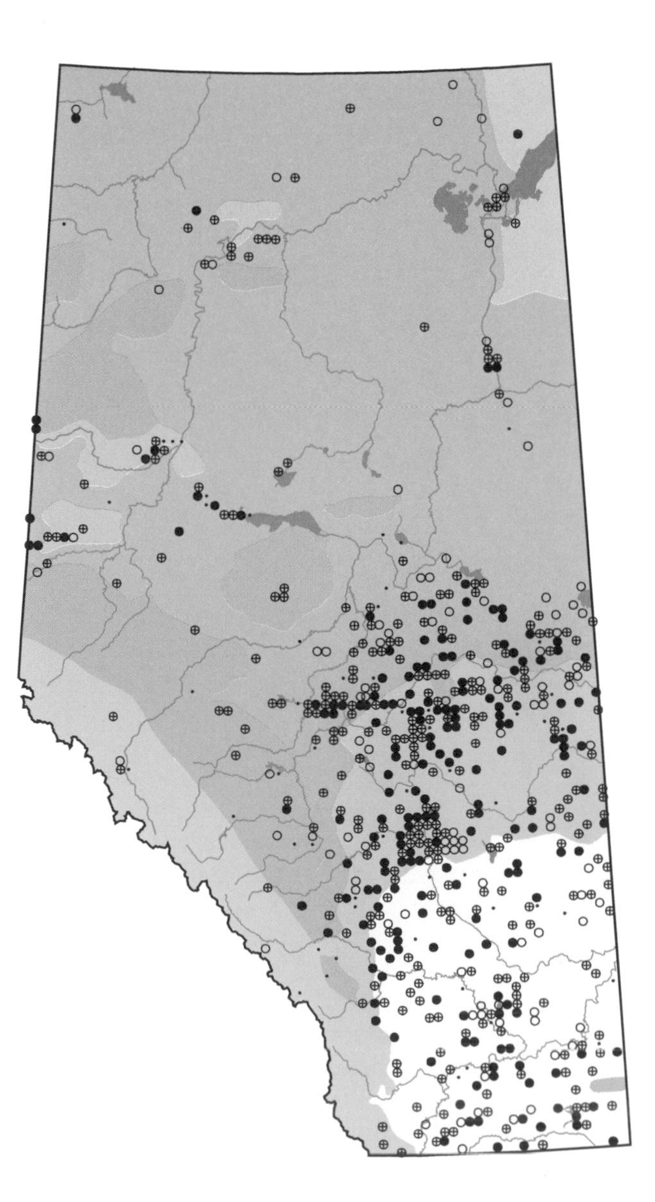

Gadwall

Anas strepera

GADW photo credit: T. Thormin

Status:

The Gadwall population is stable in comparison to other duck species. The 1991 continental population is estimated to be 23% above the 1955-1990 average. This success may be attributed to the species' tendency to nest in dense cover and to concentrate in secure nesting areas such as islands.

Distribution:

This duck breeds mainly in the Parkland and Grassland regions, where it was recorded in 38% and 26% of surveyed squares, respectively. It also utilizes the southern Boreal Forest region (20% of surveyed squares), with confirmed breeding north of Lake Athabasca and west of Bistcho Lake. Atlas data indicate that it occurs in small numbers in the Rocky Mountain, Foothills, and Canadian Shield natural regions.

Habitat:

The Gadwall utilizes marshes, sloughs, and shallow lake margins bordered by good cover where the grassy margins extend well back from the shoreline. In particular, the presence of grassy islands is of considerable significance in determining nest distribution and density (Johnsgard 1975). Throughout Alberta, alkaline waters are favored over fresh, perhaps because these lakes have fairly stable water levels and an abundance of submerged aquatics (Palmer 1976).

Nesting:

Whenever possible, the Gadwall will choose an island as the site for its nest. The nest is on the ground, usually near water, and is well-concealed by vegetation. It is a hollow that is lined with grass, reeds, weeds, down, and feathers. Preferred nesting cover consists of dense coarse vegetation, heavy grass, or brush such as that provided by shrubby willows (Johnsgard 1975).

The female incubates 8-12 dull creamy white eggs for 25-28 days. Within hours of hatching, the female leads her brood to deep water marshes and the edges of large water bodies, where they fledge in about seven weeks. Gadwalls are single-brooded.

Remarks:

Although one of the easiest species of ducks to identify in the hand, the Gadwall is perhaps the waterfowl most misidentified by hunters. This is primarily because of the species' lack of brilliant color (Johnsgard 1975). The breeding male is gray on its upper body with a white under body and black hindquarters. The female is best recognized by her association with the male. Both have yellow legs and black-gray bills.

Gadwalls are almost exclusively surface-feeders, although they have been observed diving for food. Thus, they are largely dependent on food that they can reach by tipping-up, and tend to feed in rather shallow marshes with abundant submerged plant life growing close to the surface (Johnsgard 1975). In southern Alberta, Gadwalls of pre-flight age eat surface invertebrates during their first few days. These are gradually replaced by aquatic invertebrates and plants until, by three weeks of age, Gadwalls are essentially herbivorous (Palmer 1976).

Gadwall breeding population densities in the Alberta Parkland Natural Region are among the highest in North America; some of the highest nest densities on the continent occur in Jessie Lake near Bonnyville (B. Turner pers. comm.).

In most years, the first spring migrants arrive in southern Alberta from the end of March to mid-April. In the fall, most birds have left by early October. Gadwalls have been recorded overwintering at Calgary, Wabamun Lake, and Galahad.

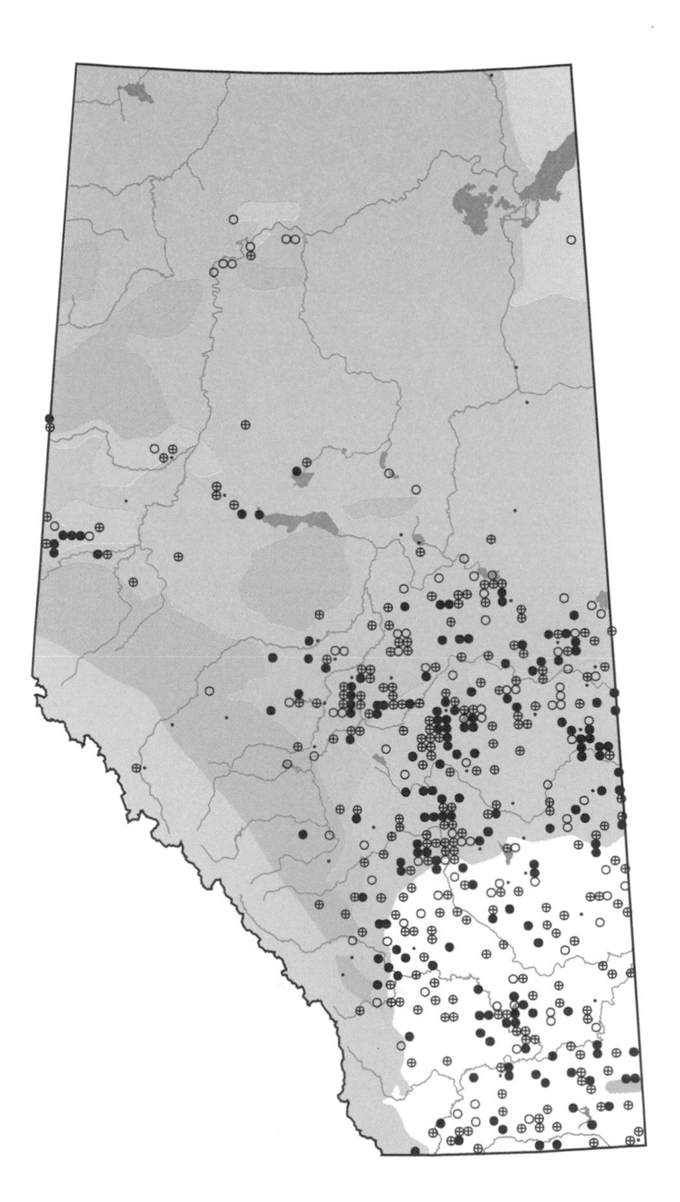

American Wigeon

Anas americana

AMWI photo credit: K. Morck

Status:

In the 1970s, the American Wigeon breeding population was estimated at three million birds. Wintering populations were reported to have increased in the Mississippi Flyway, to have stayed relatively constant in the Central Flyway, and to have declined significantly in the Atlantic and Pacific Flyways (Bellrose 1980). Recent population estimates indicate that this species is showing a long-term decline. Breeding populations in Alberta in 1991 were 67% below the long-term average (B. Turner pers. comm.). Serious habitat degradation over the past ten years is a major concern in the province (Wildlife Management Branch 1991).

Distribution:

The American Wigeon breeds across the prairie provinces and north into the Yukon and western Northwest Territories. In Alberta, it was recorded in 42% of surveyed squares in the Parkland region, 34% in the Boreal Forest region, and 22% in the Grassland region. Salt and Salt (1976) reported that the American Wigeon nests in appropriate habitat throughout Alberta but is absent from the Rocky Mountain region, where it occurs only as a migrant. The species has now been confirmed nesting in Waterton Lakes National Park, with other probable nest sites near Jasper and the Vermilion Lakes.

Habitat:

The American Wigeon prefers larger sloughs and ponds and the marshy areas bordering large lakes. Wigeons prefer to nest around lakes and sloughs that are surrounded by dry meadows (Johnsgard 1975).

Nesting:

The American Wigeon nests on dry ground in an area where the nest can be hidden in tall grass, brush, or shrubbery. In the Boreal Forest region, nests may be in leaf-litter by a tree or bush (Harrison 1978). Often the nest is some distance from water — up to 400 m has been reported. The nest is made of grass and reeds and is lined with down and feathers. It is placed in a hollow in the ground. The incubating female will usually cover the eggs with down when she leaves the nest.

The female lays 9-11 creamy white eggs. Once the eggs are laid, the male joins other American Wigeon males and leaves the incubation and care of the young to the female. Brooding female wigeons are known to be very protective of their young, using distraction displays to allow the young to flee to cover. Incubation takes 23-25 days; nestlings fledge at 45-48 days. This species is single-brooded.

Remarks:

The American Wigeon is often called the Baldpate, because of the male's characteristic white crown. In flight, a large white wing patch is conspicuous on the front of the wing in both the male and the female. While most dabblers feed on the seeds of aquatic plants or grains, the Wigeon prefers the succulent stems and leafy parts of submerged vegetation. The practice of taking food directly from the bills of diving ducks has earned this species the nickname "poacher". Most often, it will be seen helping itself to food from Canvasbacks, Redheads, and American Coots. There is one report of a Wigeon taking food from a muskrat (Fisher 1975).

The American Wigeon arrives in southern Alberta during the latter half of March and leaves the province by the end of October (Pinel et al. 1991). There are overwintering records for this duck at Lake Wabamun and at Calgary.

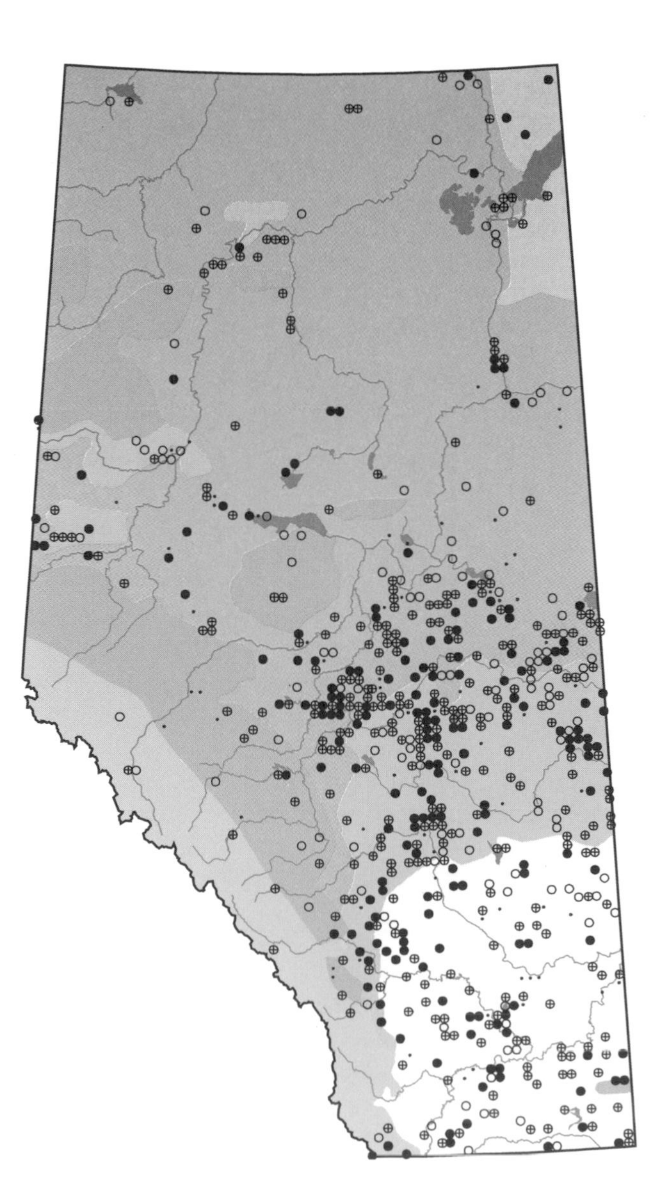

Canvasback

Aythya valisineria

CANV

Status:

The Canvasback is the least abundant of the major game duck species in North America. This species was on the Blue List from 1975 to 1981, and was designated a species of "special concern" in 1982 and 1986 (Tate 1986). Populations decreased during the 1980s and, in 1991, the continental estimate was 16% below the long term average. The harvest of this species is closely regulated because it is one of the most prized game ducks.

Distribution:

Atlas data indicate that the Canvasback prefers the Parkland region, where it was recorded in 36% of surveyed squares. The southeastern portion of the Boreal Forest region was also a favored location. As earlier reported by Salt and Salt (1976) and confirmed by the present study, this duck rarely occurs in the Rocky Mountain region and in the extreme north of the Boreal Forest region. Records show that it also breeds in the Peace-Athabasca Delta area.

Habitat:

This duck prefers lakes and ponds with emergent vegetation and vegetated margins in the Parkland and Grassland regions. In the Boreal Forest region, it utilizes open marshes. In southern Alberta, the highest incidence of Canvasback pairs was on a shallow lake with a maximum depth of eight feet, having scattered stands of bulrushes, shorelines dominated by rushes, sedges, and spike rush, and with several cattail-covered islands (Keith 1961).

Nesting:

The nest is frequently concealed in a clump of vegetation growing in water, but may be in more open sites in sedges or hidden in tall waterside plants (Harrison 1978). Several studies have shown that Canvasbacks prefer to nest in cattail cover. The female constructs the nest, which is usually a large well-built structure composed of reeds and sedges and lined with down. The female lays 7-10 grayish-olive eggs and incubates them for 24-27 days. The male deserts the female soon after egg laying commences. The Canvasback nest is frequently parasitized by both the Redhead and Ruddy Ducks.

The young are led to open water soon after hatching. The female is not as protective of her young as some other species and abandons them before they fledge, at approximately nine weeks.

Remarks:

The Canvasback is one of the largest diving ducks found in the province. The male is distinguished by its long bill, sloping forehead, chestnut-colored head and neck, and canvas-colored back. The female has the same distinctive profile and has a light brown breast and grayish sides.

They dive for the majority of their food, but they are also known to tip to feed or feed on the surface when aquatic insects are hatching. Vegetable matter makes up about 80% of the ducks diet, with pondweeds being of most importance (Palmer 1975). The remainder of the diet is made up of molluscs and aquatic insects.

This species arrives in southern Alberta from late March to early April (Pinel et al. 1991). Beaverhill and Utikuma Lakes are major moulting areas for Canvasbacks in the province. Most Canvasbacks depart major staging lakes such as Beaverhill during the first two weeks of October and leave southern Alberta by mid-November. The Canvasback has been recorded overwintering at Calgary, Edmonton, Galahad, and Wabamun Lake.

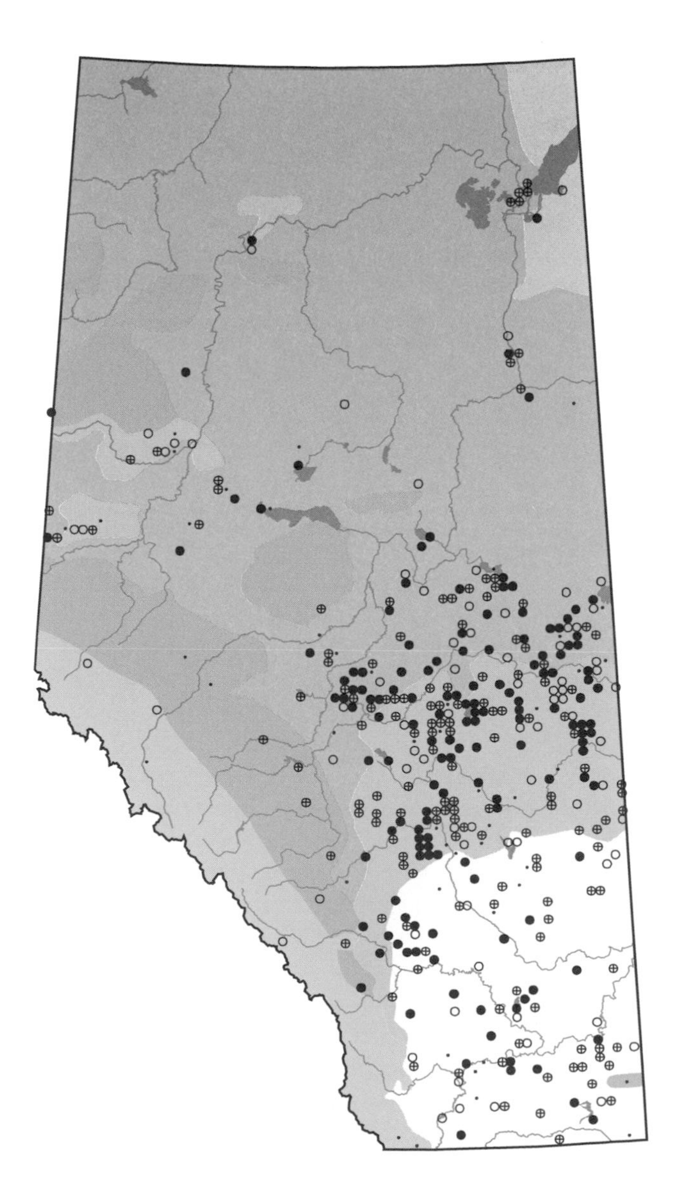

Redhead

Aythya americana

REDH photo credit: W. Burgess

Status:

The status of the Redhead is similar to that of the Canvasback across North America. Although more abundant than the Canvasback, the Redhead is scarce enough to need protection (Root 1988). In Alberta, the breeding population appears to be stable. Habitat losses have not affected diving species as much as dabbling ducks (Wildlife Management Branch 1991).

Distribution:

The Redhead breeds locally throughout Alberta. Atlas data indicate that the Redhead prefers central parts of the province, including the Parkland region, where it was recorded in 37% of surveyed squares. It was also recorded in 16% of surveyed squares in the Boreal Forest region, but was concentrated in the southern portion of this region. Although previously reported as not breeding in the Rocky Mountain region (Salt and Salt 1976), it was recorded in a small number of surveyed squares in this region, mainly along river courses. There were no records of Redheads breeding in the northeast corner of the province, although Francis and Lumbis (1979) reported southbound migrants in the Fort MacKay area.

Habitat:

Weller (1964) described the Redhead's breeding habitat as non-forested country with water areas sufficiently deep to provide permanent, fairly dense emergent vegetation for nesting cover. It breeds by fresh water, usually in the taller vegetation bordering lakes and sloughs, occasionally in more open sites in similar areas (Harrison 1978). The Redhead will nest on smaller bodies of water than the Canvasback, and females with young are often seen in reedy roadside ponds in central Alberta (Salt and Salt 1976).

Nesting:

Redheads nest on the ground or over open water, usually in cattails or similarly high vegetation. Keith (1961) reported that half the Redhead nests he found in southeastern Alberta were on land and many were very poorly concealed. The nest, made by the female, is a deep hollow of dry vegetation lined with reeds and down. The female normally lays 9-14 cream or buff eggs, which are incubated for 24-28 days. Redhead females are relatively poor parents, often deserting their brood at an early stage (Johnsgard 1975). The young fledge at 8-9 weeks.

The Redhead has the greatest tendency of all ducks to exhibit avian nest parasitism — the laying of eggs in the nest of another bird. In Alberta, Mallards had the greatest percentage of nests parasitized by Redheads, whereas Lesser Scaups had the largest number of parasitic eggs per nest (Geroux 1981). This study also showed that Redhead parasitism affected reproductive output of hosts by increasing nest desertion.

Remarks:

The adult male has a bright coppery head with a gray back and sides. It has a bluish bill with a black tip. It resembles the Canvasback, but has a different head profile than that species, with an abruptly rising forehead and a slightly concave bill (Godfrey 1986). The female is uniformly brownish, is slightly darker on the crown, and has a whitish chin.

Vegetation is the usual fare of this diving duck, preferably pondweeds, algae, and sedges. It also feeds on insects and molluscs.

Redheads arrive in Alberta in late March or early April and have migrated from Alberta in early November. There are reports of overwintering in Calgary, Edmonton, Galahad, and Wabamun Lake (Pinel et al. 1991).

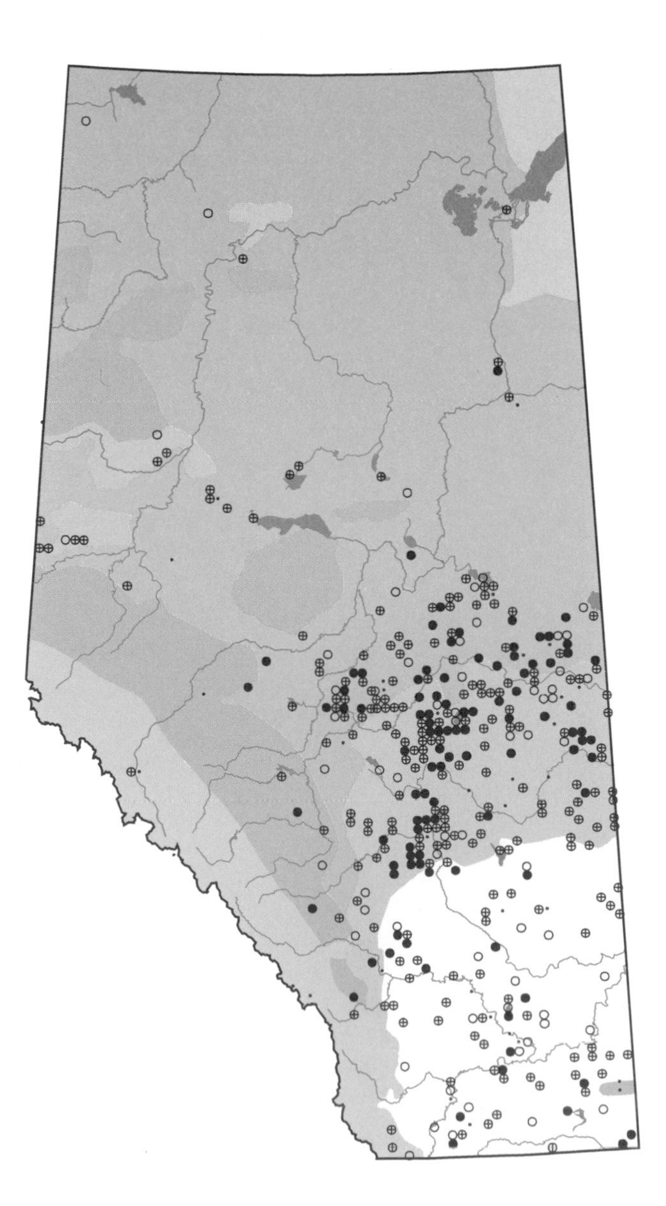

Ring-necked Duck

Aythya collaris

RNDU

Status:

The breeding population of this duck in Alberta appears to be stable throughout the province. There is no clear trend in population either in Alberta or on the continent, although it appears to be common in the southern portion of the Boreal Forest Natural Region.

Distribution:

The Ring-necked Duck breeds in north and central Alberta south to Red Deer, with some localized breeding in the Grassland region and in the Cypress Hills. Atlas data confirm reported nesting farther southwest near Bragg Creek, Turner Valley, Seebe, and Banff (Salt and Salt 1976), and in the southwest corner of the province (Sharp 1973). This species was reported as common during the breeding season in Banff National Park and uncommon in Jasper National Park (Holroyd and Van Tighem 1983). It was recorded in 38% of surveyed squares in the Canadian Shield region, 25% in the Boreal Forest region, 15% in the Parkland region, 24% in the Foothills region, and 10% in the Rocky Mountain region. It was recorded as a confirmed breeder in Jasper National Park in four separate squares.

Habitat:

These diving ducks prefer marshes, swamps, and bogs. They show a preference for fresh or acidic water rather than the brackish or alkaline water of the Grassland region. Unlike most diving ducks, Ring-necks are not often found far out on large expanses of open water (Godfrey 1986). In northern Alberta, this species prefers woodland sloughs and lakes, and muskeg regions with inaccessible ponds (Salt and Salt 1976). In the Rocky Mountain and Foothills regions, Ring-necked Ducks utilize small lakes and beaver impoundments.

Nesting:

The nest is built in damp situations and not on dry ground. The female usually selects the nest site, which is typically situated on the vegetated edge of open water, on floating islands, or on clumps of vegetation in open marshes. Ramps may be built to nests elevated above the surface, and runways to the nearest water are established (Johnsgard 1975). The nest is a hollow lined with grass, plant material, and down, and may be built up into a substantial cup (Harrison 1978). The female incubates the 8-12 olive brown eggs for 25-28 days. She normally leads the young from the nest within 24 hours, but, unlike most other waterfowl, this duck may bring her young back to the nest for brooding purposes for two to four or more days after hatching (Johnsgard 1975). The broods are not abandoned prior to fledging, which takes place at 7-8 weeks.

Remarks:

The breeding male has a glossy black head and neck, a black back and breast, gray sides, and a white band extending upward in front of the folded wing. Its bill is gray-blue with a white ring near the black tip and it has an inconspicuous chestnut ring at the base of the neck. The female is grayish brown with a white eye ring and a white ring near the tip of the bill.

Ring-necked Ducks usually feed in shallower water than most diving ducks. The diet is mainly vegetation, such as pondweeds, sedges, bulrushes, and grasses. Aquatic insects and molluscs make up the remainder of the diet.

This duck arrives in Alberta in mid-April and is gone from the province by early November. The Ring-necked Duck has been recorded overwintering in the province at Wabamun Lake and Edmonton.

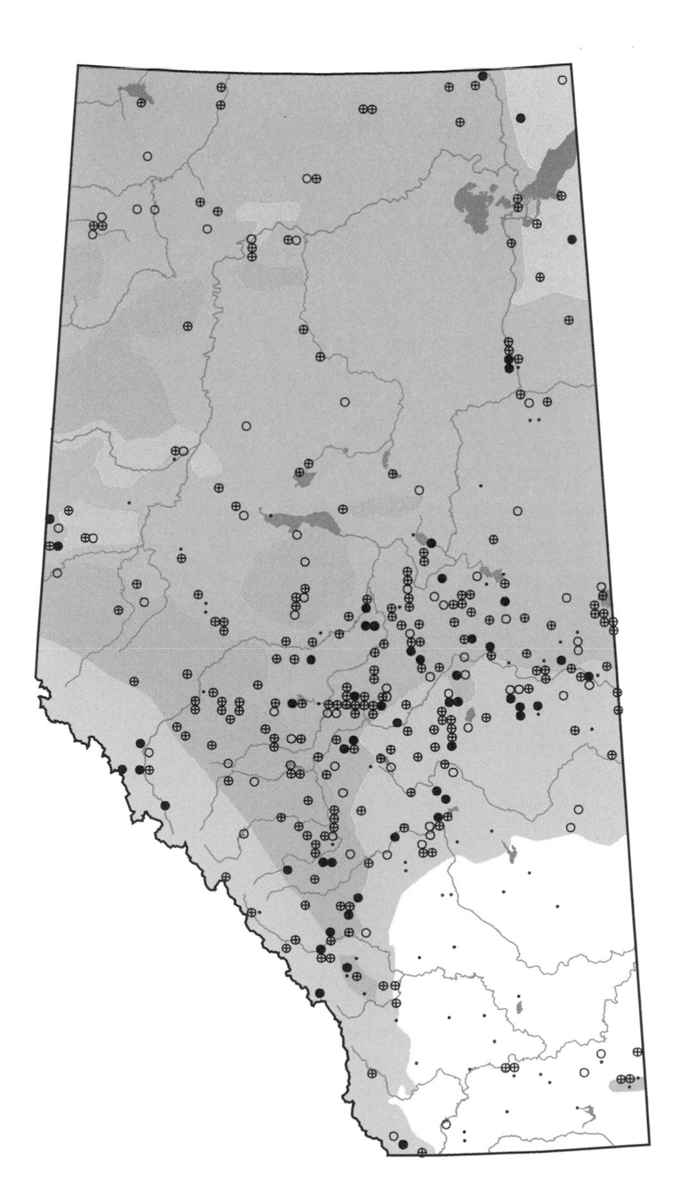

Lesser Scaup

Aythya affinis

LESC photo credit: G. Beyersbergen

Status:
The Lesser Scaup is one of the most abundant ducks in North America, but information relating to its ecology is limited (Allen 1986). The breeding population of the Lesser Scaup has decreased about 30% throughout Alberta, and there is a concern regarding the habitat losses of this upland nesting diving duck (Wildlife Management Branch 1991).

Distribution:
Salt and Salt (1976) reported that the Lesser Scaup breeds around suitable bodies of water throughout Alberta but is less common on the prairies, where many sloughs are too shallow for diving ducks (Salt and Salt 1976). Atlas data confirm this, as this duck was recorded in 40% of surveyed squares in the Parkland region, 34% in the Boreal Forest region, 31% in the Canadian Shield region, and 21% in the Grassland region. Holroyd and Van Tighem (1983) listed this as an uncommon species in Jasper and Banff national parks. Although there were 32 breeding records for this species in the Rocky Mountain region, only one was confirmed breeding.

Habitat:
Lesser Scaups in Alberta are primarily dependent upon permanent and semi-permanent wetlands. They prefer those lakes where pondweeds grow to within a foot or so of the surface and lakes that have reed beds and cattail marshes along their borders (Salt and Salt 1976). Relatively tall and dense herbaceous vegetation in close proximity to these wetlands provides preferred nest cover and, apparently, supports the highest density of Lesser Scaup nests (Allen 1986).

Nesting:
This scaup nests later than any other duck in Alberta. It prefers to nest in uplands rather than over water, but the nest is usually closer to water than that of any of the surface feeding ducks recorded in Alberta (Keith 1961). It also requires vegetation 20-60 cm high in which to conceal the nest. Nests are commonly found in sedge meadows. If available, these scaups prefer to establish nests on islands, where their reproductive success is high (Keith 1961; Giroux 1981). The nest is a hollow that is lined with grasses, feathers, and down. The female lays 9-12 olive-buff eggs and incubates them for 23-26 days. After hatching, the young are led to open water, where the female tends them. There are several reports of two broods merging, with both females present. The young fledge at approximately seven weeks.

Remarks:
The male of this species may be confused with the Ring-necked Duck, but the scaup has a pale gray back instead of black, and has a white stripe on the wing that is visible in flight. It also has a purplish gloss on the head. The female is buffy brown and has a distinctive white facial mark.
Although Lesser Scaups are good divers, they also feed in shallow water, tipping up. They normally consume about 60% plant material, consisting mainly of pondweeds, grasses, and sedges (Kortright 1943). The animal component of the diet is mainly molluscs and aquatic insects. Allen (1986) reported that, on breeding grounds, over 90% of their diet is animal food, mainly aquatic invertebrates.
Pinel et al. (1991) reported that this species first arrives in Alberta in late March or early April. The highest nest densities in North America occur on Jessie Lake, near Bonnyville (B. Turner pers. comm.). Lesser Scaups normally leave the province by late November, although there are overwintering records from Edmonton, Calgary, Wabamun Lake, and Galahad.

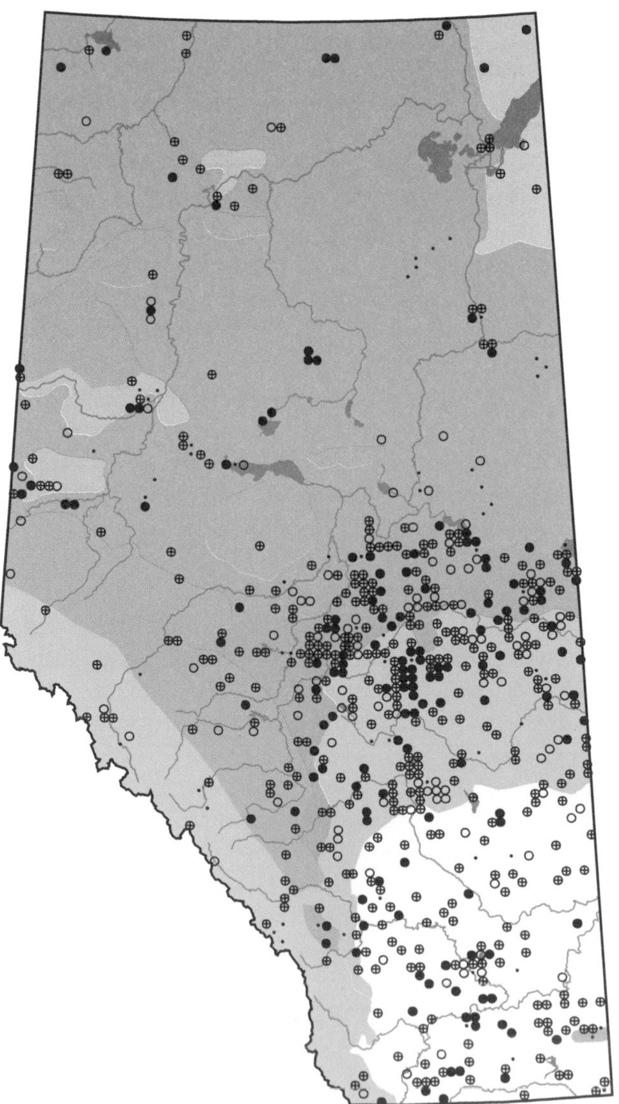

Harlequin Duck

Histrionicus histrionicus

HADU

Status:

There are two distinct widely separated populations of this duck in Canada. The Pacific population is quite large, at approximately one million individuals, and the eastern North American population is likely less than 1,000 individuals (Goudie 1991). The Committee on the Status of Endangered Wildlife in Canada has declared the eastern population as endangered. In Alberta, which is the eastern limit of the Pacific population, the Harlequin Duck appears to be stable in their limited breeding range. Harlequin Ducks breed and raise their young on fast-flowing streams, so they are susceptible to detrimental effects on stream ecology which may be caused by logging, mining and grazing (Breault and Savard 1991). Alberta populations are not as susceptible because most of their breeding range in Alberta is in national and provincial parks.

Distribution:

Atlas data confirm previous reports (Salt and Salt 1976; Godfrey 1986) which describe the breeding range of the Harlequin Duck. Of the 39 breeding records obtained during the five year Atlas survey, 31 were in the Rocky Mountain Natural Region, and the remaining 8 records were from the adjoining Foothills region. Salt and Salt (1976) reported that this duck was also recorded at Cooking Lake, Beaverhill Lake, Flemming Lake in the Caribou Mountains, and there is a wintering record from Little Buffalo River in Wood Buffalo National Park. Pinel et al. (1991) reported two late spring records from northeastern Alberta, one at Gregoire Lake and one from the Muskeg River. They also reported a winter record from Calgary.

Habitat:

Several studies on the reproductive ecology of Harlequin Ducks all point to the importance of pristine breeding habitat away from human disturbance (Breault and Savard 1991). This species is found in Alberta where there are fast flowing mountain streams providing suitable food and sheltered nest sites. Most streams used for breeding are surrounded by forests or patches of willow or alder; this duck appears to show a limited attraction to tundralike habitats (Johnsgard 1975).

Nesting:

The nest is on the ground not far from the water and it is usually protected from above by dense vegetation. Some nests have been found in crevices, under rocks, and in driftwood piles. It is a hollow that is lined with leaves, grass, feathers, and down. The female lays 6-8 creamy buff eggs, which she incubates alone for 27- 30 days. After hatching, the female leads her young to water, preferably a secluded spot. The young can fly in 40 days (Harrison 1978).

Remarks:

These are small diving ducks that appear quite dark on the water (Johnsgard 1975). The male is slate blue with bright chestnut sides and flanks. Females are dull brown, but both males and females have grayish-white areas on the cheeks, white between the eye and the forehead, and a rounded white spot between the eyes and the back of the head.

The Harlequin feeds in several different ways: wading in shallows prying among stones, swimming with the bill below water, up-ending in shallow water, diving and walking on the bottom like a dipper, or diving in deep water (Palmer 1975). The majority of its diet is of animal origin, with molluscs, crustaceans, and insects being the major components. Insect foods comprise the majority of the diet during the breeding season.

Most Harlequin Ducks arrive in Alberta in early May from the west coast (Pinel et al. 1991). The males leave the breeding areas shortly after incubation begins and return to the coast. Females remain in the province until about mid September.

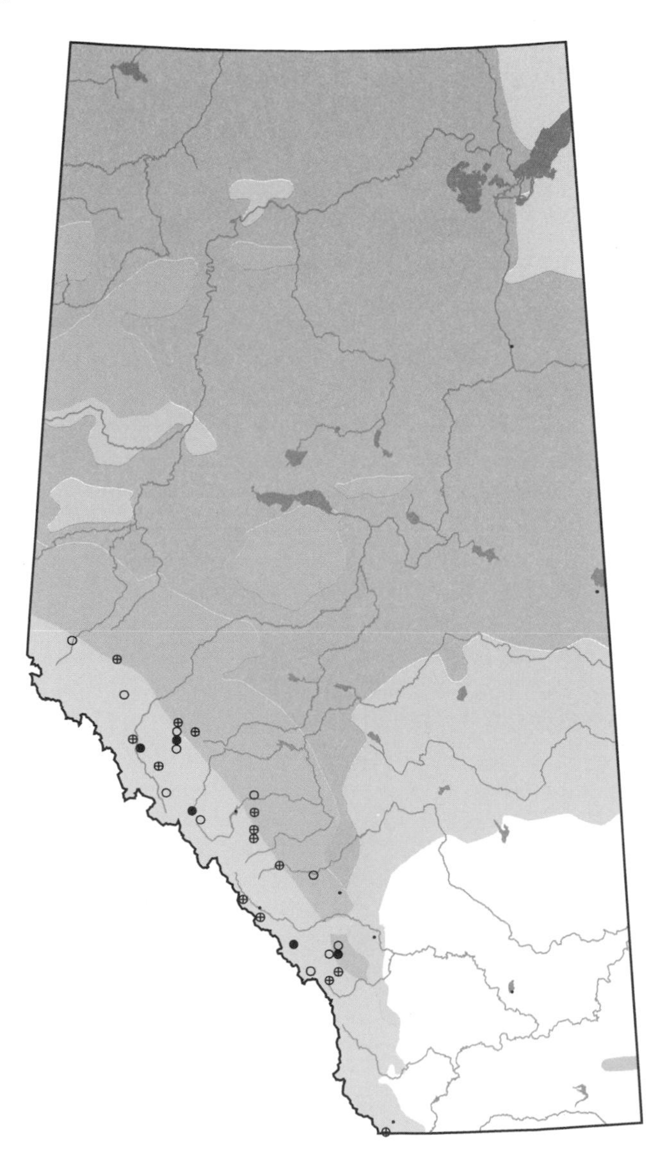

Surf Scoter

Melanitta perspicillata

SUSC

Status:
Historical records and Atlas surveys indicate that the Surf Scoter is a rare breeder in Alberta, with breeding records widespread through central and northern parts of the province. Salt and Salt (1976) reported one old breeding record from Elk Island National Park and a more recent one in Northern Alberta. Pinel et al. (1991) reported one additional breeding record from the same area as the latter. In the five year survey period of the Atlas, only six breeding records were noted and only one of these was a confirmed breeding record. Bent (1962b) reported that the nests of this species are difficult to find, because they are usually in such inaccessible places and are so widely scattered.

Distribution:
The Surf Scoter breeds in Alaska and locally in northern continental Canada (Godfrey 1986). Alberta is on the southern limit of its continental breeding range. Historical breeding records were reported from Elk Island National Park and Wentzel Lake in the Caribou Mountains (Salt and Salt 1976; Pinel at al. 1991). The breeding records from Atlas surveys come from Rocky Mountain House, Grande Prairie, Bistcho Lake, and the extreme northeast corner of the province. They have been observed in summer at Fairview and at Obed Lake, and on Barrier Lake near Seebe (Salt and Salt 1976). On migration, they have been observed in all the natural regions of the province.

Habitat:
The Surf Scoter appears to favor quiet and slow moving waters of the forest zone and semibarrens (Palmer 1975). Freshwater ponds, muskeg bogs, lakes, or streams with adjacent shrubby cover or woodland provide acceptable breeding habitat (Johnsgard 1975).

Nesting:
The nest is usually placed away from the water. In a wooded area, it is sheltered by low branches of conifers and in more open areas it is sheltered by shrubs or tall grass (Harrison 1978). The nest is a hollow that is lined with grass, other plant material, and feathers, and has an inner lining of down. The down insulates the nest from the marshy ground and is used to cover the eggs when the female leaves to feed. The nest is usually well concealed from view. The number of eggs varies from 5-9, with the typical clutch being seven pinkish or creamy white eggs. The incubation and fledging periods for this species are unknown.

Remarks:
The adult male Surf Scoter is a thickset, medium-sized diving duck. It is mainly black with a white triangle on the forehead and another on the nape, a white eye, and a black patch on the sides of the bill. The bill is heavy at the base and marked with red, orange, black, and white. The female has a brownish-black head with white cheek, ear, and nape markings. Both sexes lack the white on the wings that is present on the White-winged Scoter.

This duck dives with an awkward splash as it springs clear of the surface and lunges first forward in a high arc and then downward (Palmer 1975). It opens its wings as it goes under and uses them in its subaqueous flight in its daily pursuit of food (Bent 1962b). The main part of its diet is animal food, made up of molluscs, crustaceans, insects, and small fish. The Surf Scoter seems to be more partial to vegetable food than other scoters. These include seeds, subterranean root stems, and green vegetative matter (Cottam 1939).

Records of this species for the 1970s indicate arrival in Alberta in mid-May (Pinel et al. 1991). The Surf Scoter leaves the province in late October and early November.

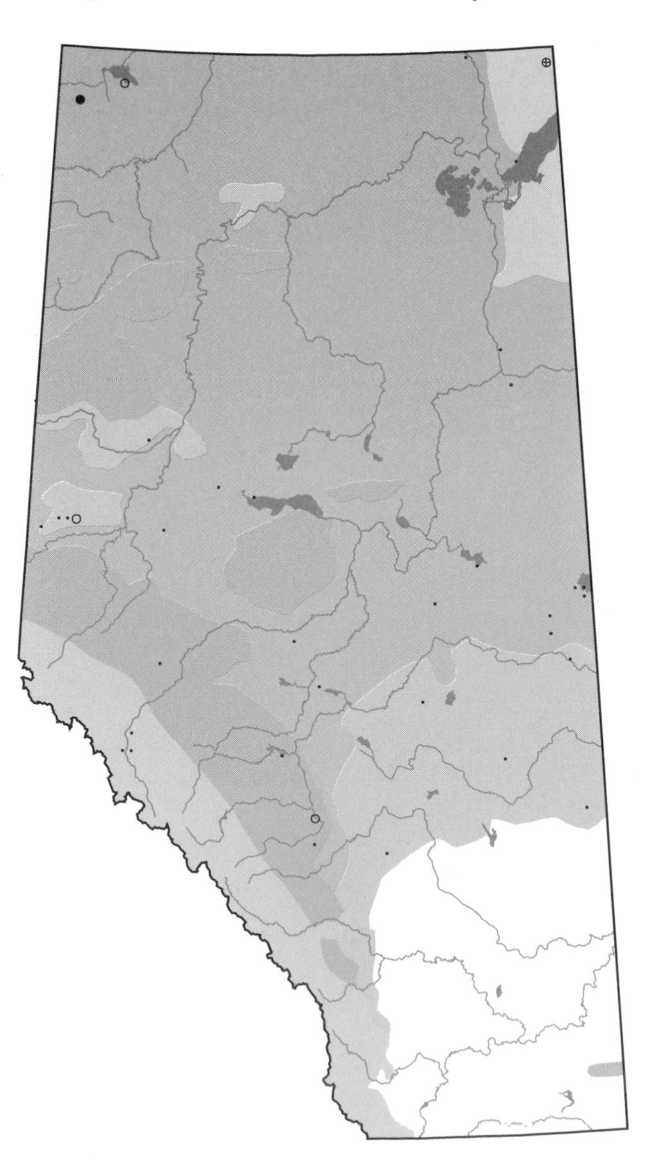

White-winged Scoter

Melanitta fusca

WWSC photo credit: G. Beyersbergen

Status:

In Canada, this is generally the most common breeding scoter species (Johnsgard 1975). It breeds across the whole of western North America, but it is not found as an abundant breeder anywhere in particular in its extensive range (Kortwright 1967). The Wildlife Management Branch (1991) reported that there is a small but stable population of White-winged Scoters in Alberta. Atlas data suggest that it is an uncommon but widespread breeder in the province.

Distribution:

Salt and Salt (1976) reported the distribution of this duck as around most of the large lakes of the prairie provinces, except in the Rocky Mountains. Atlas surveys recorded the White-winged Scoter in 3% of surveyed squares in the Grassland and Foothills regions, 8% in the Parkland and Boreal Forest regions and 13% in the Canadian Shield. Breeding was confirmed in the Cypress Hills. Distribution was recorded throughout the province, except in the Rocky Mountain region, where no breeding levels were obtained. It is observed in this region on both spring and fall migrations.

Habitat:

This scoter breeds near ponds, lakes, oxbows, and sluggish streams in treeless or relatively open country (Palmer 1975). There must be dense and low ground cover in association with these areas. Undisturbed islands in deep water lakes are preferred breeding habitat in Alberta (Salt and Salt 1976).

Nesting:

The nest site is preferably in a patch of brush or herbaceous growth. When nesting on islands, the nesting densities are high, as in the case of a small (less than 1/2 acre) willow covered island in Chip Lake where there were 20 nests (Palmer 1975). The nests are built on the ground in a scraped-out hollow that is lined with sticks, leaves, and rubbish, and lined with down. They are usually well hidden under shrubs or bushes. This late breeder lays 9-14 pinkish-white eggs in late June. They are incubated by the female for 27-28 days. After hatching is completed, the female takes the young to water, where merging of broods often occurs (Johnsgard 1975). The young are independent at about 4-5 weeks and fly at about seven weeks (Harrison 1978).

Remarks:

The male White-winged Scoter is black with a large white crescent shaped marking below the eye, a smaller one above the eye, and a bright bill. He is most easily distinguished from the similar appearing Surf Scoter by the lack of white triangles on his nape and forehead. The female and young are a dark sooty olive brown overall with small white cheek and ear patches that may be visible. They are not easily distinguished from the female Surf Scoter, although that species has no white wing patches. Both sexes display prominent white wing patches when in flight, have thick necks, sturdy heads, and a swollen region at the base of the bill.

They usually forage in water that is less than 8 m in depth and they have unusually great endurance in remaining submerged. Johnsgard (1975) reported that dives of one minute duration are not unusual. Except for when the female is nesting, these ducks normally do not come to shore. They are most frequently seen in rafts of various sizes or in low flight over water. Their diet is predominantly animal food comprised of molluscs, crustaceans, insects, and fish, with some aquatic plants also taken.

The White-winged Scoter arrives in the province in small flocks from late April to early May (Pinel et al. 1991) and leaves in late October. The western wintering distribution is from the Aleutian Islands of Alaska south to California, with a break in Oregon. Large concentrations winter around Vancouver Island (Root 1988).

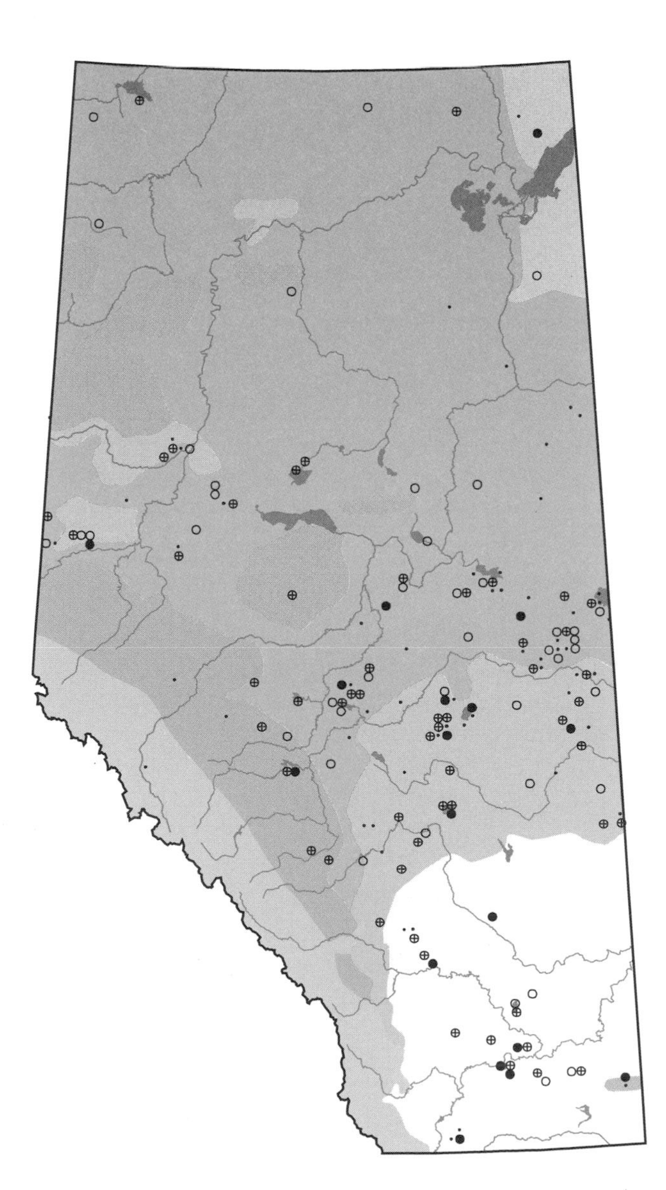

Common Goldeneye

Bucephala clangula

COGO photo credit: G. Beyersbergen

Status:

The Common Goldeneye is a common breeder in the central part of the province, uncommon in southern Alberta, and widely distributed in the northern parts of the province. The Wildlife Management Branch (1991) reported that the populations in northern Alberta have decreased by about 50% and that habitat losses in this area of the province may be cause for concern.

Distribution:

This duck was recorded in all natural regions of the province during Atlas surveys. It was recorded in only 2% of surveyed squares in the Grassland region and 6% in the Rocky Mountain region. In the Parkland region, it was recorded in 26% of surveyed squares, 22% in the Foothills, 34% in the Boreal Forest, and 25% in the Canadian Shield. Salt and Salt (1976) reported that the Common Goldeneye had been found nesting in the lower reaches of the Red Deer River and the Cypress Hills but there were no breeding records in these areas during Atlas surveys. Extralimital nesting was recorded in the Brooks area.

Habitat:

The preferred breeding habitat of the Common Goldeneye includes woodland lakes, shallow stretches of rivers, and muskeg ponds having marshy shores and adjacent stands of hardwoods to provide nest sites. In northern coniferous forest areas, aspens are apparently important for nesting (Johnsgard 1975).

Nesting:

The Common Goldeneye nests in natural cavities in trees, the abandoned holes of large woodpeckers (mainly the Pileated Woodpecker), or nest boxes. The nest site is usually 2-15 m from the ground. They have also been reported to use chimneys (Palmer 1975; Godfrey 1986). If previously used nest sites are still available from past years, these are often used, sometimes up to five years in succession (Johnsgard 1975). The female makes a rounded depression in rotted wood or chips at the bottom of the cavity and then adds down. She lays 7-12 pale bluish-green eggs, which she incubates for 27-30 days. Larger clutches, of up to 30 eggs, occur when two females use the same nest cavity. In North America, mixed clutches with Common Mergansers, Hooded Mergansers, Buffleheads, and Wood Ducks are reported (Palmer 1975). The hatchlings remain in the nest for one day before leaving the cavity. They have sharp curved nails to help them climb up the cavity to the entrance, and they jump to the ground with small wings outstretched to break the fall. They are led to water by the female, where they instinctively dive to feed themselves. The female abandons her young at a much earlier age than most other ducks. The young are able to fly at 55-60 days.

Remarks:

The male Common Goldeneye is mostly white-bodied, with a black back, an iridescent green head, and an oval white mark between the bill and the eye. The female has a gray and white body, brown head, and narrow white collar. In flight, the rapid wingbeats produce a whistling sound which accounts for its common name, whistler.

These ducks prefer water depths of 1-4 m for feeding. Their diet is mainly crustaceans, insects, and molluscs, as well as small amounts of seeds and other parts of aquatic plants (Palmer 1975). Insects and plants are more prevalent in the diet during the breeding season (Cottam 1939).

The peak of spring migration into the province is mid- to late April and fall migration takes place in October and early November. Winter records for this species come from Calgary, Banff, Stettler, Galahad, and Wabamun Lake.

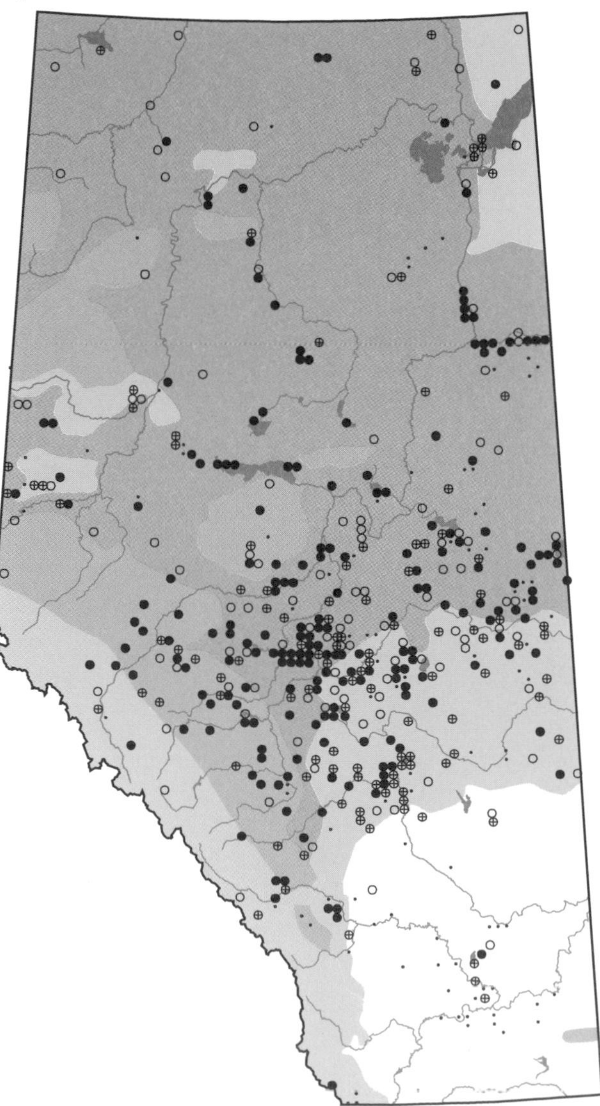

Barrow's Goldeneye

Bucephala islandica

BOGO

Status:

Barrow's Goldeneye breeds in two widely separated areas of Canada. Quebec has a small population of about 3,000 birds and there is a western population in British Columbia, the Yukon, and southwestern Alberta that numbers about 100,000 birds (Savard 1987). This latter group represents over 60% of the world's population. The overall status of this western population is undetermined, but Christmas Bird Counts do not indicate any drastic decrease (Savard 1987).

Atlas data indicate that this species is common in the Rocky Mountain region and less common in the adjacent Foothills.

Distribution:

In Alberta, the Barrow's Goldeneye was recorded breeding in 21% of surveyed squares in the Rocky Mountain region and 10% in the Foothills region. There were two extralimital breeding records from Buffalo Lake and Beaverlodge. Observations of this duck were noted from Peace River in the north, east to Cold Lake, and south to Bow Island. Salt and Salt (1976) reported migration observations from Wood Buffalo National Park and the Cypress Hills.

Habitat:

In summer, the Barrow's Goldeneye prefers alkaline lakes to freshwater ones; they are usually more productive because of the higher amount of nutrients in the water and the absence of fish which compete with the waterfowl for invertebrates (Savard 1987). Favored alpine and subalpine lakes, beaver ponds, and streamside sloughs are 1-5 m deep and have a dense growth of submerged vegetation (Palmer 1975). In flooded reservoirs with forested margins, drowned trees make excellent nesting sites.

Nesting:

This duck is typically a secondary cavity nester that uses abandoned holes of the Pileated Woodpecker or smaller Flicker cavity that have been enlarged by natural decay. It will also use natural cavities in trees, holes in the rock, and, rarely, a crow's nest (Godfrey 1986). The nest is over water or usually within 50 m of water. Nest sites are chosen by the female, and she may return to the same site as long as it remains usable. The nest is comprised of material already present in the cavity, with the addition of some feathers and down. The female lays 6-15 pale bluish-green eggs and she incubates for 30-34 days. The young soon leave the nest in a similar manner as Common Goldeneye young. Females with broods are very aggressive towards other females with broods, attacking the adult bird first and then her young. The female abandons the young when they are quite large, but before they have fledged. Fledging is at approximately eight weeks.

Hybridization is reported with the Common Goldeneye and, more rarely, with the Hooded Merganser (Savard 1987).

Remarks:

The male Barrow's Goldeneye is quite similar to the Common Goldeneye but has a purple sheen to the head and a crescent shaped white patch between the bill and the eyes. The white wing patch is divided horizontally by a black bar. The females of these two species are very similar, but they can be differentiated by their association with the male.

These diving ducks usually feed by diving 1-3 m below the surface. Small items are swallowed underwater and larger ones are brought to the surface (Palmer 1975). The Barrow's Goldeneye consumes more insects than the Common Goldeneye, but also eats molluscs, crustaceans, and some plant matter.

Pinel et al. (1991) reported that spring migration for the province begins in April and fall migration is completed in early November. They reported wintering records in the Bow Valley from Banff to Calgary, Waterton National Park, Lethbridge, Edmonton, and Wabamun Lake.

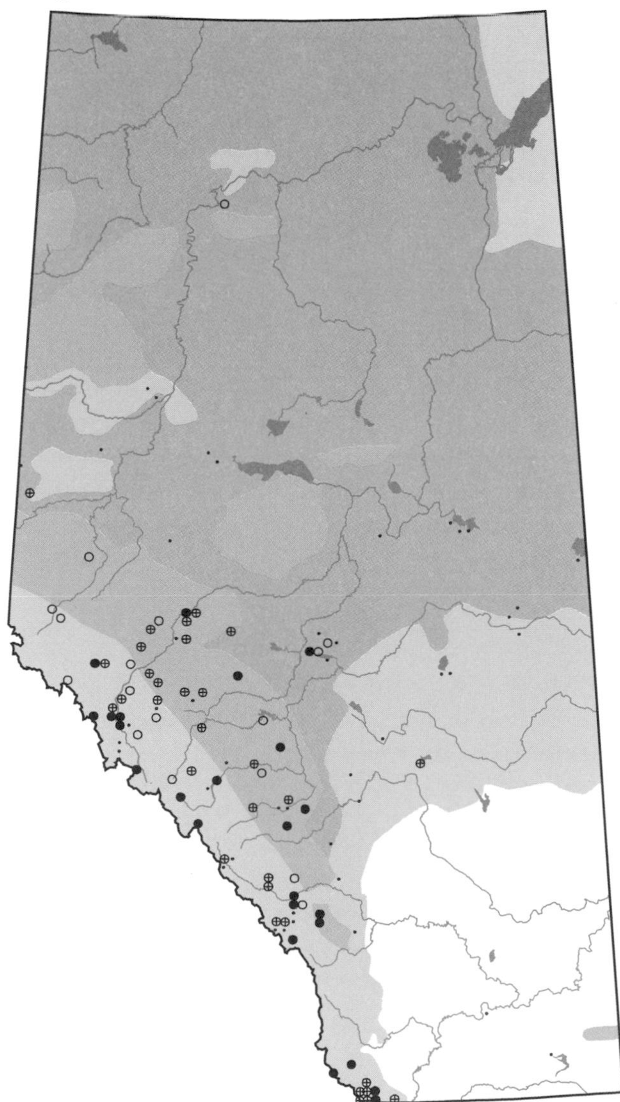

Bufflehead

Bucephalus albeola

BUFF photo credit: G. Beyersbergen

Status:
Several factors combine to restrict the numbers and distribution of the Bufflehead, which is one of the scarcest ducks in North America (Canadian Wildlife Service 1974). The use of tree holes for nesting sites excludes them from prairie and tundra habitat and the continued clearing of woodlands in the western provinces reduces excellent habitat. At present, the population in Alberta is regarded as stable (Wildlife Management Branch 1991). Atlas data indicate that the Bufflehead is common in the Parkland and southern Boreal Forest regions of the province. They are widely dispersed in northern Alberta and uncommon in the southeastern part of the province.

Distribution:
Buffleheads were recorded breeding in all natural regions of the province, but were concentrated in the central part including the Parkland region and the southern portion of the Boreal Forest region. As indicated by Salt and Salt (1976), they do breed in the southern part of the province and they were recorded in 5% of surveyed squares in the Grassland region. They were also recorded breeding in the Cypress Hills. In the Parkland Natural Region, they were recorded in 32% of surveyed squares, 31% in the Foothills, 10% in the Rocky Mountain region, 39% in the Boreal Forest, and 38% in the Canadian Shield Natural Region.

Habitat:
The breeding distribution of the Bufflehead is restricted by the distribution of woodlands required for nesting. Most lakes frequented by these ducks are at least moderately fertile, with extensive open water areas and, usually, with maximum depths of 3 m (Erskine 1971). Erskine further stated that lakes favored by breeding Buffleheads in central Alberta are small and relatively shallow, lack an outlet, have vegetated margins, and are positioned in areas where poplar communities dominate. Larger lakes and ponds, especially at higher elevations, are not favored, nor are waters broadly margined with emergent or floating vegetation (Palmer 1975).

Nesting:
Bufflehead courtship occurs through the winter, with most pairing taking place during spring migration so that birds are paired when they reach breeding areas. They start nesting soon after their arrival.
This species is the smallest of the diving ducks in Canada. It is a cavity nester that can use the former nest of flickers. The nest site is always in close proximity to water. No material is added to the nest and the eggs are laid at the bottom of the cavity. The typical clutch is 7-11 buffy to creamy white eggs, which the female incubates for about 30 days. The young are led to water about one day after hatching and are closely brooded for approximately one month. They fledge at about 50-55 days.

Remarks:
The male Bufflehead has a disproportionately large head that is dark and has a conspicuous broad white area extending back from the eyes. The body is mostly white and the centre of the back is black. The female is dark brown and has a small white cheek patch and wing patch.
This duck always feeds by diving, even in shallow water. They feed primarily on insects, shrimps, and snails during the breeding season, while the young birds feed almost entirely on insects. In the fall, plant material, mainly the seeds of pondweeds and bulrushes, becomes more important (Erskine 1971).
They arrive in the province in late March to mid-April and most leave by late October. Winter records are reported from Edmonton, Galahad, Banff National Park, Calgary, and Waterton Lakes National Park (Pinel et al. 1991).

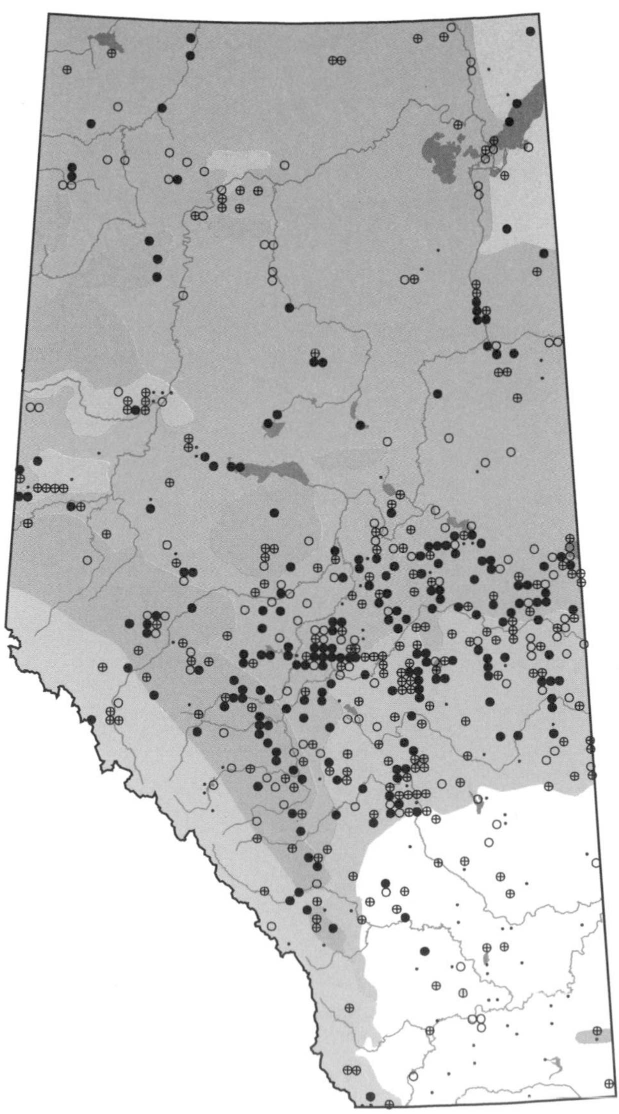

Hooded Merganser

Lophodytes cucullatus

HOME photo credit: J. Lange

Status:
The Hooded Merganser is never abundant in Alberta, even during migration. As with the other mergansers, clearing practices near shoreline or the removal of trees with nest cavities will adversely affect the local status of this already uncommon species in Alberta.

Distribution:
The Hooded Merganser has been recorded breeding, intermittently, at several locations. These have included the three mountain national parks, the Crowsnest Pass, Longview, and Barrhead (Holroyd and Van Tighem 1983; Pinel et al. 1991). Salt and Salt (1976) considered it to probably breed in the Lake Athabasca area. It is more common in southern Alberta during migration. During the five year Atlas Project, only five breeding records were confirmed, all within the historical forested range of this species. Two unusual possible breeders were recorded in the Camrose area and north of Taber.

Habitat:
The Hooded Merganser shares similar habitat requirements with the Common and Red-breasted Mergansers. It breeds in ponds, lakes, and rivers that have fish available to feed on and a woodland border to provide nest sites.

Nesting:
This duck does readily accept nest boxes with a wide range of volumes and nest hole sizes. Preferred locations are adjacent to water bodies, from 3-6 m above ground. The nest may be made of grass (Salt and Salt 1976), wood chips, leaves, or dry moss (Campbell et al. 1990). The inside is lined with down and a few feathers. The typical clutch is 5-12 white eggs. Larger clutches, as many as 36, are the result of egg dumping, when suitable nests are unavailable or when a female has returned to find her previous year's nest already occupied. Nests may also include joint clutches with Wood Ducks or the Common Goldeneye. The clutch is incubated 29-37 days (Godfrey 1986). The male leaves at this time to moult. While tending to the chicks alone, the female will use a distraction display to draw away intruders, while the young dive and swim to hide within the shoreline vegetation. Unlike the other mergansers, Hooded Merganser females do not combine broods. The young fledge at 71 days.

Remarks:
The boldly patterned male has a black head, neck, and back, reddish-brown sides, a yellow iris, and white underparts with two black bands across the breast. A large white triangular patch on the back of the top of the head forms part of a large and expandable crest. The female has a grayish-brown head and throat, dark brown upperparts with a mostly hidden white wing patch, a brown iris, and pale grayish-brown underparts with a white belly. She often wears her crest raised.

Although the Hooded Merganser is the smallest merganser, it is one of the swiftest flying ducks. It shares the same general body shape as other mergansers, which makes it awkward on land, requires it to make long takeoffs, and allows it to be an excellent swimmer and diver. Like grebes and loons, it can partially submerge its body if required.

In addition to fish, these birds will also eat crustaceans, aquatic insects, amphibians, and molluscs.

Although this species often migrates in mixed merganser flocks, their proportions within these groups are small in Alberta. They arrive in southern Alberta in mid-April and depart, in larger flocks, through the same area between September and early November. There are overwinter records for Calgary and the Sundance Cooling Pond.

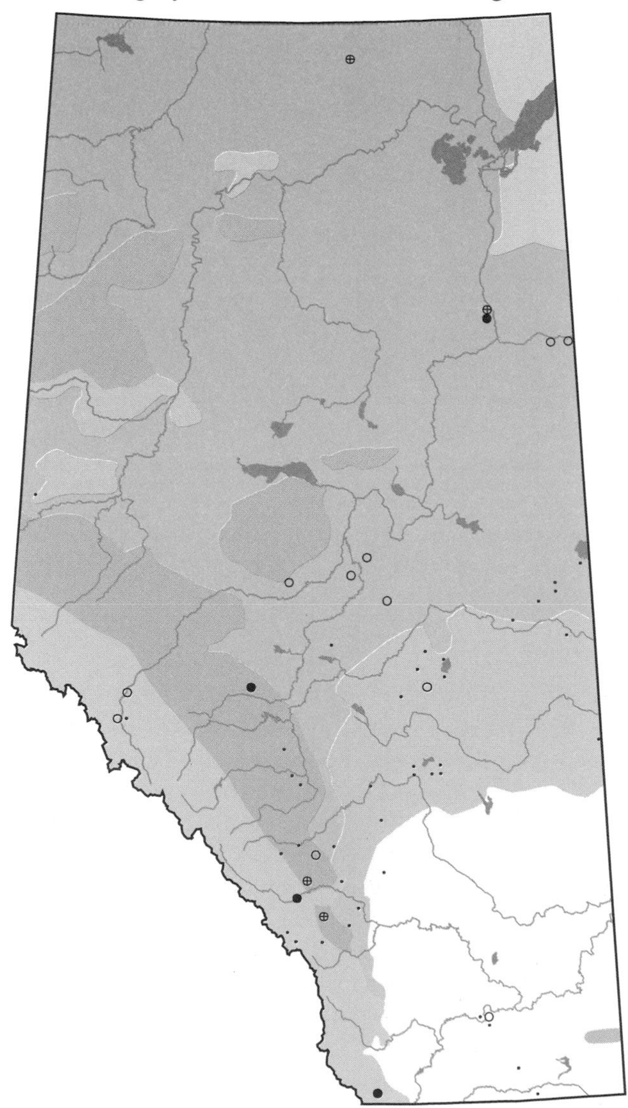

Common Merganser

Mergus merganser

COME photo credit: G. Beyersbergen

Status:
The Wildlife Management Branch (1991) does not consider the status of the Common Merganser to be at risk, because of healthy population levels and a secure habitat. However, the species is sensitive to the removal of nest sites by logging near shorelines, cottage development, and disturbance from power boats (Mansell 1978; Cadman et al. 1987).

Distribution:
Historical breeding distribution has included the northern forests, the Rocky Mountains south to the United States border, the adjacent foothills, the Bow River Valley east to Carsland, and the Cypress Hills (Salt and Salt 1976). To this, Pinel et al. (1991) added an area east of Red Deer, and Fort MacLeod. During the period of the Atlas Project, records were obtained along the St. Mary River to Lethbridge, near Taber along the Oldman River, and down the Red Deer River to Drumheller. Of the total squares surveyed in each natural region, it was most commonly recorded in the Rocky Mountain region (25%), the Foothills (19%), and the Canadian Shield (25%).

Habitat:
This accomplished diver breeds a short distance inland of the wooded shores of lakes and rivers. Probably because of feeding requirements, it shows a preference for clear water (Godfrey 1986).

Nesting:
The female selects the nest site, which is most commonly located in a tree cavity, from 5-16 m above ground, or in a nest box. If on the ground, she constructs a nest of grass; however, in any location, the nest is lined with down. The 9-16 pale buff eggs are laid on successive days. Clutches greater than this may be the result of egg dumping by a conspecific, probably due to a lack of suitable nest sites. Soon after this time, the male leaves to join rafts of other males to moult their breeding plumage. The female conducts the incubation over a 28-35 day period. The hatched young remain in the nest for only 24-48 hours, then quickly leave to join their mother on the water. They are able to swim and dive well almost immediately. When alarmed, they will seek refuge in the overhanging vegetation of the shoreline. Tending to them alone, the female feeds the young minnows, tadpoles, frogs, and insects. Young fledge after 65-85 days.

Remarks:
With a blackish head and upper neck that are darkly glossed over in green, the male Common Merganser resembles the male Red-breasted Merganser. However, the male Common Merganser is larger, has white sides, and has no crest or reddish band across the breast. The female Common Merganser is distinguished from the female Red-breasted Merganser by the Common's more distinct delineation between her red neck and her white breast. Both sexes have the characteristic merganser red bill, iris, and legs. Using their long, slender, serrated bills, Common Mergansers largely dive and feed on fish, swallowing them whole. Amphibious and water insects are also consumed. When the situation presents itself, they will cooperatively cut off and drive a school of fish into the shallow water of a bay, and engulf the prey as its movement becomes restricted.
Arriving on still largely frozen lakes and rivers between early March in the extreme south and mid-April in the north, groups of Common Mergansers feed on fish stocks available along the thawed edges of these water bodies. Fall migration is largely carried out from mid-September to the end of October. There have been several overwintering records, including Calgary, Edmonton, Fort McMurray, Lake Wabamun, Lethbridge, Seebe, and Banff and Jasper national parks (Salt and Salt 1976; Pinel et al. 1991).

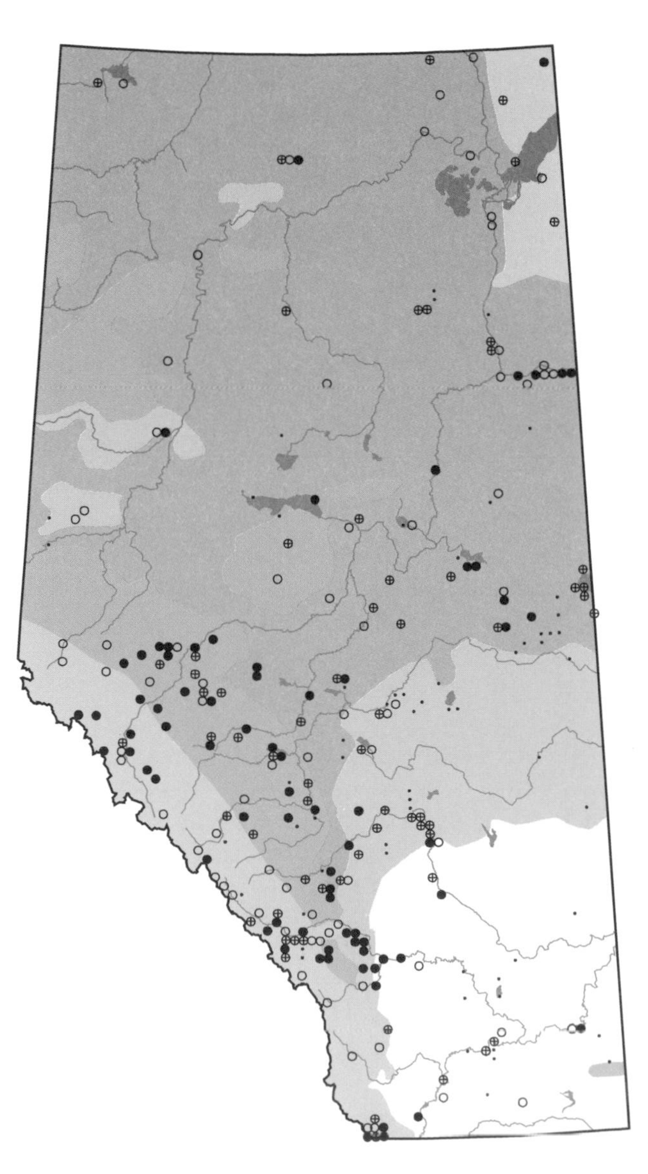

Red-breasted Merganser

Mergus serrator

RBME photo credit: G. Beyersbergen

Status:
The Wildlife Management Branch (1991) considers the present status of the Red-breasted Merganser to not be at risk. During the Atlas Project, this species' breeding status may have been underrepresented, because of the remoteness of its distribution, its well-concealed nest, and the fact that the female resembles the Common Merganser.

Distribution:
Salt and Salt (1976) reported that this species has been seen from Lake Athabasca and Wood Buffalo National Park south to Lesser Slave Lake. There is one historical nest record for the Caribou Mountains. Pinel et al. (1991) included records for Reesor, Sturgeon, Pitchimi, and Lesser Slave lakes. Although the Atlas Project was able to expand this range to include records for Bistcho Lake, Cold Lake, Muriel Lake, the Red Deer River, and Rocky Mountain House, there were only two records of confirmed breeding, both north of Lake Athabasca. It was recorded in 19% of surveyed squares in the Canadian Shield and 2% in the Boreal Forest region. Values for all other regions were less than 1%. The Red-breasted Merganser occurs throughout the province while on migration.

Habitat:
Red-breasted Mergansers feed and breed in the bays, lagoons, and estuaries of larger lakes and rivers. They are usually found in more open, deeper water than are Common Mergansers.

Nesting:
The female will usually construct her nest of grass and leaves in a depression within 8 m of the shoreline. It is lined with down, and is usually sheltered by low vegetation among tree roots or logs or in rock crevices. This species often nests in close proximity, suggesting an almost colonial lifestyle. The female lays 7-12 olive-brown eggs, which she covers with down and decaying vegetation upon departing the nest. She incubates these for 29-35 days. At this time, the male departs to moult into eclipse plumage. The female feeds her young small fish, water insects and larvae, worms, and crustaceans. Females often combine their broods, using one or several to attend to the group. Young fledge after about 60 days.

Remarks:
These mergansers have long, slender (aquadynamic), serrated bills that enable them to grasp onto fish. Their wing area is reduced, which enables strong diving but results in long takeoff runs. They are, however, strong swift fliers. With legs set toward the rear of the body to propel diving, they walk awkwardly, tilting their bodies 45 degrees to maintain balance. The male Red-breasted Merganser generally resembles the Common Merganser, having a glossy, dark green head and upper neck, a black back, and red legs, bill, and iris. Prominent differences in the Red-breasted include a double pointed crest, mottled sides, and a reddish band across the breast. More closely resembling her Common Merganser equivalent, the female Red-breasted lacks the sharp contrasting definition on the throat.
This species eats mostly fish, and hunts with its head submerged, looking for prey. It also will eat crustaceans and aquatic insects. Like the Common Merganser, it will fish cooperatively with other conspecifics, herding fish into shallower waters.
Arriving in southern Alberta at the beginning of April, they carry out a protracted migration throughout the month, moving on as water bodies thaw. At this time of the year they may be in mixed flocks with Common Mergansers. There are few fall records for the province, although some have been seen as late as November before migrating. There are overwintering records for Lake Wabamun, Sundance Cooling Pond, and Banff.

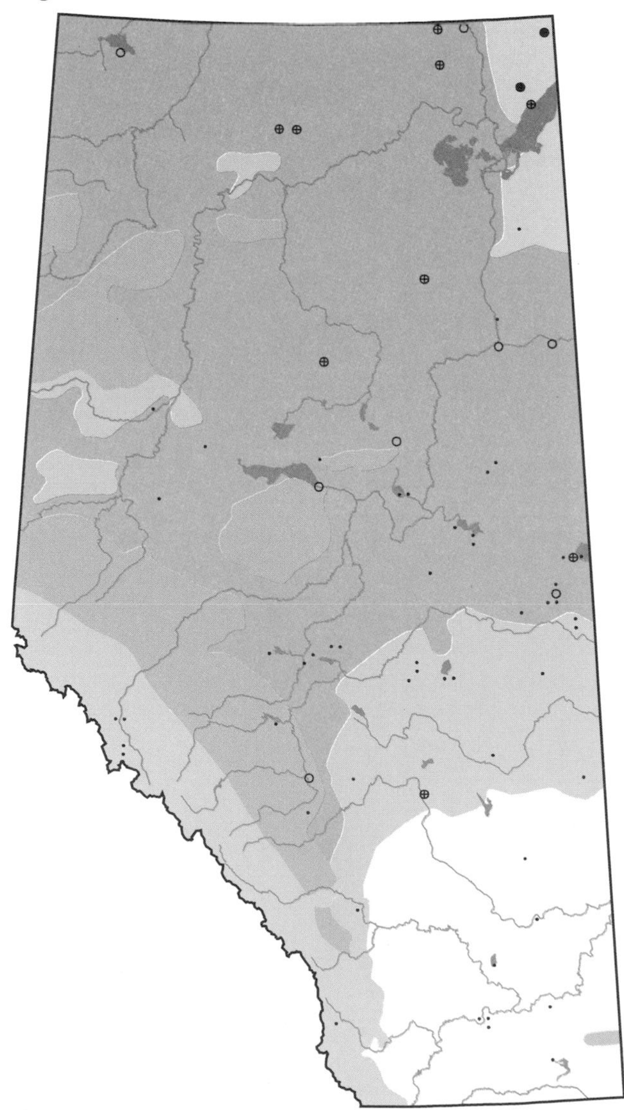

Ruddy Duck

Oxyura jamaicensis

RUDU photo credit: G. Beyersbergen

Status:
The breeding population of the Ruddy Duck in Alberta is stable or may be increasing throughout the province (Wildlife Management Branch 1991). This duck is common across the Parkland and southern Boreal Forest regions, less common in the Grassland region, and a rare breeder in the Rocky Mountain region and the extreme northern parts of the province.

Distribution:
Salt and Salt (1976) reported that the Ruddy Duck breeds across the province but is absent from the Rocky Mountains except in migration. Atlas data indicate that this species does breed in this area, with breeding records from Jasper, the upper reaches of the Highwood River, and Waterton Lakes National Park. It was recorded most often in the Parkland Natural Region, where it was found to occur in 39% of surveyed squares. In the Boreal Forest region, it was recorded in 19% of surveyed squares and 13% in the Grassland region.

Habitat:
The preferred habitat of the Ruddy Duck is permanent freshwater or alkaline marshes having emergent vegetation and relatively stable water levels (Johnsgard 1975). Suitable nesting habitat must have open water close to nesting cover, including emergent plants that provide accessability as well as adequate cover density. Cadman et al. (1987) reported that the Ruddy Duck can adapt to a wide variety of other open habitats associated with freshwater, such as ponds, lakes, and lagoons with emergent vegetation.

Nesting:
The nest site is usually in dense emergent vegetation. A nest platform is built by bending down the surrounding vegetation. The nest is a woven structure built up above shallow water 25 to 40 cm deep. It is composed of reed stems and other nearby plants and usually does not contain down (Harrison 1978). The female lays 6-10 dull white eggs, which she incubates for 22-24 days. During incubation, the male remains close to the nest. The young leave the nest within one day and swim and dive well from the beginning, but they are very clumsy on land. Unlike many other ducks, male Ruddy Ducks sometimes remain in the nesting grounds and accompany young through the breeding season. The female stays with the brood for about four weeks and then abandons them. The young ducks fledge in approximately eight weeks.

Remarks:
This is a small diving duck. The male has a reddish body plumage, a black cap on the head, white cheeks, and a bright blue bill. The female is a dark brown above and lighter below, and has a gray bill and a white cheek that is divided horizontally by a brown bar. Both sexes have broad flattened bills and long tails that are held on the water surface or variably cocked above it (Johnsgard 1975).
The Ruddy Duck is quite grebelike, in that it has large feet and the hind limbs are far back on the body. If disturbed, it will dive rather than take flight. The Ruddy usually dives and feeds on the bottom but it may also be observed dabbling. The majority of its diet is vegetable matter, with pondweed and the tubers and seeds of sedges being preferred. It also eats insects, primarily the larvae of midges and caddis flies.
The Ruddy Duck arrives in the province in mid-April to early May. Most Ruddys have left by the end of September, with some lingering into November (Pinel et al. 1991). It has been reported wintering at Wabamun Lake and Galahad.

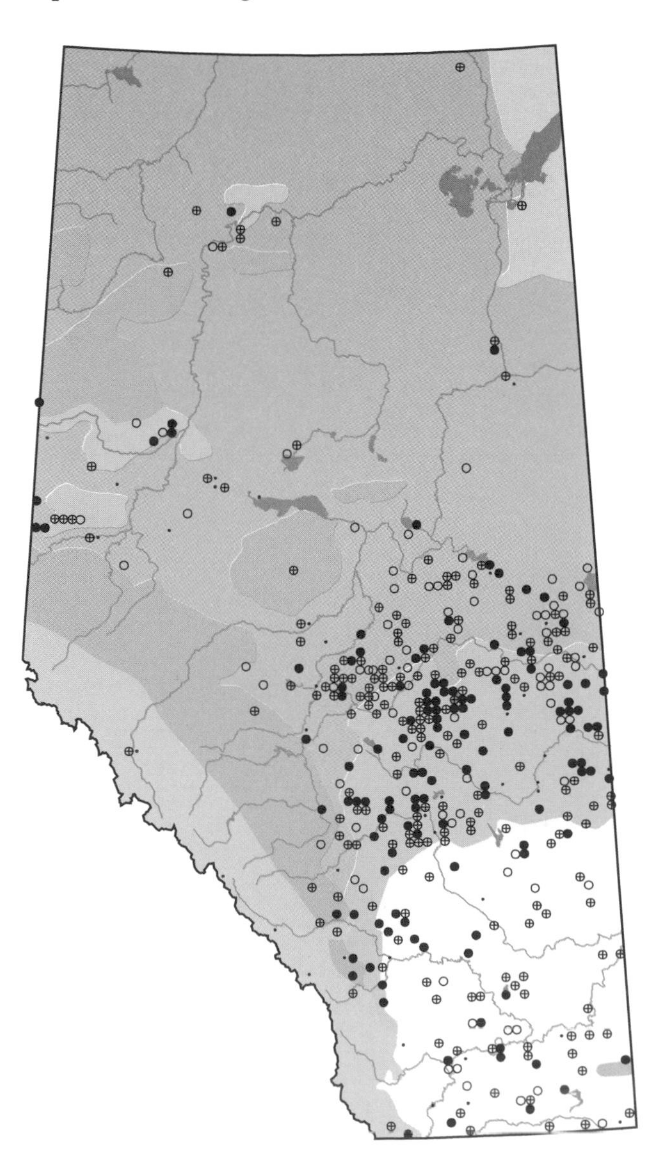

Order Falconiformes

(Family Cathartidae, Family Accipitridae, Family Falconidae)

One of the most appreciated groups of birds by even the most novice enthusiast, these raptors have strong feet for killing or grasping prey and strong hooked bills for tearing flesh. The exception to the rule is the less appreciated Turkey Vulture, which does not require talons because it generally scavenges. The Turkey Vulture has large, broad wings that permit it to soar at great heights, and a well-developed sense of smell, both of which assist it in finding carcasses. Hawks and eagles also have large, rounded wings, which make them accomplished fliers and soarers, as well as strong legs and long talons for seizing prey. Another large member is the Osprey, which has longer, pointed wings and a short hooked bill. Adapted to grasp fish, their large claws include reversible outer toes and spines on the bottom. Falcons, small to medium-sized birds, use the shallow, rapid beats of their long pointed wings to conduct swifter, more direct flights than the larger hawks. Their prey is often captured on the wing. They too have strong talons and short, hooked bills. This group tends to be more solitary. In all these birds of prey, the sexes are, or nearly are, alike and the female is larger than the male.

Turkey Vulture

Cathartes aura

TUVU

Status:

This species is uncommon in Alberta, where it is at the northern limit of its range. While populations appear to be on the increase in eastern North America, declining numbers in the west have been a cause for concern (Tate and Tate 1982). The population in Alberta is probably less than 100 pairs, and the trend is uncertain (Wildlife Management Branch 1991).

Distribution:

In Alberta, these birds are rare and local breeders found east of the foothills in the south and central parts of the province. Summer records were concentrated in the Cold Lake and Big Valley areas and, historically, the species appears to have been more common in the lower Red Deer River area than it is today. One traditional nest site at Big Valley has been used for over 50 years (C. Stoneman pers. comm.).
Atlas data indicate that the Turkey Vulture is also found in the North Saskatchewan River valley east of Duvernay, southwest of Wainwright, and in the Cypress Hills.

Habitat:

Turkey Vultures are found in various types of terrain south of the Boreal Forest region. The birds like to roost in trees near a stable food source and near water. They are rarely seen on the ground except when feeding on a carcass, and are most often observed circling on thermals.

Nesting:

Turkey Vultures nest in a variety of habitats, but tend to select undisturbed areas, preferring rocky outcrops, protected caves and crevices in cliffs, or a scrape beneath a log in mixed forest. Little attempt is made to build a nest, though the site may be cleared and trampled. Eggs are laid directly on the ground or rock. Vultures tend to be solitary nesters but, when there is a concentrated food supply nearby, several pairs may nest in proximity. If undisturbed, birds may return to a nest site for several years.
Two creamy-white eggs with brown to faint purple blotches is the usual clutch. Both parents incubate the eggs, which hatch in 38-41 days, and are involved in continuous brooding for five days. They then visit the nest with decreasing frequency, feeding the young only two to three times a day. Fledging is at nine weeks. Turkey Vultures are single-brooded, with families often staying together until the following season. Pair bonds may be lifelong for some birds.

Remarks:

The Turkey Vulture is a large bird with a wing span of up to 2 m. Its head and upper neck are red and naked, reminiscent of a turkey. Unlike other raptors, vultures have weak feet and beaks, making them obligate carrion eaters. Small-mammal carrion form the bulk of the diet, but the birds will also kill mice, snakes, and insects. Their genus name, *Cathartes*, means "purifier".
This species has a well-developed sense of smell, but mainly relies on its excellent eyesight to locate food. A vulture may skim the tree tops to locate small carrion, or observe from the heights. These birds can spot a deer carcass 6 km away. Living off of death, vultures must wait for what chance presents; sometimes they must go several days without food. They are also known for their extraordinary resistance to disease, particularly the *botulinus* toxin.
Turkey Vultures are awkward on the ground; they hop clumsily and sometimes with a sideways hitch. They are, however, masters of the air. When soaring or gliding, they hold their wings in a V, unlike hawks whose wing profile is more horizontal.
The birds arrive in Alberta in early May and most leave by September (Pinel et al. 1991).

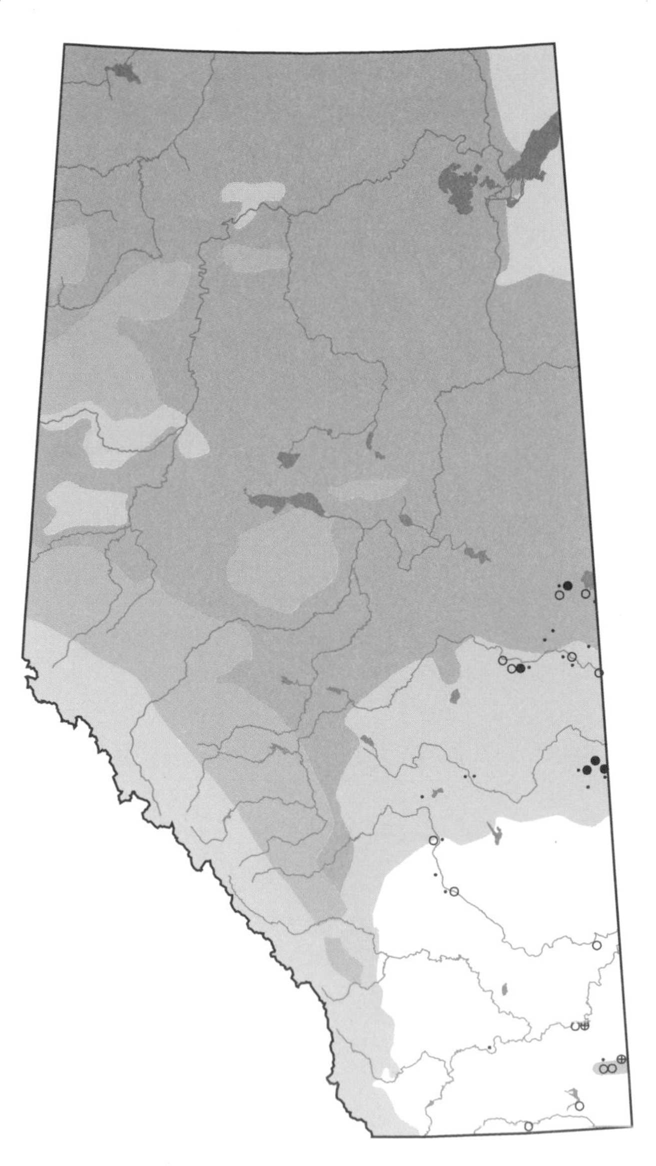

Osprey

Pandion haliaetus

OSPR photo credit: W. Burgess

Status:

The Osprey was severely persecuted by shooting during the first half of this century. Later, a serious decline in nest productivity occurred as a result of the use of chemical pesticides. The United States population was most severely affected by pesticide use; numbers in Canada have remained relatively stable. The Osprey was on the Blue List from 1972-1981. It was removed in 1981 as populations began recovering following restrictions on the use of DDT (Tate 1981).

The exact status of this species in Alberta is unknown at present, although Atlas data indicate that it is relatively common in the southern Boreal Forest region and in river valleys in the Rocky Mountain region. Continued protection of specific nest sites is necessary, as is the monitoring of the population to ensure maintenance at viable levels (Wildlife Management Branch 1991).

Distribution:

This species is found throughout Alberta, except in the arid southeast. It was recorded in approximately 20% of surveyed squares in the Rocky Mountain region, 10% in the Foothills, Canadian Shield, and Foothills regions, and less than 5% in the Parkland and Grassland regions. Almost 50% of these records were in the southern Boreal Forest region. Although Atlas surveys did not confirm any breeding north of Lake Athabasca, there are six nests between Fort Chipewyan and the Northwest Territories border (H. Armbruster pers. comm.).

Habitat:

Since its principal food is fish, the Osprey is invariably found in the vicinity of permanent lakes and rivers. It is found where water bodies contain an adequate supply of fish and suitable nesting sites occur nearby.

Nesting:

Ospreys nest near water, usually near the top of a tree, or occasionally on the pinnacle of a cliff or a man-made structure such as a wooden power pole. The nest is a massive platform of large sticks, branches, and twigs and lined with finer grasses, bark, weeds, fish bones, or down. Both sexes participate in nest building, the male bringing materials and the female weaving them in. Nests are re-used and added to each year so that many become immense structures, which can weigh several hundred kilograms. The Osprey is usually a solitary nester but, in appropriate conditions, sometimes nests in loose colonies.

The female lays 2-4 creamy white eggs, which are blotched with dark brown. Incubation is 32-33 days, mainly or solely by the female. The male feeds the female during incubation. He also brings food to the female after eggs hatch, and she tears it up into appropriately sized pieces for the young. Fledging is at 51-59 days (Harrison 1978). After fledging, the young may return to the nest for a week to feed and roost.

Remarks:

Ospreys are narrow-winged, long-legged raptors, intermediate in size between the large buteos and eagles. Their feet are equipped with long, strongly curved talons and pointed scales for maintaining their grasp on their slippery prey. The birds hunt 15-30 m over shallow water. Once a fish is sighted, the bird dives, thrusts the feet forward of the head at the last moment, and plunges into the water. The bird may be submerged for a short time, but then rises from the surface, shakes the water from its feathers, and, if successful, carries off in its talons a fish pointed head first into the flight path. Ospreys have excellent eyesight and their feathers are compact, oily, and waterproof.

This species arrives in southern Alberta in the last half of April. Fall migration is in September. There has been one winter sighting near Lethbridge.

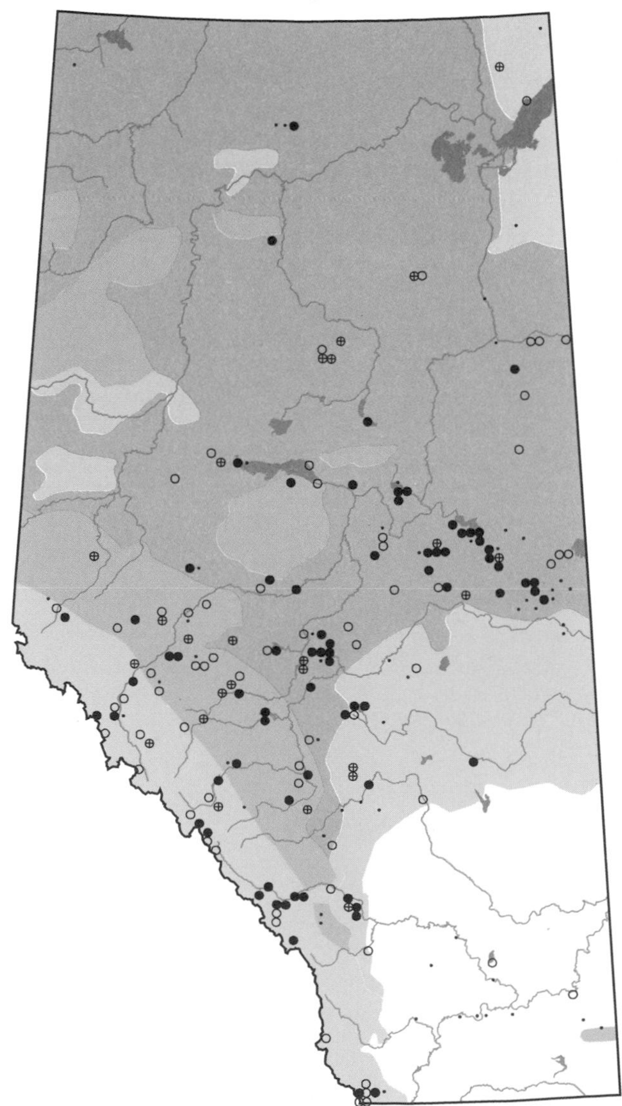

Bald Eagle

Haliaeetus leucocephalus

BAEA

Status:
There are an estimated 70,000 Bald Eagles in North America (Gerrard 1983). Its exact status in Alberta is unknown, but the species is at risk throughout much of its central North American range, and is considered endangered in the continental United States. This species is the victim of direct persecution, loss of habitat, and pesticide contamination. In most areas of the province, it nests in very low densities, except in certain lakes in northern Alberta.

Distribution:
The Bald Eagle is restricted to North America, with most of its breeding population in Canada and Alaska. The species formerly nested across Alberta, occasionally on the prairies. However, it is now confined to the northern half of the province, in the Canadian Shield and Boreal Forest regions, and the western Foothills and Rocky Mountain regions.

Habitat:
The Bald Eagle's primary nesting requirement appears to be the proximity of a large body of water—in Alberta, a large inland lake or river. Breeding areas must have suitable tall trees near shore for nesting and roosting, good fish populations, and relatively little human disturbance.

Nesting:
This raptor usually nests in mature open-crowned trees that are taller than the surrounding forest canopy. In northern Alberta, most nests are in jack pine, spruce or aspen, within 200 m of shore, and often on islands (Munson et al. 1980). The nest is placed in a crotch, or on sturdy branches near the trunk, usually near the crown. Occasionally, birds will nest on cliff ledges or, very rarely, on the ground. Bald Eagles are normally solitary nesters but, occasionally, several pairs nest in relatively close proximity on the shore of lakes with good fish populations.
The nest is a massive cup-shaped platform made of large sticks, branches, twigs, and weeds, and is lined with grasses, mosses, bark strips, and leaves. Both sexes bring material to the nest site and the female does most of the construction. The same nest site is re-used over several years or a pair may alternately occupy two or more nest sites over successive years. Since nests are added to each year, they become immense structures, some up to 3 m in height and width and weighing over a tonne.
Clutches are 1-3 eggs, usually two, dull white in color. Incubation is 34-35 days by both sexes. After hatching, food is brought to the nest by either parent, but most feeding is done by the female. Young fledge at 10-11 weeks, but may return to the nest for short periods to feed or rest.

Remarks:
With its great size and gleaming white head and tail, the adult Bald Eagle is hard to miss. Having a wingspan of over 2 m, it is North America's largest broad-winged raptor.
Bald Eagles are opportunistic feeders. Scavenged and live fish form the bulk of their diet. Eagles also hunt grounded or injured waterfowl and small mammals captured on the ground. Bald Eagles are well known for their piracy of prey items from other Bald Eagles, Ospreys, various hawk species, and Mergansers.
Like Golden Eagles, this species is only partially migratory. Birds that have access to open water will stay on the nesting grounds all year. In winters with limited open water, birds make their way south or to the coast.
In Alberta, birds arrive in early March, reaching the northern areas in early April. Fall movements begin in September and peak in November, with most birds leaving by mid-December. Most overwintering records are from the Calgary area. However, other winter sightings have occurred throughout southern Alberta, including the mountain parks, and as far north as Edmonton and Wabamun Lake.

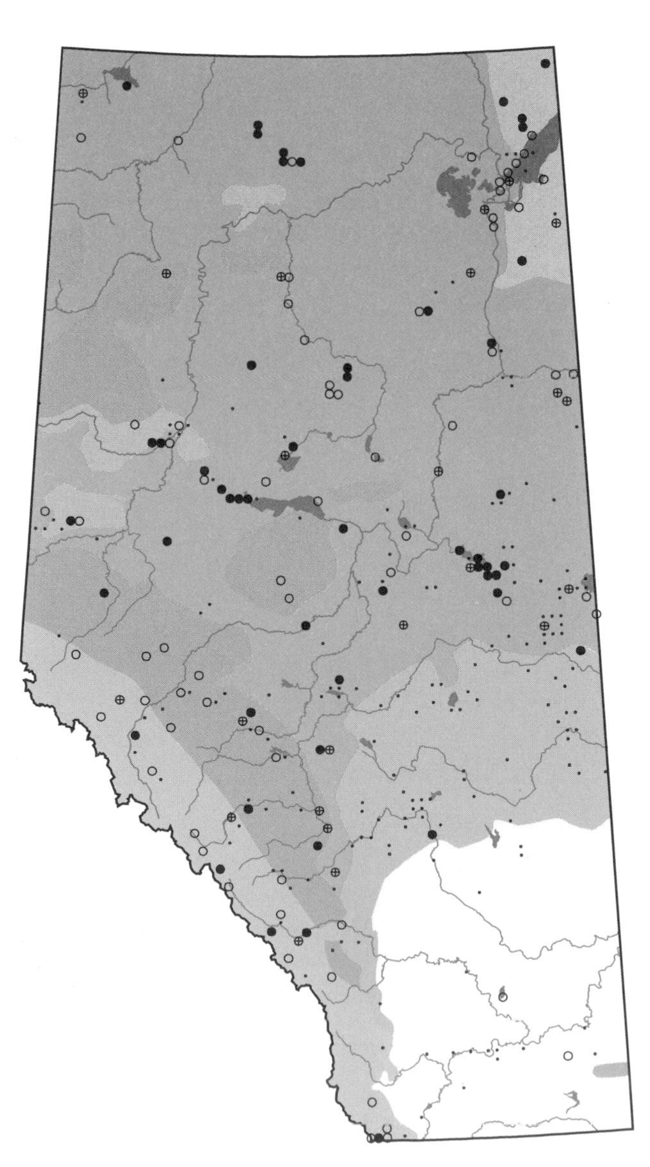

Northern Harrier

Circus cyaneus

NOHA

Status:

The Northern Harrier (formerly known as the Marsh Hawk) is widely distributed and not uncommon in southern Canada. However, its overall population has been declining, particularly in eastern North America (Robbins et al. 1986). This species is particularly affected by intensive agricultural development and destruction of marsh habitat, and it has been on the Blue List since 1972. Populations in Alberta seem to be stable.

Distribution:

The Northern Harrier is a Holarctic species breeding across North America. It breeds throughout Alberta, although Atlas data indicate that it is most abundant in the Grassland (31% of surveyed squares) and Parkland (29%) regions and less common in the Boreal Forest (22%), Foothills (17%), and Rocky Mountain (9%) regions.

Habitat:

This medium-sized hawk is a bird of the open country, where it hunts over a variety of landscapes, including marshes, meadows, and cultivated fields. It prefers damper meadows, but will also breed in drier areas so long as there is adequate food and good cover for nesting. In the mountain parks, it is found in open alluvial meadows and wetlands and is usually seen hunting over grasslands and stands of willow and birch (Holroyd and Van Tighem 1983).

Nesting:

The Northern Harrier has a spectacular aerial courtship which Savage (1985) called "skydancing". From mid-March to May, the males attract females with alternating dives and ascents, from 25 m to within 1 m of the ground, over the marshes and fields where they nest. Nesting is on the ground among low shrubs, grasses, or weeds, or in a moist meadow, cattail, or bulrush marsh. In the prairie, nests are often in clumps of rose and buckbrush. Nests are constructed of sticks, weeds, marsh vegetation, grasses, and rootlets, and lined with grasses and weed stalks. In dry areas, the structure may be quite minimal, but nests constructed over water are bulky, durable platforms. The female does most of the nest-building, though the male assists by bringing nest materials. Nests are used only once.

Clutch size is typically 4-6; the eggs are bluish white, sometimes lightly blotched with gray or brown. Incubation is 31-41 days (Sealy 1967) by the female, though the male may occasionally alight on the nest. During incubation, the male hunts for the female. After hatching occurs, he is the family's sole provider for two weeks. He does not bring food directly to the nest but calls to the female, who rises from the nest to take the food from the male's talons, or catch it in her own talons in mid-air. Fledging is at 30-35 days. In the fall the family group breaks up.

Remarks:

The Northern Harrier is North America's only harrier. Unique to this species is a partial facial disc, a ruff of feathers on the side of the head which gives the bird an owl-like appearance and focuses sound to give the bird its acute sense of hearing.

This group of hawks is known for its distinctive hunting style—a slow sail over open terrain often less than 3 m above the ground, searching for mice, small birds, frogs, and insects. Once prey has been located by sight or sound, the bird drops to the ground to claim its prize. Birds may attempt to flush prey from bushes or cattails, but if they miss on the first pounce, they rarely take up pursuit.

Northern Harriers arrive in southern Alberta in mid-March with a peak in mid-April. In the fall, the birds leave the province late in October and early November. In Alberta, overwintering has been reported in the southern part of the province with sightings at such locations as Lethbridge, High River, and Medicine Hat.

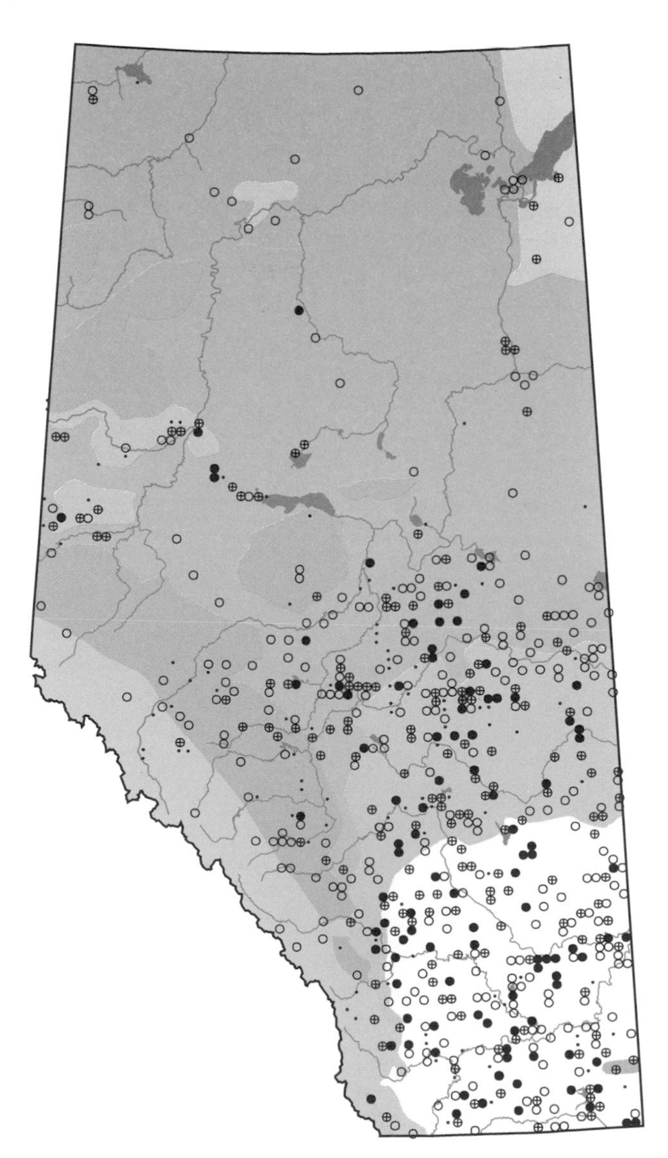

Sharp-shinned Hawk

Accipiter striatus

SSHA

Status:
The population of the Sharp-shinned Hawk is probably stable, and possibly increasing, but lower than that of the last century. This hawk was one of the main victims of the "great hawk shoots" that occurred up to the 1930s, and was also severely affected by the use of pesticides such as DDT. This species continues to be very susceptible to habitat loss, particularly through forest management practices that eliminate forests altogether or which produce large blocks of unsuitable habitat. These practices reduce potential and remove active nest sites, as well as lower the abundance of some prey species.
The Sharp-shinned Hawk has been on the Blue List since 1972 and its exact status in Alberta is unknown, although several confirmed nesting sites were recorded by the Atlas study.

Distribution:
This hawk was found throughout all of the forested areas of the province. This small hawk breeds mainly in the Canadian Shield and Boreal Forest regions of northern Alberta, as well as in the Foothills and Parkland regions of central Alberta. The lack of information in the north-central part of the province is a reflection of Atlas coverage.

Habitat:
This species prefers thick deciduous and mixed wood forests to heavy conifer growth. Birds are rarely seen above the forest canopy except in display.

Nesting:
The Sharp-shin nest, usually built annually, is a flat platform of twigs lined with finer twigs and bark. It is built in a crotch of a tree or on a sturdy branch next to the trunk. Nests are 3-21 m above the ground and are usually well-concealed by thick growth. Both sexes gather nesting materials but most of the construction is by the female.
The female typically lays 4-5 eggs, which are bluish white and heavily blotched with brown. Incubation is 34-35 days by both sexes. The male feeds the female when she is incubating. Young are tended by both parents, who are silent and unobtrusive around the nest. Fledging is at 23-25 days (Beebe 1974). After fledging, the family group remains together a few weeks near the nest site.

Remarks:
The smallest of all Alberta's accipiters, the Sharp-shin is well-equipped as a woodland predator. Its small body, long tail, and short, cupped wings allow quick maneuverability for pursuit of prey through thick woods. These birds rarely hunt in the open, but sit concealed waiting for movement. From cover, the hawk will dash out, seize its prey with outstretched feet, and disappear again. It usually hunts high in the forest canopy, but will also pursue prey on the ground. Sharp-shins primarily eat small birds (usually passerines), but small mammals, reptiles, amphibians, and insects are also taken. One brood at Beaverhill Lake was fed large numbers of House Wrens that were nesting and fledging from nest boxes (Quinn and Holroyd 1989). The pair of Sharp-shins appear to have been attracted by the large numbers of wrens that were nesting in the 200 available nest boxes.
These birds arrive in Alberta in mid-April and begin to migrate south again in August. Most birds have left the province by October. Large numbers of Sharp-shins migrate along the shores of Lake Erie and Lake Ontario, but no migration concentration sites are known in Alberta. Sharp-shins winter from southern Canada to northern South America. Overwintering birds have been observed in southern Alberta at such locations as Horseshoe Canyon, Snake's Head, Red Deer, Calgary, Banff, Lethbridge, and Edmonton.

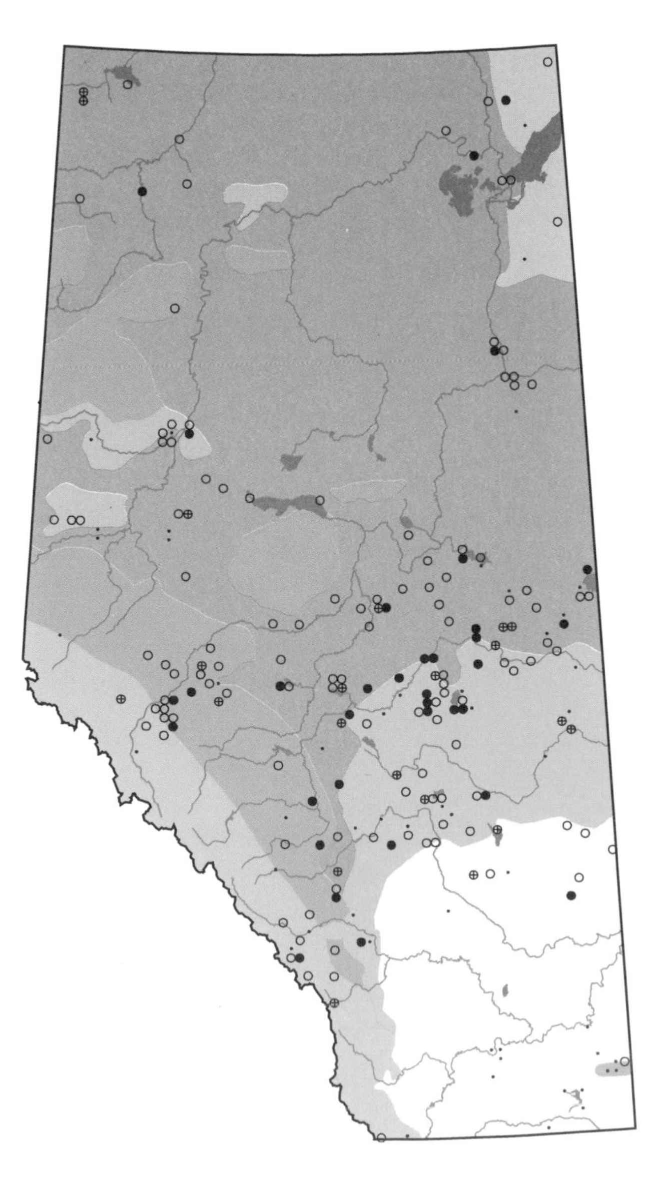

Cooper's Hawk

Accipiter cooperii

COHA

Status:

Before 1900, the Cooper's Hawk was considered a common nesting raptor. Since that time, it has suffered declines in all parts of its North American range except in western regions. The main reason for this decline is the destruction of forest habitat needed for nesting, pesticide contamination, and excessive shooting.

This hawk was on the North American Blue List from 1972-1981 and again in 1986 (Tate 1986). It is recognized as a "vulnerable" species by the Committee on the Status of Endangered Wildlife in Canada. In Alberta, population status and trend information are presently lacking.

Distribution:

The first breeding records of this species for Alberta are from the 1950s (Grant 1957). The highest concentrations of this hawk were recorded in the Parkland and southern Boreal Forest regions. It has now been observed nesting in the Rockies and foothills as far north as Jasper, then eastward along river valleys into the parklands. It was also recorded in the South Saskatchewan River Valley and there are reports of it nesting in the Cypress Hills .

The northern limit of nesting in the province appears to be south of Lesser Slave Lake. However, hawks were observed in the summer as far north as Lake Athabasca.

Habitat:

This species breeds in pure or mixed deciduous and coniferous forests. Its preference is for dense woods as opposed to the forest edge.

Nesting:

The Cooper's Hawk nests in dense cover, often near water, in the crotch of a tree or against the main trunk. Nests are usually 6-20 m above ground. A new nest is usually built each year, but sometimes nests are returned to for several years. Occasionally, birds will use an old crows nest for their nesting base. There is one report of a Cooper's Hawk making use of an artificial nesting platform (Hoffman 1988).

The nest is a well-hidden substantial mass of twigs lined with grass and shreds of bark. The male does most of the building.

The female lays 3-5 eggs, bluish white and lightly blotched with brown or violet. Incubation is 34-36 days (Meng 1951) by both sexes, primarily by the female. At first, the male brings food to the female who feeds the young; later, when she also hunts, he will also feed the nestlings. During the breeding season, these inconspicuous birds are even more shy and secretive. However, during incubation the male will "sing" to his mate using a soft, modulated version of its cackling call. Fledging is at 30-34 days. After fledging, the young will return to the nest to feed for another 10 days. The family remains together for two months after fledging.

Remarks:

In the field, the Cooper's Hawk appears much like the Sharp-shinned Hawk, except that the former is larger and has a rounded rather than a square tail. Like other accipiters, the Cooper's Hawk is unobtrusive and is rarely seen in the open. It is quick and agile in the deep woods, and is often seen flying just a few metres above the ground. It relies on a swift surprise attack to capture its prey. This hawk feeds mainly on medium-sized birds such as jays, flickers, and thrushes, and small mamals, but also takes young poultry, game birds, reptiles, amphibians, and insects. The Cooper's Hawk's predation on poultry earned it (and other hawks) the name of "chicken hawk".

These hawks arrive in southern Alberta at the end of March or beginning of April. The first fall migrants depart in August and the last birds leave by late September or early October. Overwintering has been observed at several locations in southern and central Alberta, as far north as Edmonton and Wabamun Lake.

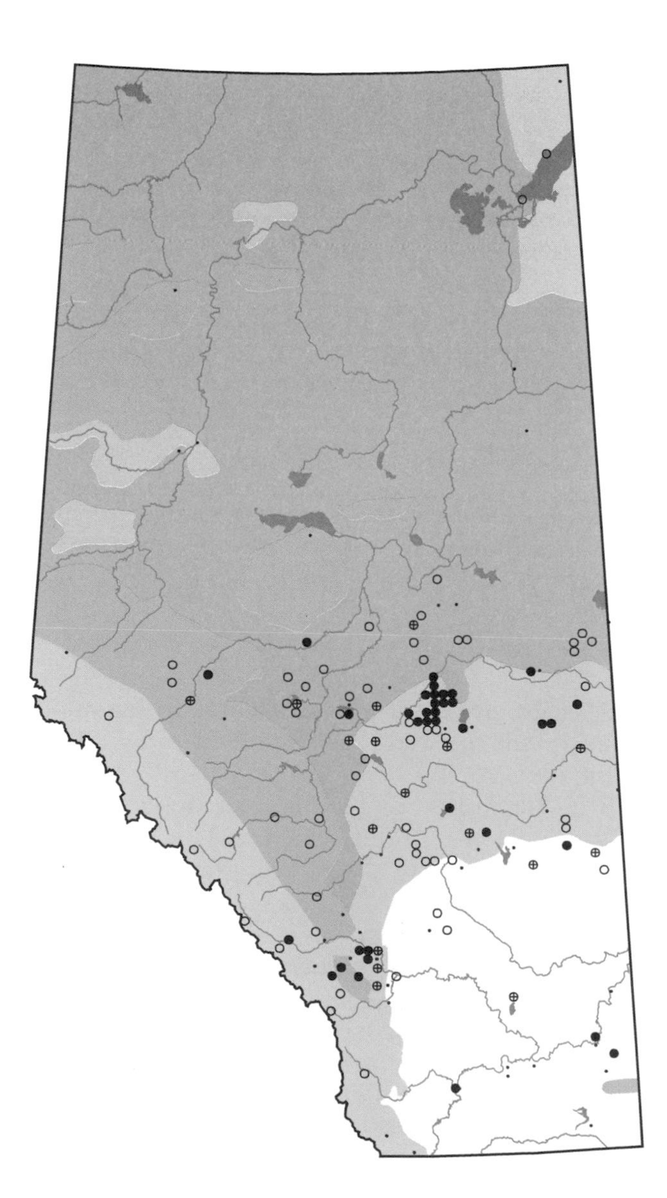

Northern Goshawk

Accipiter gentilis

NOGO

Status:

In northeastern Canada, Northern Goshawks are known to experience massive population declines when prey is scarce. In Alberta, the present status of this species is not known. There is concern over the impact of industrial development in key Boreal Forest habitats (Wildlife Management Branch 1991), and the effects of northern logging.

Distribution:

This species breeds most commonly in the densely wooded parts of northern and western Alberta. Although it was found in all the natural regions of the province, it was most prevalent in the Boreal Forest region (9% of surveyed squares), Foothills region (13% of surveyed squares), and Rocky Mountain region (7% of surveyed squares). There are extralimital breeding records from the Cypress Hills (Salt and Salt 1976).

Habitat:

Northern Goshawks are found in a variety of forested habitats, usually in mixed wood forests that are dense, but sometimes in areas interspersed with clearings or cultivation.

Nesting:

The nest of this species is a large shallow structure of dead twigs lined with shreds of bark and leafy twigs. The male builds the nest, but fresh bunches of conifer needles may be brought by the female daily after incubation. The nest is about 1 m in diameter and is built 6-23 m up in either a conifer or hardwood, usually in heavy woods. A fork in a hardwood is usually preferred. Most nests are built close to permanent water and may be used for successive years. Occasionally, the nest of another hawk will be built upon and relined.

The female lays 2-4 pale, bluish-white eggs. They are incubated 36-41 days by the female, who is fed by the male. The female feeds the young 8-10 days, then remains nearby as the male brings food. The young fledge at 40 days. Northern Goshawks are single-brooded. They are fierce defenders of the nest, and will strike human visitors who stray close to a nest.

Remarks:

The Northern Goshawk is one of the largest, strongest, and most audacious of the hawks. When hunting, it will make a quiet, concealed approach, gliding in low and fast. Often it will be involved in dramatic aerial pursuits of a Ruffed Grouse or other bird. Prey types include grouse, snowshoe hare, red squirrel, and other small mammals and birds. During the winter, when many summering birds have migrated and small mammals are hibernating, the Goshawk may turn to the barnyard for its food. This habit, plus its taste for bird game species, have caused it to be much maligned. However, Northern Goshawks are not very numerous, so their depredations are not highly significant (Salt and Salt 1976).

Generally, the Northern Goshawk is migratory, although some individuals overwinter in the northern breeding range. Other birds will move into the parklands and prairie in winter, where they are found in brushy draws, treed coulees, and river bottoms. Failures of cyclic northern food species, such as Ruffed Grouse and Snowshoe Hare, will also cause a substantial movement south. There are definite movements in Alberta in March-April and September-October. Overwintering is recorded from a number of localities throughout the province, north to Wembley and Ft. McMurray.

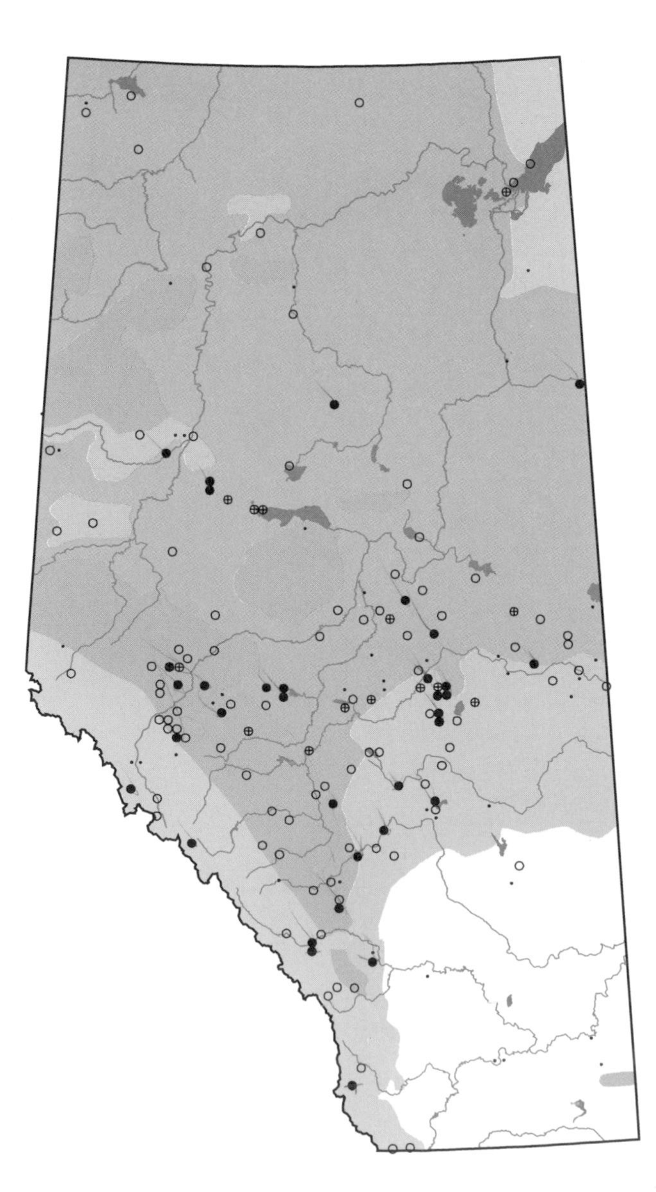

Broad-winged Hawk

Buteo platypterus

BWHA

Status:
Populations of Broad-winged Hawks in the prairie provinces have been assessed as stable with high abundance (Fyfe 1976). However, the exact status of this species in Alberta is presently unknown.

Distribution:
The Broad-winged Hawk breeds in a narrow east-west strip in the southern Boreal Forest and Parkland regions. It is found from the Miquelon Lakes north to Cold Lake, Lesser Slave Lake, and Hines Creek, with a small breeding population in the Fort MacKay area. This hawk was sighted in summer, but without breeding evidence, in the Lake Athabasca area and at Fitzgerald (just south of the Alberta/N.W.T. border). The species was first found nesting in the Cypress Hills in 1971 and the continued use of this area has been documented. The rare dark phase of this bird appears to nest only in this province (Clark and Wheeler 1978).

Habitat:
This small woodland buteo often perches in or under the forest canopy, and forages at the forest edge, near clearings and wet areas, or in the woodland itself. It rarely shows itself above the forest canopy except in courtship display.

Nesting:
Broad-winged Hawks nest in dense woodland generally 6-12 m up in a medium to large deciduous tree or, occasionally, a conifer. The nest, usually in the first main crotch of the tree, is a small shallow structure built of twigs and lined with bark chips and mosses. It is built by both sexes, but lined by the female alone. The birds generally build a new nest each year. Broad-winged Hawks do not mate for life, but do tend to return to the forest in which they bred the previous year.
The female lays 2-4 eggs, generally creamy white with markings of brown to lavender. Once the eggs are laid, the adults become very quiet, inconspicuous and hard to find. The incubation period has been variously reported, from 23-38 days. Incubation is by both sexes, but mostly by the female. When on the nest, the birds tend to sit silently, allowing an intruder a close approach.
After hatching, the female broods continuously for a week or two, and the male brings food. The female will then begin to hunt on her own again and supplement the food brought by the male. Young fledge at 24-30 days. Families remain together for several weeks after fledging.

Remarks:
The Broad-winged Hawk is an unobtrusive, crow-sized bird. Its short tail with black and white tail bars, broad, rounded wings, and melancholy cry (like a Wood-Pewee) help in its identification.
It hunts either at the forest edge or deep in the forest canopy, and generally from a perch where it watches for movement of some unsuspecting prey. Once prey is spotted, the bird pounces and carries it off to the perch. Only occasionally does this species hunt from flight or on the ground. The bulk of this raptor's diet is small mammals (voles, chipmunks, squirrels), birds, reptiles, amphibians, and a large variety of insects.
While this species is shy and retiring in nesting season, it can be observed in large numbers during spring and fall migration. The birds travel in groups called "kettles", exploiting thermals to gain altitude, then gliding on set wings in the general direction of travel.
Broad-wings arrive in Alberta in late April or early May and leave the province between mid-August and the third week of September. No overwintering has been observed, although there are November records in the south.

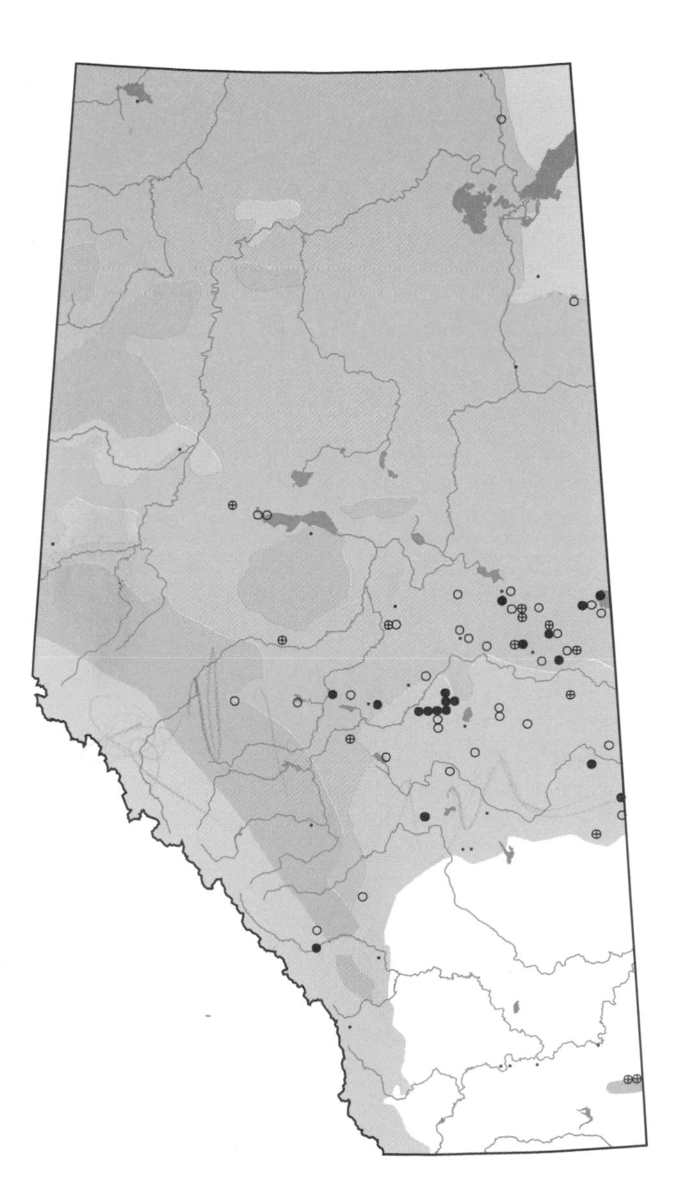

Swainson's Hawk

Buteo swainsoni

SWHA

Status:

While the status of Swainson's Hawk in Alberta is not known (Wildlife Management Branch 1991), populations of this species are thought to be stable in the Canadian prairies (Palmer 1988). However, the species as a whole has declined over much of its range, primarily because of habitat loss and excessive shooting. Because this hawk eats prey that are not major accumulators of biocides, it has been less affected by DDT and other pesticides than have other raptors. However, the effects of the continued use of pesticides in the bird's South America wintering range are unknown.

Swainson's Hawk was on the Blue List from 1972 to 1982, and in 1986 was delisted to species of "special concern" (Tate 1986). The total North American population of this species has been estimated at 300,000, give or take 50,000 (Palmer 1988).

Distribution:

In Alberta, Swainson's Hawk is found mainly south of the North Saskatchewan River in the Parkland and Grassland natural regions. It has been observed in summer north to Lesser Slave Lake, Grande Prairie and the Peace River District. Nesting outside of its more usual range is recorded for the Fort McKay area, also the Waterton Lakes, Banff and Jasper national parks.

Habitat:

Swainson's Hawk breeds in dry, open country where trees and shrubs are available for nesting. Highest densities of this hawk are found at the prairie/parkland border. The species also fares well in moderately cultivated areas where it can hunt the fields and their ungrazed, grassy borders.

Nesting:

Nesting is usually in deciduous trees, and sometimes in shrubs, bushes, or thickets. Nests are generally low to the ground, with the range being approximately 1-30 m. Rarely, birds will nest on the ground. The nest is a bulky mass of small sticks lined with bark, herbs, and grasses. Fresh green sprigs are added throughout the nesting period. Nests are re-used for several years and may reach 1 m in diameter.

Clutches are typically 2-3 eggs, pale blue to dull white, usually blotched lightly with brown. Incubation is 28 days (Schmutz 1977), primarily by the female. Young are tended mainly by the female, with the male providing all the food for the first 3 weeks. Young fledge at 42 days (Schmutz 1977) and families separate by late August. One brood is raised per year.

Birds do not seem to mate for life, but may pair again when they return to their previous nesting territory.

Remarks:

This large open country buteo is the most common hawk of the prairie ecoregion. It is extremely variable in appearance, with pale and dark phases, and every gradation between. A broad, dark band across the chest and a white area on the sides of the rump (in flight) are distinguishing marks.

Swainson's Hawks hunt from the air, or sometimes from a perch, and may chase after insects on the ground. This species' most common foods are ground squirrels, mice, voles, jack rabbits, small birds, insects, and, occasionally, carrion.

This hawk is highly migratory, travelling 20,000 km on the round trip to Argentina. Like the Broad-winged Hawk and Turkey Vulture, Swainson's Hawks use thermal updrafts and favorable winds in their migration, a technique that conserves considerable energy. The bird is thought to go without food for several weeks during the migration period.

These hawks first arrive in Alberta in late March to mid-April (Pinel et al. 1991), leaving by mid-September. A few stragglers linger until mid-October. The first winter record for the species was from Wembley in 1976. Overwintering has also been observed at Medicine Hat.

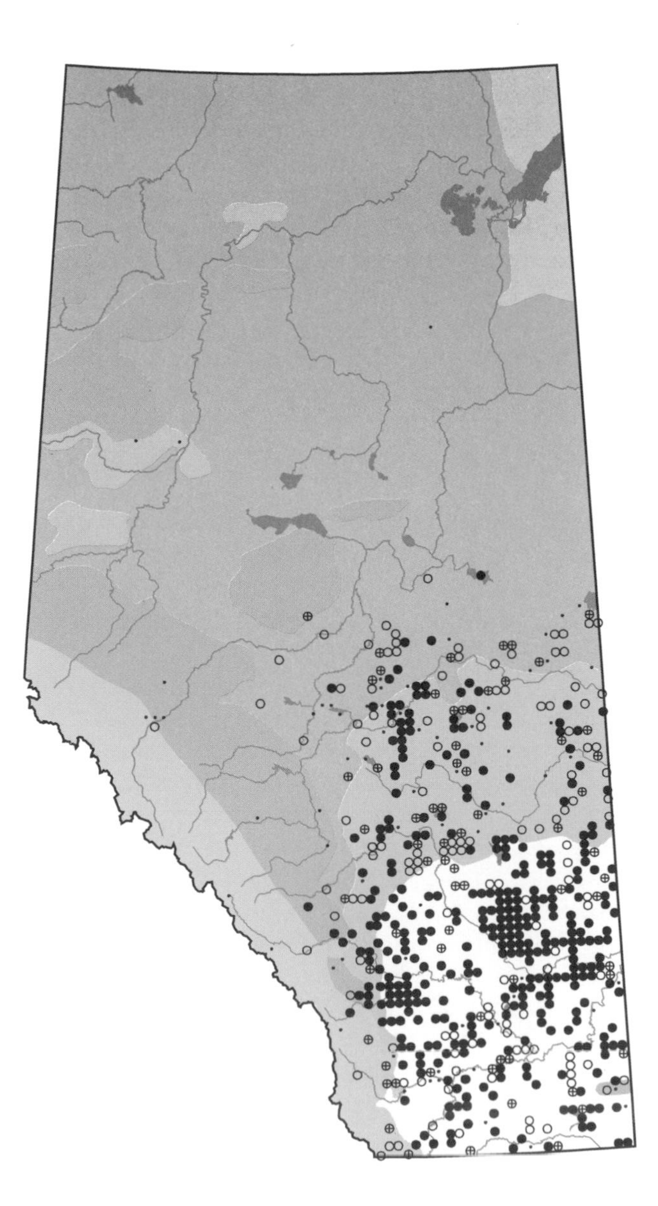

Red-tailed Hawk

Buteo jamaicensis

RTHA

Status:
This hawk is one of the most widely distributed, numerous, and commonly observed raptors in Canada. Its population in Alberta is considered healthy and not at risk (Wildlife Management Branch 1991).

Distribution:
Red-tailed Hawks breed throughout Alberta, with the majority of sightings in the central part of the province. It was recorded in over 60% of squares surveyed in the Parkland Natural Region, 47% in the Foothills, 43% in the Boreal Forest, and 30% in the Rocky Mountain region. It was also recorded in about 20% of squares surveyed in the remaining regions. It is largely absent from the southeast corner of the province, except in the Cypress Hills.

Habitat:
The preferred habitat of this species is woodland near open country, so it may be observed in diverse areas: at the edges of mixed, deciduous or coniferous woods, in agricultural areas with woodlots, along wooded streams and rivers, in woods bordering lakes, ponds or wetlands, and, occasionally, in coulees.

Nesting:
This species usually nests in tall trees, either deciduous or coniferous, most often in a main crotch or on a sturdy branch near the tree top. Small trees and, rarely, cliff ledges will be used if tall trees are unavailable.
The nest is up to 1 m in diameter and is a bulky structure of small sticks. Nests may be used in successive years and added to each year. Nest lining is usually bark, with greenery added throughout the nesting season.
The female Red-tail lays 1-4 eggs, usually 2-3, white and lightly splotched with brown. Incubation is 28-32 days, with 32 days being reported from Rochester, Alberta (Adamcik et al. 1972). The female is fed by the male while she incubates. He may also assist in some incubation duties. Young are tended by both parents and fledge at 41-46 days. Red-tails are single-brooded and mate for life.

Remarks:
This large sturdy hawk is the common soaring hawk of Alberta's parkland and northern forests. During breeding season, it can be seen circling on thermals over its nesting area, or perched high on some exposed limb or telephone pole.
Like other buteos, Red-tails are highly variable in appearance. Their most distinguishing mark is a brick-red upper tail, but this can be absent, or nearly so in many very pale or dark phase birds and immatures. Adults and immatures, except in the very dark phase, have a well-defined band across the abdomen which helps in identification. The darkest-phase birds are known as Harlan's Hawks; the very pale, as Krider's Hawks.
Red-tails are opportunistic feeders and have a variety of hunting techniques. They hunt by the perch-and-wait method, often shifting from perch to perch throughout the hunting territory. They will also search for prey from the heights, will quarter over low terrain like a harrier, hunt through trees, pursue prey on the ground or eat freshly killed carrion. The Red-tail diet consists of ground squirrels, hares, mice, voles, other small mammals, birds, and insects. These hawks are adaptable in their eating habits and will substitute one prey for another depending on availability.
This species arrives in southern Alberta in late March, with the migration peak in mid-April. Most birds have departed in fall by mid-October. There is evidence of a small passage of Harlan's Hawks in October. Overwintering has been reported at Ft. McMurray, Hinton, Wabamun Lake, Edmonton, Calgary, Snake's Head, and Lethbridge.

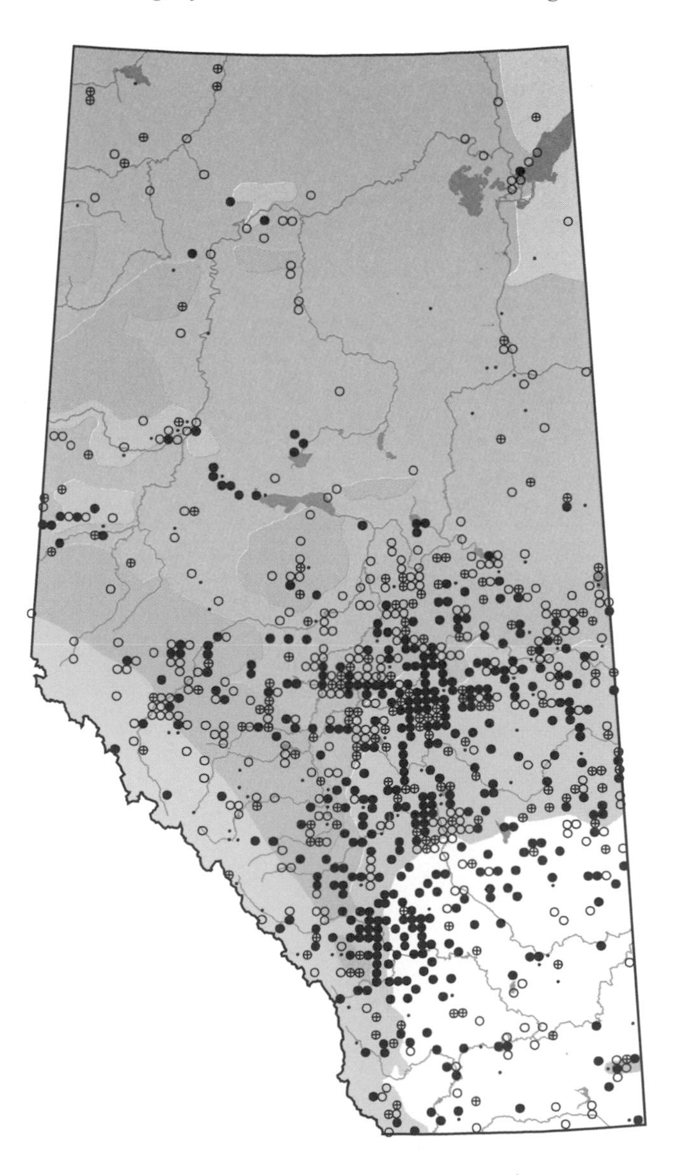

Ferruginous Hawk

Buteo regalis

FEHA photo credit: K. Morck

Status:
Historic populations in Alberta have been estimated at 4,000 pairs. The present population is estimated at 1,400 to 1,700 pairs, and is considered to be recovering (Wildlife Management Branch 1991), but the species remains on Alberta's "endangered" list.
Maintenance of this species will depend on conserving both the summer and wintering grounds, managing nest sites, and minimizing human disturbance.

Distribution:
Of Alberta's buteos, the Ferruginous Hawk is the most restricted in its breeding distribution, which includes the Grassland Natural Region, where it was recorded in 22% of surveyed squares. This hawk nests north to Trochu and Coronation, west to Calgary and southwest to Pincher Creek. One confirmed nest was north of the Battle River in the Parkland region, within its historical range. The range of this species formerly extended up to Edmonton and west to the foothills. Extralimital nesting has been reported at Waterton Lakes, Elk Island, and Banff national parks, Gooseberry Lake, Tofield, and Beaverhill Lake.

Habitat:
This hawk is a bird of the sparsely treed dry mixed grass prairie. The species tends to be most abundant in grasslands under moderate cultivation (10%-30%). With the spread of agriculture, and the encroachment of aspen parkland into the grassland ecosystem, the Ferruginous Hawk has abandoned about 40% of its former Alberta range.

Nesting:
Ferruginous Hawks prefer to nest in elevated areas such as trees, coulee ledges, rock piles, river banks, and hillsides. Traditional nesting on cliffs has drastically decreased, with a subsequent increase of nesting in trees in upland areas and farm shelter belts, but well away from human habitation.
The nest is a bulky structure of large sticks lined with sod, grasses, shredded bark, and dried cow or horse dung. Nests often contain man-made items such as fence wire, machinery belts, and small boards. Both sexes bring nest material, but the female lines the nest and moulds it. Nests are often used in successive years and can become immense.
Clutches are typically 3-5 eggs that are white with bold splotches of brown. Incubation is 36 days (Schmutz and Schmutz 1980), mostly by the female. Hatching closely coincides with the emergence of young ground squirrels. Young are tended by both parents, although the male does most of the hunting. Fledging is at 46 days (Schmutz and Schumtz 1980). Ferruginous Hawks raise one brood per year.

Remarks:
The Ferruginous Hawk is the largest and heaviest of the North American hawks. Most of these hawks (95%) are light-phased, and are identified by their white bodies and reddish-brown shoulders, backs and rumps (hence "ferruginous"). Dark-phase birds are a uniform chocolate brown.
In Alberta, this species depends mainly on ground squirrels for food, and when populations of ground squirrels increase or decline, so do Ferruginous Hawks. They also eat small numbers of hares, voles, mice, and birds.
Ferruginous Hawks hunt from a perch, from high flights, while soaring, from low flights, especially along slopes and hill sides, or while hovering. They also have the unusual strategy of laying down or crouching at a burrow.
This species arrives in late March to early April. Little is known of its fall migration, but it seems young hawks migrate on their own, leaving in August. The rest of the population leaves by early September, with some stragglers staying until late October. There is only one overwintering record, from Wrentham, in 1965.

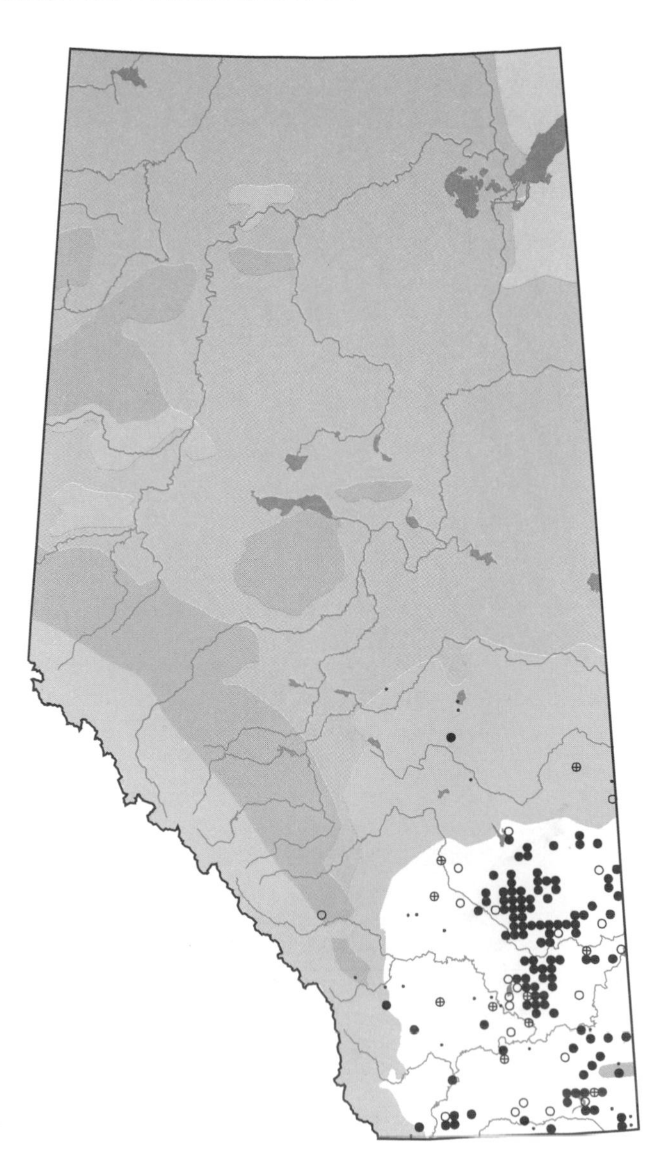

Golden Eagle

Aquila chrysaetos

GOEA photo credit: A. Carey/VIREO

Status:

Populations of this species declined during the early 1900s, primarily because of eradication programs of government agencies and ranchers. Public attitudes have shifted and eagles are now protected by legislation. Populations have begun to rebound, but are still well below former levels. In Canada, the species is showing slight increases in numbers; in the prairie provinces, populations appear to be stable, but small (Fyfe 1976). The North American population is estimated at 50,000 to 100,000 birds (Snow 1973), with over 10,000 pairs in Canada (De Smet 1987).

The Golden Eagle is scarce in Alberta, but populations appear to be stable. The provincial population has been estimated at 100 to 500 pairs (Wildlife Management Branch 1991). Because this species' population is very low and dispersed over broad areas, site-specific nest protection is appropriate.

Distribution:

This species nests in the lower reaches of the major southern river systems and the Rocky Mountain region. It is uncommon in the northern part of the province. In settled areas, this eagle is observed mainly as a transient.

Habitat:

The Golden Eagle is a bird of rocky outcrops, sparsely treed mountain slopes and grassland habitats with coulees, steep river banks, and canyons. It prefers open areas with short or sparse vegetation, especially slopes and plateaus that allow a commanding view and where there are updrafts for soaring. In forested areas, nesting territories usually contain large openings such as meadows, burns, and marshes.

Nesting:

This species nests in isolated areas, usually on high cliff ledges, escarpments, or rocky bluffs. Tree nests are common in some parts of this species' range, but not in Alberta.

Nests are large interwoven masses of sticks and twigs, and are lined with grasses, weeds, mosses, and leaves. Both sexes fetch material, but only the female works it into place. A pair may repair and tend several nests within their territory before choosing one for egg-laying. Nests may be used throughout several seasons and become enormous.

The female lays 1-3 eggs, which are white and spotted with brown. Incubation has been variously reported, from 35-45 days, and is done mostly by the female. The female also takes on most of the brooding and rearing duties, while the male does most of the hunting. Broods usually have two young, and in some cases the weaker or younger eaglet may die or be killed by its older sibling. Fledging is at 65-75 days (De Smet 1987). Golden Eagles form monogamous bonds that may be lifelong. They raise one brood per season.

Remarks:

The Golden Eagle in adult plumage is entirely dark brown except for the golden brown of the nape and hind neck.

The bulk of this eagle's diet consists of small to medium-sized mammals, particularly ground squirrels, hares, rabbits, and marmots. Hooved animals, young pronghorn, deer, elk and livestock make up no more than 5% of the diet, and much of this is in the form of carrion.

Golden Eagles hunt from high or low flight, using a low-level, high-speed glide or, occasionally, a steep vertical descent. They also hunt from an elevated perch. This species is known to pirate food from other Golden Eagles and Red-tailed Hawks.

Some eagles overwinter in Alberta, with most records from the south, but birds have been observed in winter as far north as Drayton Valley, Edmonton, Wembley and Grande Prairie. A definite movement toward nesting grounds occurs in late March, with fall migrants leaving the province by early November.

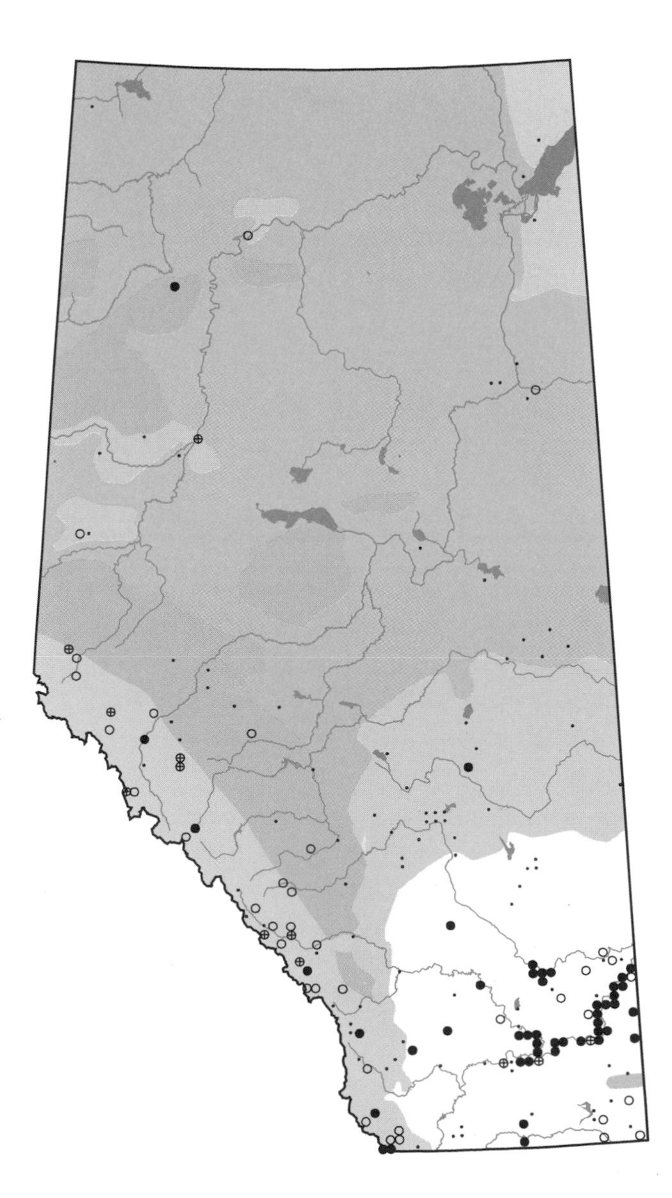

American Kestrel

Falco sparverius

AMKE photo credit: T. Webb

Status:
The American Kestrel is the most common falcon in Alberta, with a stable and secure population. The species was on the Blue List from 1972 to 1981 because of population declines in several American states (Tate 1981). Kestrel populations did not suffer the dramatic pesticide-related declines suffered by many other raptor species, mainly because of the bird's focus on a mammal and insect prey base. Cade (1982) has estimated that there are one million Kestrel pairs in North America, with numbers equally large in Central and South America.

Distribution:
This species is widely distributed throughout Alberta and Atlas data confirmed breeding in all natural regions of the province.

Habitat:
Kestrels favor semi-open to open country, breeding where trees, man-made structures, and cliffs provide cavities for nesting. Typical habitat includes grassland, farm woodlots, river bottom lands, woodland edges, burns, meadows, wooded lakeshores, and highway or railway rights-of-way. These birds are also found in urban environments such as parks, cemeteries, and residential areas. Kestrels are relatively uncommon on the prairies in comparison to other Alberta regions. In this area, they are restricted to water-courses where there are larger trees with nesting cavities, or cliffs with crevices and potholes (Oliphant 1991).

Nesting:
The American Kestrel usually nests in a natural cavity or woodpecker hole in a tree. It may also nest in crevices or holes in cliffs, bank burrows, or magpie nests. This species will also make use of a suitable recess in a building, billboard, power pole, or fence post, or use a nest box. No attempt at nest building is made and cavities are rarely lined.
Females lay 4-6 eggs, which are creamy, buff, or pinkish and marked with brown. Incubation is 29-30 days, mostly by the female, with the male providing food. The young are tended by both parents. Initially, the male does all the hunting; after a week or two, the female assists. Fledging is at 29-31 days. The young stay with the family for up to a month after fledging, but, in future years, they do not return to the territory where they were reared. Adults, however, do return to previous territories. Kestrels raise one brood per season.

Remarks:
The American Kestrel, formerly known as the Sparrow Hawk, is a jay-sized falcon, the smallest diurnal raptor in North America. The black-and-white face markings and the red plumage on its back distinguish it from the Merlin; long pointed wings prevent confusion with the Sharp-shinned Hawk.
The most important prey of the Kestrel are mice, voles, and insects, primarily grasshoppers. Occasionally, it will hunt birds (mainly nestlings), small reptiles, and amphibians. Generally, it still-hunts from a perch or hovers in the air with its tail spread and wings beating rapidly. It dives head first to take mammal prey and feet first for an insect, but it always uses its feet for the capture. Kestrels will also hunt insects on the ground.
Male Kestrels arrive in southern Alberta in the last half of March, and in mid-April in the north. Females arrive a week or two later. Fall migration begins in late August. Most birds leave by the end of September, but a few stragglers linger into November. Winter records are mainly from southern Alberta, with reports as far north as Devon, Edmonton, and Grande Prairie.

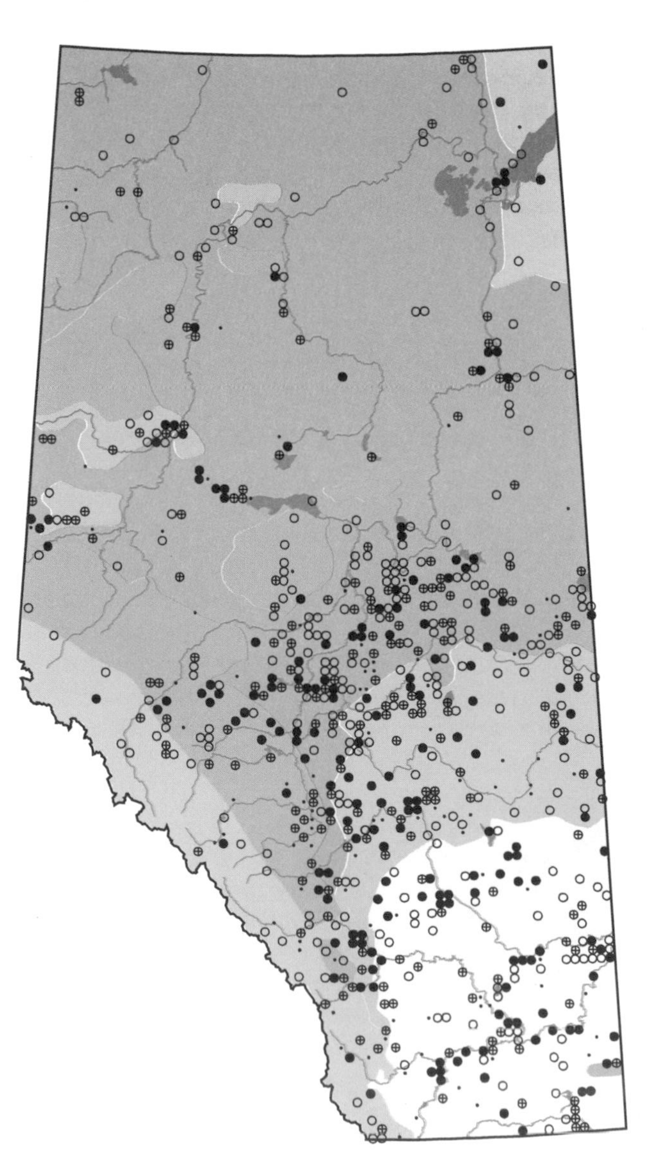

Merlin

Falco columbarius

MERL

Status:
The status of North American Merlin populations has not been adequately documented. Migration and Christmas Bird counts indicate that Merlin populations are healthy and increasing. The Merlin was on the Blue List from 1972 to 1981. It was delisted when population levels began to increase, but it remains a species of "special concern" (Tate 1986). Between 40 and 80 pairs nest in Edmonton alone (G. Holroyd pers. comm.)

Distribution:
This species is found throughout Alberta. The Richardson's Merlin (*F. c. richardsonii*) breeds in the Grassland and Parkland regions and the Taiga Merlin (*F. c. columbarius*) breeds locally in the Boreal Forest region.

Habitat:
In the Grassland and Parkland regions, Merlins breed in farm shelterbelts, small stands of trees adjacent to mixed grass prairie, and wooded river valleys or coulees. Boreal birds are found in sparse woodland bordering forest openings, burns, rivers, lakes, or bogs. These birds need an adequate avian prey base, open to semi-open habitat for hunting, woodlands or cliffs that provide inaccessible nest sites, and a regularly breeding corvid population for nests. Some southern birds are now nesting in prairie cities and towns, a shift made possible by a preceding movement of crows and magpies into the urban environment. Urban nest sites are usually in parks, cemeteries, ravines, or river valleys.

Nesting:
This small falcon usually appropriates the abandoned nest of a crow or magpie set in a tree or bushes. Occasionally, it nests in a tree cavity, on a cliff ledge, or, rarely, on the ground. Magpie nests are considered the most productive nest sites, because their overhead cover prevents predation by crows. Nests may be relined with bark. Nesting territories may be used several years in succession, but it is unclear whether there is true nest site fidelity. Pair bonds are monogamous, but this species does not mate for life. Clutches are usually 4-6 creamy white eggs with profuse blotching of reddish and dark brown. Incubation is 28-34 days, primarily by the female. The male does all the hunting in courtship, incubation, and early brooding. He calls the female off the nest to give her food, which is transferred in the air. Fledging is at 25-30 days, but the young remain dependent on their parents for up to a month after fledging.

Remarks:
Merlins are rapid and agile in flight. When flying, they look much like fast-flying domesticated pigeons, hence the vernacular "pigeon hawk". Male Merlins are slaty blue and females are a dusky brown. As is true of all hawks, the female is larger and heavier than the male.

Merlins prey on a variety of small- to large-sized birds, generally whatever passerines are locally abundant. In the Grassland region, Merlins depend primarily on Horned Larks, Chestnut-collared Longspurs, and native sparrows. They may also take a few small mammals and grasshoppers. Urban-dwelling Merlins take mainly House Sparrows, while those that winter in cities also prey on Bohemian Waxwings.

Migrating males arrive a few days to a month before migrating females, the first arriving in March. In some cases, the females remain on the territory all year and wait for the males to return. Increasing numbers of birds are overwintering in the prairies, often in urban areas. Most reports are from southern Alberta, with birds reported as far north as Wabamun Lake, Edmonton, Fort Saskatchewan, and Wembley. Overwintering has been reported rarely in Banff and Waterton Lakes national parks. Most migrating Merlins have left the province by November.

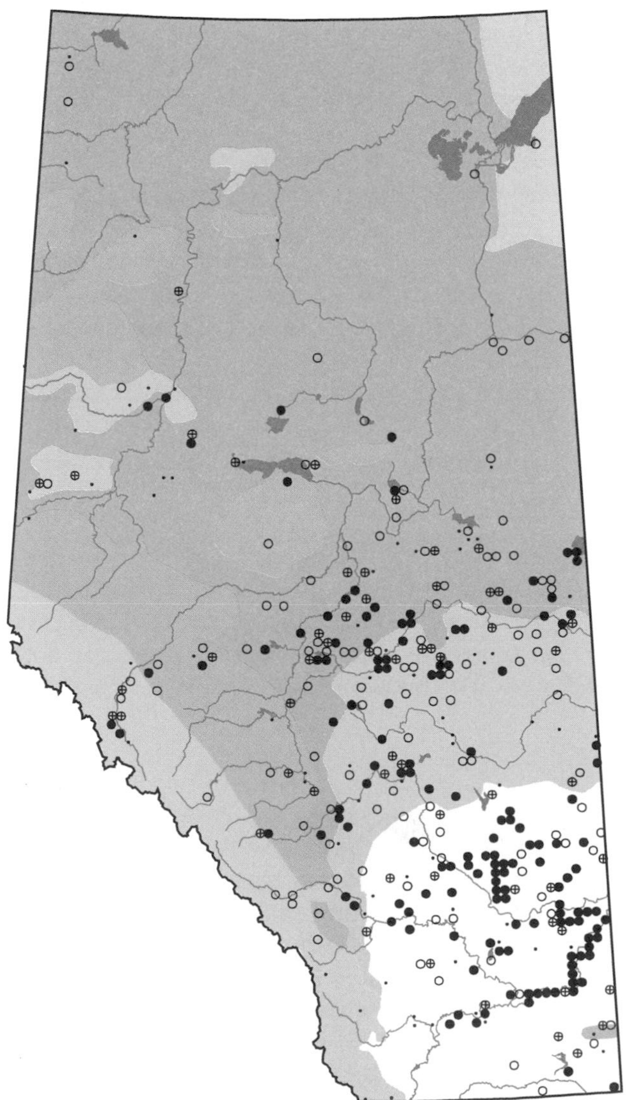

Peregrine Falcon

Falco peregrinus

PEFA

Status:

Of the three subspecies of Peregrine Falcon that breed in Canada, only one, *F. p. anatum*, is reported to breed in Alberta. This subspecies has been on Canada's "endangered" list since 1978, and is considered endangered in Alberta. Subject to the devastating effects of agricultural pesticides, particularly DDT, the *anatum* subspecies was extirpated from most of its range in southern Canada and United States, including southern Alberta. In northern Alberta, the number of occupied sites declined significantly.

Distribution:

The Peregrine Falcon has the most widespread occurrence of any falcon in the world, breeding on every continent except Antarctica. In Alberta, a small remnant population exists in the northeast corner of the province (Poston et al. 1990). Some sites in more southern locations, mainly on cliffs along major rivers, are also occasionally occupied by falcons. Summer records from the Rocky Mountains may indicate a few breeding pairs in that area. There is one pair in Calgary and two pairs nesting in Edmonton as of 1991.

Habitat:

The Peregrine seeks out cliffs near water for nesting sites and open fields, swamps, and marshes for hunting. In Alberta, the Canadian Shield region provides ideal Peregrine habitat. Birds are occasionally found in urban settings, where tall buildings take the place of cliffs.

Nesting:

Peregrines do not build a nest. Rather, they lay their eggs in a hollow scrape in the ground or directly on a building ledge, and they use very little nesting material. Pairs will often use the same nest site for several years, or may have several alternates in a nesting territory. Nesting Peregrines are intolerant of human disturbance.

The female typically lays 3-4 eggs, which are creamy or buff with heavy red or chestnut speckling. Incubation is 33-35 days, by both sexes, but primarily by the female. The female closely broods and feeds the young for about two weeks, while the male hunts. When he brings food, he calls the female off the nest and food is passed in the air. Fledging is at 35-40 days and the young depend on their parents for food for up to two months after fledging.

Remarks:

This falcon is considered to be one of the swiftest birds in the world when stooping or diving at prey at nearly 300 km/h. Peregrines eat songbirds, waterfowl, and shorebirds and, rarely, mammals; young Peregrines may chase insects on the ground. The species is specialized for hunting by direct aerial pursuit, usually from a high vertical stoop. It may hunt from a perch or a low glide, or from soaring.

Captive-raised and release programs have reintroduced *anatum* birds into part of their former breeding range in Canada. Of approximately 600 captive birds released in Canada since 1977, 30 breeding pairs have become established. Several young birds are fostered to wild parents in northern Alberta each year, to increase nest productivity.

The greatest threat to the *anatum* subspecies continues to be exposure to pesticides in winter in Central and South America. Recovery of the subspecies in Alberta depends on releases of captive-reared chicks into southern areas (Wildlife Management Branch 1991).

These birds are first seen in Alberta in mid-April, with the migration peak in early May. In the fall, birds leave between early September and mid-October. Overwintering has been reported in southern and central Alberta at Calgary, Cochrane, Longview, Bruce, Crowsnest Pass, Monarch, Pigeon Lake, and Edson.

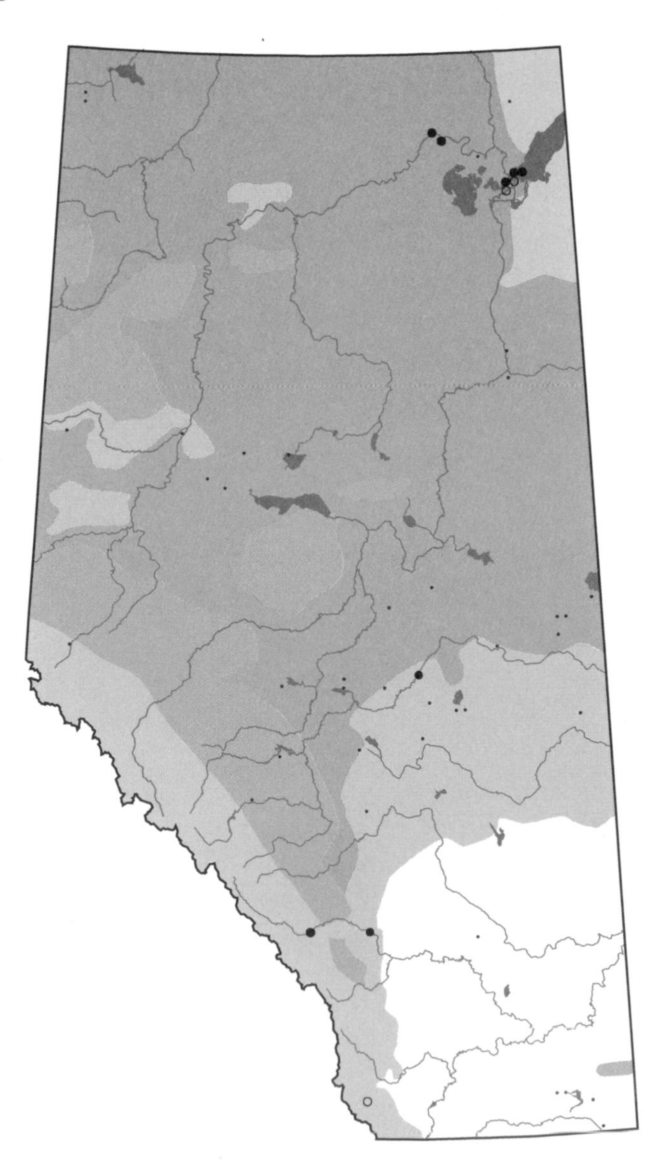

Prairie Falcon

Falco mexicanus

PRFA

Status:

Like many other raptors, the Prairie Falcon has been the victim of shooting, human disturbance, the taking of eggs and nestlings, and pesticides. In the 1920s and 1930s in Alberta, *Falco mexicanus* was a species for which a bounty was paid.

Recent surveys show a small but stable, and possibly increasing, population of Prairie Falcons in Canada. Total numbers of this species have been estimated at 5,000 to 6,000 pairs (Palmer 1988), with no more than a few hundred pairs breeding in Canada (Oliphant 1991). One estimate has determined that 250 breeding pairs occur in Alberta (Woodsworth and Freemark 1982). The range of this species in Alberta's northern areas has been reduced from historic levels. Its core range in southern Alberta is dependent on availability of secure nest sites and an adequate ground squirrel prey base (Wildlife Management Branch 1991).

This species was on the Blue List from 1972 to 1980. It was removed from the list in 1981, despite concerns about some local populations because most respondents thought the species was "doing well" (Tate 1981).

Distribution:

The Prairie Falcon breeds locally in the Grassland Natural Region (recorded in 12% of surveyed squares), west into the Rocky Mountain region and north to Calgary, Red Deer, and Rumsey. There are historic nesting records for the North Saskatchewan River 50 km southwest of Edmonton, and one nest site near Devon, where the young were banded (H. Armbruster pers. comm.).

Habitat:

A bird of dry, open country, the Prairie Falcon is found in the vicinity of the canyons and coulees of the badlands, or about the cliffs of southern river valleys. It hunts the adjacent grasslands.

Nesting:

This species nests in cliffs, using potholes, shallow caves, or ledges with overhangs. Man-made excavations in otherwise unsuitable cliff faces are also accepted. Eggs are normally laid in a scrape late in April and early in May. Occasionally, this falcon will appropriate the old stick nest of another cliff nester such as a Golden Eagle, Red-tailed Hawk, or Common Raven. Birds return to previous territories, but may change the nest site from one year to another.

Clutches are 3-6 cream or buffy eggs splotched with brown. Incubation is 33-35 days, mostly by the female. The male brings the female food during incubation and through the period of brooding, after which she hunts again. Fledging is at 36-41 days. Before they scatter, the young may stay near the nest to be fed for another two to three weeks. Prairie Falcons raise one brood per year.

Remarks:

The Prairie Falcon is as large as the Peregrine Falcon, but is much paler in color. Its favorite prey are ground squirrels, mice, insects, and birds. It may hunt from a perch or from low level flight where it flushes prey close up, or it may make a rapid, vertical swoop. Most prey is taken on the ground. Birds may range over 25 km from the eyrie in search of food.

This falcon winters throughout much of its breeding range, but most northern birds move south in winter. In spring, migrating falcons arrive in breeding areas in March. Either sex may return first to establish the nesting territory (R. Fyfe pers. comm.). In July and August, post breeding falcons hunt in the Rockies where ground squirrels are still active (Dekker 1984). In the fall, most birds have departed by late October. They migrate with other hawks and may soar in thermals like the buteos. Overwintering has been reported for various locations in southern and central Alberta as far north as Edmonton.

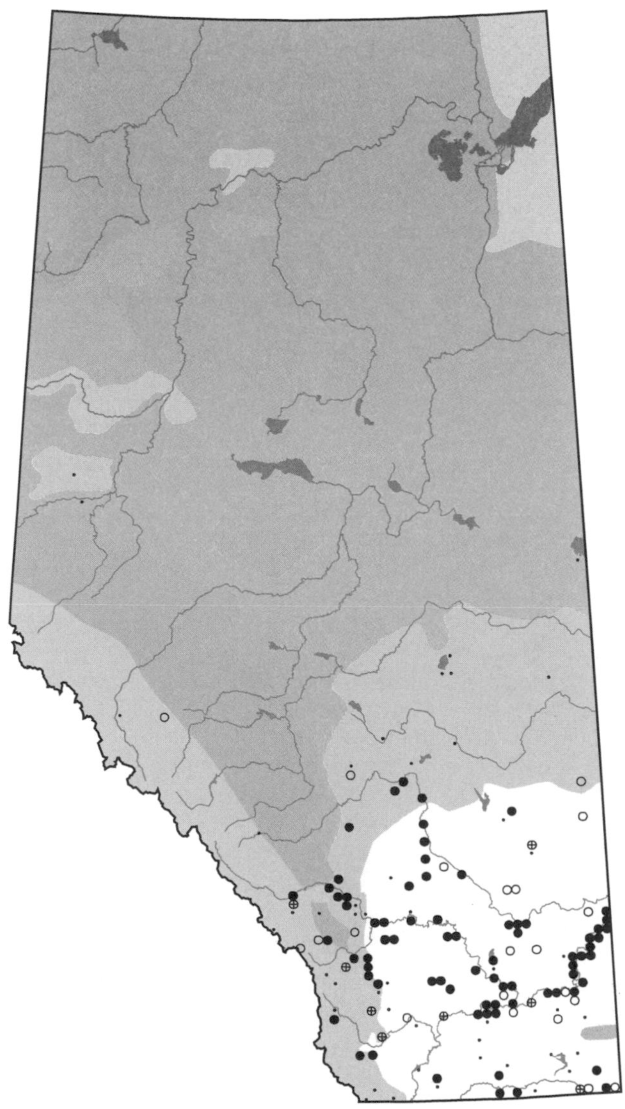

Order Galliformes

This order of upland game birds includes all grouse and ptarmigan, the Ring-necked Pheasant, the Gray Partridge, and the Wild Turkey. Largely terrestrial, all birds of this order have large feet, rounded wings for short, explosive flights, and small bills to feed on seeds and other vegetation. Although the grouse, ptarmigans, and Gray Partridge are all plump, compact birds that resemble domestic fowl, only the grouse and ptarmigans have, in varying degrees, feathered legs. Despite their cryptic coloration, they are favored prey items of both raptors and carnivorous mammals, including man. Grouse species range from gregarious to solitary, although all conduct some kind of elaborate courtship display. In the same family as domestic fowl, male Ring-necked Pheasants are the most brightly colored of any member of this order in Alberta. In contrast, the larger Turkey is quite uniformly colored, although his long neck bears a bold red wattle. Both species have naked, spurred legs. Where there are sufficient numbers, Turkeys are gregarious. The sexes of the species in this order vary in appearance, with the exception of the Turkey.

Gray Partridge

Perdix perdix

GRPA

Status:
This game bird of Alberta peaked in population in the late 1970s (Pinel et al. 1991). Although present population levels appear to be stable (Wildlife Management Branch 1991), these levels are susceptible to sharp drops during severe winters. This is somewhat mitigated by the species' high reproductive potential. During the five year survey period of the Atlas Project, the Gray Partridge was fairly common in the Grassland and Parkland regions, where it was recorded breeding in 23% of surveyed squares.

Distribution:
The Gray Partridge is a resident throughout its range, which includes the Parkland and Grassland natural regions and the Peace River district. It is a very rare visitor to the Foothills and Rocky Mountain regions, where it was recorded in 5% and 2% of surveyed squares, respectively. Although Salt and Salt (1976) and Godfrey (1986) reported the distribution of the Gray Partridge as continuous from Mayerthorpe northeast to the Peace River district, Atlas data support Pinel et al. (1991) who reported that the Peace River population is disjunct.

Habitat:
This bird resides in areas of open grassland and agricultural land which have adjacent areas of woody cover. It is most frequently observed around the border between scrub and cultivation.

Nesting:
Prior to courting, males engage in prolonged battles to establish breeding territories. During courtship, the male will attempt to attract prospective mates with a lateral display, exposing his barred flanks. Paired in April, the Gray Partridge constructs a nest of grass over a scraped hollow on the ground, concealed in surrounding grasses or low bushes. Often positioned in roadside ditches, these nests are rarely above ground.
A clutch may include from 10-22 unmarked olive or olive-buff eggs, which are laid at 1-2 day intervals. Although larger broods may be the efforts of two females, hatching success is normally unaffected. Upon completion, the female begins incubating her single brood for 23-25 days in late May or early June, while her monogamous mate defends from a close distance. The young hatch over a short period, often leaving the nest on their first day. Both adults brood and tend the young, which grow quickly and fledge in 16 days. Family units are seen travelling with the female in front while her mate acts as rearguard. Adults are known to use both distraction and aggression in defence of their young. Family groups remain together until the following year.

Remarks:
This bird was first imported from Hungary in 1908, when 70 pairs were released near Midnapore. From this came the local name Hungarian Partridge. Alberta's smallest game bird is generally gray with a pale brown eyeline, cheeks, and throat. This bird is also distinguished by its unmarked, chestnut outer tail feathers in flight and a chestnut "horseshoe" on the male's breast. Flushing warily, this bird explodes with a clatter of wings and a rapid cackle, assuming a low and fast flight path.
The Gray Partridge feeds on leaves, grass and clover shoots, insects, grasshoppers, and weed seeds. In winter, it can often be seen seeking the protection and waste grain around farm buildings. In colder weather, coveys of up to 30 birds will huddle in a circle with heads outwards and tails toward the middle. Digging into deep snow, a Gray Partridge will escape the wind, bask in the sun, and feed on stubble. However, unlike the Sharp-tailed Grouse, this species does not tunnel in the snow.

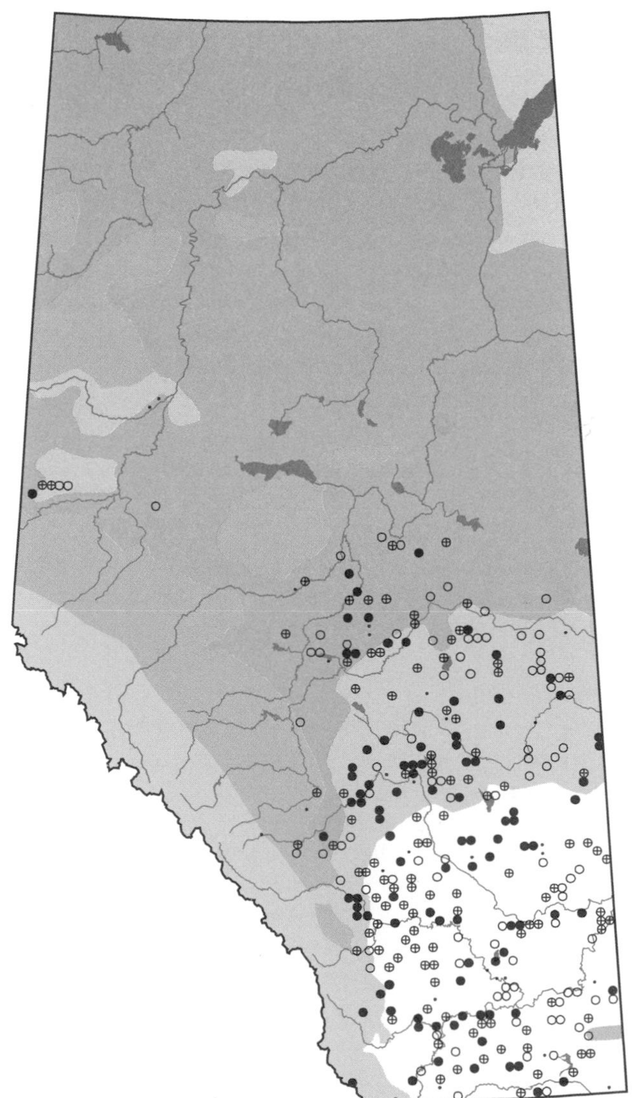

Ring-necked Pheasant

Phasianus colchicus

RGNP photo credit: T. Thormin

Status:

The population of this colorful game bird has decreased from a maximum of approximately 1.5 million birds in the early 1960s to 250,000 in the 1980s, and even fewer today (Fish and Wildlife 1983). Despite an increase in range because of further land clearing, local population levels are decreasing as a result of habitat degradation, mainly the reduction of cover habitat and small water bodies. This species is also prone to high mortality rates in severe winters, although this is somewhat mitigated by hatchery releases.

Distribution:

Since being introduced into the province from Eurasia in 1908, the Ring-necked Pheasant has become a resident of the Parkland and Grassland natural regions of central and southern Alberta, where it was recorded breeding in 20% of surveyed squares. This bird was later released into the Peace River district, and it is considered only a very rare visitor in the Foothills and Rocky Mountain regions.

Habitat:

The Ring-necked Pheasant is most abundant on farmland and/or grassland, which has adjacent areas of suitable cover. This may include local reedbeds, hedgerows, willow scrub, or woodland borders, or similar dense cover along irrigation canals. It may occasionally be seen in city parks or backyards, especially near centres of annual releases.

Nesting:

In early spring, the male establishes a breeding territory that he is tenacious in defending, although persistent challenges and other demands result in flexible boundaries. Practicing polygamy, the male will protect a harem of up to six females. Females start their solitary nesting duties in mid-April to early May, constructing a shallow hollow that is sparsely lined with nearby grasses and dead leaves. This ground nest is concealed by tall grasslands or by adjacent shrubs and is invariably near water. Clutches consist of 6-15 unmarked olive or buff eggs, which are laid on consecutive days. Occasionally, females are known to parasitize the nests of Ruffed Grouse and Blue-winged Teal. Incubation is carried out by the female for a period of 22-25 days, at which time she will leave to forage only a few short times a day. Young are led to food and brooded only at night or during inclement weather. Fledging takes 12-14 days and, when half-grown, the young will roost in trees at night.

Remarks:

The brillant coloration of the male makes this bird very distinctive from other species in the same habitat. This large ground-dwelling bird has a long tapering tail, iridescent green head, bronze body, and naked legs, and often sports a white neck band. The female is smaller and is generally buffy and marked with black and brown. She is distinguished from Sharp-tailed Grouse females by her buffier appearance, lack of white patches, long tail, and naked legs. Naked legs and no black abdominal patch separate it from any Sage Grouse. Relying on her coloration, the female will remain motionless when approached while alone or on the nest. However, when accompanied by her brood, she may either noisily flush or use a distraction display to draw attention away from the scattering young. Perhaps because of its brillant color, the male is more easily flushed when disturbed.

The Ring-necked Pheasant feeds diurnally on insects, weed seeds, waste grains, wild fruits, berries, and small vertebrates. In areas of deeper snow, it may be forced to forage on the buds of trees and bushes and in the thin snow under trees.

Loosely communal winter flocks are formed based on sex and a strong dominance hierarch.

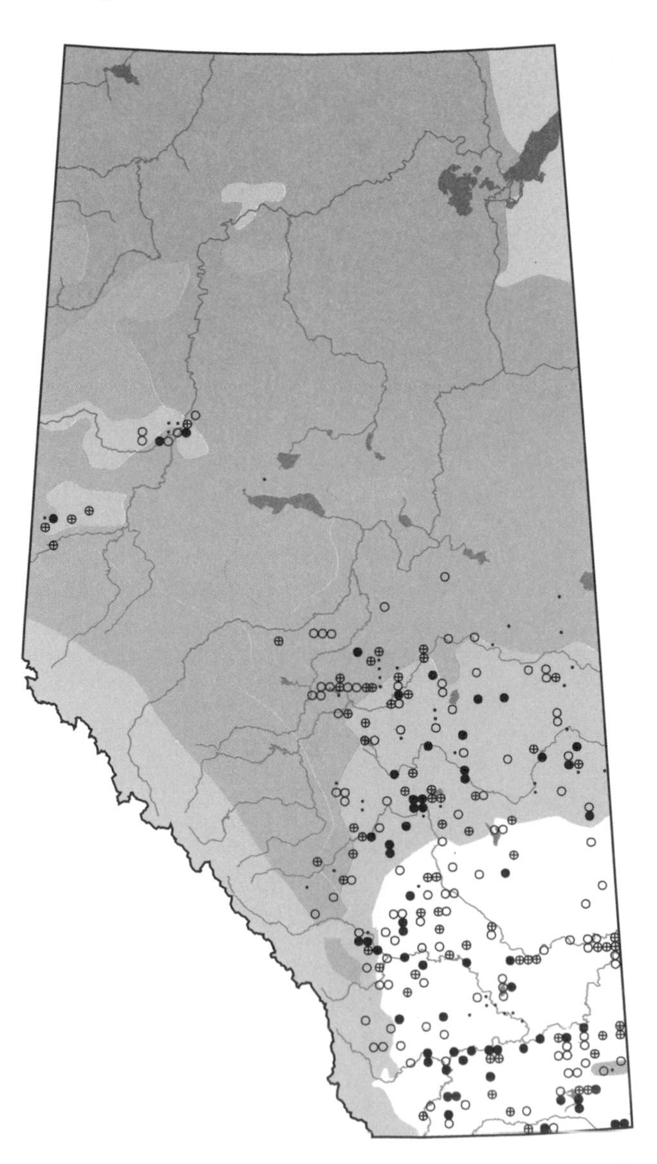

Spruce Grouse

Dendragapus canadensis

SPGR photo credit: A. Yuill

Status:

Although the status of the Spruce Grouse is described as healthy and not at risk (Wildlife Management Branch 1991), land clearing and fragmentation have reduced available habitat in central Alberta (Pinel et al. 1991). Holroyd and Van Tighem (1983) stated that population fluctuations in the mountain parks can vary its status from common to uncommon. Although small pockets remain in central Alberta, land clearing has restricted the largest proportions into the more remote peripheries of western and northern Alberta.

Two subspecies of the Spruce Grouse are found in Alberta, *D. c. canadensis* is found in central and northern Alberta, except in the Rocky Mountain Natural Region, where *D. c. franklinii* is found.

Distribution:

Salt and Salt (1976) reported that the Spruce Grouse formerly bred in all of northern, central, and western Alberta south to about Red Deer, but the destruction of coniferous woods has forced it to retreat farther north. Atlas surveys recorded this grouse in 50% of surveyed squares in the Canadian Shield, 20% in the Foothills, and 24% in the Rocky Mountain region. It was only recorded in 7% of surveyed squares in the Boreal Forest Natural Region, but it was widespread.

Habitat:

This species prefers coniferous and mixed wood forests with muskegs and small openings (Godfrey 1986). Holroyd and Van Tighem (1983) reported that the coniferous forests in the montane and subalpine were used, with the majority of observations in closed lodgepole pine or lodgepole pine/spruce forest.

Nesting:

Johnsgard (1973) described in detail the territorial and courtship displays of the male Spruce Grouse. The male may flutter from a low bough, producing a Ruffed Grouse-like drumming sound that is punctuated by a wing clap as it alights. A strutting display is performed in front of the female, with the male partially spreading his tail and erecting his red combs.

In May, the hen builds a sparsely lined nest of leaves and grass over a depression, either on the ground or in moss under a low hanging conifer bough. Open coniferous forest is selected and a wet bog is often nearby. The typical clutch is 7-8 eggs, which are buff and mottled with various shades of brown. The female incubates for extended periods during the required 21-24 days, hatching her only brood of the year. Boag et al. (1983) reported that Spruce Grouse are relatively vulnerable to egg predation, with the major predator being the red squirrel. The young birds feather quickly and fledge in 10 days. The female broods and tends to the chicks alone for the summer, sometimes accompanied by a male in the fall.

Remarks:

The Spruce Grouse male is generally barred in black and brown with a black throat, breast, and tail. The female, being quite different, is generally barred brown, gray, and rust. The conspicuous black and white markings of the underparts of males distinguish Spruce Grouse from Blue Grouse (Johnsgard 1973). The lack of a ruff and a shorter tail with no subterminal band sets this species apart from the Ruffed Grouse. Leaves, insects, and berries are eaten in summer while conifer needles are the staple in winter (Salt and Salt 1976). Spruce Grouse are renowned for their tameness, unless the chicks are alarmed, in which case the female may make aggressive challenge.

Winters are spent in the protection of denser coniferous stands.

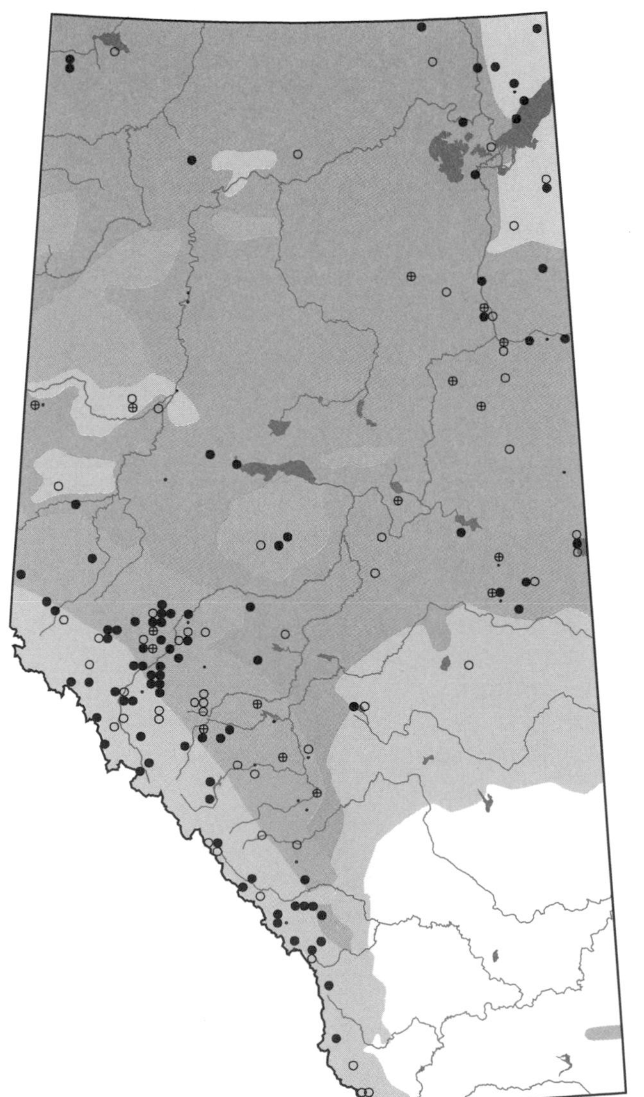

Blue Grouse

Dendragapus obscurus

BLGS photo credit: S. Gunsch

Status:
No major changes have occurred in the range of the Blue Grouse since historical times. However, there have been local population fluctuations, such as a decline since the 1950s in the Kananaskis/Sheep River area.

Distribution:
The Blue Grouse is a permanent resident of the Rocky Mountain Natural Region and adjoining Foothills Natural Region. Its range is from the headwaters of the Smokey River in the north, south to the international boundary, east as far as the lower foothills, and west into British Columbia. It was recorded in 11% of surveyed squares in the Rocky Mountain Natural Region, and 3% in the Foothills Natural Region.

Habitat:
This species occupies a fairly broad vertical range, breeding at low elevations and spending fall and winter close to the timberline. In the spring and summer, the Blue Grouse is mainly terrestrial, living in a variety of treed habitats in or near openings such as forest burns or logging slashes. In winter, the Blue Grouse moves to higher elevations, inhabiting thick coniferous forests where it becomes more arboreal.
In Banff and Jasper national parks, most records are from the dry slopes and burns in the montane/subalpine region of the drier, lightly wooded valleys of the Front Ranges (Holroyd and Van Tighem 1983). In Kananaskis Provincial Park, the Blue Grouse inhabits the upper 100-300 metres of the forest, along the interface of subalpine fir or mixed conifers with alpine larch or larch/fir (Pinel et al. 1991). In Waterton National Park, they frequent a variety of elevations, ranging from prairie poplar mosaic to the subalpine zone. The species is most common in coniferous types (Pinel et al. 1991).

Nesting:
The male establishes his territory and announces his presence by emitting a gutteral hoot produced by filling the yellow or plum-colored (as pictured above) nuchal sacs with air. The females are not fixed to territory and wander through. They are pursued by any male whose territory they enter (Salt and Salt 1976).
The Blue Grouse nest is a depression in the ground, usually well hidden under a rock or some type of vegetation, such as a small tree, shrub, or fallen log. Nest materials are taken from close at hand and include grasses, twigs, roots, leaves, needles, moss, herbaceous vegetation, and feathers.
The female lays 5-10 eggs, typically 6-8, which are buff with fine spots of reddish or light brown (Harrison 1978). She will partly cover the eggs when she leaves the nest. Incubation is 24-25 days by the female. Young can feed themselves shortly after hatching, but are brooded by the female for 8-10 days. The broods move from the immediate nesting area to thick cover. The young begin to fly at 6-7 days and they are fully fledged by two weeks (Harrison 1978).

Remarks:
The adult male Blue Grouse is a large dark slate-colored grouse that is similar to the smaller Spruce Grouse, but lacks the black breast patch of that species (Godfrey 1986). Females of these two species are similar, but the female Blue Grouse lacks the dark breast patch of the female Spruce, and is less extensively barred on the underparts than is that grouse.
In its summer range, the Blue Grouse feeds primarily on flowering plants or insects. The young feed heavily on insects for the first few days and then include more plant material in their diet.

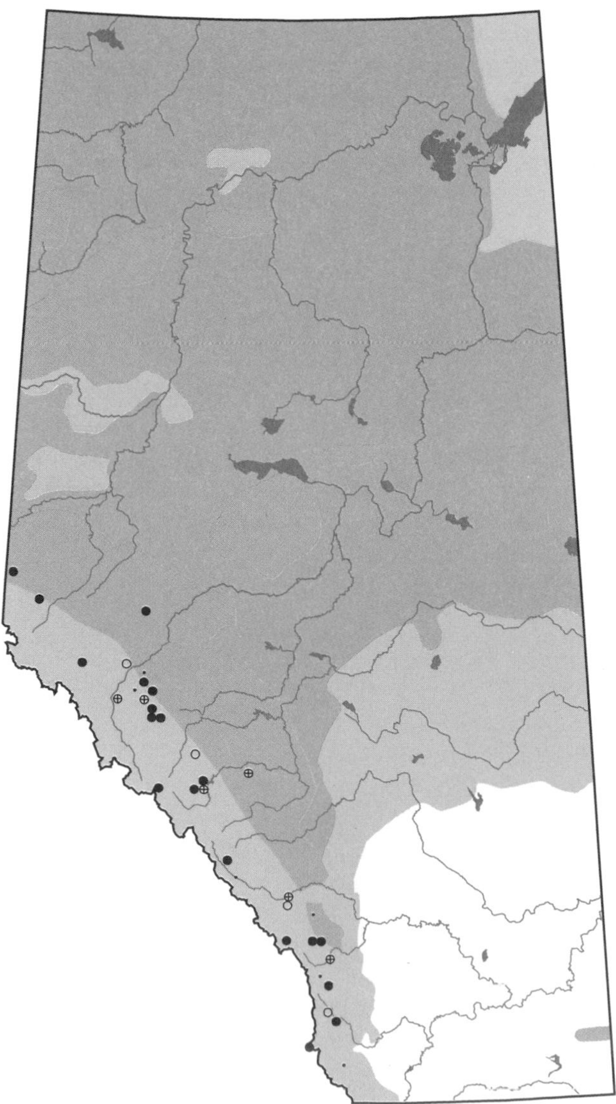

Willow Ptarmigan

Lagopus lagopus

WIPT

Status:
The Willow Ptarmigan is a rare breeder in a very localized part of the province. Holroyd and Van Tighem (1983) determined them to be an uncommon resident year-round, experiencing yearly population fluctuations.

Distribution:
Salt and Salt (1976) described the breeding range of the Willow Ptarmigan as the northern part of Jasper National Park as far south as the Tonquin Valley. Pinel et al. (1991) suggested that these ptarmigan may be resident as far north as the Kakwa area southwest of Grande Prairie. In winter, flocks from the Northwest Territories move south into the northern parts of the province. Salt and Salt (1976) reported that they have been recorded as far south as Sullivan Lake. During the five year survey period of the Atlas Project, there were only three confirmed breeding records and all were in the same square.

Habitat:
Breeding only in the northern part of the Rocky Mountain Natural Region, the Willow Ptarmigan prefers willow/dwarf birch meadows and willow bordered stream bottoms near or just above timberline (Godfrey 1986). It is more frequently seen, while breeding, in the open forests and shrubby meadows of the upper subalpine than is the White-tailed Ptarmigan (Holroyd and Van Tighem 1983). Bent (1962e) reported that they move to interior river bottoms and creek beds in winter, where there is food available.

Nesting:
The male begins the breeding season with territorial display. This includes plumage display, hooting, booming, and descending flight spirals. High annual mortality provides a turnover of territorial dominance, permitting a high number of yearlings to breed. In late May or early June, the female constructs a nest of grass, moss, and feathers over a small scrape in the ground. Other vegetation, usually woody, overhangs this, concealing all but a small open entrance. A nest from a previous year may be repaired and reused. The 7-10 yellow-brown eggs are spotted and blotched with rich reddish-browns. These are incubated by the female for 22 days. They are led away from the nest soon after hatching. Although only the female broods the chicks, the male is the most attentive of any grouse, and will assist in brood defence. The well camouflaged chicks find their own food, insects and plants, and concentrate in wet areas. They fly at 12-13 days. The brood remains together until late autumn and may group with others to form winter flocks (Harrison 1978).

Remarks:
Although this arctic grouse shares the same general plumage as that of the White-tailed Ptarmigan, it can be distinguished from that species by its black tail and by more white on its wing in summer. The male is more reddish brown on his head, back, and breast and the female is grayer overall and more heavily barred on her breast and flanks. Adapted for the cold and snow of winter, this species also has feathered legs and toes and assumes a white plumage by October, except for its black bill, eyes, and tail. At this time, the female's tail is more brown.
Adults forage on the leaves of willow, birch, and alder and on berries, seeds, and some insects. Single-species flocks form in the fall and migrate to valley lowlands where they browse on the buds and twigs of willow, alder, and birch. The northern population migrates into frozen muskegs, lake and stream margins, woodland openings, and, in some years, into cultivated fields.

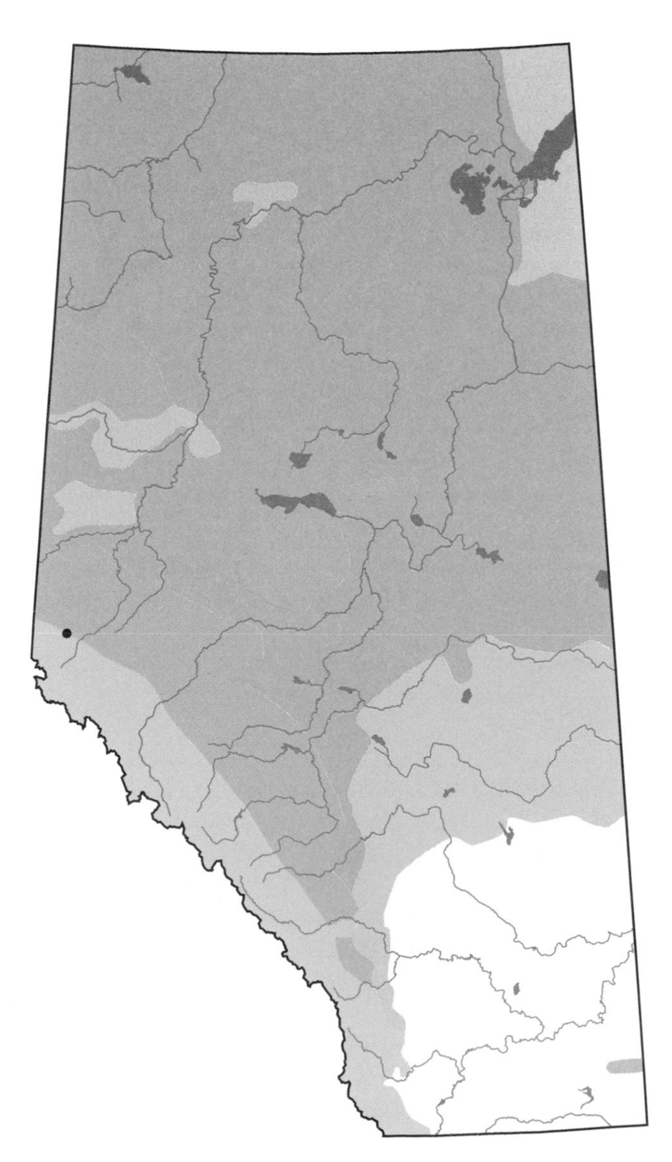

White-tailed Ptarmigan

Lagopus leucurus

WTPT photo credit: G. Beyersbergen

Status:
A fairly common year-round resident, our smallest ptarmigan is considered widespread throughout the Rocky Mountain Natural Region.

Distribution:
Holroyd and Van Tighem (1983) reported that the White-tailed Ptarmigan is widespread, occurring in suitable habitat in both Banff and Jasper national parks. Salt (1984) included Kananaskis Country in the breeding range. Atlas surveys indicate that the breeding range extends from the Willmore Wilderness Area south to Waterton Lakes National Park.

Habitat:
The White-tailed Ptarmigan frequents open rocky areas and both moist and dry alpine meadows in summer (Holroyd and Van Tighem 1983). Common features of this landscape include open scree, rock slides, boulder fields and nearby water. Selected terrain should provide snow cover, escape cover, and a stable substrate for new vegetation. In winter, a short downslope migration is made into the open forests and willow meadows of the subalpine. This windblown landscape, with its irregular features, creates the deep snow needed for snow burrowing, as well as exposing the willow buds which form a major component of the winter diet. Herzog (1980) reported that winter habitat use by the sexes differs in that males are found at or above the treeline (cinque and krummholz types), while females are found below the treeline in subalpine and stream course habitat.

Nesting:
The male on territory will court a prospective mate with a courtship chase and procession of quick and slow strutting and a display of swollen red eye combs. Nesting is delayed by the female until she has attained her summer camouflage. A loose nest of grasses, leaves, and lichens is built over a shallow depression, and it is lined with feathers. It is positioned on dry ground in an open meadow of short grasses or stunted willow, or surrounded by stone. Southerly slight or moderate angled slopes are preferred. In mid-June, the female begins her single brood, laying 4-8 eggs. The buff eggs, which are spotted with brown, resemble the eggs of the Spruce Grouse. The female covers the eggs with lichens when she leaves the nest, and she incubates them for 22-23 days, while the male remains nearby. When disturbed at the nest, the female will perform a distraction display with her wings dragging and her head held low. The female and young will remain on the nesting territory if the food quality is good. The male leaves the nest locality by the last week of incubation and is gone from the nesting territory by the time the young are ready to feed with the female (Salt 1984). The young can fly at 10 days. The broods remain together until the following spring (Harrison 1978).

Remarks:
An all white tail distinguishes this ptarmigan from the Willow Ptarmigan, with its black tail. White wings are evident in flight. The summer plumage of mottled black, brown, and white blends in well with the surrounding rocks and vegetation, allowing it to avoid detection. In summer, adults will feed on willow, leaves, sedges, heaths, berries, wildflowers, seeds, and insects. Chicks consume droplets of dew, small insects, and larger insects broken up by the female (Salt and Salt 1976). In winter, the White-tailed Ptarmigan feeds on willow and may be seen in close proximity to flocks of Willow Ptarmigan (Holroyd and Van Tighem 1983). In late summer, these ptarmigan will slowly retreat to winter on the willow streamside courses of the subalpine. Flocks range from an average of six up to 35 individuals. However, large flocks of ptarmigan are rare in the Alberta Rockies, which is due, in part, to small populations (Salt 1984).

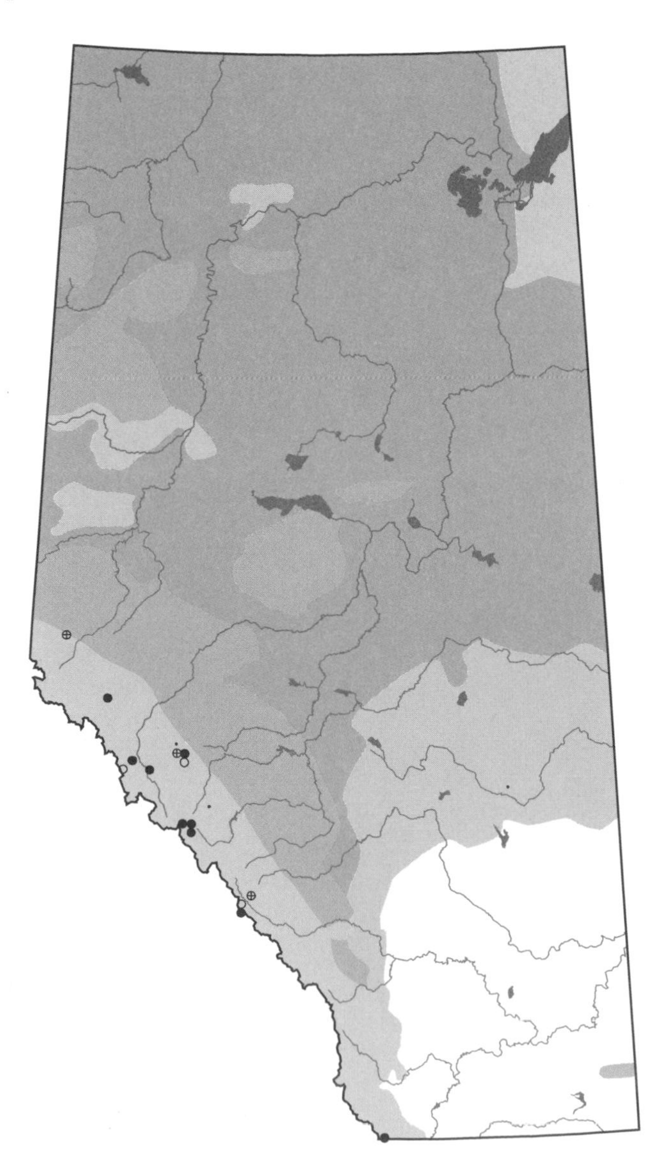

Ruffed Grouse

Bonasa umbellus

RUGR photo credit: E. Bruns

Status:
Alberta's most common grouse enjoys a healthy and widespread population base (Wildlife Management Branch 1991), despite high winter mortality due to raptors and weather severity. Overall levels fluctuate along a 10 year cycle. Godfrey (1986) indicated that two of the nine subspecies occur in Alberta.

Distribution:
The Ruffed Grouse is found mainly in the central part of the province, with a widespread distribution in the northern part of the province. It is very scarce in southeastern Alberta, with the exception of Cypress Hills, where it was introduced. It was recorded in 31% of surveyed squares in the Parkland region, 26% in the Boreal Forest and Rocky Mountain regions, and 42% in the Foothills region.

Habitat:
In Alberta, Ruffed Grouse are most numerous in aspen-dominated and mixed wood forests. Small openings in the deciduous forest are an important part of preferred habitat and, for brood use, a heavy understory is needed for drumming sites (Johnsgard 1973).

Nesting:
The familiar drumming of the male Ruffed Grouse heralds, for many, the coming of spring. The promiscuous male positions himself atop a prominent log in some shrub cover. In courtship display, he extends his ruff, erects his crest, trails his wings, and erects and fans his tail. To drum, the wings are brought forward and upward slowly at first, accelerating into a frenzied whirring motion and then quickly ceasing. The dull, muffled sound carries far in the still woods.
Once mated, the female constructs a ground nest of grass and leaves, which is lined with feathers, down, and pine needles. This is situated under or near a fallen log, in a ground hollow, or at the base of a tree, rock, or brushpile. In early May, 8-14 buff-colored eggs, which occasionally have fine spots, are laid at day and a half intervals. Should this clutch be lost, the female will replace it with a smaller clutch. The female, a very close sitter, will incubate for 24 days. The chicks soon follow their mother through brushy areas while feeding themselves. If approached, the mother will distract the intruder with hisses and mock attacks, while the young quickly scurry into surrounding cover. Young are quite vulnerable to the effects of late spring rains. The young fledge in 10-12 days, and the brood breaks up after 12 weeks.

Remarks:
This native upland game bird appears in two phases in the province, gray and red, indicating the various dominant shades of brown. The gray phase is typically associated with northern areas or higher altitudes, while the reddish-brown color phase is more characteristic of southern and lower altitude populations (Johnsgard 1973). Each has a crest, an umbrella-like ruff, a mottled back, lightly barred underparts, feathered legs, naked toes, and a longish tail with a broad, dark subterminal band that is most noticeable in flight. The female is smaller, has a shorter ruff and tail, and has an incomplete tail band.
Summer foraging is on seeds, leaves, fruits, berries, and some insects. Insects make up a small proportion of the adult birds' diet, but they are the basic food of young chicks for the first two weeks. In winter, this normally solitary grouse may be seen in small flocks, eating the flower buds of aspens and poplars, as well as the buds and twigs of willow and other bushes. Afterwards, they will fly off to roost in a dense coniferous tree or burrow into a snowbank. Fall drumming may be hatched-year males assuming previously vacant territory.

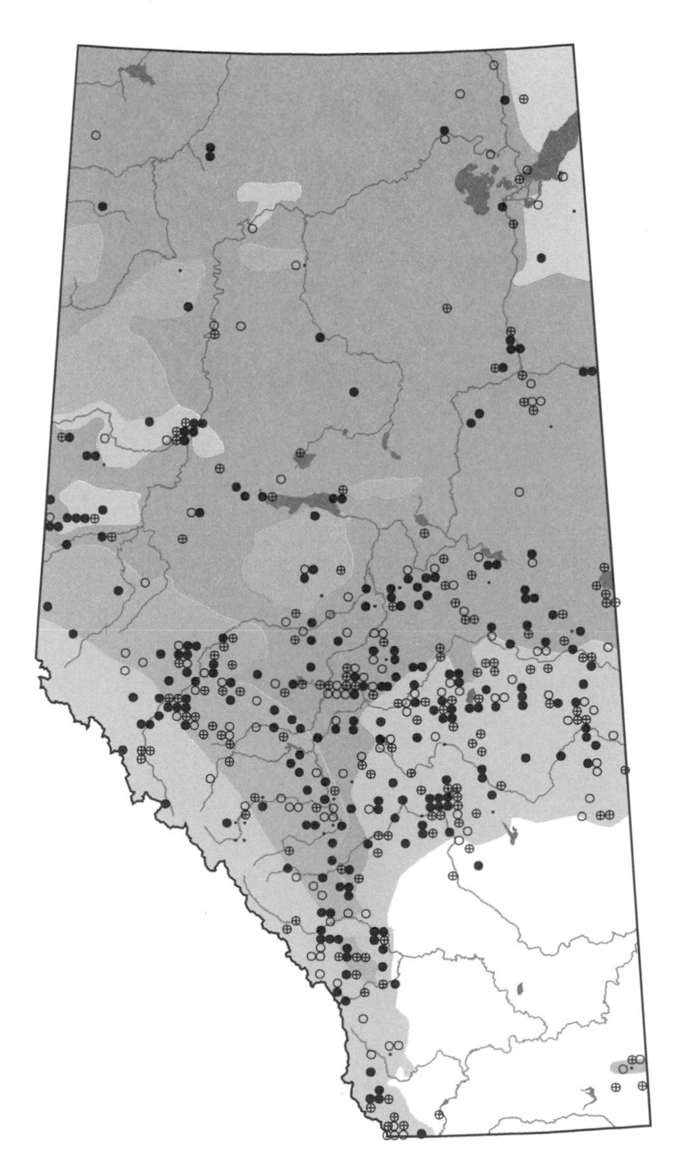

Sage Grouse

Centrocercus urophasianus

SAGR photo credit: M. Preston

Status:
The Sage Grouse is a scarce permanent resident (Salt and Salt 1976) and Alberta shares Canada's only population with Saskatchewan. It was on the National Audubon Society's Blue List from 1972 to 1981 and was listed as a species which merited Special Concern from 1982 to 1986 (Wildlife Management Branch 1991). Its population appears to be stable in a very restricted area of unique habitat. Population maintenance depends on the availability of this habitat, and any alteration to it will be detrimental to population survival. Population levels have decreased since the 1960s, apparently because of habitat deterioration resulting from vehicular disturbance and grazing (Fish and Wildlife 1984).

Distribution:
The extremely limited provincial distribution of this species was confirmed by Atlas surveys. Habitat constraints and land management have restricted the Sage Grouse to breeding in the extreme southeast corner of Alberta east of the Milk River, and south of the Cypress Hills.

Habitat:
The Sage Grouse requires the large stands of sagebrush which sometimes exist in the shortgrass prairie. Wet meadows, river bottoms, or green areas are required by broods for insect foraging (Ehrlich et al. 1988).

Nesting:
In April, males congregate to form territorial leks on traditional mating grounds or arenas. Each day at dawn the promiscuous males defend favored mating spots and passively attract prospective mates with ceremonial displays of strutting, hooting, snorting, and grunting. A sharp popping or booming sound is generated by quickly releasing air from the brightly colored balloon-like nuchal sacs through the mouth. Females move through the leks searching out dominant males. Selection may be based as much on the mating spot location as on a particular male's physical attributes. Population maintenance depends on the availability of this habitat, and any alteration to it will be detrimental to population survival.

Dancing ceases by early May and the females leave to build loosely constructed nests of grass and leaves. These are placed over a shallow depression in the ground, usually in the shelter of sagebrush or a tuft of grass (Godfrey 1986). The typical clutch consists of 6-9 buff-colored eggs, which are evenly and finely spotted with brown. The female incubates alone for 25-27 days, hatching and tending to the chicks. Almost as quickly as their down dries, they leave the nest and begin to feed themselves. One study suggests that the quantity of insects taken in during the first three weeks may strongly determine a chick's growth rate and survival (Johnson and Boyce 1990). While brooding, the chicks are arranged in a circle around their mother, their heads pointing outwards. The young are able to fly in 7-10 days and become relatively independent at about 10 weeks.

Remarks:
This large grouse has upperparts that are finely marked in buff, brown, dull white, and black, and rather long tapering tail feathers (Godfrey 1986). Each sex has a black abdominal patch, although this covers more area on the larger male. In summer, insects, grasshoppers, tender shoots, and leaves are eaten, but sage leaves are the sole dietary item in winter. Thus, the Sage Grouse is totally dependant on the limited distribution of sagebrush in the province.
Reluctant to fly when alarmed, it prefers to run, but will resort to a grouse-like flutter and cackle when pressed. Voluntary flights are toward water bodies. These very social birds form large gender-specific flocks in fall and winter.

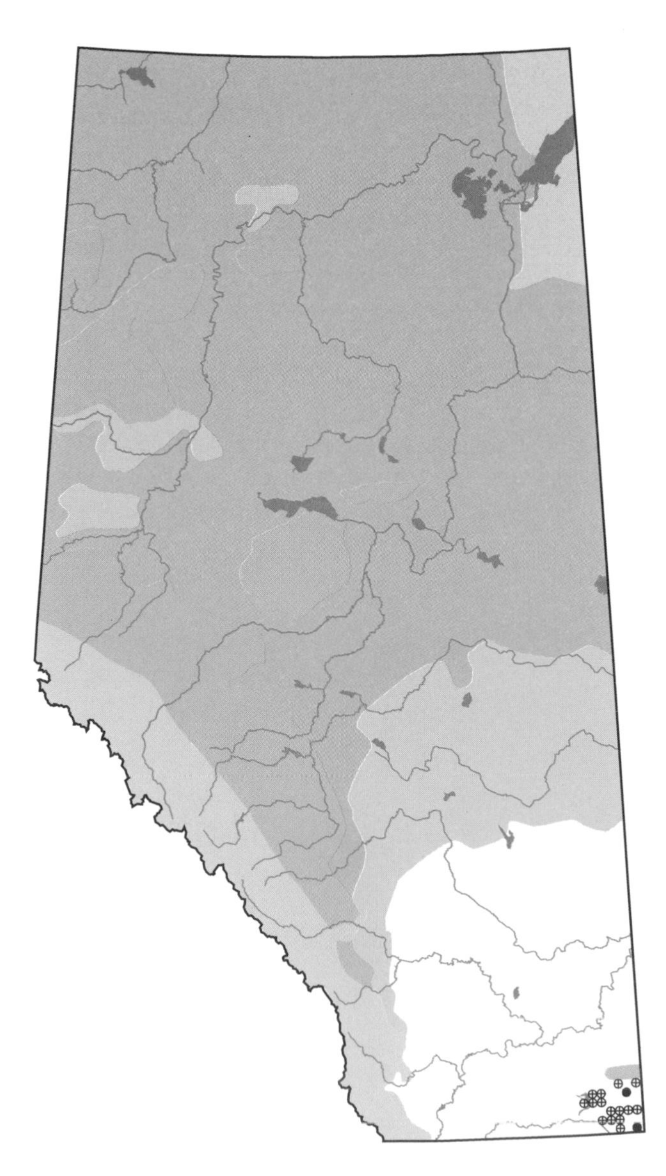

Sharp-tailed Grouse

Tympanuchus phasianellus

STGR photo credit: G. Horne

Status:

The Sharp-tailed Grouse is relatively common throughout the Grassland, central Parkland, and Peace River Parkland regions, although farming intensification has decreased habitat availability in central Alberta (Wildlife Management Branch 1991). Population abundance is dependent on open grassland/shrubland during mating season, and the retention of undisturbed grassland habitat is essential for continued population welfare. High winter mortality is common to this species (Fish and Wildlife 1984).

Distribution:

The Sharp-tailed Grouse appears throughout the province where suitable habitat exists. It is only a rare fall and winter visitor in Jasper and Banff national parks and Holroyd and Van Tighem (1983) reported no breeding records. Atlas surveys resulted in four breeding records for the Rocky Mountain region in the five year study period. One of these records was in Jasper and the other three were south of the Highwood River. This grouse was recorded in 9% of surveyed squares in the Grassland and Parkland regions, 7% in the Boreal Forest, and 5% in the Foothills region. Within the Grassland Natural Region, densities are greater where wooded escape cover still remains.

Habitat:

The relative success of this grouse species' distribution is created by its more generalized habitat requirements. In the Grassland region, it is found in open prairie, shrubby sandhills, coulees, and margins of water courses. In the Parkland region, it favors farmland and open woodland. In the Boreal Forest region, it utilizes openings made by fire and man, muskegs, and bogs (Godfrey 1986).

Nesting:

Beginning in April, males will gather at dawn on ancestral dancing grounds and defend territory within the lek with a mixture of gobbles, strutting, and fighting (Salt and Salt 1976). Within this area of sparse vegetation, the males carry out dancing duels for nearby females by rapidly stamping their feet, lowering their heads, and ruffling their plumage. Booming sounds are produced by the release of air through the mouth from inflated purple neck sacs on the side of the neck (Godfrey 1986). A shrill "chilk" is used to attract mates. Females, appearing passive and disinterested, move to the centre of the lek where most are mated by dominant males. These activities also take place at dusk during the height of the breeding season. The females then disperse to carry out the solitary duties of their single brood.

A nest of grass is constructed over a shallow scrape in the ground and concealed within the grasses or brush of grassland or open woodland. The female lays 10-14 olive-brown eggs, which are each finely dotted with reddish brown. The eggs are incubated for 23-24 days. The young leave the nest within hours and are led to moister areas, where they will be brooded, protected, and shown food. The initial diet of the chicks is insects but, later, they feed on seeds, leaves, flowers, and fruit. A pecking order develops within each sex of the brood, which prepares them for later adult social behavior. The young, weak flyers by the tenth day, disperse after 6-8 weeks.

Remarks:

The upperparts of this medium-sized grouse are grayish and mottled with brown, black, buff, and white.The underparts are mostly white with black V marks. The yellowish-orange eye comb is often hidden and the short pointed tail has long dark middle feathers and short white outer feathers. Fall dancing is carried out by young males to replace vacancies. Large flocks form in the fall and move into birch and aspen stands to feed on berries, buds, and catkins. Shelter is taken in snowdrifts.

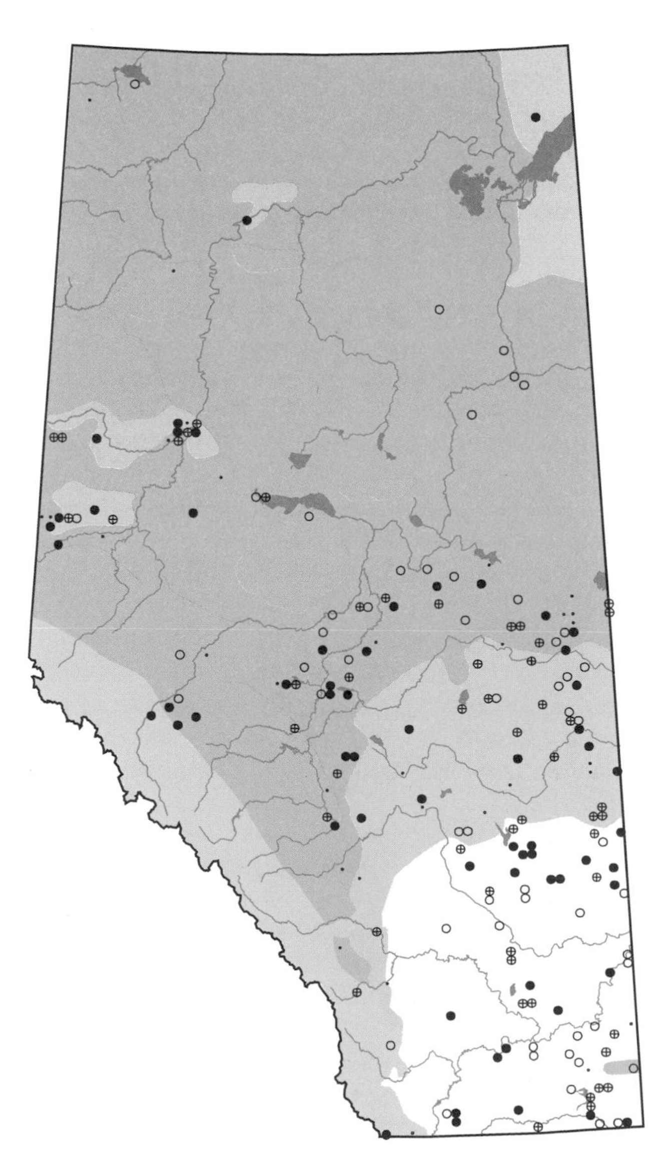

Wild Turkey

Meleagris gallopavo

TURK photo credit: W. Burgess

Status:

Alberta's Wild Turkey is an introduced species whose original range extended from southern Ontario through the eastern and southwestern United States to Mexico. The species was extirpated from Ontario and many other areas by the early 1900s, mainly as a result of hunting, habitat alteration, and disease spread by domestic poultry. The Wild Turkey has since been reintroduced into much of its former range in Alberta and elsewhere in southern Canada and the United States.

The province currently has two populations of Wild Turkeys with a total of about 350 birds. No specific management efforts are thought to be required for this species.

Distribution:

Wild Turkeys were first introduced into Alberta in the Cypress Hills in 1962. This population is now thought to number about 150-175 birds. The birds were also introduced into the Porcupine Hills in 1966 and 1973; this population is thought to number about 150 birds, but it has not grown substantially since 1984 (G. Ericksen Pers. Comm.). In 1971, Turkeys were released near Forestburg and near Stettler; a few young were raised in the following two years, but, by 1973, very few could be found in these areas (Salt and Salt 1976). In 1988, birds from the Porcupine Hills were introduced into the Lees Lake and Todd Creek areas. In 1990, Lees Lake had about 125-175 birds; there may be about 40-60 birds in two groups at Todd Creek. Turkeys sighted in the Canmore, Cochrane, and High River areas are probably birds released by locals and are not thought to be breeding.

Habitat:

This species favors open deciduous forest, forest edges, and agricultural fields. It remains on the breeding grounds over winter, but is quite dependent on supplemental feeding during harsh winter weather, when it frequents cattle feeding sites and barnyards.

Nesting:

The nest of the Wild Turkey, built by the female, is a hollow lined with grasses, weeds, and leaves. It is usually well-hidden in tangled undergrowth, bushes, or grass. Nests are generally located relatively close to water.

Clutches are about 10-18 eggs; the eggs are pale buff and are spotted with browns or purplish gray. Incubation is 27-28 days by the female alone. Two to three females will sometimes share a nest and share incubation. Males are polygamous and take no part in nesting duties (Harrison1978).

Chicks brood beneath the body, tail, and wings of the female for about four weeks, and fledge at about two weeks. After fledging, they can be seen roosting on branches with the female. One brood is raised per season, and the brood stays together until winter.

Birds generally flock together from early autumn to early spring, with the males flocking separately from the females and young.

Remarks:

The Wild Turkey, sometimes called Merriam's Turkey, is a large bronze and brown bird with a naked head and long brown tail barred with black and tipped with rusty brown. The domestic turkey, which originated from a Mexican subspecies, is lighter in color and has tail coverts tipped with white or pale buff (Godfrey 1986).

The main foods of the Wild Turkey are seeds, grains, wild grasses, tubers, flower buds, berries, insects, and worms, and livestock feed from farmsteads in adverse weather conditions.

This species roosts in trees at night. When alarmed, it is more likely to run than to fly.

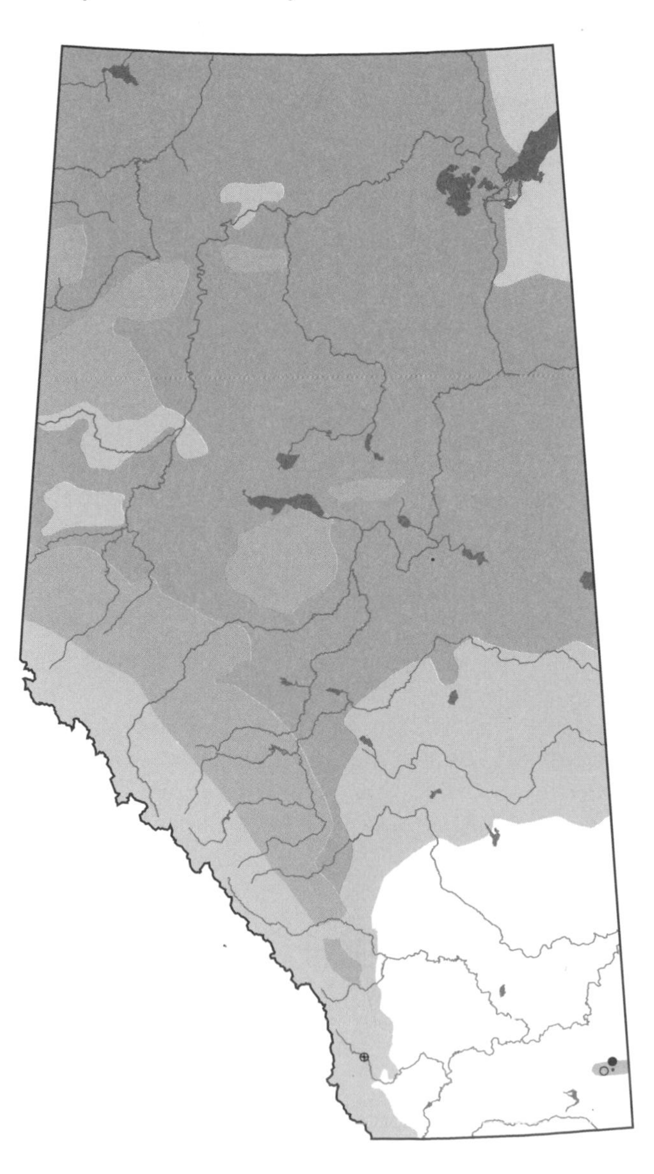

Order Gruiformes

(Family Rallidae, Family Gruidae)

This order hosts the rails, coots, and cranes. In Alberta, this includes the Sora, the Yellow and Virginia Rails, the American Coot, and the Sandhill and Whooping Cranes. The rails are adapted to life within a forest of reeds, preferring to walk about the vegetation rather than swim or fly. They all have relatively long legs and long toes, short wings, and short tails. Although all have strong bills, these vary in length. A characteristic walk includes head-bobbing and a flirting tail. Solitary and secretive, they prefer to remain inconspicuous, flushing only when given no other option. The American Coot also has a laterally compressed body to navigate about the vegetation, but is larger, has lobed toes, and has a white bill which meets with the forehead. Using its lobed toes to swim effectively, including in open water, it is often mistaken for a duck by the uninitiated. Vastly different in appearance, Cranes are our tallest birds. They have long, straight bills, long legs and toes, long necks, and short tails. Although they spend most of their time on the ground, their large wings enable accomplished flight and soaring. Sometimes mistaken for herons while flying, cranes fly with both the neck and legs extended. These two species are gregarious out of the breeding season. The sexes of all species of this order are alike.

Yellow Rail

Coturnicops noveboracensis

YERA photo credit: C.R. Sams II and J.F Stoick/VIREO

Status:
This most secretive of our Alberta rails is believed to be locally common, though censusing efforts have been few and without significant success. The Yellow Rail is rarely seen, even in those localities where it is relatively common (Salt and Salt 1976).

Distribution:
The historical range of the Yellow Rail in Alberta has been concentrated in east-central Alberta, with occurrences as far north as Fort MacKay and south to Ribstone Creek. The only records for the Rocky Mountain Natural Region are of two specimens taken at Jasper early in the century (Salt and Salt 1976). Atlas data for this species were scant, but indicated that this rail may be breeding as far south as Calgary. Not one confirmed nesting was recorded in the five-year field survey.

Habitat:
This species shares similar habitat requirements as the Sora and Virginia Rail, sometimes experiencing territory overlap. In particular, the Yellow Rail prefers larger, dense grass or sedge marshes with little or no standing water. It has been suggested (Anderson 1977) that only wetlands large enough to support a small group of territorial birds are likely to have Yellow Rails on a long-term basis.

Nesting:
The nest of the Yellow Rail is constructed of grasses and other dead emergent vegetation into a shape ranging from a depression to a deep cup. It is built within a tussock or on a mat of dead grass and situated on the ground in or near the marsh, often in drier habitat than other rail species. The nest is well concealed by a canopy of bent-over grass and is very difficult to find, paralleling the birds secretive lifestyle.
The 6-10 eggs are buff-colored and sparsely covered by fine reddish-brown spots which sometimes form a wreath on the large end. The eggs' shapes, narrower than for other rails, range from sub-elliptical to oval, and appear slightly glossy and smooth. Incubation is approximately 18 days. Whether both sexes participate in incubation is unknown, as is the brood number. The young are led from the nest within two days of hatching, and they fledge at 35 days.

Remarks:
The Yellow Rail is our smallest rail and is distinguishable by its short, yellow bill, overall buffy plumage, and white tips on the secondaries. Fine white crossbars on the back separate this bird from the similarly colored immature Sora, for which the white bars are longitudinal. This species is heard more often than seen and may be distinguished from the Sora's "whinny" call by its own nuptial call. This is comprised of a series of five metallic notes, most often heard at night: "tik-tik-tik, tik, tik".
This shy bird shares the same retiring habits as other rails. When confronted alone, it prefers to freeze and rely on its coloration, but will quietly slip off the nest when alarmed and hide in a maze of tangled stems. Its long toes and laterally compressed body aid it in this activity. It is reluctant to flush and when forced to fly, its short round wings carry it awkwardly a short distance over the reeds and back down into the security of vegetation. This rail forages for snails and other invertebrates in the drier parts of large grass and sedge growing in freshwater marshes (Ripley 1977).
Spring migration is later than most migrants, coming in late May. Fall migration is believed to be in September, although records are incomplete on this point. It is known to be more widespread at this time, passing through such other habitats as dry meadows and croplands.

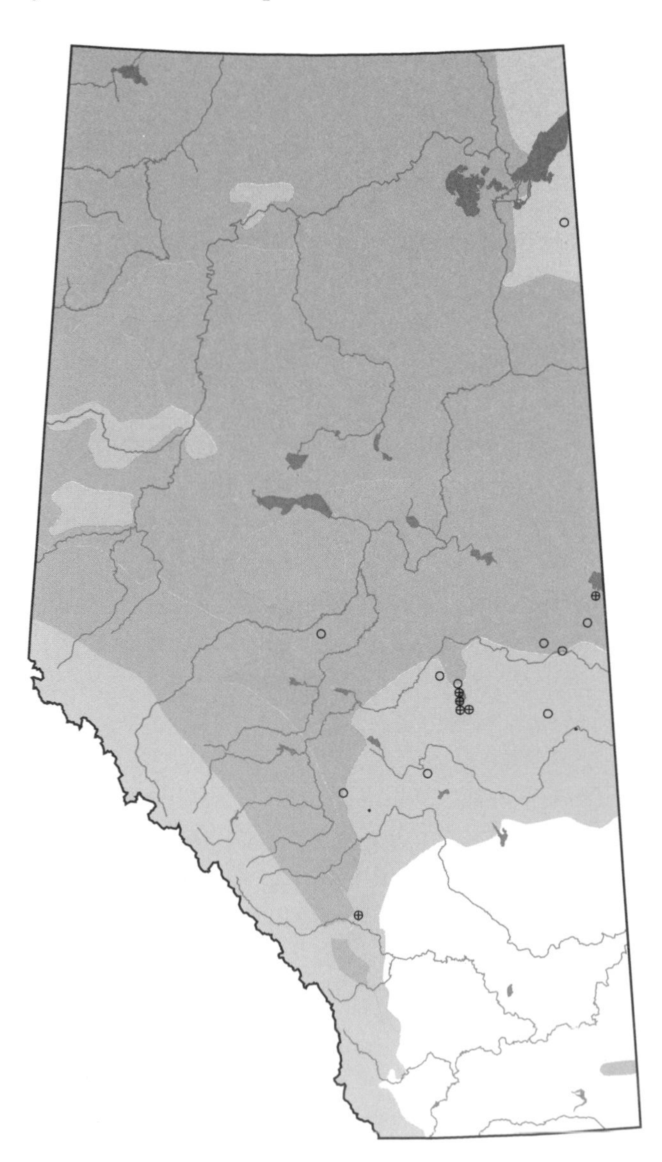

Virginia Rail

Rallus limicola

VIRA photo credit: B. McGillivray

Status:
The status of the Virginia Rail in Alberta is relatively unknown. Few historical or Atlas accounts exist and certainly none on a yearly basis. Like other rails, censusing of these secretive birds is very difficult, leading to a cautious status rating of uncommon to rare.

Distribution:
Historically, the distribution of this species has been in eastern Alberta, with individual accounts ranging from Fort Chipewyan in the north to Brooks in the south. The westernmost sighting was in the Valleyview area, east of Grande Prairie. Atlas observations are concentrated in the Parkland and Grassland regions of east-central Alberta, with one confirmed record of breeding in each region.

Habitat:
The Virginia Rail can be found in breeding habitat that is often occupied by the more common Sora. This includes areas of freshwater lakes, ponds, marshes, sloughs, and bogs, which host an extensive coverage of bulrushes, cattails, and sedges. Where they coexist, the Virginia Rail will often be found locally in drier areas than the Sora.

Nesting:
The pair will participate in a number of displays, including mutual preening and courtship feeding. A variety of vocals are used on the breeding grounds, the most prominent being a repetition of loud metallic notes, given day or night. The nest is constructed of a loose mat of surrounding vegetation, occasionally lined with grass and forming a general depression. This structure is attached within a tuft of vegetation that rises well above the ground, mud, or shallow water. The surrounding stalks are pulled over to form a well-concealed canopy. Rising water levels will sometimes infiltrate the nest, after which the breeding pair will add materials to elevate the eggs.
The usual clutch is 7-12 eggs, which are white or buffy with small spots of reddish brown, often forming a wreath. They are generally paler and less spotted than those of neighboring Soras. Both sexes participate in incubation, which begins before the final egg is laid and continues for 19-20 days. Almost immediately after their hatching, the young have the ability to run, swim, dive, and climb, the latter aided by a tiny claw on the outer digit of their wings. They are led away from the nest site to one of several brooding platforms by the male. These young fledge in 25 days and appear mostly black by late summer.

Remarks:
The Virginia Rail shares the same general adaptations to marsh life as do our other Alberta rails. These include long toes, short round wings, and a laterally compressed body. However, its long, down-curved, reddish bill is a departure from that of other rails, being adapted to probe the mud for earthworms and insect larvae, as well as to capture slugs, snails, beetles, and caterpillars. Occasionally, vegetable matter and fish are consumed as well.
If this rail is approached while on the nest, it will, like its cousins, quietly slip off before being even seen or heard. If the intruder continues toward the nest, the adult will endeavor to distract him by splashing with its wings. Off the nest, this species will freeze in position and rely on its coloration, generally brown with black-and-white banded flanks, to elude the observer.
This low-flying migrant is especially secretive during migration, furnishing little detail on its migration periods in Alberta. Data from Pinel et al. (1991) suggest that spring arrivals enter southern Alberta early in May and leave the province in early September.

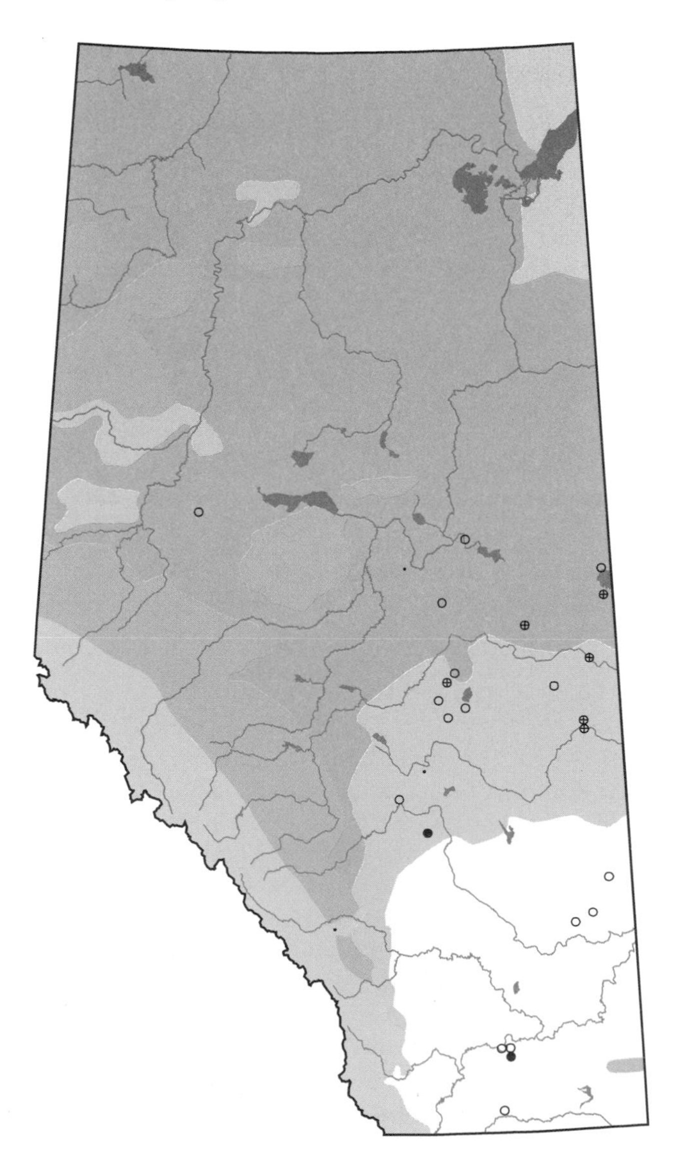

Sora

Porzana carolina

SORA photo credit: W. Burgess

Status:
The Sora is the most common rail species in Alberta. The status of the population of this species is unknown, but appears to be stable within the province.

Distribution:
The Sora is found in all regions of Alberta where suitable habitat is available. It was recorded in 25% of surveyed squares in the Canadian Shield, 32% in the Boreal Forest, 38% in the Parkland, 17% in the Grassland, 21% in the Foothills and 15% in the Rocky Mountain Natural Region.

Habitat:
Alberta's most common and least secretive rail has adapted to a variety of wetland habitats. Wherever there are freshwater ponds, meandering streams, marshes, sloughs, or wet meadows, with at least a partial margin of extensive sedges, rushes, or cattails, a breeding pair has likely taken up residence. Flooded willow swamps are less favored, but also occasionally chosen.

Nesting:
The make-up of the Sora's nest reflects the variety of the dominant vegetation of its nest site. Usual constituents include a woven basket of available sedge species, bulrushes, cattails, and occasionally willow twigs. When forced to select a nest site with lower surrounding water levels, a greater mix of construction elements is chosen. The nest is anchored within a tuft of rushes or, rarely, in a clump of willow, and positioned 15-30 cm above water. This deeper water and a more common nest lining of finer grasses and sedges may distinguish this nest from a similar one of a neighboring Virginia Rail. The surrounding vegetation above is bent over for concealment and a runway is often constructed to provide access. Nest-building is often completed after egg-laying has begun.
The Sora lays 7-13 buff-colored eggs, more heavily spotted with reddish brown than those of the Virginia Rail. Incubation duties are shared by both sexes and continue for 16-20 days. The chicks depart the breeding nest shortly after hatching. Young Soras develop a generally buffy appearance with a white throat and longitudinal streaks on the back. These birds fledge in 21-25 days. Adults move off their territories once the broods have been raised because they are single-brooded.

Remarks:
The Sora has the common traits of many rails, including long unwebbed toes adapted to displace their weight over floating vegetation. Other similarities include short round wings that provide weak flight and a laterally compressed body to navigate tangled vegetation. Distinguishing features include a gray breast, a black throat and black in front of the eyes, a stout and pointed yellow bill, and a habit of walking with its tail erect. The Sora is more often heard than seen. Its descending series of loud notes, often described as sounding like a whinny or maniacal laugh, can be heard by day or night.
Common to other rails and coots, the Sora is very hesitant to take flight when alarmed, preferring to slip away quietly into its wetland jungle. This bird forages in shallow water, feeding on seeds, molluscs, mosquitoes, and dragonflies.
Soras arrive in Alberta in late April or early May and depart by mid-September for their wintering grounds. In the fall, their distribution is more widespread and conspicous.

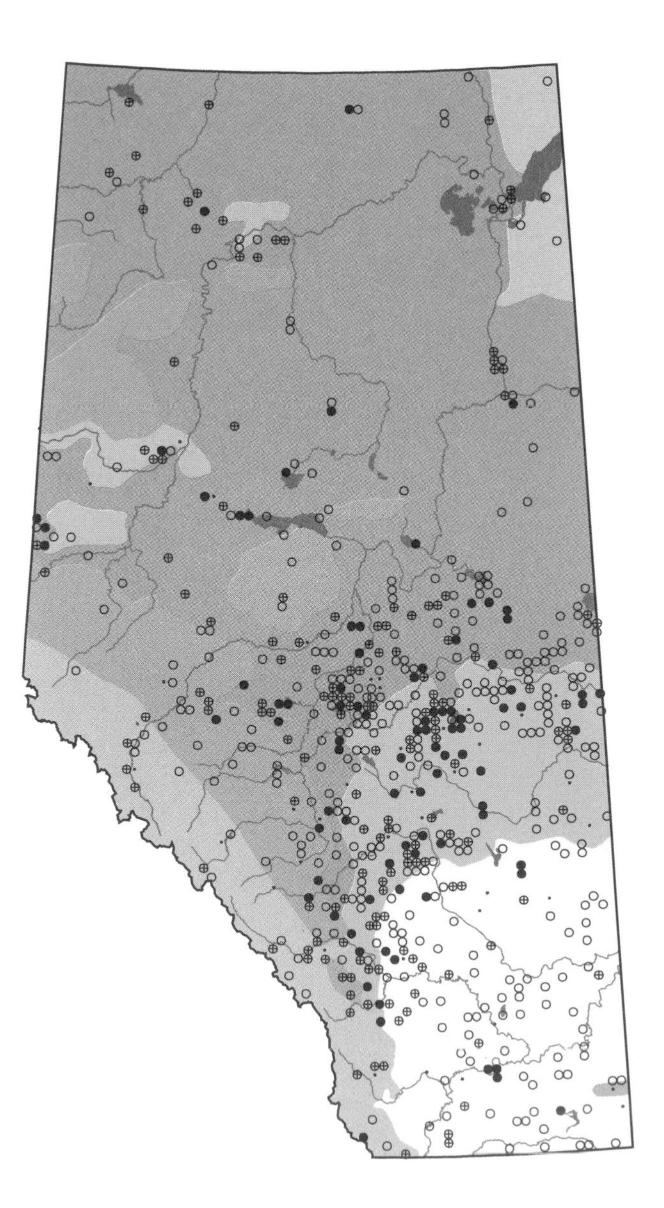

American Coot

Fulica americana

AMCO photo credit: G. Beyersbergen

Status:
This waterbird is common in marshes across the province. There is an abundant breeding population throughout Alberta and the key habitats are generally secure (Wildlife Management Branch 1991).

Distribution:
The American Coot breeds throughout Alberta where appropriate habitat may be found. They are most prevalent in the Parkland Natural Region, where they were recorded in 50% of surveyed squares. In the Boreal Forest region, they were recorded in 32% of surveyed squares and 20% in the Grassland region. The coot was previously reported as not known to nest in Banff, Waterton Lakes, and Jasper national parks (Holroyd and Van Tighem 1983). Atlas data records confirmed breeding in Waterton Lakes and Banff national parks and probable breeding in Jasper National Park. They are less frequent breeders in the Foothills and Rocky Mountain natural regions, where they were recorded in only 10% and 8% of surveyed squares, respectively.

Habitat:
Similar to other members of the Rallidae family, coots breed in freshwater bodies that host margins of bulrushes, cattails, and sedges. Being more duck-like in their habits than other members of this family, the American Coot prefers an area of open water adjacent to its breeding grounds.

Nesting:
The nest of the coot is a bowl or saucer-shaped construction of decaying emergent vegetation lined with finer grasses and sedges. It is anchored in or near local rushes but designed to move in response to water level changes. Most often, the top of the nest is 10-30 cm above water and is reached along a ramp.
This species is a solitary breeder, but nesting pair densities will increase in relation to the density of available vegetation. The female constructs the nest as the male supplies the material. Several additional nests are constructed for displaying, copulating, and brooding purposes.
The female lays 8-12 smooth and slightly glossy eggs which are buff-colored and spotted with dark brown. Incubation is shared by both sexes for a 23-24 day period. The young fledge at 49-56 days. This species may renest to take advantage of a longer breeding season or to replace an earlier nesting that failed.

Remarks:
The coot is generally a dark slate color with a black head and neck and a whitish bill. Although more closely related to the rails, with which it shares the characteristic long toes and short round wings, the coot is sometimes mistaken for a duck.
Broadly scalloped lobes on its toes permit this bird to swim and forage in open water, as does the buoyancy created by a rounder body. However, close observation will highlight differences between it and other waterfowl. On land, this species has longer legs, distinctive toes, a shorter bill, and a more upright posture. In both elements, the coot nods its head in relation to its leg movements and is smaller and noisier than ducks. Like rails, it is difficult to flush out, preferring to hide or dive. When forced, the coot will use a long, pattering take-off to attain a low-level flight of 3-5 m.
Being omnivorous, this species will scavenge the shoreline for snails, insects, and dead fish, but will dabble or dive for shellfish, crustaceans, and submergent vegetation. It is also known to pirate food from nearby waterfowl.
These birds migrate to Alberta in late April or early May and depart in late September. Individuals have been known to overwinter on Wabamun Lake (Pinel et al. 1991).

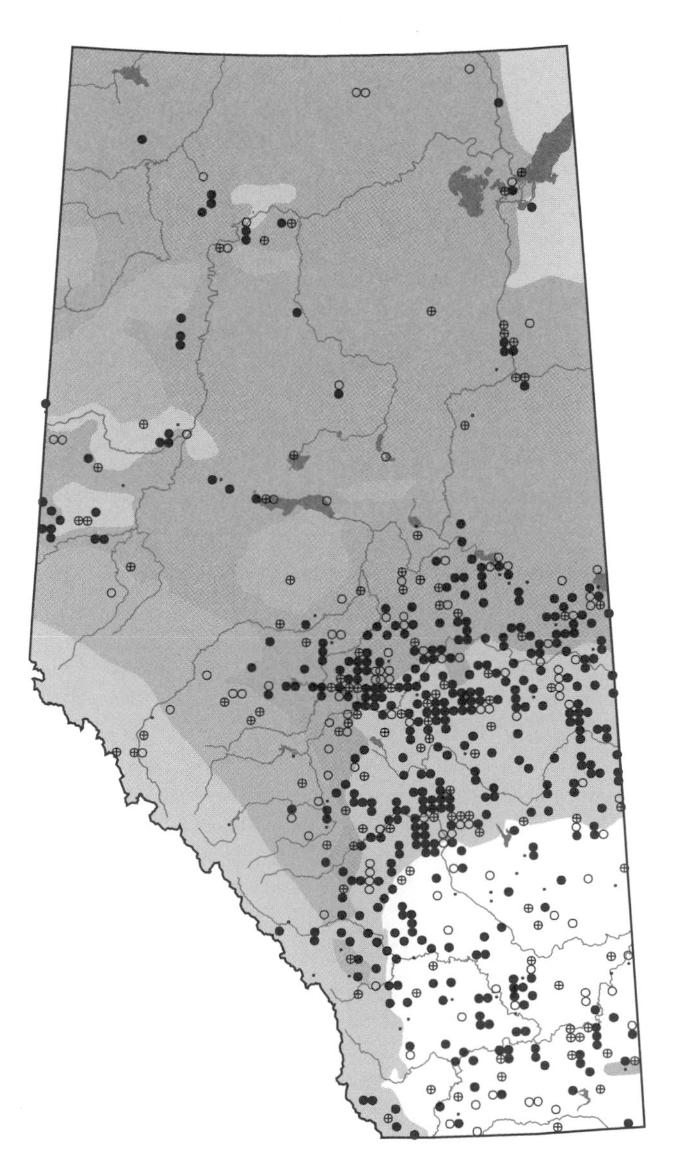

Sandhill Crane

Grus canadensis

SACR

Status:
As a result of settlement and agricultural practices, the Sandhill Crane no longer breeds in the Grassland Natural Region. It is more common in the Canadian prairies during migration. Successful research into their age of maturity, mating, and territory procurement will promote future conservation strategies.

Distribution:
As suggested by Salt and Salt (1976), the Sandhill Crane appears to be expanding southward into part of its former range. Several records were made south of the species previous limit, Rocky Mountain House, terminating in the Bottrell area. Atlas findings also confirmed breeding at Police Outpost Provincial Park. In the Canadian Shield, Foothills, and Boreal Forest natural regions, it was recorded in 19%, 10%, and 9% of surveyed squares, respectively.

Habitat:
Breeding habitat includes marshes, bogs adjacent to ponds, and large marshes with some open water and tall grasses and rushes. An area must be secluded and undisturbed.

Nesting:
This species migrates to the same breeding grounds each year. It carries on a courtship display of a lowered head, wing raising, and hopping. The nest may be constructed on land or over water. In a dry area, the nest may be a simple hollow with a thin grass lining. Next to a marsh or muskeg, it may be a depression in a hummock or on a muskrat house. Over shallow water, a larger nest is constructed of loose aquatic vegetation, and lined with grass, willow sticks, and roots. It is generally concealed by dead rushes and weeds. The typical clutch consists of two olive buff eggs, which are spotted and blotched with brown. The eggs are incubated by both parents for 29-32 days. Adults are very secretive on the nest, calling only at dawn and dusk. The precocial and downy young leave the nest soon after hatching, but are fed by both parents for about one week. Young fledge at about 70 days, but remain with adults until the following spring.

Remarks:
These metre-tall birds are brownish slate overall, with some staining of rust, black wing tips, and a bald red upper face and forehead. Juveniles are similar to adults, with many tawny-tipped wing and body feathers. Seen on migration, Sandhill Crane flocks may appear in different formations, but are most often seen in linear formation. The fully extended neck and quicker wing upstrokes distinguish it from the Great Blue Heron. Trailing legs, soaring circles, and a rolling call differentiate it from migrating Gray Geese. Foraging is done on land, and this species mainly consumes roots, insects, crustaceans, frogs, and mice. Occasionally, the bill is used for digging. Thermal updraughts are used for soaring while migrating from the southern U.S. and Mexico. Rest and foraging take place in marshes or corn stubble fields late in the day. Small numbers group together at night and larger numbers congregate after dawn to feed and dance.

Individual Sandhill Cranes begin arriving in late April and peak in early May. The subspecies *G. c. rowani* nests predominantly in Alberta, but the subspecies *G. c. canadensis* only nests here occasionally. Reverse migration occurs from August to mid-October.

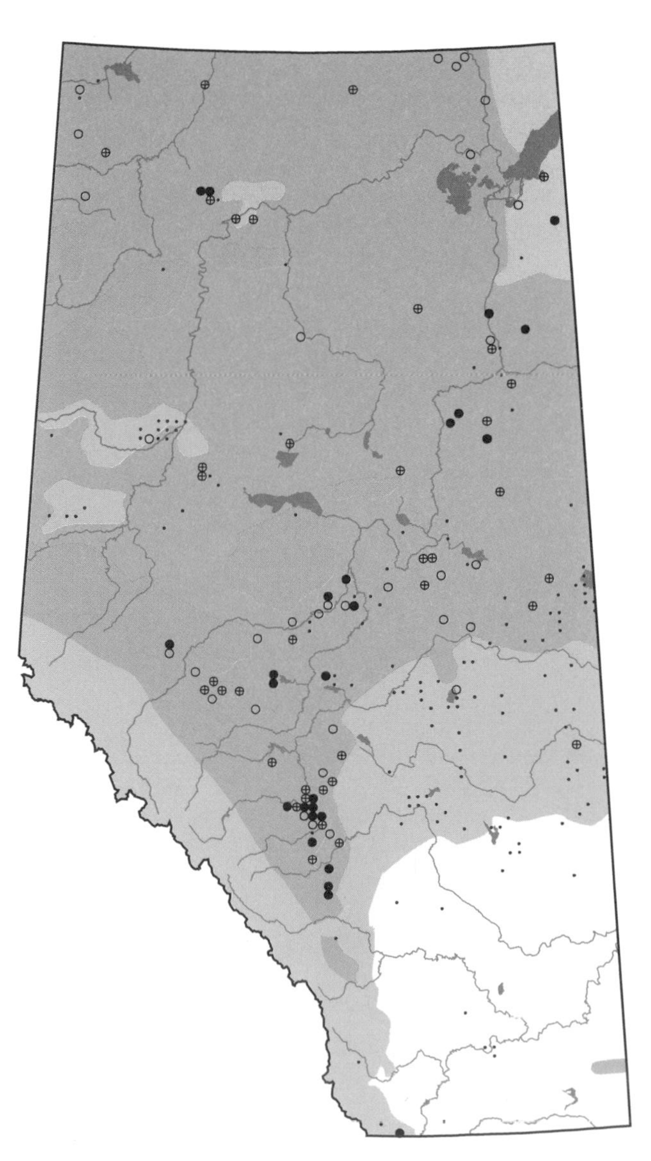

Whooping Crane

Grus americana

WOCR

Status:
Hitting a low of only 21 birds in 1941, the population of this endangered species has increased marginally, despite intensive management through cooperative Canada and United States protection efforts. Hunting, egg collecting, and the destruction of nesting habitat had created pressures from which this species could not naturally recover. This deficit is compounded by young that only start breeding at 4-6 years of age, and the practice of only raising a single young. The wild population was estimated at 146 in the fall of 1989 (Kuyt 1990).

Distribution:
This species previously bred south to the Battle River (Salt 1976), the Whooper has more recently been restricted to the marginal habitat of Wood Buffalo National Park, straddling the border between northeastern Alberta and the Northwest Territories. It winters along the Gulf coast of Texas.

Habitat:
The large, relatively open marshy areas of Wood Buffalo National Park provide the isolation required for successful nesting for this species. Poor drainage and spring rains can cause nesting delays and reduced nesting success. Excessively dry weather will require adults to travel further to forage and increase the fire hazard.

Nesting:
Although pairs return to their same nesting territories in late April or early May, their old nests are not reused in raising their single brood. The nest is a large mass of bulrushes, cattails, and sedges, and may be up to 1.5 m in diameter. It may be built in shallow water, on ground surrounded by water, or on a muskrat house. If the nest is destroyed before mid-incubation, renesting will take place. The two large olive or buff eggs are blotched with brown. Only one young is raised. If the young dies early, the second egg will be used. Incubation duties are shared by both for 29-31 days. Both parents tend to the young whooper, which fledges in about 14 weeks.

Remarks:
Standing over 1 m in height, the heron-like Whooping Crane is the tallest Canadian breeder. It has long black legs and a very long neck. It is distinguished in flight by drawn back legs and an extended neck. Adults are all white with black-tipped primaries and coverts that show in flight. A red facial patch bifurcates from the bill over the crown and from the bill and extending under the eye. Immatures, which adopt adult plumage by their second fall, are a general mix of white and rusty brown, and have no facial patches. The relatively long and heavy bill probes the mud for insect larvae, crustaceans, and small molluscs.
Fall migration (mid-September) is done singly, by family unit, or in small groups, sometimes among Sandhill Cranes. Although this species may live 22-24 years (Smith et al. 1986), increases in the wild population have been slow, although greater successes have occurred since 1982. The surplus second egg of each clutch is gathered by the Whooping Crane Recovery Team and placed among clutches of foster Sandhill Crane adults in Idaho, as are some eggs from a captive population of Whoopers. Continued management success requires the protection of winter staging and breeding grounds for this wary and highly territorial Crane.

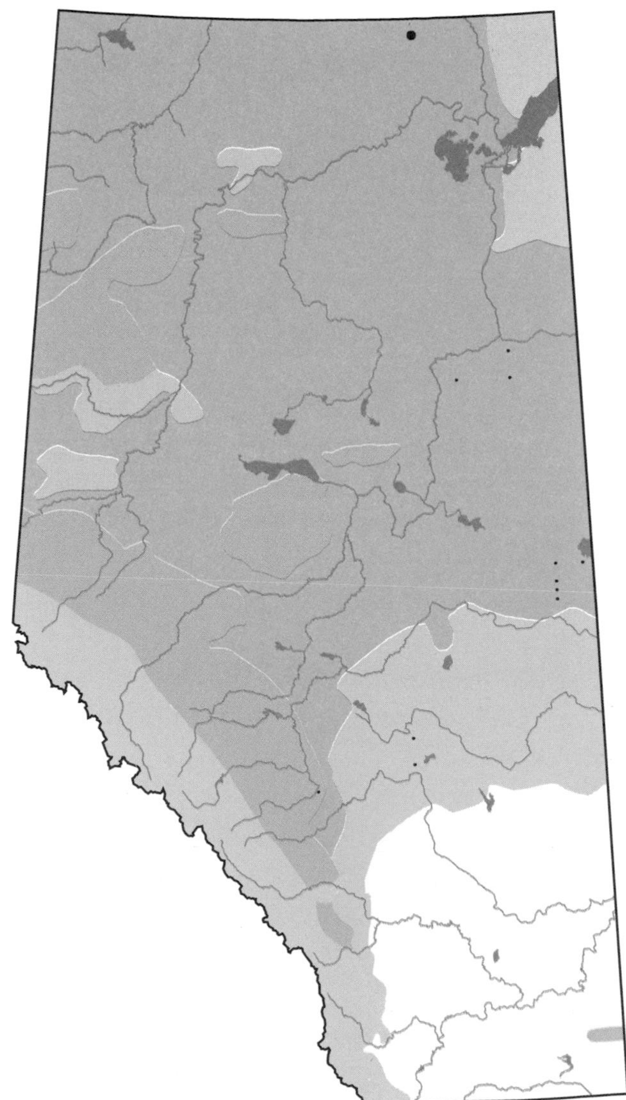

Order Charadriiformes

(Family Charadriidae, Family Recurvirostridae, Family Scolopacidae, Family Laridae)

This large order includes plovers, shorebirds, gulls, and terns. Like the members of the Order Anseriformes, these members have developed various leg, neck, and bill structures in response to their feeding strategies. Avocets have relatively long necks and legs and long upwardly curved bills. They sweep their bills sideways through shallow water when they are feeding. Plovers are smaller and more terrestrial, swiftly running about as they feed with their short, somewhat thick bills. Sandpipers share a similar body shape with the plovers, but spend more time near water, feeding with their longer, although variable, bills. The above groups share a few similarities: all are, at some time, waders, all have long wings enabling strong, swift flight, and each sex has a similar, or nearly similar, plumage. Conversely, phalaropes, equipped with flattened lobes on the sides of the toes, swim when foraging, using their sandpaper-like bills to capture stirred-up crustaceans. Also, each sex is brightly colored, although the intensity of this coloration varies. Distinct from the other members of this order are the gulls and terns. These birds are stout-bodied, and have long pointed wings, medium to long tails, and webbed feet. The terns tails are forked, and many have crests. The slender, stout bills are pointed on the terns and usually hooked on the gulls. Both groups have variously arranged gray, white, and black coloration, and all are gregarious. All fly, glide, and soar well, although the wingbeats of terns are shallower and more rapid.

Semipalmated Plover

Charadrius semipalmatus

SEPL

Status:
The numbers of this species dropped dramatically in the late 1800s as a result of uncontrolled hunting, but, with the enactment of the Migratory Bird Convention Act in 1918, by both Canada and the United States, populations were soon able to recover (Root 1988). The Semipalmated Plover is considered to be a relatively abundant species in Arctic North America, and is not presently in any danger (Johnsgard 1981). It is a rare breeder in Alberta, where it is at the southern periphery of its breeding range.

Distribution:
This species has been confirmed breeding in Alberta at Grosbeak Lake in Wood Buffalo National Park (Kuyt 1982). There have also been probable breeding observations in the Birch River Delta. The Semipalmated Plover is most often observed as a migrant in Alberta and, in the five year survey period of the Atlas Project, there were no breeding records for this species.

Habitat:
During migration, the Semipalmated Plover frequents mudflats, sandy or muddy beaches, and the flat open margins of ponds, lakes, and rivers (Godfrey 1986). For breeding, it prefers open sandy, gravelly, or pebbly shores, including sandbars and river gravel bars (Johnsgard 1981).

Nesting:
This small shorebird is semi-colonial and nests on dry ground near water. It lays its eggs in a hollow scrape in sand or gravel that is sheltered from the wind but is in full sun. The scrape, built by the male, may be unlined or sparsely lined with stone chips, plant materials, and debris. Some pairs of this plover exhibit strong nest site fidelity (Campbell et al. 1991).
The clutch typically consists of four buff eggs, or rarely three or five, with blackish-brown blotches (Harrison 1978). Incubation lasts 23 days and is undertaken mostly by the male. After hatching, young are guarded and brooded by both parents. Parents will give a broken wing display if intruded upon at their territory and will run ahead of an intruder for long distances before being pressed into flight. The chicks leave the nest shortly after hatching, are soon independent and fledge between 22-31 days. One brood is raised per season.

Remarks:
The Semipalmated Plover looks somewhat like a Killdeer, but is smaller, has a single rather than double breast band, lacks the Killdeer's rufous rump, and has an orange bill with a black tip. It also resembles the Piping Plover, but its plumage is much darker, and its face is dark rather than light. The feet of this plover have a greater amount of webbing between the three toes, giving it the name "semipalmated".
This species eats insects, crustaceans, worms, other small invertebrates, and occasionally seeds. When feeding, the bird runs with rapid little steps, then pauses, and may use foot trembling to entice prey to move (Johnsgard 1981).
On migration, this shorebird occurs in small flocks on mudflats and beaches of lakes (Salt and Salt 1976). Individuals scatter when the flock alights, and often mingle with other shorebirds.
Semipalmated Plovers first appear in southern Alberta in the last week of April and the first week of May. Fall passage is extended, starting in early July and continuing to late October. Juveniles do not migrate until mid-August, and stragglers, likely birds of the year, are reported into mid-September (Salt and Salt 1976).

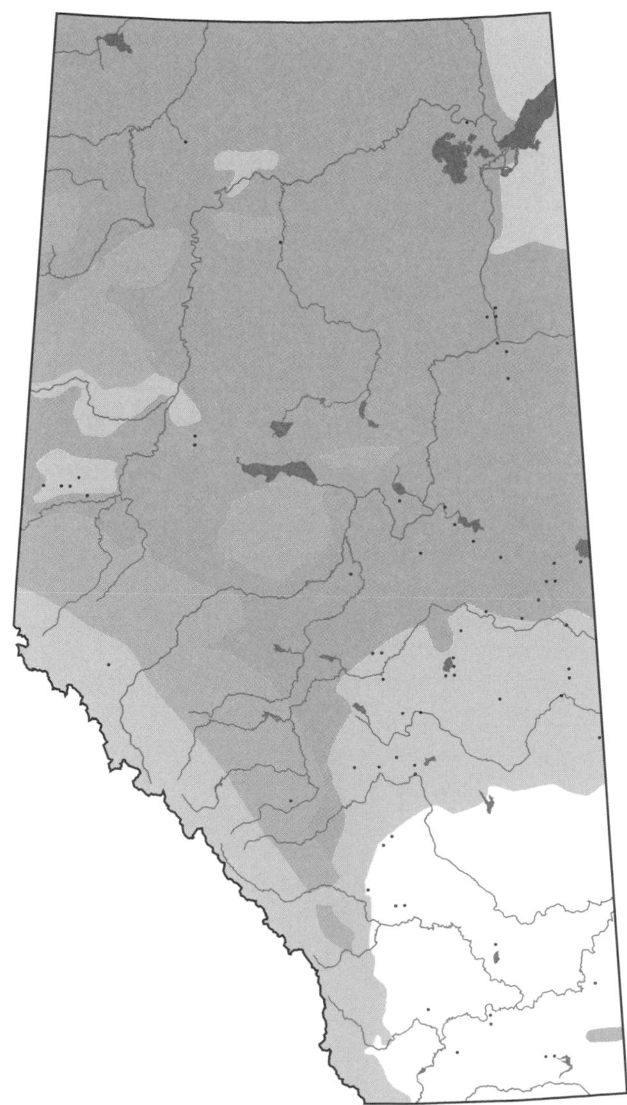

Piping Plover

Charadrius melodus

PIPL photo credit: R. Gehlert

Status:

The Piping Plover was classified as "threatened" by the Committee on the Status of Endangered Wildlife in Canada in 1978 but, because of continued population declines, its status was upgraded to "endangered" in 1985 (Haig 1985). The species is also considered endangered in Alberta and, in the United States, it is federally listed as endangered in the Great Lakes area and threatened in the remainder of its range (U.S. Fish and Wildlife Service 1985). It has been on the National Audubon Society's Blue List since 1973 (Tate 1986).

Historically, the Piping Plover has undergone dramatic fluctuations in its population numbers. It was hunted close to extinction in the late 1800s and early 1900s prior to the Migratory Birds Convention Act. The plover appeared to recover under this protection but began to decline again in the 1940s (Haig and Oring 1985). The present decline is linked mainly to increased recreational use of beaches required by the plover as nesting habitat. Other problems include fluctuating water levels, disturbance by cattle, and predation (Wershler and Wallis 1986; Goossen 1990).

The first nesting record for this species in Alberta was in 1930 (Farley 1932) and it has been regarded as a rare and local summer resident since that time. The present world population of the Piping Plover is nearly 5,500 (Goossen and Johnson 1992), of which 180 adults are in Alberta (Hofman 1992). Both Canada and the United States are implementing recovery plans to assist this species.

Distribution:

It breeds in the Grassland and Parkland natural regions north to the North Saskatchewan River, south to the Oldman and South Saskatchewan rivers, and east of Red Deer to the Saskatchewan border. The species no longer nests in areas on the periphery of its Alberta range, probably because of fluctuating water levels, and small populations have been displaced on some lakes because of recreational activity (Wershler and Wallis 1986).

Habitat:

Typical Piping Plover habitat is an extensive sandy, gravelly beach on the shore of a saline lake or large saline pond. The species requires dry, open backshore areas for nesting and wet shoreline close to water's edge for feeding (Wershler and Wallis 1986).

Nesting:

The Piping Plover typically nests on dry, gravelly, unvegetated or sparsely vegetated beaches of saline lakes and large ponds (Wershler and Wallis 1986). The higher portion of a gravelly point is a favored location. The nest is a shallow scrape lined with pebbles. Nesting may be loosely colonial and at least some birds return to the breeding territory of the previous year.

The female lays 3-5 eggs, usually four, which are light buff to whitish in color and have blackish-brown spots. Incubation is 27-31 days by both sexes. Young leave the nest within a few hours of hatching to forage for themselves. Adults brood and protect the young, which fledge at four to five weeks. The female abandons the family before the young fledge, leaving parental duties to the male.

Remarks:

This species is similar in appearance to the Semipalmated Plover, but is much paler and lacks any black or dark brown on its cheeks. It is named for its melodic piping calls.

The Piping Plover eats insects and other invertebrates, which it obtains by gleaning or gently probing the sand and gravel of the lake shoreline.

The earliest spring date for this species is April 25 for the Calgary and Hanna area, and the latest fall reports are for the first week of August.

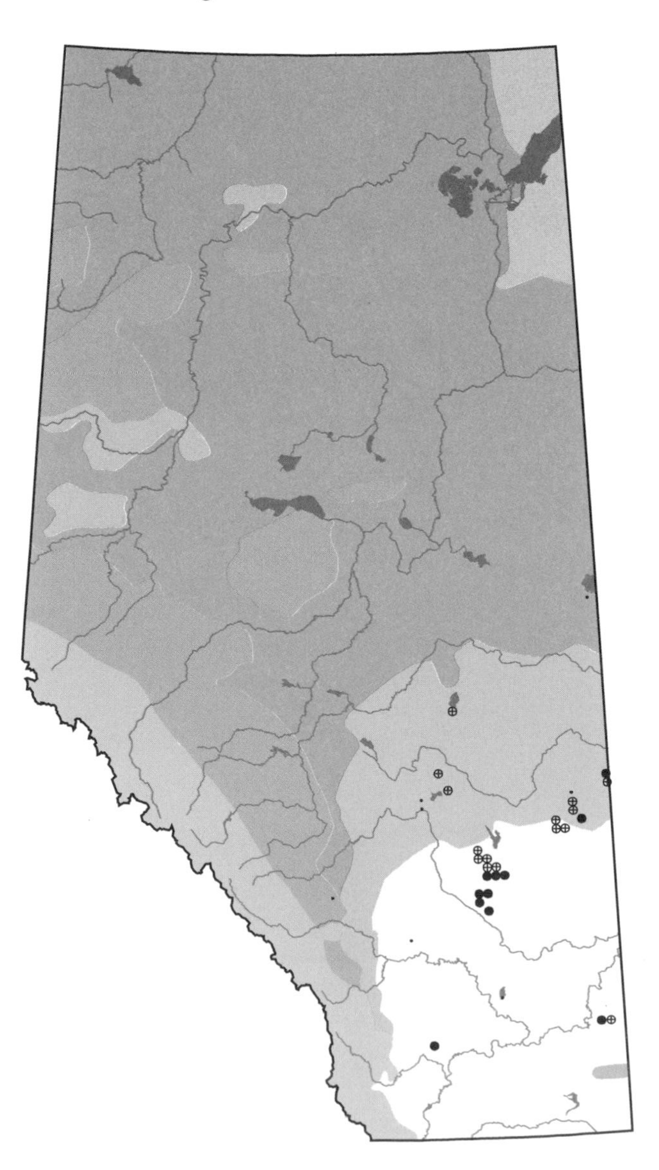

Killdeer

Charadrius vociferus

KILL photo credit: T. Webb

Status:
The Killdeer is a common summer resident in much of Canada and throughout Alberta. This species' capacity for exploiting agricultural and other human-altered habitats is the key to its present success.

Distribution:
The Killdeer breeds throughout Alberta where suitable habitat is available. It was recorded in 40% of surveyed squares in the Grassland region, 53% in the Parkland, 43% in the Boreal Forest, 25% in the Canadian Shield, 38% in the Foothills, and 23% in the Rocky Mountain region.

Habitat:
This plover breeds in open areas with minimal vegetative cover, not necessarily close to water. Its natural habitats include open grassy uplands, lakeshore clearings, river banks, and woodland clearings. The Killdeer also breeds in human-modified habitat such as pastures, cultivated fields, roadsides, gravel pits, golf courses, parking lots, lawns, and rooftops. After nesting, it is more likely to frequent the margins of ponds and lakes, and other muddy, moist places. In the Rocky Mountain region, Killdeer frequent gravelly stream and river channels, sedge and willow meadows with ponds and streams, lake shores, and disturbed areas such as sewage lagoons, landfills, and borrow pits (Holroyd and Van Tighem 1983).

Nesting:
The Killdeer nests in areas with gravelly substrates. In vegetated areas, the nest site will be a bare patch often next to a rock, piece of wood, or other conspicuous object. The nest is a fully exposed shallow scrape in the ground, which is either unlined or sparsely lined with pebbles, grass, wood chips, or twigs. During pair formation, the male will dig a succession of scrapes before one is chosen as the nest site. Birds often return to previous nesting territories.
The clutch is typically four eggs with a buffy background and dark brown or black blotches. Incubation is 24-26 days, by both sexes. The embryo inside the egg is very sensitive to high temperatures, and the Killdeer will soak the feathers of its belly and use them to wet the eggs, cooling them through evaporative heat loss (Ehrlich et al. 1988). This species commonly raises two broods. The young leave the nest soon after hatching and are able to forage for themselves. They are guarded and brooded by both parents. Fledging is at about 30 days.

Remarks:
The Killdeer, named after its loud, persistent call, is one of the most widespread and familiar of all North American shorebirds. It is easily identified by its twin black breast bands and cinnamon-colored rump and tail coverts. Other "ringed" plovers have only one breast band, are smaller, and are shorter-tailed. The Killdeer is well known for its elaborate broken wing display, which it uses to lure intruders away from the nest or young.
Insects, mostly beetles, and other invertebrates form the bulk of this species' diet. It forages in the usual plover fashion, by running quickly forward, stopping, then seizing its prey from the surface of the ground. It does not probe for its food. The Killdeer's large eyes enable it to forage at dusk and at night, as well as during the day.
This species is an early migrant, usually arriving in Alberta in early to mid-March in the south, and mid-April farther north. Birds arrive singly or in small groups and soon disperse over the breeding grounds. In the fall, some birds linger until late November or early December before migrating south (Pinel et al. 1991). Overwintering is reported from Banff, Jasper, and Waterton Lakes national parks, Wabamun Lake, Calgary, and Lethbridge.

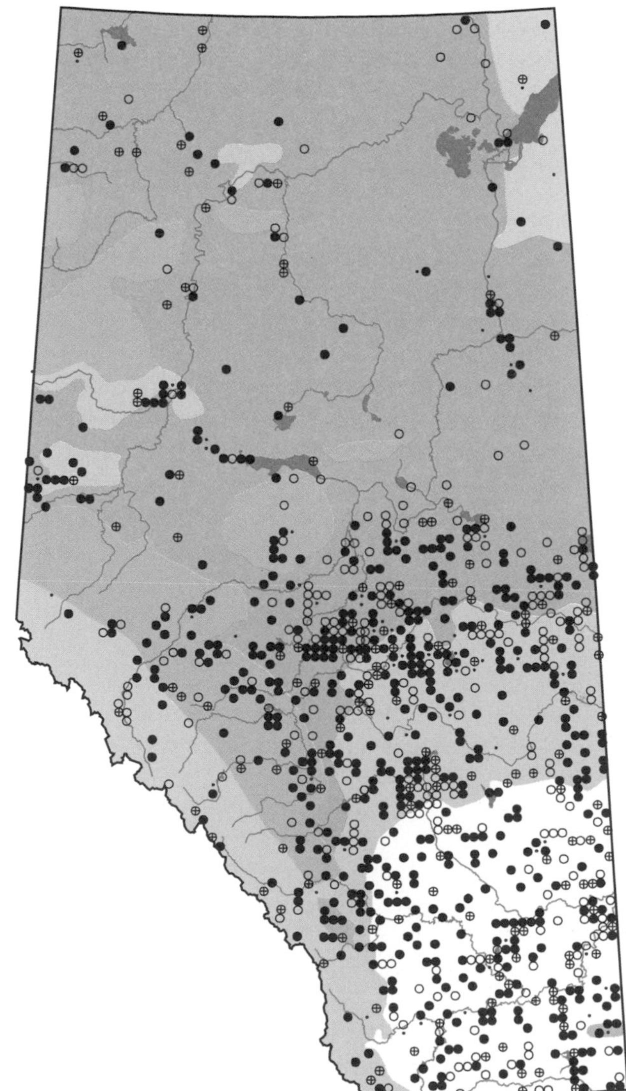

Mountain Plover

Charadrius montanus

MTPL

Status:
Over the last century, this grassland species has experienced declines in distribution and/or population size estimated at 50% to 89% (Leachman and Osmundson 1990). It suffered from overhunting prior to the Migratory Birds Treaty of 1916, and has been a victim of large-scale habitat destruction throughout much of its range. It has been designated as an "endangered" species in Canada (Wershler and Wallis 1986) and in Alberta (Wildlife Management Branch 1991). Management of this species' nesting habitat and protection of nesting birds is required to ensure its continued success in Alberta (Wershler and Wallis 1986; Wershler 1991).

Distribution:
In Alberta, the Mountain Plover is at the extreme northern periphery of its range. The first nesting of this species in Canada and Alberta was documented in 1979 (Wallis and Loewen 1980). Since that time, several nests and broods have been documented within an area of about 500 ha in the Lost River area of southeastern Alberta, with the maximum number of birds and nests observed being 11 adults and six nests in 1981 (Wershler 1991).

Habitat:
Contrary to its common name, this plover does not include mountains in its breeding range or wintering areas. In Alberta, its breeding habitat is heavily grazed or recently burned and grazed native, mixed grassland in flat upland areas (Wershler 1987). Preference is shown for land that has been used as winter feeding pasture. The Mountain Plover has a narrow range of habitat requirements and avoids areas of cultivation, tall grasses or shrubby vegetation, and poorly drained soils (Wershler and Wallis 1986). In Saskatchewan, this species has been observed in a Black-tailed Prairie Dog town and, in Montana, prairie dog towns constitute its major nesting habitat (Wershler and Wallis 1986).

Nesting:
The nest of the Mountain Plover is a shallow scrape lined with lichens, club mosses, rootlets, grass, dried cow manure chips, and rabbit droppings. The birds construct several scrapes before nesting, but only the actual nest is lined. Adults and first year plovers may return to a nesting site if habitat conditions remain stable. Mountain Plovers nest singly or in loosely associated groups.
The female lays 1-4 eggs, usually three. Eggs are olive-buff with black markings. Under suitable conditions, the female will lay a second clutch, which she incubates while the male incubates the first set. The female will sometimes switch mates before laying her second clutch. Incubation averages 29 days. Chicks leave the nest within a few hours of hatching, but are brooded regularly for the first two weeks. Most chicks stay within 300 m of the nest until they are fledged at 33-34 days.

Remarks:
This pale-bodied, inconspicuous, dryland plover has a slender bill and long legs. It approaches the Killdeer in size, but lacks the black breast bands and rust-colored rump of that species. When approached, the Mountain Plover may crouch motionless on the ground or run ahead of the intruder until pressed into flight.
Its diet consists mainly of insects, including beetles, grasshoppers, crickets, ants, and flies, plus a few seeds. It searches for prey by running or walking and pausing periodically to look around. When it stops, it may bob up and down, a behavior characteristic of other shorebirds.
The Mountain Plover probably arrives in the second to third week of April and leaves by the first week of September.

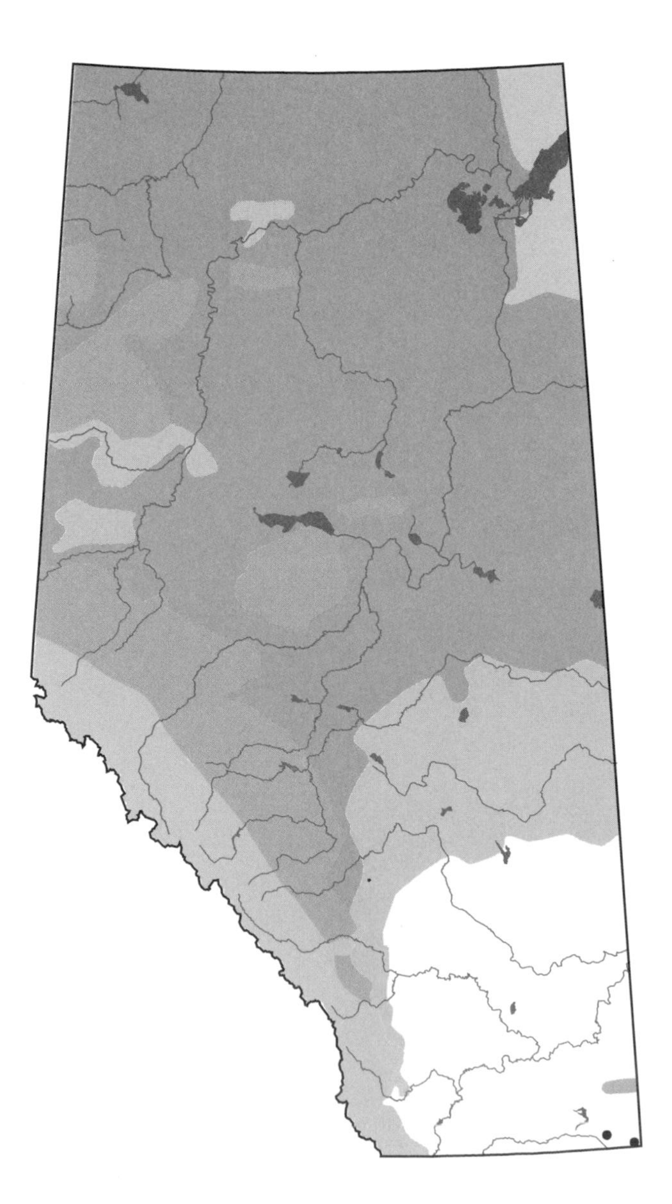

Black-necked Stilt

Himantopus mexicanus

BNST

Status:
This species is a very rare, erratic and local breeder in southern and central Alberta. All nesting records for Canada are in southern Alberta. On the periphery of its normal distribution, the Black-necked Stilt enjoys a secure habitat in the province (Wildlife Management Branch 1991). Continued sporadic appearances may be expected.

Distribution:
The Black-necked Stilt breeds in Alberta, south of Edmonton. In the past, nesting had been confirmed at Beaverhill Lake, New Dayton, and near Calgary. Other Alberta sightings have been at Pakowki Lake, Lethbridge, Airdrie, Stirling Lake, St. Albert, Irricana, Longdon, and Brooks. The Atlas Project was able to secure five breeding records for the province, including four in the Grassland Natural Region, and a single record in the Parkland. Nest sites often vary from year to year.

Habitat:
Black-necked Stilts frequent wet pastures and the grassy shorelines of shallow and brackish pools, marshes, and lakes, especially where there are extensive areas of mudflats.

Nesting:
Courtship display consists of the male alternating between pecking at the ground or water and preening his breast. He also circles the female, flicking water with his bill. During this ceremony, she assumes an elongated posture.
These long-legged waders generally nest in small, loose colonies of less than 20 pairs, usually in the open and close to foraging areas. Nests are usually on the ground of a dry mudflat or on a hummock in a wet pasture or marsh. Both sexes participate in building the nest, which is generally a shallow hollow lined with stems, grass, pebbles, and other debris. On wetter sites, nests may be built up with vegetation and may be added to again during incubation, if the water level should rise further.
The female usually lays four buffy eggs, which are spotted with brown and often nest-stained. Incubation, by both sexes, lasts 25-26 days. When adults leave the nest, they often throw bits of vegetation over a shoulder. To distract nest predators, adults may use aerial flight, mock incubation, or feign injury. The young, which are tended by both parents, run and feed themselves soon after hatching. Adults perform strong distraction displays when their nest or young are approached. Fledging is at 28-32 days and family groupings may stay together through migration. Black-necked Stilts are single-brooded.

Remarks:
The Black-necked Stilt is a large shorebird with contrasting black upper parts, white under parts, geranium-red legs, and a very long slender bill. This bird "talks" a lot, is sharp-eyed, and emits a loud alarm call.
It feeds mainly on aquatic and terrestrial insects and their larvae, as well as crayfish, shrimp, snails, small fish, and seeds of aquatic plants. It picks up its prey from shallow water or from muddy shores, sometimes pecking and sometimes plunging its head into the water. It does not use the side to side scything motion of the American Avocet, which shares the same habitat.
The earliest spring records are from the third week of April and the latest summer records are from early August. This species does not overwinter, but migrates, instead, to southern South America.

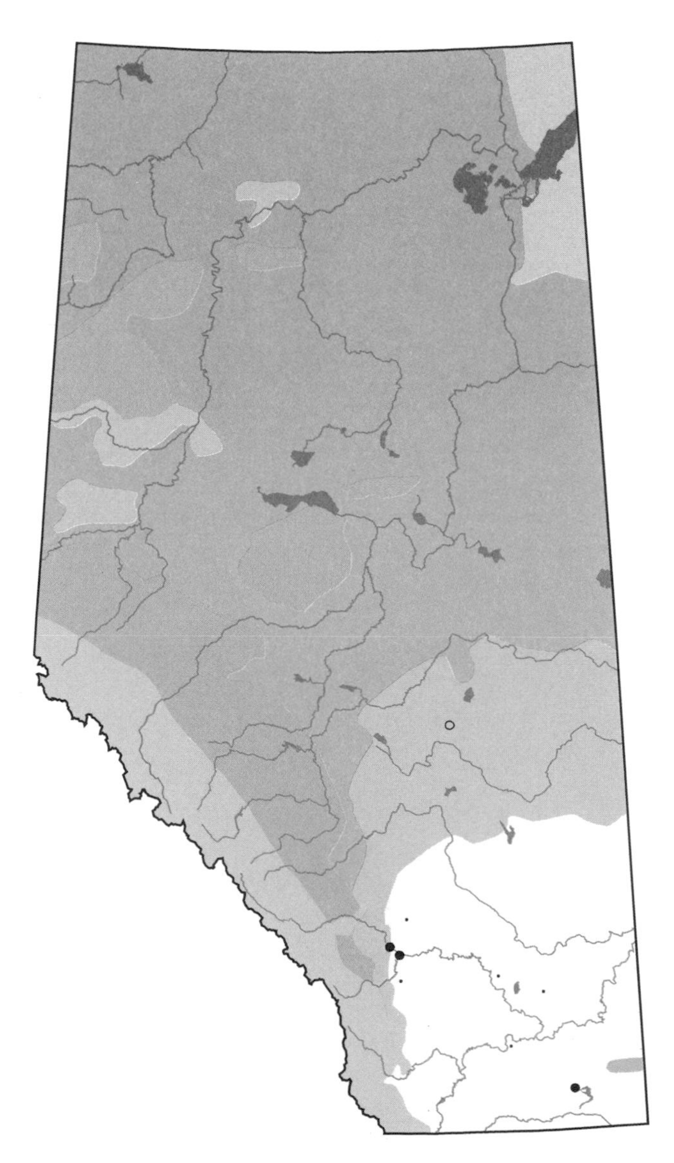

American Avocet

Recurvirostra americana

AMAV photo credit: T. Webb

Status:
Atlas data indicate that this species is a common breeder in the Parkland and Grassland regions. Salt and Salt (1976) reported that, formerly, the range extended much farther north to Fort Chipewyan and north of the Alberta boundary.

Distribution:
This shorebird breeds in appropriate habitat throughout central and southern Alberta, east of the Rockies and north to Edmonton, Beaverhill Lake, and Minnie Lake. It was recorded in approximately 20% of surveyed squares in the Parkland and Grassland regions. It was also recorded in 4% of surveyed squares in the Boreal Forest region, with the majority of those records coming from the southeastern fringe of this region. Extralimital breeding was recorded east of Lesser Slave Lake in the McLennan area, and observations were recorded in the Grande Prairie, Peace River, and Fort McMurray areas. Extralimital breeding was recorded by Francis and Lumbis (1979) in the Fort MacKay area.

Habitat:
The American Avocet favors the shallow muddy borders of saline or alkaline lakes or sloughs, but it is also found in sparsely vegetated marshes or broad wet meadows with shallow pools. Receding pools of standing water attract flies, making this type of area ideal for feeding. When nesting, however, the Avocet prefers drier, more open habitat. Sun-baked mudflats or low gravelly or sandy islands with sparse vegetation furnish the desired breeding habit for this species (Bent 1962c).

Nesting:
Avocets nest in small, loose colonies, usually on dried-out mud shores or on islands of shallow waterbodies. They generally build their nests in areas that afford them 360 degree vision. If an intruder is seen close to the nests, these birds will display communal distraction techniques; other Avocets will join the parent to mob the intruder and drive it away. The nest is a shallow scrape in the ground, which is usually lined with grasses and weeds, but also, sometimes, dry mud chips, pebbles, twigs, feathers, or bones. Some nests may be built up, especially in areas subject to flooding. The female lays 3-5 eggs, usually four. The eggs are light brown to dark olive, and are much spotted with dark browns, and an occasional splash of lavender. Some nests have been found with up to eight eggs, but these are likely the product of two females. These "superclutches" occur more frequently in the northern part of the Avocet breeding rang (Giroux 1985).
Incubation is by both sexes and takes 22-24 days. Young are tended by both sexes and fledge after 28-35 days.

Remarks:
With its striking black and white plumage, the American Avocet is one of our most handsome and showy shorebirds. In breeding plumage, this bird's head and neck are light cinnamon.
It is often seen foraging in shallow water or on the barren mud flats along the edges of lakes. Its recurved bill is admirably suited to scraping up food from both mud and water. The avocet eats some vegetation, generally seeds of aquatic plants, but the bulk of its diet includes insects, crustaceans, fish, brine shrimp, and carrion. Unlike many shorebirds, the avocet will swim when it gets beyond its depth. While swimming, it feeds by tipping up like a surface-feeding duck (Bent 1962c).
Avocets are found in central Alberta in early May, converging in large flocks. They are rather noisy at this time, putting on courtship displays and squabbling. They soon settle down and break up into smaller nesting groups. They migrate early in the fall and are seldom seen in Alberta after the end of September.

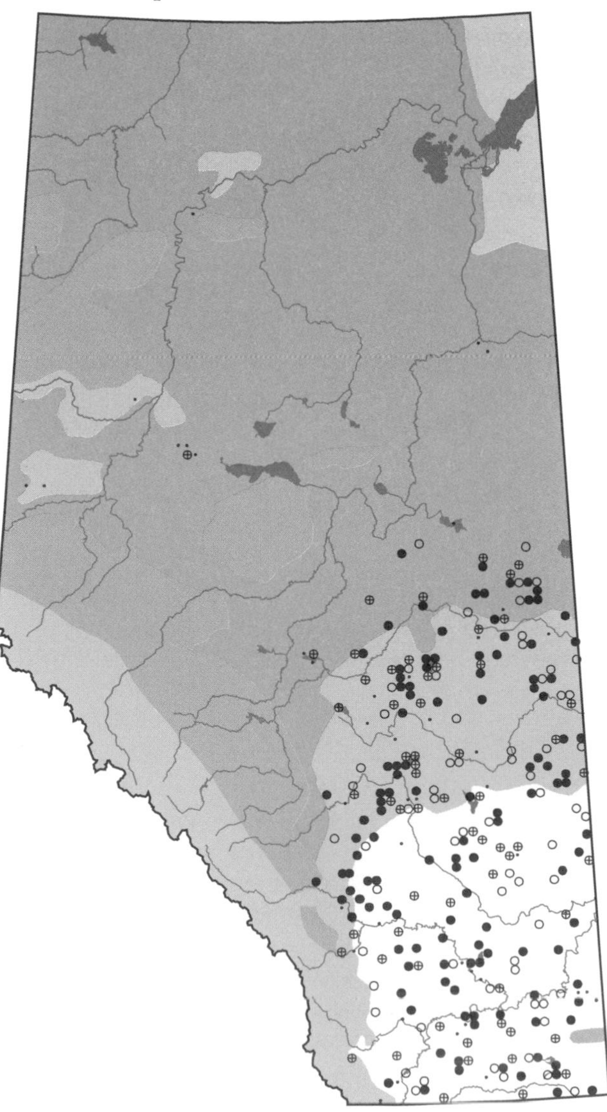

Greater Yellowlegs

Tringa melanoleuca

GRYE

Status:
The first nesting record for this species in Alberta was near Fawcett in 1930 (Salt and Salt 1976). It is less common than the Lesser Yellowlegs, but its population is considered healthy and not at risk (Wildlife Management Branch 1991).

Distribution:
The Greater Yellowlegs breeds throughout the boreal regions of Canada. It was recorded breeding most often in the Boreal Forest and Foothills regions of the province. Salt and Salt (1976) and Godfrey (1986) both included the northwest corner of the province within the known breeding range of this species, but there were no breeding records for the Greater Yellowlegs in this area during Atlas surveys. Although reported as rare (Salt and Salt 1976) and uncommon (Holroyd and Van Tighem 1983) in the Rocky Mountain region, there are several breeding records from the Athabasca River valley in Jasper National Park.

Habitat:
This long-legged wader breeds in wooded muskeg country, burned-over areas, and open, mixed forests with low sparse undergrowth (Godfrey 1986). It favors those areas with open bogs, marshes, and small ponds. In the northern part of the province, muskegs with tamarack and spruce seem to be important areas of breeding (Salt and Salt 1976). In the mountain parks, it is found in wet shrub meadows, sedgy wetlands with scattered trees, and alluvial wetlands (Holroyd and Van Tighem 1983).

Nesting:
Greater Yellowlegs usually nest on drier ground, often on a ridge or hummock, generally near water. Eggs are laid in a depression or scrape lined with grasses or leaves. The nest is often screened to some extent by vegetation or debris.

The female lays four eggs, which are buffy and blotched with browns. Incubation is 23 days, probably by both sexes. Greater Yellowlegs are vigorous and noisy defenders of their nests. Young are tended by both parents, which take the chicks from the vicinity of the nest to water within a day or two of hatching. Young can usually fly in 18 to 20 days.

Remarks:
The Greater Yellowlegs is one of Alberta's largest shorebirds. It is generally gray and brown above with white underparts and long, bright yellow legs. It is similar to the Lesser Yellowlegs, but stands much higher and has a longer and slightly upturned bill than that species. Both species nod their heads as they walk about, and are vigilant and quick to raise the alarm when approached by intruders.

This species eats insects, small crustaceans, small fishes, tadpoles, worms, and berries. It usually forages in shallow water, picking food off the surface or sometimes "ploughing" with an open bill. Occasionally, it forages for insects on the shore.

Spring migrants begin to arrive in early to mid-April, with last spring migrants observed in southern Alberta in mid-May (Pinel et al. 1991). Fall migration begins as early as late June or July, with birds congregating on lakes in central Alberta. Peak fall passage is in August, with the last stragglers leaving in late October and early November (Pinel et al. 1991).

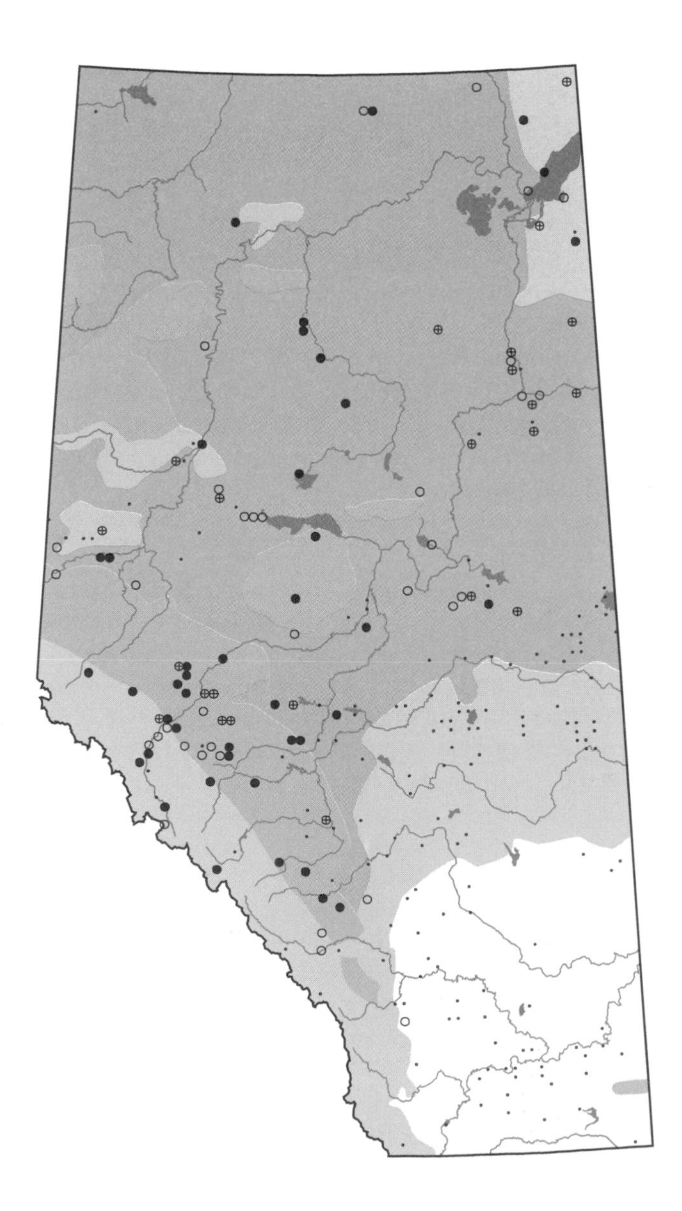

Lesser Yellowlegs

Tringa flavipes

LEYE

Status:
This species is one of the most common medium-sized shorebirds in North America (Johnsgard 1981). Its population in Alberta appears to be relatively secure.

Distribution:
The Lesser Yellowlegs breeds most commonly in northern and central parts of the province east of the Rockies. It was recorded in 50% of surveyed squares in the Canadian Shield region, 23% in the Boreal Forest, 15% in the Parkland, and 9% in the Foothills region. There is only one confirmed breeding record from the Rocky Mountain region, which confirms Salt and Salt (1976), who reported this species as a rare transient in this region.

Habitat:
It inhabits open woodlands and burns with sparse low undergrowth, especially areas interspersed with muskegs, and marshy or grassy ponds and lakes. In Alberta, Lesser Yellowlegs have been found nesting among broken hills that are covered with burned and fallen timber and have a second growth of low poplars (Johnsgard 1981). Wet areas are used for feeding and raising young. After the young have fledged, birds appear on larger lakes, where they forage along sandy beaches and mudflats close to shallow water.

Nesting:
Lesser Yellowlegs nest on the ground, usually in dry, lightly wooded areas close to water. The nest is a shallow scrape lined with dry grasses, leaves, or twigs. It is often placed by a stump, log, small tree, or in the open. Several scrapes may be made before one is chosen for nesting. Several Lesser Yellowlegs may share an area of good nesting habitat and gather at the same feeding pond, but this species does not form colonies (Salt and Salt 1976).
The clutch typically consists of four eggs, which are buffy and blotched with brown. Incubation is 22-23 days, by both sexes. The young leave the nest soon after hatching, and are tended by both parents, though the male assumes most of the rearing duties. The young are strongly protected by the parents, which persistently dive at human intruders, but never actually strike them (Johnsgard 1981). Fledging is at approximately 23 days.

Remarks:
The Lesser Yellowlegs is a dusky brown sandpiper with long yellow legs. It shows a white tail and rump in flight. It is smaller than the Greater Yellowlegs, and has a more slender and straighter bill than that bird. It is a noisy bird, quick to sound an alarm should an intruder appear.
This species feeds in shallows, often with water up to its breast, gleaning its food from the surface of the water or mud. It seldom probes the mud and rarely skims its bill from side to side like the Greater Yellowlegs does. It uses a more deliberate pecking action in mud and shallow water. This species' main foods are insects and their larvae, small crustaceans, worms, and even the occasional small fish.
Spring migrants arrive from early to late April, with fall movement beginning in late June or early July. Large numbers of birds then congregate on larger lakes from late July through September and the last fall birds leave by mid-October (Pinel et al. 1991).

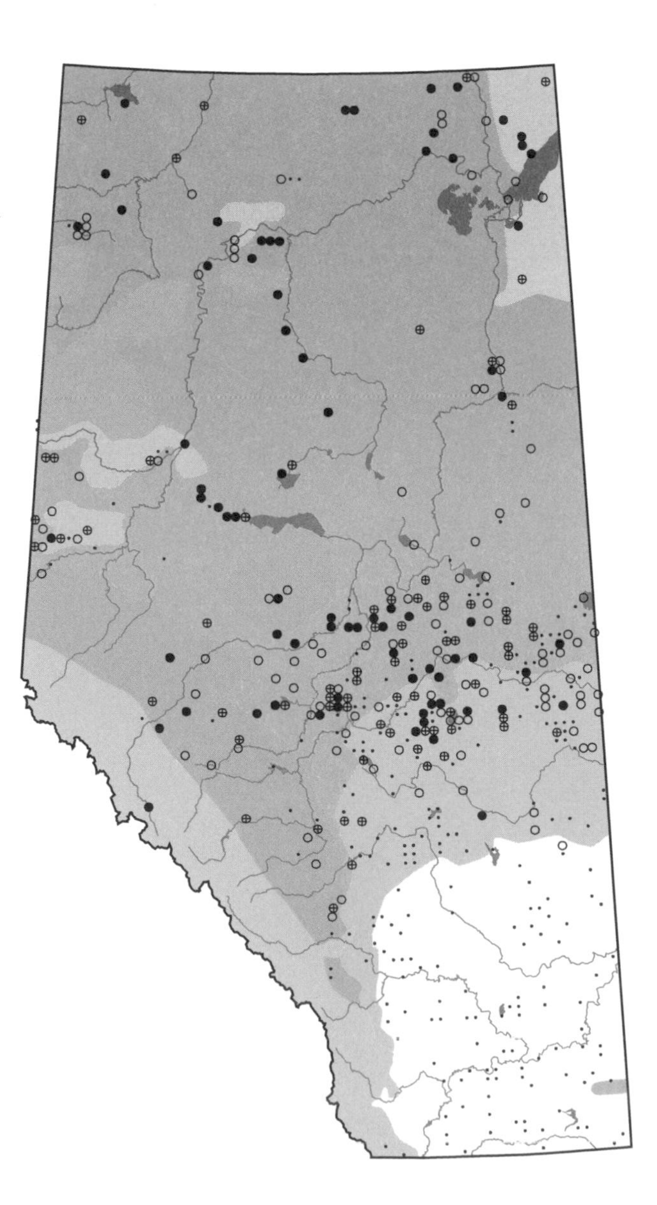

Solitary Sandpiper

Tringa solitaria

SOSA

Status:
It is impossible to judge this species' status, because it is so elusive and non-gregarious throughout the year. Therefore, the status of this species in Alberta has not been determined. However, it appeared to be relatively common in the southern part of the Boreal Forest and Foothills regions, and widespread in the northern part of these regions.

Distribution:
This medium sized sandpiper breeds through the subarctic and boreal zones of North America. The Solitary Sandpiper breeds within suitable habitat in all of Alberta's natural regions, but rarely in the Grasslands region, where fewer than 1% of squares surveyed recorded this species. It was recorded in 30% of surveyed squares in the Foothills region, 19% in the Boreal Forest, and 13% in the Rocky Mountain region. Atlas data confirm the distribution of this sandpiper as reported by Godfrey (1986), with the exception of one extralimital confirmed breeding record southeast of Hanna.

Habitat:
The Solitary Sandpiper prefers open, wet woodlands, nesting near wetlands, ponds, and lakes. It is often seen feeding at the edges of shallow pools and muskeg ponds, or the muddy margins of sloughs and rivers. It also forages in wet fields, and meadows.

Nesting:
This species nests in the old nest of a Robin, Rusty Blackbird, or similar-sized bird, at any height in a coniferous or deciduous tree (Godfrey 1986). Nests are usually near a stream or small pond. Both sexes are involved in seeking out the nest, though the male is probably the main nest prospector, with the female making the necessary adjustments to the nest (Johnsgard 1981). The nest lining may be rearranged, but no new lining is added. This nesting strategy is unique among North American shorebirds.
The female typically lays four eggs, which are greenish or buffy and spotted with browns or a little bluish gray. The incubation period is about 23-24 days for this species. The relative roles of the parent birds in incubation are not confirmed, though both probably participate (Johnsgard 1981). A few hours after hatching, the downy young are encouraged by the female to tumble down from the nest. Both parents tend the hatchlings. The main defense of the young is their coloration and their ability to crouch motionless on the ground. The fledging period of this species is unknown and only one brood is raised per season.

Remarks:
The Solitary Sandpiper is a slender, straight-billed sandpiper that is dark brown above with dark green legs and a white eye ring. Like the yellowlegs species, it has the habit of nodding. Birds are usually observed alone, but they may, occasionally, be observed in groups of two or three during migration.
This species forages at the waters edge or in shallow water, often vibrating the leading foot as it walks. This activity stirs up the bottom enough to disturb insects. Aquatic and terrestrial insects and their larvae, worms, molluscs, and small crustaceans form the bulk of this sandpiper's diet. It also captures winged insects in the air, as well as spiders, grasshoppers, and even small frogs (Bent 1962c).
Early spring migrants arrive in the province in late April or early May, and birds may leave the breeding grounds as early as late June or early July (Pinel et al. 1991). The main fall passage is in August, with a few birds lingering to mid-September.

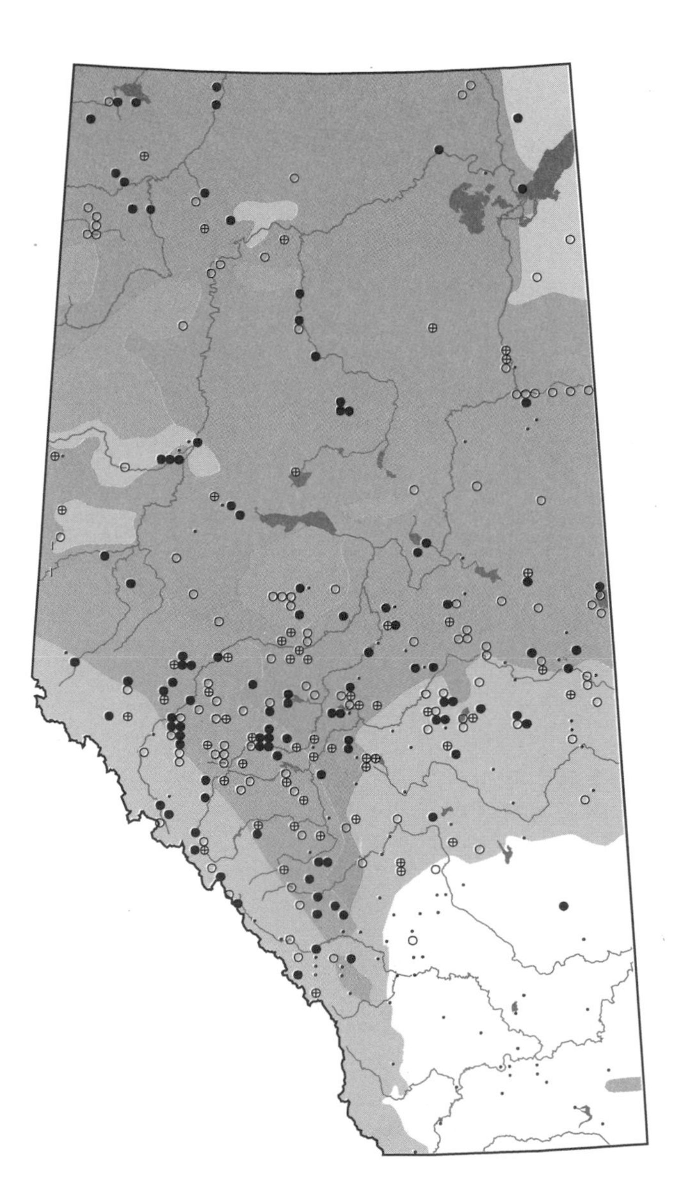

Willet

Catoptrophorus semipalmatus

WILL photo credit: T. Webb

Status:
This species is considered to be moderately abundant over much of its range (Johnsgard 1981). Salt and Salt (1976) reported that it was formerly a common summer resident but had become relatively scarce throughout much of its range. Robbins et al. (1986) reported that Breeding Bird Survey data indicated a significant increase in continental populations, with a high count coming from Alberta and Saskatchewan. Atlas data confirm that the Willet is common in the Grassland and Parkland regions, where it was recorded in 30% and 20% of squares surveyed, respectively.

Distribution:
The Willet breeds in central and southern Alberta west to the foothills and north to about the North Saskatchewan River. Atlas records outside of the Grassland and Parkland natural regions were minimal (less than 2% of surveyed squares) and were close or adjacent to these regions. Pinel et al. (1991) reported an extralimital summer record for Fort McMurray.

Habitat:
This species frequents moist and wet meadows and the grassy edges of prairie sloughs and lakes, rarely wandering far from water. On the prairies, a small weed-covered pond or a barely discernible creek meandering through a shallow valley may have sufficient water to attract a pair (Salt and Salt 1976).

Nesting:
The Willet is semi-colonial, with pairs maintaining separate nesting and feeding territories. They usually nest on the upper edges of beaches or on drier grassy areas near marshes. The nest is generally found in the proximity of a piece of wood, dried cattle dung, a stone, or other such objects (Higgins and Kirsch 1979). It is a shallow scrape lined to varying degrees with grasses, sedges, or twigs. It is often well-concealed in grass, but is sometimes in the open and almost completely exposed. The nest site is chosen by the female.

The female typically lays four eggs, which are olive-buff and blotched with browns. Incubation is reported at 21-29 days (Johnsgard 1981) by both sexes, with the male taking over at night and occasionally at midday. The female abandons her mate and brood two to three weeks after the young hatch. The male then attends the brood for two more weeks. Adults often leave the breeding area before the young have fledged. The fledging period for this species is 30-35 days. This species raises one brood per season, and has a strong nest site and mate fidelity between seasons.

Remarks:
The Willet is a large, grayish, straight-billed shorebird, suggesting a somewhat robust Greater Yellowlegs but with gray-blue legs (Godfrey 1986). In flight, it can be recognized by a broad white band on the wing which contrasts with the black tips, producing a showy pattern.

This wader hunts primarily by sight; it captures prey by walking along, pecking or probing and turning over rocks and debris with its bill. Its main foods are insects, worms, crustaceans, and other small invertebrates.

Birds arrive in southern Alberta in the last two weeks of April, and most have left by the end of August, though a few stragglers carry on into October (Pinel et al. 1991).

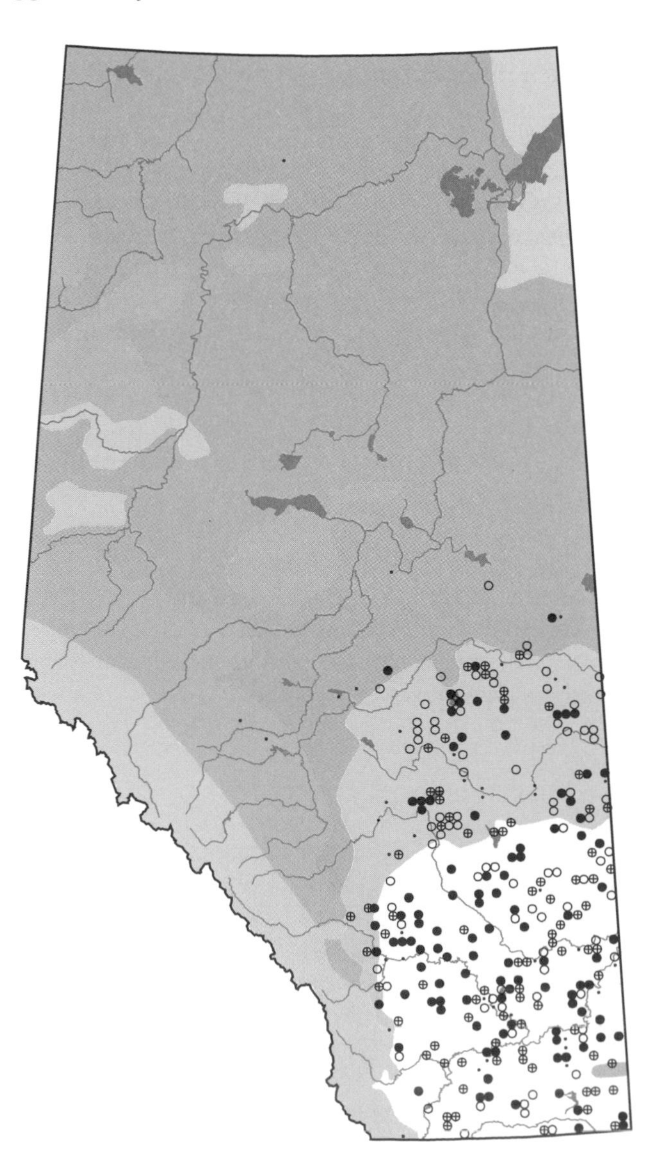

Spotted Sandpiper

Actitis macularia

SDSA photo credit: T. Webb

Status:
This very successful and adaptable species is the most widely distributed and best known sandpiper in Canada. In Alberta, it is a widespread and abundant species.

Distribution:
The Spotted Sandpiper breeds from the Arctic south to the southern United States. It is found in suitable habitat throughout Alberta. It was recorded in approximately 40% of surveyed squares in the Canadian Shield, Rocky Mountain, Foothills, and Boreal Forest regions. It was also recorded in 30% of surveyed squares in the Parkland region and 18% in the Grassland region.

Habitat:
This species occurs in all the natural regions of the province. It breeds by freshwater, frequenting ponds, lakes, rivers, and streams, usually in open or fairly open places. Typical breeding habitat would be a clearing near water, an open river bank or a weedy roadside. Foraging occurs in diverse areas including shorelines, gravelly beaches and gravel bars, mudflats, wet meadows, wetlands, grazed grasslands, and cultivated fields.

Nesting:
The nest of the Spotted Sandpiper is a hollow in the ground, which is sparsely lined with grasses, leaves, stems, moss, twigs, or other vegetation. Nests are generally close to water and at least partly concealed by grass, shrubbery, or a log or similar object. Nest prospecting is done by both sexes. These birds sometimes nest in colonial fashion, especially on small islands (Johnsgard 1981).
Clutches are usually four eggs, which are buffy with blotches of brown and spots of bluish gray. The female Spotted Sandpiper is polyandrous (one female mating with more than one male), moving between nesting territories and possibly producing several clutches with different males. When more than one clutch is laid, the earlier clutches are incubated by the male alone and the last clutch by both sexes (Harrison 1978). Incubation is 20-24 days. Young may be tended by the male alone, or by both sexes. Newly hatched chicks leave the nest once their down is dry. Young fledge at 16 to 18 days and the brood scatters with fledging. Adult birds return to previous nesting territories each year.

Remarks:
The Spotted Sandpiper is a small, greenish-brown shorebird that has distinctive breast spots in the spring. It is well-known for its rapid, quivering flight, constant teetering motion when on the ground, and clear, repetitive "peet-weet" call. This sandpiper might be confused with the Solitary Sandpiper, but the Spotted is smaller, has a flight entirely different, and is distinguished by a white stripe along the spread wing (Munro 1963).
This species is very active, foraging by pecking food items from the ground or water surface, by stalking, or by snatching its prey in mid-air. Its most common foods are terrestrial and aquatic insects and their larvae, small molluscs, and crustaceans.
The earliest spring migrants arrive in the first two weeks of May, and most birds have left the province by the first week in September. Some stragglers remain until mid-October (Pinel et al. 1991).

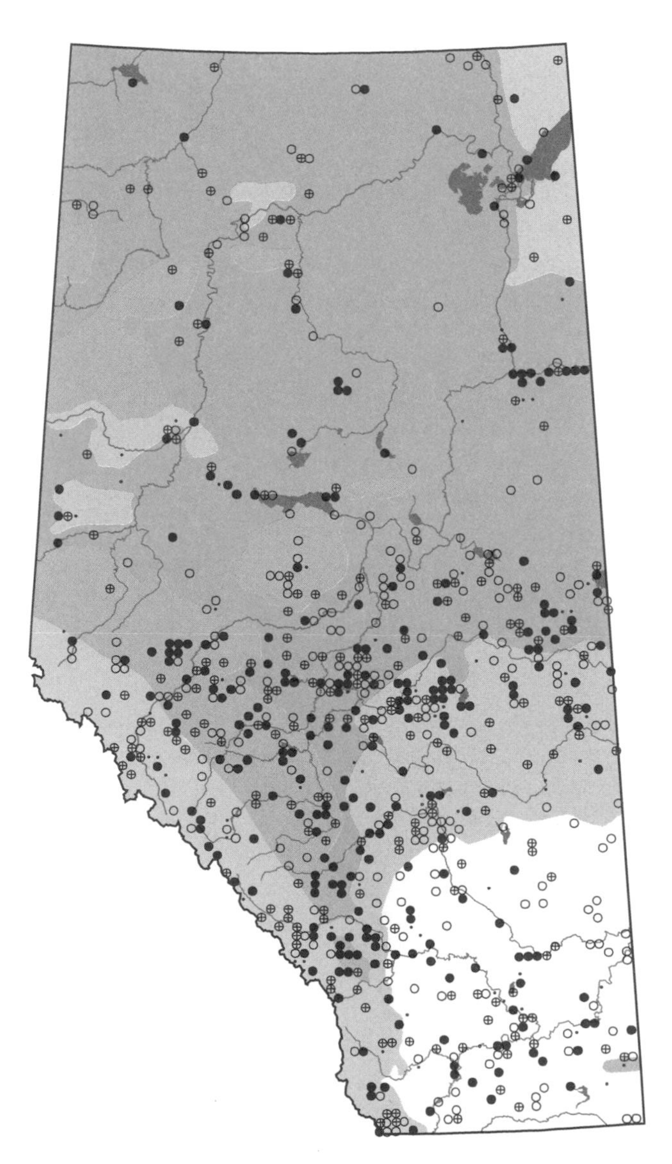

Upland Sandpiper

Bartramia longicauda

UPSA

Status:

This species used to be a fairly common resident of the prairies, but has suffered from the loss of its natural prairie grassland habitat. Its current status in Alberta is unclear. No population figures are available, but it appears to have disappeared from some areas of the province (Wildlife Management Branch 1991; Pinel et al. 1991).

The Upland Sandpiper has been on the North American Blue List since 1975, but Breeding Bird Survey data indicate increasing populations on the North Amercan continent as a whole since 1968 (Chandler et al. 1986) and Robbins et al. (1986) reported that increases in Canada were significant.

Distribution:

Atlas data agree with Pinel et al. (1991), that the Upland Sandpiper is a regular breeder in the Grassland region, with extremely local nesting distribution in northern Alberta. This contrasts with the extensive range reported by Salt and Salt (1976). Over 70% of breeding records obtained from Atlas surveys were from the Grassland Natural Region. Extralimital breeding reported from Waterton Lakes National Park in the south and Grande Prairie in the north (Pinel et al. 1991) was confirmed by Atlas data. The Upland Sandpiper was also recorded as occurring in the extreme northwest corner of the province south and west of Bistcho Lake.

Habitat:

Historically, this sandpiper is an inhabitant of mixed prairie, tall-grass prairie, and aspen parkland areas of Canada, but it also utilizes permanently open areas of the Boreal Forest (Erskin 1977). It is primarily found in open, grassy uplands, hay fields, pastures, wet meadows, and old fields with minimal shrub or tree growth. It rarely uses cultivated fields. Upland Sandpipers primarily utilize upland areas for nest selection and foraging.

Nesting:

Upland Sandpipers nest singly or in small groups. The nest is usually a depression in the ground lined with grasses, leaves, or small twigs, and generally well-hidden in grass or other vegetation along the edges of clearings, sloughs, and ponds. Both the male and female participate in nest scraping behavior.

Clutches are typically four eggs, sometimes three; eggs are pale buff and spotted with browns and a little bluish gray. The eggs are incubated by both sexes for 24 days. The adult birds are very secretive while incubating, and are difficult to flush. Young leave the nest soon after hatching and are then tended by both parents. The adults are very alert and vocal in defending their young and will feign injury if the young are threatened. Fledging is at 32-34 days. One brood is raised per season.

Remarks:

This species was called the Upland Plover until recently, but it is a true sandpiper, though it is a bird of the dry uplands. It is without any conspicuous markings, but may be distinguished from other sandpipers by its light median stripe on the crown, rounded head, slender neck, and dark Vs on the breast (Salt and Salt 1976). Another notable feature is its habit, when alighting, of extending the wings high over the back before folding them into place. Its wolf-whistle call, (described as "whip-whee-ee-you"), extraordinary display flights, and habit of sitting visibly on fence posts are also diagnostic.

Its diet is primarily terrestrial insects, spiders, snails, and earthworms (Johnsgard 1981). It is also known to eat weed seeds and waste grain (Bent 1962c).

This species first arrives in the province from early to mid-May, and departs again by late August or early September.

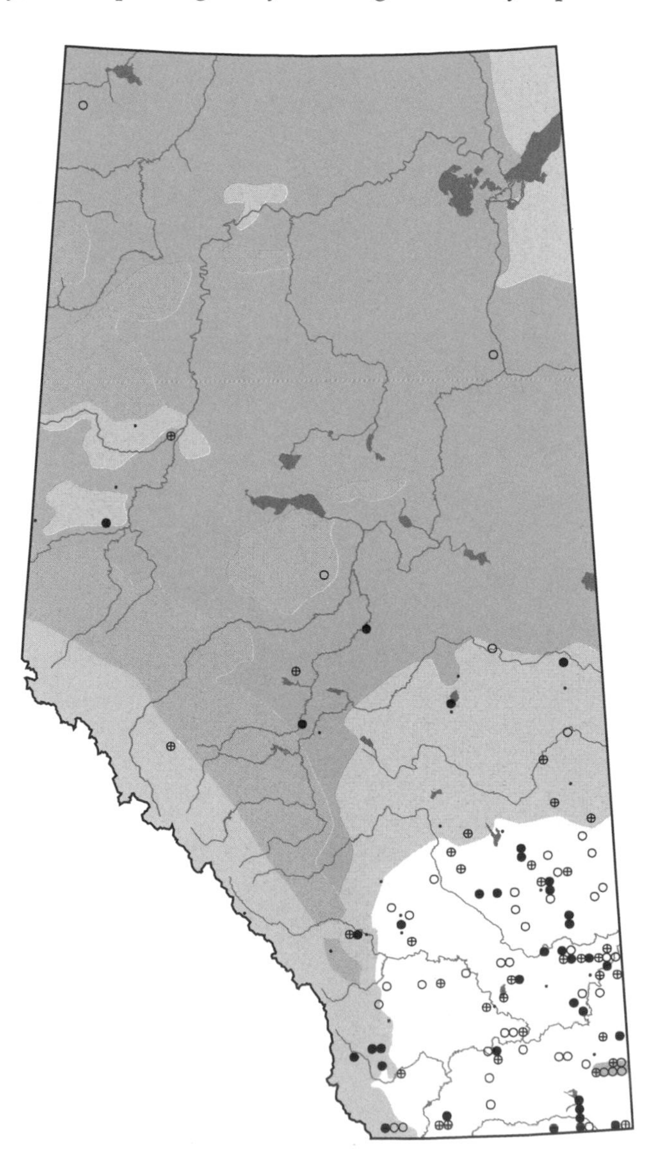

Long-billed Curlew

Numenius americanus

LBCU

Status:
Since the late 1800s, the Long-billed Curlew has undergone considerable reductions in numbers and breeding range, mainly beacause of over-hunting and loss of its grassland habitat. In Alberta, Long-billed Curlew numbers are low and probably declining and, at present, there is insufficient information on the species in the province to allow its efficient management (Wildlife Management Branch 1991).
The Long-billed Curlew was Blue-listed in 1981 and 1982, but was down-listed to a species of "special concern" in 1986 (Tate 1986).

Distribution:
This species was once common over most of the prairie regions of southern Canada, but it is presently found only in isolated populations in south and central British Columbia, southern Alberta, and southwestern Saskatchewan.
The northern edge of the Long-billed Curlew's breeding range in Alberta closely parallels the Grassland/Parkland border. This bird was recorded in 22% of squares surveyed in the Grassland Natural Region and in less than 2% of squares in other regions. It nests as far north as Elnora and Castor and west almost to the foothills. Extralimital breeding has been recorded at Glenbow Lake and Beaverhill Lake.

Habitat:
This species requires large tracts of open grassland with low vegetative cover and no visual barriers for its nesting habitat. Nesting occasionally occurs in fallow and stubble fields, or in forage and grain crops. In Alberta, maximum breeding densities occur in moderately grazed grassland on sandy loam (de Smet 1989).
Brooding and rearing generally take place in areas that have more cover and are wetter than the nesting site. In spring migration, birds may be found on dry upland prairies; in the fall, they are more likely to frequent mudflats, lakeshores, and river valleys.

Nesting:
Long-billed Curlews nest singly or in small, loose colonies, usually returning to the same nesting territory year after year. The nest is a depression in the ground, and may be lined with grass, plant stems, or straw. It is usually placed on flat terrain in short grass cover or stubble, and near a conspicuous object such as a rock.
The female lays four eggs, which are olive-buff and spotted with browns and olive. Both sexes incubate, for 27-30 days. The male sits at night, and the female during the day. Breeding adults are very vocal and aggressive in defending the nest and will be joined by other Long-billed Curlews in the area should an intruder appear.
Young leave the nest within a few hours of hatching, even before the down is dry, but will return at night for a week or so to be brooded. The female abandons the brood after two to three weeks, leaving the care of the chicks to the male. Young fledge at 41-45 days.

Remarks:
The Long-billed Curlew is North America's largest shorebird. Its size, long legs, long down-curved bill, and buffy color are distinctive.
This species has opportunistic feeding habits. Its main foods in breeding season are terrestrial insects, caterpillars, and seeds, but it is also known to eat nestlings, eggs, earthworms, berries, and, probably, the occasional reptile, amphibian, and small mammal. During migration, birds frequent wetter environments, where they take a variety of insects, crustaceans, molluscs, and other invertebrates. This species forages by probing and pecking.
Long-billed Curlews arrive in southern Alberta in the third week of April, and leave the province again by the end of August.

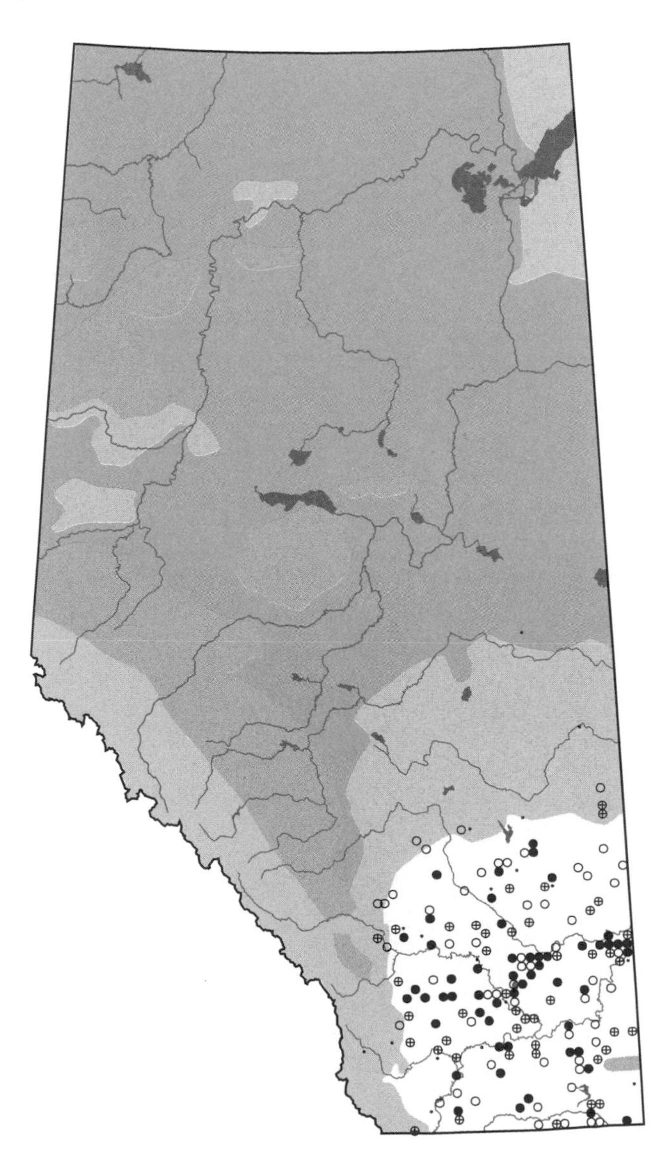

Marbled Godwit

Limosa fedoa

MAGO photo credit: T. Webb

Status:
Available breeding habitat for this prairie-dependent species has declined over the last several decades. Its status in Alberta remains undetermined, although it is fairly common in the Grassland Natural Region.

Distribution:
The Marbled Godwit breeds throughout the prairie regions of western Canada. In Alberta, it is found north to Lac La Biche and High Prairie and southwest to the base of the foothills. This species is primarily found in the province's Grassland and Parkland regions, where it was recorded in 28% and 16% of squares surveyed (respectively). It was also recorded in the southern portion of the Boreal Forest Natural Region, which is a northern extension of the range reported in Godfrey (1986). There is one record of nesting in Jasper National Park (Holroyd and Van Tighem 1983).

Habitat:
This species frequents prairie wetlands and sloughs, close to streams and shallow parkland lakes. Grasslands of low to medium stature seem to be preferred over tall cover, and native prairie is apparently preferred over cropland cover (Johnsgard 1981).

Nesting:
Marbled Godwits nest in the vicinity of water, often back on the grassy prairie or on the low meadowland bordering a lake. Nesting may be semi-colonial. They usually nest in short grassy cover, where their clutches are poorly concealed by adjacent vegetation (Higgins et al. 1979). The nest, formed by the male, is a shallow depression in the ground sparsely lined with grass, often near a rock or clump of weeds. Nest scraping is a common display between paired birds.
The female typically lays four pale buff eggs, which are sparingly blotched and spotted with browns. Incubation is 23 days, by both sexes, with the male incubating throughout the night. The birds stand very close to the nest and allow a close approach. Young leave the nest shortly after hatching and are tended and vigorously defended by both parents. They fledge in about three weeks, and the male leaves the family at this time.

Remarks:
The Marbled Godwit is a large, noisy, and conspicuous shorebird, typical of the prairie slough. It has dark buffy brown and black upper parts which are spotted and streaked pinkish-cinnamon, giving it a "marbled" effect. It has a long slender slightly upturned bill which is two-tone pink and black. It is similar in size and plumage to the Long-billed Curlew, but the Marbled has a shorter, upturned bill.
This species eats mainly aquatic and terrestrial insects and molluscs, crustaceans, and worms, feeding along the shoreline or in shallow water, or in nearby grassy flats. Dekker (1976) reported Marbled Godwits eating sticklebacks at Beaverhill Lake. The birds capture their insect prey by pecking along the surface of the substrate, by probing into the mud, or by picking victims off the short grass. They have been observed in Saskatchewan eating tubules of aquatic vegetation, a food source identified in studies of other godwits in Europe (H.L. Dickson pers. comm.).
The first spring migrants arrive late in April. Fall migration is early, with few birds remaining after August, though stragglers are occasionally recorded into mid-October. No overwintering of this species has been reported.

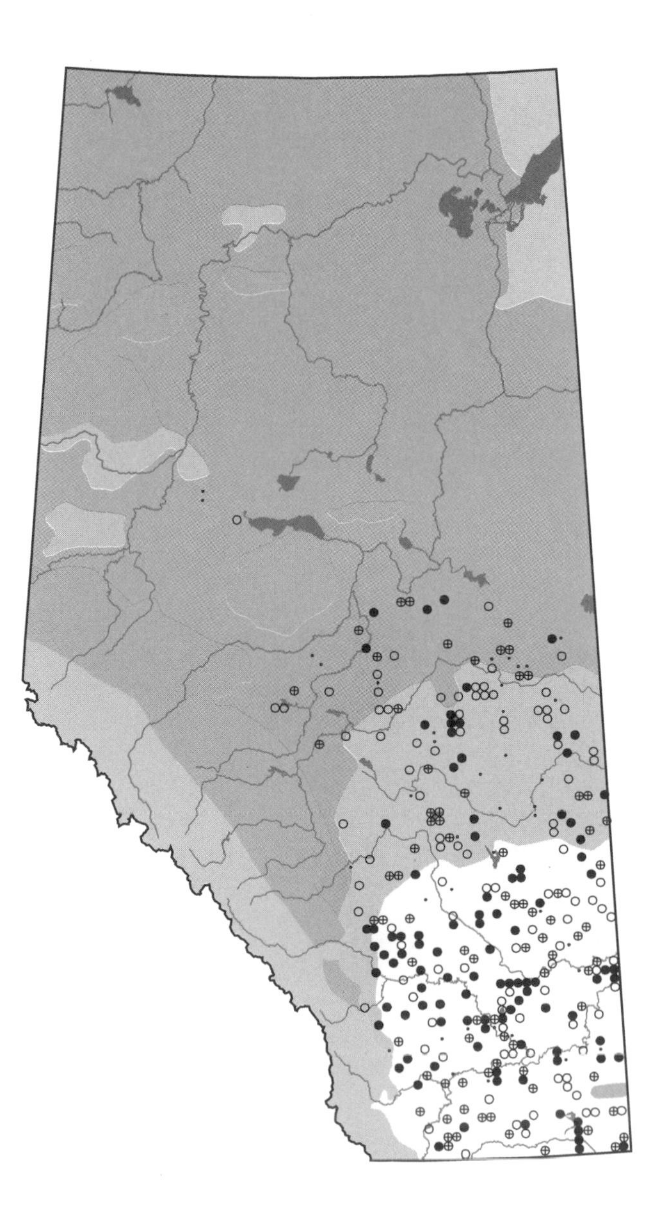

Least Sandpiper

Calidris minutilla

LESA

Status:
The Least Sandpiper is a common to abundant breeder over most of its range, although hunting up through the early 20th century decimated its numbers considerably. It is reported as a regular spring and fall transient in the province (Salt and Salt 1976). In Jasper, this species is a fairly common autumn migrant, especially in the lower Athabasca River valley (Holroyd and Van Tighem 1983). It is a very rare breeder in Alberta, where it is at the periphery of its range.

Distribution:
This species breeds from western Alaska across subarctic Canada. There were several observations of the Least Sandpiper during Atlas surveys. However, there was only one record of possible breeding in the five years of the project. This was recorded in the Rocky Mountain House area. Alberta observations from Lake Athabasca and Lillaboo Lake indicate possible nesting in these areas (Höhn 1972).

Habitat:
Subarctic and northern Boreal Forest areas are the usual breeding habitat of this species. In breeding season, it is usually found near boggy places with sedges, mosses, or grass, or on higher ground with low vegetation. During migration, it frequents the muddy margins of lakes, mudflats, and wetlands.

Nesting:
Least Sandpipers nest in open areas, usually in sedges or moss, in wet, swampy ground. Birds sometimes nest on drier ground in sedges, but always near water. The nest is a shallow depression sparsely lined with dry leaves, grass and moss, frequently at the base of a small willow. The dead leaves seem to be placed very purposefully in the nest, rather than accidentally, and are often of the same color and shape as the eggs (Johnsgard 1981).
Clutches are usually four eggs, which are pale or olive buff and blotched with browns. Incubation is 20-23 days, mainly by the male. Young may spend a day near the nest, then wander widely. Both sexes tend the young, with the male taking most responsibility. Young are reared in well-vegetated, swampy grounds (Campbell et al. 1990). Fledging occurs at about 18 days. One brood is raised per season, although several instances of renesting in the same season have been recorded (Johnsgard 1981).

Remarks:
This species is the smallest of our native sandpipers; it is under 15 cm in length. In breeding plumage, this sandpiper has a streaked and light brown head and underparts. It has a short thin dark bill which curves down gradually, and pale legs. The Least Sandpiper is colored very much like the Baird's Sandpiper. However, the former is smaller and has yellowish-green legs, while the latter has black legs (Godfrey 1986). It is also easily confused with the Semipalmated Sandpiper, but is browner and has lighter colored legs than that species (Salt and Salt 1976).
This bird is gregarious and is usually found in flocks. In flight, these flocks twist and turn, alternately flashing white bellies and dark backs. This species forages mainly by making pecking, rather than probing, movements (Johnsgard 1981). Its main foods are insects, their larvae and pupae, other invertebrates, and seeds.
Early spring migrants arrive in southern Alberta during the last two weeks of April, arriving in the north in mid-May. They usually leave the province by the end of May to continue north. Fall passage is extended, with migration from late June or early July to mid-October. No overwintering has been reported.

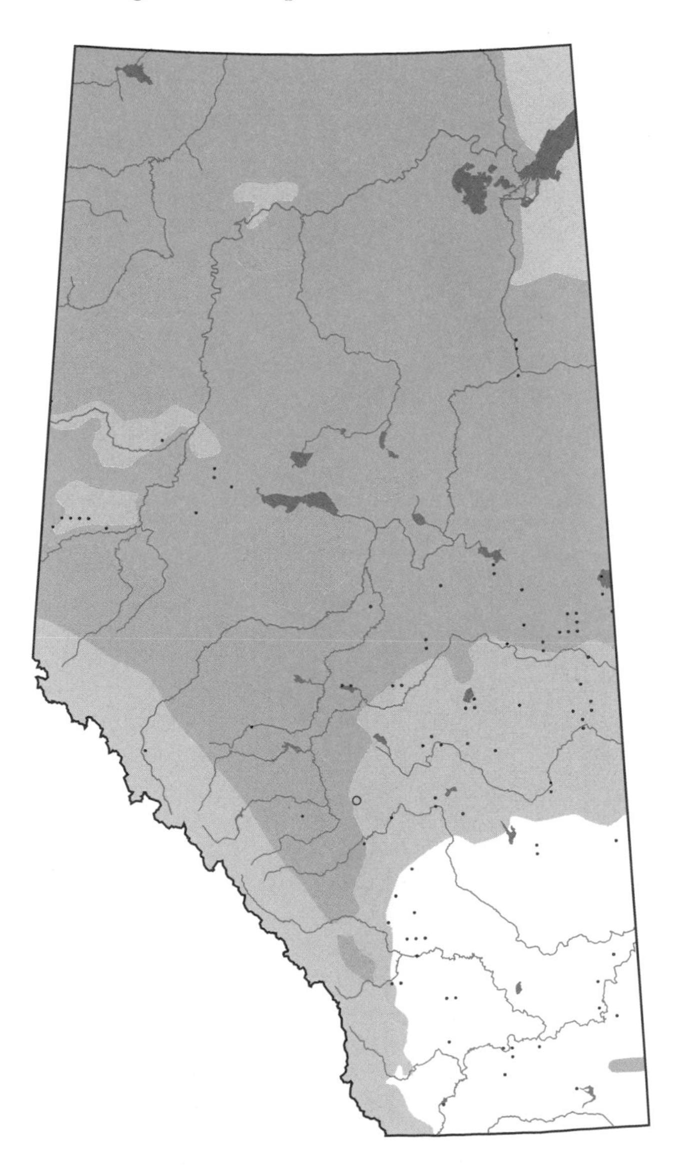

Short-billed Dowitcher

Limnodramus griseus

SBDO

Status:
This species has a widespread breeding range in Canada, but its total population is completely unknown (Johnsgard 1981). The status of the Short-billed Dowitcher in Alberta has not been determined. However, atlas data indicate that it is an uncommon breeder in the province, with 17 breeding records in the five year survey period and only four of these were confirmed breeding. Salt and Salt (1976) stated that the habitat preferred by this species in central and northern Alberta has been greatly reduced by drainage of marshes and small lakes for agricultural and other uses.

Distribution:
The subspecies found in Alberta, *L. g. hendersoni*, breeds from eastern British Columbia to Manitoba and north to the southern Mackenzie. In Alberta, birds breed very locally from about Edmonton north to the 60th parallel (Salt and Salt 1976; Godfrey 1986) and are found primarily in the Boreal Forest region. Atlas data concur with Salt and Salt (1976), who reported that this species is a transient elsewhere in the province. They further reported that there were no records for the Rocky Mountain region, but Pinel et al. (1991) reported that there are now several migration records for this region.

Habitat:
This species inhabits muskegs and similar boggy and marshy places with low vegetation (Godfrey 1986). In migration, they are usually seen at the soft, muddy margins of lakes and sloughs.

Nesting:
The nest is usually built on a dry hummock in muskeg, marsh, or a grassy meadow, but may also be found well-concealed in a tuft of grass or sedge cover. It is usually quite close to water, if not actually surrounded by it (Johnsgard 1981). The nest is a cup-like hollow or depression lined with moss, leaves, grass, and small twigs.

Clutches are typically four eggs, rarely five, which are greenish to olive-buff and spotted with browns. Incubation is 21 days, by both sexes, and the young are cared for mostly by the male. Nesting pairs remain close together and they will defend their nesting territory against intruders in concert (Johnsgard 1981). No fledging time is recorded for this species. One brood is raised per season.

Remarks:
This sandpiper has a straight bill that is twice as long as its head. While similar to the Common Snipe, it is distinguished from that species by its reddish breast, lack of buffy stripes on the back, and white patch on rump and back. It is so similar to the Long-billed Dowitcher that many individuals cannot be safely distinguished in the field (Godfrey 1986). Bill lengths are so similar between these species that only extremes can serve as useful indicators. Dowitchers with extensive white on the belly, no bars on the sides of the breast, and short to moderate bills are Short-bills (Godfrey 1986). Only the Short-billed Dowitcher nests in Alberta.

The diet of this species consists primarily of insects and their larvae, worms, and molluscs. It forages by probing with rapid bill movement while standing in shallow water.

First spring migrants arrive in early May. The Short-billed Dowitcher is an early migrant, with fall migration beginning in July and lasting, in some years, to late October. Breeding birds move south in the early part of migration; juveniles move at a later date. This species does not overwinter.

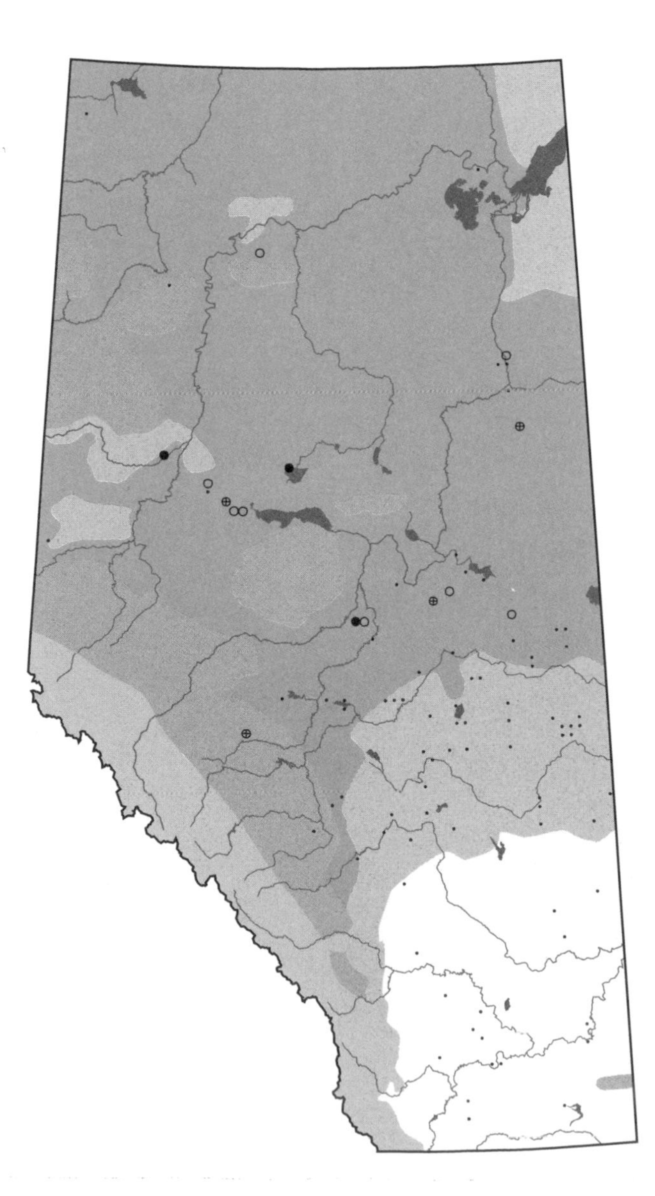

Common Snipe

Gallinago gallinago

COSN

Status:
This is the most abundant shorebird in North America (Tuck 1972). Alberta populations of the Common Snipe are thought to be stable. Atlas data indicate that it is a common breeder in the Parkland, Foothills, southern Boreal Forest regions and breeds locally in the other regions of the province. It is the only hunted species of shorebird in Alberta.

Distribution:
The Common Snipe has a holarctic distribution, breeding in temperate northern regions in Alaska, all of subarctic Canada, the northern United States, and subarctic Eurasia. It breeds in suitable habitat throughout Alberta and was recorded in 31% of surveyed squares in the Canadian Shield region, 38% in the Boreal Forest, 40% in the Parkland, 13% in the Grassland region, 46% in the Foothills, and 27% in the Rocky Mountain region.

Habitat:
In Alberta, this species is typically found in woodland bogs, fens, alder and willow swamps, grassy margins of sloughs, moist meadows, and similar places that provide soft mud and low cover. In the Rocky Mountain region, sedge and willow wetlands on stream floodplains are the most important habitat. In migration, the species also frequents ploughed and wet fields, damp stubbles, ditches, streams, and creek edges.

Nesting:
The nest is a shallow depression lined with grass, reeds or sedge, situated in tufts of grass or sedge, or on a hummock, usually in the drier parts of a bog or similar wet terrain. Nests are usually well-hidden in vegetation, and birds may weave a canopy of grass or sedges over the nest. The female builds several scrapes before selecting one for laying.
The clutch is usually four eggs which are greenish buff to brownish and are strongly marked with dark browns. Incubation is 18-20 days and is performed mainly, if not solely, by the female. The birds are very secretive during incubation, and the female sits very close to the nest. If flushed, she will perform an injury-feigning display. This bird is more sensitive to nest disturbance than any other nesting shorebird in Alberta, and often abandons the nest after a single visitation by humans (H.L. Dickson pers. comm.).
Young leave the nest when the down is dry, but they initially take food from their parents' bills (Harrison 1978). Chicks are tended by both parents, who may divide the brood between them, with the male normally taking care of the older chicks (Tuck 1972). The young fledge at 19-20 days.

Remarks:
The Common Snipe, formerly known as Wilson's Snipe, is a long-billed shorebird which can be recognized by its erratic zig-zag flight when flushed and by the hollow drumming sound of its "winnowing" aerial display in the spring.
The species uses its long, sensitive, and flexible bill to probe the mud for worms, insects, and other small invertebrates which form the bulk of its diet. Foraging is done in small areas with little walking; the bird often probes in a small semi-circle by pivoting its legs (Johnsgard 1981).
Spring migrants generally arrive in southern Alberta in the first three weeks of April, and towards the end of the month in northern Alberta (Pinel et al. 1991). In the fall birds may linger in the province until mid-November. Overwintering has been reported in Banff, Canmore, Waterton Lakes National Park, Calgary, Caroline, and Snake's Head.

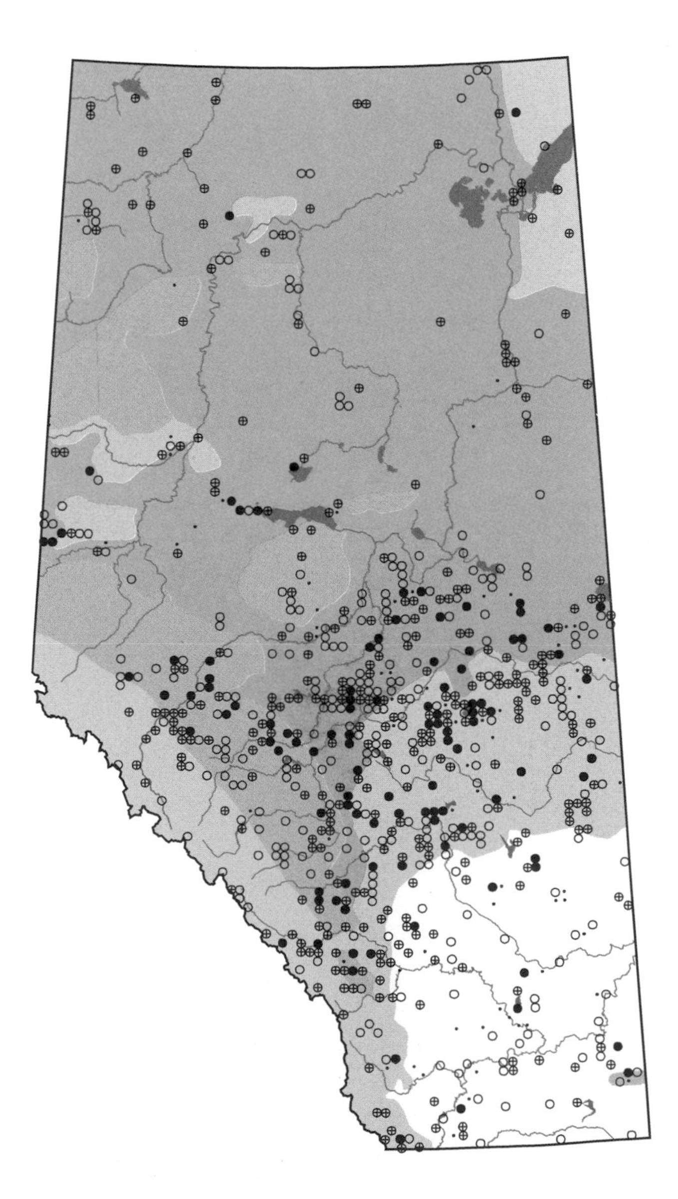

Wilson's Phalarope

Phalaropus tricolor

WIPH photo credit: G. Beyersbergen

Status:

This species is relatively common over much of its range and, unless drainage of its prairie marsh becomes much more serious, it does not seem to be in any immediate danger (Johnsgard 1981). Wilson's Phalarope, the largest of the three phalaropes in Canada, is a common breeder in the Grassland and Parkland natural regions of Alberta.

Distribution:

These shorebirds breed in the interior of North America, primarily in prairie areas. Salt and Salt (1976) described their Alberta distribution as nesting regularly in southern Alberta west to the foothills and north to Lesser Slave Lake and fairly commonly northwest to Hay Lake and Zama Lake. It was observed in Wood Buffalo Park and the delta of the Athabasca River. Godfrey (1986) expanded the breeding distribution to include the Foothills and Rocky Mountain regions and probably the Athabasca delta. Atlas surveys confirmed nesting in the Rocky Mountain region and the area north of Lake Claire. However, there was no evidence of breeding recorded in the northwest corner of the province.

Habitat:

This species favors sloughs and shallow lakes where wet meadows, grassy marshes, or short sedges are present.

Nesting:

Wilson's Phalarope nests on the ground, usually in wet meadows and grasses around the perimeter of a marsh or pond, sometimes in a hay meadow or pasture near water, or on an islet in a lake. The nest is a depression that may be either sparsely or well-lined with grass or sedges, and is often well-concealed by over-arching grasses. The male may make several scrapes before one is finally chosen by the female. This species is often loosely colonial and several pairs may nest in close proximity.

The female usually lays four eggs, which are buffy green and blotched with dark brown. Incubation is by the male alone, and lasts 16-21 days. He also tends the young, keeping them in dense cover until fledging. The fledging period for this species is not recorded. One brood is raised per season.

Remarks:

The genus name for this species, *Phalaropus*, means "coot-footed", referring to the lobes of skin along the sides of the toes (Höhn 1965). This is a trait the phalaropes share with the coots which assists their swimming. This species was named after Alexander Wilson; an early 18th century American ornithologist. Phalaropes' needle-like bills and slender necks distinguish them from sandpipers and the lack of a white stripe on the spread wing separates the Wilson's Phalarope from other members of this group.

The sex roles are reversed in the phalaropes, with the female being larger and having more colorful plumage than the male (female pictured above). She has a black stripe from the eye down the side of the neck merging into chestnut (Godfrey 1986). She initiates courtship, and then defends her mate from the intrusions of other females. After the female lays the eggs, the male takes on all major nesting duties, and the female goes off to find another mate.

Phalaropes are the only shorebirds that habitually swim but, unlike other phalaropes, the Wilson's is inclined to forage on land. It will be seen wading at the edge of a slough or in a nearby field, probing in the mud or grass for food. Its diet consists primarily of insects, plus a few crustaceans and the seeds of aquatic plants.

The first spring migrants of this species arrive in southern Alberta during the last week of April and the first week of May. The females arrive first, followed by mixed flocks of males and females. In the fall, the birds usually depart in August and September.

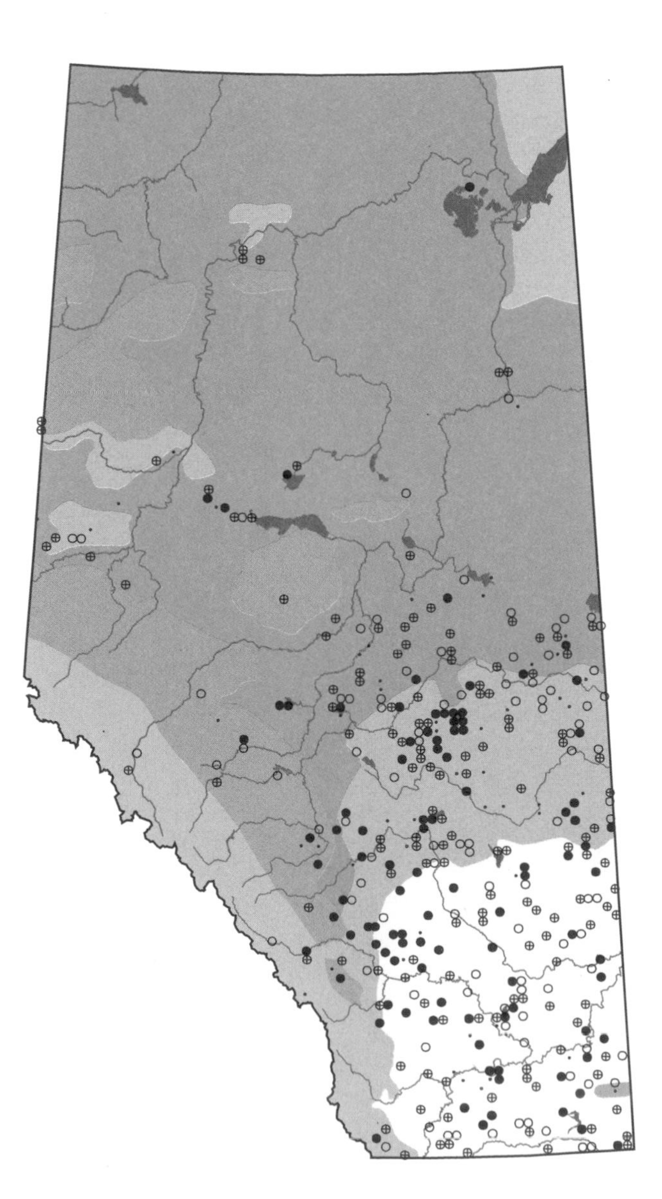

Red-necked Phalarope

Phalaropus lobatus

RDNP

Status:
This species has an extremely broad breeding range that is beyond the range of most human activity, and its wintering areas are primarily oceanic, so it probably has a large, though inestimable, total population (Johnsgard 1981). The first confirmed nesting record of the Red-necked Phalarope in Alberta was in 1979 (Höhn and Mussel 1980). Worldwide, this is the most abundant and widely distributed of the phalarope species.

Distribution:
The Red-necked Phalarope has a circumpolar distribution with the northern edge of the province at the southern extremity of its breeding range. It is a regular spring and fall migrant in the province. The only confirmed nesting record in Alberta is of a male with egg, plus nest, near Rock Island Lake in the Caribou Mountains (Höhn and Mussel 1980). Other sightings in this area (Höhn and Marklevitz 1974; Höhn and Burns 1975) also suggest breeding. There were no breeding records for this species in the five year survey effort of the Atlas Project.

Habitat:
This species breeds in tundra and tundra-forest transition zones, preferring small lakes, pools, bogs, and marshes where adjacent sedges and mossy hummocks provide suitable nesting sites. During migration in Alberta, they prefer areas of shallow water far out from shore (Höhn 1981).

Nesting:
The nest of the Red-necked Phalarope is a small hollow lined with grass and leaves. It is usually built in a grassy tussock, on moss or among sedges, and is found near water. Several scrapes may be made by male or female before nesting.
The female lays four eggs, which are olive buff and spotted with brown. Incubation is variously reported, 17-23 days, by the male alone. Chicks leave the nest within a few hours of hatching, are tended by the male alone, and fledge at 20 days. The male abandons the chicks at about 14 days. Serial polyandry, where females lay two successive clutches for two different males, is a regular feature of this species, provided excess males are available (Johnsgard 1981). Hilden and Vuolanto (1972) reported that females may lay two clutches in a single season of an early nest is destroyed and the male is able to remate either with its original mate or a new one, or if there are excess males in the population available for incubating a second clutch.

Remarks:
This species, formerly known as the Northern Phalarope, is a small, graceful shorebird whose breeding plumage includes a chestnut band across the breast and side of the neck. The lack of extensive red on the underparts and a very slender, needle-like bill distinguish this species from the Red Phalarope, which is an uncommon migrant in Alberta. A white wing stripe visible in flight separates it from the Wilson's Phalarope. The female is pictured above.
Like other phalaropes, the Red-necked does much of its foraging while swimming, catching its prey while spinning and rotating. Its primary foods are invertebrates, including flies, mosquito larvae, aquatic insects, and small crustaceans. The spring passage for this species is early May to early June, with numbers peaking during the last two weeks of May. During this period, birds may often be observed on lakes in east central Alberta. Fall migration takes place from early July to mid-October (Pinel et al. 1991).

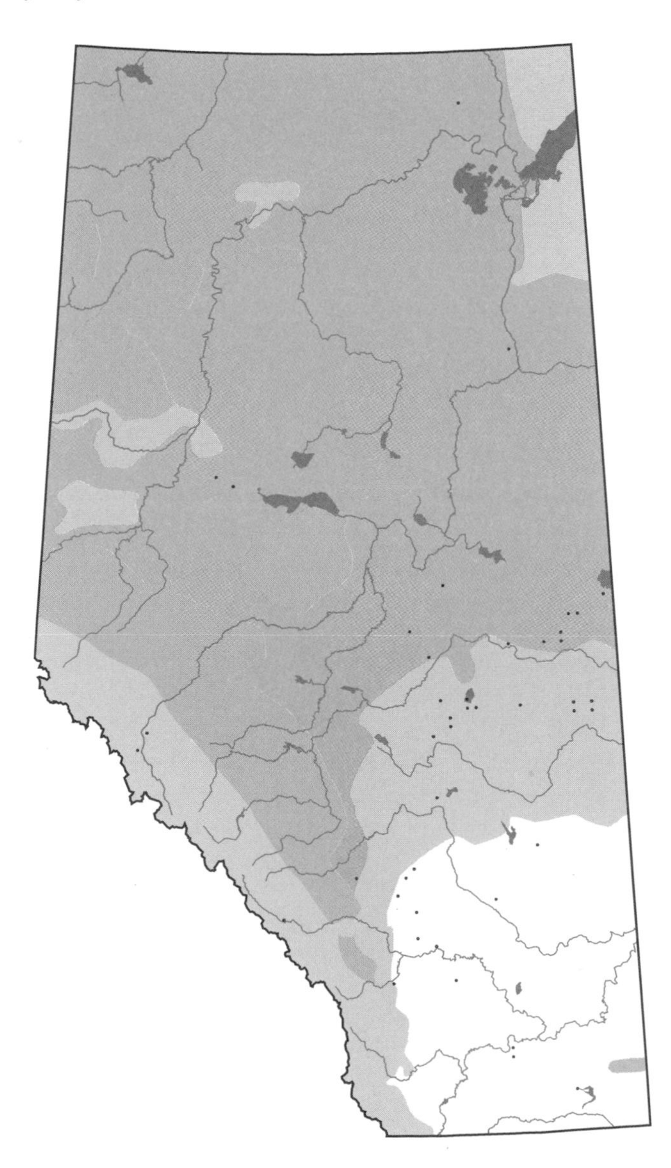

Franklin's Gull

Larus pipixcan

FRGU photo credit: R. Gehlert

Status:

Pesticide use is blamed for the considerable reduction of this species' population on the prairies in the 1930s and 1940s (Koonz and Rakowski 1951). Its present status in Alberta is unclear, but it may be in decline (Wildlife Management Branch 1991). As a marsh-nesting species, Franklin's Gull is susceptible to agricultural practices which affect wetlands, including marsh drainage, cultivation to the shoreline and stabilization of water levels resulting in marsh stagnation. Human disturbance at critical times during the nesting period will result in abandonment of the nest (Guay 1968).

Distribution:

Franklin's Gull breeds commonly in Alberta east of the mountains. It is found primarily in south and central Alberta, but, since the 1950s and 1960s, it has been expanding its range and is now becoming more numerous in areas of northern Alberta. It was recorded in 17% of surveyed squares in the Parkland region, 13% in the Canadian Shield, 11% in the Boreal Forest, and 6% in the Grassland region. The most northerly nesting colony of this species in North America was reported at Loutit Lake, near Lake Athabasca (Francis and Lumbis 1975). This was confirmed by Atlas records for this area.

Habitat:

This species frequents larger, reedy lakes and marshes, and forages over water, grassy fields, meadows, ploughed land, and other open areas. During migration, the Franklin's Gull has been observed at landfill sites and other sites of human refuse (McNicholl 1978).

Nesting:

Franklin's Gulls nest in large marshes or along the marshy margins of lakes, usually near open water. They nest close together in large colonies that often number several thousand birds.

The location of these colonies often shift annually according to water levels. The nest of this gull is a floating mass of old reeds, usually with a well-formed cup, and anchored to nearby emergents. The nest is added to throughout the nesting period.

The female lays 1-3 eggs, usually three (Guay 1968). These eggs are olive with blotching of dusky brown to black. Incubation is conducted by both sexes and is variously reported. The common incubation period in central Alberta is 21-28 days and, farther south, the typical period is 21-25 days (Guay 1968).

Young birds leave the nest after two days, and can swim at three days. Very young vagrant chicks may be retrieved and taken in by surrogate parents, but older ones are not tolerated and may be pecked to death by adult birds. The young are tended by both parents, and are fed by them for 10 days, after which they increasingly feed themselves. Birds fledge at 28-30 days. This species raises one brood per season.

Remarks:

This is a common black-headed white-bodied gull of the Canadian prairies. It has a gray wing with a white band bordering the black wing tips, and a dark red bill. The Franklin's Gull feeds primarily on insects, including some regarded as pests by prairie farmers. Flocks of wheeling gulls often follow the plough, hawking airborne insects, or searching for larvae on the ground.

This species arrives in southern Alberta in the first two weeks of April and later in the month in the north. From late July to late September, these gulls are observed in high numbers on lakes in southern Alberta. Most birds have left the province by the end of October (Pinel et al. 1991). No overwintering is recorded for this gull.

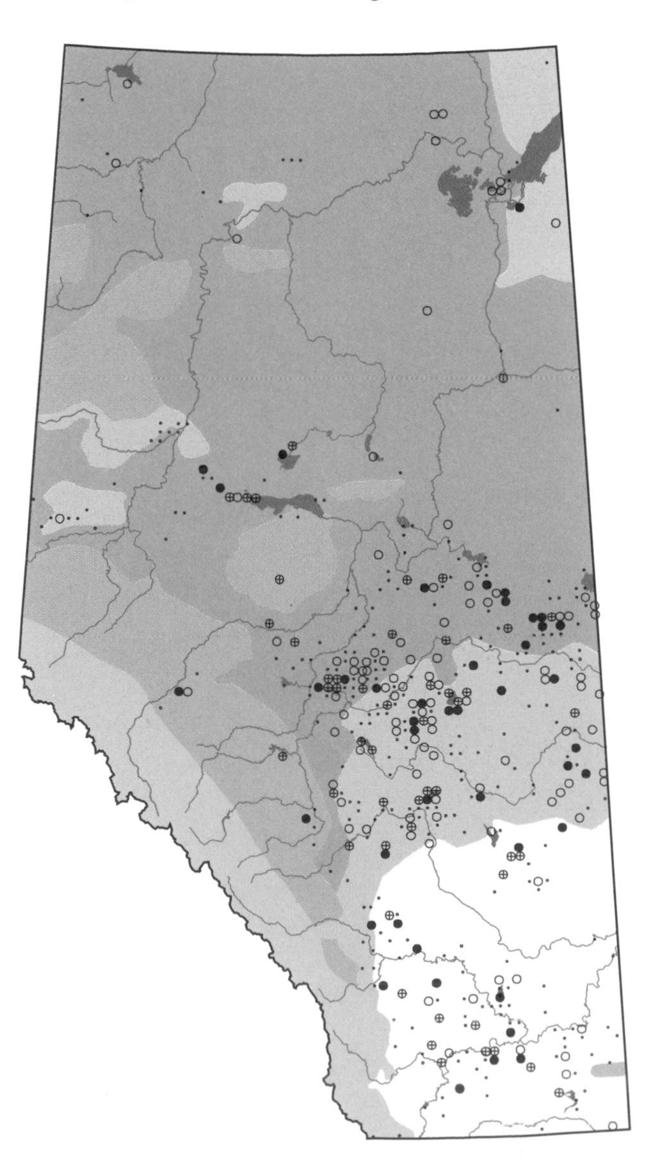

Bonaparte's Gull

Larus philadelphia

BOGU

Status:

The status of this species in Alberta is presently unknown, though it is possibly declining in some areas (Wildlife Management Branch 1991). The small muskeg lakes upon which Bonaparte's Gull breeds are the kind of habitat that will disappear with the cutting of northern forests.

Distribution:

Salt and Salt (1976) reported that the breeding distribution of Bonaparte's Gull was widespread but local in the forested areas of central and northern Alberta, from Battle Lake and Edmonton in the south, north to the 60th parallel. Atlas data confirm this report, with the addition of breeding records in the southwestern portion of the Boreal Forest region south to Caroline. This species was recorded in only 4% of surveyed squares in the Parkland Natural Region, and was not recorded as breeding in the Grassland region. There also were no breeding records for the Rocky Mountain Natural Region. Holroyd and Van Tighem (1983) reported this gull as a rare migrant and a very rare summer visitor. Extralimital nesting has been reported from Waterton Lakes and Miquelon Lakes (Pinel et al. 1991) and Camrose (Farley 1931).

Habitat:

Bonaparte's Gull breeds in the vicinity of lakes, ponds, and muskeg in the coniferous woodlands.

Nesting:

Nesting is in conifers near water, with small muskeg lakes preferred. Normally, there is only one nesting pair per lake. The nest is made of twigs, mosses, and lichens, and is lined with bark, moss reeds or grass. It is placed on a sturdy branch close to the trunk, 2-7 m up from the ground. Both sexes build the nest and strenuously defend the nest site against intruders (Twomey 1934). Salt and Salt (1976) also reported nests of reeds in the reeds and rushes of a marsh. These birds may nest together in colonies of 3-4 pairs, or birds may nest singly, with one pair per pond. Bonaparte's Gulls have been observed in a mixed colony with Franklin's Gulls at Loutit Lake (Francis and Lumbis 1979).

Clutches are 2-4 eggs, usually three. The eggs are olive-gray and are marked with browns and lavender. Incubation is 24 days by both sexes. The chicks tumble to the ground a few days after hatching and are taken to water. Fledging time is uncertain. By the end of July or early August, most birds have left their breeding ponds and have assembled on larger lakes. Bonaparte's Gulls raise one brood per season.

Remarks:

This small, graceful gull is likely to be confused only with Franklin's Gull in its breeding area. Bonaparte's Gull has a black bill (the Franklin's is red), and the triangle formed by the extended primaries is white in the Bonaparte's, dark in the Franklin's. In the fall, the black ear spot of the Bonaparte's is diagnostic. The flight of the Bonaparte's Gull is light, buoyant, and very tern-like.

Main food items for this gull are insects, worms, other small invertebrates, and small fish. Birds will pick up food while swimming, hawk insects delicately from the air, or drop onto insects and fish from a hover. This is one gull that is not found at landfills.

Bonaparte's Gulls arrive in Alberta in late April or early May, and return to southern Alberta in late July and August (Pinel et al. 1991), gathering with other waterbirds in preparation for fall migration. Most stragglers have left the province by November. No overwintering is reported for this species.

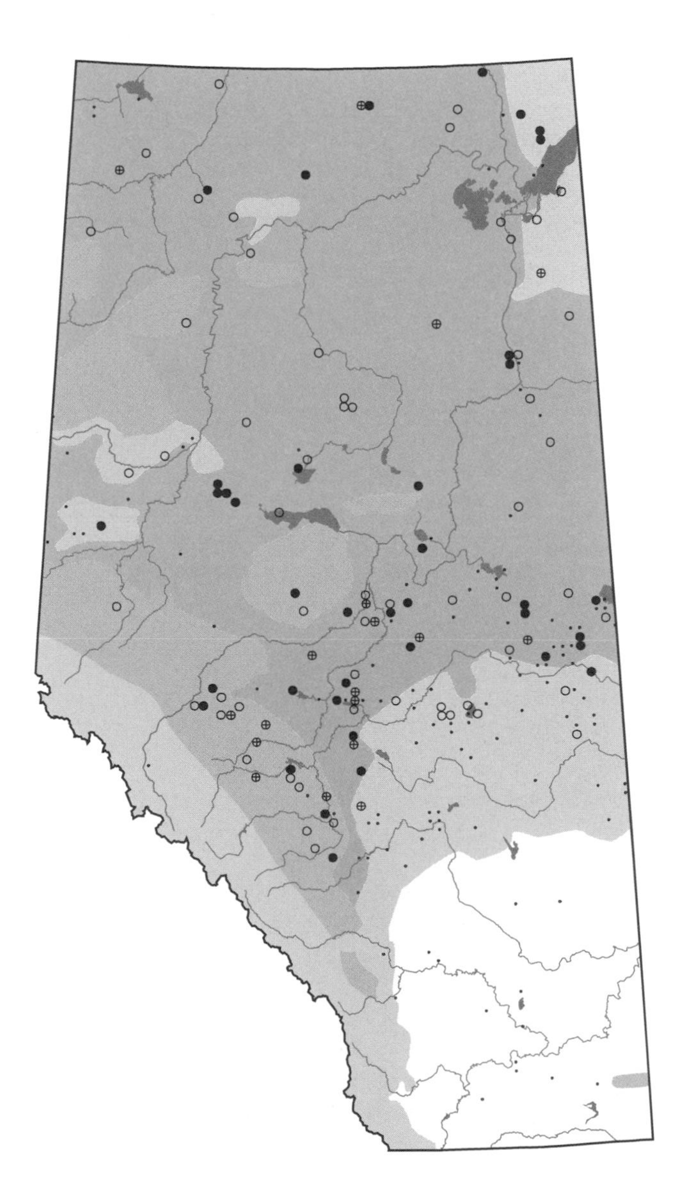

Mew Gull

Larus canus

MEGU

Status:
Mew Gulls likely have a stable or increasing population in their North American range. In Alberta, they are very rare and local breeders, being at the periphery of their range.

Distribution:
This species breeds from the Mackenzie River drainage south into northern Alberta and Saskatchewan. The birds are common summer residents on many remote northern Alberta lakes, but most individuals seen are subadults. There is breeding evidence from Margaret Lake in the Caribou Mountains, Fort McMurray, Bocquene Lake, Rock Island Lake, and Eva Lake (Pinel et al. 1991). Holroyd and Van Tighem (1983) reported this gull as a very rare migrant and very rare summer resident in the Rocky Mountain region. In southern Alberta, this gull is a casual transient that can be observed in migration at Calgary and Edmonton landfill sites. During the five year survey period of the Atlas Project, the Mew Gull was recorded on five occasions, with one confirmed breeding record from the Canadian Shield Natural Region in the extreme northeast corner of the province. The other four records were of possible breeding and were also all in the northeast corner of Alberta. Three of these records were from the Boreal Forest region.

Habitat:
In breeding season, Mew Gulls frequent marshy areas, shallow lakes, ponds, and rivers in Alberta's northern Boreal Forest and Canadian Shield regions. During migration, they are likely to be seen at urban landfill sites.

Nesting:
This small, white-headed gull breeds singly or in small colonies. Nests are usually built on the ground, on the marshy or rocky shores of lakes and ponds and their islands. Occasionally, some birds may nest on tussocks or in trees near the shoreline. The nest is built of twigs and coarse rushes or grasses, mainly by the female. In some areas, the Mew Gull shares its nesting site with other gulls.
The female lays 2-5 eggs, typically three. The eggs are buffy to green-brown and are blotched with browns and grays. Incubation is 23-26 days by both sexes. Chicks leave the nest after a day or two, but remain close by, where they are tended by both parents. Young fledge at four to five weeks. One brood is raised per season and young often return to breed at the natal colony.

Remarks:
Named for its distinctive "mewing" cry, the Mew Gull is sometimes difficult to distinguish from other white-headed gulls, particularly the Ring-billed Gull. Unlike the latter, the Mew Gull has a short, unmarked bill which is entirely yellow, and it shows more white on its wing tips. The Mew Gull was formerly known as the Short-billed or Common Gull.
The prey of this gull includes insects, other invertebrates, and small fish. It dives for fish more than other gulls, but it rarely, if ever, plunges below the surface. In some parts of its range, the Mew Gull is known for its habit of dropping molluscs on the rocks from high in the air in order to crack them open. This gull also feeds at Alberta's garbage dumps, mostly during migration.
Records show the Mew Gull arriving in the Fort McMurray area in late April, and in fall migration birds are seen in southern Alberta as late as early November (Pinel et al. 1991). There are no records of overwintering.

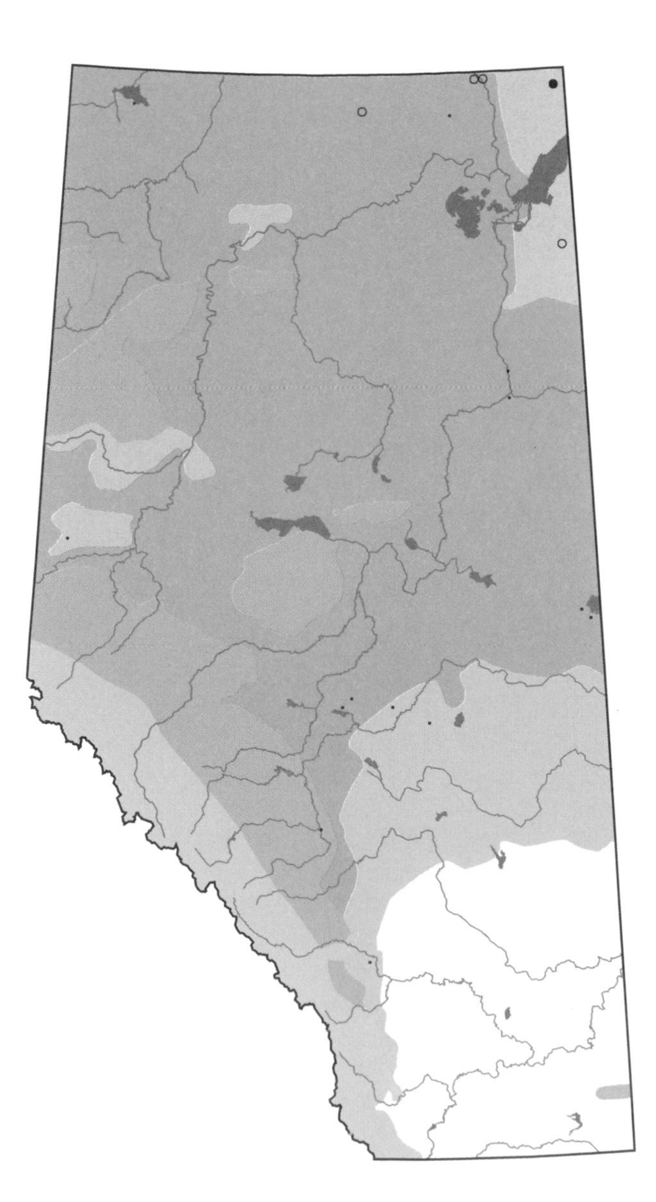

Ring-billed Gull

Larus delawarensis

RBGU photo credit: T. Webb

Status:

In the late nineteenth century, the Ring-billed Gull suffered a population decline because of shooting for its feathers. Since protection in 1916 by the Migratory Birds Treaty, the species has reoccupied much of its former range in North America, and is now increasing rapidly and expanding its range. The success of this species is the result of its adaptability to humans, their agricultural practices, refuse disposal, and urban habitat, plus the species' long life span and high reproductive success. It is now the most abundant gull in North America and is causing serious management problems in some areas.

Distribution:

Ring-billed Gulls breed locally throughout eastern Alberta, from Pakowki Lake in the south, north to Athabasca, and west to Brooks, Pigeon Lake, Lac la Biche, and the Caribou Mountains (Salt and Salt 1976). Large colonies with more than 1,000 pairs exist in central and southern Alberta at Miquelon Lake, Lake Newell, Keho Lake, Frank Lake, Irricana Reservoir, Dowling Lake, and Buffalo Lake. It is found in suitable habitat throughout the Grassland, Parkland, Boreal Forest, and Canadian Shield regions.

Habitat:

This species breeds on the shores and islands of larger freshwater and alkaline lakes and impoundments, and forages over fields, mudflats, water, and landfill sites.

Nesting:

Ring-billed Gulls are social birds which nest in dense colonies on isolated and sparsely vegetated islands and beaches of large waterbodies. Colonies can become very large, with over 10,000 pairs. These birds often share nesting sites with a variety of colonial birds. Ring-billed Gulls return to previous nesting territories but will also colonize new areas.

The nest of this species is on the ground, usually on higher land, and is a simple scrape made with a lining of small sticks, weeds, grasses, feathers, and debris. Most nests are in the open, but some are partially hidden in vegetation or rocks. Both sexes build the nest.

Clutch size is 2-4, typically three. Eggs are buffy to gray and are marked with dark brown and lavender. Both sexes help with incubation, which lasts about 25 days (Vermeer 1970). Young remain in the nest a few days after hatching, but soon learn to run about and hide in nearby cover. Chicks wandering far from their parents' territory may be pecked to death by neighboring adults. The young learn to swim at an early age and are fed until they fledge at 37 days (Vermeer 1970). One brood is raised per season.

Remarks:

This common, white-headed gull is very similar to the Herring Gull and California Gull, but differs from these by being smaller and having a black ring around its yellow bill. It also has yellowish legs and feet, while the Herring Gull's feet are pinkish.

An opportunistic feeder, the Ring-billed Gull eats insects, rodents, grains, the eggs and nestlings of other birds, (usually shorebirds and waterfowl), road-kills, and various items scavenged from landfill sites. In late summer and autumn, it is a common sight at garbage dumps and other urban locations such as parking lots, drive-ins, parks, and golf-courses. These birds are also known to congregate where there are high populations of grasshoppers.

Ring-billed Gulls arrive on the breeding grounds in late March and early April, often in the company of California Gulls. The last fall migrants are seen in the province late in November. Overwintering has been reported from Edmonton and Waterton Lakes.

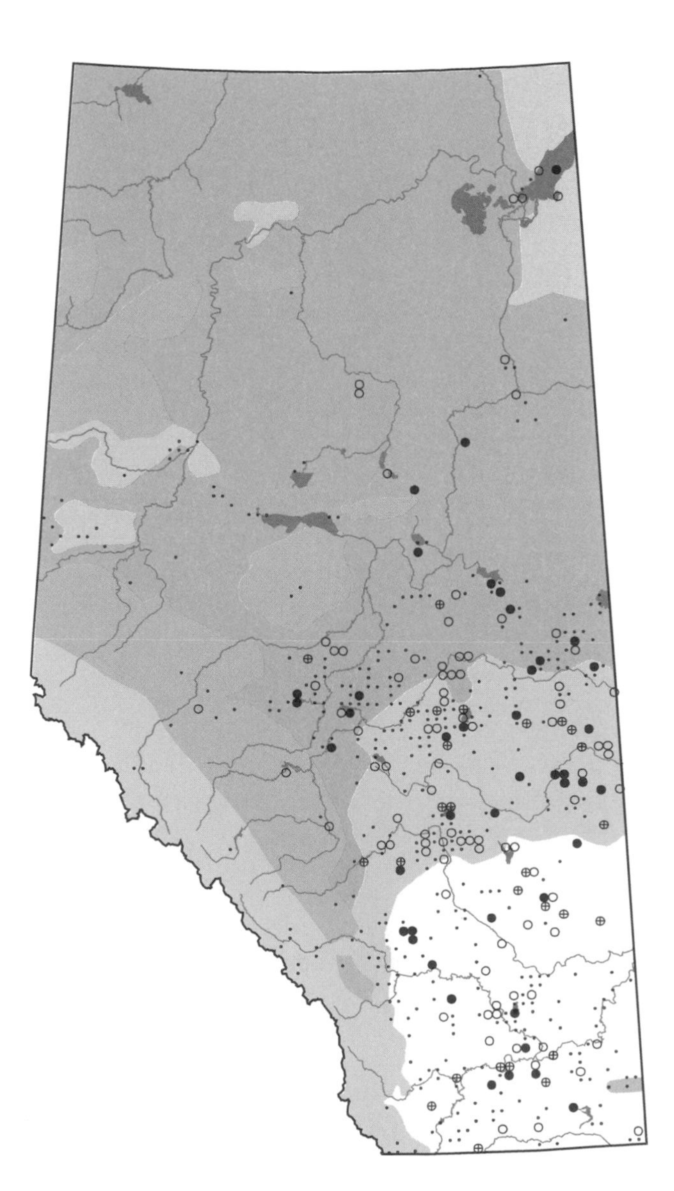

California Gull

Larus californicus

CAGU

Status:

Populations of this species in Alberta are healthy and widespread, and their key habitat is relatively secure (Wildlife Management Branch 1991).

Distribution:

Salt and Salt (1976) reported that the California Gull breeds locally throughout eastern Alberta from Lake Athabasca in the north to Pakowki Lake in the south, and west to Brooks, Calgary, and Miquelon Lake. Nesting evidence beyond this range has been reported northwest at the Wapiti River south of Grande Prairie, west to Edson and the Brazeau Canal Reservoir, and south to Shanks Lake, Ross Lake, and St. Mary's Reservoir (Pinel et al. 1991). Colonies of over 10,000 pairs can be found at Irricana Reservoir, Lake Newell, Miquelon Lake, St. Mary's Reservoir, Keho Lake, and Chip Lake. Atlas data confirmed breeding in all natural regions except for the Foothills and Rocky Mountain regions. It was recorded in 18% of surveyed squares in the Canadian Shield region, 9% in the Boreal Forest, 11% in the Parkland, and 5% in the Grassland region. Approximately 50% of the confirmed breeding records were in the Boreal Forest Natural Region.

Habitat:

This gull breeds in the vicinity of freshwater and alkaline lakes, marshes, and rivers, but may also be seen around freshly ploughed fields and city dumps.

Nesting:

Preferred nesting habitat for the California Gull is a treeless island of a lake, marsh, or river where spring vegetation is sparse. At Miquelon Lake, ridges of boulders and debris formed by windblown ice were favored by nesting California Gulls. This gull usually nests in colonies in the company of other water birds, particularly the Ring-billed Gull. The nest is on the ground, either in the open or in low vegetation. It is a scrape with sparse to substantial lining of sticks, grass, weeds, feathers, and rubbish. Both sexes help with nest building.

Two to three eggs are laid. They are white to buff, and are marked with brown and black or violet. Incubation, by both sexes, averages 25-27 days (Vermeer 1967; Harrison 1978). Young are also tended by both male and female, and hatchlings leave the nest and swim after a few days (Harrison 1978). Fledging of this species is at approximately 40 days (Vermeer 1970). The California Gull is single-brooded.

Remarks:

The California Gull is intermediate in size between the Herring Gull and the Ring-billed gull, all of which have a generally similar breeding plumage (Salt and Salt 1976). The most important diagnostic features of the adult California Gull are a dark eye color, as opposed to the light eye color of the Ring-billed and Herring gulls, red (or sometimes black) on the lower mandible of the bill, and a yellow leg color (Weseloh 1972). First year California and Herring gulls are very similar in appearance, and do not attain adult plumage until the fourth year (Godfrey 1986).

The California Gull is an opportunistic and omnivorous species. During the breeding period it feeds on insects, rodents, nestlings and eggs, grains, refuse and the occasional fish and amphibian. It may also follow the plough with Franklin's and Ring-billed Gulls.

This species arrives in Alberta the last two weeks of March, or early April, often before the snow has left the ground. Fall migration begins in August, with the last migrants recorded in late October or early November (Pinel et al. 1991). Overwintering has been recorded at Edmonton.

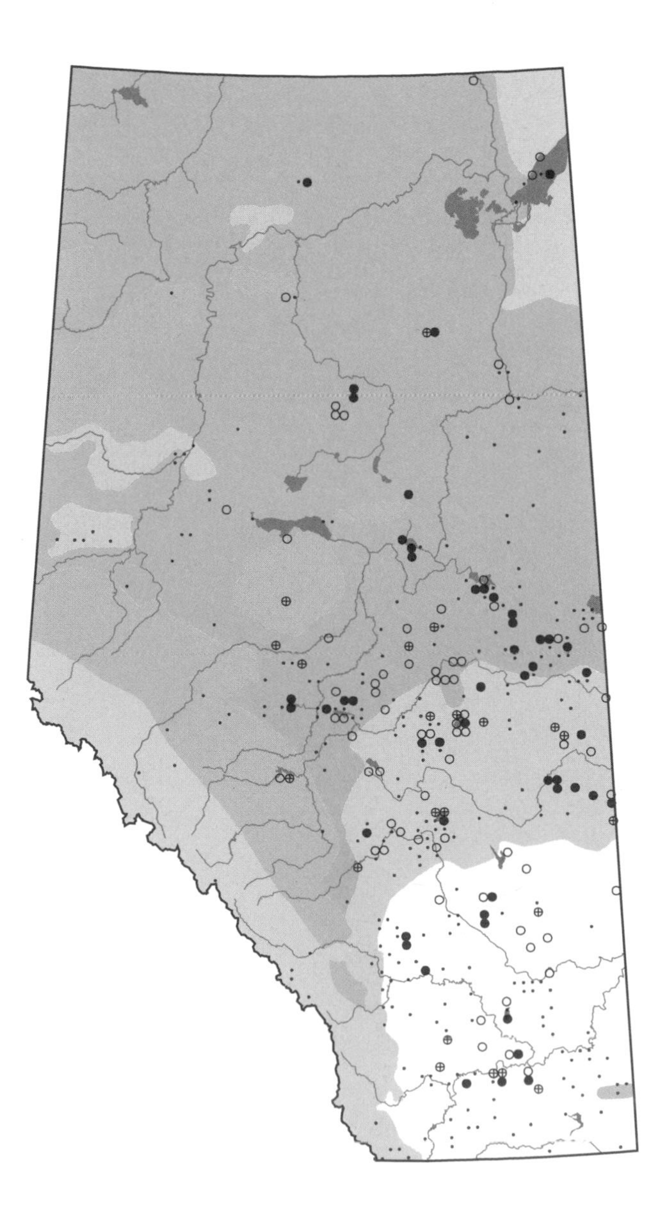

Herring Gull

Larus argentatus

HGGU

Status:

Egg collecting and the milliner trade had a profound effect on this species in the late 1800s, and populations suffered major declines at that time. Protection under the Migratory Birds Treaty of 1916 allowed the bird to subsequently recover, and to expand its range and breeding numbers. The species is now the most widespread gull in North America and is presently the most numerous, next to the Ring-billed Gull. It has profited greatly from human activity and waste and, in some areas, this speciesis undergoing a major population expansion.

In Alberta, Herring Gull populations are stable and likely increasing, and this species is possibly expanding its range (Sadler and Myers 1976). Its breeding habitat is restricted in the province, so protection of nesting colonies is key to its management (Wildlife Management Branch 1991).

Distribution:

Herring Gulls breed only in northern and northeastern Alberta, with nesting birds or colonies known at Bistcho Lake, Rock Island Lake in the Caribou Mountains, Wentzel Lake and Lake Athabasca south to Namur Lake and Lower Therien Lake. Extralimital breeding has been reported at Lesser Slave Lake, Lac la Biche, Calling Lake, Cold Lake, Beaverlodge, and the Bow River in Banff National Park. In south and central Alberta, these gulls are seen most commonly as transients in spring and fall at local landfill sites.

Habitat:

This species breeds in the vicinity of large freshwater lakes and rivers, and forages over lakes and ponds, open areas, farmland, and garbage dumps.

Nesting:

Herring Gulls will nest as single pairs or in colonies of various sizes, often in the company of other colonial species including gulls, terns, and cormorants. Nesting is usually on islands, rocky promontories, or large boulders in larger lakes and rivers. Birds return to previous nesting territories when possible.

Nests are most often built on the ground in areas with or without vegetation, and usually in the open. Occasionally, these gulls will nest in trees, but always near water. The nest is a sparse to bulky mound of twigs, moss, grass, weeds, and feathers, with a shallow to deep cup and a large base. Both birds build the nest.

Clutches are generally 2-3 eggs which vary in color from a blue gray to brown. The eggs are blotched with dark brown and lilac. Incubation is conducted by both sexes, but primarily the female, and is variously recorded at 25-28 days. Herring Gulls will take in wandering chicks up to 4-5 days old; after this, hatchlings receive a ferocious response from neighboring adults. Young are tended by both parents and fledge at 35 days. They are fed by the adults for up to 40 days after fledging. Herring Gulls are single brooded and mate for life.

Remarks:

This large, conspicuous gull is difficult to tell from the California Gull or the Ring-billed Gull, unless the Herring's pink legs and the absence of black on the bill can be seen. Herring Gulls are mainly scavengers, but will take fish, mice, frogs, insects, crustaceans, molluscs, and berries, plus the eggs and young of other birds.

First spring migrants arrive in the province from mid-March to mid-April, with a peak in the third week of April. In the fall, birds migrate south from early August to the third week of November. Overwintering is reported from Waterton Lakes.

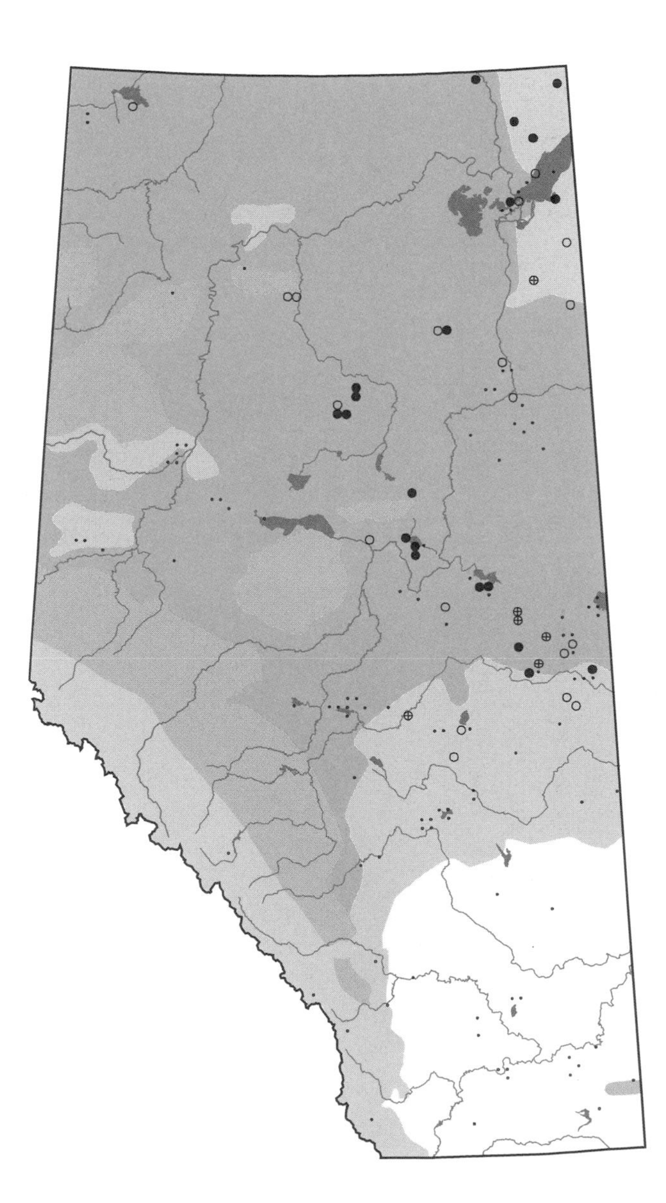

Caspian Tern

Sterna caspia

CATE photo credit: R. Gehlert

Status:

The Committee on the Status of Endangered Wildlife in Canada (COSEWIC) considers the Caspian Tern to be a "vulnerable" species in this country. The species is sensitive to water pollution, changes in water levels in nesting habitat, and disturbance of the nest site, particularly in the early stages of breeding.

Available historical data indicate that there has been no reduction in the breeding range of the Caspian Tern in Canada, but there have been population fluctuations in some areas. The tern is a rare breeder in Alberta, with approximately 50 pairs. This represents about 1% of the total Canadian population of Caspian Terns (Martin 1979). Only two breeding colonies are known to exist in the province, so protection of these sites is necessary to ensure a viable population (Wildlife Management Branch 1991).

Distribution:

The Caspian Tern has a worldwide, but highly disjunct breeding distribution. In Canada, most of the population exists in a few large colonies in Manitoba and Ontario. In Alberta, two small colonies are known, in two widely separated areas. One colony is at Lake Newell in the south, and the other is at Egg Island in the northwest part of the province. A single pair with young was observed on Lost Lake, northwest of Vauxhall, and there have been other sightings of this species in this general area.

Habitat:

This species frequents large lakes rich in small fish. Nesting habitat is usually a small, isolated island with little or no vegetation.

Nesting:

Caspian Terns nest on sandy or cobbly beaches on higher ground, most often on islands. They may nest singly, although they usually do so in large, tightly-packed colonies, often in the company of other terns or gulls. The nest, built by both sexes, is a depression in the ground, unlined or with a scanty lining of grasses.

When a female has reached at least 4-5 years of age, she lays 2-3 eggs, which are buffy and lightly spotted with browns. Incubation is 26 days (Martin 1978), by both sexes. Young are tended by both parents and fledging is at 37 days. Juveniles may be fed for up to seven months after fledging, the longest duration for parental care among the terns. Caspian Terns retain their mates between years and are single-brooded. Young stay on the wintering grounds for their first year and, where the population is stable, the terns will return to their natal colony to breed.

Remarks:

This tern can be distinguished from other Alberta terns by its large size (that of a medium-sized gull), slow wing beats, slightly forked tail, and fierce temperament. Despite breeding colonially, it is, otherwise, the least gregarious of the terns. It feeds almost exclusively on fish, which it takes by hovering and diving. Birds are most often seen hunting over deep water with their beaks pointing down. Caspian Terns may forage up to 70 km away from their nesting colony. They will occasionally steal food from others.

It is thought that Caspian Terns return to the province in late April or early May, though the first sightings of this species have been in June. The birds leave the province by mid-September. This species is a summer resident only, and no overwintering has been recorded. Winters are primarily spent around coastal and inland waterbodies in Columbia and Venezuela.

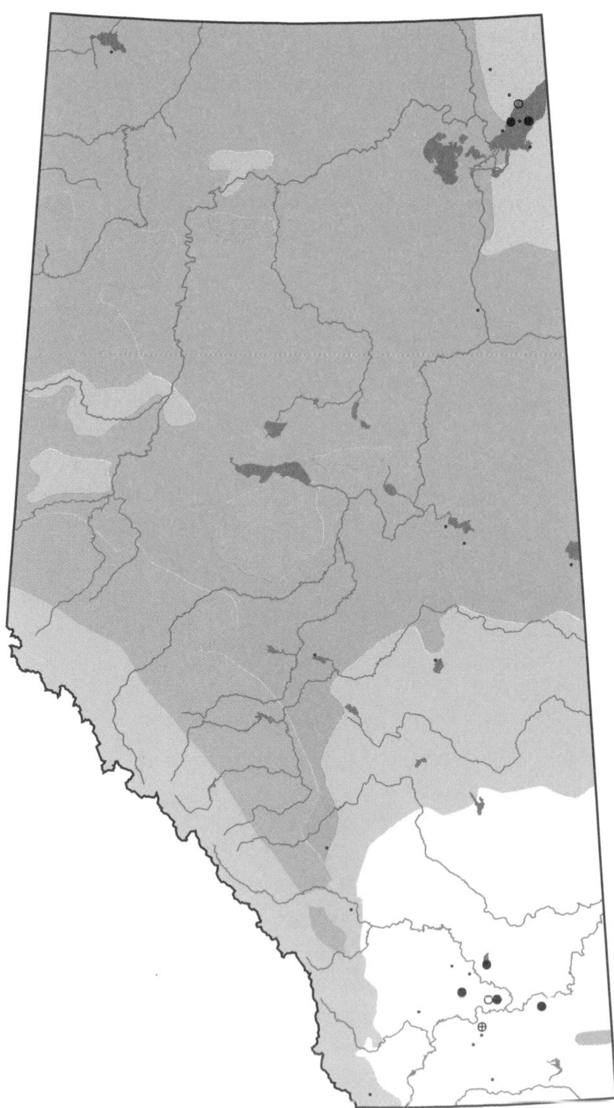

Common Tern

Sterna hirundo

COTE photo credit: R. Gehlert

Status:

In the 19th century, the North American population of Common Terns declined dramatically when the species' eggs were exploited for food, and its feathers for the millinery trade. The Migratory Birds Treaty Act (1918) in the United States and the Migratory Birds Convention Act (1917) in Canada protected terns and allowed their recovery.

While these terns are commonly observed in southern and central Alberta, they are still under pressure, and are considered to be in decline in some parts of their range. They must compete with rising populations of gulls for nesting sites and face problems of human intrusion, shoreline development for recreation, water pollution, and water level fluctuations. They are also susceptible to the effects of chemical contamination (Fox 1976), particularly in their wintering range in South America.

This species was on the Blue List from 1978 to 1981; was down-listed to a species of "special concern" in 1982, and was considered of "local concern" in 1986.

Distribution:

In Canada, the Common Tern breeds from the foothills of the Rocky Mountains in Alberta east to the Atlantic, except for the Arctic. It breeds locally throughout much of Alberta, with most colonies in the central and southern parts of the province. Colonies have been recorded at Buffalo Lake, Namaka Lake, Eagle Lake, Lake Newell, and Pakowki Lake in the south, at Lac Ste. Anne, Beaverhill Lake, and Miquelon Lake in Central Alberta, and at Barstall Lake and St. Agnes Lake in the north. Atlas data confirm historical distribution, as well as adding Bistcho and Andrew lakes as breeding locations. It was recorded with similar frequency in the Grassland (6%), Parkland (7%), and Boreal Forest (7%) natural regions. Seemingly quite common in the Canadian Shield Natural Region (50%), this may be more attributable to a small sample size or habitat bias.

Habitat:

This species is found in the vicinity of large bodies of open water, rivers, irrigation canals, and creeks, which it patrols for food. It breeds on islands, peninsulas, and shorelines of lakes, and, occasionally, in marshy situations.

Nesting:

Common Terns are ground nesters that prefer the open edge of a sandy or gravelly beach, or a flat area on an island, often near an edge of sparse low vegetation. The nest, built by both sexes, is a shallow depression in sand or gravel, which is lined with bits of grass or reeds. These birds occasionally nest in marsh habitat, on platforms of aquatic vegetation, or on muskrat houses. Common Terns may nest singly, but they are usually found in colonies, often in association with Ring-billed Gulls, California Gulls, or other terns.

The female lays 2-4 eggs, usually three. The eggs are pale buff, brown, or olive and are blotched with browns and purplish gray. Incubation is by both sexes, mostly the female, and lasts 21-30 days. The young are tended by both parents. They may leave the nest after a few days, but will return for brooding. They swim at an early age and fly at 28 days. Common Terns are single-brooded.

Remarks:

This graceful and elegant species feeds primarily on small fish. During the early phase of courtship, the male displays his fishing skill by holding a fish in his beak and flying back and forth over the colony, uttering a "fish" call. When he is joined by a female, the two will fly in parallel, banking and turning in synchrony, and often soaring to great heights (Severinghaus 1983).

Common Terns arrive in spring in late April to early May. Most leave again by mid-September, with a few lingering to October. No overwintering has been reported.

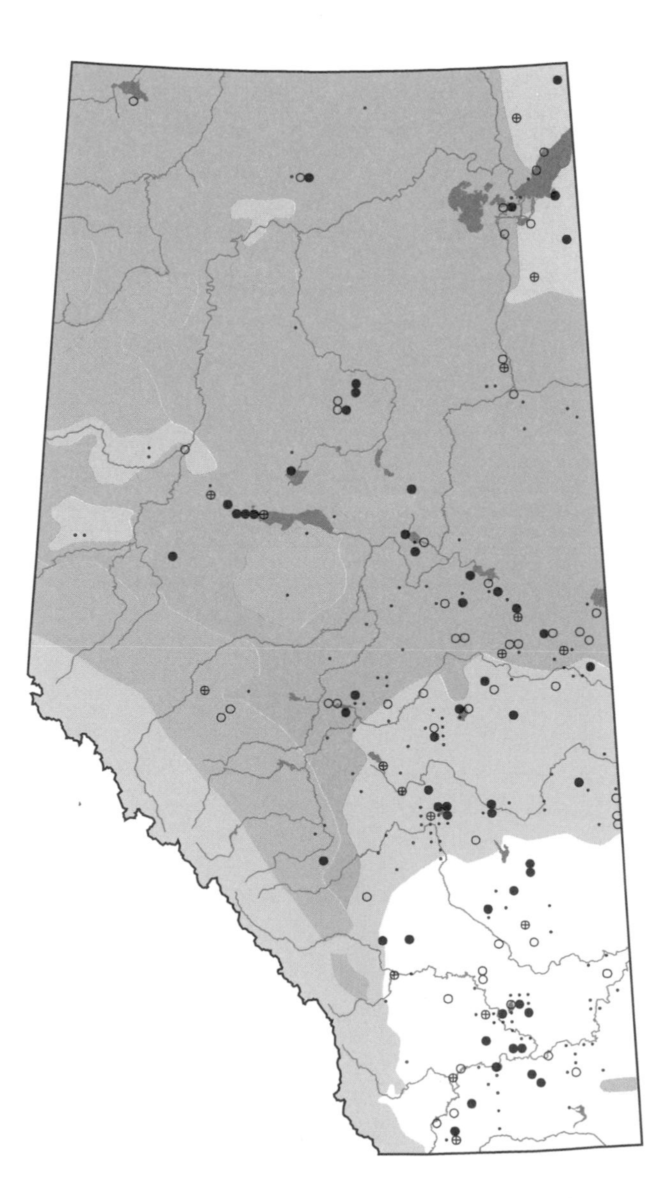

Forster's Tern

Sterna forsteri

FOTE photo credit: G. Beyersbergen

Status:
The Forster's Tern is an uncommon breeder in Alberta, where it is at the northern limit of its range. Its status in the province is unknown (Wildlife Management Branch 1991), but loss of marsh habitat may be affecting this species.

Distribution:
In Alberta, this species historically bred locally in central and southern parts of the province, north to the Edmonton area. Colonies were known from Pakowki Lake, Stobart Lake, Eagle Lake, Namaka Lake, Buffalo Lake, Driedmeat Lake, Lake Isle, Wabamun Lake, Dillberry Lake, and Tofield. Extralimital breeding has been recorded at Lesser Slave Lake, Fawcett Lake, and the Birch River Delta. Atlas findings include all of the previously known range, as well as a number of records extending further north. This includes the areas in and around Moose Lake, Muriel Lake, Cold Lake, Lac la Biche, Winagami Lake, and Fairview.

Habitat:
Forster's Tern inhabits marshes and, often, the marshy bays of bordering lakes. Suitable habitat is found in the Grassland, Parkland, and southern Boreal Forest regions of the province. It is occasionally associated with Yellow-headed Blackbirds (Ehrlich et al. 1988).

Nesting:
For nesting, this tern prefers the deeper portions of large cattail marshes where there are abundant stands of emergent vegetation. Nests are built by both sexes. They are usually constructed over water on floating mats of aquatic vegetation lodged in standing vegetation. Some terns will also use a muskrat house, an old grebe's nest, or a marshy island. The nest is dry and high enough to be protected from wave action. It may be a scantily lined deep hollow or a compactly woven neat cup. Forster's Terns usually nest in loose colonies.
Clutches are 2-4 buff to olive-buff colored eggs, which are marked with dark brown, gray, and purples. Larger clutches are the result of two females. Incubation is 23-25 days, by both sexes. Although Forster's Terns exhibit only weak site tenacity, they will defend their nest sites vigorously. The young remain in the nest for several days after hatching, but will escape to water if disturbed. They are tended by both parents. The fledging period for this species remains unreported. Forster's Terns raise one brood per season. They are occasionally parasitized by American Coots and Red-Necked Grebes, and are hostile to the presence of other species. Pair bonds in this species are monogamous.

Remarks:
This medium-sized, black-capped tern is very similar in habit and appearance to the Common Tern, and the two are often confused. The Forster's Tern can be identified by its gray tail (not white as in the Common Tern), its white breast and belly plumage, and its harsh call. In winter, the Forster's Tern has black ear patches, while the Common Tern sports a black bar extending around its head.
This species subsists mainly on small fish. When diving for food, it will fold its wings on entering the water and submerge completely. Emerging with a fingerling in its bill, the tern will shake off the water in mid-flight before carrying off the prey or eating it. It occasionally eats flying insects and frogs.
Forster's Terns arrive in the province in early May and depart by the third week of September. No overwintering has been recorded. Winters are spent along a range from western Guatemala to the Bahamas and extending to the eastern Greater Antilles.

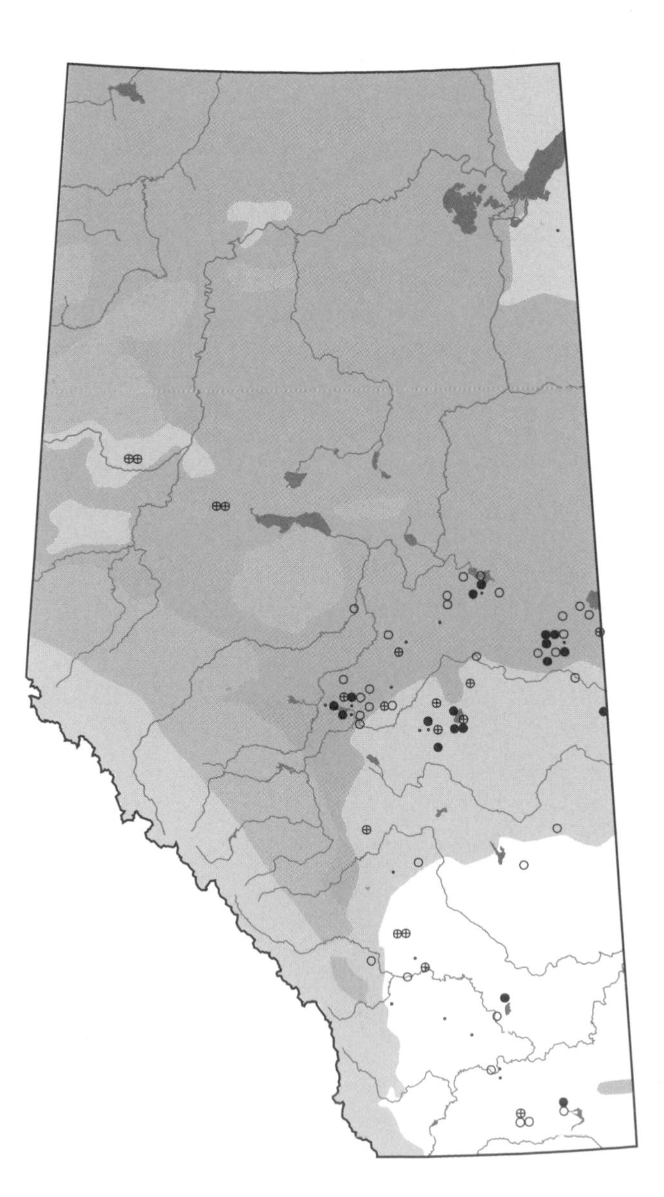

Black Tern

Chlidonias niger

BLTE photo credit: G. Beyersbergen

Status:
Despite its apparent abundance in Canada, Breeding Bird Survey analysis shows that Black Tern populations declined by 8% of their total population annually in Canada and across the continent from 1966 to 1985 (Gerson 1988). This trend is most likely due to habitat loss; the marshes inhabited by Black Terns are often drained for agricultural purposes or developed for residential and industrial use. The accumulation of environmental contaminants on this species' wintering grounds and migration routes may also be a factor in the decline. Because of declining populations, this species has been on the Blue List from 1978 to 1986. It is not considered threatened in Canada at present, and its exact status in Alberta is unknown (Wildlife Management Branch 1991).

Distribution:
Black Terns breed in all the natural regions of Alberta, with the greatest occurrence statistics coming from the Parkland region (44% of surveyed squares) and Boreal Forest (31%) natural regions. In the drier habitat characteristic of the Grassland Natural Region, the Black Tern was recorded in 10% of the surveyed squares. It bred with similar frequency in the Foothills (10%), and less so in the Rocky Mountains (4%). There were records for 19% of the surveyed squares in the Canadian Shield Natural Region.

Habitat:
This graceful and conspicuous little tern inhabits shallow lakes, marshes, sloughs, ponds, and wet meadows, where there are extensive shallows and moderate amounts of emergent vegetation. The species requires large open areas of water in the period just prior to nesting and after the young have fledged.

Nesting:
Black Terns usually nest in small colonies, though pairs may, occasionally, nest singly. Nests are built on rafts of aquatic vegetation, floating and often anchored to emergents, or on marshy hummocks. Other sites include muskrat houses, old grebe's nests, or raised mud patches. The nest itself is a depression in the nest substrate, which is lined with pieces of marsh vegetation. Because nests are only a few centimetres above the waterline, there is often egg loss from wave action or rising water levels. Birds return to the same nest area year after year unless emergent vegetation becomes too dense or water levels change markedly. Black Terns are noisy and fierce defenders of the nest and will divebomb intruders.
Clutches are 2-4 eggs, usually three. Eggs are buffy or olive and are marked with brown or black. Incubation is 21-22 days, by both sexes, and young are tended by both parents. Chicks feed near the nest for about two weeks, hiding when disturbed. Fledging is at 18-21 days. Black Terns raise one brood per season.

Remarks:
This small dark tern feeds mainly on dragonflies, damselflies, mayflies, and other insects, as well as taking in small fish. It spends much time hovering over open water picking up floating insects or hawking flying insects in mid-air. Black Terns have also been observed hunting over farmers' fields.
This species arrives in Alberta by mid-May, and leaves the province by late August or early September. Juveniles remain on the wintering grounds in South America for their first year. No overwintering has been reported for Black Terns in Alberta.

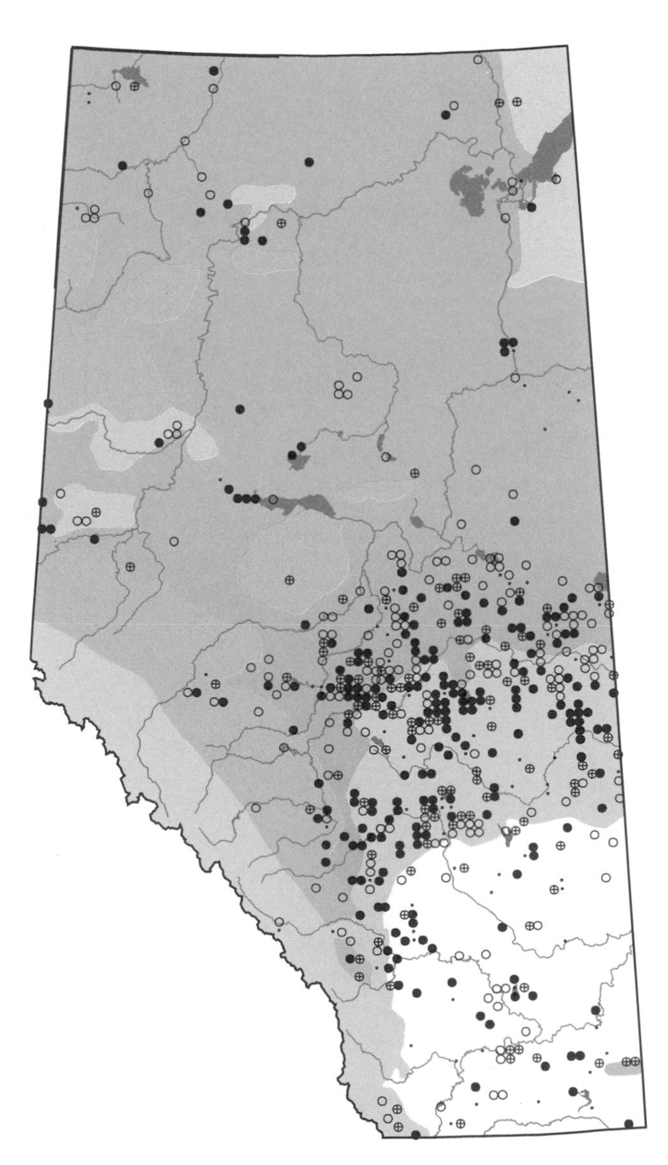

Order Columbiformes

Although larger members of this order are usually called pigeons and smaller ones are usually called doves, the two names are often interchangeable. Only two species from this order breed in Alberta, the native Mourning Dove and the introduced Rock Dove or common pigeon. They can be recognized by their generally compact, medium-sized bodies, medium to long tails, short legs, stout, medium-sized bills, small heads, and dense, soft plumage, which is highly variable among individual Rock Doves. On the ground, they are distinguished by their jerky, chicken-like walk. The more gregarious Rock Doves are much maligned for spoiling outdoor human structures, while, in parts of its North American range, the Mourning Dove is a commonly sought-after game bird. However, its high reproductive success and increasing habitat may allow the Mourning Dove to withstand hunting pressures, unlike its extinct cousin, the Passenger Pigeon. This species was not thought to be common in Alberta, although it was abundant in other parts of the continent. Both existing species have the unique ability to produce "pigeon's milk" for their young, a viscous milky substance produced in the crops of the adults.

Rock Dove

Columba livia

RODO photo credit: K. Morck

Status:
A year-round resident, the common pigeon is an old world species introduced into North America. North American populations are derived from several Old World subspecies and, consequently, are considered a mongrel stock (Godfrey 1986). A secure habitat ensures continued widespread distribution, and healthy population levels for this familiar bird.

Distribution:
Feral Rock Doves that have left farms where they were introduced many years ago are now established in most settled areas of the prairie provinces (Salt and Salt 1976). Present in the Parkland and Grassland natural regions, the pigeon completely mimics human settlement patterns. This strategy has also brought it into the Boreal Forest Natural Region, where breeding was recorded in the Peace River district as far north as Manning.

Habitat:
Familiar to all city dwellers, the pigeon will use a variety of human structures to nest. It uses semi-open, open, and bare areas to forage. Occasionally, it will nest in cliffs. It may also appear in some rural areas, but only if there is human habitation nearby. It usually avoids extensive forest, but occasionally perches in trees (Godfrey 1986).

Nesting:
A very prolific species, the Rock Dove will have multiple broods in a year. In rural areas, breeding may take place from March to September. However, given mild weather conditions, pigeons will irregularly breed in urban areas at any time of the year. They were recorded nesting in Edmonton in January (McGillivray 1988). On the ground, the monogamous male will court the female by inflating his neck, spreading his tail, bowing, and cooing. Pursuit and circling the female are incorporated in flight display. Nest locations include dark areas inside and outside buildings, ledges, and bridges. Cliff crevices are occasionally used. This species is usually colonial, although it will nest independently.
The nest consists of a scant collection of straw and twigs shaped into a saucer and without lining. The male usually brings nest material as the female builds. Both adults share in incubating the two white eggs, which hatch in 17-19 days. The nestlings are initially fed "pigeon's milk" (a substance produced within the crop of the parent from predigested food) from the crop of the adults and, later, switch to regurgitated grain (Godfrey 1986). Young fledge after 25-26 days, and become self-reliant 5-10 days later.

Remarks:
The domestic pigeon was initially introduced into rural areas before moving into settled regions. Varied plumage, from the white of a dove to pale brown to the original blue-gray, is the result of selective breeding. Usually, a pigeon will have black wing bars, a broad black terminal band on the tail, a white rump, and red legs. Short legs, a small head, and a dark bill are common traits, as is a quick flapping flight, an uneasy dihedral glide, and a chunky appearance. Head bowing, used mostly by the male, is used for assertive and defensive posturing. Pigeons will visit open areas such as cultivated fields, barnyards, dumps, railway yards, streets, and sidewalks, to forage on seeds, insects, human scraps, and, occasionally, green leaves. Individuals exhibit different feeding preferences, leading them to visit various sites. Pigeons are deemed pests by some, because their excrement accumulates in areas over years. Any history of pest control has been largely ineffective. Preble and Heppner (1981) found some evidence of population regulation by nest assertion. Pigeons now form a dependable prey base for nesting Peregrine Falcons in Edmonton and Calgary.

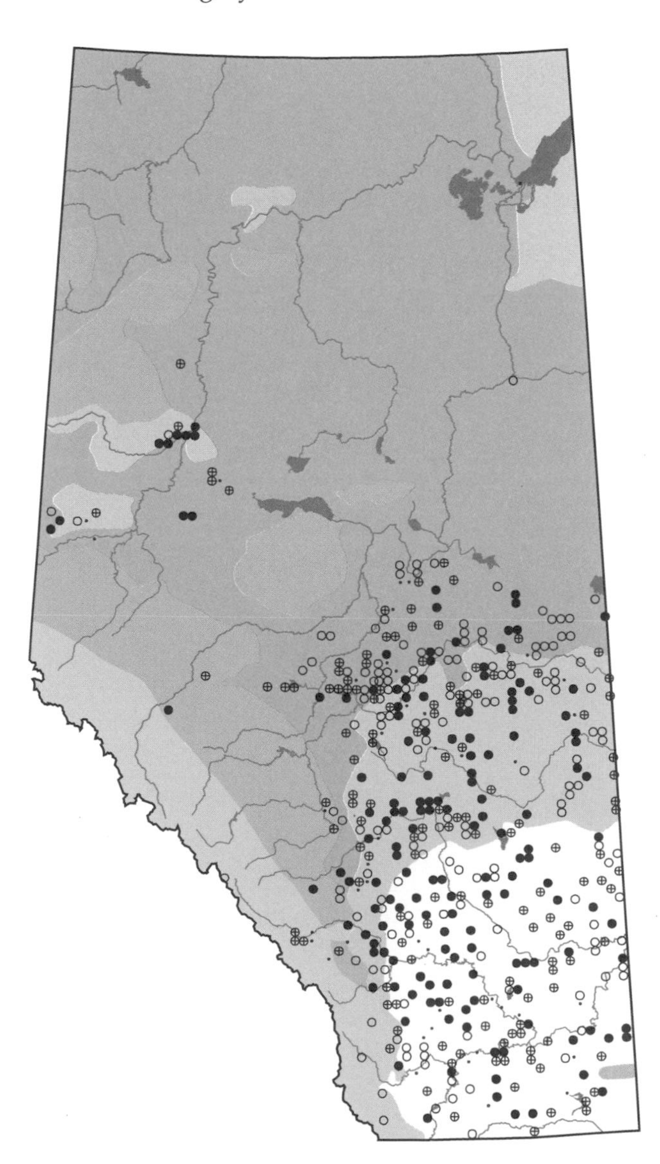

Mourning Dove

Zenaida macroura

MODO

Status:
Once thought not to occur in Alberta (like the American Crow and the Barn Swallow), the Mourning Dove has benefitted from expansive human settlement. With its present status not at risk, this species enjoys a healthy and widespread population and secure habitat (Wildlife Management Branch 1991).

Distribution:
The Mourning Dove breeds mainly in central and southern Alberta. It was recorded in 27% of surveyed squares in the Grassland and Parkland regions, 13% in the Boreal Forest, and 7% in the Foothills and Rocky Mountain regions. It was recorded in northwestern Alberta in Fort Vermillion and Meander River. Salt and Salt (1976) also reported it breeding in the Peace River district and beyond Lesser Slave Lake, and listed a nesting record for Banff, where it is usually an uncommon summer visitor. Unlike the Rock Dove, the Mourning Dove prefers rural areas.

Habitat:
This species will breed in a variety of habitats, ranging from open woodland to semi-arid grassland. However, any chosen habitat must be within reach of water. It requires coniferous or deciduous areas, such as coulees, river valleys, aspen groves, shelterbelts, and woodlots for nesting. Adjacent cultivated fields or meadows provide foraging area.

Nesting:
The monogamous male includes displays of high flapping flight and long spiralling glides in his courtship efforts. Mutual preening, head bobbing, and billing are also used. The male collects nesting material as the female builds. The placement of the nest may be in bushes, trees, or on the ground. Normally in conifers, the frail platform of twigs and dried grasses will be placed at any height and position. When on solid ground, the nest will be a scrape lined with grass and leaves, in proportions based on their availability. They may also re-use an old nest or nest on top of the nest of another species. The two white eggs laid are incubated by the male by day and by the female at night for 13-16 days. The chicks are tended to by both parents and one parent is usually present at all times. The nestlings are initially fed "pigeon's milk", a secretion produced in the crop of the adults. Soon, soft insects and plant food constitute fewer and larger feedings. A distraction display of wild fluttering may be used to ward off intruders. Fledging in 12-15 days, the young will soon flock with other young and wander far and wide, where they may be seen in the northern areas of their range in late summer or early fall. The prolific adults will go on to produce a second and possibly a third clutch.

Remarks:
Pigeon-like in shape and size, the Mourning Dove is dull brown with a buffy brown head, neck, and underparts. Some black spots are visible on the wings, as well as one under the ear. A long tail is distinguishable. The female is duller and has a shorter tail. Juveniles lack the ear spot and also have a shorter tail. A squeaky take-off and alighting bracket strong swift flight. These birds are most often seen along dusty country roads, where they may be basking or dust-bathing. Being ground feeders, they forage on waste grains, weed seeds, berries, and insects.
Mourning Doves usually arrive in Alberta in single numbers in April or early May. Flocks of young may join adult flocks in fall or they may migrate earlier. Most of both age groups have departed for central Panama by mid-October. There have been winter records for Wabamun and St. Albert.

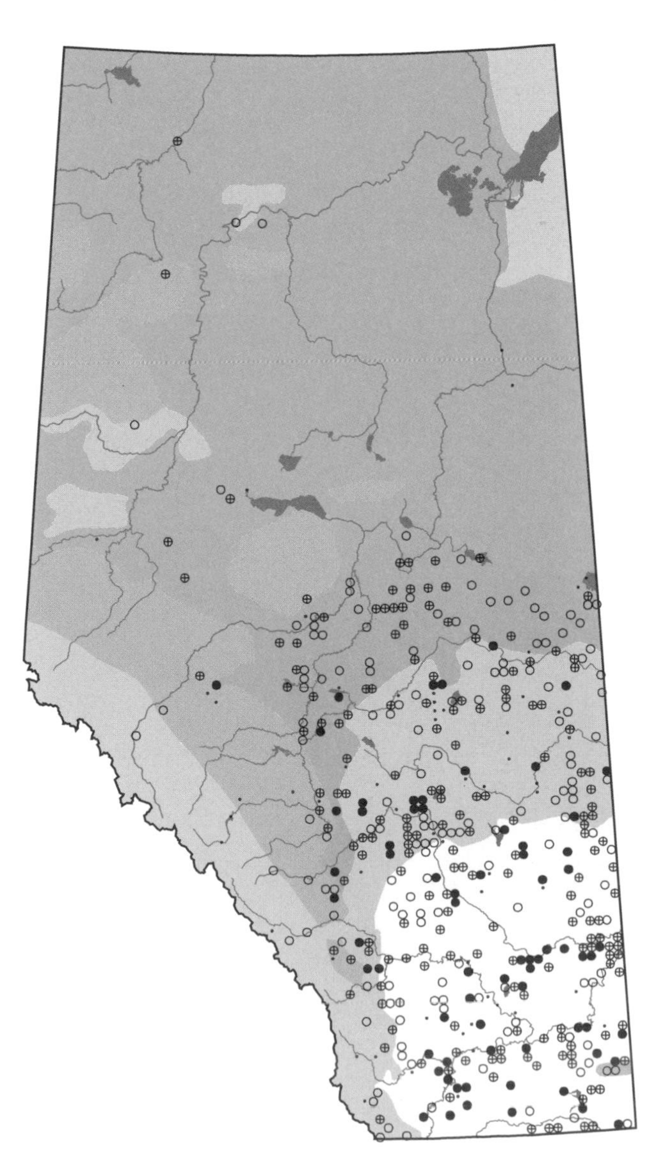

Order Cuculiformes

Only one species of these pigeon-sized birds breeds in the province, the Black-billed Cuckoo. A rather long, sturdy bill, a large head, and long wings and tail give this bird an elongated appearance. It also has short legs and zygodactylous toes. The sexes share the same coloration. This species conducts its breeding within dense shrubbery, but is a good flier nonetheless. However, this type of habitat is conducive to the species' solitary, wary tendencies. A late migrant into our region, the Black-billed Cuckoo builds a flimsy nest, which is sometimes destroyed in a heavy storm. This may explain occasional late renesting dates and, sometimes, nest parasitism of other species' nests. However, the latter practice is much less frequently used by the Black-billed Cuckoo than by the cuckoos of Europe.
After the downy phase, chicks of this species produce hard sheathed feathers that are quill-like, and may protect the young from nest robbers. Once fully feathered, these sheaths are quickly lost. Of particular interest is the fact that these birds consume injurious hairy caterpillars, such as the Tent Caterpillar, that most species will bypass. In general, the secretive nature of the Black-billed Cuckoo has limited our knowledge of this species and forced speculation on other scanty information.

Black-billed Cuckoo

Coccyzus erythropthalmus

BBCU

Status:
The Black-billed Cuckoo appears to be uncommon in the province.

Distribution
Sightings during the Atlas Project confirm previous literature findings (Salt and Salt 1976), that this species is locally and irregularly distributed in central and southern Alberta and in Waterton Lakes National Park. Breeding was only confirmed in central Alberta in the transition area between the Parkland and Boreal Forest natural regions. Athabasca marks the northernmost known nesting location for this species (Salt and Salt 1976), although there were no confirmed breeding records in the far north during Atlas surveys.

Habitat:
When found, the cuckoo appears within brushy thickets along the roads and streams of the Parkland Natural Region. In the Grassland Natural Region, it may be found in the dense bush of some coulees. In general, it prefers open woodland with tangles of willow, alder, and vines (Godfrey 1986).

Nesting:
Courtship feeding is the only example of courtship display known. Composed yet wary, Black-billed Cuckoos may nest near humans. A loose and fragile platform of twigs, dead leaves, and grass is built on the branch of a dense coniferous or deciduous shrub, bush, or low in a tree. It is lined with soft plant material. The typical clutch is 2-4 eggs that are pale bluish green. Larger clutches are apparently associated with caterpillar outbreaks. Within its North American range, it may rarely parasitize the nest of a Yellow-billed Cuckoo. Hatching asynchronously, beginning 10-13 days later, the nestlings are fed insects, beetles, grasshoppers, and hairy caterpillars from the throat pouch of each adult. The chicks have disks of soft papillae in their mouths that aid in grasping the parents' bills and food. About one week after hatching, the young are able to leave the nest and move about the branches of the nest tree. Their voracious appetite permits them to grow quickly. They will fledge 14-17 days later. The various ages of the young may overextend the adults, reducing the overall success of the clutch.

Remarks:
This secretive, rarely seen bird has brown upperparts, white underparts, a sturdy black down-curved bill, red eye ring, zygodactylous toes, and a graduated long tail with narrow white tips. Adding to the difficulty of its detection, this species does not sing much, usually producing only strange guttural notes that resemble hollow wooden clucks. However, when found at the nest, it is very approachable, eventually fluttering off and scolding the intruder from a lower bush. Nestlings will feign death.
Almost entirely insectivorous, this species feeds on insects, larvae, and hairy caterpillars, such as the tent caterpillar, proving it to be highly beneficial to the landscape. Some small vertebrates, fruits, and berries are also eaten.
Local migration data is not precise. Arrival in mid-June coincides with the densest profusion of deciduous foliage. Departure in August to northern South America is thought to occur shortly after the young become independent.

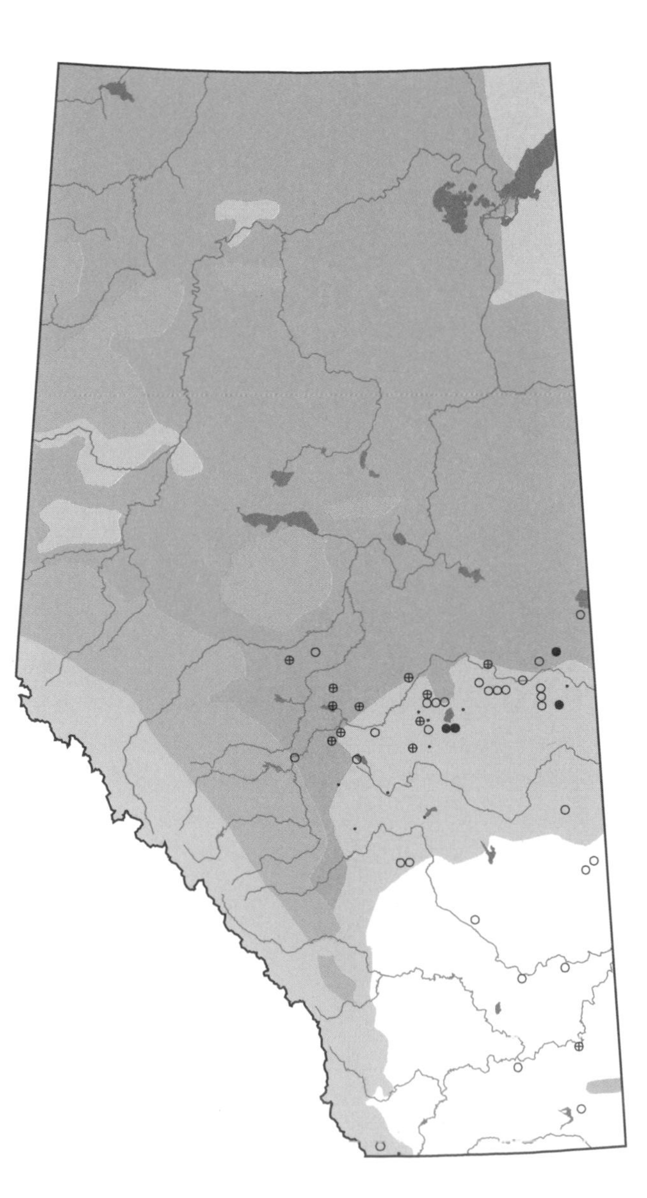

Order Strigiformes

Most members of this order of raptors are nocturnal, thus filling the niche occupied primarily by the Order Falconiformes by day. Their physical adaptations complement their sit-and-wait hunting strategy. These stocky birds of prey have large heads and short necks, which can rotate almost completely around. This assists in the use of the large, non-moving eyes, which are capable of binocular vision. The eyes are deeply set in round or oval facial discs. The short hooked bills of these birds allow adults to tear flesh to feed their young. Large auditory openings and an acute sense of hearing allow these birds to locate their prey. The long, strong toes are used to pounce on and kill quarry. The plumage is similar between the sexes. The broad wings and rounded tails, like the rest of the plumage, are equipped with thick, soft feathers that are adapted to promote the silent flight they require as they approach their prey. Females are usually larger than males, which is common among raptors. All species in Alberta are arboreal, except the Burrowing Owl. However, none of the owls build nests, rather they usually occupy old nests or cavities of other species.

Great Horned Owl

Bubo Virginianus

GHOW photo credit: D. Wiggitt

Status:
The Great Horned Owl, the official provincial bird, is common in the Grassland and Parkland regions, and the southern portion of the Boreal Forest region. It is widely dispersed throughout the rest of the province. It is resident throughout its breeding range (Salt and Salt 1976).
Godfrey (1986) reported that there are two subspecies present in the province, *B. v. subarcticus* in all parts of the province except in the southwest part of the Rocky Mountain Natural Region where *B. v. lagophonus* occurs.

Distribution:
This owl was recorded breeding in all the natural regions of the province. It was recorded in 27% of surveyed squares in the Grassland, 46% in the Parkland, 22% in the Boreal Forest and Foothills, and 13% in the Canadian Shield and Rocky Mountain regions. The widespread distribution recorded in the northern part of the province is reflective of the survey effort in these areas during the early breeding season of this species.

Habitat:
Probably no other owl in North America lives in so many habitats and under so many climatic variations and, thus, it is very difficult to characterize habitat requirements for this species (Johnsgard 1988). Godfrey (1986) described the habitat as both deciduous and coniferous woodlands from extensive heavy forests to isolated groves, woods of city parks, and, in the Grassland region, wooded coulees and river valleys.

Nesting:
Great Horned Owls chose their nesting territories in February and early March (Salt and Salt 1976). The nest is almost always the abandoned nest of a hawk, usually a Red-tailed Hawk, or a crow. Nothing is added to the nest and Eckert (1974) reported that the nest is sometimes in such a state of disrepair that eggs or young will fall through holes and perish. The female incubates the clutch of 2-3 eggs for 30-35 days. The young are brooded by the female for about three weeks, and the male brings all of the food at this time. The nest is vigorously defended by both adults. Young leave the nest at approximately five weeks, but they do not fly very well until 9-10 weeks (Harrison 1978). They are dependent upon the parents for a considerable period while they slowly acquire hunting skills (Johnsgard 1988).

Remarks:
Eckert (1974) described the Great Horned Owl as the fiercest, most aggressive, and most impressive owl of North America. Its large size and raised ear tufts are quite distinctive. It has a wide variety of vocalizations that are described in detail by Eckert (1974), but the most common is the hoot, a familiar "whooo-whooo-whoooooo-whoo-whooo".
Pinel (1978), in conducting a food habit study of this owl near Calgary, reported that the most abundant prey species for the Great Horned Owl were Varying Hares, Richardson's Ground Squirrels, and Pocket Gophers. Of the 60 prey species found in nests, 31 were mammal and 29 were birds. Several sources report the variety of prey species taken (Bent 1961b; Rusch et al. 1972; Eckert 1974).
The Great Horned Owl is nonmigratory but will move in response to shortages of prey species. Salt and Salt (1976) also reported that, in fall and winter, it may appear in regions where it was not seen in summer as young birds wander about looking for new territories.

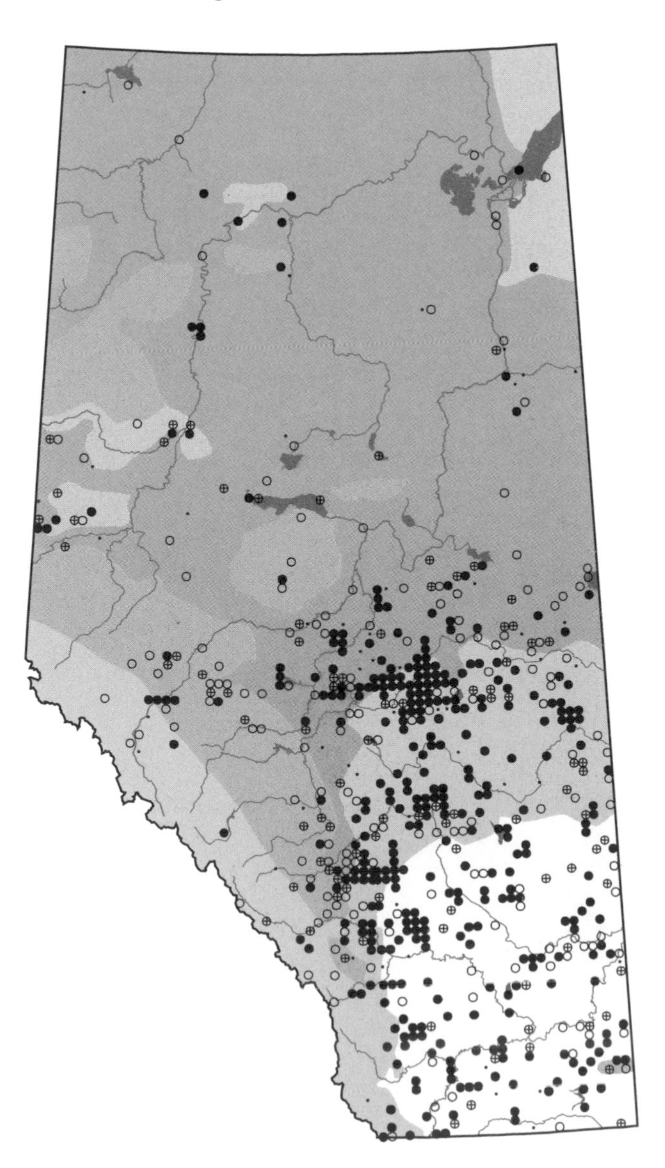

Northern Hawk Owl

Surnia ulula

NHOW

Status:

It is difficult to assess the population status of this species, because of the remoteness of its breeding range. Atlas results indicate that the Northern Hawk Owl is uncommon and widely dispersed in suitable habitat in Alberta. There is a potential for this species to be adversely affected as boreal forest habitats are altered (Wildlife Management Branch 1991).

Distribution:

Salt and Salt (1976) reported the breeding range of the Northern Hawk Owl as the northern half of Alberta, southwest into the Rocky Mountains and the Banff area. Atlas surveys resulted in a total of 36 breeding records for this owl over the five year survey period, 22 of which were in the Boreal Forest region, 11 in the Foothills, 2 in the Rocky Mountain region and 1 in the Peace River Parkland.

Habitat:

The Northern Hawk Owl is a bird of the northern coniferous forests. It frequents open coniferous or mixed woodlands, muskeg, brushy edges of clearings, and old burns.

Nesting:

It may nest in a natural tree cavity, the hollow top of a dead tree stub, a woodpecker hole, or sometimes an abandoned crow or hawk nest (Godfrey 1986). However, unlike most other owls, this owl will often construct its own nest—a crude affair of interlocking twigs and leaves, sometimes with a lining of grasses, bits of moss, and plumage from the female's breast (Eckert 1974).

The Northern Hawk Owl lays 3-7 glossy white eggs, which are incubated by the female for 25-29 days. The male brings food to her during incubation and early brooding. Young are tended by both parents and nestlings fledge at 23-27 days. Family groups tend to remain together until the following spring, when the adult birds begin nesting activities (Eckert 1974).

Remarks:

This is a medium sized owl that has dark brown upperparts that are spotted with white, a buffy white facial disc with a black outer border, sharply barred underparts, and a long tail. Its hawklike posture and flight give this bird its common name.

This owl is much more active during the day than most woodland owls, hunting both in the morning and late afternoon. It will perch on top of a tall tree or old snag watching for movement below. Swift swoops down, followed by a return to its elevated perch, are apparently typical (Johnsgard 1988). The Northern Hawk Owl never carries food in its beak, but it can carry animals much larger than itself in its talons. Its usual prey are mice, lemmings, ground squirrels, other small mammals and birds, and insects. It is sometimes killed by Great Horned Owls when it roosts at night.

The Northern Hawk Owl is not regularly migratory in its habits. However, in late fall, it may move south to set up a winter territory in a more settled area. Here, it may be found in a stand of tall trees near a cultivated field where it can hunt. In March, the owl will return to its more northerly breeding grounds. In severe winters or when prey is scarce, the owl will move far south of its breeding range into southern Canada and the northern United States.

Pinel et al. (1991) reported wintering observations from Edmonton, Edson, Stettler, Seebe, Calgary, Priddis, and Turner Valley.

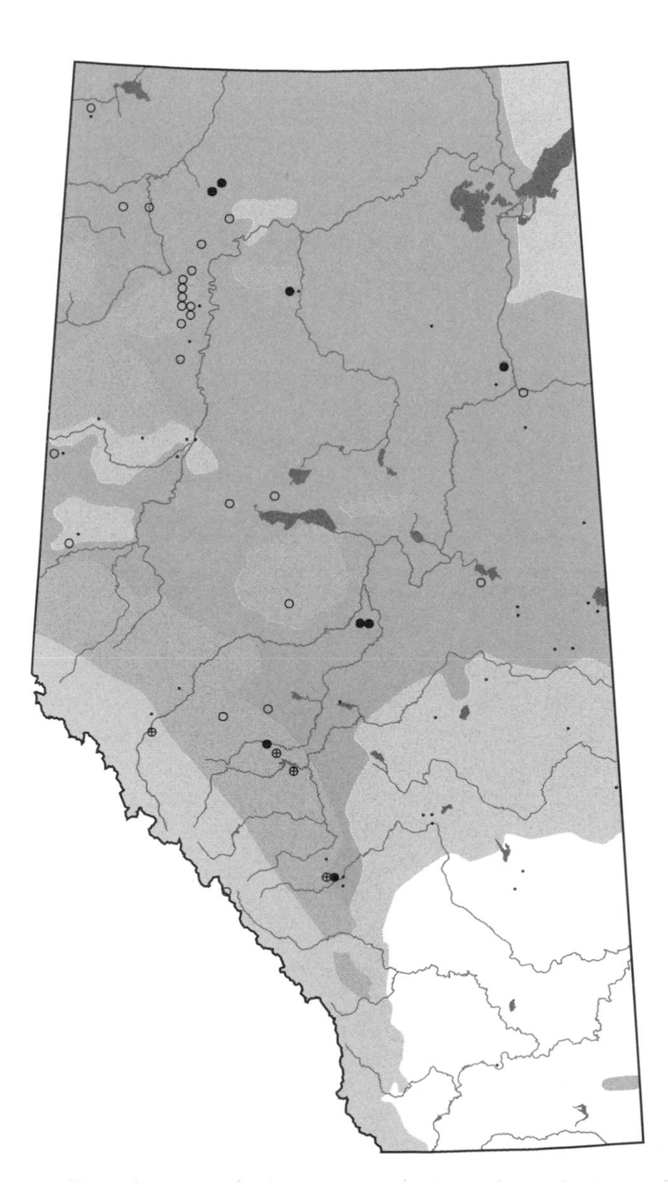

Northern Pygmy Owl

Glaucidium gnoma

NPOW

Status:
This is an uncommon species in Alberta. The first confirmed nesting in the province was in 1971 and the next, subsequently, in 1977 (Pinel et al. 1991). Johnsgard (1988) stated that this species is tolerant of a broad range of habitats in western North America and is not harassed by humans or seriously affected by their activities. Partial forest clearing may improve hunting opportunities for it.

Distribution:
There were a total of 25 breeding records for this species in the five year study period of the Atlas, with nine of these being in the confirmed category. The majority of these records were in the southwest portion of the Boreal Forest (11) and adjacent Foothills (8) region. It was also recorded breeding in the Rocky Mountain region. Holroyd and Van Tighem (1983) reported nesting on Tunnel Mountain near Banff, as well as spring observations near Jasper.
This owl is regularly observed in the Parkland and Grassland regions in winter.

Habitat:
Salt and Salt (1976) described the breeding habitat of the Northern Pygmy Owl as heavy stands of coniferous trees broken by small clearings. Mixed wood forests are quite acceptable, providing that there is a good proportion of spruce, pine, or fir. These owls are usually found in the vicinity of meadows or other sizeable openings in the forest, and they avoid areas of dense unbroken forest (Bent 1961b).

Nesting:
The nest is usually the abandoned hole of a Hairy Woodpecker or Northern Flicker, and it may be used for several years. No improvement is made to the nest and the eggs are laid upon the bare wooden bottom of the nest hole (Bent 1961b). The typical clutch is 3-6 white glossy eggs, which are incubated by the female for 28 days. The female broods the young and feeds them on food brought entirely by the male (Harrison 1978). The young leave the nest and are able to fly at 30 days, but remain with the parents for an additional 20-30 days (Johnsgard 1988).

Remarks:
Godfrey (1986) described the Northern Pygmy Owl as a bluebird-sized owl with a small head, no ear tufts, and a rather longish tail that is often held quite high. It may be confused with the slightly larger Northern Saw-whet Owl, but the Pygmy Owl has darker stripes on the underparts and a black and white half collar on the hind neck. The female is larger and her general coloration is more reddish than that of the male.
The primary call of this species is a monotonous repetitive series of "hoot-hoot" notes (Johnsgard 1988).
The calling is usually performed at night, but this bird hunts during the day. Its small size and, consequently, high metabolic rate, requires it to hunt almost constantly. It is most active just before dusk and soon after dawn. Northern Pygmy Owls use the technique of surprise attack on their prey, gliding and diving down from an elevated perch after first locating their prey visually (Johnsgard 1988). The favored prey are mice, small birds, and large insects, with small birds taken mainly when there are young in the nest.
The Northern Pygmy Owl tends to remain solitary or in highly dispersed pairs or family groups. It is essentially non-migratory, although there is some movement to lower elevations in winter.

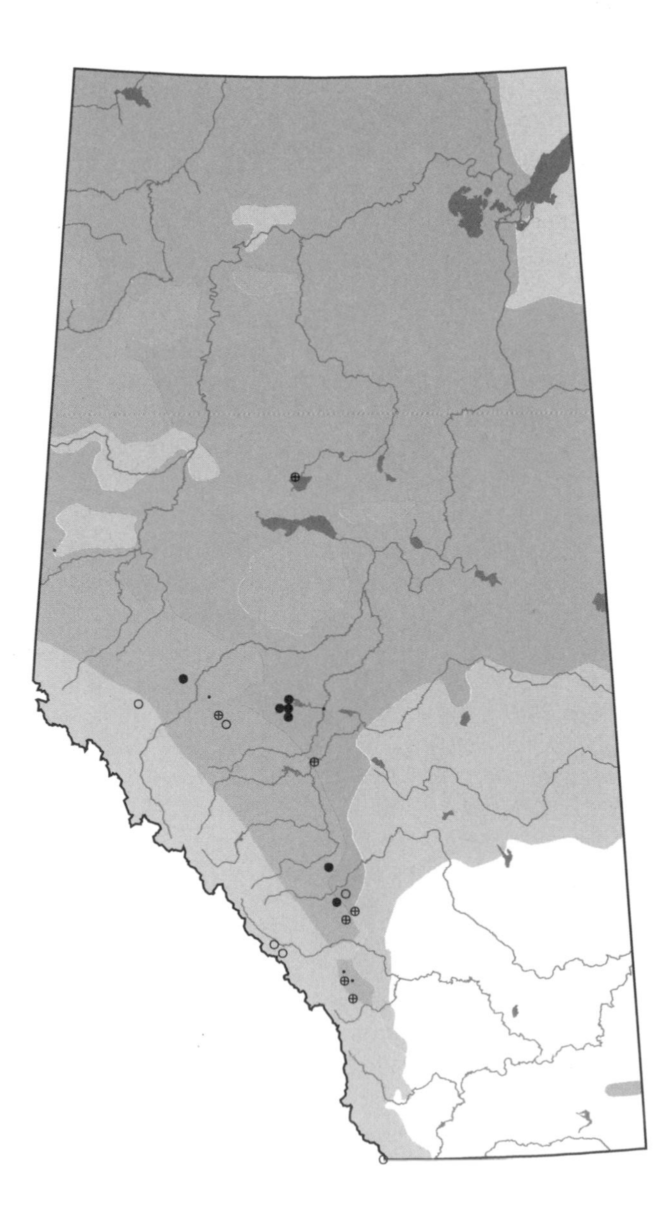

Burrowing Owl

Athene cunicularia

BUOW photo credit: R. Gehlert

Status:

This species has suffered over most of its historical range in western North America as colonial rodent populations have been controlled or eliminated by poisons, insecticides have reduced its food supplies and have perhaps directly poisoned it, and traditional rangelands have been converted to agricultural purposes (Johnsgard 1988). The number of owls that are killed by vehicles is also a significant mortality factor. After surveys in the late 1970s indicated a decrease in Burrowing Owl populations in the western provinces, The Committee on the Status of Endangered Wildlife in Canada declared that the species was "threatened". In 1986, the species' status was changed from "threatened" to "endangered". It was listed as a "species of special concern" in 1986 on the National Audubon Society's Blue List of apparently declining species. In Alberta, it is designated as an endangered species (Wildlife Management Branch 1991) with 700-900 breeding pairs in the province.

Distribution:

Erickson (1987) described the range of the Burrowing Owl as including all of the Grasslands Natural Region and extending north and west into the Parkland region. Atlas survey results show that all but three breeding records were recorded in the Grasslands region, where it was recorded in 30% of the squares surveyed. Historically, this owl has been observed west to Banff and north to Jasper and Cooking Lake (Pinel et al. 1991).

Habitat:

The basic criterium for optimum Burrowing Owl habitat are openness of site and availability of nest burrows. The habitat of this species is level open short grass areas that are typical of heavily grazed or low stature grassland (Johnsgard 1988).

Nesting:

Burrowing Owl pairs may nest singly, in loose association within an area, or in tightly knit colonies of a dozen or more pairs in a single field (Savage 1985). The nest is usually located in the unoccupied burrow of a ground squirrel, prairie dog, or badger. On rare occasions, the nesting burrow may be excavated by the owl but, normally, the only excavation undertaken is to modify and enlarge an existing burrow. The nest is located in a chamber at the end of a burrow 2-5 m long. Both male and female bring in nesting material to line the nest chamber. Favored materials are grass, shredded cow dung, feathers, bits of bone, and rubbish (Eckert 1974; Harrison 1978).

The typical clutch is 5-7 glossy white eggs. Incubation is usually 28-29 days, primarily by the female. During this time, the male supplies her with food. As the young mature, they spend increasing time at the mouth of the burrow, retreating into it when alarmed (Harrison 1978).

Remarks:

The Burrowing Owl is generally light brown with spots, a very rounded head with no ear tufts, long, lightly feathered legs, and a stubby tail. The most common call of this species is a deep melodius "Coo-coo-o-o". It also issues a defensive call when an intruder disturbs or enters the mouth of the burrow, which is an accurate vocal mimicry of the buzzing rattle of the rattlesnake (Eckert 1974).

This owl hunts by direct animal chase, by hovering flights, by running down prey on the ground, and by hawking insects during short sorties from perches (Johnsgard 1988). The diet is varied, with insects and mice making up the majority of the fare (Eckert 1974). It also takes worms, frogs, toads, ground squirrels, and songbirds up to its own size.

The Burrowing Owl arrives in Alberta in late March to mid-April and leaves the province from mid-September to mid-October (Pinel et al. 1991).

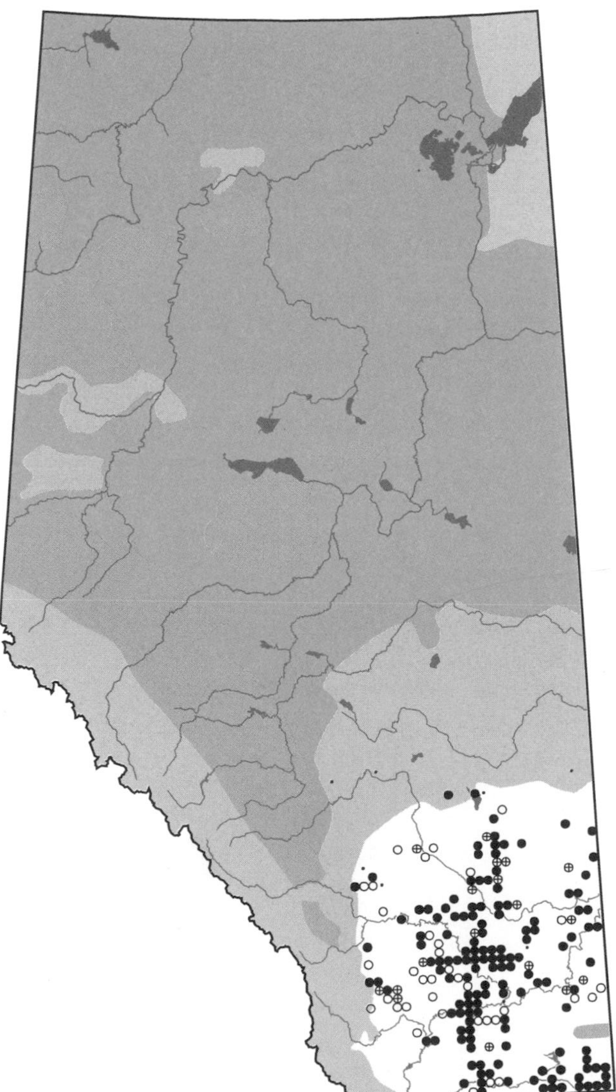

Barred Owl

Strix Varia

BAOW photo credit: R. Gehlert

Status:

The first provincial breeding record of the species was not made until 1949 (Boxall and Stepney 1982) and the second not until 1966 (Jones 1966). Boxall and Stepney (1982) reported that only eight definite nestings had been recorded. There is no doubt that this is one of the rarest owls in the province, despite its large size and characteristic vocal behavior (Boxall 1986).

Distribution:

The Barred Owl has been seen at Sturgeon Lake, Lesser Slave Lake, Calling Lake, and Fort McMurray, south to Edmonton, Rocky Mountain House, and Calgary, and west to the Smoky River and Jasper (Salt and Salt 1976). The northernmost observation was at the confluence of the Peace and Notikewin rivers and the southernmost at Beauvais Lake Provincial Park (Boxall and Stepney 1982). Atlas data concurred with Boxall's summary that records of sightings since 1959 were concentrated in the Boreal Forest region northwest of Edmonton and the primarily coniferous foothills and montane forests west of Calgary and in Jasper National Park. There is one extralimital possible breeding record in northern Alberta near La Crete.

Habitat:

The typical breeding habitat of the Barred Owl consists of relatively heavy mature woods, often with nearly open country for foraging, but with densely foliaged trees for roosting, and the presence of enough large trees with suitable cavities to allow for nesting (Johnsgard 1988). It has been stated that Barred Owls in Alberta preferred the mixed wood boreal forest, but have adapted to the forests of a predominantly coniferous character (Boxall and Stepney 1982). The preferred nesting habitat in Alberta is mixed woods with large deciduous trees, particularly along lakeshores and stream valleys (Pinel et al. 1991).

Nesting:

This owl nests in natural cavities in large trees, usually not over 10 m from the ground, or in the old nest of a crow or hawk. When the nest is a natural cavity, no nesting material is taken in, but the cast off downy feathers of the owl are used; when an old hawk's nest is used very little is done to it (Bent 1961b).

Breeding is in late February or early March. The female lays 2-3 white eggs, which she incubates herself for 28-33 days. The young are brooded for about three weeks and are tended by both parents. They usually leave the nest at 4-5 weeks and perch on an adjacent branch. Young Barred Owls fly at approximately six weeks.

Remarks:

The Barred Owl is a large gray owl with a large, rounded head, large dark eyes, and no ear tufts. It is barred laterally on the throat and upper breast with vertical striping on the lower body.

Without doubt, this is the most vocal of all owls of North America and the owl with the widest range of calls. The most common and often heard call is the resonant and far-carrying hoot for which this bird is nicknamed "Hoot Owl" (Eckert 1974). This call is said to sound like "Who cooks for you, who cooks for you all".

The diet of this owl consists of a wide variety of small mammals, including: mice, voles, chipmunks, and squirrels; a wide variety of birds including sparrows, juncos, jays, warblers, woodpeckers, robins, and swallows; fish and insects.

It has the habit of drinking water frequently and bathing often, usually with seeming delight, even in the midst of winter (Eckert 1974).

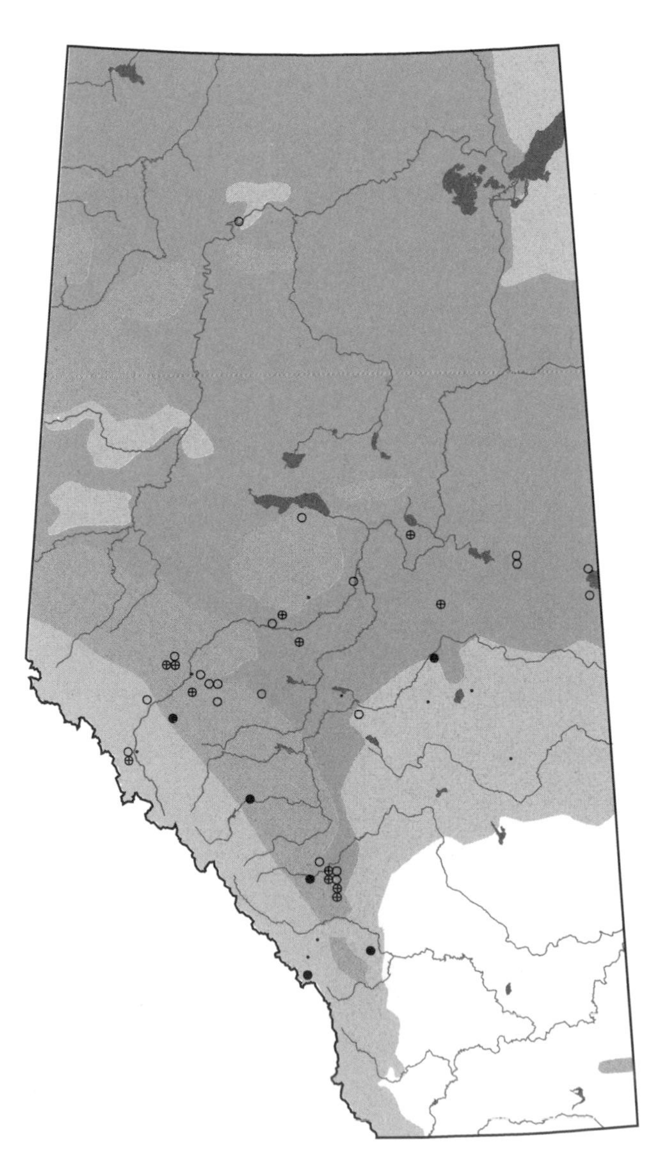

Great Gray Owl

Strix nebulosa

GGOW photo credit: T. Webb

Status:
The Great Gray Owl is an uncommon inhabitant of northern and western Alberta. The Committee on the Status of Endangered Wildlife in Canada (COSEWIC) considers the Great Gray Owl to be a "vulnerable" species, that is, a species at risk because of low or declining numbers. Nero (1980) estimated the North American population of Great Gray Owls at 50,000 birds.

Distribution:
Salt and Salt (1976) reported that this owl breeds in northern Alberta south to Jasper, Rocky Mountain House, and Miquelon Lake, and perhaps to Sundre and Calgary. Breeding records obtained from Atlas surveys generally concur with this range, although nest sites were reported in the Calgary area at Sarcee Butte and the Brown Lowery Natural Area.

Habitat:
The Great Gray Owl is a bird of undisturbed boreal forest. It is found in coniferous, deciduous, and mixed woodlands, usually near water sources such as muskegs, marshes, and wet meadows. It hunts the forest margin in brushy clearings and natural forest openings.

Nesting:
The Great Gray Owl uses the abandoned nest of a crow, raven or hawk, which is sparsely refurbished with a few twigs, leaves, feathers, or strips of bark. The owl may nest in a hardwood or conifer, but preferred nesting sites appear to be stands of mature poplar in muskeg country. Nests are usually 2 to 15 m above the ground. The Great Gray Owl lays 2-5 dull white eggs, which the female incubates for about 30 days. The male begins to feed the female in courtship and continues this through development of the nestlings. Once the young have hatched, the adults may become fierce defenders of the nest. Young fledge at 21-28 days, but do not fly well until 35 days. Generally, one brood is raised per year, but there may be a second laying if the first nest is destroyed.

Remarks:
While the Great Gray Owl appears to be the largest of our owls, it only weighs about a kilogram under its fluffed-out feathers. Despite its size it is elusive and is most often observed in non-breeding season, when it comes out of the deep forests and into more settled areas.
The most conspicuous features of the owl are its large round head and very dark feathers. Unlike the Great Horned Owl, the Great Gray Owl has no feathered tufts on the top of its head. Its perfectly circular facial disc indicates an advanced sense of hearing. As the owl moves from perch to perch watching and listening for prey below, the incoming sound waves are caught and focused by specialized feathers of the facial disc. It is the only species among the eleven in genus *Strix* that is not primarily nocturnal, its particularly small eyes being an adaptation for daylight hunting. When prey is heard or seen, the owl will take flight, hover over the prey, and dive towards the prey with talons outstretched. This owl hunts most often at dawn and dusk.
This owl shows very little fear of man and, except when there are nestlings in the nest, it is quite easy to approach. Its lack of suspicion and quietude once made it an easy target for trophy hunters.
Migrations of Great Gray Owls are sporadic and dependent on cyclic levels of small mammals in the northern breeding range. In some years there is little movement, while in other years owls will move south and be found in a variety of wooded habitats in more settled parts of the province.

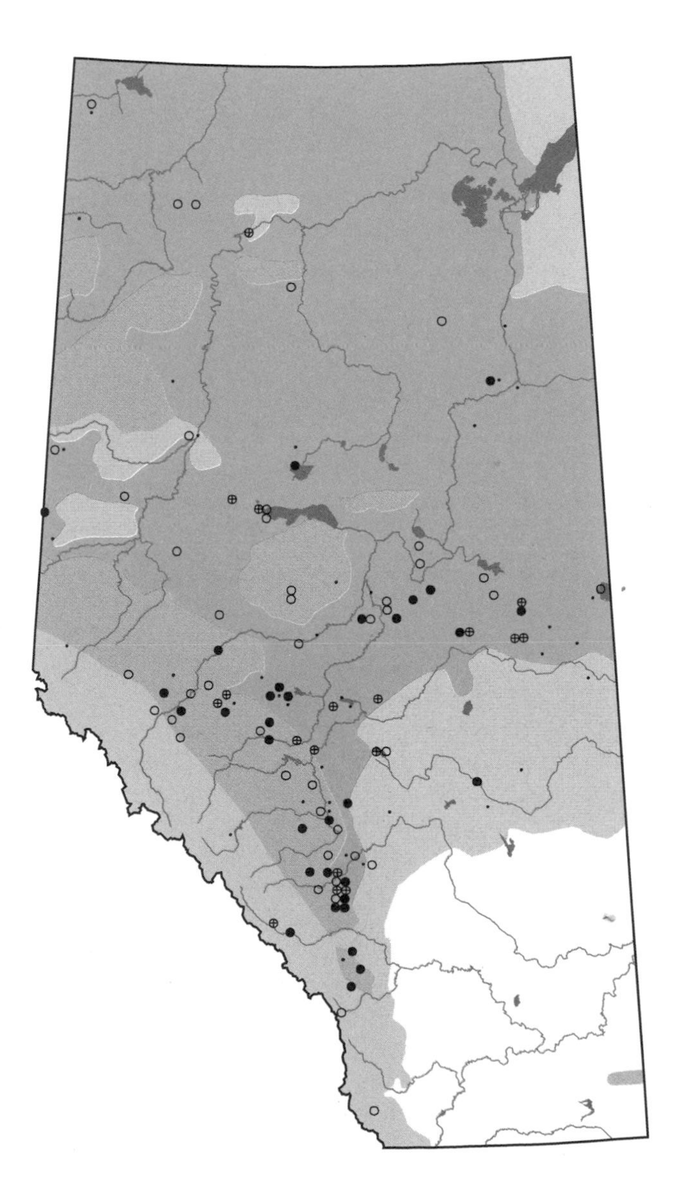

Long-eared Owl

Asio otus

LEOW photo credit: T. Webb

Status:

The Long-eared Owl appears to be widespread and locally common in the central and southern parts of the province. Cadman et al. (1987) suggested that because it is well hidden by day, is exclusively a nocturnal hunter, and calls infrequently, it is the most difficult owl to locate during the breeding season.

Distribution:

The breeding range of this owl in Alberta includes the Grassland, Parkland, and southern Boreal Forest regions, as well as the southern portions of the Foothills and Rocky Mountain Regions. Salt and Salt (1976) suggested that the Long-eared Owl may breed in the northern part of Alberta, as it was recorded breeding in the southern Mackenzie area. Atlas surveys found no evidence of this; there was only one breeding record north of Athabasca, and that was just west of Grande Prairie. Pinel et al. (1991) reported an extralimital breeding record near Ft. Chipewyan.

Habitat:

During the breeding season, the Long-eared Owl is associated with woodlands, forest edges or patches, or similar partially wooded habitats of coniferous, deciduous, or mixed forest (Johnsgard 1988). It prefers these well-timbered areas to be in close conjunction with water including streams, marshes and lakes.

It is more easily found and observed in the aspen groves of the Parkland region, and in the wooded coulees and river valleys in the Grassland region (Salt and Salt 1976).

Nesting:

This owl very seldom builds its own nest, but prefers to use the abandonded nest of some other species such as the American Crow or the Red-tailed Hawk. Unlike other owls, the Long-eared does improve the nest with strips of bark, leaves, moss, and breast feathers. The nest is usually more than 6 m from the ground near the trunk of a dense tree in heavy cover (Eckert 1974). It has been reported, rarely, nesting on the ground sheltered by the base of a tree, or among low shrubby growth (Harrison 1978). The female lays 3-6 white glossy eggs, which she incubates for about 28 days. She is fed by the male during incubation and, after hatching, she feeds the young with food provided by the male. The young leave the nest at 23-24 days and can fly well by the 35th day. The young birds continue to be fed by the adults until the ninth week and the family tends to stay together until the onset of winter (Eckert 1974).

Remarks:

The distinctive long ear tufts, the reddish-brown facial disc with narrow black markings around the eyes, and stripes on the abdomen are distinctive features of this owl. It is much smaller than the Great Horned Owl and its ear tufts are positioned more to the centre of its head. The most common call is an extended series of soft "Kwoo-kwoo-kwoo" and, on some occasions, it will produce a series of rapid piercing shrieks (Eckert 1974).

The majority of hunting activity takes place in open areas on the edges of woodlands. Rodents, mainly voles and mice, make up the majority of its diet, but, during nesting, the Long-eared Owl will take other birds up to the size of grouse.

Pinel et al. (1991) recorded the arrival of this bird in Alberta in late March and early April. In the fall, The Long-eared Owl leaves the province from mid-October to mid-November. It has been observed wintering in Calgary.

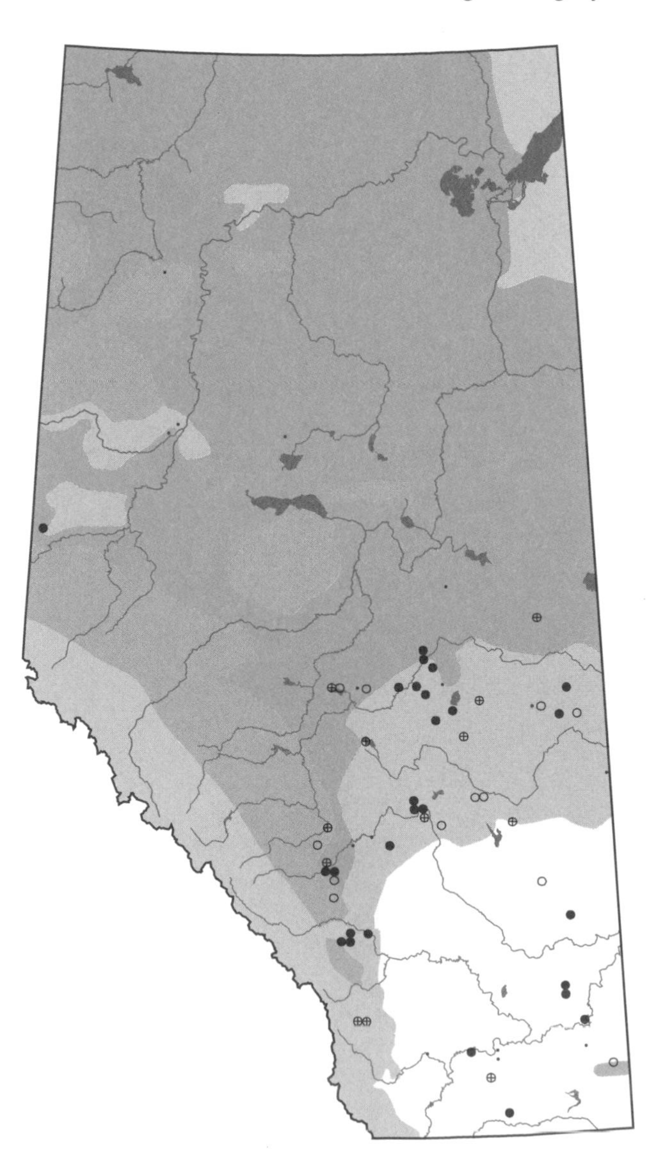

Short-eared Owl

Asio flammeus

SEOW photo credit: T. Webb

Status:
This owl has been on the Audubon Society's Blue List of declining species from 1976 to 1986, but it has not been recognized as a species of special concern by authorities (Johnsgard 1988). In Alberta, it is fairly common in the Grassland and Parkland regions and uncommon in the Boreal Forest region.

Distribution:
Salt and Salt (1976) reported the Short-eared Owl as breeding throughout the province with the exception of the Rocky Mountain Natural Region. Atlas data indicate that this owl breeds in the south and central parts of the province, with the northern limits of breeding being Peace River, Lesser Slave Lake, and Cold Lake. It was recorded in 8% of surveyed squares in the Grassland region, 10% in the Parkland, and 2% in the Boreal Forest region. There were only two breeding records in the Foothills region and none in the Rocky Mountain region.

Habitat:
This owl prefers to breed in relatively open country such as grassland, grassy or brushy meadows, marshland, pastures, stubble fields, croplands, and previously forested areas that have been cleared. A combination of areas of suitable resting and nesting cover, with adjacent hunting areas containing an abundance of small mammals, is a dominant factor in selecting breeding habitat (Johnsgard 1988).

Nesting:
The nest is usually found on the ground. It is no more than a slight depression in a little rise of ground, well hidden by heavy surrounding reeds or grasses. The depression is lined with some dried grasses, weed stalks, and, occasionally, feathers from the female's breast (Eckert 1974). It is often in the vicinity of water.
The typical clutch is 5-7 eggs, which are white with a faint bluish tinge. Larger clutches, up to 14, may be laid when food supplies are abundant. The female incubates the eggs for 24-28 days and is fed by the male during this time. The young may begin to leave the nest at 14-18 days, walking and running in the vicinity, before they actually fledge at 24-27 days (Johnsgard 1988). The parents continue to bring the young food until nestlings can hunt for themselves, usually at 25 days. To gain the parents' attention, a young bird will perform a wing fluttering, food begging display (Clark 1975). When the nestlings are endangered, one or both parents will feign injury to lead the intruder away. This species is usually single brooded, but double broods are reported when food is plentiful (Harrison 1978).

Remarks:
The Short-eared Owl is medium-sized and buffy white in color with broadly streaked upperparts. It has small facial discs, short ear tufts, and a distinctive patch of black on the underwing near the wrist. This owl is relatively silent in comparison to other owls. The most common call associated with this bird is a barking sound which Eckert (1974) compared to the yapping of a very small dog.
This owl flies low over its territory and pauses often to hover before swooping down on its prey. Small rodents make up the majority of the diet, primarily the Meadow Vole. Other prey species include small mammals (ground squirrels, shrews, rabbits, and young muskrats), small birds (sparrows, Red-winged Blackbirds, Horned Larks, Western Meadowlark, and cowbirds), larger insects, and caterpillars.
Most Short-eared Owls overwinter in their breeding range. Clark (1975) suggested that some may occasionally migrate. When migration does take place, the first migrants return to the province during March and early April (Pinel et al. 1991).

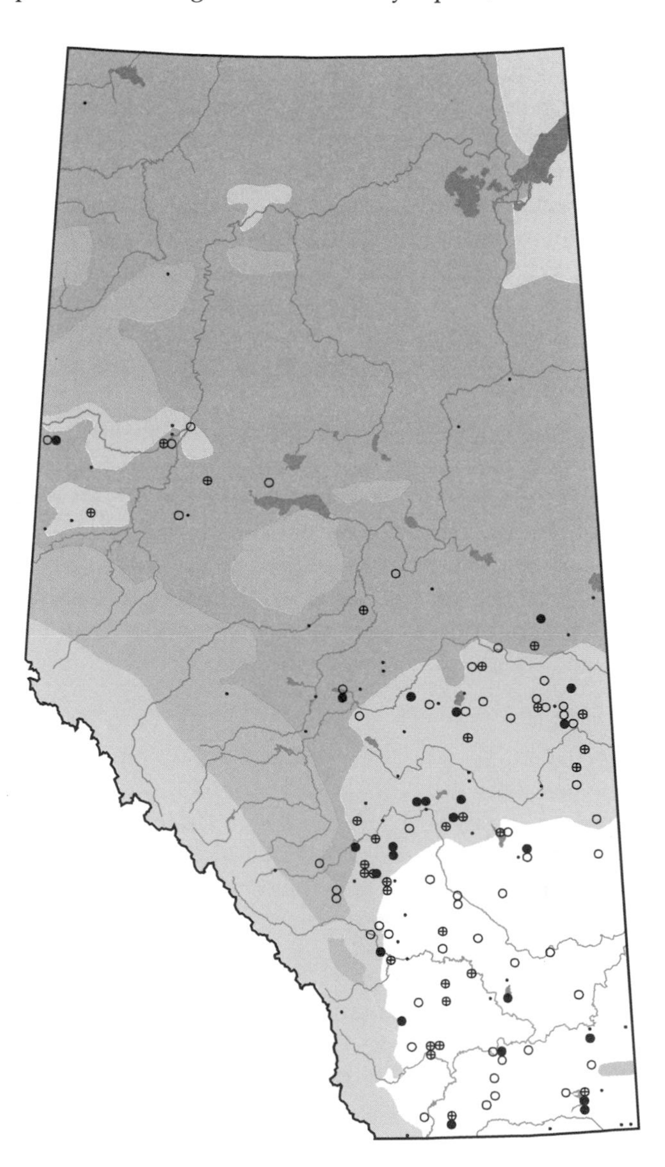

Boreal Owl

Aegolius funereus

BOOW

Status:
The Boreal Owl was found to be uncommon and widely distributed during the Atlas survey period. There were only 36 breeding records during the five year survey period, and only five of these were confirmed breeding records. This may be a result of the birds elusive, mostly nocturnal nature and its primary association with relatively inaccessible areas of coniferous forest (Johnsgard 1988). Pinel et al. (1991) reported that there were more records of the Boreal Owl in southern Alberta in the 1970s than there were in the 1960s.

Distribution:
The breeding range of this owl as delineated by Atlas surveys includes the Boreal Forest Natural Region (14 records) south to the North Saskatchewan River and the Foothills (14 records) and Rocky Mountain regions (8 records) south to Canmore. Pinel et al. (1991) reported breeding south to Bragg Creek and Priddis and indicates that the breeding range in the Rocky Mountains may extend from Canada south to northern New Mexico.

Habitat:
Coniferous and mixed wood forests are the preferred breeding areas for the Boreal Owl. They tend to avoid large unbroken stands of pine. In winter it wanders into settled regions where it may roost in isolated deciduous trees as well as coniferous woodland, sometimes buildings, and even on a haystack (Godfrey 1986).

Nesting:
Most nesting occurs in abandoned Northern Flicker or Pileated Woodpecker holes in both dead and living trees, usually at a height of 3-8 m (Eckert 1974). Occasionally, the birds build a crude nest of sticks and twigs, which is lined with leaves (Reed 1965). Eckert (1974) reported that the nest of a Rusty Blackbird or Gray-cheeked Thrush is sometimes used, especially in the more northerly limits of the bird's range. The typical clutch is 4-6 glossy white eggs. The incubation, 26-28 days, is by the female and the male provides food for her during this time. The female broods the young for the first three weeks (Harrison 1978) and they fledge at 30-35 days.

Remarks:
The Boreal Owl closely resembles the Northern Saw-whet Owl, but is larger than that species. Its upperparts are deep brown with white spots on the crown and its underparts are white with grayish-brown streaking. It has grayish-white facial discs with a black outer border. The bill of this owl is yellow, distinguishing it from the saw-whet, which has a black bill. The call of this owl is heard only during courtship and the nesting season. It resembles the winnowing noise of the Common Snipe, being a trill of essentially constant pitch and lasting 1-2 seconds (Johnsgard 1988).

This owl usually uses low perches, from which it dives to take ground-dwelling prey. Voles make up the majority of the diet, but mice, squirrels, sparrows, frogs, and larger insects are also eaten.

There is no regular annual migration, although periodic irruptions do occur, bringing large numbers of these owls into southern Canada and the northern United States (Johnsgard 1988). These migrations may occur in response to scarcity of prey species.

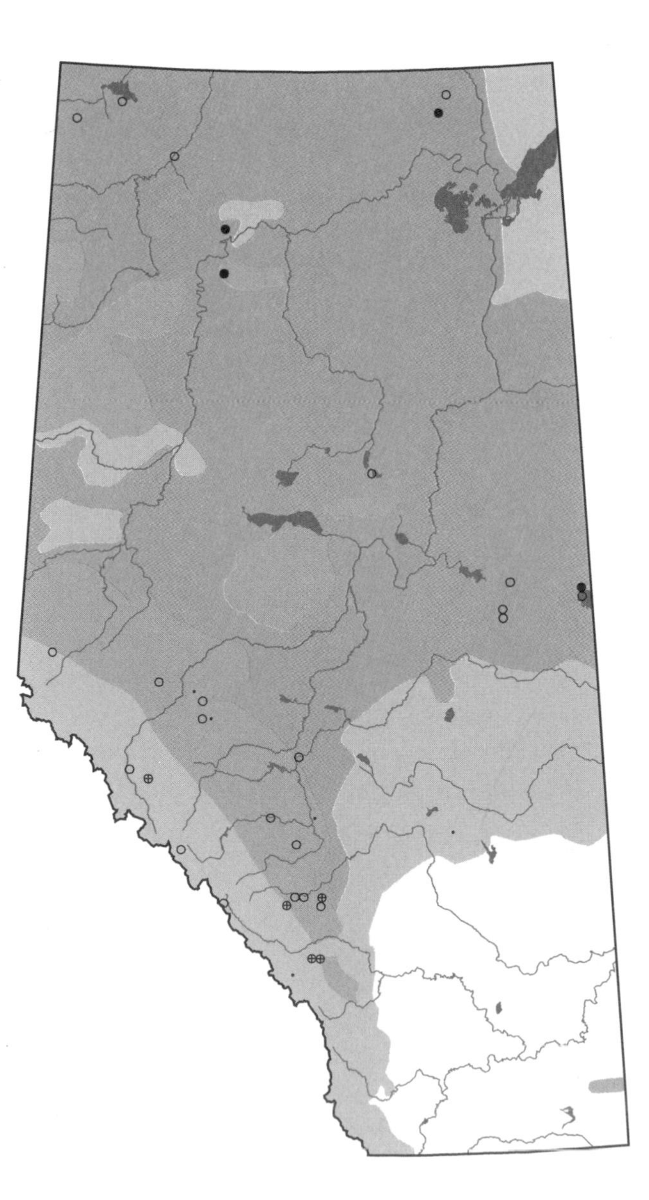

Northern Saw-whet Owl

Aegolius acadicus

NSWO photo credit: T. Webb

Status:

There is no detailed information available on the status or possible population trends of this species in North America (Johnsgard 1988). In Alberta, it is fairly common locally in the central part of the province.

Of the two subspecies that occur in Canada, *A. a. acadicus* is the one that occurs in Alberta (Godfrey 1986).

Distribution:

Salt and Salt (1976) reported the breeding range of this owl as the southern half of the province west into the Rocky Mountain region and north to Valleyview and Big Mouth Creek on the Athabasca River. Pinel et al. (1991) reported that this is misleading as there were only three historical records of breeding in the Grassland region. Atlas surveys only recorded one possible breeding record for this region.

It was recorded in 12% of surveyed squares in the Parkland and Foothills regions, and 7% in the Boreal Forest and Rocky Mountain regions. The majority of confirmed breeding records (75%) were from the Parkland region. The northern limits of the breeding distribution as shown on the map agree with those described by Salt and Salt, with the exception of one northern record in the Twin Lakes area.

Habitat:

Breeding habitats are typically mature forests with a mixture of large trees, both living and dead, and with medium sized woodpecker cavities present (Johnsgard 1988). The Saw-whet prefers moister areas, alder thickets, and tamarack bogs to drier deciduous areas.

Nesting:

The most important factor in selecting a nesting area is the presence of woodpecker cavities, especially those of the Northern Flicker. The nest may have wood particles and "sawdust" as a lining, but may also contain dry grass, moss, feathers, and pieces of bark (Campbell et al. 1991). The typical clutch is 5-6 glossy white eggs, which the female incubates for 26-28 days. The male provides food for the female through the incubation and early brooding stages. The young fledge at approximately 34 days. Double brooding is not unknown to the Saw-whet, although it is primarily single brooded (Eckert 1974).

Remarks:

The Northern Saw-whet Owl is the second smallest of the owls found in western Canada with a length of about 20 cm. It has no ear tufts but a relatively large head with small facial disks and whitish streaking on the crown. The underparts are white with broad, irregular chestnut streaks. It is distinguished from the Boreal Owl by the streaking on its crown and a black beak instead of that species' yellow beak. The somewhat metallic call, like the filing of a saw, for which the bird has been named, is not its most common call. The more common call heard during the breeding season is a more melodious whistled note, "whook", that is uttered three times every two seconds (Bent 1961).

Prey for this owl include mice, voles, bats, shrews, young rabbits, frogs, and some small birds. Because of its high rate of metabolism, this small owl must eat often during its active period (Eckert 1974).

Salt and Salt (1976) reported that this owl is usually considered to be a permanent resident in its territory, but winter sightings in Alberta are rare. Pinel et al. (1991), in reporting the scarcity of winter records, suggested that perhaps the majority of Northern Saw-whet Owls in Alberta are migratory.

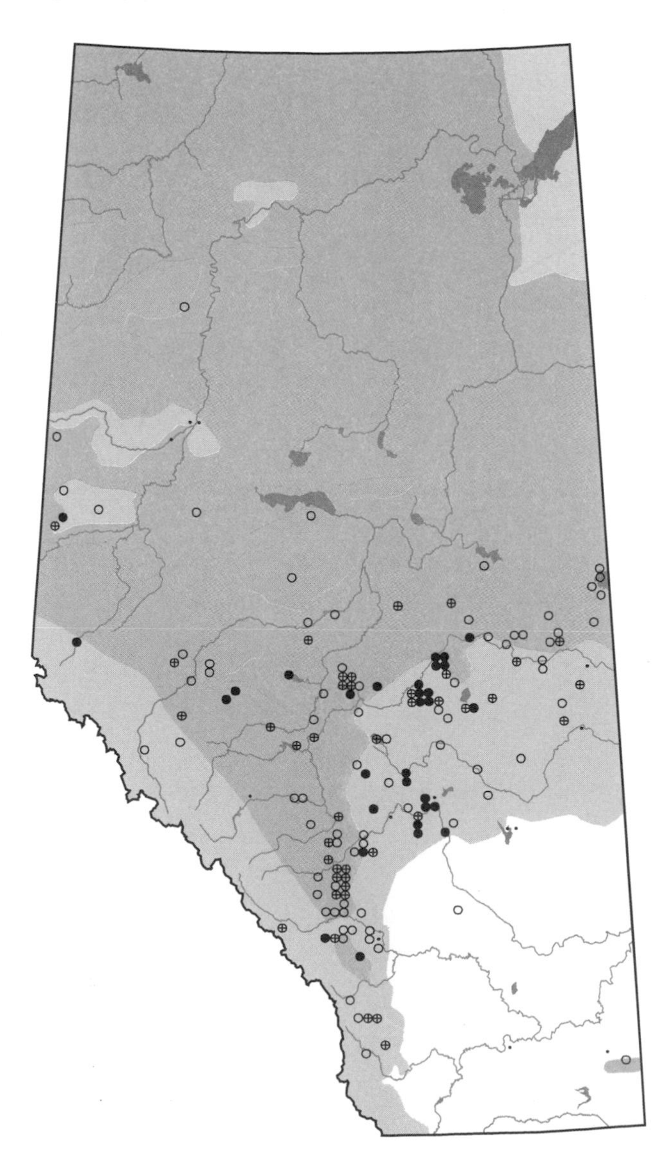

Order Caprimulgiformes

In Alberta this order of goatsuckers (or nightjars) includes the Common Nighthawk and the Common Poorwill. Their name was derived from the ancient belief that these species fed by suckling goats. The shapes of their mouths was partially responsible for this belief. However, their wide gape is an adaptation, much like that of the swallow family, and is used to scoop up insects while in flight. Other adaptations which aid in this activity include a large head, medium to long pointed wings, and tail, and a small bill with bristles on the sides to prevent insects from escaping. Other features, such as small legs, feet, and claws, have been reduced because of their small role in these species' dichotomous lifestyle. By day, these solitary crepuscular (or nocturnal) creatures perch on the ground or lengthwise on a branch, resting before their nightly activity. Remaining motionless, they are very well camouflaged by their soft cryptic coloration. Eggs are usually placed directly on the ground without any use of a nest. In the evening or at night, they take flight, using their swift and acrobatic flight to scoop up insects in almost bat-like fashion. The common Poorwill has been recorded elsewhere hibernating over winter in the crevices of canyon walls.

Common Nighthawk

Chordeiles minor

CONI photo credit: T. Webb

Status:
Once common, populations of Common Nighthawk may be slowly decreasing, or even undergoing local declines as seen in the Calgary area in the 1970s (Pinel et al. 1991). The Wildlife Management Branch (1991) reported the status of this species as undetermined with insufficient information available.

Distribution:
Atlas surveys found this nighthawk in every natural region within the province. It was recorded in 62% of squares surveyed in the Canadian Shield Natural Region, approximately 15% in the Boreal Forest, Parkland, Grassland, and Foothills regions, and 6% in the Rocky Mountain region. It's excellent camouflage on the nest resulted in breeding confirmations being very low, in proportion to sightings.

Habitat:
This species breeds in or near open or semi-open habitat in a variety of areas, including forest clearings, burnt-over areas, cultivated fields, gravel pits, barren rock, and beaches. In cities and towns, tar and gravel roofs have often been used.

Nesting:
A male will initiate the breeding season by using a display to court a mate. After a preliminary aerial display, he lands, then stands with his tail fanned and wagging, and rocks his body. Puffing out his throat, the male exposes a white patch and emits guttural croaks. Nesting is very simple, no construction is involved. The Nighthawk nests on sand, rock, gravel, leaf litter, and forest duff. This may be on flat areas or slopes and may be sheltered by nearby boulders or shrubs. A solitary nester, conspecifics may occasionally nest in close proximity. The two pale olive-buff eggs are speckled with brown, black, and gray and are nearly elliptical. These eggs may be shifted short distances away from water or into shade, as required. Intruders may be received with a distraction display. A single brood is raised. The female incubates the eggs for 19 days, perhaps with assistance from the nearby male. The semi-precocial and downy chicks are brooded by the female and fed by the male. The young are fledged in 23 days, becoming independent after 30.

Remarks:
The highly cryptic adult of this species has a barred belly, a white wing bar, and blackish upperparts marbled in gray, whites, and buffs. The male has a white tail bar and a white throat band which appears buffy on the female. This bird has a large head and a small weak beak with a large gape, adapted for scooping up insects in flight. These birds are normally seen in flight at dusk or dawn, readily identified by their dark appearance, slightly forked tail, and long, pointed wings with broad white patches on the primaries. Their flight call, a nasal-sounding "peent", is often imitated by the Starling. During the nesting period, the male will create a localized sonic boom near the nest area. This territorial display is produced by air rushing through the flight primaries as the male rounds out of a steep dive. The observer is less likely to view a bird while it's perched motionless, eyes closed, lengthwise on a branch, on a roof, post, or on the ground. Not related to true hawks, these twilight feeders scoop up insects such as mosquitoes, beetles, flying ants, moths, and grasshoppers, which have risen with the warm air currents. This acrobatic flight can be seen over pastures and forest openings.
Arrival in late May may be timed with higher daytime temperatures and insect populations. Rarely seen in flocks, most have left by late August to overwinter from Mexico to central South America.

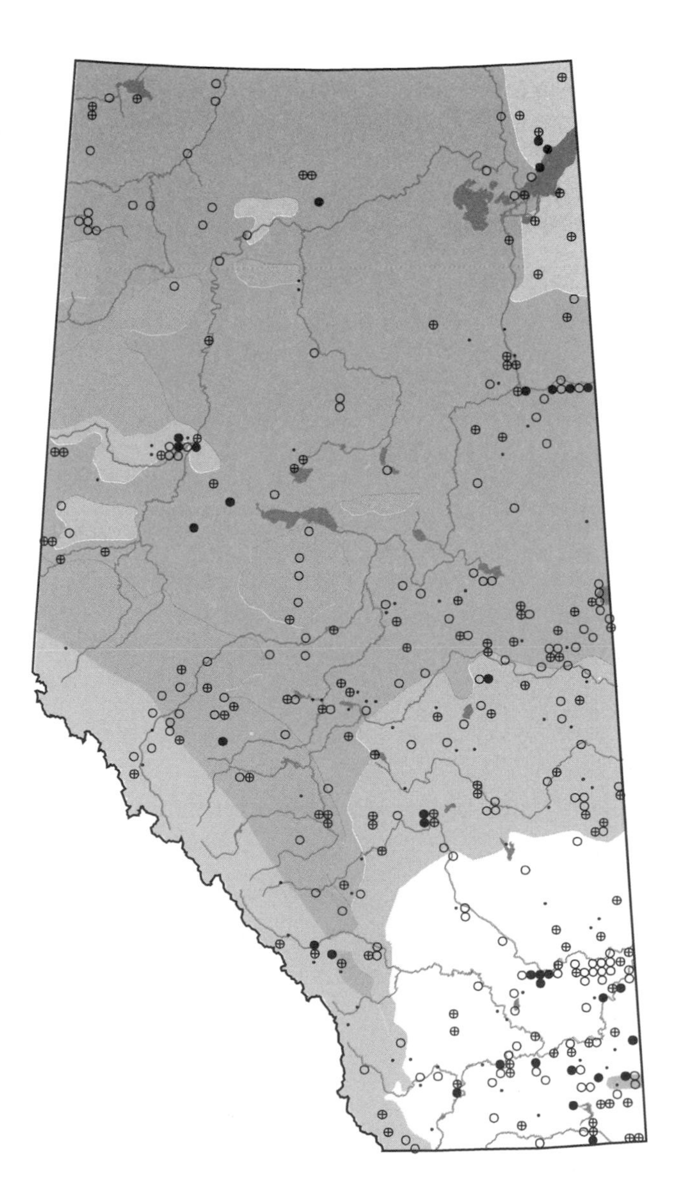

Common Poorwill

Phalaenoptilus nuttallii

COPW

Status:
The provincial status of the Common Poorwill still appears to be in question. Although Salt and Salt (1976) reported that there have been several sightings since 1945, there has never been confirmation of breeding. Two further sightings in the 1970s (Pinel et al. 1991) and one possible breeder during the Atlas Project failed to clarify this status. However, suitable habitat does apparently exist around the Cypress Hills of southeastern Alberta. Based on the above information, it is not possible to accurately determine this species' status, with Alberta marking the periphery of its normal distribution.

Distribution:
Several sightings have centred about the Cypress Hills, where Salt and Salt (1976) considered it to probably breed. However, other observations have been located in Lindbergh (Boag 1972), Calgary (Pinel and Riddell 1986), and, during the Atlas Project, in the Red Deer River area northeast of Dinosaur Provincial Park. The Cypress Hills of Saskatchewan and northeast and northwest Montana mark adjacent areas of Poorwill distribution.

Habitat:
This species breeds in semi-arid sagebrush benchlands and the grassy openings of dry open woodland (Godfrey 1986). Such habitat is available in and around the Cypress Hills.

Nesting:
Within its known breeding range, the Common Poorwill has been reported to return to the same locality each year (Bent 1964c). No nest is constructed. Instead, the female lays her two pinkish-white eggs directly on the ground. This substrate may be of rock, leaves, or gravel and usually contains a small depression or hollow, often shaded by overhanging brush or grass. The periods of incubation and of fledging are unreported, although both adults are known to incubate and care for the clutch. Like its roosting habits, while adults are on the nest they remain motionless. This, coupled with their cryptic coloration and nondescript surroundings, makes the Common Poorwill very difficult to find on the nest. If disturbed, it will raise the wings above its head and, with a wide open gape, will utter a snakelike hiss.

Remarks:
Adult Common Poorwills, both sexes similar in appearance, have short rounded tails and upperparts distinctly marked in black and finely mottled in gray and brown, white throat patches and blackish throats and upper breasts. Very similar to our more abundant Common Nighthawk, it may be distinguished by its smaller size, lack of white on the wings, rounded wings and tail, the absence of any white tail band, and the presence of white-tipped outer tail feathers. Also, unlike the high-flying Nighthawk, this species will more frequently be seen in bat-like flight at levels below the tree tops. Roosting motionless on the ground, often under grass or a shrub, or lengthwise on a tree limb, the Common Poorwill remains inconspicuous. More often heard than seen, its repertoire includes a whistled "poor-will",which at closer proximity sounds like "poor-will-uck". Sung most often at dusk and dawn, this whistled call is a far-carrying rendition.

A nocturnal feeder, it takes flight in order to hawk insects on the wing. This bird has also been observed making short jump-flights from the ground to hawk insects (Bent 1964c). Their diet includes mosquitoes, flying ants, beetles, grasshoppers, and moths. It also drinks on the wing.

The few site records available have not provided the migration dates for this species locally.

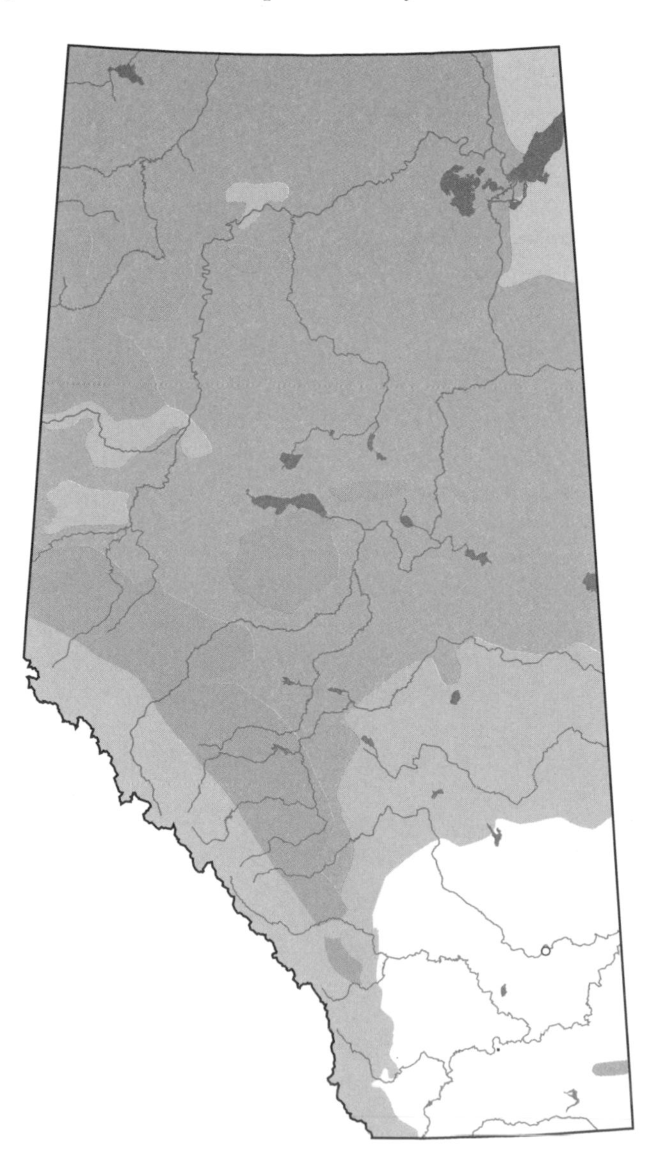

Order Apodiformes

(Family Apodidae, Family Trochilidae)

Literally meaning "legless", this order derives its name from the very small legs and weak feet of its members, the hummingbirds and the swifts. The visual similarities of the two families ends here. Although all have small, compact bodies with long wings, hummingbirds are much smaller. In fact, they are the smallest birds in the world. Hummingbirds have long slender bills and protrusible tongues, which they use to feed on nectar producing flowers. Because their rapid and agile, bee-like flight requires a high metabolic rate, they must carry out a high rate of feeding, hovering about wildflowers, cultivated flowers, and feeders. They have variable and iridescent plumages, and males are more brightly colored. Swifts, catching insects on the wing with their small bills and large gapes, may resemble nightjars or swallows. However, the tip of this species' upper bill is more decurved, it forages at great heights and over great distances by day, never alights upon trees, and sometimes breeds colonially. The sexes are similarly colored and, at rest, when not on the nest, they cling to vertical surfaces. As well, swallows and nightjars are much more common in this province than is the Black Swift.

Black Swift

Cypseloides niger

BLSW photo credit: G. Holroyd

Status:
The Wildlife Management Branch considers the provincial status of the Black Swift to be at risk. An extremely rare summer resident, this species occupies nesting habitat that is vulnerable to disturbance. The availability of its prey base, flying ants, is also uncertain.

Distribution:
The Black Swift is found only in the Rocky Mountain Natural Region of Alberta. Only two nesting areas have been confirmed in Alberta. One long-term site is located in Johnston's Canyon in Banff National Park. A more recently confirmed site is in Maligne Canyon, Jasper National Park. Additional historical sightings have been made at Sunwapta and Athabasca falls, Jasper, Canmore, Kananaskis, and Bragg Creek. The Atlas Project confirmed breeding in Johnston's Canyon, Maligne Canyon, and Bow Lake, as well as a possible breeder in the Alexandra River Canyon.

Habitat:
The Black Swift breeds in areas with cliff faces in canyons, in a moist situation near some seepage or by or behind a waterfall (Harrison 1978). Complementing these are the mountain slopes, over which the Black Swift forages.

Nesting:
The monogamous pair conduct their courtship and copulation efforts on the wing (Bent 1965a). Returning to the same site each year, Black Swift pairs select dark, damp, and out-of-the-way cliff crevices near and behind waterfalls (Harrison 1978). If conditions permit, a location may host a colony of birds. A simple nest of grass, moss, pine needles, and ferns is lined with fine rootlets and pine needles. It is often built over top of an old nest, and it is not known if only one or both contribute to the nest construction. Other swifts utilize a saliva and mud plaster when building nests, but this is not true of the Black Swift.
In late June or early July, the female lays the single dull white egg for her only brood, possibly timed to coincide with a later flying ant eruption. Nestlings have been observed as late as September, elsewhere. Both adults participate in the incubation, which lasts 24-27 days. The young have very weak feet, and are unable to leave the nest until they fledge about one month later. They are fed by regurgitation, mostly at dusk and dawn. Unlike most young birds, the young swift goes long hours without food.

Remarks:
The Black Swift is Canada's largest swift. It has a very small bill and very large mouth, small but strong feet, and a hind toe that is reversible (Godfrey 1986). The male is dark sooty above and below with a small black area in front of the eyes and a slightly forked tail. The female has less of a forked tail and the feathers of the abdomen and undertail coverts are tipped in white. Unlike the Vaux's Swift, there are no spines at the tips of the tail feathers. As their name would imply, these birds fly very quickly, aided by their elongated wings. Resembling the Purple Martin, the Black Swift's wings are longer and its head, body, and tail are of equal width.
Entirely insectivorous, this swift forages at great heights during fair weather, when it's barely visible, and lower during dull weather. It also ventures far from the nest in search of food, especially in periods of inclement weather.
Little is known of this species' migration patterns. Arriving in the mountain parks in mid-June, their nesting activities in Banff in the past often went unnoticed. The dates of fall migration are uncertain but are suspected to be in mid-September.

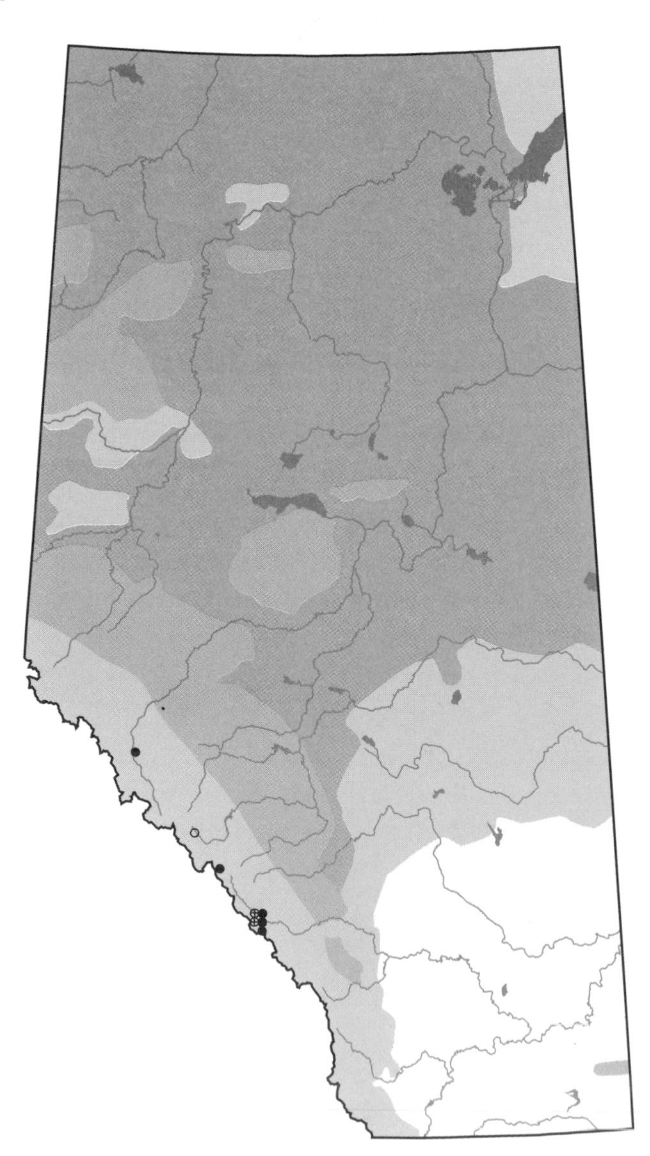

Ruby-throated Hummingbird

Archilochus colubris

RTHU

Status:
The status of the Ruby-throated Hummingbird in Alberta appears to be healthy and the species distribution is widespread where suitable habitat is available. Continual clearing activities northward for agriculture and forestry may increase the proliferation of this species.

Distribution:
Appearance through central and southern Alberta marks the most northerly and western distribution of the only hummingbird species native to eastern North America. In Alberta, this extends north to the Peace River area, Lesser Slave Lake, and Fort McMurray, and west into the foothills of the mountain parks. It was most commonly recorded in the Parkland region and the southern portion of the Boreal Forest region.

Habitat:
The high number of sightings in the Parkland region of central Alberta delineate the optimal habitat for this species as woodland clearings, edges, and gardens. The relatively few sightings in the Grassland region are the result of both fewer treed or shrubby areas for nesting, perching, and shelter, and fewer nectar-providing cultivated flowers or wildflowers.

Nesting:
The male Ruby-throated Hummingbird arrives in advance of the female locally, and establishes a territory whose size depends on the quality and density of nectar-producing flowers. The female constructs a nest in a woodland edge near water, saddles it on a limb 3-7 m above ground and lines it with plant down for warmth. Integrity is provided by the use of mosses and rootlets, upon which lichen flakes are placed for camouflage. The whole structure is secured with spider and insect silk. The female will lay two bean-sized, white eggs a day apart, sometimes even before the nest has been completed. At this point, the male becomes disinterested and returns to his feeding and territorial requirements. The female incubates the eggs for 16 days and also tends the young alone. The nestlings are fed aphids and spiders, which she gleans from nearby plants. The female thrusts her bill deep into the young bird's throat, placing the food directly into its crop. The young leave the nest at about 19 days, and are shown what to eat before separating from the female. They are known to raise second broods.

Remarks:
The male is characterized by his iridescent throat, which varies from orange to red to black depending on the angle of view. The female has a plain white throat, a lack of brown on her sides, and a dark forked tail. Like other hummingbirds, this species is highly territorial, often even guarding artificial feeders while standing watch from a nearby perch. The humming noise is created by its wings, which regularly beat at 78 beats per second, accelerating to 200 beats per second in courtship flight. Hummingbirds are capable of flight in all directions. This places high energy demands on the bird, which must take in half its weight in food and 16 times in weight in water daily. A long, slender bill and extensile tongue permit these species to feed on the nectar of various flowers. Some insects are consumed.
These tiny birds complete a long migration to and from the southern U.S. and central America; many cross the Gulf of Mexico. Arriving in mid-May, the Ruby-throated is known to feed on the running sap furnished by the drilling of sapsuckers. More visible as they wander in the late fall, these birds have departed our province by late August or early September.

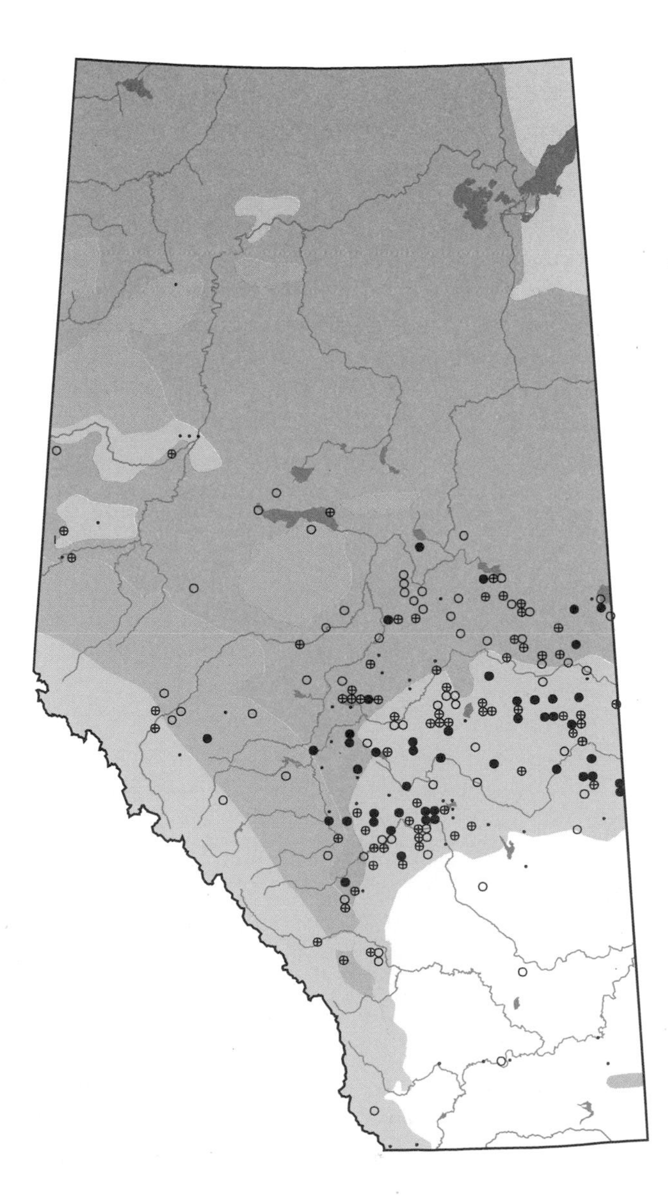

Calliope Hummingbird

Stellula calliope

CAHU

Status:

Little is known of the provincial status of this species. Extreme difficulty separating Calliope from Rufous Hummingbird females has led to poor breeding confirmation. Of 28 records for this species during Atlas surveys, only four were confirmed breeding records. Salt and Salt (1976) categorized them as scarce in the Rockies from Waterton to Jasper and east to Longview and Exshaw.

Distribution:

Alberta marks the eastern range of the Calliope. Breeding was confirmed in the upper Wapiti River, in the upper Red Deer River, at Canmore, and in Waterton Lakes National Park. It is thought to breed in Jasper National Park (Salt and Salt 1976) and east within the Foothills Natural Region. The confirmed breeding in the Wapiti River area is the northernmost breeding record for the province.

Habitat:

In the Rocky Mountain Natural Region, this hummingbird frequents avalanche slopes, burns, shrubby meadows, and open timberline forest, and other disturbed areas where flowers and berry shrubs are abundant (Holroyd and Van Tighem 1983). It nests in open woodland, in either coniferous, deciduous, or mixed wood.

Nesting:

Female Calliope Hummingbirds base their mate selection on the quality of food on his territory. Courtship is accomplished with the use of flight displays. The male will fly back and forth like a pendulum, with wings humming and gorget fanned (Salt and Salt 1976). The female then constructs a nest of moss, lichens, and plant down, all bound together by cobweb silk. The nest is saddled on a low, sloping tree branch, generally facing eastward and protected by branches above. Brunton et al. (1979) reported a female nesting on top of an old nest in Kananaskis Provincial Park. Bent (1964c) reported similar examples, as well as the use of old nests. Two very small, white eggs are laid up to three days apart (Brunton et al. 1979). The male becomes disinterested after the eggs are laid, and wanders off to defend a territory of open area, having no ability to produce fluids from his crop for the young. The female will vigorously defend her wooded breeding territory alone. The female incubates the eggs for 15-16 days, sometimes completing nest construction during this time. The young fledge in 18-22 days, although one or both may leave the nest prematurely to perch in the immediate vicinity, where they will be fed and preened.

Remarks:

The male appears iridescent green above, and dull white below. On the sides of the throat, he has long iridescent red feathers, which are fanned out during display. The female, easily confused in the field with the Ruby-throated Hummingbird, is iridescent green above and dull white below. The Calliope is the smallest North American hummingbird, but has the most controlled flight. In territorial display, the male swoops in a deep "U", emitting a loud whistle or squeak at the bottom. It may also hover about 10 m above the intruder and descend slowly. Less noisy than the Ruby-throated, it is still a vigorous defender of its territory, challenging all other birds and animals. Attracted to brightly colored flowers, particularly red, it uses its long, slender bill and extensile tongue to procure nectar. It also consumes tiny insects and spiders and often flycatches on the wing.

Spring arrival occurs in the last two weeks of May, and autumn migration to Mexico is completed in late August.

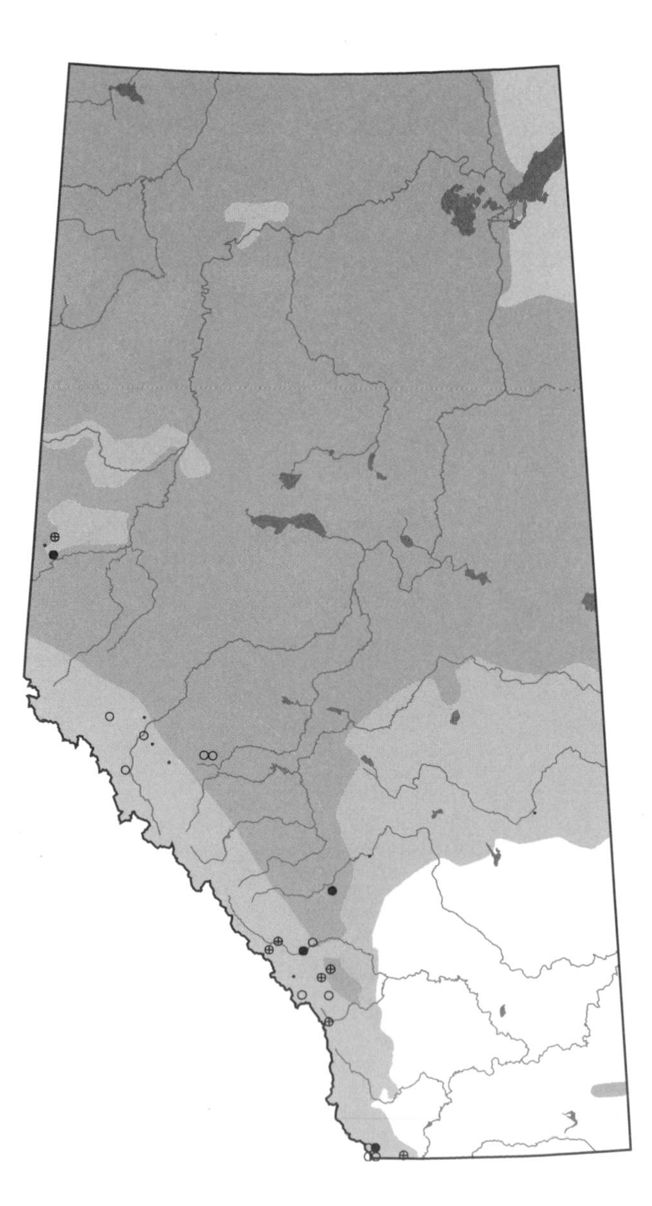

Rufous Hummingbird

Selasphorus rufus

RUHU photo credit: T. Webb

Status:

The current status of the Rufous Hummingbird in Alberta has not been well researched and is therefore considered undetermined. Holroyd and Van Tighem (1983) considered it to be a fairly common summer resident of the mountain parks.

Distribution:

This species breeds in the Rocky Mountain and Foothills natural regions, where it was recorded in 23% and 5%, respectively, of squares surveyed by the Atlas project. Salt and Salt (1976) reported this hummingbird breeding from Waterton Lakes National Park in the south northward to Jasper, wandering eastward, where it was seen at Pigeon Lake, Edmonton, Red Deer, Little Fish Lake, Strathmore, and Lethbridge. Atlas data records probable breeding north of Jasper National Park and the Wapiti River, and there was one confirmed breeding record near Peace River. Like the Ruby-throated Hummingbird, its northern distribution may be restricted to areas of nectar producing plants and Yellow-bellied Sapsucker distribution.

Habitat:

Within its range, the Rufous Hummingbird breeds in areas that provide nectar-producing flowering plants. This may include mountain wildflower meadows, valley bottoms, old burns, open Douglas-fir and treeline forests, montane shrub meadows, forest edges, shrubby stream margins, avalanche slopes, and cultivated flower gardens.

Nesting:

The female begins construction of her nest in mid- to late June. The firm, cup-shaped nest is constructed of mosses, shreds of bark, lichen flakes, and plant down, and is bonded with cobwebs. It is saddled on a tree bough or positioned in a bush or vine. In early spring, it is usually located in the low branch of a conifer. Later nesters may construct the nest in the higher branches of a deciduous tree. It is protected from above by higher branches. A female may re-nest on top of a previous nest. Neighboring conspecifics may nest in close proximity. The female incubates her two small, white eggs for 12 days and feeds her chicks regurgitant after they hatch. As the nestlings' feathers grow fully, their diet switches to nectar and tiny insects. Fledging in three weeks, the young may return to the nest area for a short time for feeding. Otherwise, they disband shortly and wander, becoming more visible to observers.

Remarks:

The male carries its namesake color on the back, sides, flanks, and tail. His crown and back are glossy green, while his iridescent throat varies from scarlet to greenish to brown, based on the viewer's perspective. The female (pictured above on the nest) and young are difficult to distinguish from female Calliope Hummingbirds. The Rufous has more contrast between its reddish sides, flanks, and undertail coverts and its white breast and belly. In flight, the wings create a "z-e-e-e" noise and their voice is a soft "tchup". Typical of other hummingbirds, this species is aggressively territorial, selecting prominent perches to conduct his surveillance. He may use a variety of territorial displays, including gorget spreading, shuttle flight, and a tilted oval-shaped dive with wings buzzing, whining, and rattles. Both sexes may use tail fanning in flight or while perched. Feeders, flowers, and sapsucker holes are tenaciously defended, even on migration.

The Rufous Hummingbird arrives in small flocks in the last two weeks of May. Similar to the Ruby-throated, this bird uses Sapsucker holes to obtain late spring nectar and tiny insects before flowers begin blooming. Individuals depart, males first, in late August, to overwinter in southern California, Texas, and Mexico.

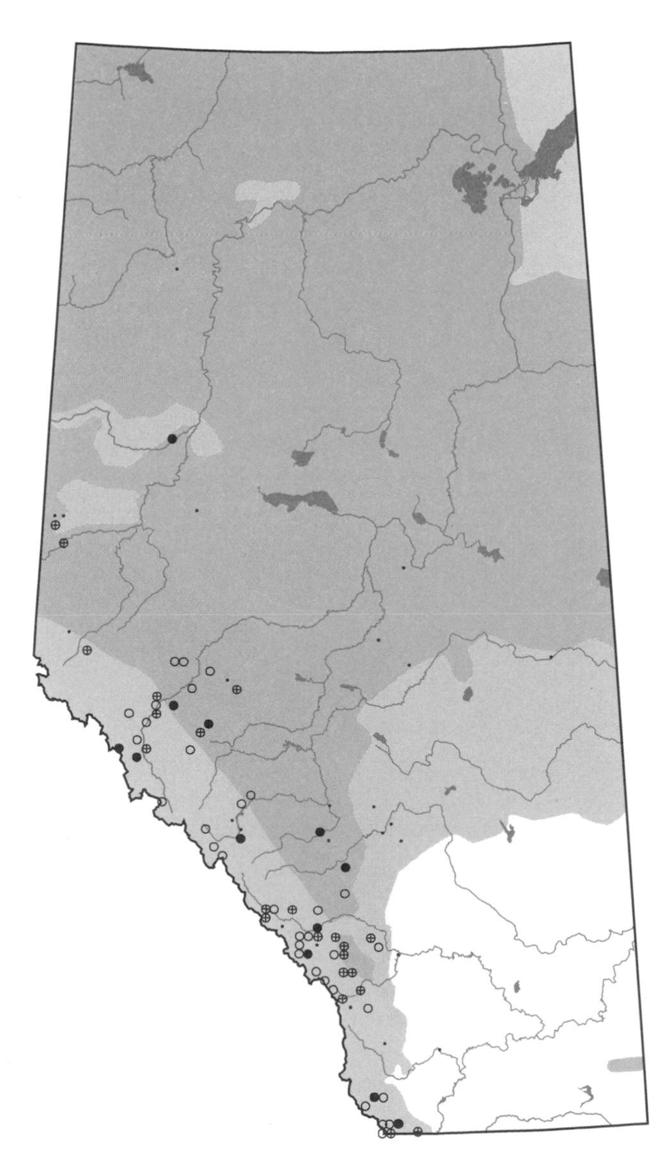

Order Coraciiformes

Not only in Alberta, but throughout most of Canada and the United States, the Belted Kingfisher is the only breeding member of this order. Its stocky appearance includes a short neck, a large head, a heavy, pointed bill, short wings and tail, a thick body, and small legs and feet, which have syndactyl toes. A crest adorns the head. The thickly oiled plumage, needed to repel water, is generally blue above white, with a prominent white collar. Quite solitary, these birds are seen along shorelines, where they perch on dead tree limbs. Hunting primarily for small fish, as well as shallow water crustaceans and amphibians, Kingfishers pounce on their prey by making a sharp-angled dive into the water. Occasionally, they may hover before attacking. When disturbed, the Kingfisher swoops down from its perch, making a short, strong flight over the water before alighting farther down in its territory. This is accompanied by an unmistakeable harsh rattle. Among its own species, it is an assertive territorial defender. When available inside the territory, the preferred nest site is a burrow near the water's edge with a connecting tunnel beyond the accessibility of mammalian predators.

Belted Kingfisher

Ceryle alcyon

BEKI

Status:
The Belted Kingfisher is a relatively common spring and summer resident in Alberta.

Distribution:
This species breeds in suitable habitat throughout Alberta. It was recorded in over 20% of squares surveyed in the Rocky Mountain, Foothills, and Canadian Shield regions, 15% in the Boreal Forest and 10% in the Parkland region. In Banff and Jasper national parks, the Belted Kingfisher is most common around the Athabasca, Bow, lower Cascade, and North Saskatchewan rivers (Holroyd and Van Tighem 1983). It is rare and irregular along the lower reaches of prairie rivers and other water bodies in southern Alberta (Pinel et al. 1991). It was recorded in only 4% of surveyed squares in the Grassland region.

Habitat:
Kingfishers are found in the vicinity of lakes, ponds, rivers, and streams where there are good populations of small fish, the water is clear and shallow, and there are suitable perching and nesting sites.

Nesting:
To nest, this species digs a burrow in a steep vertical bank, preferably one close to water and fishing areas. These birds will use cliffs along rivers or lakes, or sometimes road and railway cutbanks or gravel pits. Sawdust piles, tree cavities, and the earth-filled roots of upturned trees have also been used. The burrow, excavated by both sexes, has an oval opening and twin ruts made by the birds' feet. It is up to 2 m in length. It is usually dug close to the top of the bank, offering protection from predators. While nesting is often solitary, Kingfishers will sometimes be found in the vicinity of Bank or Rough-winged Swallow colonies. Burrows are often returned to over successive years.
The eggs are laid in an enlarged circular chamber at the end of the burrow, usually on bare ground. Nests are occasionally lined with fish bones, grasses, or other debris. Clutches are typically 6-8 glossy white eggs. Incubation, mostly by the female, lasts 23-24 days. The young, tended by both parents, fledge at 30-35 days. After the young fledge, parents teach them fishing to perched young by dropping dead meals, into the water below the perched young for retrieval (Ehrlich et al. 1988). When young can fish for themselves, they are forced from the older birds' territory. One brood is raised per season.

Remarks:
With its blue-and-white plumage, distinctive head crest, and harsh rattling call, the Belted Kingfisher is readily identifiable. It is a solitary bird, often seen near water, where it will sit motionless on a perch watching for prey in the water below. The above picture is that of a female.
The kingfisher hunts from the air as well as a perch, and may hover in the air for a moment before hitting the water with a splash. It will re-emerge with a fish in its mouth, which it takes to a perch, pounds if it is still alive, and swallows head first. Small fish are the bulk of the diet, along with crayfish, frogs, tadpoles, and insects when fish are scarce.
Belted Kingfishers arrive in Alberta in April, and depart by mid-October, although stragglers have been reported into early December. This species frequently overwinters where water remains open, with reports from Banff, Canmore, Hinton, Calgary, Medicine Hat, Lethbridge, Red Deer, Snake's Head, the Clearwater River near Rocky Mountain House, and Waterton Lakes.

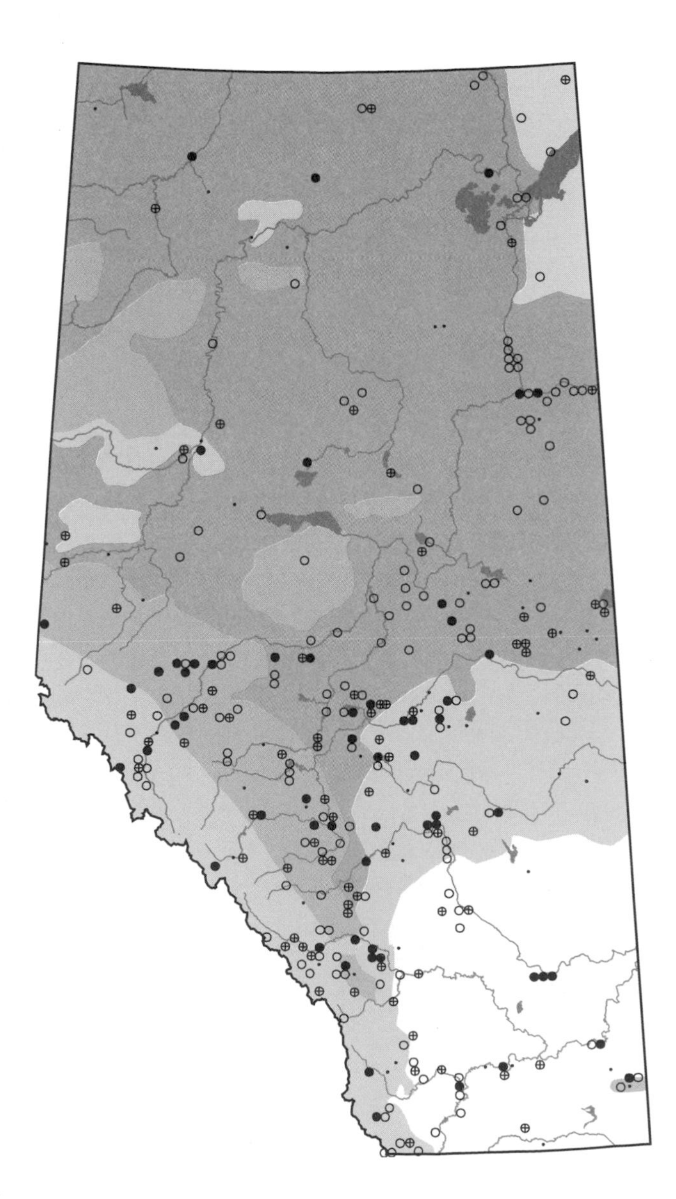

Order Piciformes

Although a few songbirds may glean the bark on the trunks of trees, only the woodpeckers chisel and excavate with such vigor and success. To accomplish this, they have made several adaptations. Species bill length may vary, but the bills of all are pointed and sturdy. When feeding, a long extrusible tongue unfurls, capturing insects on its sticky or barbed surface. The heavy skull includes an air membrane that encircles the brain. Massive neck muscles thrust the head forward and the stiff pointed feathers of the round or wedge-shaped tail act as a prop. Zygodactylous toes (two pointing forward and two pointing back), provide a sure grip on the vertical bark surface. Two species, the Three-toed and Black-backed Woodpeckers, have only three toes, two of which project forward. Movement along the surface is a short, jerky hopping motion. A drumming of the bill may be used by most species for courtship or territorial purposes. Observed in their treed habitat, most exhibit some combination of black, white, red and/or yellow coloration, and distinctive, deeply undulating flight. The sexes usually differ to a small degree in their coloration. Obligate cavity nesters, woodpeckers excavate cavities that are eventually used by many other birds and some animal species.

Yellow-bellied Sapsucker

Sphyrapicus varius

YBSA photo credit: T. Webb

Status:
The Yellow-bellied Sapsucker is a relatively common summer resident in Alberta.

Distribution:
This species breeds from Alaska and northern Canada east of the Rockies, south to the central and eastern United States. It is found in most wooded areas in Alberta, but nesting in the Grasslands Natural Region is uncommon (it was found in less than 3% of surveyed squares). It was found in 36% of surveyed squares in the Boreal Forest region, 31% in the Parkland region, 28% in the Foothills region, and only 7% in the Rocky Mountain region. In the mountains and foothills, its distribution overlaps with that of the Red-naped Sapsucker. There is evidence that these species may hybridize, making delineation of ranges difficult (B. McGillivray pers. comm.).

Habitat:
In the breeding season, they frequent deciduous or mixed woodlands, especially where birches and poplars are prevalent. Most nest sites are at the forest edge adjacent to water bodies or near other forest openings.

Nesting:
The Yellow-bellied Sapsucker nests in tree cavities, preferring deciduous trees to conifers. Trees are often dead, or partly so, and an aspen with rotting heartwood is the most-favored site. The male chooses the nesting site and both male and female excavate the cavity. Nest trees are often returned to repeatedly over successive years. The birds will usually excavate a new nest cavity, but will occasionally use an old one or the excavation of another species of woodpecker or a Starling. Cavities are at various heights, usually in the trunk, and are often lined with wood chips.
The female may lay 4-7 white eggs, though clutches are typically 5-6 eggs. Both sexes incubate over 12-13 days. Young are tended by both parents and may be fed regurgitant for a few days after hatching and then insects brought in the bill. Young are taught sap-sucking when they fledge, at 14 days (Ehrlich et al. 1988). After leaving the nest, the young depend on their parents for another 1-2 weeks, after which the family scatters. One brood is raised per season, but a lost clutch is usually replaced (Harrison 1978).

Remarks:
The adult male has a red crown and throat patch, the adult female a red crown (occasionally black) and white throat. Both sexes have a black breast patch and yellowish-white abdomen.
Sapsuckers feed a great deal on the sap or cambium of trees. They drill rows of holes in horizontal lines or a checkerboard pattern through the outer bark. Other food items include insects and larvae, fruit, berries, and some seeds.
This species arrives in the province in late April or early May and leaves the province again by early September. No overwintering has been reported.

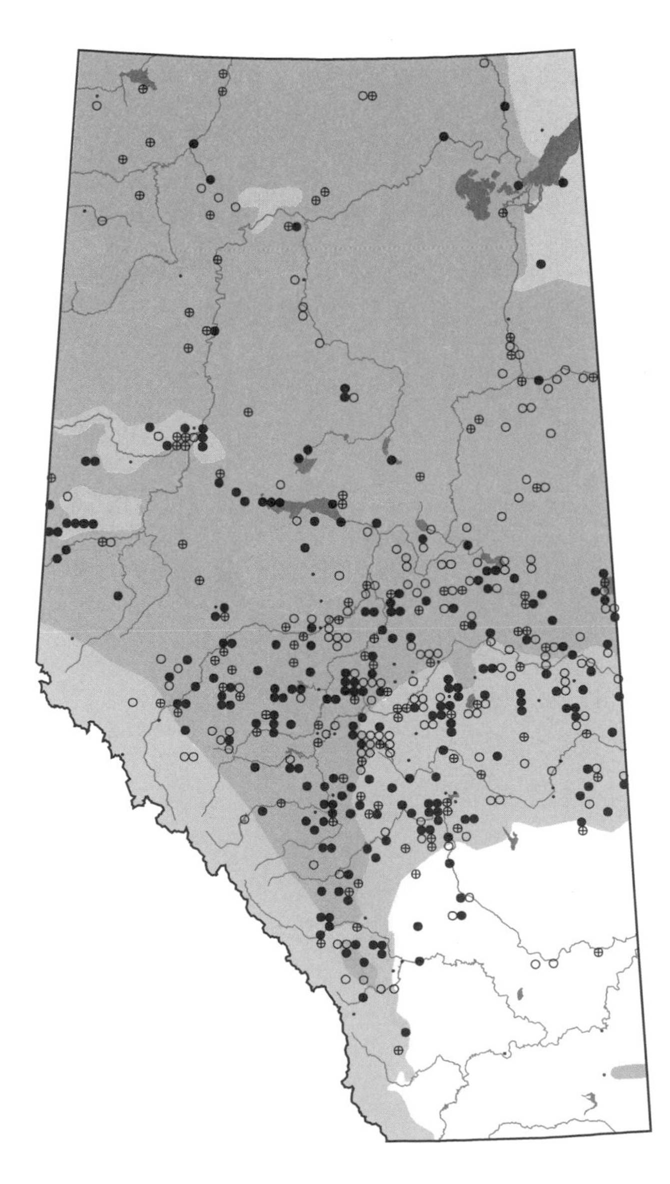

Red-naped Sapsucker

Sphyrapicus nuchalis

RNSA

Status:
Until 1985, this species was considered to be a subspecies of the Yellow-bellied Sapsucker (*S. varius*). It is common in southwestern Alberta from Banff to the United States border.

Distribution:
The Red-naped Sapsucker breeds from southern British Columbia, southern Alberta, and western Montana south to California. In Alberta, it is found in the foothills and mountains of southwestern Alberta, north to at least Banff National Park and Sundre. There were 34 confirmed nesting records in this area during the five years of the Atlas surveys. There were also two confirmed records in the Parkland region near Calgary. Pinel et al. (1991) reported a population in the Cypress Hills and this was confirmed by records at the Provincial Museum of Alberta, and by Atlas records. Birds have also been observed nesting in the lower Milk River and lower Red Deer River areas (Pinel et al. 1991), and in Edmonton (B. McGillivray pers. comm.)..

Habitat:
Preferred breeding habitat for this species is deciduous or mixed woodland, particularly where aspen, other poplar, and birch are important components.

Nesting:
Red-naped Sapsuckers nest in tree cavities preferring deciduous to coniferous species, and particularly aspen and birch. Nests are at various heights, usually in living trees. Nest trees are often at the edge of woodlands adjacent to water bodies such as streams, ponds, sloughs, and lakes, and other open areas such as road edges, logging slashes, transmission line rights-of-way, and mountain meadows (Campbell et al. 1990). Both sexes help to excavate the cavity, which is lined with wood chips from the digging process. Birds may re-nest in the same tree over successive years, but not necessarily in the same nest hole.
Clutches are 3-7, typically 4-5, white eggs. Both male and female incubate, the male at night and the female during the day, for 12-13 days. Young are tended by both parents, with fledging at 25-29 days. Young are taught sapsucking at fledging, and then remain dependent on their parents for only a short additional period (Ehrlich et al. 1988). One brood is raised per season.

Remarks:
This species very closely resembles the Yellow-bellied Sapsucker, but has more red on its head and on the nape of its neck. In adult males, the red throat extends farther posteiorly onto the black breast patch and it has less white on the back (Godfrey 1986). Its diet includes insects and larvae gleaned from under tree bark, with flying insects sometimes hawked from the air. It also eats tree sap and cambium, fruit, and berries. It will guard its sapwells from other species, including hummingbirds, warblers, and chipmunks (Ehrlich et al. 1988).
It is unknown whether the timing of this species' migration differs greatly from that of the Yellow-bellied Sapsucker. The latter arrives in Alberta in late April and early May, and leaves by late September. No overwintering has been reported.

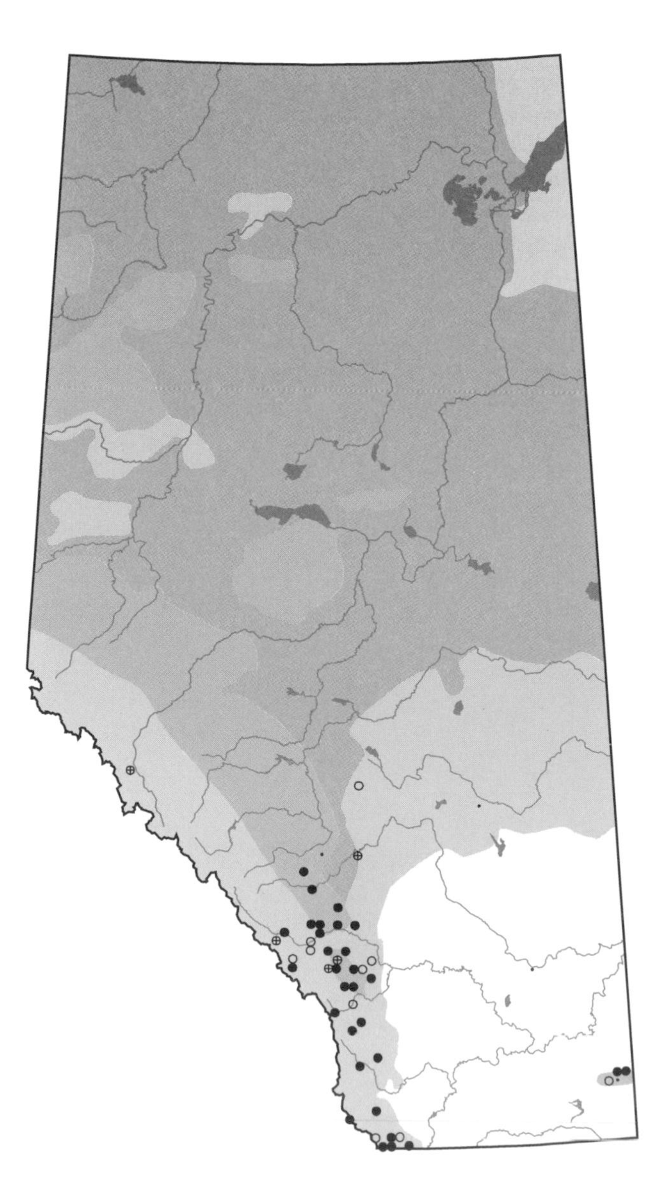

Downy Woodpecker

Picoides pubescens

DOWO

Status:
This woodpecker is a common year-round resident of Alberta. Its populations are healthy and stable throughout its range.

Distribution:
The Downy Woodpecker breeds in woodlands from southeastern Alaska across Canada south of the treeline. Though the species wanders widely in winter, it is not migratory. It breeds throughout Alberta, but there are few records for the northern Boreal Forest region. The majority of records were from the Parkland region of central Alberta, where it was recorded in 32% of surveyed squares. There were only sporadic nesting records in the southeastern part of the province, where this woodpecker is restricted to heavily wooded valleys and coulees. Subspecies *P. p. nelsoni* is found through most of Alberta south to Banff, while *P. p. leucurus* is found in the extreme southwest of the province.

Habitat:
This species prefers deciduous and mixed forests to conifer forests. It nests in a variety of situations, including wilderness, farm shelterbelts, city parks, and golf courses. In the mountain parks, it breeds in aspen stands and tall deciduous shrubbery, and in winter it is usually seen in shade trees and at feeders in Banff and Jasper townsite (Holroyd and Van Tighem 1983).

Nesting:
For nesting, Downy Woodpeckers excavate cavities in dead trees, or dead parts of living ones, with deciduous species (particularly poplars) preferred. The female chooses the nest site, and both sexes excavate the cavity, with the male doing the bulk of the work. The excavation generally takes two to three weeks to complete. Nest cavities are usually excavated anew each year, and are most often in the main trunk of a tree. The tree may be shared with other nesting woodpeckers and starlings. Eggs are laid on a bed of wood chips left over from the excavation process.
The typical clutch is 4-5 glossy white eggs. Both sexes incubate, but predominantly the male, for about 12 days. Young are tended and fed by both parents; initial food is caterpillars and soft insects. Young are at the nest entrance for feeding after 12 days, and fledge at 21-25 days. Adults are highly secretive during the nesting period, but nests can often be located by the noise of the nestlings. Young depend on their parents for another three weeks subsequent to fledging. One brood is raised per season.

Remarks:
The Downy is the smallest of Alberta's woodpeckers. It closely resembles the Hairy Woodpecker, but can be distinguished from that species by its smaller size, short stubby bill, and black spots on white outer tail feathers. It is rather tame and easy to approach and will sometimes feed from the hand. It is often encountered feeding on young saplings, hanging onto branches in chickadee style.
Like many other woodpeckers, it searches for wood-boring beetle larvae, chipping away at a tree to reach the insect's tunnel, then impaling the occupant on its barbed tongue. Other insects are sought on the tree bark and in crevices. While up to 85% of the Downy's diet is insects, it also eats small fruits, seeds, and sap from sapsucker holes.
This species disperses widely in the fall and, in winter, is a common visitor to household feeders. For winter comfort and protection, it usually excavates a roost hole in a dead tree.

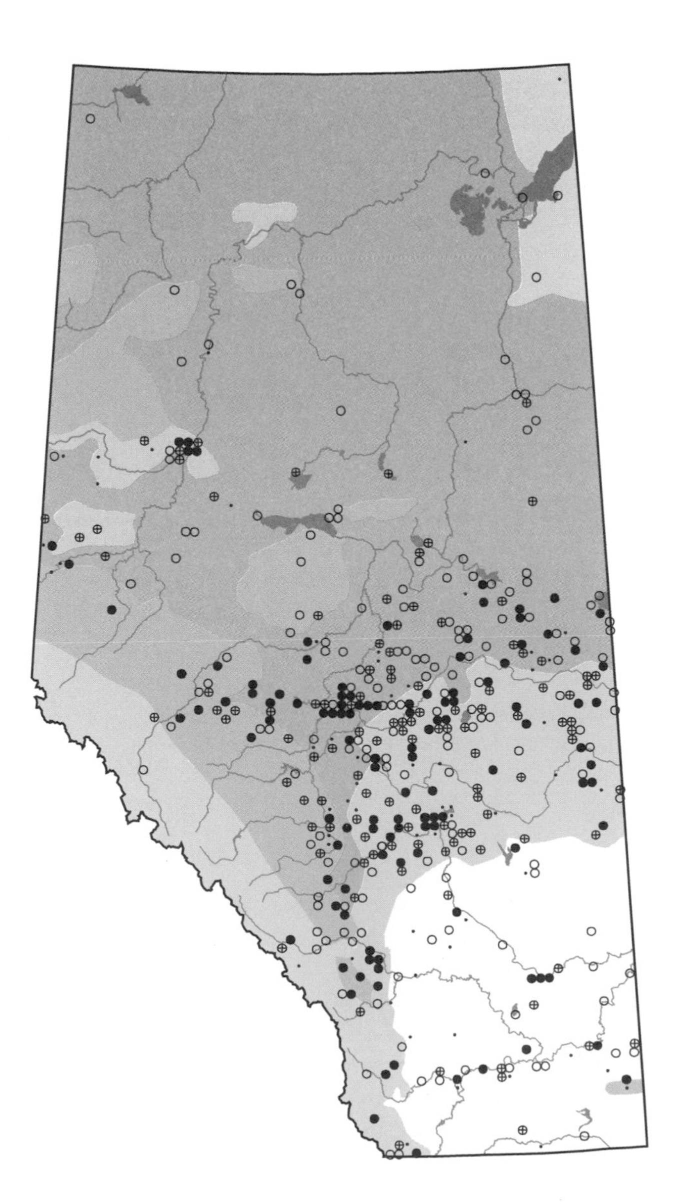

Hairy Woodpecker

Picoides villosus

HAWO

Status:

Populations of the Hairy Woodpecker are considered to be healthy and stable in the province. This species was on the Blue List from 1975 to 1982 because of population declines, but was upgraded to a species of "special concern" in 1986 (Tate 1986) as populations stabilized. Atlas data indicate that it is a common breeder in the Parkland region and in southern portions of the Boreal Forest region. To maintain nesting habitat for the Hairy Woodpecker and other hole-nesting birds, forests subject to logging must be managed so that clumps of snags in all successional stages are retained (Quinlan et al. 1990).

Distribution:

This species is a permanent resident throughout the woodlands of North America. It is found in suitable habitat in all natural regions of the province, with subspecies *P. v. septentrionalis* nesting east of the Rockies, and *P. v. monticola* in southwestern Alberta. Records in the Grasslands Natural Region are mainly along wooded river courses.

Habitat:

The Hairy Woodpecker inhabits deciduous, mixed, and, occasionally, coniferous forests, often where openings such as burns, logged areas, meadows, or marshes occur. In winter, these woodpeckers frequent a wide variety of trees, often visiting ornamental trees or tall shrubbery, even in urban areas (Godfrey 1986).

Nesting:

This species excavates nest cavities in trees, generally preferring deciduous species. The trees can be dead or living; living trees often have heart or sap rot. Patches of snags are favored nesting areas. Birds also occasionally use fence posts or wooden power poles. Nest holes average 2-6 m above ground, and are excavated by both sexes, with the male doing most of the work. Excavation takes about three weeks. Wood chips are the only nest material.
Clutches are 3-6 glossy white eggs, with incubation by both sexes for 11-15 days. This species nests earlier than most other woodpeckers, and eggs usually hatch before the end of May. Young climb to the nest hole entrance for food at 17 days, and fledge at 24-30 days. Both parents tend the young, although the female does most of the brooding. Young are fed directly by food carried in the bill. Parental care continues for several weeks after fledging. Once the young are self reliant, they leave the family and wander widely. One brood is raised per year.

Remarks:

The Hairy Woodpecker is black and white with a broad white stripe down the middle of the back. The adult male also has a small bar of red on the nape. This species closely resembles the Downy Woodpecker, but is larger, has a much longer bill, and, in most areas, has white outer tail feathers (Downys have black spots or bars). Hairys are also much more wary than Downys and will generally fly away when approached. This species provides many of the cavities used by non-excavating birds and mammals.
Main food items for the Hairy Woodpecker are ants, caterpillars, and beetle larvae, which it captures by gleaning and drilling snags, live trees, and downed logs. It also eats sap at sapsucker holes. Seeds retrieved by these birds from winter feeders will often be jammed into bark crevices and then hammered open with the bill.
This species is found throughout Alberta in winter.

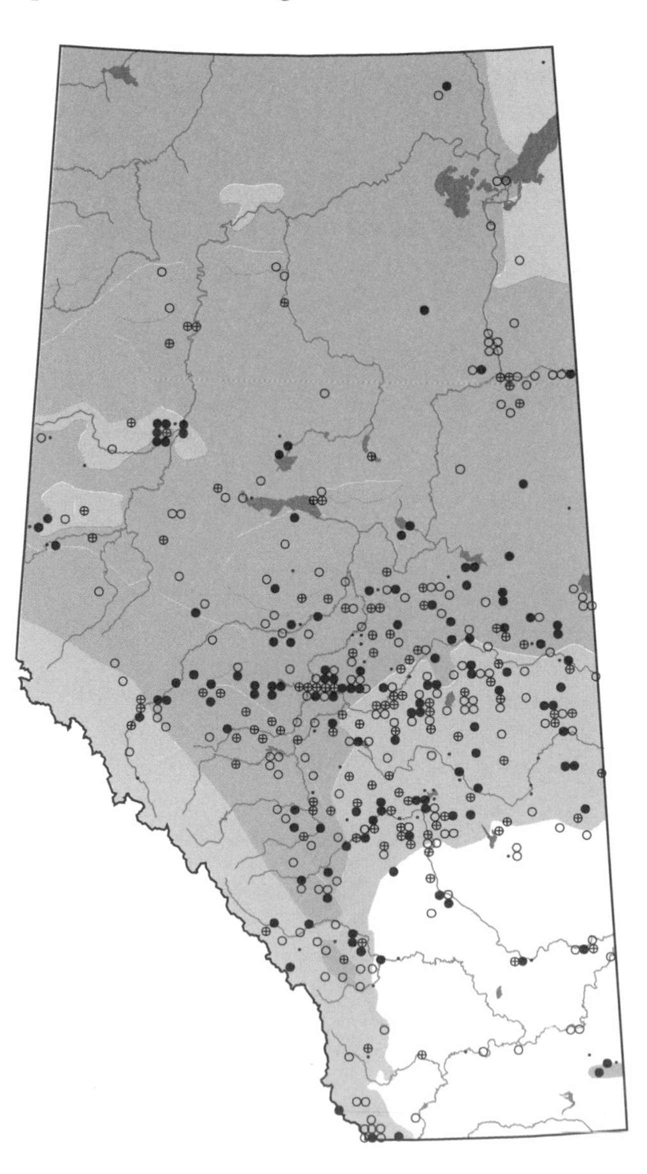

Three-toed Woodpecker

Picoides tridactylus

TTWO photo credit: R. Gehlert

Status:

The Three-toed Woodpecker is not abundant anywhere in its range, although numbers may increase locally during and after forest insect outbreaks. In Alberta, it is an uncommon resident for which the current status is unknown. Its reliance on mature and burnt-over conifer stands make it a species sensitive to northern logging. The subspecies of this woodpecker found in Alberta is *P. t. fasciatus*.

Distribution:

This woodpecker is a circumboreal species that breeds throughout northern Canada, the northeastern United States, and the Rocky Mountains. It was recorded in 20% of the surveyed squares in the Foothills region, 13% in the Rocky Mountain region, and 6% in the Boreal Forest region. Over 50% of the confirmed nesting records came from the Foothills Natural Region. Extralimital nesting is reported from Waterton Lakes, Edmonton, Elkwater, and Keoma (Pinel et al. 1991).

Habitat:

The Three-toed Woodpecker is a bird of mature coniferous forest with open areas near burns and clearcuts. In Alberta, it utilizes areas of spruce, fir, and pine, with a preference for mature spruce forest (Salt and Salt 1976; Holroyd and Van Tighem 1983). Foraging is often in burnt-over stands or elsewhere where there are stands of dead trees.

Nesting:

This species nests in cavities excavated in dead or living trees, with conifers preferred, particularly spruce trees. Nesting trees are usually near forest openings made by burns, logging operations, ponds, lakes, bogs, or muskeg. Male and female both excavate the nest hole, which is lined with wood chips. Cavities are in the main trunk of the tree, at various heights.

Clutches are typically 4-5 eggs, sometimes 3-6, the eggs a glossy white. Incubation, by both sexes, is 14 days, with the male incubating at night. Young are tended by both parents, even after fledging, which is at 18-23 days. Chicks are fed on food brought in a parent's bill. The pair bond of this species may be maintained all year, and in successive years. One brood is raised per season.

Remarks:

This rather inconspicuous woodpecker was formerly known as the Northern Three-toed Woodpecker. It has three toes, which distinguishes it from all other woodpeckers except the Black-backed. The male Three-toed, with its yellow crown, is similar to the Black-backed, except that the former has white bars on the back. The female Three-toed lacks the golden crown patch but also has a white back. Both male and female are heavily barred on the sides.

This bird is solitary and quiet and is relatively shy of humans. One is more likely to hear its drumming than any vocalization, except perhaps when parents are with young. It eats insects of various kinds, but a large percentage of the diet consists of the larvae of wood-boring insects. These it obtains by scaling the bark of conifers. Other foods include cambium and sap. Pairs forage separately, with the female foraging higher in trees than the male.

The Three-toed Woodpecker overwinters in the province, with sightings in such locations as Fort McMurray, Wembley, Edmonton, Wabamun Lake, Elk Island National Park, Fort Smith, Forestburg, Snake's Head, Horseshoe Canyon, Calgary, Red Deer, Banff, Jasper, and Waterton Lakes (Pinel et al. 1991).

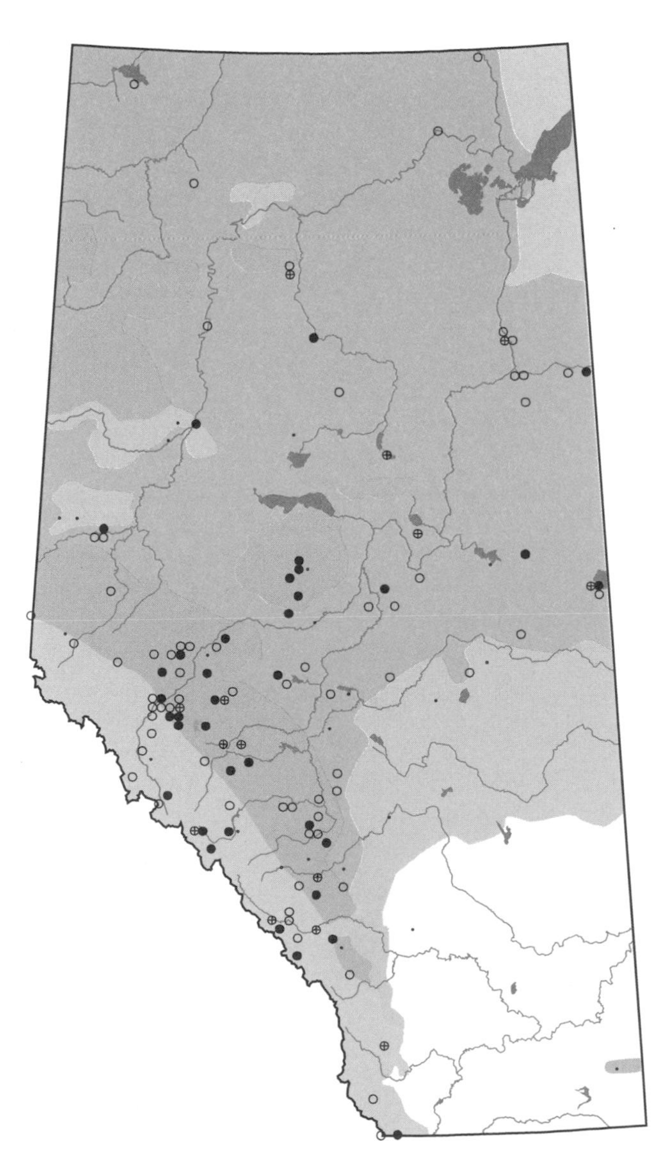

Black-backed Woodpecker

Picoides arcticus

BBWO

Status:
The current status of this species in Alberta is unknown, but it is not particularly abundant anywhere in its range. Maintenance of mature coniferous forests with standing dead trees is required to maintain populations of this bird in the province. Holroyd and Van Tighem (1983) considered this species a very rare resident in the Rocky Mountain Natural Region. This species is opportunistic, and may become temporarily abundant during and after forest insect outbreaks, but is usually uncommon and relatively inconspicuous (Bock and Bock 1974).

Distribution:
The Black-backed Woodpecker is a local resident in the coniferous forests of Alaska, Canada, and the northern United States, south in the mountains to California. There were only 33 records of this species during the Atlas Project. The records were divided almost equally between the Foothills Natural Region and the Boreal Forest Natural Region, with only two records from the Rocky Mountain Natural Region south to Waterton Lakes National Park.

Habitat:
This species breeds in dense mixed or coniferous forests, often choosing nest sites in stubs or decaying trees in burns, logged areas, windfalls, or other openings such as bogs, swamps, and lakeshores.

Nesting:
Nesting is in excavated cavities in the main trunk of dead or living trees, with conifers strongly preferred. Utility poles are occasionally used in some areas. Cavities are excavated by both sexes, but mostly by the male, and are usually less than 5 m above ground. The bark is often chipped away from the cavity entrance and the nest is lined with wood chips.

The female typically lays 3-4 eggs, sometimes 2-6; eggs are glossy white. Both sexes incubate for approximately 14 days. Both sexes also tend the young, feeding them soft insects and regurgitant. Fledging is probably at about 25 days. One brood is raised per year, but this species is known to renest if the clutch is lost (Harrison 1978).

Remarks:
This bird was formerly known as the Arctic Woodpecker and the Black-backed Three-toed Woodpecker. Its plumage is entirely black above and whitish below, but the male also has a yellow crown patch. This is a difficult woodpecker to observe, because of its coloration, which blends in with its dark surroundings, and its retiring nature. Movement of the bird from tree to tree is often the only means by which it can be discovered (Cadman et al. 1987).

The species mainly eats wood-boring insects, which it obtains by scaling bark from trees. It also eats other insects, spiders, cambium, mast, and wild fruit. It prefers to forage in trees with easily peeled bark, and commonly works low down on the trunk. Evidence of this woodpecker's activity is a dead conifer with large areas where the bark has been scaled.

The Black-backed Woodpecker winters throughout its range, moving south in winter. It has been reported at Edmonton, Calgary, Bow Valley Provincial Park, Fort Smith, Waterton Lakes and Elk Island national parks, the Peace River area, Wembley, Wabamun Lake, Rocky Mountain House, Lacombe, and Fort Saskatchewan.

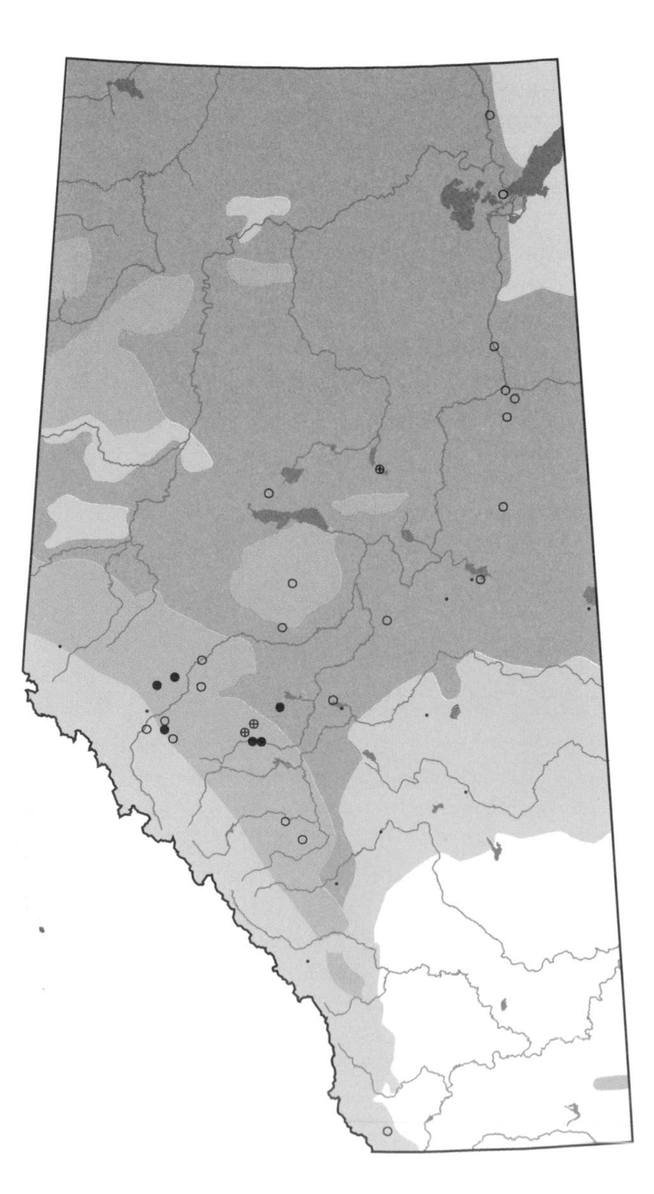

Northern Flicker

Colaptes auratus

NOFL photo credit: T. Webb

Status:
Yellow-shafted Flickers and Red-shafted Flickers, both of which breed in Alberta, are now considered to be two subspecies of the Northern Flicker. Populations of this species are considered to be healthy and stable in the province.

Distribution:
The Northern Flicker is the best known and most widely distributed of Alberta's Woodpeckers. It breeds throughout Alberta, but is scarce in the Grasslands region. It was recorded in 63% of surveyed squares in the Canadian Shield, 48% in the Boreal Forest, 45% in the Parkland, 52% in the Foothills, and 38% in the Rocky Mountain region. The Yellow-shafted Flicker (*C. a. borealis*) breeds in most parts of the province, except the mountain regions of the southwest, where the Red-shafted Flicker (*C. a. colaris*) occurs. Interbreeding is common, with hybrids found all the way to the British Columbia border. Very few pure Red-shafted flickers are now found in Alberta (B. McGillivray pers. comm.) Intergrades also commonly occur in the Cypress Hills.

Habitat:
This species breeds in a variety of habitats, including mixed, deciduous, and coniferous forests. The preference is for moderately open situations. It is often found at forest edges on logged areas, burns, wetlands, and shelterbelts. In the Grassland region, it is restricted to wooded valleys and coulees.

Nesting:
Flickers are cavity nesters that excavate their own nest holes in deciduous trees that are often dead or have decaying heart and sap wood. They also nest in hydro poles, fenceposts, nest boxes, and sometimes clay banks. Both sexes excavate the nest cavity, with the male generally choosing the site. Cavities are at various heights, but are often low to the ground (less than 3 m). No nesting material is introduced into the nesting cavity. Pair bonds in this species are for life and birds tend to return to their previous nesting territory.
Females lay large clutches of 3-12 eggs, typically 5-8. The eggs are a glossy white. Both sexes incubate for 11-13 days and both also tend the young. Young birds appear at the nest hole at 17-18 days and fledge at 25-28 days.

Remarks:
The Northern Flicker is a large woodpecker with brown underparts, a spotted breast, and a black crescent on the throat. In flight, this species' white rump is quite conspicuous. The two subspecies are separated by recognition of yellow or red under the tail and wing linings and the presence or absence of red on the nape. In Alberta, the subspecies' interbreed so that various combinations of these characteristics occur. The name of the species was derived from the "flicka, flicka, flicka" call the male gives when courting the female.
Its diet consists mainly of ants and their pupae and larvae, which it gleans from trees or by foraging on the ground. The birds also eat beetles, caterpillars, worms, and berries.
The first spring migrants arrive in late March and early April and most leave by the end of September, with some remaining into December. Birds regularly overwinter throughout the province, with records from Wembly south. Winter records have increased significantly since the 1970s (Pinel et al. 1991).

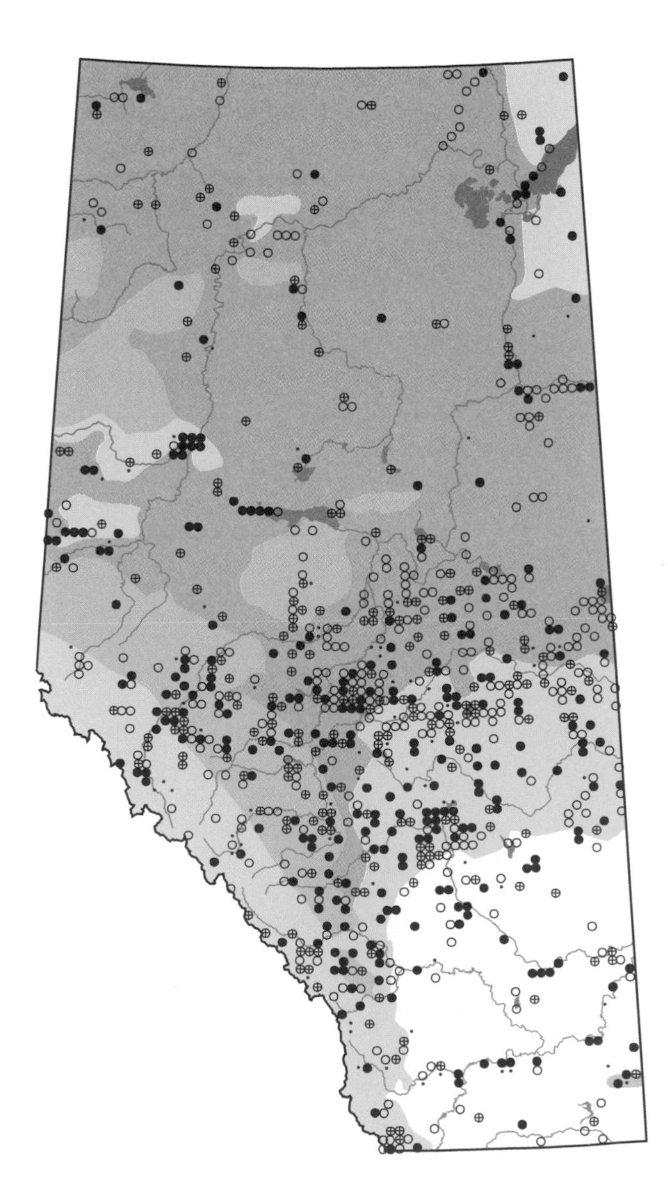

Pileated Woodpecker

Dryocopus pileatus

PIWO photo credit: T. Webb

Status:
The Pileated Woodpecker population in Alberta is thought to be stable. The relative frequency of breeding data from Atlas records supports this status. Because of its need for large mature and old growth trees for nesting, this species is likely to be severely harmed by intensive forest management aimed solely at timber production (Bull and Meslow 1977).

Distribution:
This species is resident in Canada from the Mackenzie District and Great Slave Lake south to the American border. In Alberta, it breeds mainly in the Boreal Forest, Foothills and Rocky Mountain natural regions. It was also recorded in 15% of surveyed squares in the Parkland region. Breeding records in the Grasslands region were restricted to wooded river valleys. It has continued to extend its range east along the Red Deer River, having been observed at Duchess, Dinosaur Provincial Park, and Jenner (Pinel et al. 1991). In winter, the species tends to wander farther south and east.

Habitat:
The Pileated Woodpecker prefers older, mature dense-canopied forest, particularly mixed and deciduous woods where there are large dead or dying trees for nesting and downed woody material for feeding. It is occasionally found in immature forests, woodlots, and parks. Unlike other woodpeckers, it rarely occurs in burns or drowned timber. It is becoming more prevalent in urban areas as suitable habitats mature. Birds defend a territory year-round.

Nesting:
This species is a cavity nester, excavating its nest hole in dead or partially dead trees, preferring deciduous to coniferous species. Nest trees are generally mature, with a minimum diameter of 40-50 cm, and close to water. Most nests are excavated in the main trunk, although large tree limbs are occasionally used. Both sexes help in the excavation. Adults tend to remain in a nesting territory year round, but usually build a new nest cavity each year. The nest is lined with wood chips.
The female lays 3-4 white eggs which are then incubated by both sexes for 18 days. The young are tended by both parents and fledge in 22-26 days. Nestlings are fed by regurgitation in the nest and post-fledging until they can feed on their own. Families may stay together until the fall, when the young disperse and wander widely. This species is single-brooded and has a year round pair bond.

Remarks:
The Pileated is our largest woodpecker. It is a crow-sized bird with mainly black plumage, a white stripe on the side of the head and neck, and a prominent flame-red crest (female is pictured above). Its loud call and slow, heavy drumming are distinctive.
This woodpecker forages on snags, live trees, and dead downed trees, feeding mainly on wood-boring insects and their larvae, carpenter ants, other insects, fruit, and nuts.
There are many winter records for this species, from Wembley south. In winter, birds will excavate separate cavities for roosting.

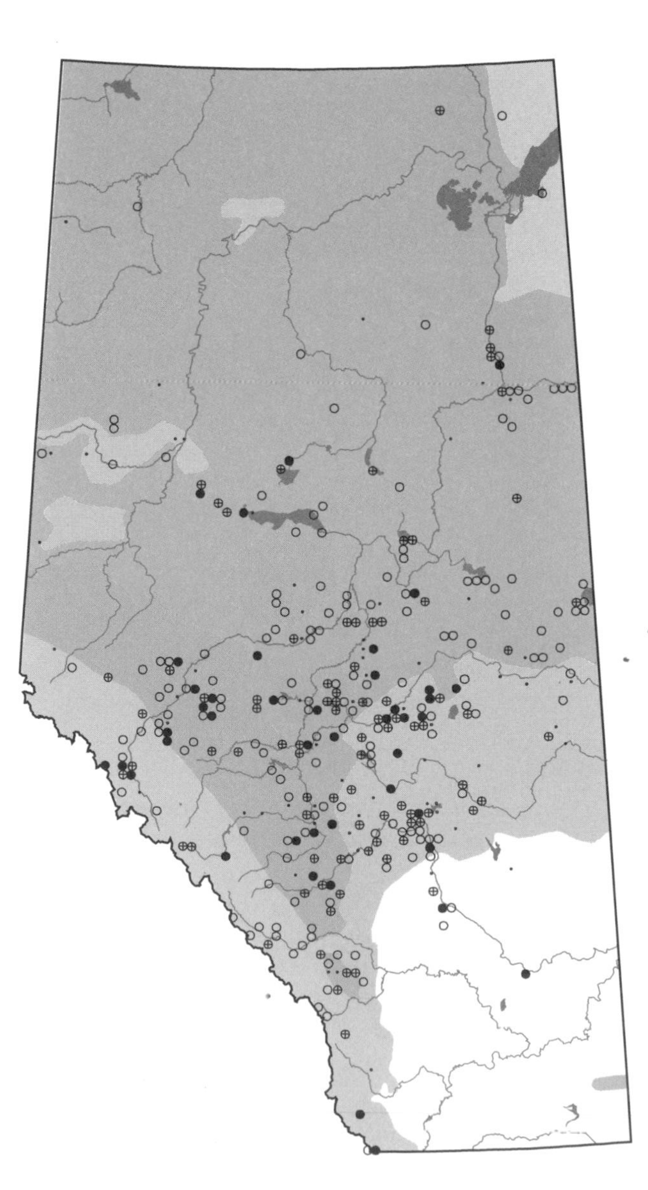

Order Passeriformes

(Family Tyrannidae, Family Alaudidae, Family Hirundinidae, Family Corvidae, Family Paridae, Family Sittidae, Family Certhiidae, Family Troglodytidae, Family Cinclidae, Family Muscicapidae, Family Mimidae, Family Motacillidae, Family Bombycillidae, Family Laniidae, Family Sturnidae, Family Vireonidae, Family Emberizidae, Family Fringillidae, Family Passeridae)

The songbirds comprise the largest and most diverse order of birds in the avian world, including 60% of all living bird species. In Canada in spring, their spirited singing marks the end of winter. However, because most migrate south in late summer, their departure foretells of the coming shorter, cooler days of fall. The two most common characteristics of songbirds are their perching and singing abilities. Equipped with three forward toes and a single strong one pointing backward, these birds are able to lock tightly around a stem or branch, assuring their stability in the face of strong winds or other loss of balance. Vocalizations are broken into two categories, calls and songs. Calls are short, simple notes given by all ages and sexes to elicit a call response from another. Songs are longer and more complex, and are often repeated over prolonged periods. These are usually only given by males and only during the breeding season. Generally considered as a symbol of masculinity, songs serve two purposes: the proclamation of territory and the attraction and stimulation of a mate. While learning songs, juveniles will sing subsongs, a garbled mixture of random notes including a few from the completed rendition. Passeriformes are split into two suborders—Tyranni, which perch but do not sing, and Passeres, which do both. Kingbirds, peewees, and flycatchers make up the former group, and all others make up the latter suborder.

Olive-sided Flycatcher

Contopus borealis

OSFL

Status:
This species is reported as an uncommon resident in the mountain parks (Holroyd and Van Tighem 1983). Atlas data indicate that it is uncommon across its range, but common locally in the southern Boreal Forest.

Distribution:
The Olive-sided Flycatcher breeds in wooded areas across North America, from near the tree line to the southern United States. In Alberta, it is found in the northern and central parts of the province south to about the North Saskatchewan River in the east, and in the west south through the Rockies and their foothills to the Waterton Lakes. It was recorded in 31% of surveyed squares in the Canadian Shield, 28% in the Foothills, 25% in the Rocky Mountain region, 21% in the Boreal Forest, and 8% in the Parkland region.

Habitat:
Preferred breeding habitat includes semi-open coniferous and mixed forest near water, often where there are standing dead conifers. Favored sites are bogs and muskegs, lakes with water-killed trees, burned or logged areas with snags, and other woodland clearings.

Nesting:
The nest of this flycatcher is built well out on a horizontal branch, usually in a conifer, less often a deciduous tree. It is geneally well-concealed and at a considerable height but, in rare cases, may be as low as 2 m (Godfrey 1986). Probably constructed by the female, it is a bulky structure of twigs, mosses, lichens, and rootlets, lined with fine grasses and rootlets. Clutches are typically 3-4 eggs, which are a buffy color with browns and purplish gray wreathing the larger end. Incubation is 16-17 days by the female. Both parents tend the young, which are brought insects in the bill. Young fledge at 21-23 days but still depend on their parents after this and may stay with the family group until fall migration. Pair bonds in this species are monogamous and they are vigorous defenders of the nest and territory.

Remarks:
This stoutly built, short-necked flycatcher is entirely olive-gray except for some white on throat and belly, and white rump patches that show best in flight. It is wary and difficult to approach, but can often be identified at a distance by its bulky appearance and its tendency to perch conspicuously on the tops of trees. It is well-known for its emphatic "quick-three-beers" spring song.
Olive-sided Flycatchers are almost exclusively insectivorous, taking bees, flying ants, flies, beetles, grasshoppers, moths, and dragonflies. They catch their prey on the wing, often with a loud snap of the bill, and return with it to their perch. This species is a late migrant, generally arriving around the third week of May. Birds move southwards again in late August.

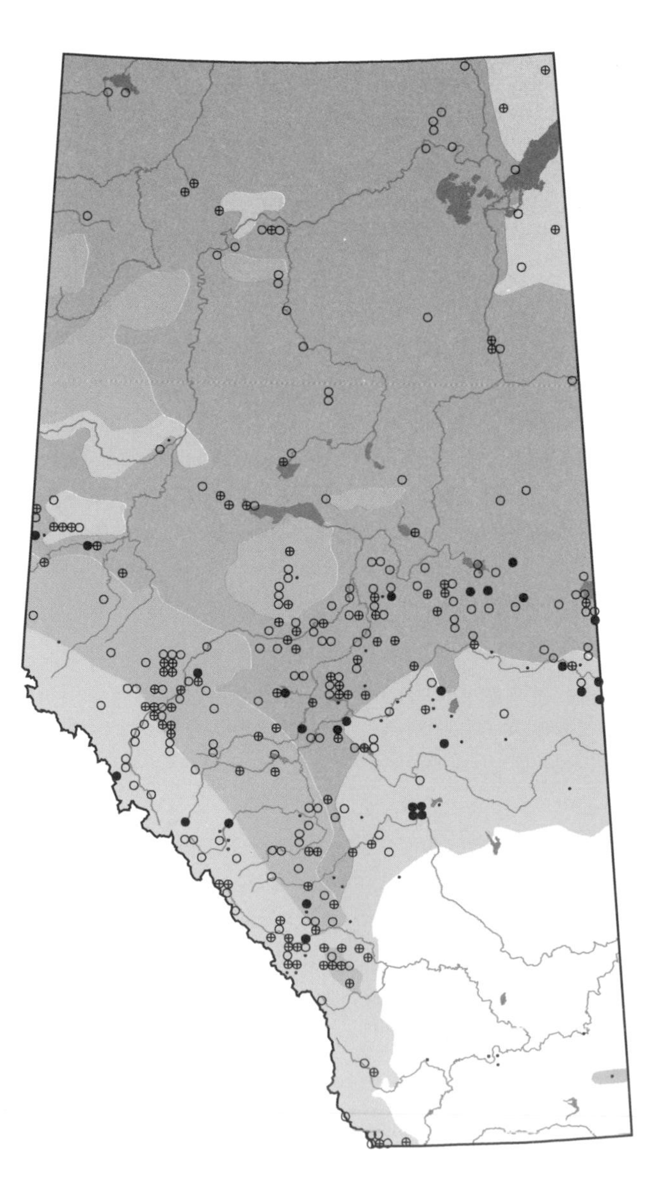

Western Wood-Pewee

Contopus sordidulus

WWPE photo credit: R. Gehlert

Status:
This species is a fairly common summer resident in Alberta. However, its exact status in the province is unknown (Wildlife Management Branch 1991).

Distribution:
The Western Wood-Pewee is widely distributed in Alberta, but nests only sparsely in the southern prairies and is rare and local in extreme northern Alberta (Pinel et al. in prep.). It is documented as nesting in the Cypress Hills, but is not widespread there (Pinel et al. in prep.). Atlassers recorded this species in 38% of surveyed squares in the Boreal Forest, 33% in the Parkland, and 25% in the Canadian Shield region. It was also found in 19% of surveyed squares in the Rocky Mountain region, and 9% in the Grassland region.

Habitat:
This species is found in open mixed, deciduous or coniferous woodland, often at the forest edge adjacent to lakes, rivers, wetlands or other openings. On the prairies, it nests only in deep, well-wooded coulees, in cottonwoods along rivers (Salt and Salt 1976), or in farm windbreaks (Pinel et al. in prep.). In the Cypress Hills, it has been found at lower elevations in deciduous-dominated mixed woods, but it is apparently absent from higher elevations, despite an abundance of suitable habitat (Pinel et al. in prep.). In the mountain parks, it is observed most commonly in deciduous forest (Holroyd and Van Tighem 1983) and, in the north, conifer-dominated forests are preferred (Francis and Lumbis 1979). The species can tolerate moderate human activity, so is sometimes found near human habitation.

Nesting:
The nest of this flycatcher is generally saddled on a horizontal limb, sometimes in an upright crotch. Nesting trees can be coniferous or deciduous and nests are, on average, 3-15 m above ground. The female builds the nest, a well camouflaged shallow structure of grasses, plant down, weed stems, bark scale, and lichens. It is bound with spider webs and lined with fine grasses, hair, and fibres.
Clutches are 2-4 eggs, which are creamy white with a ring of brown speckles near the large end. Incubation is 12-13 days, by the female alone. This species is not shy or wary. The female will return to the nest with no attempt at deception, even if an observer is in plain view close by (Salt and Salt 1976). The young are tended by both parents, though they are brooded by the female for most of the first four days. Fledging occurs at 14-18 days. Nestling pewees are initially fed by regurgitation, then whole insects are brought in the bill. One brood is raised per season.

Remarks:
The Western Wood-Pewee is dark gray and intermediate in size between the Phoebe and the Least Flycatcher. It is much darker, particularly on the underparts, than the small *Empidonax* flycatchers, and it lacks their eye-ring. In the field, this bird is most likely to be identified by its call, a distinct "pee-wee" or "pee-err".
The diet of this flycatcher is almost entirely insects, along with the occasional berry. The bird sits on a limb waiting for an insect to fly by, hawks it in the air, and returns to its perch. Insects are hunted both in the woods and at the forest edge.
The species arrives in central Alberta around mid-May and departs by mid-September. It does not overwinter.

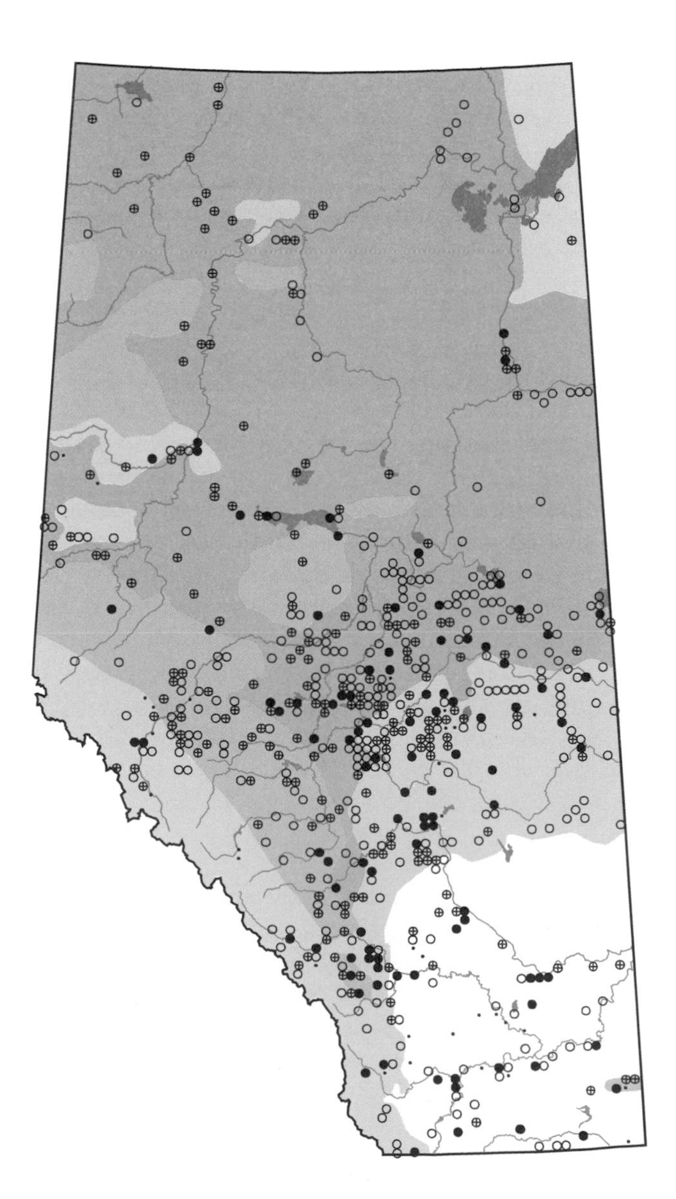

Yellow-bellied Flycatcher

Empidonax flaviventris

YBFL

Status:
The status of this species in Alberta has not been determined, but it is generally scarce in the province (Pinel et al. in prep.). Atlas data indicate only two confirmed breeding records for this species in the five year survey period.

Distribution:
Salt and Salt (1976) reported the Yellow-bellied Flycatcher as breeding in northern Alberta south to about Grande Prairie, Glenevis, and Rochester. Atlas data confirm this breeding range, but also record breeding in the Hinton area. Pinel et al. (in prep.) reports that, during the nesting season, there are records from a few places as far south as the Turner Valley area.

Habitat:
This shy and elusive flycatcher is found primarily in the Boreal Forest Natural Region. In northern Alberta, it favors receding muskegs and the edges of any coniferous and mixed wood forests where thick brush shades a moss covered floor (Salt and Salt 1976). Those authors further reported that tangled alders in swamps bordering woodland lakes and streams also provide nesting territories. Francis and Lumbis (1979) reported that the preferred nesting habitat in the Fort MacKay area was black spruce open above 2 m. Cadman et al. (1987) reported that, in the Ontario Atlas surveys, the many gaps in the Yellow-bellied Flycatcher distribution likely reflected the difficulty in locating this species, rather than its absence, because the areas where it is found are not easily penetrated.

Nesting:
The Yellow-bellied Flycatcher nests on or near the ground, often in the side of a mossy hummock or among the roots of a fallen tree, usually in wet or swampy places. The nest, built by the female, consists of mosses, plant stems and fibres, rootlets, and grasses, and has a lining of grasses and fine rootlets. It is so well concealed that it is rarely found. The female typically lays 3-5 eggs, which are white with brown speckling, and incubates them for 12-14 days. Both parents tend the young, which fledge at 13-14 days. The female sits very tight to the nest while incubating, and both adults are highly secretive when feeding the young (Cadman et al. 1987). The number of annual broods raised by this species is unknown.

Remarks:
Like other *Empidonax* flycatchers, this species is difficult to identify. It is the most brightly colored of these flycatchers, being greener above and yellower below than the others and having a yellow throat. The male and female have similar plumage. The Yellow-bellied's call notes — "cheleck" and "pe-wheep"— are helpful for identification in the spring, but it is a less vocal species than most other flycatchers.
This is a retiring bird that keeps to the dense foliage of the forest edge, rarely perching openly in trees or bushes. It is considered to be the most elusive of the *Empidonax* group (Salt and Salt 1976). Like other flycatchers, it picks insects and spiders off foliage, and complements this diet with a few berries. It forages mainly in the shady lower statum of the woodlands in which it hides (Godfrey 1986). The Yellow-bellied Flycatcher is not a fast flier, although it is strong and manoeuvreable.
Little information is available on migration dates for this species in Alberta. Spring migrants probably arrive in late May, with fall migration taking place in late July and August (Pinel et al. in prep.). The juveniles leave a few weeks after the adults (Hussel 1982). No overwintering is recorded for this species.

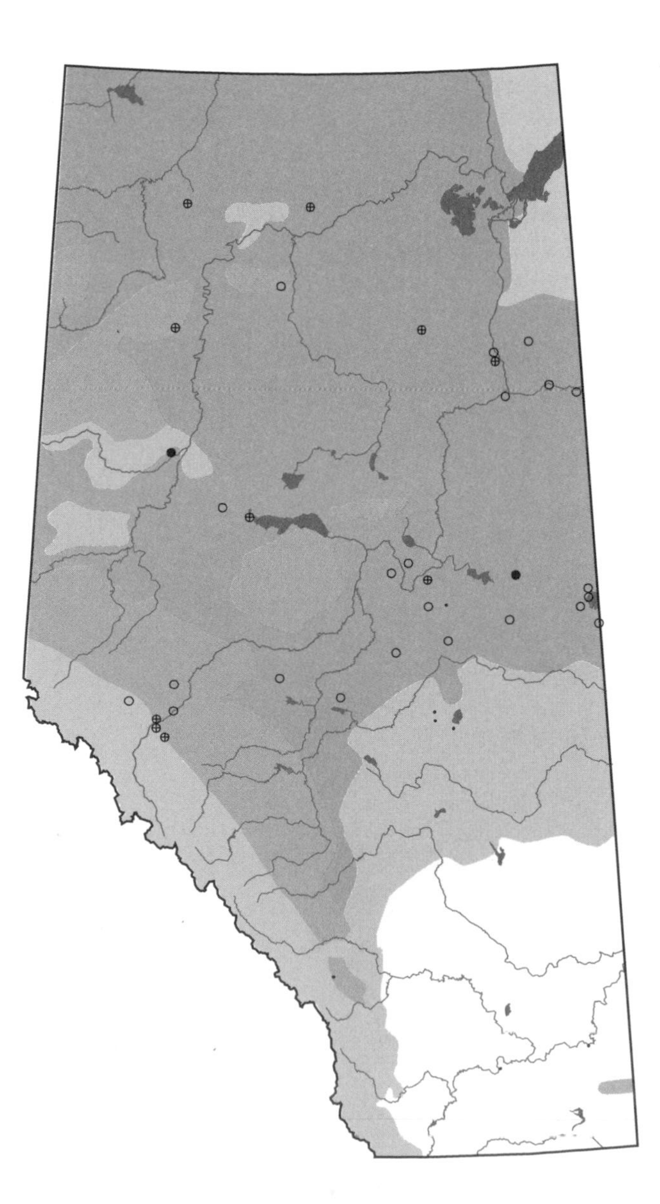

Alder Flycatcher

Empidonax alnorum

ALFL photo credit: T. Webb

Status:
Until 1973, the Alder Flycatcher and the Willow Flycatcher were treated as one species, the Traill's Flycatcher (American Ornithologists' Union 1973). The status of the Alder Flycatcher in Alberta has not been accurately determined (Wildlife Management Branch 1991), but it is a relatively common breeding bird in the province.

Distribution:
This species breeds across northern North America from Alaska and the Yukon east to Newfoundland and south to central British Columbia and the central United States. Its nesting range is not completely understood at this time, but Atlas data indicate it is found throughout Alberta except perhaps in the extreme south. It is also uncommon in the mountains, where its range overlaps with that of the Willow Flycatcher. It is more common in Jasper than Banff National Park (Holroyd and Van Tighem 1983). The species was recorded by Atlassers in 40% of surveyed squares in the Foothills, 37% in the Boreal Forest, 25% in the Canadian Shield, and 21% in the Parkland region. It was found in 14% of surveyed squares in the Rocky Mountain region, and 3% in the Grassland region.

Habitat:
The Alder Flycatcher breeds in alder, willow, and similar thickets in moist, but sometimes drier situations (Godfrey 1986). In northern woodlands, it nests in thickets near muskegs, bogs, marshes, streams, lakeshores, forest margins, or roads. In the south, it is found in willow-birch thickets in prairie coulees and along prairie creeks and rivers (Salt and Salt 1976).

Nesting:
The nest of this flycatcher is built low in bushes, shrubs, or small saplings, usually in an upright or slanting fork or around supporting stems. It is likely built by the female. Nests are compact and usually well-concealed. They are made of grasses, weed stalks, plant fibres, bracken, and moss, and have a lining of soft plant fibres, plant down, and fine grasses. Trails of nesting material often hang down from the nest.

Clutches are typically 3-4 eggs, which are finely-spotted in browns, mostly at the larger end. Incubation, by the female alone, is 12-13 days. Male and female both tend the young, with fledging at 13-14 days. The nests of this species are occasionally parasitized by cowbirds. The number of broods raised per season has not been determined.

Remarks:
The male Alder Flycatcher is conspicuous in the spring as he calls his "fee bee o" from prominent perches in the breeding territory. When birds become silent it becomes nearly impossible to distinguish this species from the Willow Flycatcher. The Alder Flycatcher is more green above and yellow below then the Least Flycatcher; not as green above or as yellow below as the Yellow-bellied Flycatcher. The alder is one of the larger members of the *Empidonax* group.

Most often, this flycatcher perches low in dense bushes, dashing out to catch an insect, then retreating again to its perch. It eats mainly flying insects, plus a few berries. These birds are shy and disappear into dense foliage when approached. They are also very secretive around the nest.

Spring migrants arrive in southern Alberta in the last two weeks of May and birds depart again mainly in August (Pinel et al. in prep.).

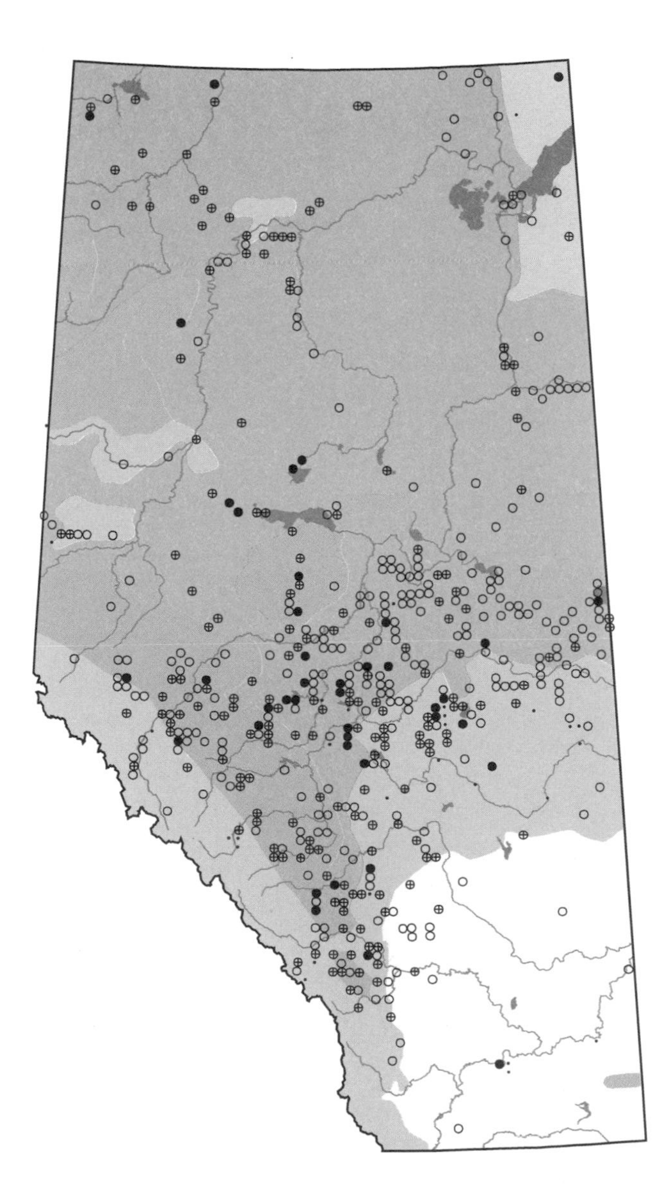

Willow Flycatcher

Empidonax traillii

WIFL

Status:

This species is locally common throughout its range in southern Canada and the northern United States (Cadman et al. 1987). Holroyd and Van Tighem (1983) considered it be a fairly common summer resident within the mountain parks. It was on the North American Blue List from 1980-1982 because of apparent declines in west coast populations (Tate and Tate 1982), but was down-listed to a species of "special concern" in 1986 (Tate 1986) as populations stabilized.

Distribution:

The Willow Flycatcher breeds in continental United States northward into southern Canada. In Alberta, it is known to frequent the mountains and foothills, with more records in Banff than in Jasper (Holroyd and Van Tighem 1983). Atlas sightings in central and northern Alberta may be evidence of a northward extension of this species' range (as is occurring in eastern North America) or perhaps of difficulties distinguishing this species from the Alder Flycatcher. The Willow Flycatcher was recorded in 9% of surveyed squares in the Rocky Mountain region and 1% in the Parkland and Foothills regions.

Habitat:

Primary habitat of this flycatcher is shrubbery along streams or lake edges and secondary shrub growth in open areas. This species does not appear to breed in as wide a variety of habitats as the Alder Flycatcher. In some areas, it particularly prefers willow and rose to other shrub types (Godfrey 1986) and, in the mountains, only birch-willow thickets are used (Holroyd and Van Tighem 1983).

Nesting:

This species nests in thickets and low shrubby woods, in the crotch of a shrub or small tree, usually less than 2 m from the ground. The nest, a compact cup, sometimes with dangling material, is built of grasses, plant down, plant fibres, bark strips, weed stems, fine rootlets, and similar material. The lining is fine grasses, plant down, weed stems, mosses, and conifer needles. Nests have a cottony appearance and look similar to those of the Yellow Warbler and the Goldfinch. The female selects the nest site, accompanied by the male, and builds the nest.

Three to four white eggs, spotted with browns, are incubated by the female for 12-13 days. Both male and female tend the young, and the female broods for 7-8 days. Youngsters fledge at 12-14 days. This species raises one brood per season.

Remarks:

The Willow Flycatcher and Alder Flycatcher are practically identical in appearance and were considered one species until 1973 (American Ornithologists' Union 1973). It is easiest to distinguish them by voice, though even the songs are close enough that mistakes can be made. The Willow Flycatcher's song is a "fitz-bew", compared to the Alder's "fee-bee-o". The Willow is similar to the Yellow-bellied Flycatcher, but has less green on the upper parts and less yellow below. It can also be mistaken for a Least Flycatcher, but has more green above, and yellower underparts.

This species is an energetic little bird which is often seen flitting in and out of bushes as it hunts for its winged insect prey. A few berries and seeds complement its insect diet.

The Willow Flycatcher probably arrives in the province by mid-May and departs again by early August. No overwintering is reported.

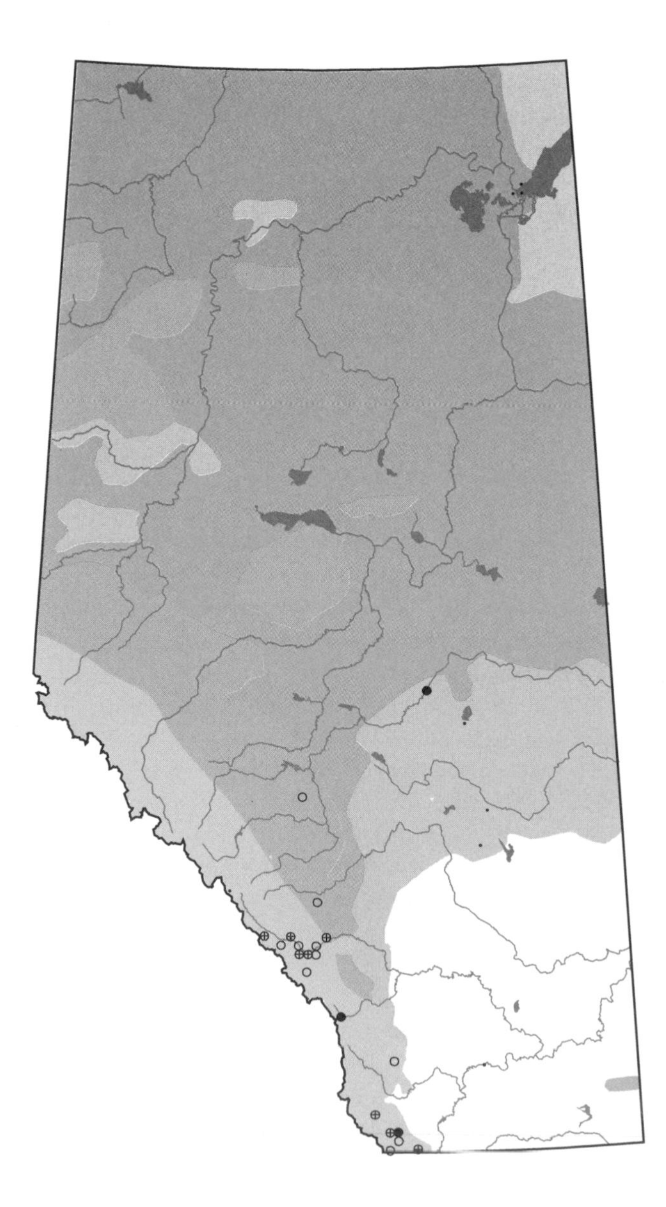

Least Flycatcher

Empidonax minimus

LEFL photo credit: T. Webb

Status:

The Least Flycatcher is the most common and most widely distributed of the *Empidonax* flycatchers, and its populations in Alberta are believed to be healthy and stable (Wildlife Management Branch 1991). However, this species was on the North American Blue List in 1980, was a species of "special concern" in 1981-1982, and "local concern" in 1986; there have been continuing declines in some parts of its range (Tate 1986).

Distribution:

The range of this species extends from the Yukon and southern Canada (excluding much of western British Columbia) south to Montana and North Carolina. It breeds in suitable habitat in all of Alberta's natural regions. It was recorded by the Atlas in 49% of surveyed squares in the Boreal Forest region, 43% in the Parkland, 38% in the Foothills, and 21% in the Rocky Mountain region. It was also found in 19% of surveyed squares in the Canadian Shield and Grassland regions.

Habitat:

The Least Flycatcher is commonly observed in more open deciduous or mixed woodland (occasionally in conifer groves), particularly near forest edges and openings such as burns, swamps, and bogs. In the Parkland region, aspen and poplar scattered along roads and through fields are popular areas; in the Grassland region, poplars and cottonwoods in coulees and river valleys are favored (Salt and Salt 1976). In the Rocky Mountain region, willow and aspen are preferred haunts (Holroyd and Van Tighem 1983), and, in the north, open deciduous woods or deciduous portions of mixed forests are used (Salt and Salt 1976; Francis and Lumbis 1979). Like the Western Wood-Pewee, the Least Flycatcher can tolerate some human activity and sometimes nests in urban residential areas.

Nesting:

The nest of this species is a small, deep compact cup set into the upright crotch of a tree branch or on a horizontal branch and supported by small lateral branches. Deciduous trees are preferred and usually a small tree or sapling is favored rather than a large one. The female builds her nest of bark fibres, grasses, weed stems, plant down, and other vegetation, and binds it with spider's webs or cocoons. The lining is fine grass, plant fibre, plant down, and sometimes hair or feathers. There is one report of a dragonfly wing lining (Briskie 1988). Nests are usually 2-8 m above ground. Clutches are 3-6, usually four, creamy white eggs. Incubation is variously reported, 12-16 days (Bent 1965a; Harrison 1978; Godfrey 1986), and is by the female alone. Young are tended by both parents, and young birds fledge at 13-16 days. They continue to be fed by their parents another 10 days. Once nesting duties are complete, adult birds leave the breeding grounds and begin their southward migration, although Bent (1965a) reported that, in the southern portions of this species' range, two broods are raised in a season.

Remarks:

The Least Flycatcher is the smallest of Alberta's flycatchers. It can be distinguished from others in its family by its paler underparts and more conspicuous eye ring, but particularly by its distinctive call, a loud, emphatic "chebec". This little bird is highly pugnacious around the nest.

Like other flycatchers, it catches most of its prey on the wing, but it also searches under leaves for caterpillars, ants, and other insects (Salt and Salt 1976). Berries and other small fruits are occasional complements to its insect diet.

This species arrives in Alberta in mid-May and departs again in August, with some stragglers staying into September (Pinel et al. in prep.). No overwintering has been reported.

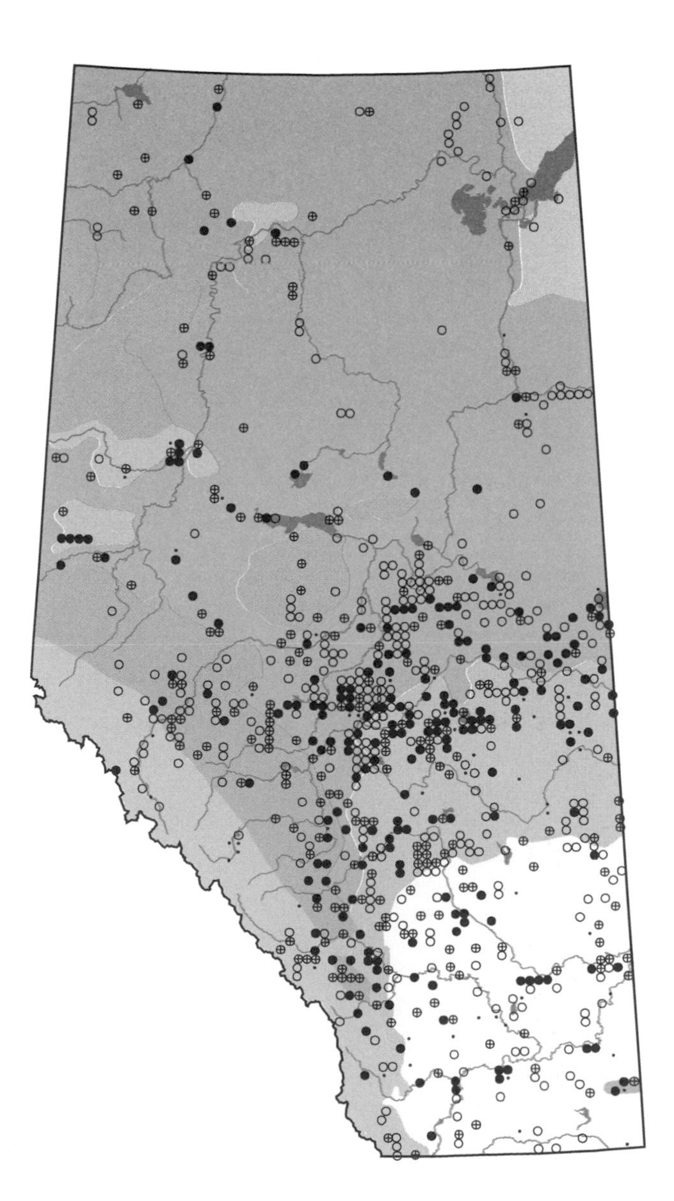

Hammond's Flycatcher

Empidonax hammondii

HAFL

Status:
Hammond's Flycatcher is not common anywhere within its range. Its abundance and distribution in Alberta are difficult to determine because of problems in the field distinguishing it from the Dusky Flycatcher (Holroyd and Van Tighem 1983).

Distribution:
This species breeds only in western North America, from Alaska south to California and east through the Rockies. In Alberta, its range is restricted to the Rocky Mountains, from Jasper National Park south to Waterton Lakes (Salt and Salt 1976). It is more common in the Jasper area and northern Banff National Park than in the mountains to the south. Atlassers recorded this species in 8% of surveyed squares in the Rocky Mountain region, and in less than 1% in the Foothills region.

Habitat:
This flycatcher is a bird of mature coniferous forests, also inhabiting mixed woodlands and forest edges, usually at higher altitudes. It is generally active in the upper canopy of these forests. In the national parks, it is found most often in Douglas fir forests, open white spruce stands on dry slopes, shrubby cinquefoil/bearberry — northern bedstraw, and in lower densities in dry lodgepole pine (Holroyd and Van Tighem 1983). Hammond's Flycatchers occasionally nest at lower altitudes, where Dusky Flycatchers are found, but these two species inhabit quite different habitats. The Dusky prefers deciduous scrub, and the Hammond's prefers coniferous woods (Salt and Salt 1976).

Nesting:
The nest of the Hammond's Flycatcher is usually in the fork of a horizontal branch, most often in a conifer, generally 8-18 m above ground. A few nests have been reported at much lower levels (Godfrey 1986). Probably built by the female, the nest is a loose cup of plant fibres, rootlets, grass, and weed stems, and is lined with grass and hair.
This species' 3-4 eggs are white and are unmarked or have small brown spots. Incubation is 12-15 days, by the female alone. The young are tended by both parents, with fledging at 17-18 days. Hammond's Flycatcher is probably single-brooded.

Remarks:
This small, grayish *Empidonax* flycatcher is extremely difficult to distinguish from the Dusky Flycatcher, particularly outside of breeding season. These species' plumage is very similar, although the Hammond's appears darker, especially on the throat, and it has a slightly shorter tail. The Hammond's bill is a little shorter and the lower mandible is darker than the Dusky's. Also, their spring songs are hard to tell apart. Both have a three part song, the Hammond's being a "sweep, tsurp, seep", with the last note rising (Godfrey 1986). This song is quite different from that of the Alder, Willow, or Least Flycatcher. In breeding season, habitat differences are probably the best guide in identifying the Hammond's and Dusky flycatchers.
Hammond's Flycatcher is exclusively insectivorous, hawking flying insects from higher branches. It arrives in the Banff and Jasper area in mid- to late May (Holroyd and Van Tighem 1983) and departs again in August (Pinel et al. in prep.). In migration, it is usually seen in pairs or singly. No overwintering is recorded for this species.

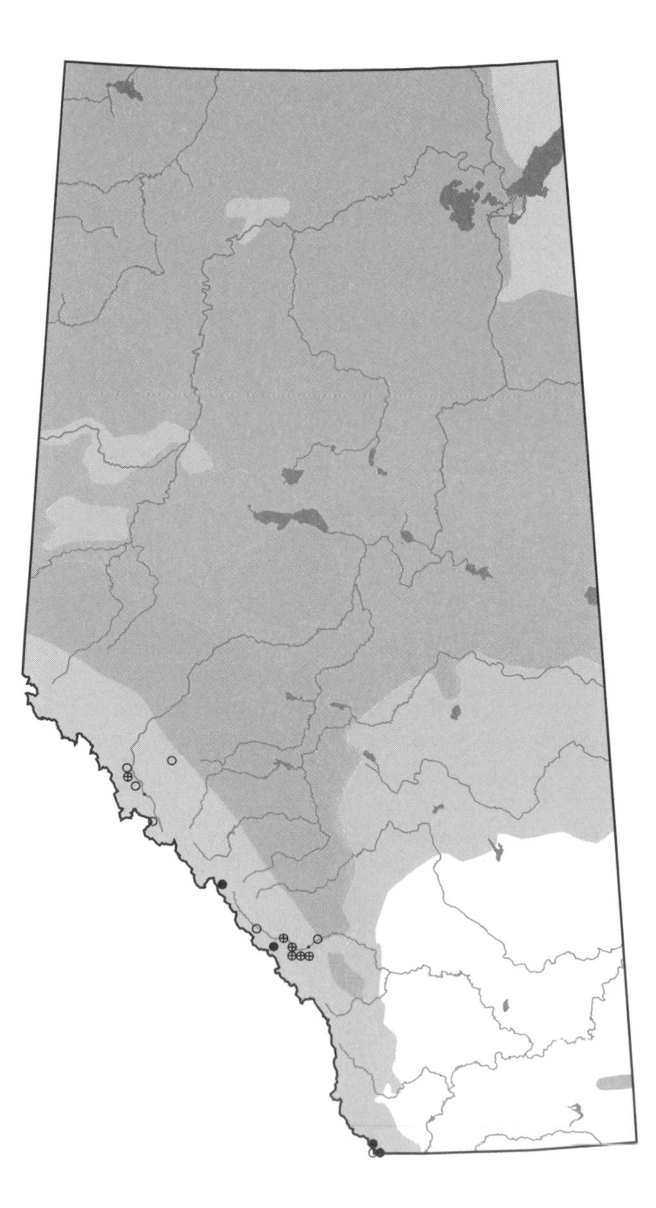

Dusky Flycatcher

Empidonax oberholseri

DUFL

Status:
The status of the Dusky Flycatcher in Alberta has not been fully determined, but Holroyd and Van Tighem (1983) considered it to be an uncommon summer resident of the mountain parks. Similarities to the Hammond's Flycatcher make abundance estimates and distribution difficult to define.

Distribution:
This species frequents the Rockies and their foothills from Jasper National Park south to Waterton Lakes, and east to Beauvais Lake, Longview, Bragg Creek, and Silver Creek (Salt and Salt 1976). Nesting also occurs in the Cypress Hills. In the mountains, this flycatcher is more common in Banff during breeding season than in Jasper, the reverse pattern of that for the Hammond's Flycatcher (Holroyd and Van Tighem 1983). It was recorded by Atlassers in 15% of surveyed squares in the Rocky Mountain Natural Region, 12% in the Foothills, and less than 1% in the Parkland region.

Habitat:
The Dusky Flycatcher is usually found in deciduous habitats at low altitudes, particularly open deciduous woods, alder and willow thickets, and other low scrubby growth. Aspens and shrubby burns are highly preferred habitats. This flycatcher is sometimes found in mixed woods, but rarely enters the coniferous forests preferred by Hammond's Flycatcher. It forages and nests at lower levels than the Hammond's, but sometimes perches on top of good-sized trees (Godfrey 1986). In the mountains, this species is most likely to be found in aspen forests, other deciduous forests and on sub-alpine fir/willow avalanche slopes (Holroyd and Van Tighem 1983).

Nesting:
The Ducky Flycatcher nests in the upright crotch of a willow, alder, aspen or similar tree (rarely a conifer), at heights averaging 1-2 m (Godfrey 1986). The nest is a neat, compact cup of plant fibres and grasses, lined with similar, but finer material, plant down, and feathers. The outside and base are loosely constructed and odd ends sometimes hang down. Spider webbing is often used to bind the outside of the nest. The female builds the nest with some assistance from the male.

The clutch is typically 3-4 dull, white eggs, incubated 12-15 days, mostly by the female, with, possibly, some help from the male. The female sits tight to the nest when incubating. Both sexes tend the young, which fledge at 18 days. This species may raise more than one brood per season. Pair bonds are monogamous.

Remarks:
Formerly known as Wright's Flycatcher, the Dusky Flycatcher is very similar in plumage and voice to the Hammond's. The underparts of the former are paler and less uniform (the throat is whitish instead of gray, contrasting more with the gray of the breast) and the outer web of the outer tail feathers is more whitish. The bill of the Dusky is a little longer and the lower mandible lighter and more variable in color than that of the Hammond's (Godfrey 1986). The call of the Dusky Flycatcher tends to be softer and less burred than the Hammond's, but is not sufficiently distinct to be reliable in the field. Outside of the breeding season, these species are nearly impossible to distinguish.

Like other flycatchers, this species feeds almost entirely on insects, which it hawks in the air. It often flicks its tail while perched.

Spring migration of this flycatcher in Banff and Jasper occurs in late May, with autumn migration occurring in August. No overwintering has been reported.

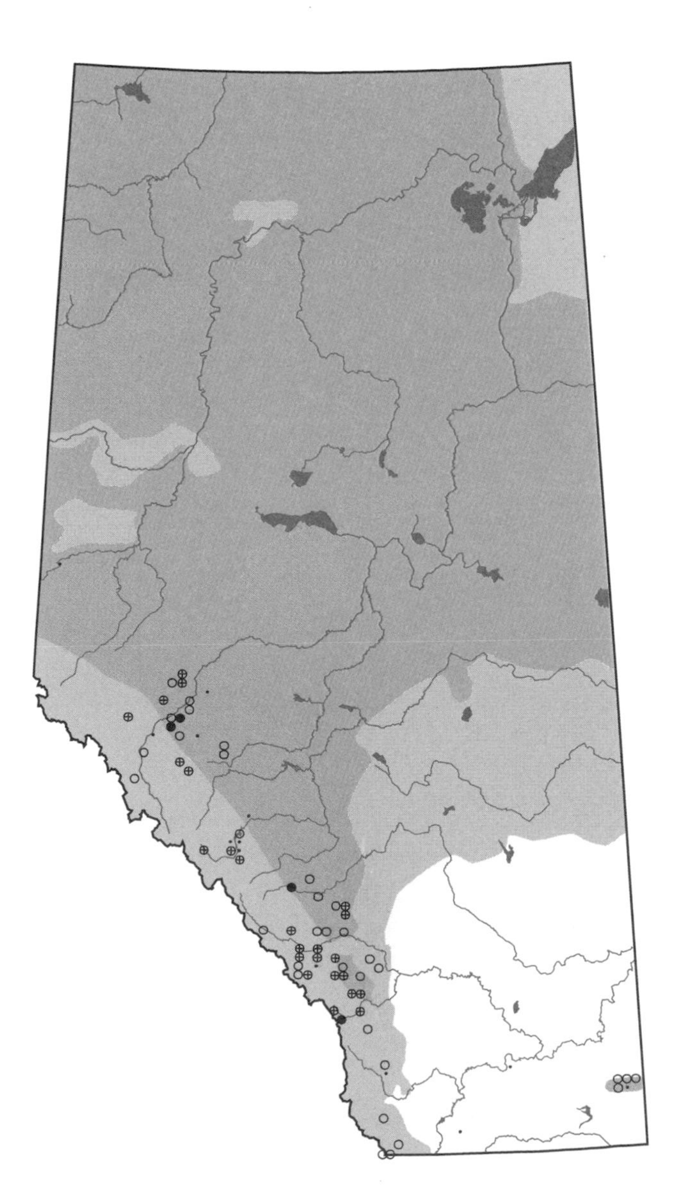

Cordilleran Flycatcher

Empidonax difficilis

COFL

Status:
The status of this species in Alberta has not been determined, but it is considered to be a rare summer resident in the mountain parks, which constitute a significant part of its provincial breeding range (Holroyd and Van Tighem 1983). It was first discovered in Alberta at the R.B. Miller Biological Research Station west of Turner Valley in 1954 (Salt and Salt 1976).

Distribution:
The Cordilleran Flycatcher breeds mainly west of the Rocky Mountains from Alaska south to Baja California, west Texas and Belize. In Alberta, it is found in the southwestern part of the province, primarily in the mountains and foothills north to Jasper. Small numbers of this species have been recorded in several valleys of the mountain parks in the summer (Holroyd and Van Tighem 1983). Atlassers recorded the species in 15% of surveyed squares in the Rocky Mountain region, 2% in the Foothills, and less than 1% in the Parkland and Boreal Forest regions.

Habitat:
This small flycatcher inhabits shady but more open open deciduous or coniferous woodlands, and is usually found near streams, ravines, seepages, or other moist areas. The dense thickets of tall shrubbery preferred by Alder and Willow Flycatchers tend to be ignored by this species (Godfrey 1986).

Nesting:
The nest of the Cordilleran Flycatcher, built by the female, is a finely woven deep cup of plant fibre, mosses, rootlets, and weed stems lined with finer materials and hair. Nests are placed in a variety of locations — in or under the banks of a stream, in the upright fork of a small tree, in an old stump, or among tree roots. The cavity of a tree trunk, a woodpecker hole, a crevice, cavity, or ledge in a rock face or in an unoccupied building may also serve as a nest site. Birds may return to favored nesting sites over successive years, repairing the old nest or partially rebuilding it (Bent 1965a).

The female lays 3-4 eggs, which are white and lightly or moderately spotted with reddish browns. Incubation is 14-15 days by the female alone, although the male stands guard nearby and drives away any other birds that venture too near the nest (Bent 1965a). Both sexes tend the young, with the female brooding. Fledging is at 14-18 days, and the young are fed another 10-11 days after leaving the nest (Harrison 1978). There is no conclusive information for thisspecies on the number of annual broods.

Remarks:
Except for the Yellow-bellied Flycatcher, the Cordilleran Flycatcher is the only *Empidonax* flycatcher in its range with both yellow throat and underparts. The Cordilleran is less green above than the Yellow-bellied, but song and nesting habits are the most accurate ways of separating these species in the field. The song of the Cordilleran is a "pseet, ptsick, seet", with the last note highest (Godfrey 1986).

Like all flycatchers, the Cordilleran Flycatcher hawks on the wing, often over streams. Bent (1965a) reported that the majority of its diet is insects and their larvae and that this flycatcher appears to eat more ladybird beetles than does any other flycatcher. Seeds make up a very small portion of the diet.

Information on migration is scant, but birds likely arrive in the last half of May and leave by September (Pinel et al. in prep.). No overwintering is recorded.

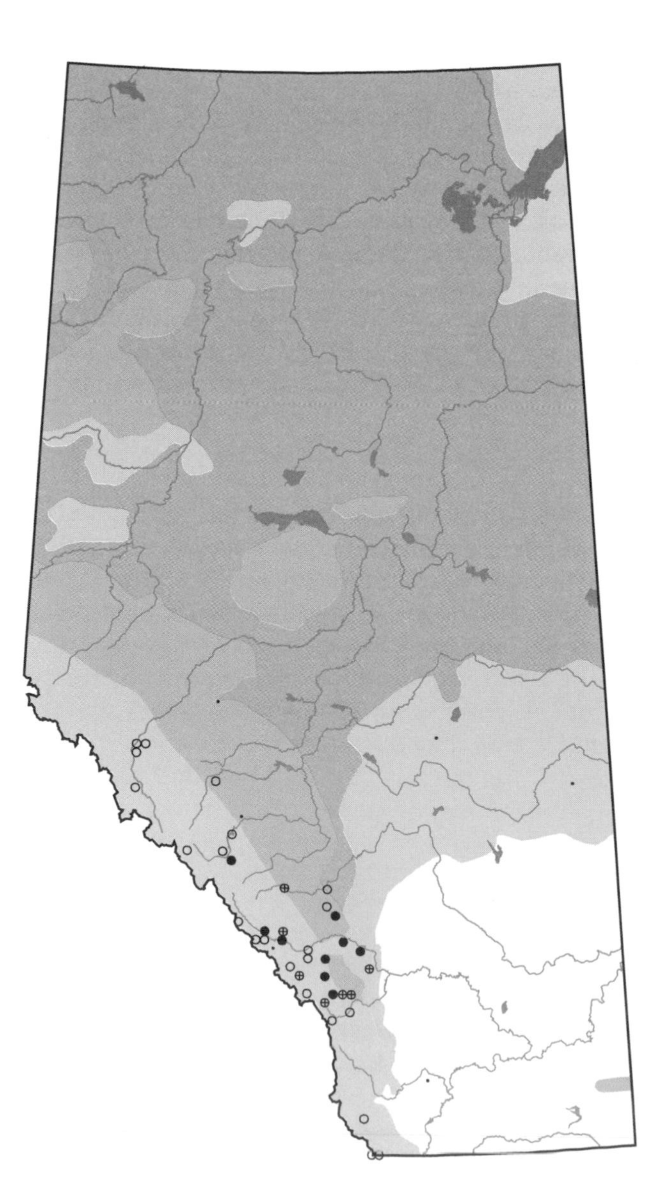

Eastern Phoebe

Sayornis phoebe

EAPH photo credit: T. Webb

Status:
The Eastern Phoebe is a fairly common summer resident in Alberta, particularly in central and southern areas, but its status in the province has not been accurately determined. This species was on the North American Blue List in 1980, and was a species of "special concern" in 1981-1986, primarily because of population declines in eastern parts of its range and the Midwest (Tate 1986).

Distribution:
This phoebe breeds from the Mackenzie area east across southern Canada and south to New Mexico and South Carolina. It is found in most parts of Alberta, but only very locally in the Rocky Mountains and southern prairies (Salt and Salt 1976). It was recorded by Atlassers in 35% of surveyed squares in the Parkland, 34% in the Boreal Forest, 25% in the Canadian Shield, and 18% in the Foothills region. In the Rocky Mountain region and the Grassland region, it was recorded in 3% and 2% of surveyed squares, respectively.

Habitat:
Eastern Phoebes breed in open wooded areas, including mixed, deciduous and coniferous forests, often in the vicinity of streams and lakes, and most usually at the forest edge or near open areas. They also frequent areas of human habitation, including farmyards and cottage properties.

Nesting:
Natural sites chosen for nesting include cliff ledges, caves, earth-bank overhangs, tree-roots and shrubs. However, this species, having adapted somewhat to human habitation, also seeks out man-made structures such as ledges and beams of buildings and bridges. The nest, built by the female, is a rather bulky structure of mosses, mud, grass, rootlets, plant stalks or other vegetation, and, occasionally, hair and plant fibres. It is rarely concealed. Nests situated against a vertical surface generally contain more mud. This species returns to the same nest site over successive years, often building up tiers of nests as it does its annual repairs.

The female typically lays five eggs, which are white and are, very occasionally, speckled with reddish-brown. Incubation, by the female, is variably reported from 13-20 days. Young are tended by both parents for 15-17 days until fledging, and are then fed for another 2-3 weeks. Females often produce two broods, and occasionally three. This species is often parasitized by cowbirds.

Remarks:
This sparrow-sized, dull-colored flycatcher is similar to the Western Wood-Pewee, but is lighter below, has inconspicuous wing bars and a completely black bill. Its call note, "fee-be", and habit of wagging its tail when perched are its two most distinguishing features in breeding season. It shows little of the excitability common to the flycatchers and is quite unafraid of humans. Like other flycatchers, the Eastern Phoebe is primarily insectivorous, though its diet consists of about 10% vegetable matter throughout the year, a much higher percentage than other phoebes (Root 1988). Most of the fruits, berries, and seeds it takes are eaten in the winter months.

This species is the first flycatcher to arrive in the spring, appearing about mid-April. It leaves the province again in August, with the last birds generally observed in late August or early September (Pinel et al. in prep.).

The Eastern Phoebe was the first species ever banded in North America, by John J. Audubon, in Pennsylvania, in 1840 (Root 1988).

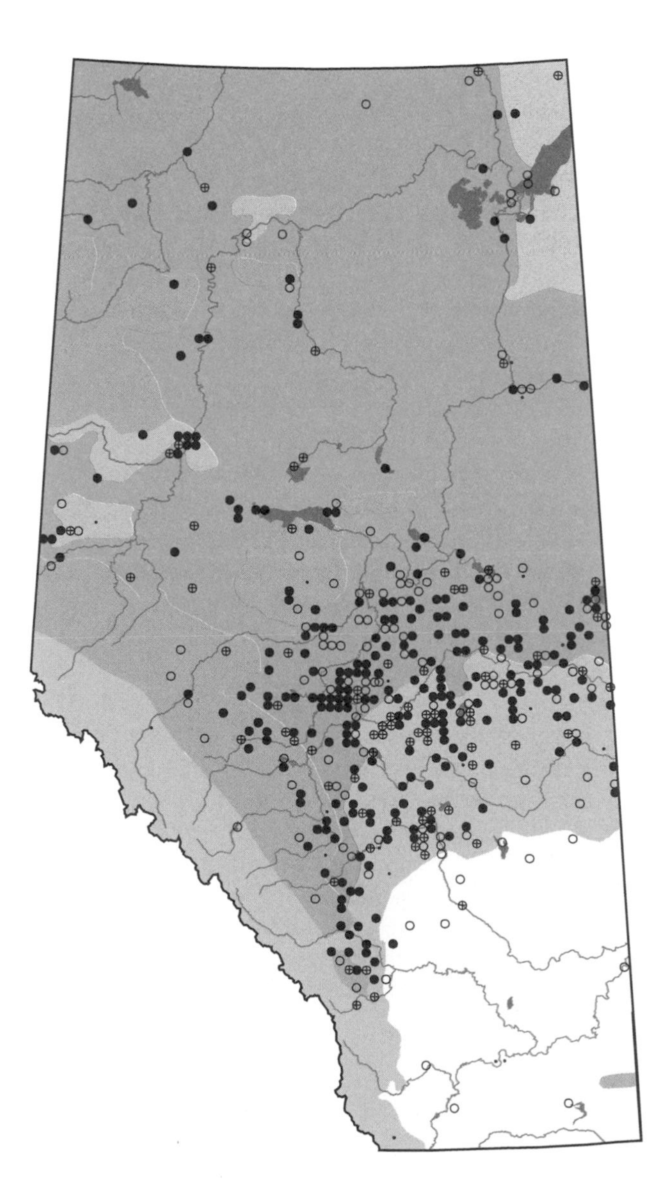

Say's Phoebe

Sayornis saya

SAPH

Status:
The status of this species in Alberta has not been determined (Wildlife Management Branch 1991), but Atlas data indicate that it is an uncommon breeding bird in the province. It is considered to be a rare summer visitor to the mountain parks (Holroyd and Van Tighem 1983).

Distribution:
Say's Phoebe is found in western North America from Alaska south to northern Mexico. Salt and Salt (1976) reported that, in Alberta, it breeds most commonly in the southeastern prairies, but also in the parklands and foothills. It was recorded by the Atlas in 8% of surveyed squares in the Grassland region, 3% in the Rocky Mountain region, 2% in the Foothills, and about 1% in the Parkland and Boreal Forest regions.

Habitat:
This is a bird of open dry country, particularly the badlands, prairie coulees, river banks, and prairie farms. It is often seen perching on a hoodoo, a low bush or weed stalk, a fence post, wire, or building. In Banff and Jasper, it has been observed in open areas from the montane to the upper subalpine (Holroyd and Van Tighem 1983).

Nesting:
Nesting is generally in a sheltered area with some overhang, including shelves or crevices of rocky cliffs, holes in banks, ledges in caves, or natural cavities in trees. A sheltered ledge in a building, under a bridge, or in other types of man-made structures may also be used. Bent (1965a) reported that burrows of Bank Swallows are occasionally occupied. The nest, probably built by the female, is a flat bulky cup of grasses, plant fibres, weed stems, moss, wool, hair, and spiders' webs. It is lined with wool and hair (Harrison 1978).
Clutches are 3-7 eggs, typically 4-5. The eggs are usually white, but sometimes have a few brown spots. The female alone incubates for 12-14 days. Young are tended by both parents, and are fed insects, initially by regurgitation. When the young fledge at 14-16 days, the female leaves to re-nest and raise a second brood, and the male takes over care of the first brood. This species is occasionally parasitized by cowbirds.

Remarks:
Say's Phoebe is a meduim-sized flycatcher (the largest of the phoebes), and is easily identified by its reddish-brown belly and black tail, and the habit of flipping its tail. Its song is a plaintive "pee-urr" (Godfrey 1986).
To capture its prey, it will hover American Kestrel-style or sit on a low perch, catching most insects on the wing. Its diet includes bees, wasps, flies, grasshoppers, crickets, and dragonflies, plus a few berries. This species regurgitates pellets containing indigestible portions of its prey (Bent 1965a).
Say's Phoebe arrives in southern Alberta the last week of April, usually alone or in two's or three's. Extralimital spring migration records from late May and early June suggest a minor migration, route east of the Rocky Mountains, of birds heading for nesting grounds in northern British Columbia, Northwest Territories, Yukon, or Alaska (Pinel et al. in prep.). The species departs again from the province in late August, or early September, with just a few stragglers remaining after mid-September. There are no winter records for this species.

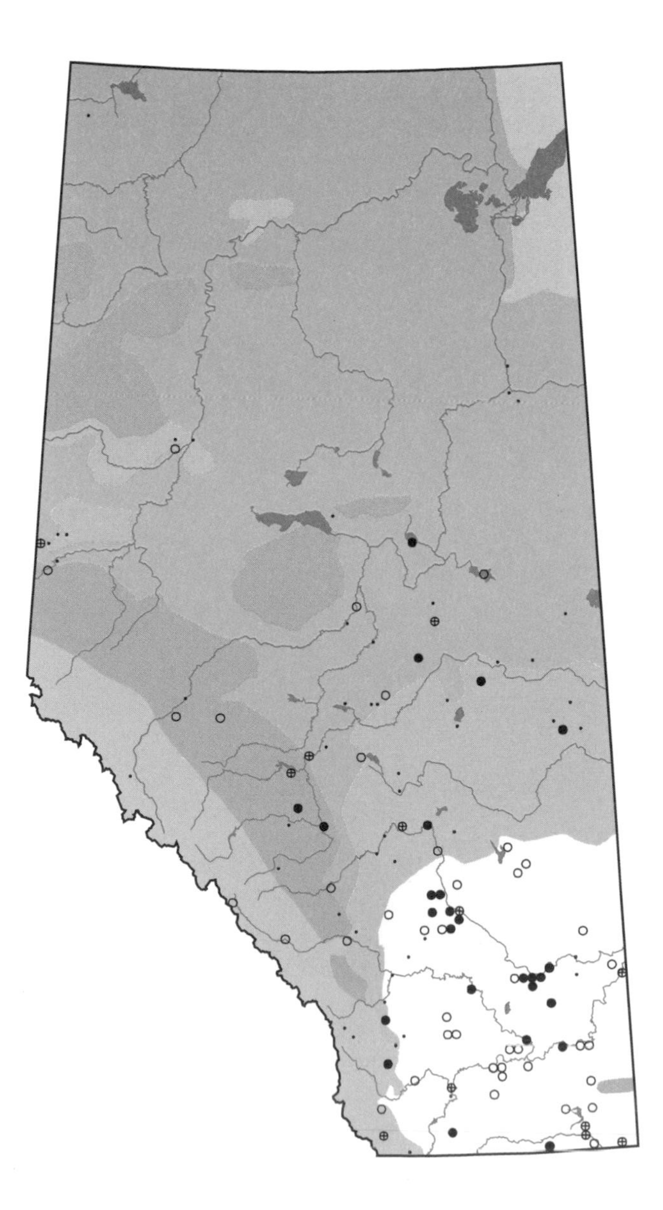

Great Crested Flycatcher

Myiarchus crinitus

GCFL

Status:
The status of the Great Crested Flycatcher in the province is undetermined, but it is not a common breeding bird in Alberta. This species was first sighted in Alberta in 1939 (Ussher 1940), and more recent records indicate nesting in the east central area of the province. Salt and Salt (1976) suggested the appearance of this flycatcher in Alberta is the result of the gradual destruction of suitable breeding habitat farther east.

Distribution:
Since its initial sighting in Alberta at Elk Island National Park in 1939, there have been a variety of sightings near Two Hills, Elk Island, Bonnyville, Cooking Lake, St. Paul, Miquelon Lake, Botha, Mann Lake, and as far northwest as Lesser Slave Lake and Dunstable. The first confirmed nesting of this species in the province was in 1976 near Miquelon Lake (Trann 1976,1977), where it also nested in 1978 and 1979 (Pinel et al. in prep.).

Habitat:
The Great Crested Flycatcher is partial to Alberta's Parkland Natural Region, where there were 17 breeding records over the Atlas study period. There were 7 records from the southern fringe of the Boreal Forest region, and one extralimital record from the Canadian Shield region north of Lake Athabasca.
It breeds in mature deciduous and mixed woodlands, usually in more open parts, close to clearings or the woodland edge. It is also found in scattered trees in farmland, and sometimes in fence rows, at roadsides, and in residential areas. Most foraging is done in the upper branches of deciduous trees.

Nesting:
This species nests in cavities in dead or living trees, usually deciduous, often choosing a natural cavity or woodpecker hole. It will also nest in bird boxes, hollow logs, open pipes, fence rails, and other artificial sites. Most nests are 3-7 m above ground. The male and female both construct the nest, which is a bulky structure of leaves, grass, moss, hair, feathers, other litter, and, often, cast snakeskins (or onion skins, plastic, or cellophane). Nests may be used over successive years.
Clutches are 4-5 eggs, which are creamy white or buffy and are streaked with browns. The female incubates for 12-15 days. Fledging is at 12-21 days, and young are fed by their parents for some time after fledging. This species is intolerant of intrusion and defends its nest vigorously, particularly against small mammals and small birds (Salt and Salt 1976). It is single-brooded.

Remarks:
The Great Crested Flycatcher is a large flycatcher with yellow belly, gray throat and upper breast, and much rufous on the tail. It is distinguished from the Western Kingbird by its darker upper parts, white wing bars and rusty-colored tail and wings (Salt and Salt 1976). The Great Crested's main call note , "wheep", is distinctive.
This species hawks insects high in the upper canopy of trees, taking flies, bees, wasps, beetles, grasshoppers, crickets, butterflies, moths, and dragonflies. It captures flying insects by dashing out on quivering wings, much like a kingbird, with its tail spread to show the rufous coloration (Salt and Salt 1976). It also eats spiders and a few berries.
Birds probably arrive in the province in late May and depart by late August or early September. A summation of records up to 1980 show that this species has been observed in Alberta from May 22 to September 9 (Pinel et al. in prep.). No overwintering is reported.

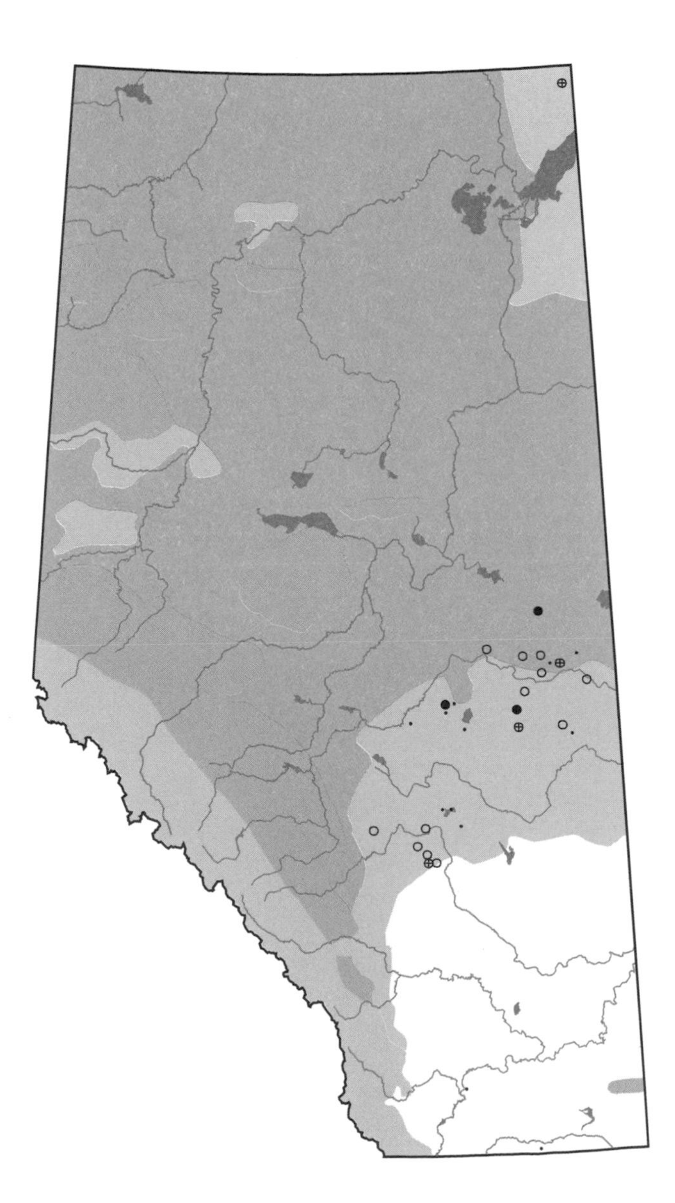

Western Kingbird

Tyrannus verticalis

WEKI

Status:
The status of the Western Kingbird in Alberta has not been accurately determined, but it is a relatively common species in appropriate habitat.

Distribution:
The Western Kingbird breeds in west central Canada from southern British Columbia south to Baja California and northwestern Mexico. It is found in southern Alberta primarily in the Grassland Natural Region, where it was recorded by the Atlas in 31% of surveyed squares. It was also recorded in 5% in the Parkland and less than 1% in the Rocky Mountain and Foothills regions. Since the late 1800s, the range of this species has expanded north and eastwards as suitable nesting sites have become available with settlement and agricultural development.

Habitat:
This kingbird is commonly found in open situations such as prairies, but generally near trees, telephone poles, or buildings (Godfrey 1986), where it can perch and maintain a clear outlook. It nests in tall cottonwoods along river valleys, in wooded coulees, and in farm shelter belts (Salt and Salt 1976), and also in shade trees along streets. In the mountains, this kingbird frequents open areas among scattered trees (Godfrey 1986).

Nesting:
The nest of this species, built probably by both male and female, is a bulky untidy structure of grasses, hair, wool, plant down, and other fibrous material. It is lined with similar but finer material felted together into a tight cup. The nest is built in a shrub, on a horizontal branch or in the crotch of a tree, or on utility poles, buildings, and other human structures. Occasionally, several pairs will nest close together in a grove of trees (Bent 1965a).
Clutches are 3-5 eggs, usually four, and the eggs are white and strongly marked with browns. They are very similar to those of the Eastern Knigbird, though slightly smaller. Incubation is 12-14 days by the female, with fledging at 16-17 days. Both parents tend the young and are noisy and aggressive defenders of the nest. This species is probably single-brooded.

Remarks:
The Western Kingbird is a large flycatcher distinguishable by its yellow belly, pale gray head and breast, and black tail with narrow white stripes down the sides. Its voice is usually lower pitched than that of the Eastern Knigbird, and includes various shrill twitters and squeaks, a series of harsh notes, or a single "kip" (Godfrey 1986).
This species is primarily insectivorous, ready to hawk from its perch or pounce on prey on the ground. It always returns to its perch after a hunting foray. Small fruits and berries are occasional additions to its diet. The young are fed insects, first by regurgitation, then torn to pieces.
Spring migrants arrive in mid-May or shortly thereafter, and fall migration is generally complete by September. This species does not overwinter.

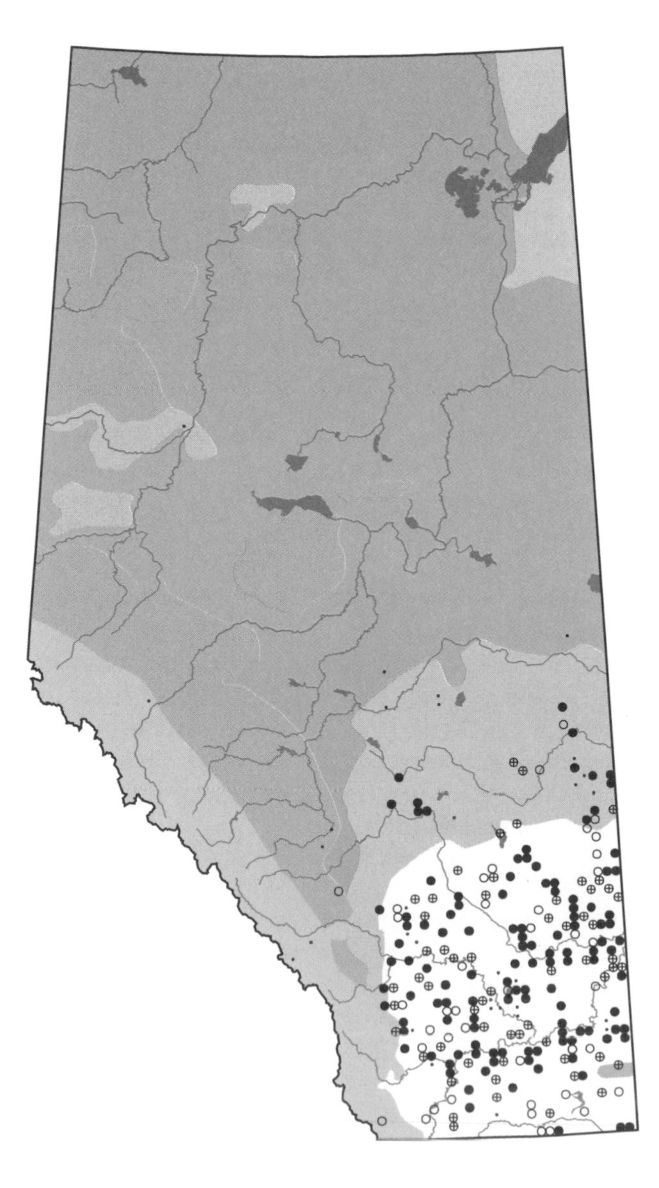

Eastern Kingbird

Tyrannus tyrannus

EAKI photo credit: T. Webb

Status:
The Eastern Kingbird is a common and abundant bird in Alberta.

Distribution:
This large and conspicuous flycatcher occurs throughout the province although it is scarce in heavily forested areas in the north and in the Rockies (Salt and Salt 1976). Atlassers recorded this species in 45% of surveyed squares in the Parkland region, 41% in the Grassland, 38% in the Canadian Shield, and 30% in the Boreal Forest region. It was also recorded in 15% of surveyed squares in the Rocky Mountain region and 14% in the Foothills region.

Habitat:
This species prefers open country with scattered trees, as well as woodland edges and shrubby fields and pastures. It frequents the farm shelterbelts and wooded coulees of the Grassland region, and the light woods with fields and meadows of the Parkland region (Salt and Salt 1976). In the more densely forested areas of the north, it is found near clearings such as roadsides, burns, beaver ponds, bogs, or other wetlands, particularly where there are standing dead trees. In the mountains, these birds are most likely to be found in willow or birch thickets, or in sedge meadows (Holroyd and Van Tighem 1983).

Nesting:
The nest of the Eastern Kingbird is a bulky and untidy structure built of grasses, plant stems, twigs, plant down, rootlets, string, or feathers. It is lined with fine grasses, plant down, and hair, and is placed on the branch of a tree or bush (usually deciduous), often over water. Sometimes, a stump, fence post, or upended tree root will also serve as a nesting site. Most nests are less than 6 m from the ground. Both male and female construct the nest, and may return to the same nesting site over successive years.
The female lays 3-4 eggs, which are white and spotted with browns. Incubation, primarily or solely by the female, is 12-13 days. This species is a common cowbird host, but it usually ejects or damages the cowbird's eggs. Young Eastern Kingbirds are tended by both parents and fledge at 16-18 days. The parents continue to feed the young for about 35 days after fledging. One brood is raised per season.

Remarks:
This flycatcher is easily identified because of its contrasting black and white plumage, rapid quivering flight, and habit of perching conspicuously well out on a branch, fence post, or telephone wire. From these perches, it makes its forays to capture insects, picking them off with a loud snap of the bill. This species also eats berries and other small fruits.
The Eastern Kingbird has a fearless disposition and is unafraid of humans. It is also highly aggressive around the nest and does not hesitate to attack crows, hawks, or magpies who approach too near to the nest. The intruder is pursued until well away from the nest. During an attack, the Kingbird's crest will be raised, revealing a bright red or orange patch on the crown.
The call of the Eastern Kingbird is a buzzing "dzeet", although, in attack or mating, it becomes a loud and rising "kit kit kit" (Bovey 1988). The male advertises his territory by display flights in which he flutters and swoops about 6 m above the ground giving his harsh, shrill calls.
This species is a relatively late spring migrant, with birds arriving from mid- to late May. Fall migration occurs in August, with the last birds usually observed late in August or early in September (Pinel et al. in prep.). There are no reports of overwintering.

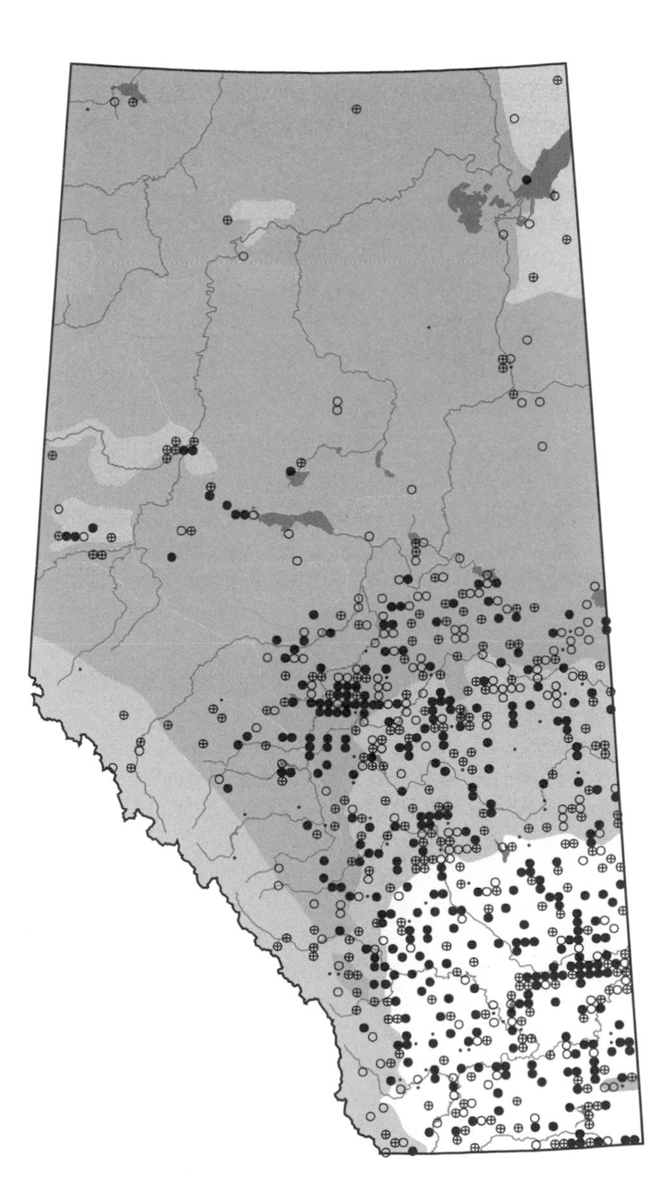

Horned Lark

Eremophila alpestris

HOLA

Status:
Of the 75 species of Larks, 50 are found in Africa, and 11 in Europe, but only the Horned Lark is native to North America. Four subspecies breed in various parts of Alberta. The Horned Lark is considered to have a healthy population and secure habitat, and is, therefore, not at risk (Wildlife Management Branch 1991).

Distribution:
The Horned Lark breeds throughout most of the Northern Hemisphere (Holroyd and Van Tighem 1983). Atlas data agree with Salt and Salt (1976), that this species breeds in southern and central Alberta north to Athabasca and west into the mountains. One of the subspecies may breed north around Lake Athabasca. In Alberta, it is found throughout the Grassland Natural Region, where it was located in 47% of surveyed squares, and less commonly in the Rocky Mountain (10%) and Parkland (17%) regions.

Habitat:
The Horned Lark breeds where open ground can be found. The species uses grassland, savannah, and even deciduous forests, but avoids coniferous woods. This preference is also found in the species' mountain habitat (Holroyd and Van Tighem 1983).

Nesting:
Males establish their territory and attract females with a strutting display, which includes dropping their wings, fanning their tails and erecting their horns. In territorial disputes, chasing one another and ground pecking are common behaviors (Bent 1963). The nest of the Horned Lark is a cup constructed of dry grass, twigs, and leaves. It is built in a hollow in the ground, often near a clump of grass, a stone, or a stump, and is lined with fine grass, down, and feathers. The nest is built by the female and will often have an area on one side "paved" with pebbles.
Clutch size is variously reported, but 3-4 eggs, which are grayish white and speckled with reddish brown, are typical. The female alone incubates the eggs for 10-12 days. During the early nesting period, excrement is removed by the adults by consumption, later on it is simply removed and discarded (Bent 1963).
The nest is defended by abandonment if the intruder is detected while far from the nest, otherwise a distraction display is used (Bent 1963).
The young leave the nest in 9-12 days and fledge 3-5 days later (Harrison 1978).

Remarks:
The Horned Lark has a whitish to yellowish face and underside with a black patch on the upper chest and a narrow black streak extending from the bill to under the eye and down the cheek. A second black stripe across the crown terminates in two "horns" on the side of the head, which are actually erectile feathers. Cadman et al. (1987) reported that the subspecies color variation of this species can be correlated to geographical variation, resulting in the dorsal surface of the bird blending closely to the nesting substrate.
Horned Larks eat mainly plant seeds, but also eat insects and worms. Most foraging is carried out on the nesting territory (Bent 1965a).
The bird's love of open ground has resulted in a terminal habit of staying on roads in ignorance of vehicles (Craig G. and E. Craig 1978).
The Horned Lark is one of the earliest spring migrants, with some arriving in mid-February. However, most do not arrive until April (Salt and Salt 1976). In fall, they move south in September, but stragglers have been recorded as late as November. Overwintering has been recorded in the mountains, southern Alberta and Edmonton.

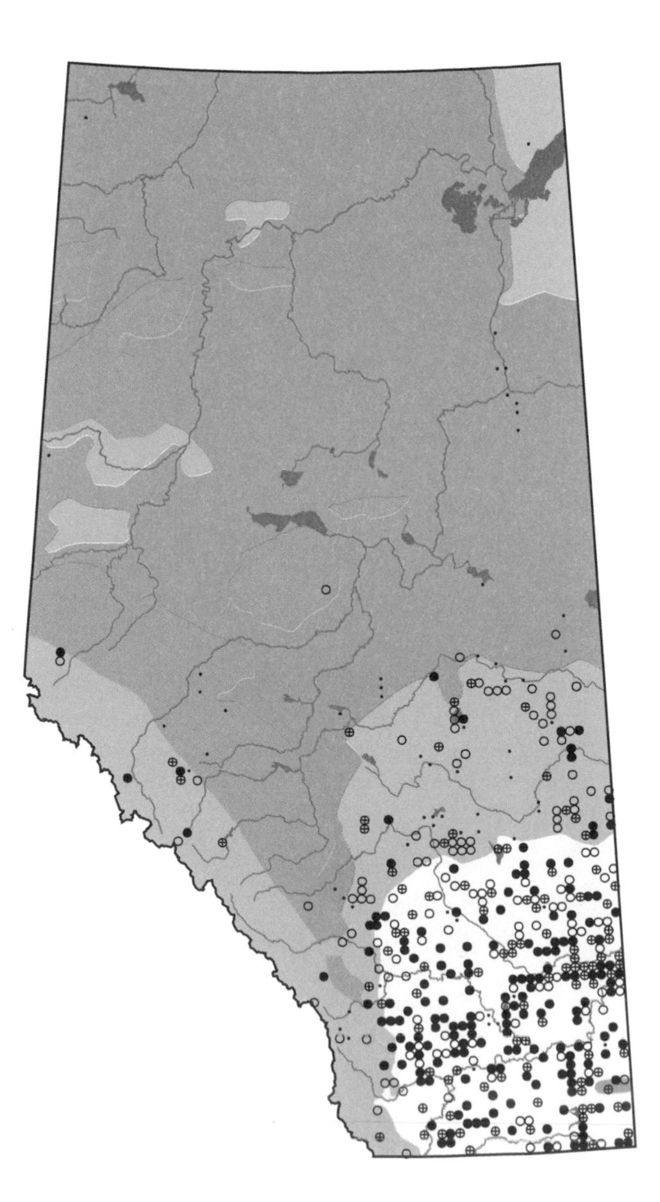

Purple Martin

Progne subis

PUMA photo credit: C.R. Sams II & J.F. Stoick/VIREO

Status:
The population of the largest North American swallow is decreasing over its total range, partly as a result of competition with House Sparrows and European Starlings for nest cavities. Dragonflies and other insect prey may be reduced because of pesticide use. Wintering grounds in South America are also threatened, creating further pressure. The increased number of communal nest boxes provided for martins has somewhat ameliorated the population decline.

Distribution:
Very local distribution of the Purple Martin is found in east-central Alberta, bordered on the west and north by Wabasca-Desmarais, on the south by Calgary, and east to the Saskatchewan border. Breeding colonies are also found in the Peace River region.
As acreage subdivisions proliferate, there is, to some extent, a lateral migration of colonies from more populated centres. The first Purple Martins to nest in man-made colony boxes in Alberta were at Camrose in 1918 and Red Deer in 1920 (C. Finlay pers. comm.). Edmonton had its first nesters in 1937 but it was not until 1946 that the first colonies began. Calgary had its first recorded nest in 1971.

Habitat:
This species was present in 16% of the Parkland and 6% of the Boreal Forest squares surveyed by the Atlas. In other natural regions, the Purple Martin was sighted in fewer than 2% of surveyed squares. Under natural conditions, the Purple Martin nests in mature woodlands and burns spotted with lakes, meadows or marshes. Human settlement in central Alberta has largely removed these stands, limiting the species' nesting to artificial nest boxes around the open areas of Parkland human settlements.

Nesting:
Very few observations are now made of nesting in this species' original habitat, such as in cliff crevices, woodpecker holes, and hollow trees. The Natives of eastern North America first lured these birds to nest nearby by providing hollowed-out gourds. This has given way to individual and, more commonly used, colonial nest boxes. The nest, an accumulation of grass, feathers, and twigs, is built by both sexes, but mainly by the female. It often includes fresh leaves as a lining. A small mud wall in front, to prevent the eggs from rolling out the entrance, is sometimes present. Typically, 4-6 smooth white eggs are laid and primarily incubated by the female for 15-17 days. Once hatched, the nestlings are tended by both parents and fledge in 24-28 days, returning to roost in the nest for 7-10 days afterwards.

Remarks:
The most domestic of our swallows, these birds are well-regarded for their approachability and insectivorous habits. However, they do not eat the large numbers of mosquitoes that some martin enthusiasts maintain (C. Finlay pers. comm.). The two-year-old and older male is entirely dark, glossy purplish bluc, while the female and yearling male are gray-brown below. The yearling may have some dark blue spots extending below. In flight, the Purple Martin resembles the Starling, but can be distinguished by its forked tail, longer wings, wider bill, and swallow-type flight.
They forage on the wing over open land, water, and marshes for larger insects, including mayflies, flying ants, dragon flies, butterflies, and moths.
The Purple Martin has a distinctive loud call, "chew-chew", as well as a pleasant liquid gurgling song during breeding.
Adult birds arrive in late April or early May. The yearling birds return around the third week of May. Once the young have become strong fliers, all birds will flock together for their journey to South America, resulting in few Alberta sightings after mid-September.

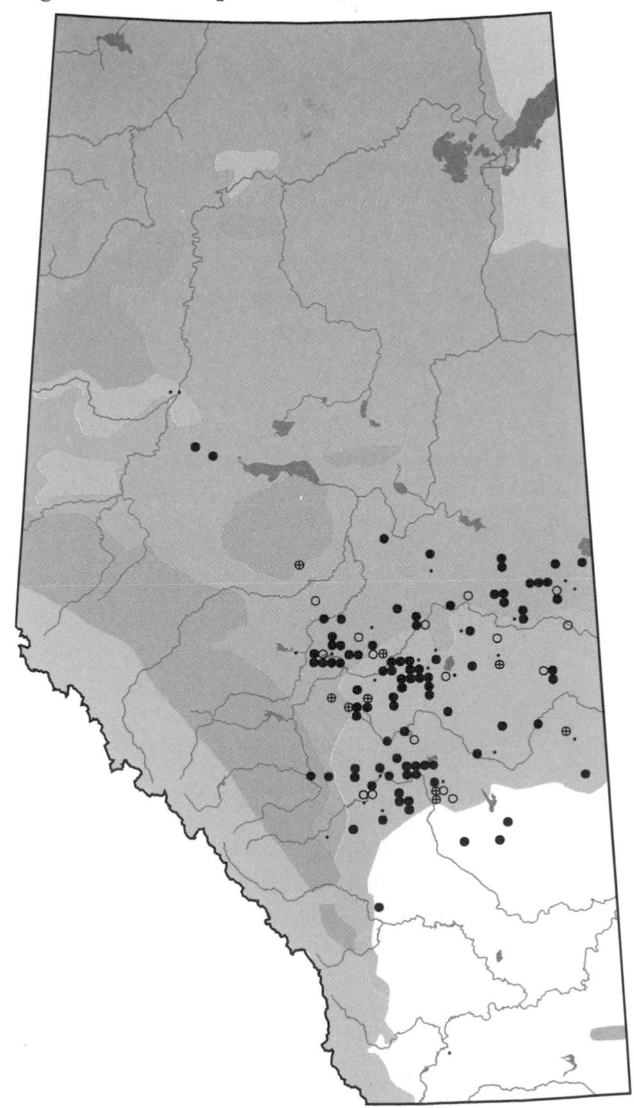

Tree Swallow

Tachycineta bicolor

TESW photo credit: R. Gehlert

Status:
The status of this species in Alberta is presently undetermined, but Atlas data indicate that it is a relatively common bird across Alberta.

Distribution:
This swallow appears throughout the province. Atlas records show that it is a common breeder in the Parkland and southern Boreal Forest natural regions, and less common in the Grassland region, where it must rely heavily on the use of nest boxes. As more boxes are put out this swallow increases in number. The tree swallow is present in all six natural regions, with concentrations ranging from 13% of Grassland squares surveyed to 60% in the Parkland region. The visibility of nest cavities and boxes provided high levels of breeding confirmation.

Habitat:
The natural breeding habitat of mature woodland provides dead or dying trees in which woodpeckers can excavate nesting cavities. Tree swallows may occupy these in following years. Adaptation has been made to nest boxes positioned on roadsides and near lakes, ponds, sloughs, marshes, wet meadows, bogs, or other flooded areas. Local populations may concentrate where numerous standing trees are present.

Nesting:
A well-researched species, the following observations have been made: polygamy occurs and, possibly, double brooding (Stiles 1982); females can brood alone, given an adequate food supply (Leffelaar and Robertson 1985); lower natural cavities are more frequently preyed upon and larger floor area results in larger clutches (Rendell and Robertson 1989). The female Tree Swallow may return to her previous nest site for up to five years, sometimes in a new pair bond (Collister 1990). An obligate cavity nester, the tree swallow may use single or colonial nest boxes, natural tree cavities, or, occasionally, buildings as possible sites. There is also heavy use of nest boxes on the various bluebird trails in the province. They have been known to place their nests over Bufflehead eggs. Given the opportunity, they may nest in relatively close quarters, although, when too close, aggression may prevent some pairs from breeding. Nest sites near water host larger clutches and more fledged young (Holroyd 1987). Inside the nest, the female builds a cup-shaped nest of dry grass and a few feathers. The 4-6 white eggs are incubated by the female for 13-16 days. The male will perch nearby for long periods and provide some courtship feeding. Both parents tend to the young, feeding them every few minutes. Once fledged (16-24 days), all leave the area to flock and feed over water or bask on power lines.

Remarks:
The male is a dark, glossy greenish blue above and white below. The adult female is similar, but duller above. Yearling females are brownish in color. Juveniles are grayish black above and can be distinguished from young Bank Swallows by the latter's breast band and from young Rough-winged Swallows by their pure white throats. Tree Swallows respond to human intruders by silent circling or by diving and loud chatter.

In times of inclement weather, these insect eaters are known to eat berries. A couple of weeks after spring migration, the males separate, search out a nest and twitter constantly to attract a mate. The Tree Swallow also has "chee-veet" and "silip" calls.

An early spring migrant, these are usually the first swallows to arrive, doing so in numbers beginning in mid-April. However, this leaves them susceptible to late winter storms. These swallows depart by late August. Stragglers linger until the third week of September in some years.

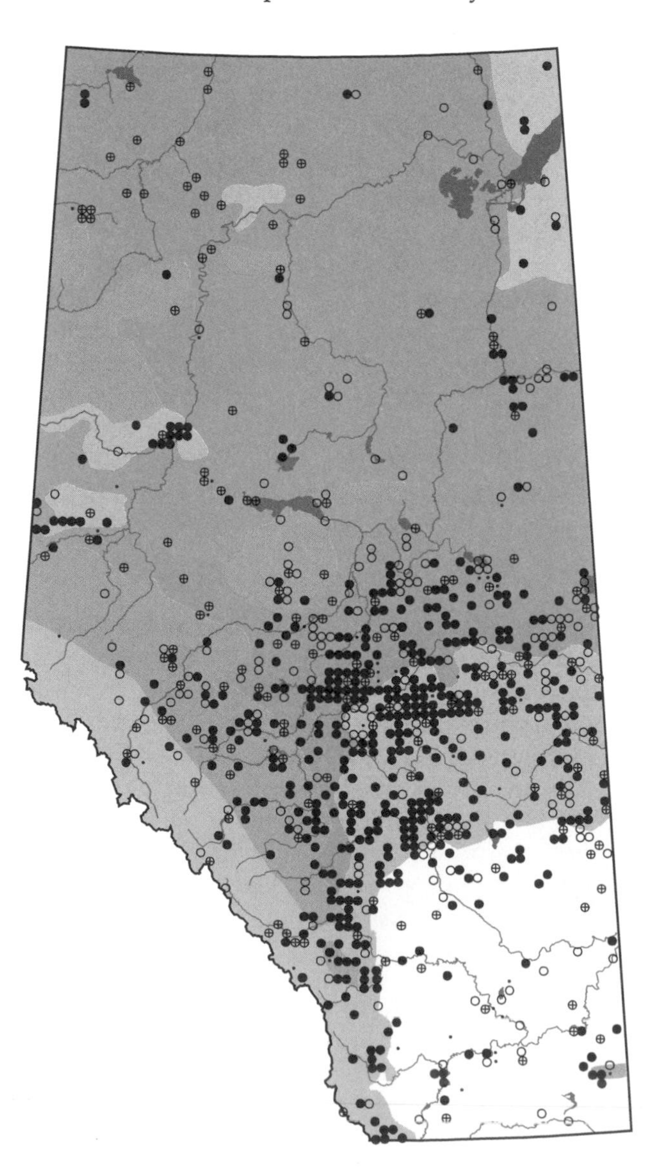

Violet-green Swallow

Tachycineta thalassina

VGSW

Status:
The status of this species has not been adequately studied and remains undetermined (Wildlife Management Branch 1991). Salt and Salt (1976) reported it to be fairly common in the Rocky Mountains, locally common along the Eastern Slopes, and often abundant within the human-inhabited areas of the Rocky Mountain region. However, Holroyd and Van Tighem (1983) considered it to be uncommon in Jasper and Banff national parks in the fall.

Distribution:
In addition to breeding in the mountain parks from Jasper through to Waterton, the Violet-green Swallow was found to be breeding in the Red Deer area and in southeastern Alberta along the Milk and South Saskatchewan rivers. This eastward movement concurs with past sporadic extensions east of the mountain areas (Salt and Salt 1976). This species was recorded in 12% of the surveyed squares in the Rocky Mountain Natural Region. It was also recorded in the Grassland and Foothills regions, but only in 1% of surveyed squares. It was even more rarely encountered in the Parkland and Boreal Forest natural regions.

Habitat:
Primarily a bird of the Rocky Mountain region, this species keeps to mountain valleys, only occasionally venturing into the lower subalpine. Foraging where flying insects are numerous, the Violet-green Swallow frequents beaver ponds, lakeshores, open country, and sometimes forest. Willing to live in close proximity to man, it can often be seen in townsites. Occasional irruptions bring this species east of the Foothills Natural Region and into habitats of similar variety.

Nesting:
Remote natural cliffs, crevices, and rock faces are this species' most commonly used nest locations. Human structures, birdhouses, and tree cavities may also be used. Social and tolerant, especially of Tree Swallows, this species will nest in colonies where adequate nest sites exist. The female builds most of the nest, using grass and small twigs, and lines it with feathers. Four to five smooth and non-glossy white eggs are laid. Incubation will begin before the clutch is complete . Chicks hatch after 15 days. Both parents tend to the young, which open their eyes in 10-11 days and fledge after 23-25 days. Adults may occasionally feed young on the wing. Only one brood is raised.

Remarks:
At first glance, an observer may mistake this species for the more common Tree Swallow. However, when well illuminated, the male is a dark glossy green above with a bronze tint. It is also distinguished by white on the earpatch, above the eye, and on the sides of the rump. The female is similar, but paler above. Juveniles are brownish above with brownish wash across the breast. Typical of the swallows, the Violet-green Swallow can often be seen carrying out various aerobatic manoeuvres with wide sweeps and short turns. It does this while foraging for flying insects, seemingly using little caution around animals or buildings. It will hawk these insects high above water or flutter along the surface. This may be one of few birds seen while hiking high mountain trails.
This species arrives in Alberta in late April to early May, in mixed flocks with other swallows, and has departed the province by the end of August. Difficult to distinguish from adjacent Tree Swallows during this time, they may be separated by their slightly earlier migration from the province.

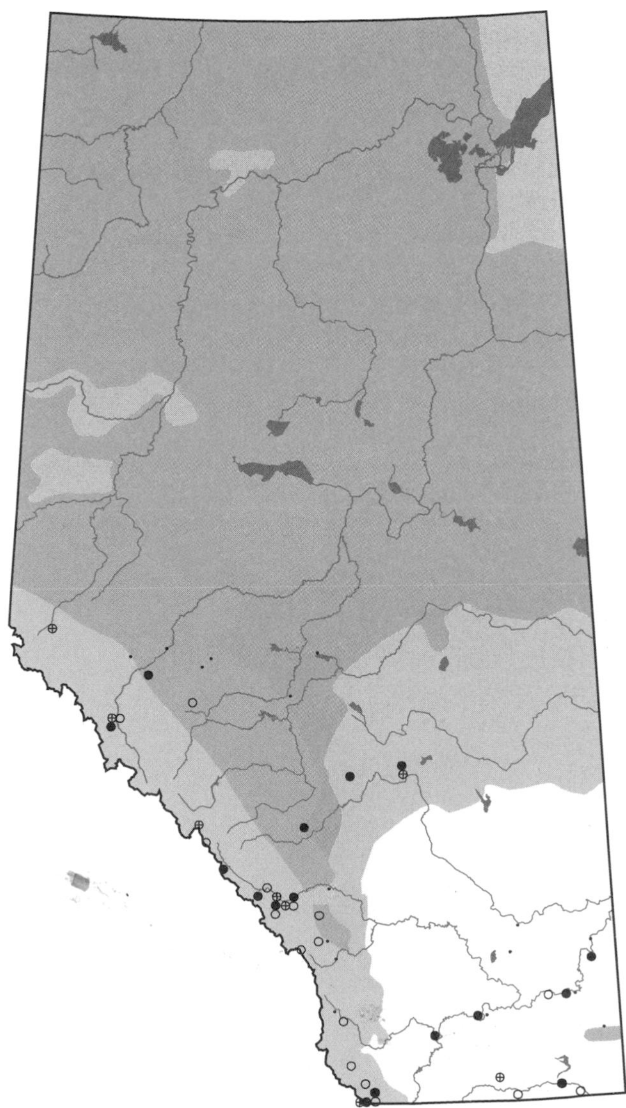

Northern Rough-winged Swallow

Stelgidopteryx serripennis

NRWS

Status:
The Wildlife Management Branch (1991) has considered the status of the Northern Rough-winged Swallow as undetermined, although not immediately considered at risk. Salt and Salt (1976) reported it to be common in the south and Holroyd and Van Tighem (1983) considered it an uncommon summer resident of the mountain parks and formerly more common in Jasper. Cadman et al. (1987) suggested that its range may be expanding with further human settlement.

Distribution:
Historical and Atlas Project sightings consider the breeding territory of this species to be south of a line which connects Empress (near the Saskatchewan border), Edmonton, and west to the mountains. This species was recorded in 8% of the Grassland squares that were surveyed, as well as 10% of the Rocky Mountain squares and 6% of the Parkland squares. Although it uses soil banks for nest sites, as does the Bank Swallow, its solitary nature made actual nest confirmations less numerous.

Habitat:
The soft earth banks of creeks and irrigation canals in the Grassland Natural Region are natural locations, as are the low stream banks of the Parkland region. Human landscaping has added many artificial sites, such as road and rail cutbanks and embankments. Water and/or open land, even semi-arid, must be nearby in order for the birds to forage. Mountain and, occasionally, lower subalpine areas are used.

Nesting:
Courtship is stimulated by the male's pursuit and tail fanning. Selecting banks of sand, clay, or gravel, this swallow excavates a chamber and tunnel of variable length. Depending on the texture of the substrate, the entrance to a Rough-winged Swallow's tunnel will appear rough and elliptical, whereas that of the Bank Swallow may be neat and more circular. The crevices in rocks or cliffs, holes in cement walls and bridges, drainpipes, and deserted Kingfisher burrows may also be occupied. Although not colonial, 2-3 pairs of these swallows may nest in the same bank. The female builds an untidy nest of dry grass, feathers, rootlets, and leaves in the chamber at the end of the tunnel. The female, predominantly, incubates the 5-7 white eggs of her single brood for 16 days.
The chicks are tended to by both adults. Feathered after 12-13 days, the young depart the nest after 18-21 days. The nesting habits of this species are less known than are those of our other swallows.

Remarks:
The least domestic of Alberta's swallows, the Northern Rough-winged Swallow derives its name from the serrated edge of the primary feathers of the adult male. Male and female plumage are the same—brown above with a grayish wash on the chin, throat, and upper breast. This is distinguishable from the white throat and dark breast band of the Bank Swallow. Juveniles are similar, but have cinnamon wing bars. Sharing similar habits, this species is commonly mistaken for the Bank Swallow, although the Bank Swallow's wingbeats are quicker and more shallow. Rough-winged Swallow nest sites host less activity and are harder to find, although an intruder can expect a loud and aggressive challenge.
This species arrives in mixed flocks with other swallows in late April to early May, though they are difficult to detect. Little information is available on fall migration, although it appears that most birds leave the province by late August.

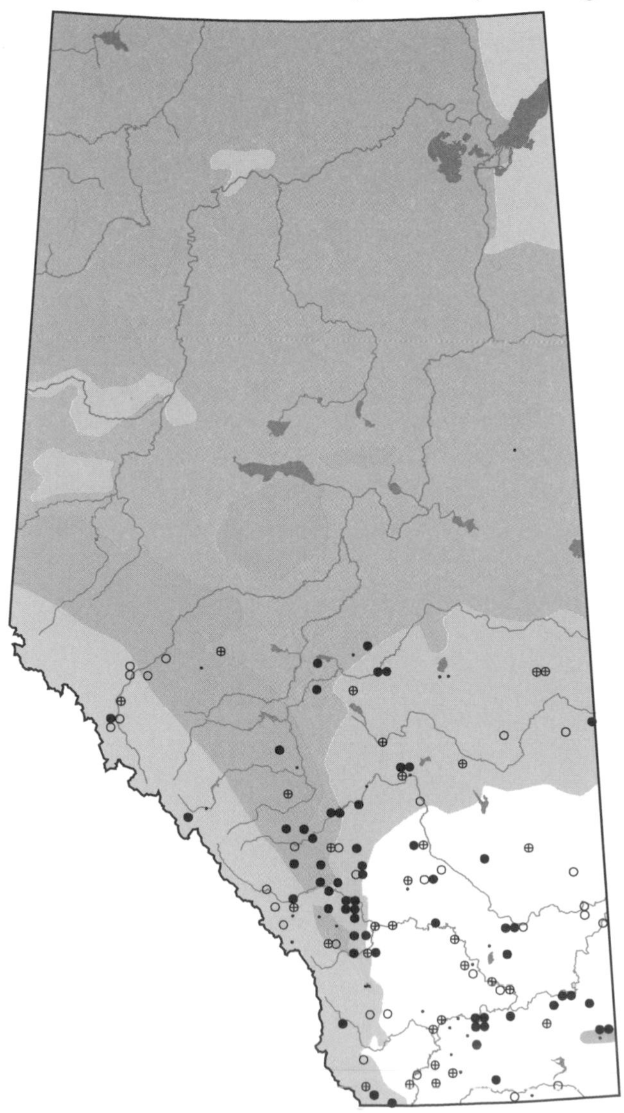

Bank Swallow

Riparia riparia

BKSW

Status:
The status of the Bank Swallow has not been well studied in the province and is undetermined. Holroyd and Van Tighem (1983) reported this species as a fairly common migrant and an uncommon summer resident of the mountain parks. Colonies are fairly common in cutbanks and gravel pits in central Alberta (C. Finlay pers. comm.). Local populations may have increased because of these man-made cutbanks and excavations.

Distribution:
The Atlas Project found this species to be widely distributed in each natural region of the province, with the exception of the Canadian Shield, where no sightings were reported. The Bank Swallow was recorded in 15% of surveyed squares in the Grassland and Rocky Mountain regions, 22% in the Parkland region, and 14% in the Boreal Forest and Foothills regions. The high percentage of breeding confirmation is attributed to the visibility of nesting colonies. Local distribution is determined by the availability of nesting habitat and nearby water.

Habitat:
Natural features providing soil banks, such as the sides of rivers and lakes, are used for nest burrows. Similar features exposed by human landscaping may also be colonized. The presence of nearby water for foraging is an additional requirement.

Nesting:
Returning to the same nesting location each year, this colonial species may reuse old burrows or dig new ones. Rarely, existing artificial holes may be used, though this is more likely to be true of Rough-winged Swallows (Ajertoas 1988). Use of the vertical face of an aged and stable sawdust pile in Hinton has been recorded (Wilde 1987). Colony size depends on dimensions of the bank. Burrows are located near the top of steep embankments, generally in open areas. Both sexes tunnel the burrow, using their bills to excavate and their claws to kick the sand out. A simple nest of some grass and feathers is placed in the small chamber at the end of a 60-90 cm tunnel. Four to six white eggs are laid and both share in the incubation duties. Pale gray downy chicks hatch in 14-16 days. While the male will do some brooding, he primarily forages for the young. The female shares in this activity once brooding is complete. Young are initially fed in the cavity, graduating to its entrance later. Juveniles instinctively begin burrowing in the nest chamber, creating cavities that are used for resting. Nest predation is low, but is occasionally caused by snakes, mice, skunks, and foxes. Despite being able to fly after 14 days, young do not fly out of the burrow until after about 19 days. Families soon desert the nest site to wander before migration.

Remarks:
The similarly appearing sexes are grayish brown above with white behind the earpatch, are white below, and have a distinctive brown breast band with a short central stripe trailing beneath it. This distinguishes them from the pale brown throat and breast of the less widespread Northern Rough-winged Swallow. The Bank Swallow also has shallow, rapid wingbeats. Locally, the Bank Swallow may be seen above cliffs, flying around and emitting soft twitters. Being very social birds, aggression only occurs when neighbors appear at the same nest entrance.
During migration, these widespread transients may be seen within mixed flocks, hawking insects over shallow waters. Spring arrival is in late April to mid-May, and up to one month later in the mountain areas. Autumnal migration to South America occurs by the third week of August, with some lingering into early September in some years.

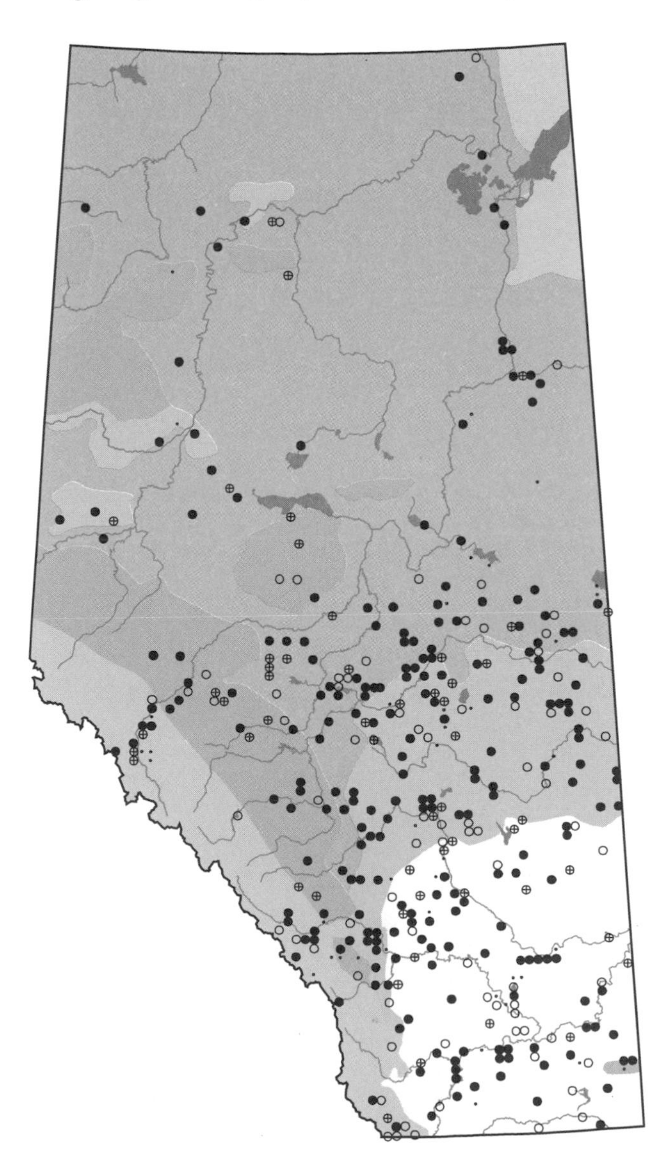

Cliff Swallow

Hirundo pyrrhonota

CLSW

Status:
The provincial population of Cliff Swallows appears to expand as human habitation expands. The building of human structures has provided more opportunities for this species to nest, thus increasing its population. Further activity should bode well for this species. Where distribution overlaps with that of House Sparrows, nests may be parasitized and usurped, as was reported by Lang (1974) in the Calgary area.

Distribution:
Widely distributed within the province, the Cliff Swallow breeds in every natural region. It was recorded in 35% of surveyed squares in the Foothills region, 27% in the Parkland region, 25% in the Rocky Mountain and Grassland regions, and 20% in the Boreal Forest region. As a result of their colonial tendencies and, often, use of human structures, a significantly large proportion (77%) of breeders were confirmed.

Habitat:
A variety of sites provide the vertical and overhanging surfaces required for nesting. Concrete bridges, rural settlements, cliffs, high ledges, culverts, steep clay riverbanks, and, rarely, the inside of a building are suitable. Nearby water should be available for gathering mud, drinking, and foraging. It will also forage over open land and marshes. In the mountains, Cliff Swallows normally breed from the montane to the upper subalpine areas, usually near rivers or lakes. However, two colonies have been reported breeding in the alpine ecoregion (Holroyd and Van Tighem 1983).

Nesting:
Bent (1965a) suggested that courting and mating activities centred around nest building. Ehrlich et al. (1988) suggest that courtship flight is followed by copulation on the ground. The gourd-shaped nests are constructed of mud pellets with the entrance placed at the edge of a down-curved neck. Previous nests may be repaired and reused. Nests are plastered between a vertical surface and an overhang of less than 90 degrees. Being very colonial, sometimes in the hundreds or more, Cliff Swallows stack their nests very close together. Constant activity gives the area the appearance of a beehive. Despite the close proximity of nests, adults recognize their own. On natural surfaces, the colony will be confined by areas exposed to rainfall. Inside the nest, 4-5 white eggs spotted with brown are laid on a few feathers and bits of grass. The female does most of the incubating (12-14 days). Both adults tend to the young, which fledge in 21-24 days.

Remarks:
Most similar to the Barn Swallow in appearance, the Cliff Swallow is distinguished by its shorter squarish tail, white forehead, and buffy rump. Juveniles are gray-brown in the dark areas of the adult with a paler throat and darker forehead. The soft voice of the Cliff Swallow is squeakier than other swallows. These swallows can be seen skimming over rivers, lakes, and sloughs, scooping up insects. They also feed on airborne insects. An advantage of a colonial situation is that conspecifics observe birds that have foraged successfully and then follow them out.
Cliff Swallows arrive in the province in early May and begin to nest in colonies by the end of the month. Once young have fledged these swallows leave to gather in flocks, and leave the province by the end of August, with a few remaining into September.

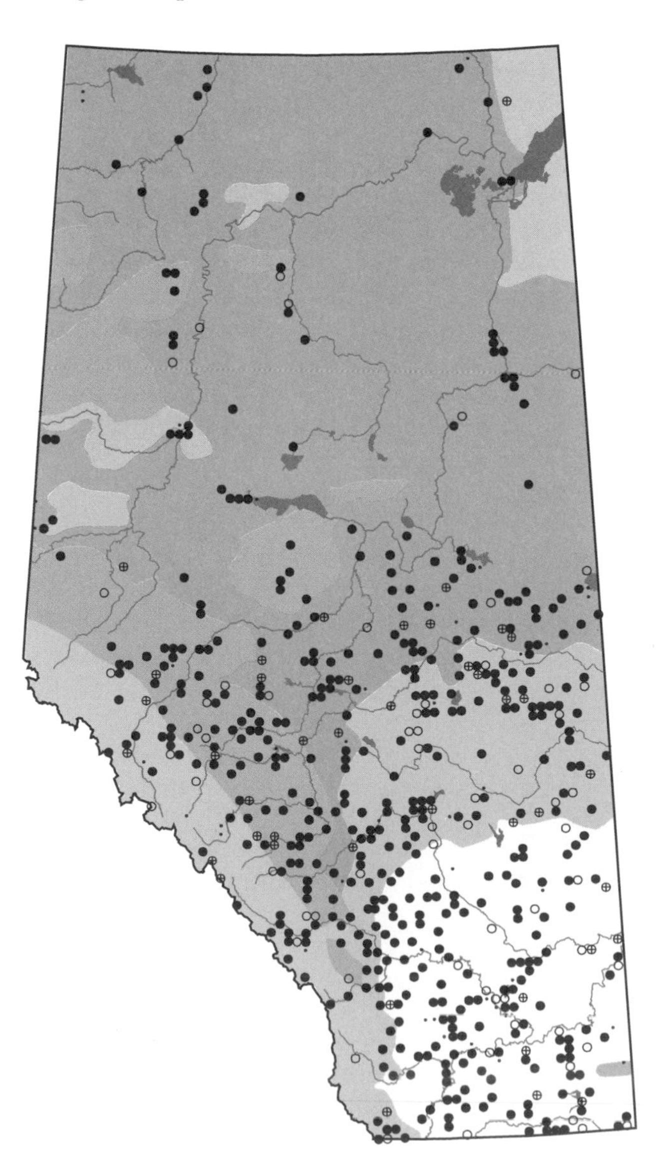

Barn Swallow

Hirundo rustica

BRSW

Status:
The Barn Swallow appears be relatively stable, with its distribution becoming increasingly widespread as more of the province is settled. In areas of greater wilderness, particularly the Rocky Mountain, Boreal Forest, and Canadian Shield regions, this species is less common, although established.

Distribution:
This species breeds in every region of Alberta, although it is more common in the heavily settled areas of the south. Distribution in the north will undoubtedly increase as further human structures are erected and more land is cleared, as was the case with its expansion of numbers in the Peace River region in the early 1960s.

Habitat:
The Barn Swallow uses human structures to provide suitable nest sites in rural and suburban areas. Nearby water bodies, gardens, and livestock are also preferred for foraging, drinking, and mud gathering. In the mountain parks, it primarily nests in the montane and lower subalpine ecoregions, though it may range into alpine regions during migration (Holroyd and van Tighem 1983).

Nesting:
The generally monogamous male (Wolinski 1985) courts a prospective mate by carrying out lengthy pursuits, rubbing necks, interlocking bills, and mutual preening. The female's mate selection is influenced by the male's tail length (Moller 1990). Nests are built by both sexes and located on the horizontal or vertical surfaces of rafters or eaves of barns and older buildings, under bridges, in crevices, or in hollow trees. Construction takes 6-8 days and involves countless pellets of mud and grass transported in the bird's mouth. The resulting shallow cup-shaped nest is moulded by the bird's body and lined with grass and feathers. The nest is always sheltered from above, typically under the eaves of a building. Pairs will often nest in the same area as the previous year, even rebuilding a nest if possible. They are usually single-brooded. A typical clutch is 4-6 smooth white eggs marked with shades of brown and reddish brown. These eggs are very similar to those of the Cliff Swallow. The female predominantly incubates these eggs for a period of 14-15 days, occasionally assisted by her mate. Each mate alternates feeding the young by regurgitation, leading to fledging in 17-24 days. The fledglings will be fed and defended for a week or two afterwards. If the young hesitate to fly once able, the parents may withhold food and coax by calling and fluttering their wings.

Remarks:
The male of this commonly seen swallow is dark blue above, has a longer tail than any other local swallow, and can be distinguished from others by white on its tail. It also has a reddish-brown forehead and throat and cinnamon or buffy underparts. Females and juveniles are similar but with shorter tails and paler underparts. The song is a pleasant liquid trill, sometimes punctuated by a slightly rising, grating trill. Its flight call can be notated as "kvick, kvick". Usually foraging low, the Barn Swallow can be seen around farm yards, over water and open land, and near marshes. On warm evenings it will fly higher, following the rise of flies, gnats, and flying ants. Of benefit to the farmer, they also may be seen buzzing around cattle and tractors as insects are kicked up. Drinking, bathing, and courting are also taken up on the wing.
The Barn Swallow is usually the last swallow to arrive in the spring, with the first ones coming in late April and peaking by the end of May (C. Finlay pers. comm.). They are also the last species of swallow to depart in the fall, often still feeding nestlings in early September and departing in late September.

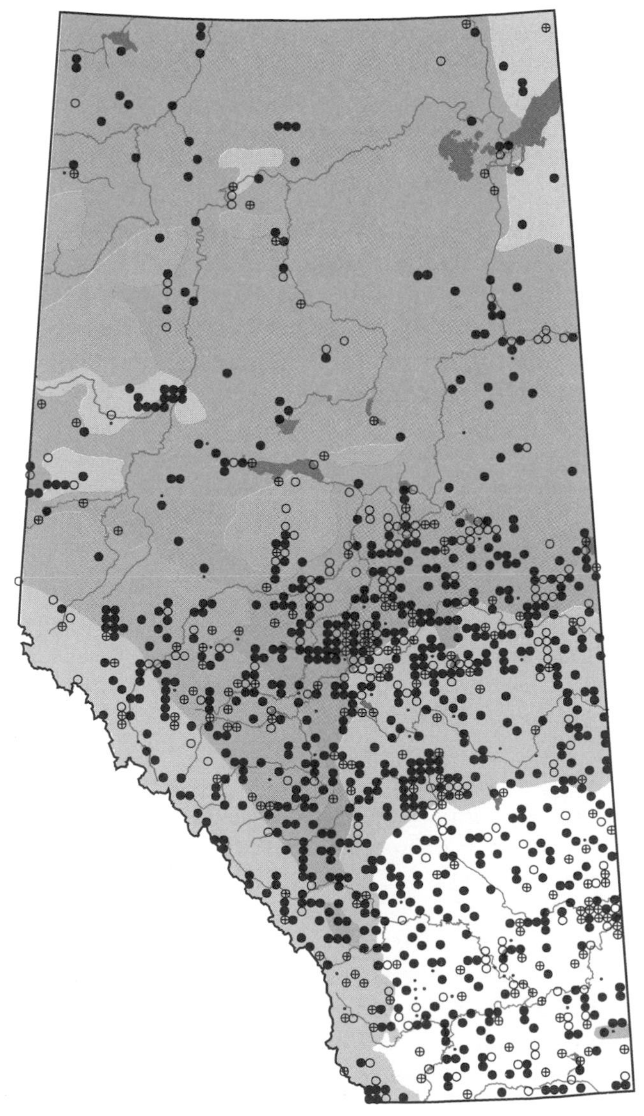

Gray Jay

Perisoreus canadensis

GRJA photo credit: T. Webb

Status:
With its status not in danger (Wildlife Management Branch 1991), this permanent resident ranges from common (Salt and Salt 1976) to very common (Holroyd and Van Tighem 1983) in the Rocky Mountain and Foothills natural regions of the province. Atlas data indicate that its status in northern Alberta is expected to be secure.

Distribution:
The Gray Jay is found in the boreal regions of Canada from the Yukon to Newfoundland. In Alberta, it was found in over 35% of squares surveyed in the Canadian Shield, Boreal Forest, Foothills,and Rocky Mountain natural regions. There were very few sightings in the Parkland region (4%) and no sightings in the Grassland region.

Habitat:
This species can be found in a variety of woodlands, including mixed wood, black spruce and tamarack lowland bogs, dense white spruce, and forest openings. It is most common in dense coniferous stands. In winter, it may wander into townsites, the Parkland region, and, occasionally, into the coulees of the Grassland Natural Region. In the mountain parks, it prefers the closed forests from montane to subalpine.

Nesting:
These life-long monogamous pairs begin the nesting activity in February. Both adults help build the bulky mass of twigs and grass and line the deep cup with moss, feathers, hair, and, occasionally, fur. This well-insulated structure takes two weeks to build in fair weather. Well-camouflaged and difficult to find, the nest is usually positioned within a crotch of horizontal branches and against the trunk of a spruce or fir. It is left partly exposed to the sun's rays and from 1.5-4.5 m above ground. In March, 2-6 greenish-gray eggs blotched with browns and lavenders are laid. The female begins incubation with the first egg and continues for 16-18 days. She is a very close sitter. The male will courtship-feed the female as she broods the young for the first few days. When weather conditions allow, the female will leave the nest for periods to assist the male in foraging for the nestlings. Young fledge after 17-20 days. This species will re-nest if the initial brood fails as a result of extreme weather or predation from squirrels or ravens. Otherwise, it is single-brooded.

Remarks:
Also known by some more colorful names—Canada Jay, Whiskey Jack, and Camp Robber—this highly sociable bird can often be seen around campsites in the mountain parks competing for handouts with the Clark's Nutcracker. Male and female adults have the same appearance: gray nape and upperparts, white crown with no crest, white throat and undertail, short black bill, longish tail, and pale gray underparts. The juvenile is a sooty gray overall. Their distinctive flight from tree to tree involves short periods of flapping, alternating with long glides. Sometimes a noisy bird, its repertoire includes whistles, harsh chatter, cackles, and an almost introspective soft babble.
Some small groups may wander in winter south and east of their usual range (Salt and Salt 1976). The Gray Jay usually maintains a year-round territory in the province, enabling it to horde food for the winter. Scraps, insects, young fir needles, spiders, berries, and seeds are all part of its omnivorous diet.

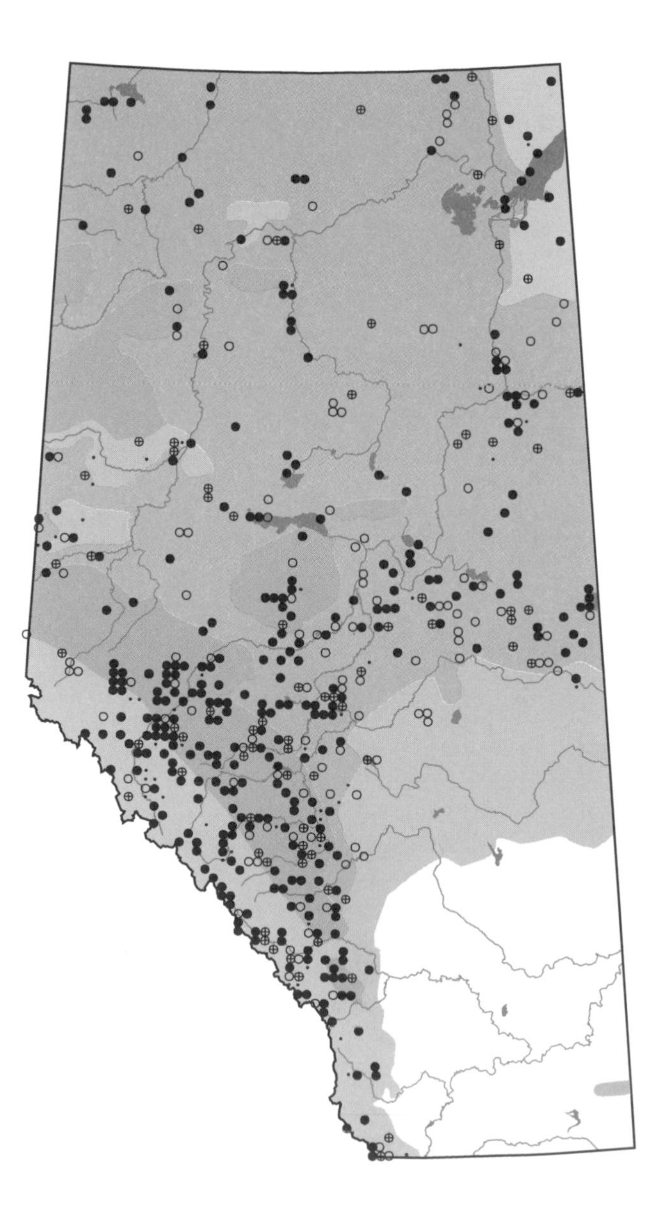

Steller's Jay

Cyanocitta stelleri

STJA

Status:
Accidental or peripheral to the province, the local distribution of the Steller's Jay is not well known (Wildlife Management Branch 1991). Atlas findings concur with past analysis, that it is a rare but permanent resident of the mountain parks (Holroyd and Van Tighem 1983). Salt and Salt (1976) also suggested it to be an uncommon resident of Jasper and Waterton Lakes national parks, and a rare visitor to Banff.

Distribution:
Breeding in Alberta is restricted to the Rocky Mountain and Foothills natural regions, with one confirmed record in each region. It has been observed to sporadically wander eastward into central Alberta.

Habitat:
Breeding in the montane and lower subalpine, the Steller's Jay prefers coniferous woods above 1200 m (Salt and Salt 1976). It may sometimes be found within the dense shrubbery of avalanche slopes (Holroyd and Van Tighem 1983).

Nesting:
The Steller's Jay is wary and secretive as it begins to nest in late April or early May. Normally in a coniferous tree, often a Douglas-fir, the nest is built on a horizontal branch or in a crotch 2.5 - 8 m up the tree. Late in the season, a deciduous tree is occasionally used (Bent 1964d; Ehrlich et al. 1988). The well-made nest is a bulky mass of twigs, bark, and leaves, moulded with some use of mud and lined with fine rootlets, grass, and pine needles. Larger than the Blue Jay's, the nest is constructed by both male and female. The female lays her single brood of 3-5 bluish-green eggs, spotted with brown, and begins an incubation period of 16-17 days. The male will courtship-feed his mate at this time. Subsequently, the chicks are tended by both parents.
When approached at the nest, the female gives off a scolding alarm call and slips quietly off. It is not known when the young fledge. The adults and young travel as a family group until fall.

Remarks:
Exhibiting the general size and shape of a Blue Jay, including the crest, the Steller's Jay has a black head, throat, shoulders, and nape, and is dark blue otherwise. Blue streaks may extend from the base of the crest on some. The young generally resemble an adult, with blackish back, rump, and underparts. The adult female is smaller than the male (Bent 1964d). Habitat preference also distinguishes this species from Blue Jays. However, intergrades have occasionally been recorded where the species' distributions overlap in Colorado (Ehrlich et al. 1988). The commonly used call is "shaak, shaak, shaak, shaak". Various other calls are used or mimicked, including a warbled song and a Red-tailed Hawk imitation. Displaying typical jay-like curiosity, it can be attracted by an imitation owl call.
Foraging on insects, small vertebrates, eggs, nestlings, seeds, nuts, and fruit, the Steller's Jay will horde and cache excess quantities into crevices and holes dug into the ground.
Small wandering flocks move down into lower elevations in winter, sometimes using the seeds, waste grain, and garbage available in mountain towns and foothill ranches. This movement has resulted in fall and winter records as far east as Beaverlodge, Edmonton, and Red Deer (Pinel et al. in prep.).

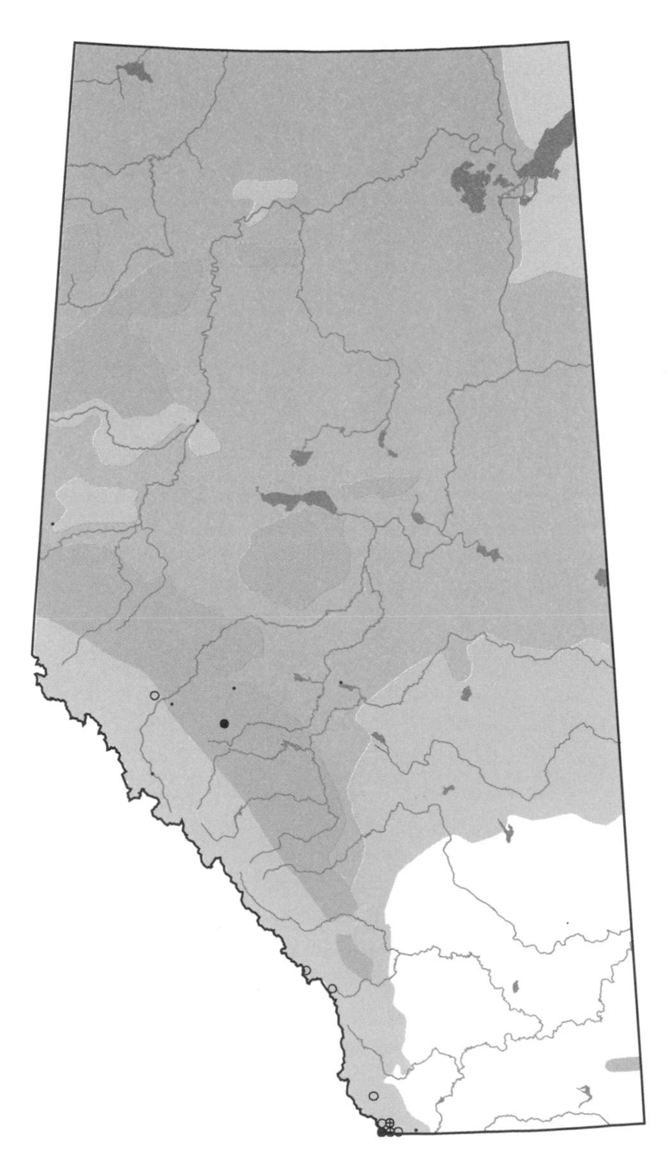

Blue Jay

Cyanocitta cristata

BLJA photo credit: T. Webb

Status:

Familiar to everyone, the Blue Jay experiences a healthy provincial population and secure habitat. Present status is not considered at risk.

Distribution:

Provincial distribution has been historically focused on central Alberta. The Blue Jay has moved northward with expansive human settlement, reaching the Grande Prairie, Peace River, Lesser Slave Lake, and Fort McMurray regions (Salt and Salt 1976). The normal southern boundary of its breeding range corresponds with the southern limit of the Parkland Natural Region. However, as the urban forests of prairie cities proliferate, sporadic incursions are made by Blue Jays. This has been documented in Lethbridge and in areas east and northeast of that city (O'Shea 1980). From Atlas data, there are confirmed breeding records in the Brooks area and probable breeding records east of Lethbridge.

Habitat:

Mixed and deciduous forest is preferred habitat of this jay. This includes semi-open areas of ornamental shade trees. Adapted to the urban landscape, the Blue Jay may breed within cities and towns of the Parkland, Boreal Forest, and, to some extent, Foothills natural regions.

Nesting:

These birds are year-round residents of Alberta and begin preparing for courtship in late winter. The 3-6 potential male suitors of a flock will intimidate, fight, and bob at each other before a single individual is finally chosen. This male will courtship-feed his potential mate, as well as present twigs to her for a mock nest (Savage 1985). A conifer tree is usually chosen for nesting. The nest is built near the trunk and elevated 2.4 m to 7.5 m above ground. Both adults construct the bulky nest of twigs, bark, moss, rags, and lichens, and line it with fine rootlets, grass, and, occasionally, feathers. The 4-6 olive or buff eggs are variably marked with browns. The following period of 17-18 days is quietly spent by the female incubating, while the male feeds her and loosely guards the location. The single brood will be replaced if lost (Harrison 1978). Although only the female broods the chicks, both adults tend to the young, feeding them caterpillars and worms. The young fledge after 17-21 days, and quickly become as vocal as their parents.

Remarks:

Considered by many as the most attractive North American Jay, the Blue Jay is blue above and grayish white below, with white barring on the wings, a crest, a longish tail, and a black necklace. It may be mistaken for only the Steller's Jay, which is a darker blue overall, but distribution of these two jays rarely overlaps. The Blue Jay uses a characteristic level flight and has a wide vocal repertoire. A loud "jay, jay" call, often discharged by a sentinel, warns of intruders. Other calls, ranging from soft to harsh, often mimic other birds, particularly the Red-tailed Hawk. This behavior is unlike that exhibited during the breeding season, when it is quiet and retiring. Becoming more visible in the late summer, these somewhat gregarious birds frequent gardens and farmyards, exhibiting the noisy, boisterous, and curious nature we expect of jays.

The Blue Jay has an omnivore's diet, including fruits, insects, grains, and the eggs and nestlings of songbirds.

Only partially migratory, young disperse a few hundred kilometres, and adults remain near their breeding territory (Savage 1985). The proliferation of winter feeders has reduced the requirement to migrate. An expandable throat pouch and aggression permit gorging at and domination around feeders.

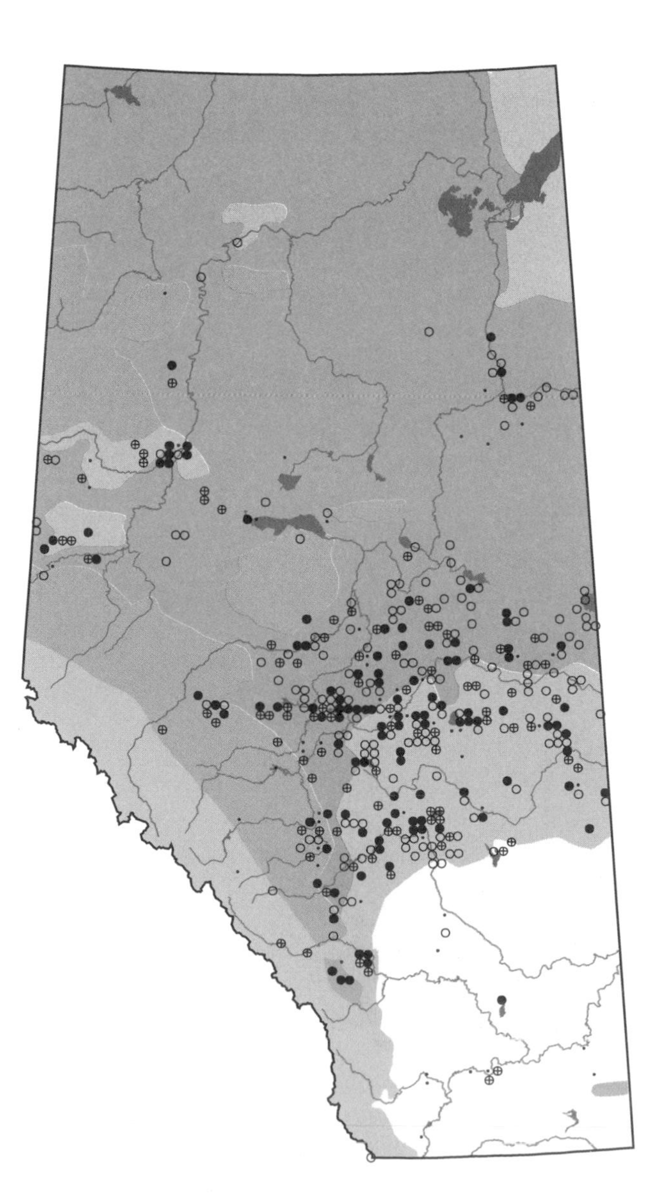

Clark's Nutcracker

Nucifraga columbiana

CLNU photo credit: T. Webb

Status:
The low natural population and limited distribution of this year-round resident present special management concerns for biologists. Not considered at risk at this time, the status of this species may be sensitive to change (Wildlife Management Branch 1991). A common summer resident of the mountain regions, it is fairly common at other times of the year (Holroyd and Van Tighem 1983).

Distribution:
This is a bird of the Rocky Mountain Natural Region, where it was recorded in 21% of surveyed squares. There are also some records in the adjacent Foothills region, as well as sporadic eastward irruptions of small groups, usually 2-5 birds, as far as the Cypress Hills. Theories vary on the cause of these movements, including the failure of local cone crops and assistance from passing weather systems (Fisher and Myres 1980b).

Habitat:
The Clark's Nutcracker predominantly breeds in the coniferous forests of the subalpine, with one additional population commonly breeding around Banff townsite. Preferring open or broken stands of conifers, this bird also uses edges, clearings, and burns up to timberline. The montane ecoregion is frequented at other times of the year.

Nesting:
Typical of our corvids, the normally vocal and boisterous Clark's Nutcracker becomes silent and secretive as it begins nesting in March. The nest, built mainly by the female, consists of a platform of twigs lined with conifer needles, grass, shredded bark, and hair. This is positioned on a horizontal limb 3-17 m up a coniferous tree. The female lays a single brood of 2-4 greenish eggs sparsely marked with small brown spots. Both adults incubate for a period of 16-17 days, well-hidden in the large deep cup. Tended and brooded by both parents, the nestlings are fed regurgitated, hulled seeds. Fledging in 22 days, the young continue to be fed by the adults for some time afterwards.

Remarks:
The name of this species is derived from William Clark, of the famed Lewis and Clark expedition, and from the bird's innate ability to crack open nuts and cones. A stout gray bird, the nutcracker has black wings and central tail, large white wing patches and white outer tail feathers, a long and pointed heavy black bill, and a shorter tail than that of the Gray Jay. Juvenile appearance is similar, with some brown tinging. Travelling singly or in small groups up to 10 birds, this can be a typically conspicuous and noisy corvid, occasionally using a "kaar-kaar" call that belies its presence. An omnivore, the nutcracker will eat insects, juniper berries, small vertebrates, and the eggs and nestlings of small birds. It also has a preference for whitebark and limber pine seeds. It has become accustomed to hand feeding and human scraps. Natural food sources in winter are buds, carrion, and conifer seeds. When a food source is abundant, this species will use a unique lower mouth pouch to horde the item and cache the excess for harsh winter days and late spring. Caches are established on snow-free, south-facing slopes for easy future access.
In winter, nutcrackers carry out an altitudinal migration to lower ski hills, valleys, and townsites. They compete with the Gray Jay for human scraps and bird feeders at these locations. The metabolic heat produced during the digestion of such food boosts survival chances during extreme cold weather. Some birds may wander out of the mountain parks in winter.

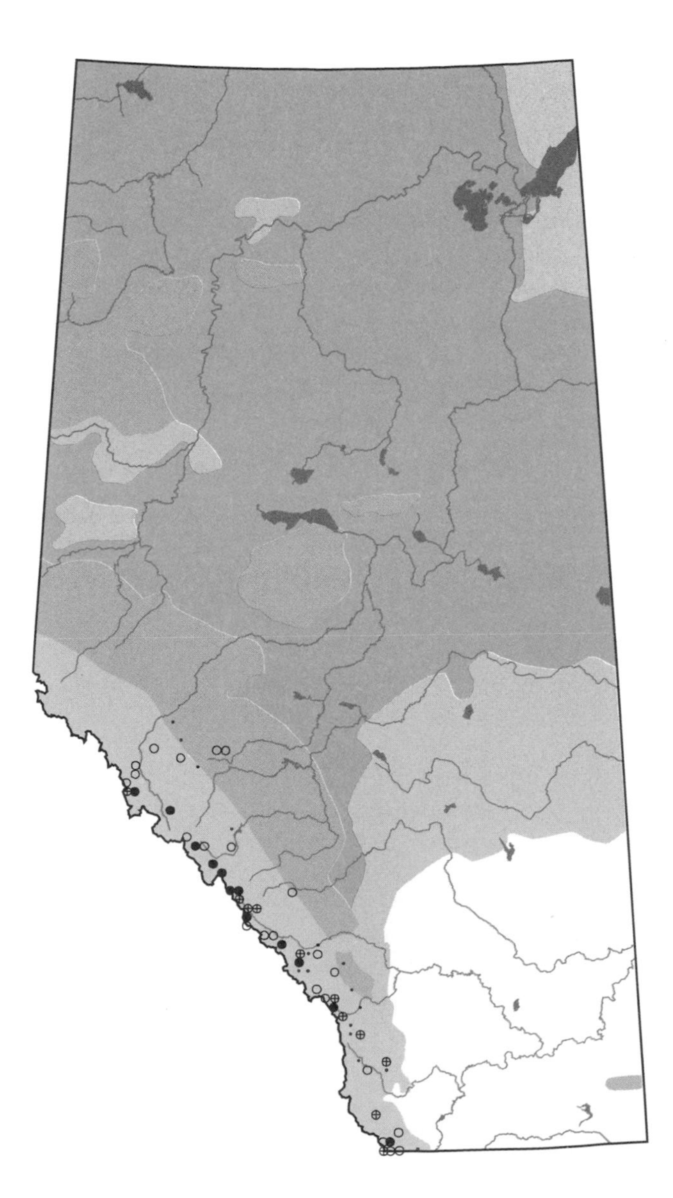

Black-billed Magpie

Pica pica

BBMA photo credit: G. Beyersbergen

Status:
This much-maligned resident is common in central and southern Alberta west to the mountain parks, where its status is fairly common (Holroyd and Van Tighem 1983). This was not always so. The magpie was considered to be very rare in Alberta until the turn of the century, when it began to appear in response to human settlement. It may also have been common in the days of buffalo herds (Bent 1964d). Culling programs featuring bounties have proven ineffective.

Distribution:
Magpie distribution reaches north to Athabasca from the American border and locally around the settled areas of Jasper, Banff, Peace River, Grande Prairie, High Level, and Fort McMurray. Pinel et al. (in prep.) predict range expansion northward into the Boreal Forest region as clearing and settlement continues.

Habitat:
The magpie may be found in all natural regions except the Canadian Shield. It frequents a variety of local areas, including thickets and scattered trees in open areas, shrubbery and trees along coulees, shelter belts, tree patches, edges, and woodland openings. It is more common around human activity.

Nesting:
The monogamous pair begin building their domed nest in early April. Both mates help to construct a large platform of heavy twigs, after which the female completes an inside cup of grasses, mud, rootlets, and hair. Meanwhile, the male completes the canopy of heavy twigs, allowing for one or more side entrances. Magpies are also known to repair and use old nests. The nest may be situated in a tree (deciduous or coniferous) or a tall bush. It is usually located below 7 m. The female lays 6-9 greenish-gray eggs, which are heavily blotched with brown, incubates them for 16-18 days, and broods the young alone. The male feeds and defends the nest occupants, exhibiting the highest level of parental care known among the passerines (Dunn and Hannon 1989). A dominance hierarchy quickly begins, where larger siblings receive more food and smaller ones become subordinate. Young leave the nest at 22-28 days.

Remarks:
This Alberta bird is easily recognizable from all others locally, being generally black and white with iridescent blue on the primaries and green on the long tail, and having a heavy black bill. Recently fledged young appear the same, but with a shorter tail. The familiar "yak, yak, yak, yak, yak" call is recognized by many, being only one of several calls used. Left undisturbed, an individual may be heard to utter a soft "mumbling". Like other corvids, the magpie is omnivorous, feeding in summer on grains, insects, eggs, nestlings, carrion, mice, and human food scraps. It adds fruits and seeds to its diet in winter. Although physiologically adapted to winter (Root 1988), flocks of all ages roost in conifers or dense thickets to reduce the effects of wind and the risk of predation. Whether a magpie is part of a winter group, a family group, or a juvenile group, a dominance hierarchy is always maintained, usually headed by the largest (alpha) male. Territoriality is exhibited year-round, although there is more tolerance for intruding conspecifics outside of the nesting season. In the late fall and early spring, ceremonial gatherings may take place, where numerous magpies congregate to form new pair bonds as required and to compete for new nest sites and territies (Scharf and Clover 1983).

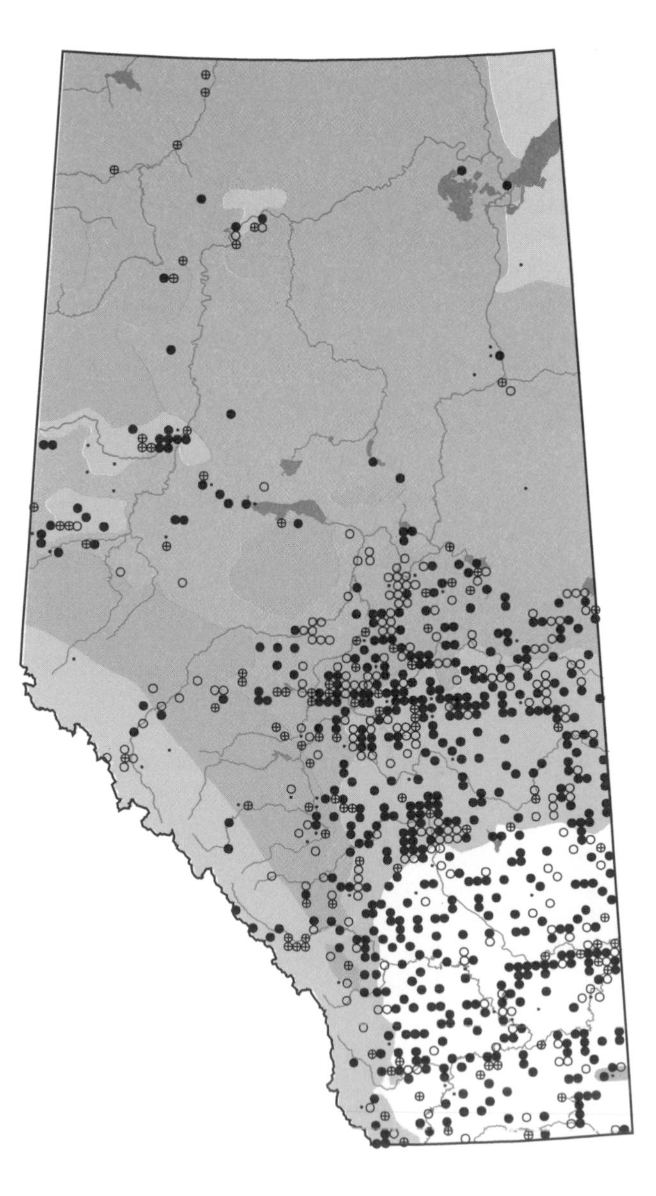

American Crow

Corvus brachyrhynchos

AMCR photo credit: G. Beyersbergen

Status:
The crow has maintained its common status throughout southern Alberta and ranges farther northward with continued human settlement. Increased population levels are due in part to the accessibility of roadkills and landfills. Efforts at persecution have had limited short-term results. Holroyd and Van Tighem (1983) reported it as a very common summer resident and a rare winter visitor to the mountain parks.

Distribution:
Flexibility of habitat selection has permitted the crow to establish itself throughout southern and central Alberta. In addition, as areas in the northern half of the province have been settled, the northward expansion of the breeding range has continued.

Habitat:
Appearing in all natural regions, the crow generally requires open areas for foraging and wooded areas to nest and roost. This edge-type habitat is common and comes in many forms: prairie coulee, aspen grove, forest edge, woodlots, shelterbelts, dumpsites, townsites, and open coniferous forest. Though common in the Grassland Natural Region, it is largely absent from large tracts of native grassland, preferring areas of human habitation and cultivation. It is resident in alpine areas only at lower elevations. In all areas, the availability of nearby water is preferred.

Nesting:
Normally noisy and highly visible during other parts of the year, the crow, like other local corvids, is quiet and unobtrusive during the nesting period. Only when a conspecific encroaches upon their territory will nesting crows become intolerant. This species will nest at various heights in both coniferous and deciduous trees and bushes. Occasionally, nests will be in low bushes or on the ground and sometimes in the proximity of others. The nest, built by both sexes, is a flat bulky structure of sticks, lined with bark, grass, hair, and rootlets. An old nest may be rebuilt. The 4-7 eggs are greenish or bluish and variously spotted and blotched with browns. These are incubated by both adults for a period of 17-20 days, during which the female is fed by the male (Ehrlich et al. 1988). Tended by both adults, the nestlings take 20-30 days to feather out and fledge about 35 days after hatching. The family group then bands with other crows to feed on swarming insects in grain fields and hay meadows.

Remarks:
Likely only confused with the Common Raven, the American Crow is all black with some a faint greenish-blue gloss above, which is visible only in ideal lighting. The crow is 16-17 cm smaller than the raven, and the black bill of the crow, though relatively long and thick, is not as thick as that of the raven. In flight, the crow has a straight takeoff and a fan-shaped tail, and does not soar. The familiar "caw-caw" is higher and more grating than that of the raven. Occasionally, a soft breeding song of musical chuckles and gurgles can be heard.
Like the other corvids, which possess the largest brain relative to body size of all birds, the crow is resourceful and cunning. A sentinel is often used when foraging. An omnivore and ground feeder, the crow includes in its diet waste grain, insect pests, human scraps, weed seeds, carrion, waterfowl eggs, and nestlings of many species.
Crows begin arriving before most passerines, in early to mid-March. In fall, crows may flock in great numbers as they roost and are pushed on southward by storms. The majority have departed by late September, with small bands lingering into early December. Wintering numbers are increasing to the extent that they are no longer considered extralimital in Calgary and Banff (Pinel et al. in prep.).

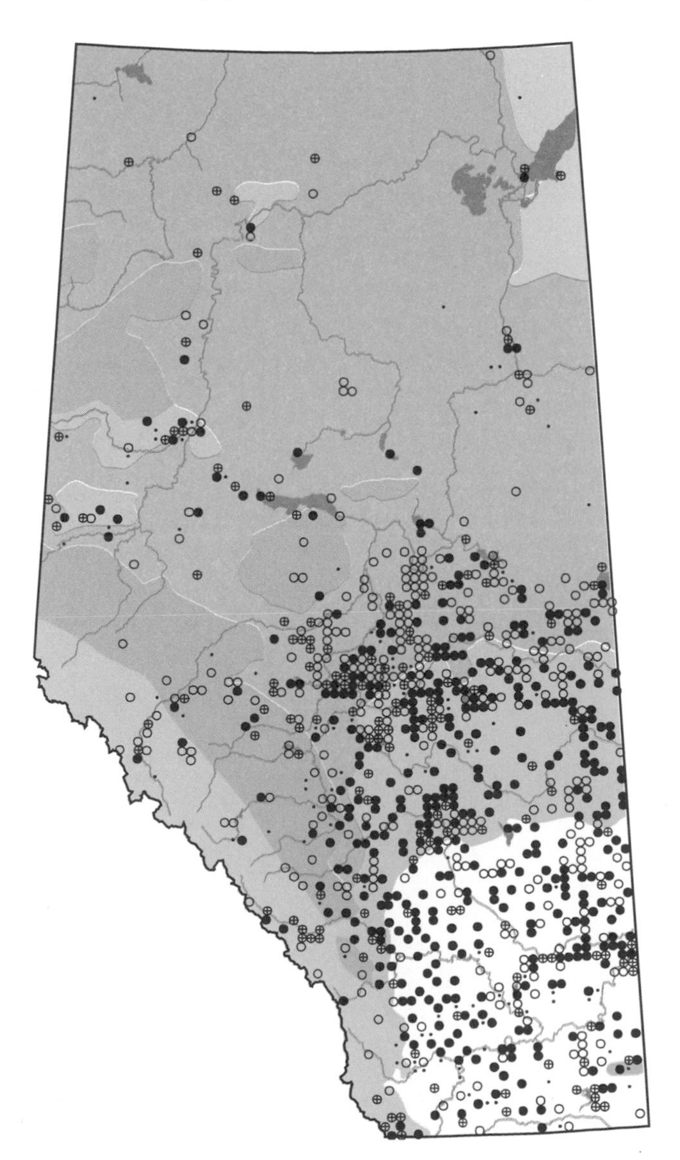

Common Raven

Corvus corax

CORA

Status:
North America's largest passerine enjoys healthy local population levels and widespread distribution. Holroyd and Van Tighem (1983) reported this species to be a common permanent resident of the mountain parks, growing in numbers since the 1940s. Over its North American range, the raven may have been more numerous at one time, when it lived in association with the bison (Root 1988).

Distribution:
The Common Raven breeds in all provincial natural regions. It is most common in the Canadian Shield, Boreal Forest, Foothills, and Rocky Mountain regions. There is only one confirmed breeding record for the Grassland Natural Region. Atlas data indicate that, since the *Birds of Alberta* (Salt and Salt 1976) was published, the breeding range has extended south to fill much of central Alberta down to a line connecting Edmonton and Lloydminister. Wandering far and wide in winter, their distribution is heavily influenced by the availability of food scraps.

Habitat:
As long as a forested region is available nearby for nesting, the raven can be found in all types of habitat, attempting to maximize its generalist habits. In the mountain parks, it is most common in the montane and lower subalpine ecoregions.

Nesting:
Mated for life, raven pairs carry out courtship displays annually, to reinforce their bond. In late winter, acrobatics, soaring, preening, billing, and vocalizing satisfy this requirement. Pairs begin nesting anytime from early April to mid-May. They are quiet and solitary at this time. A conifer tree normally hosts the nest, which is a mass of sticks and twigs lined with grass, bark, moss, and hair, and is built by both adults. The nest is positioned near the treetop and in a fork for strength. Sticks that drop from the nest are not retrieved, revealing the Common Raven's presence to observers below. An old nest or sheltered cliff may be used. Larger than those of the crow, the 4-7 eggs are greenish and blotched with brown, though markings may be variable within the same clutch. The female, while being fed by the male, will incubate her single brood for 20-21 days. The chicks are brooded by the female and attended by both. The adults transport water in their throats to nourish the nestlings. Finally fledging after 5-6 weeks, the young leave the nest with their parents to roam extensively.

Remarks:
This long-lived bird is similar in appearance to the crow. Close observation will note these differences: the raven is 16-17 cm larger, has a heavier bill that is more noticeable when open, and has pointed and elongated throat feathers. Unlike the crow, the raven sometimes soars. This species has a rounded or wedge-shaped tail and a lower voice than the crow. The song is characterized as a wooden "kwawk". Other diagnostic characteristics include flight acrobatics, a bell-like note, and 2-3 hops are sometimes taken before flight. Juveniles are noisier, calling for longer periods.
Ravens are normally seen singly or in small groups (up to five), except at carrion or at a landfill. A generalist by food and habit, this ground feeder will forage over wide areas, concentrating along lakeshores, rivers, roadsides, and landfills. In summer, it will eat seeds, insects, berries, small vertebrates, grasshoppers, worms, eggs, nestlings, and carrion. In winter, it is restricted to scavenging off winter-killed animals and human garbage. It is dominant over scavenging gulls, crows, and magpies, and caches excess food. The raven ejects pellets. In winter, it may concentrate around human habitations and garbage, or it may wander sporadically over large areas.

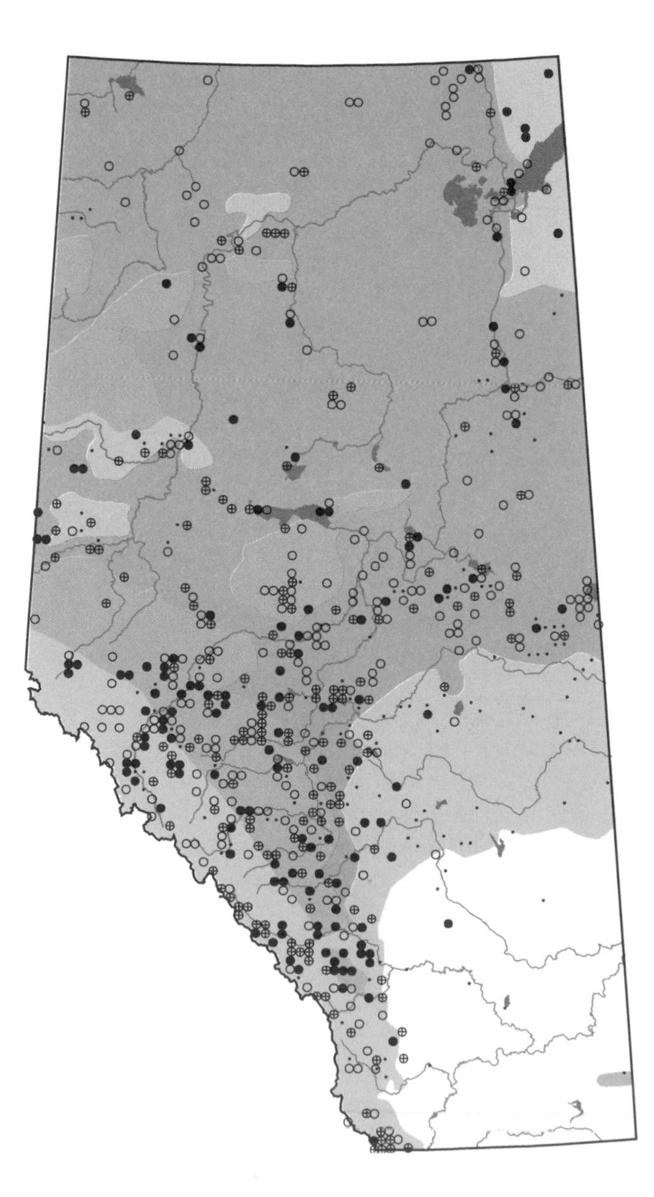

Black-capped Chickadee

Parus atricapillus

BCCH photo credit: T. Webb

Status:
In Alberta, the Black-capped Chickadee is considered to have a healthy population and secure habitat (Wildlife Management Branch 1991). In Banff and Jasper, Holroyd and Van Tighem (1983) reported this species to be a fairly common resident year-round.

Distribution:
In Alberta, this species is found in suitable habitat in all natural regions. One of the most common songbirds in the province, the Black-capped Chickadee was found in 52% of Parkland squares surveyed, as well as 50% in the Foothills, 44% in the Boreal Forest, 26% in the Rocky Mountain, and 31% in the Canadian Shield region. In the Grassland Natural Region, where it was recorded in only 10% of squares surveyed, it is restricted to wooded coulees and river valleys when nesting.

Habitat:
The Black-capped Chickadee prefers deciduous or mixed woods. In the Rocky Mountain Natural Region, Holroyd and Van Tighem (1983) found there to be a higher usage of coniferous trees in winter, as opposed to during the breeding season.

Nesting:
Trees chosen by these birds for nesting are usually deciduous, and the bark will generally be broken, revealing the soft inner wood. The majority of trees chosen by Black-capped Chickadees are dead, and there is a preference for snags with broken tops (Runde and Capen 1986).
Pairing appears to be gradual and territorial behavior begins once pair bonds are established. In disputes, the males call to one another loudly, and the defending male will often chase the intruder and even grapple with him in the air; females will occasionally join in the fray (Odum 1941a).
The Black-capped Chickadee is a cavity nester, usually excavating its own nest hole of around 25 cm in depth, but occasionally using a woodpecker hole or nest box. The hole is excavated by both birds, which carry the chips away and scatter them about, thereby not betraying the nest location by a heap of woodchips at the base of the tree.
The nest, constructed by the female alone, consists of moss, grass, plant fibres, and bark. It is lined with fur, feathers, and plant fibres.
Some mate-feeding of the female occurs during the excavation and nest building, but this behavior is more commonly seen during nesting (Odum 1941b). Usually, 6-8 white eggs, spotted with reddish brown, are laid, and these are incubated for 11-13 days by the female alone. She will incubate for regular periods, with shorter periods away from the nest for foraging (Odum 1941b). Both parents attend the young, bringing insects in the bill. Fledging is at 16 days and the young will remain with the parents until fall, but are only fed by the adults for a short period after leaving the nest.

Remarks:
The Black-capped Chickadee is a well-known bird at bird feeders in winter. It is stocky in appearance, with slate-colored wings and tail and white underparts. Its white cheek between the black crown and throat makes it easily identifiable. It may be distinguished from the Mountain Chickadee by the lack of a white eyebrow line. It has a whistled "phee-bee" call, as well as a "chicka-dee-dee-dee-dee-dee" call that is longer and clearer than it's neighboring conspecifics.
This species eats mainly insects and insect eggs, but also other invertebrates, seeds, and berries. Food is collected while the bird clambers over branches at any angle necessary to get at its food.
This species overwinters in small flocks in the same area as the nesting territory, although there may be some withdrawal from the far north and higher altitudes of the mountains (Salt and Salt 1976).

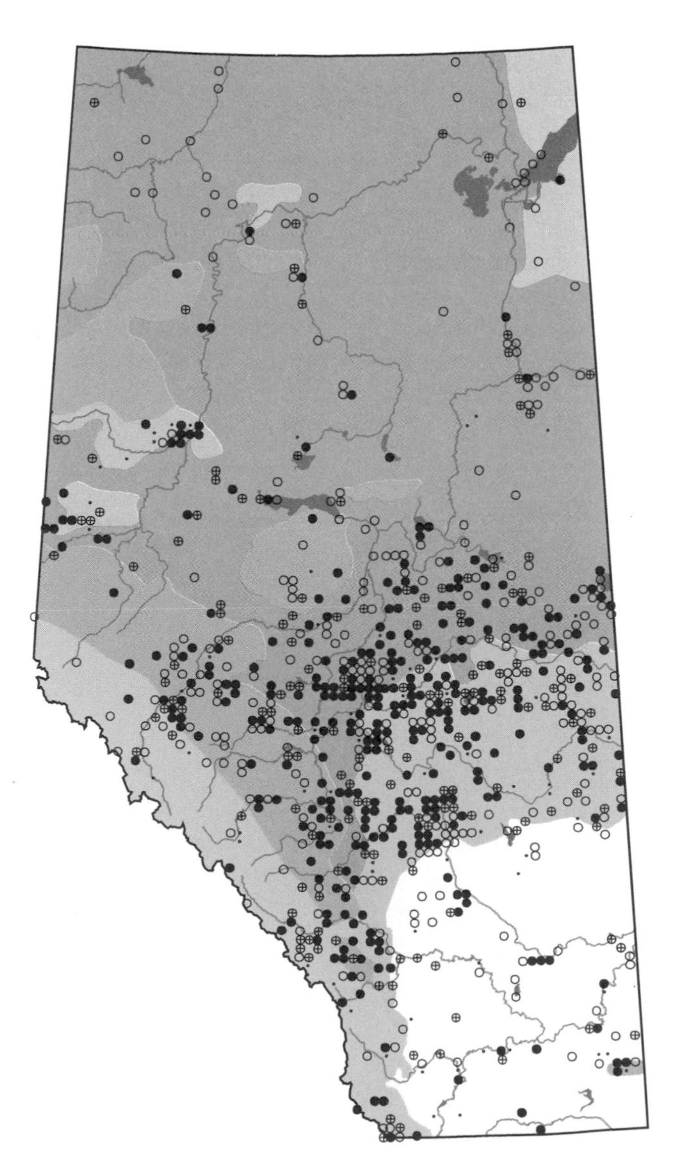

Mountain Chickadee

Parus gambeli

MOCH

Status:

The current status of the Mountain Chickadee in Alberta is unknown. In Banff and Jasper, it is a permanent resident considered to be fairly common in winter and common during the rest of the year (Holroyd and Van Tighem 1983), including the winter season.

Distribution:

The Mountain Chickadee is found as a permanent resident in mountainous areas from northern British Columbia south to California and Texas. In Alberta, this species is found mainly in the Rocky Mountain Natural Region, where it was recorded in 28% of the squares surveyed. It was also found eastward in the Foothills Natural Region, but with less frequency (4% of squares surveyed).

Habitat:

Mountain Chickadees, as the name suggests, are found in mountainous habitat, usually between 800 m and 3,300 m elevation (Root 1988), or near timberline. They prefer open coniferous forests as well as mixed woods for foraging. This includes more frequent use of tree crowns than the Black-capped Chickadee, a neighbor in some areas.

Nesting:

Mountain Chickadees are secondary cavity nesters, using old nesting cavities of woodpeckers and Black-capped Chickadees, natural cavities, or even nest boxes. These cavities are usually low to the ground, most commonly reported to be below 3 m.

Around mid-May, a nest of moss, plant fibres, fur, and feathers is constructed. It is lined with like material. The female typically lays 6-9 white eggs, usually plain, but occasionally spotted with reddish brown. The female incubates the eggs for 14 days, being fed by the male at this time. Once hatched, young are tended by both parents, which feed them insects brought in the bill. They leave the nest at 20 days (Harrison 1978).

There are records of double-brooding for this species.

Remarks:

The Mountain Chickadee is similar in appearance to the Black-capped Chickadee. Each has a black crown and nape and dark gray tail and wings. The former also possesses whitish underparts, a white cheek, and a white eyebrow line to distinguish it from the latter. Another possible neighbor, the Boreal Chickadee, is distinguished by its brown cap.

The Mountain Chickadee's primary call is shorter, harsher, and lazier than the Black-capped Chickadee's equivalent call, and is more reminiscent of the Boreal Chickadee. The whistled call, "fee-bee-bee", is usually 3-4 syllables, although it may be only two, thus causing confusion between it and the Black-capped Chickadee.

The Mountain Chickadee, although energetic, is somewhat more reserved than the Black-capped Chickadee. It is often found in close proximity to that species. It tends to forage higher in the forest canopy and is more likely to be found in conifers (Hill and Lein 1988).

Mountain Chickadees feed mainly on insects, but will also consume seeds and berries.

These chickadees will move to lower elevations in winter and join small flocks of Black-capped and, perhaps, Boreal Chickadees (Salt and Salt 1976). There are records of this species overwintering in Elk Island National Park (Carroll 1989).

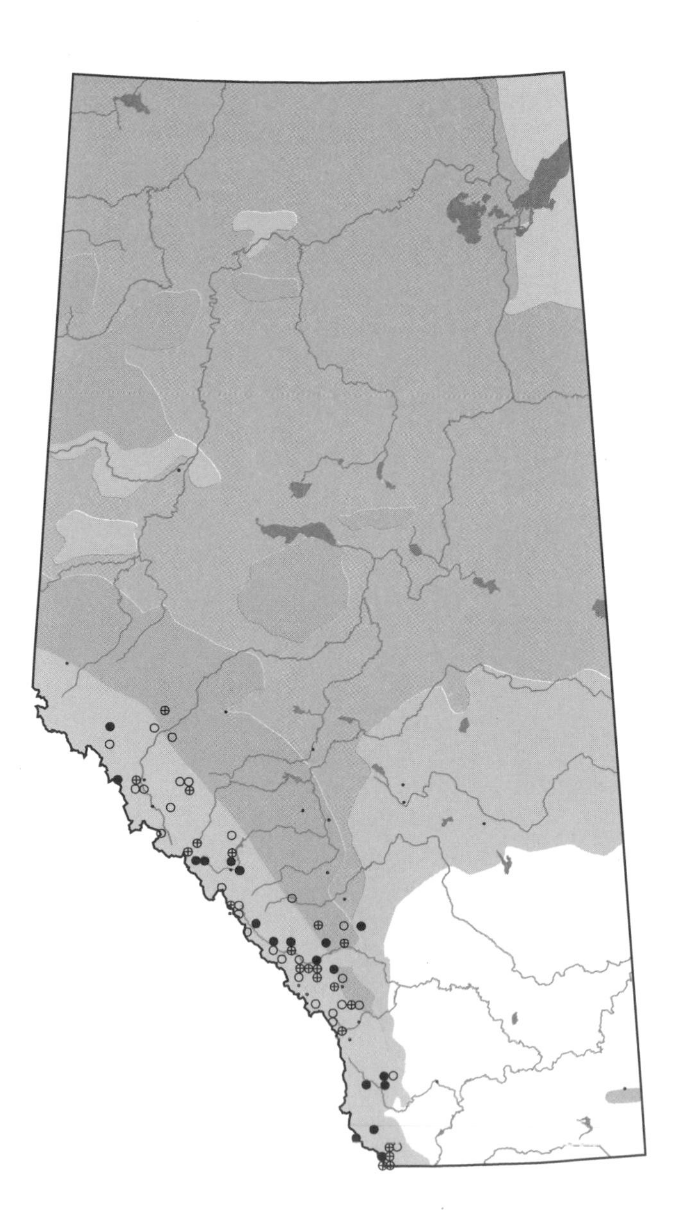

Boreal Chickadee

Parus hudsonicus

BOCH photo credit: D. Wiggett

Status:

The Boreal Chickadee is considered to have a stable population and habitat. It is a permanent resident in Banff and Jasper national parks (Holroyd and Van Tighem 1983).

Distribution:

This species is found in suitable habitat from Alaska to Labrador and south into the northern United States. The Boreal Chickadee is found from the north of the province through the Boreal Forest region, where it was recorded in 19% of squares surveyed. It was most commonly recorded in the Foothills (35%) and Rocky Mountain (33%) regions, as well as in the Canadian Shield region (19%). Boreal Chickadees were found in only 9% of the Parkland squares surveyed. No records were tabulated in the Grassland Natural Region.

Habitat:

The Boreal Chickadee is a bird of coniferous woods, preferring spruce and fir habitats over others. It will, however, use mixed woods. It is a permanent resident of dense coniferous and mixed wood forests throughout Alberta, with the exception of the Cypress Hills. Salt and Salt (1976) reported usage of small stands of conifers in central Alberta.

Nesting:

McLaren (1975) reported that Boreal Chickadees form pair bonds gradually as their winter flocks break up. It is further reported that territory establishment follows soon after the bonding has been established. The courtship display consists of the male chasing the female in a downward spiral from the treetop until she emits a solicitation call. After pairing, the female begs the male for food. These birds do not advertise their locations, although they will actively defend territories with calls and chasing (McLaren 1975). This species will often choose nesting areas that are near bogs or muskeg.

Boreal Chickadees nest in cavities located in stumps, trunks, and branches of whole trees. They generally excavate their own cavities, with the female taking the lead role (McLaren 1975), but will also use woodpecker holes or natural cavities. These holes are generally 1-4 m above the ground (Cadman et al. 1987). Although usually unoccupied, they may aggressively take over another songbird's prepared cavity. Unique to this species is a tendency to place the excavated wood chips in conifers, rather than simply dropping them on the ground. The nest, constructed by the female alone, consists of moss, lichens, bark, hair, and fur. Typically, 4-7 white eggs with spots of reddish brown are incubated for 15 days by the female. She is fed by the male for 2-3 weeks after pair formation and through the incubation period. After hatching, both parents tend to the young, bringing them insects in the bill. Fledging is at 18 days, and young remain with the adults for two weeks. The male assumes the majority of this responsibility. Young are fed progressively less frequently, until independence is reached. A single brood is raised in a season (Ehrlich et al. 1988).

Remarks:

A Boreal Chickadee is distinguished from a Black-capped or Mountain Chickadee by its brown cap, black throat, and brownish body. The Boreal Chickadee is rather quiet and reserved, and is more difficult to spot than his more boisterous relatives. This species is primarily insectivorous and climbs along the ends of branches in search of food. It also eats some vegetable matter.

This species generally overwinters in the same area as it breeds, although it ranges over a larger area and moves from higher to lower elevations (Salt and Wilk 1966). In some years, however, there are irruptions southward that are probably caused by food shortages (Livingston 1966; Root 1988).

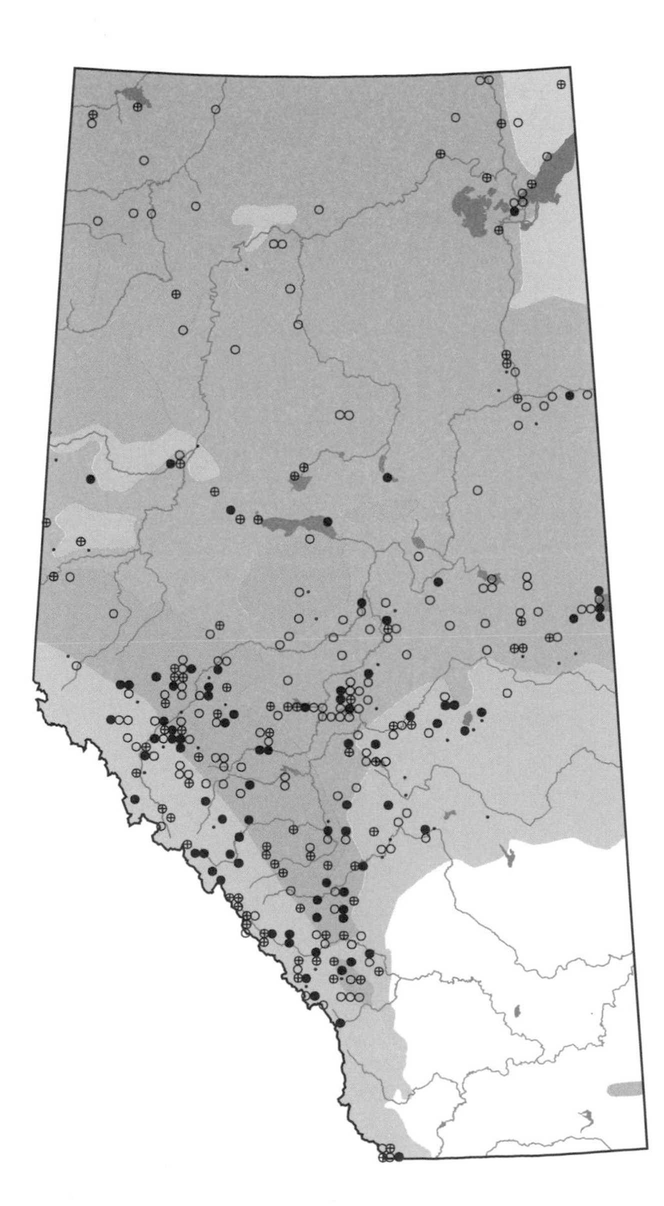

Red-breasted Nuthatch

Sitta canadensis

RBNU photo credit: T. Webb

Status:
The Red-breasted Nuthatch is rated as a fairly common resident in the mountain parks from spring to autumn (Holroyd and Van Tighem 1983). Atlas data also indicate that it is a common breeder in the southern portions of the Boreal Forest region and in the Foothills region.

Distribution:
In Alberta, this species is found in suitable habitat in the Rocky Mountain (40% of surveyed squares), Foothills (48%), Boreal Forest (25%), and Parkland (12%) natural regions. It is also found in river valleys in the Grassland Natural Region and in the Cypress Hills.

Habitat:
This active but inconspicuous little bird is found primarily in coniferous and mixed wood forests. Holroyd and Van Tighem (1983) reported that, in the mountain parks, the Red-breasted Nuthatch is found in coniferous forests where there are a variety of conifers as opposed to uniform lodgepole pine stands.

Nesting:
In late April or early May, the pair begin chiselling out their nest cavity in a decaying tree, stump or post 2-12 m above the ground. It has been suggested that unmated adults will occasionally assist a mated pair throughout the nesting period (Harrison 1978). Characteristic of the Red-breasted Nuthatch is its habit of smearing pitch around the outside of the nest hole or bird box throughout the nesting period. The purpose of this behavior is not clearly understood.
The nest is usually a cup made of grass, rootlets, and hair, but wood chips are occasionally the only nesting material.
The female lays 4-7 white eggs, which are spotted with reddish browns. Incubation is about 12 days and the young, tended by both parents, leave the nest 18-21 days after hatching. Young are fed insects, which are brought in the bill by the adults. The nestlings' faeces are removed throughout the nesting period. The family will leave the nesting area by early July (Salt and Salt 1976).

Remarks:
The Red-breasted Nuthatch has a white face and a black crown and eyestripe, which create a characteristic white eyebrow line. The underside is reddish brown, but is variable in intensity.
A characteristic habit of all nuthatches is the tendency to forage downward on tree trunks, with head down, relying entirely on their strong claws for support. This is distinct from woodpeckers, which climb trunks upright and use their tail for support. The Red-breasted Nuthatch also climbs on to the ends of branches to forage, often upside down.
This species' diet consists largely of vegetable matter, although young are fed mostly animal foods, some of which is hawked, as well as being collected in typical nuthatch fashion (Ehrlich et al. 1988).
This species arrives in Alberta in late April to mid-May and leaves the province again in late August and September, with some lingering in southern Alberta into November. They migrate in both directions in the company of warblers and other small birds (Salt and Salt 1976).
Red-breasted Nuthatches are known to remain on the breeding territory, if there are sufficient food resources (Ehrlich et al. 1988). There are overwintering records of this species from southern Alberta to Fort McMurray (Slater 1989, 1990). Numbers and locations of wintering birds vary greatly, influenced by weather and availability of food (Pinel et al. 1991).

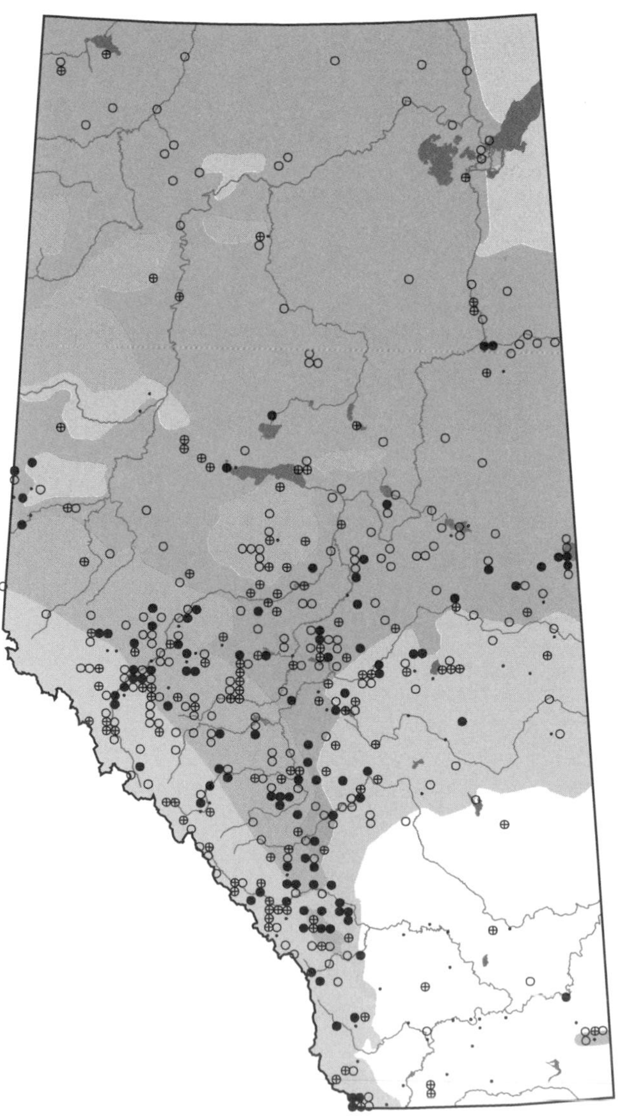

White-breasted Nuthatch

Sitta carolinensis

WBNU photo credit: R. Carroll

Status:
Salt and Salt (1976) suggested that, until 1950, the White-breasted Nuthatch was rarely seen in the province, but that it has become more common since that time. It was first recorded in Edmonton in the late 1950s and has since become a common species in central Alberta (C. Finlay pers. comm.). Holroyd and Van Tighem (1983) recorded it as a very rare visitor to the mountain parks, with no records from Jasper and very few from Banff, where it may stay year round. All records in Banff have been since 1970.

Distribution:
This species is found in suitable habitat across southern Canada. In Alberta, it is found mostly in the Parkland Natural Region, where it was recorded in 11% of surveyed squares, and in the southern part of the Boreal Forest and Foothills natural regions (10% of surveyed squares). Atlas data indicate that the breeding range of this species has extended farther north, east, and west than the range reported by Salt and Salt (1976). There are now breeding records as far north as Grande Prairie and Lesser Slave Lake, east to the Frog Lake area, and south to Lethbridge.

Habitat:
This species breeds in mature deciduous forest, mixed woods, and conifer woodlands, and also in trees around cultivated areas and houses (Harrison 1978).

Nesting:
During courtship, the male feeds the female and will display, bowing and singing, with head feathers erect and tail spread (Ehrlich et al. 1988). White-breasted Nuthatches may use natural cavities or those made by other birds, or they may excavate their own cavity for nesting. The cavity is usually 2-10 m above ground. It is lined with bark, twigs, grass, hair, and feathers. A layer of sap is placed below the nest hole, presumably to discourage squirrels and other predators.
The typical clutch consists of 5-8 white eggs, which are spotted with browns or a hint of light purple. They are incubated for 12 days by the female alone, which is fed by the male during this period. Young are tended by both parents and are fed for two weeks after leaving the nest. This species is single-brooded.

Remarks:
Nuthatches were originally named in Europe by observers who watched the birds cram nuts and seeds into crevices and strike them with their bills to break them open (Shortt 1977). Hatch is probably a derivative from the German word "hacke".
The White-breasted Nuthatch is black or dark blue-gray on the top of the head and back of the neck with white on the underparts and on the side of the face to above the eye. The lack of a black eye stripe distinguishes it from the Red-breasted Nuthatch.
Unlike creepers and woodpeckers, which brace against their tails for support when climbing a tree trunk, the nuthatch depends entirely upon its claws as it moves up and down the trunk (Wetmore 1964). This species descends tree trunks head down, stopping and looking around with the head held out at a 90 degree angle. It searches the bark crevices for insects, although the majority of its diet consists of plant matter such as seeds and nuts (Root 1988). The young are fed small insects, caterpillars, and spiders. White-breasted Nuthatches, like chickadees, will store excess food in cracks and holes in the bark. Not being particularly sociable birds, they are rarely seen in groups of greater than two (Salt and Salt 1976). This species is nonmigratory (American Ornithologists' Union 1983).

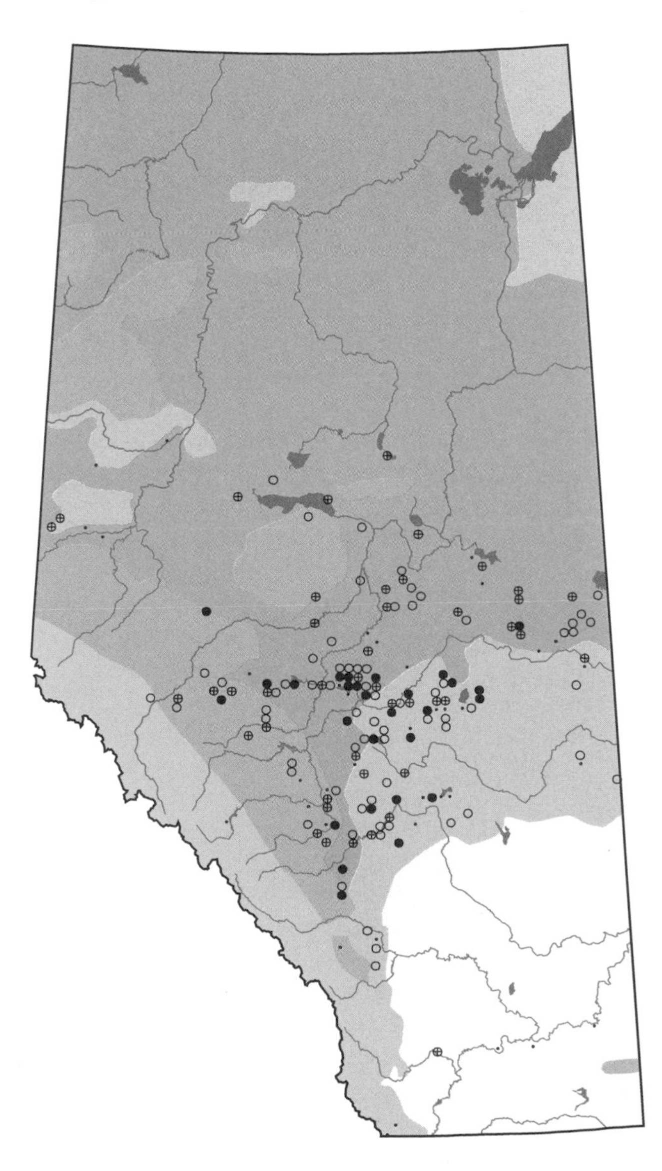

Brown Creeper

Certhia americana

BRCR photo credit: D. Farr

Status:
The status of the Brown Creeper in Alberta is ambiguous and the security of its habitat is uncertain because of its preference for mature forests (Wildlife Management Branch 1991).

Distribution:
The Brown Creeper is found from Alaska across southern Canada to Newfoundland and south through the mountains into Central America.
Provincially, it was reported across central Alberta, south to Calgary and through the Rockies (Salt and Salt 1976). It is a permanent resident of Banff and Jasper national parks, although it is less common during winter (Holroyd and Van Tighem 1983). Atlas data indicate that the Brown Creeper does breed farther north into the Boreal Forest Natural Region, with records north of Fort McMurray. There is also one confirmed breeding record in the extreme southwest corner of the province.

Habitat:
Brown Creepers are birds of mature forests, showing a preference for coniferous woods or mixed woods. In the mountain parks, it inhabits mature coniferous forests, particularly Engelmann spruce/fir and lodgepole pine/spruce stands (Holroyd and Van Tighem 1983).

Nesting:
During courtship, the Brown Creeper gives up its habit of short utilitarian flights and is seen in spiralling display flights and chasing; this creeper's complete song is also heard at this time of year. It soon reverts back to clambering on the tree trunks and even mating there (Shortt 1977). The nest of the Brown Creeper is usually well concealed under the loose bark of a tree from 1-15 m above the ground (Ehrlich et al. 1988). The nest is an untidy crescent-shaped structure of twigs, roots, moss, and grass on a base of twigs. A cup-shaped depression at the centre of the nest is lined with feathers and bark shreds. The nest is constructed by the female with the assistance of the male over a period of 6-30 days (Ehrlich et al. 1988).
The female lays 5-6 white eggs, marked with reddish browns, and incubates them for 14-15 days. The male feeds the female during this time. Once the young are hatched, they are tended by both parents for 14-16 days before leaving the nest.
Whether Brown Creepers have more than one brood per season has not been definitively documented.

Remarks:
The Brown Creeper is a more slender-looking bird than a nuthatch and has a long tail, a slightly down curved bill, and long curved claws. Its upper parts are brown and streaked with grayish white, and it has a light eyebrow line and buff underparts. The tail is made of stiff pointed feathers and looks ragged.
This small solitary bird is difficult to see as it creeps up tree trunks, fastidiously searching for insects, spiders, and other invertebrates. Seeds are eaten on occasion. Like woodpeckers, the Brown Creeper holds itself erect by pressing its long, stiff tail feathers against the trunk. After it has gone up some distance, it will make a short swooping flight to the base of a nearby tree and again begin its upward climb in search of food.
This species winters in the province.

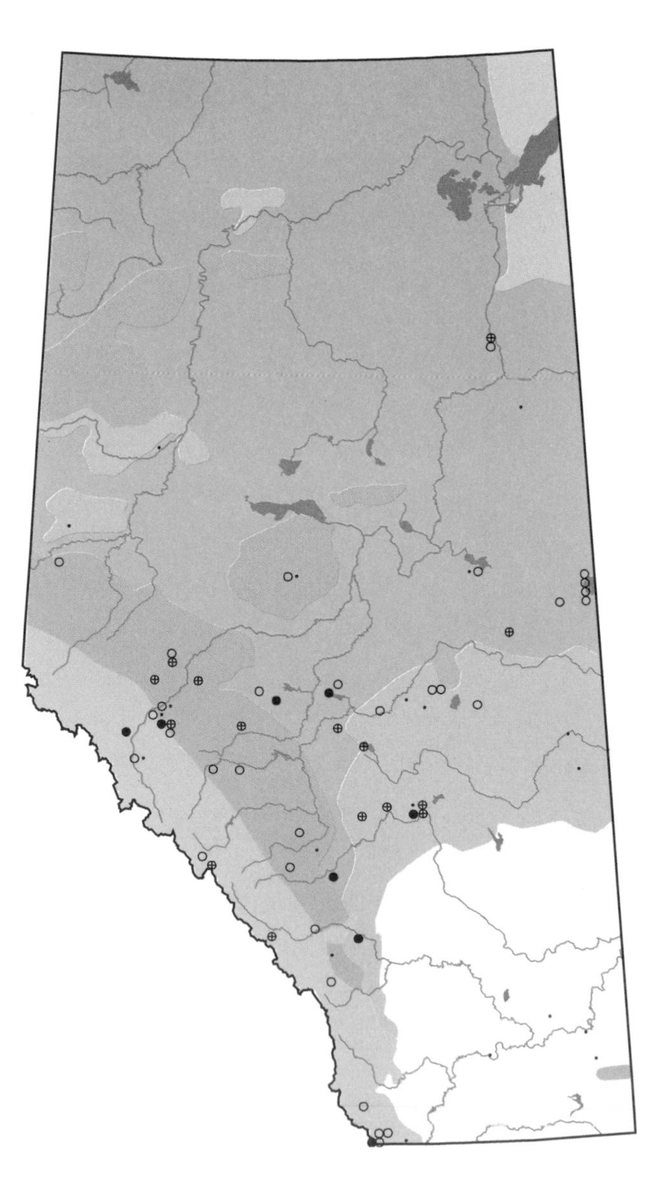

Rock Wren

Salpinctes obsoletus

ROWR

Status:
The status of this species in Alberta is undetermined, although it was commonly found in river valleys in the Grassland Natural Region.

Distribution:
The Rock Wren nests in southern Alberta north to the Red Deer River, where it was recorded in 7% of the squares surveyed in the Grassland region. It was also recorded in the Rocky Mountain region as far north as Jasper. It has been reported extralimitally at Fort Chipewyan (Salt and Salt 1976).

Habitat:
In Alberta, the Rock Wren inhabits areas featuring large rock outcrops or the sparsely vegetated walls of valleys or canyons, often in hot dry situations. The important characteristics appear to be a substrate with little or no vegetation, over which the birds can forage, and crevices where they can nest and escape the heat (Renaud 1979). Rock Wrens are found in both the mountains and the coulees of southern Alberta. In the Rocky Mountain Natural Region, they occur in the montane and lower subalpine areas, where they are found on rock slides and bluffs exposed to the sun. They also use man-made habitats such as gravel pits, road cuts, or spoil piles from mines.

Nesting:
The Rock Wren selects crevices in and among rocks or holes in cliffs for its nest (Salt and Salt 1976). The nest's base and entranceway are often paved with small flat stones. The cup-shaped nest is made of grass, weeds and rootlets, and is lined with fine grass, hair and feathers.
The female typically lays 5-6 eggs and, rarely, 9-10 (Harrison 1978). The eggs are a pure, glossy white, lightly speckled with reddish brown. Incubation is by the female. However, the male helps build the nest, feeds the female on the nest, and helps to tend the young. The number of days to fledging is unknown. Two, perhaps three broods are raised per season.

Remarks:
The Rock Wren is gray-brown with a light eye stripe and dusky striping on a gray breast. It has a plump body, short rounded wings, tapering tail, and thin, sharp bill. All these features are typical of most wren species.
This is an active bird which moves in and out of the nooks and crannies of its rocky environment probing for its food. Rock Wrens forage for insects by digging and probing. They also eat ants and, thus, are not dependent on day-flying insects (Smyth and Bartholomew 1966). In the northern extremes of this species' range, it has been observed gleaning small insects from small poplars and willows (Seutin and Charter 1989). The Rock Wren seems rarely, if ever, to drink water, probably getting the fluid it needs from its insect diet.
Its clear, rollicking song is a cheerful surprise in the barren landscape.
The Rock Wren arrives in Alberta in late April or early May and is thought to leave in September. However, fall migration data are limited.

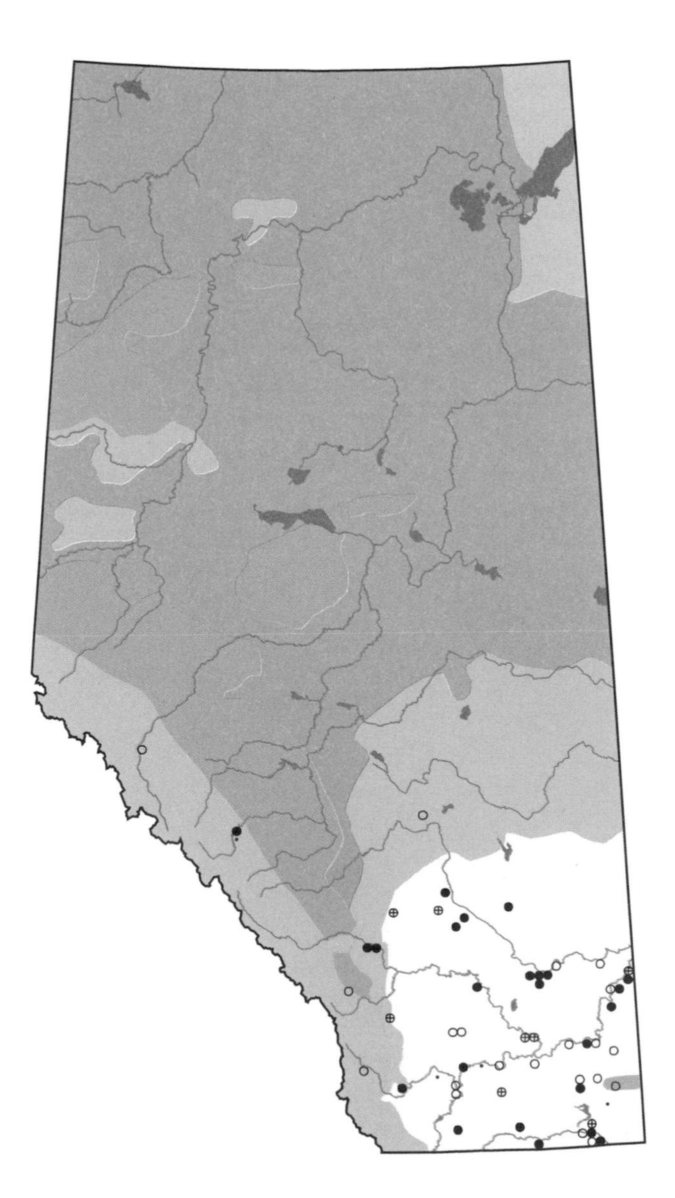

House Wren

Troglodytes aedon

HOWR photo credit: T. Webb

Status:

In 1948, Bent reported that House Wrens in eastern North America had recovered somewhat in population numbers, after a long period of scarcity. However, the wrens were still much less common than they were before the introduction and spread of the more aggressive House Sparrow, which competes with House Wrens for nest holes. Its recovery in the 1930s and 1940s was related to a decrease in House Sparrows. Tate and Tate (1982) reported the House Wren to be a species of special concern because it showed population reductions in its United States range, specifically from the mountains across the Midwest and into the western Great Lakes area.

In Alberta, this species is considered rare in the mountains (Holroyd and Van Tighem 1983) but, for the province as a whole, it seems to have secure habitat and a stable population (Wildlife Management Branch 1991).

Distribution:

The House Wren is found throughout southern Canada. In Alberta this species occurs from the southern part of the Boreal Forest Natural Region, south to the border with the United States. It was recorded in only 15% of surveyed squares in the Rocky Mountain and Foothills regions.

Habitat:

House Wrens are adaptable birds, living in open woodlands, coulees, thickets, and cultivated areas around human settlements. They are often found in areas with dense shrubbery, but they prefer to nest where there is sparse vegetation.

Nesting:

House Wrens will nest in natural cavities, including those made by other birds. They will also use bird boxes and holes or ledges in buildings.

The male usually arrives in the territory about 10 days before its mate. It builds a number of dummy nests in deciduous trees or snags and then attracts a mate. The female will then line one of the nests, or help to build another for ultimate use. The nest is composed of twigs, stems, and leaves. It is lined with feathers and plant material and is located 0-5 m above the ground.

The female lays 6-8 pinkish white eggs, spotted with brown, and incubates them for about 14 days. Once hatched, the young are tended by both parents for 12-18 days before fledging. The young may continue to be fed and to roost at the nest for a few days, until independence is reached.

This species is known to change mates between the first and second brood. Furthermore, some males may practise polygamy. House Wrens have two and occasionally three broods per season.

Male House Wrens are highly territorial, sing throughout the nesting period, and will engage in vocal contests. During the nesting period House Wrens will destroy the nests of other birds. The purpose of such attacks is not clearly understood (Quinn and Holroyd 1989).

Remarks:

The House Wren, like all wrens, has a short, slightly down-curved bill, is squat-looking, and often carries its tail pointed straight up. This species has a grayish brown underside, a brown tail and wings with dark bars. The side of the head is pale brown with an almost indistinguishable eyebrow line and a dark crown and back. This bird's diet is made up entirely of insects. House Wrens are curious birds. Observers can easily attract them in appropriate habitat by making a "pishing" sound.

Age and sex of wrens may affect their date of return to the breeding ground (Drilling and Thompson 1988), though most arrive in southern Alberta in mid-May. Most wrens have left the province by mid-September.

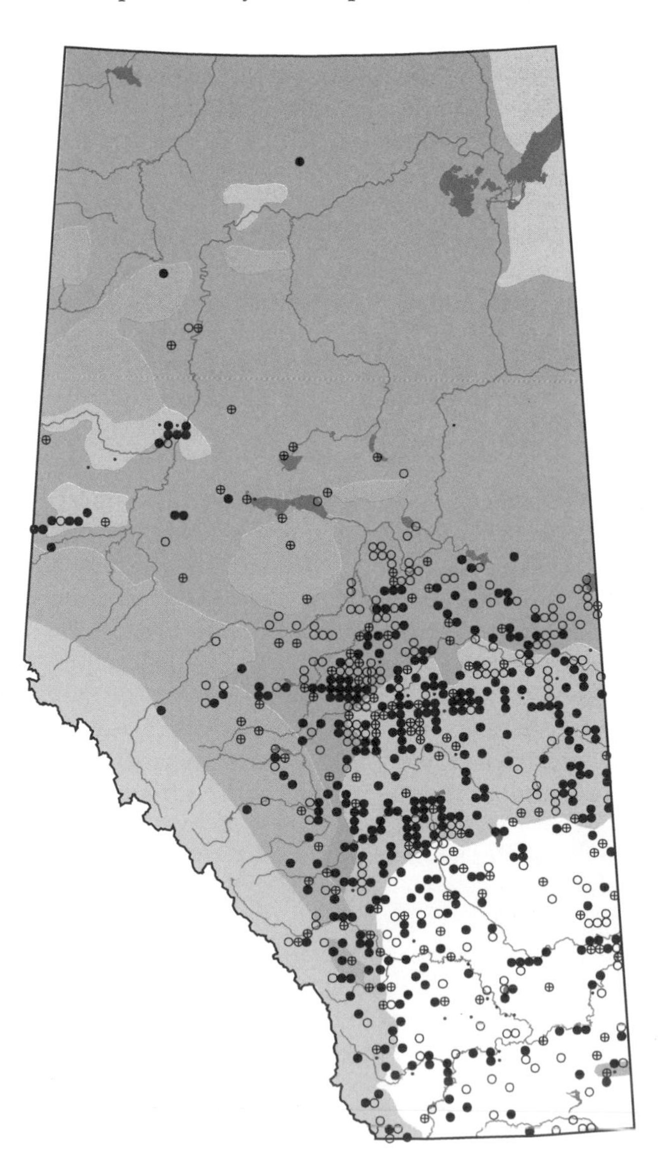

Winter Wren

Troglodytes troglodytes

WIWR

Status:

The status of the Winter Wren in Alberta is unknown, but there are indications that its eastern population may be declining (Wildlife Management Branch 1991). It is considered a fairly common resident in the mountain parks, in spring and summer (Holroyd and Van Tighem 1983).

Distribution:

In Alberta, this species occurs predominantly in the Rocky Mountain region, where it was recorded in 16% of surveyed squares, Foothills region (9% of surveyed squares), and the Boreal Forest region (4% of surveyed squares). In the Rocky Mountain region, they are found south to Banff National Park, and in Waterton Lakes National Park. There were, however, no records in the intervening mountain areas during the Atlas surveying period. This may be the result of poor access in this area and may not indicate the absence of the species. Salt and Salt (1976) and Pinel et. al. (in prep.) both indicate a dijunct population of the Winter Wren in northeastern Alberta. Salt and Salt (1976) reported an absence of this species in the southwest portion of the Boreal Forest Natural Region, including the Grande Prairie area and south and west of Lesser Slave Lake to the Rocky Mountains. Atlas surveys recorded several occurrences of the Winter Wren, including two confirmed records, in this area.

Habitat:

Winter Wrens are birds of the deep forest, selecting moist areas with thick underbrush and fallen trees in coniferous and mixed woods. Dense undergrowth at the edge of the woods, tangles roots, the crowns of fallen trees, and slash piles from lumbering activities are also used (Salt and Salt 1976). Sightings of this species in the mountain parks have been almost exclusively in Engelmann spruce-subalpine fir forests (Holroyd and Van Tighem 1983).

Nesting:

The nest of the Winter Wren is a spherical structure, constructed of moss, twigs, grass, plant fibres, bark shreds, and hair, and features a side entrance. The male usually builds a number of dummy nests, one of which is lined with feathers by the female (Harrison 1978). These nests are usually located close to the ground, in roots of upturned trees, in or under fallen logs or stumps and in crevices of trees, where they are hidden and shaded.

The female typically lays 5-6 white eggs, which are sparsely speckled with reddish brown. She incubates them for 14-16 days, during which time she may be fed by the male. The young leave the nest at 19 days (Ehrlich et al. 1988). The role of the male during the raising of the young is not clear. The male is often polygamous.

Remarks:

The Winter Wren has reddish brown upperparts, barred flanks, a light buff eyebrow line, and a rather stubby cocked-up tail.

This tiny wren, the smallest of the wren family, is difficult to spot because of its habitat choice and mouselike behavior of scurrying along the ground and crawling under logs, stumps, and brush in search of insects.

This species' response to human intruders varies, but "pishing" is likely to elicit a response (Cadman 1987). It sings from the ground or a treetop, with a ringing volume that is altogether out of proportion to the size of the bird.

Winter Wrens arrive in Alberta in mid-April to May and leave again in August and early September (Holroyd and Van Tighem 1988). In the late 70s and early 80s there were a few overwintering records for this species (Slater 1985, 1987), but it has been recorded in winter at Calgary (Salt and Salt 1976).

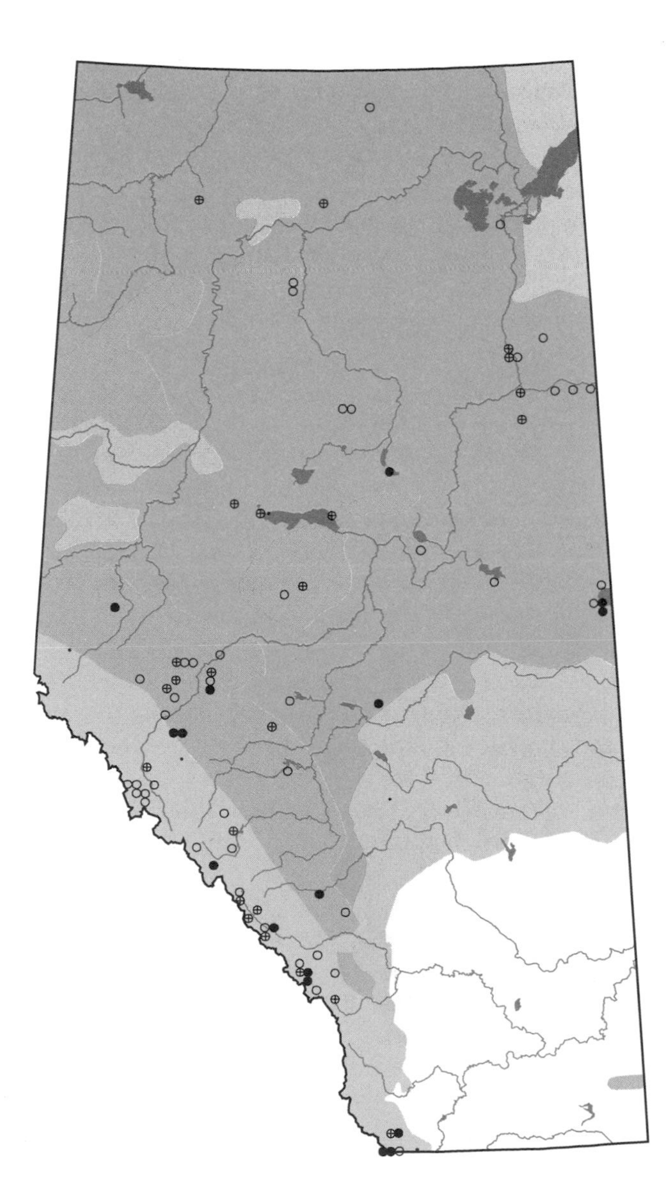

Sedge Wren

Cistothorus platensis

SEWR photo credit: W. McGillivray

Status:
The Sedge Wren, formerly known as the Short-billed Marsh Wren, was on the Blue List in 1979 and 1981. It was subsequently downlisted but kept as a species of special concern from 1982-1986 (Tate and Tate 1986) because of its declining numbers in the northeast parts of the American Midwest. The Wildlife Management Branch (1991) considered the Sedge Wren to be a nonbreeding migrant whose status was undetermined.

Distribution:
The Sedge Wren has a centrally located breeding population in Alberta. The range as determined by the Atlas Survey extends from Sangudo and Rocky Mountain House in the west to Cold Lake and Provost in the east. Records were restricted to the southern Boreal Forest and Parkland natural regions. Pinel et al. (in prep.) counts reports from Ft. MacKay and Cereal in 1976-77 as extralimital.

Habitat:
This species uses the sedge- and grass-dominated areas of marshland, usually avoiding the wetter emergent areas. It will also use uncultivated and cultivated tall grass fields, bogs and meadows with willow and alder thickets (Peck and James 1983b; Salt and Salt 1976).

Nesting:
The Sedge Wren nests both in small colonies and singly. Its nest is ball-shaped with a side entrance, and is constructed of sedges, grasses and other plant fibres. It is located 0.2-0.5 m above the ground among sedges and grasses. The male builds a number of these nests, and the female will choose one, which she will line with feathers, down, fine grass and hair. The nest is concealed in upright grasses or sedges (Godfrey 1986). The female lays 5-7 white, unmarked eggs, and incubates them for 12-14 days. Both parents tend the young, which leave the nest at 12-14 days. They may be attended for a few days after fledging until independence is reached. Sedge Wrens will destroy the neighboring nests and eggs of other Sedge Wrens, along with those of similarly sized birds. They do not show much fidelity from year to year in their selection of nesting grounds (Ehrlich et al. 1988). This species may be double-brooded.

Remarks:
The Sedge Wren is a small brownish wren with a streaked crown and back. It lacks a distinct eye line. The streaked crown and short bill are good field marks to identify this bird. It is difficult to find and confirm the Sedge Wren as a breeding species, both because of its small size and elusive nature, and because its habitat is somewhat inaccessible (Cadman et al. 1987). Similar to the Atlas Project in Ontario, the high proportion of possible and probable records during the Alberta Atlas Project (only 2 confirmed) indicates the difficulty of obtaining confirmed nesting records for this species.

This species' diet consists of spiders and insects such as ants, beetles, caterpillars, and grasshoppers (Bent 1964e).

Sedge Wrens will roost in their nests during the nonbreeding season. The migration dates of this species are not documented.

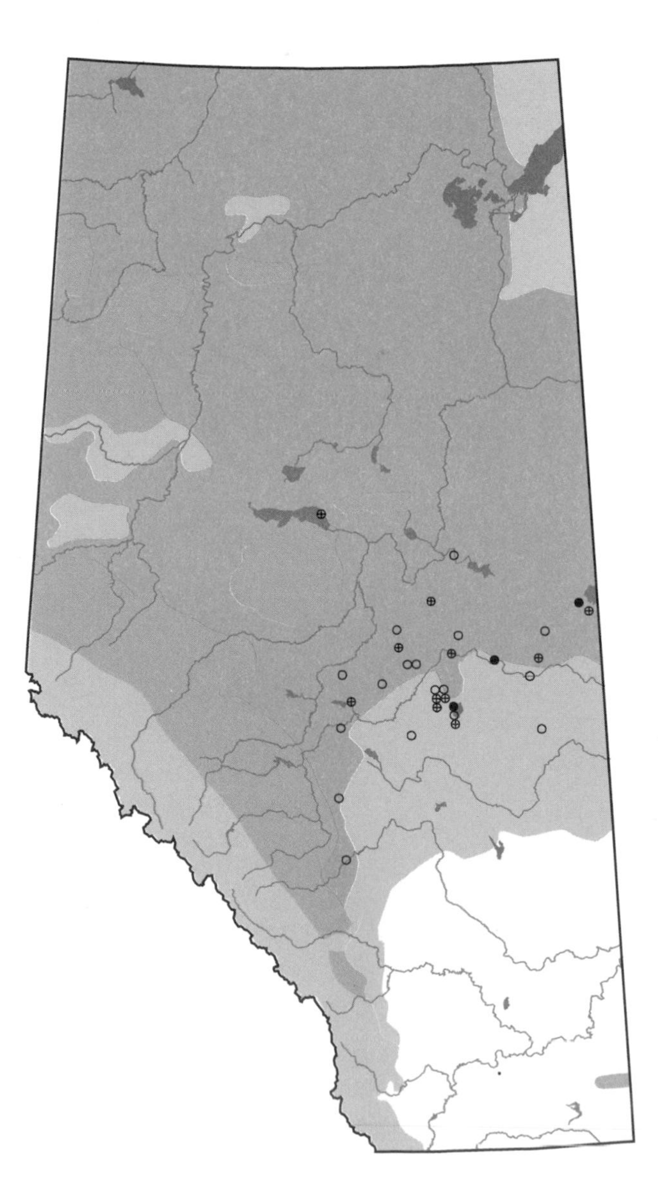

Marsh Wren

Cistothorus palustris

MAWR

Status:
In Alberta, the Marsh Wren is common in the Parkland Natural Region and in the extreme southern portions of the Boreal Forest Natural Region. It does nest sporadically throughout the Boreal Forest region. It is more localized in the Grassland region, where it is found predominantly around man-made bodies of water. In the mountain parks, it is considered a very rare visitor from spring through autumn and has not been recorded in winter (Holroyd and Van Tighem 1983).

Distribution:
The Marsh Wren is found in suitable habitat across most of southern Canada. In Alberta it is found locally throughout much of the province, but it is noticeably lacking in the Jasper area. It was most abundant in the Parkland region, where it was recorded in 14% of surveyed squares. In the Grassland region, it was only recorded in 5% of surveyed squares.

Habitat:
Marsh Wrens primarily occur in marsh habitats with cattails or bulrushes. They have also been recorded in the sedge and cattail stands of montane wetlands in the mountain parks (Holroyd and Van Tighem 1983).

Nesting:
Within the marsh habitat, this species most often chooses nest sites over water, commonly among the cattails. The nest is a woven oval or oblong structure of grass, sedge and cattail leaves and is attached to the supporting vegetation, usually 0.5 to 0.9 m above the water (Peck and James 1983). The male, as with other wrens, builds a number of dummy nests, one of which the female may pick and line. She may, instead, build a new one with her chosen mate (Leonard and Picman 1987a,b). The nest has a side entrance and is commonly lined with cattail down and feathers. The female lays 4-6 pale chocolate brown eggs, spotted with darker shades. These are incubated by the female for 12-14 days. Young are tended by both parents and leave the nest at 13-15 days. They are then helped by the male for up to seven days, while the female renests (Harrison 1984). This species is commonly polygamous. During the breeding season, the Marsh Wren is highly territorial and will chase other Marsh Wrens from its nest area. The male continues to build nests and sing throughout the breeding season. This species, like the House Wren and Sedge Wren, destroys the nests of other birds in its vicinity. This behavior is thought to play a role in reducing intraspecific and interspecific competition (Picman 1977a,b).

Remarks:
The Marsh Wren has a brown crown, a white eye line, a triangular black/brown patch on the back streaked with white, and a brown tail and wings barred with dark brown. Its underparts are whitish to buff from chest to coverts. This species is difficult to spot because it spends much of its time near the water, among the emergent vegetation.

This bird is a compulsive vocalist and can be noted by its song, which is a vigorous bubbling succession of notes, louder and more musical than that of the House Wren (Salt and Salt 1976).

This wren finds most of its food near or on the water surface. Its diet consists primarily of insects and spiders (Welter 1935).

This species arrives in mid- to late April and the majority have left the province by late September to early October.

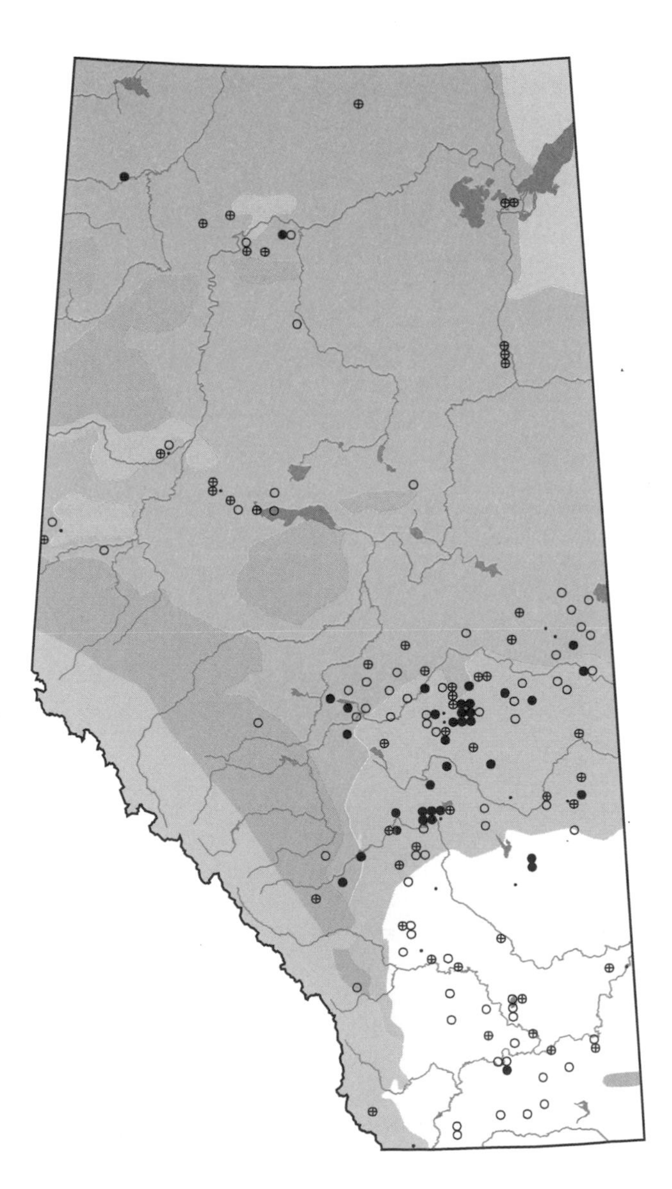

American Dipper

Cinclus mexicanus

AMDI photo credit: R. Gehlert

Status:
The status of the American Dipper in the province has not been accurately determined (Wildlife Management Branch 1991), although this bird breeds fairly commonly within its Alberta range.

Distribution:
Dippers are found in appropriate habitat in the Rocky Mountain and Foothills regions of southwestern Alberta, with records from the mountain national parks and Kananaskis Country. Dippers were recorded in 27% of surveyed squares in the Rocky Mountain region, and 3% in the Foothills region.

Habitat:
The dipper is usually associated with clear, fast-flowing, Rocky Mountain streams but may use the shorelines of clear mountain lakes or ponds. Birds will often overwinter where streams remain open in cold weather.

Nesting:
The dipper nest is usually placed on a stable rock ledge below an overhang, or in a rocky niche, generally above rapid-running water. It occasionally nests behind a waterfall. The roots of upturned trees or the crossbeam under a bridge are also used. Both members of the pair build the nest. It is a bulky, dome-shaped structure with an outer shell of moss and an inner cup of dry grasses. The entrance is on the side of the nest that faces the water. Nesting territories and nesting sites are often used over successive years. Dippers also show considerable fidelity towards their mates. Returning males mate with new females only when their previous mate has died (Ealey 1978). Polygyny is occasionally reported.
The female lays 3-6, typically 4-5, white eggs and incubates them for about 18 days (Ealey 1978). Both adults care for the young, which fledge at 24-26 days. After leaving the nest, the young are fed for up to two weeks, during which time the brood may be divided between the parents. Juveniles wander widely once they have reached independence. Through much of its continental range this species is occasionally double-brooded, but two broods are very rare in Alberta (Ealey 1977).

Remarks:
The American Dipper is a chunky, short-tailed, slate gray bird with the habit of rapidly "dipping" its body. This species' coloration makes it a difficult bird to observe against rocks along streams. Its song is long and melodious, like that of a wren, and can sometimes be heard in winter.
The American Dipper is the only truly aquatic passerine, obtaining most of its food by searching around and under stones on the bottom of rushing mountain streams. Dippers are able to swim at the surface or under water, where they use their wings for propulsion. They are also able to walk submerged along the bottom of streams. Two adaptations which permit this foraging behavior are flaps that close the bird's nostrils under water, and a nictitating membrane covering the eyes. A very large oil gland also enables this bird to keep its feathers waterproofed, and dense, compact plumage provides insulation and also sheds the cold water in which the bird is frequently submerged (Ealey 1977).
The dipper's main foods are adult aquatic insects and their larvae, other invertebrates, and small fish.
Migrating dippers arrive in early April and depart when winter ice takes over the stream or river they have claimed as territory. Overwintering individuals have been reported along open water stretches at Jasper, Canmore, Banff, Sundre, Calgary, and the Sheep River near Okotoks. However, where the majority of the dippers winter is still unknown (Pinel et al. in prep.).

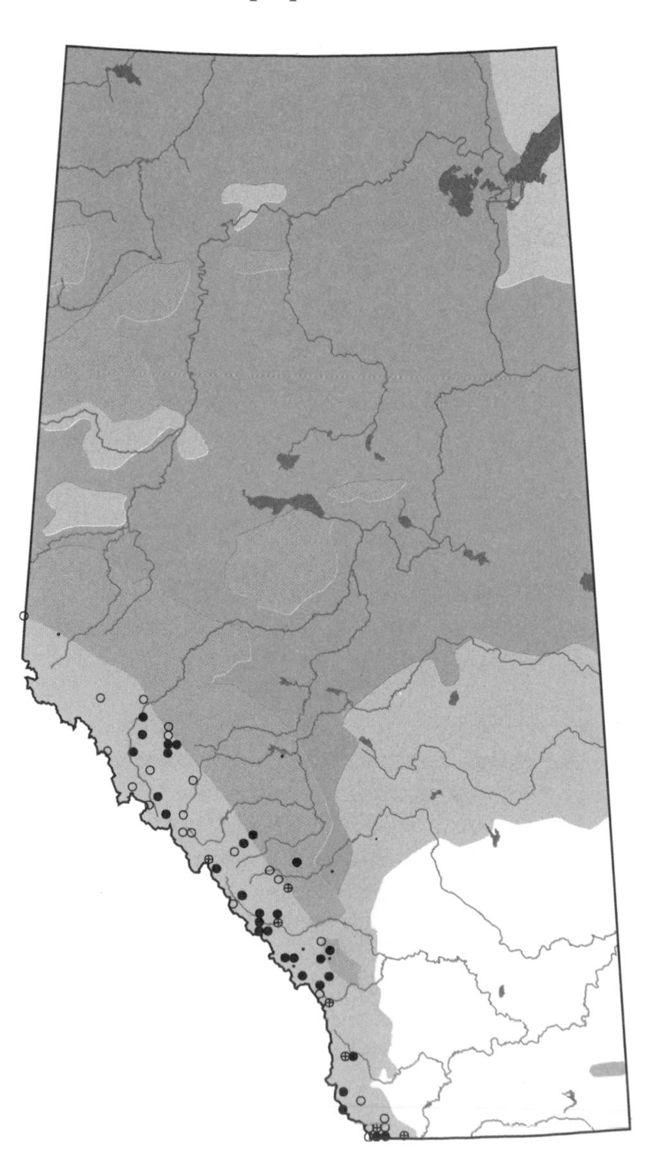

Golden-crowned Kinglet

Regulus satrapa

GCKI

Status:
Holroyd and Van Tighem (1983) reported the Golden-crowned Kinglet as a common migrant and a very common summer resident of the mountain parks. It was found to be common in the Foothills Natural Region and widespread but uncommon in the Boreal Forest region. Local abundance of this species may decline following logging activities (Quinlan et al. 1990).
Godfrey (1986) reported three subspecies for Canada, of which two breed in Alberta. *R. s. satrapa* breeds in eastern Canada and west to northeastern Alberta, and *R. s. amoenus* breeds in the interior of British Columbia and the mountains of Alberta.

Distribution:
Breeding was recorded in 35% of survey squares in both the Rocky Mountain and Foothills natural regions, and 59% in the Boreal Forest region. It is likely that the lower number of records in the Boreal Forest region reflect the low density of atlas coverage in that area. Although not recorded by Atlas surveys, it has historically bred in Fort Chipewyan and was suspected to breed in the Cypress Hills (Salt and Salt 1976). It is a transient and winter visitor to the Grassland Natural Region.

Habitat:
Although primarily a bird of the mountains and foothills, the habitat preferred by the Golden-crowned Kinglet is found throughout the Boreal Forest region. This species breeds in mature and old coniferous forests and groves (Quinlan et al. 1990) of all types, with higher densities in stands dominated by spruce (Holroyd and Van Tighem 1983). It will breed in aged mixed wood stands that are dominated by conifers.

Nesting:
The nest is a spherical mass of moss, lichens, and leaves, and it is suspended purse-like along the tip of a high branch. It may be placed on a limb or positioned as low as 1.5 m above the ground. It is very similar to the nest of the Ruby-crowned Kinglet. The deep thick-walled cavity is lined with hair, feathers, and fine rootlets and is bound and supported with spider's web. The 5-10 creamy eggs are spotted with pale brown, and are laid in two layers. They are incubated by the female for 14-15 days. The young are tended by both adults, and are fed insects and spiders before fledging after 14-19 days. Ehrlich et al. (1988) reported that two broods are raised in at least part of this bird's North American range.

Remarks:
This kinglet is the smallest songbird in Canada. It is often confused in general appearance with the Ruby-crowned Kinglet. The Golden-crowned has a gray nape, olive-green upper parts, whitish underparts, dusky wings and tail, a short pointed black bill, and two thin white wing bars. Other distinguishing marks are a yellow crown bordered in black, except at the rear, a white line over the eye, and a black line through it. The male has a red medial stripe in the yellow crown.
The food of this species consists almost entirely of insects, their eggs and larvae, and spiders taken from leaves, branches, and bark of trees (Bent 1964f). These birds are also expert at taking insects on the wing.
Arriving singly or in small groups of up to 10 from late March to early May, they may be loosely associated with chickadees and creepers. Although most depart from late August to early October, small numbers of these hardy birds may overwinter in the south or in the lower valleys of the mountain parks (Holroyd and Van Tighem 1983), and in coniferous stands in valleys and coulees of the Grassland region (Pinel et al. in prep.).

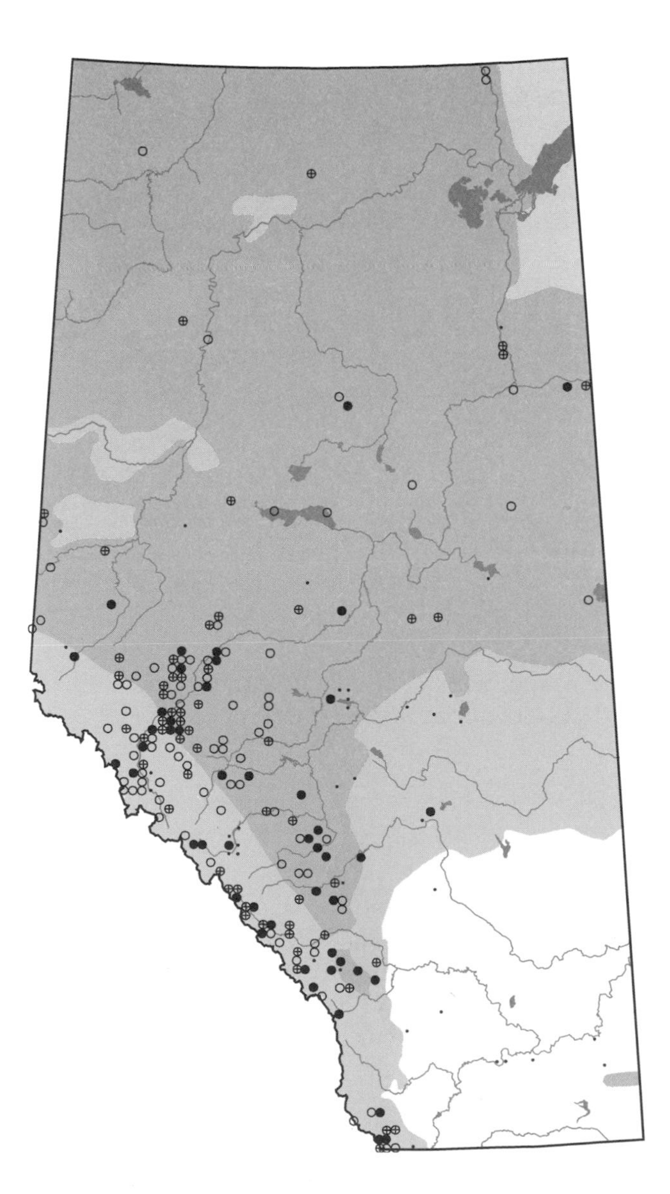

Ruby-crowned Kinglet

Regulus calendula

RCKI

Status:

There are three subspecies of the Ruby-crowned Kinglet in Canada, of which two breed in Alberta. *R. c. calendula* breeds in all parts of the province except the southwestern mountains, where *R. c. cineraceus* is found (Godfrey 1986). Holroyd and Van Tighem (1983) found this specis to be a common migrant and a very common resident of the mountain parks. This was confirmed by Atlas data, which also indicated that it is common throughout the Foothills, Boreal Forest, and Canadian Shield natural regions. It is not considered to be at risk.

Distribution:

The Ruby-crowned Kinglet was found to breed in the coniferous and mixed forested areas of all regions, including the Cypress Hills. It was not recorded as a breeder in the Grassland region, but it is a common transient in that region. It was recorded in 55% of surveyed squares in the Rocky Mountain and Foothills regions, 30% in the Boreal Forest and Canadian Shield, and 10% in the Parkland region.

Habitat:

This small forest bird may be found in any type of woods during migration. In the breeding season, it prefers coniferous forest, although it will tolerate some mixed wood containing large stands of evergreen (Salt and Salt 1976). It is most common in the black spruce and tamarack stands around muskeg, where the habitat is more open or has edge areas. It is also common in lodgepole pine-dominated forests in the Foothills region (B. McGillivray pers. comm.). The montane and upper subalpine are used in the mountain areas.

Nesting:

In courtship, the monogamous male erects his red crest and sings to his prospective mate (Ehrlich et al. 1988). The female is then thought to construct the spherical nest, a mass of lichens, plant down, and moss, bound together with spider's web. The deep cup is lined with hair and feathers and completely conceals the incubating female. The nest is similar to that of the Golden-crowned Kinglet and is very difficult to find because it is usually hung under the high branch of a tall conifer. There are usually 7-8, sometimes 5-11, eggs which are white and spotted with brown (Harrison 1978). The clutch is incubated for 12-15 days, and is rarely parasitized by cowbirds. Thought to fledge 12 days later, the young continue to be tended by both adults for some time afterward (Bent 1964f). Juveniles leave in August to wander widely before migrating south.

Remarks:

This plump and very small greenish bird is similar to the Golden-crowned Kinglet, but may be distinguished by its light eye ring, lack of any stripes over the eye, two white wing bars, and the male's red crown patch. The Ruby-crowned is easily agitated around the nest.

This species is less restricted by food and habitat than the Golden-crowned because it forages in both coniferous and deciduous trees, especially when accompanied by fledged young. The Ruby-crowned prefers the middle and upper portions of tall coniferous trees, and can be observed constantly examining needles for small insects, aphids, insect eggs, tiny moths, grubs, and some vegetable matter. They also hover around branches, and are skilful flycatchers. Arriving in Alberta from late April to late May, the subspecies *R. c. calendula* inhabits all but southwestern Alberta, where *R. c. cineraceus* dominates. Leaving from late August to late October, they may be seen in any type of woods within mixed flocks of kinglets, chickadees, nuthatches, and warblers. There are winter records from Jasper and Banff.

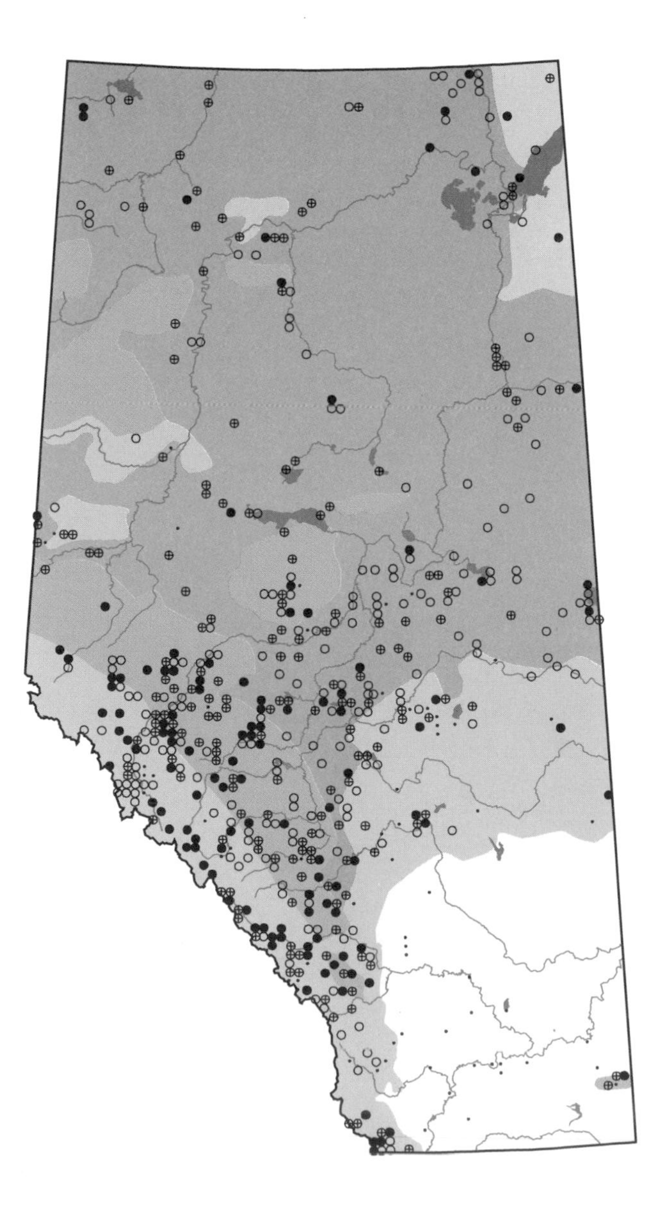

Eastern Bluebird

Sialia sialis

EABL photo credit: W. Horn

Status:
Previously considered an extralimital visitor to the province, the Eastern Bluebird is a rare breeder at this time. Within its North American range, it has encountered the same threats as those faced by the Western Bluebird, as well as occasional large winter die-offs, the destruction of native berry plants, and the removal of vital perch sites. However, like the other bluebirds, local populations have been assisted by the establishment of bluebird trails in appropriate habitat.

Distribution:
This species breeds west into the Cypress Hills of Saskatchewan (Godfrey 1986) but, in Alberta, Eastern Bluebird observations have been few and scattered, occurring mainly in the southern half of the province since 1974. These areas include the Cypress Hills, Botha, Vermillion Lakes, Calgary, Millarville, and one observation in Edmonton (Pinel et al. in prep.). There was one Atlas record of confirmed breeding in the Battle River area northeast of Wainwright.

Habitat:
In its Canadian range, this species breeds in open and sparse woods, wood edges, burns with old standing trees, and farmland (Godfrey 1986). Each area must have a nearby nest cavity, as well as bare patches or short grass for feeding.

Nesting:
Usually arriving a few days earlier than potential mates, the male establishes a territory that he will vigorously defend. Drawing attention to the nest site, he approaches the cavity in full song. He will then enter the nest and poke his head out intermittently. An interested female will enter the nest and, later, force the male out. The male will preen his mate and bring her food during courtship and incubation. Although the male will handle nest materials, he does not participate in the construction of the nest, a loose cup of grass and twigs that the female lines with hair and feathers. Typically, four pale blue eggs with light brown spots are laid in this nest. The male does assist in some of the incubation duties, which last 12-16 days. However, he spends more time guarding the nest site. Both adults tend to the young, supplying them with an insect diet. In the case of a second brood, the young of the first brood may assist in this duty. Young fledge 15-20 days after hatching. If the female renests, often with the same mate and in the same territory, the male will assume total parental care for the fledged young. Harrison (1978) reported two, or occasionally three, broods raised in a season.

Remarks:
The Eastern Bluebird male is a bright deep blue above, including the sides of the head and most of the wings and tail. The reddish-brown throat, breast, sides, and flanks contrast with a white abdomen and undertail coverts. The female resembles the male, but is much duller overall. Similar but distinguishable from its cousins, the Eastern sports a reddish-brown breast that separates it from the Mountain Bluebird and a same-colored throat that separates the Western Bluebird's blue throat (male) and gray throat (female). The song is a cheery, low pitched, and short warble that is frequently used during courtship and nest building, but is largely absent during nesting (Bent 1964f). Unlike that of other thrushes, the Eastern's song is not loud. Gleaning from foliage and ground, the Eastern Bluebird prefers to hunt from a low perch and pounce on its prey or capture it in low flight. Grasshoppers, crickets, beetles, ants, caterpillars, and wild fruit make up its diet.
As a result of the lack of sightings in the province, migration dates are not available.

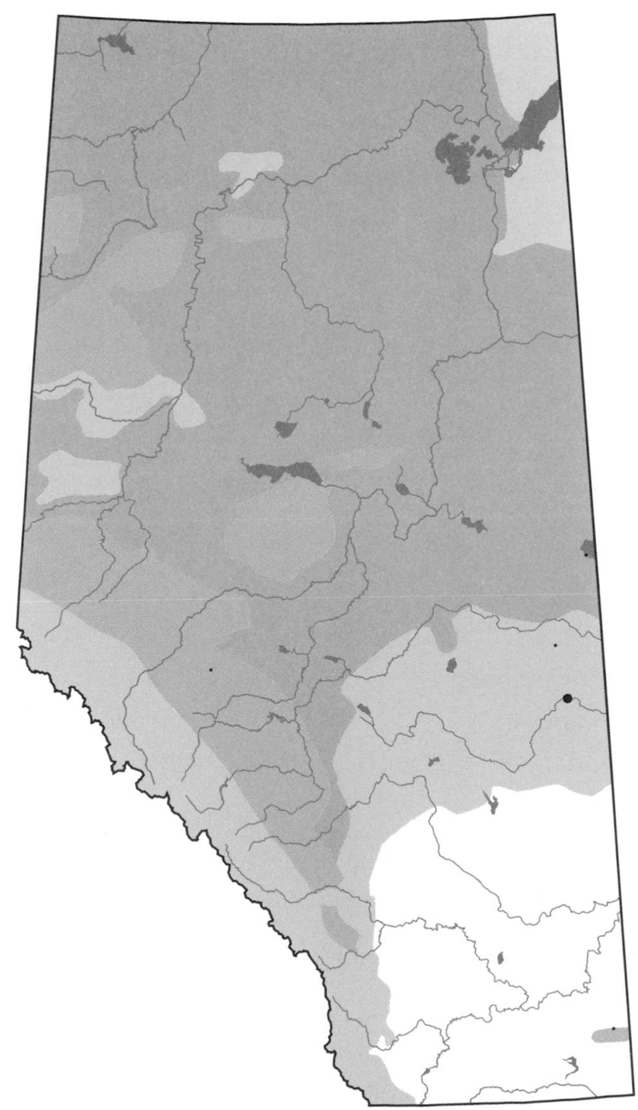

Western Bluebird

Sialia mexicana

WEBL

Status:
Previously considered, at best, a casual visitor (Godfrey 1986) to the province, the Western Bluebird is now confirmed as a breeder in Alberta. In other parts of its range, the Western Bluebird has suffered habitat destruction, the introduction of aggressive nest competitors, and the reduction in food supply that follows the use of herbicides and pesticides (Savage 1985). The provision of nest boxes has improved some local populations.

Distribution:
This "new" breeder has a scattered distribution locally. The three confirmed breeding records from Atlas surveys are from the House River in Banff National Park and between Wainwright and Czar in the Parkland region. In 1991 and 1992, three additional confirmations of nesting were reported west of Nordegg, two of which were in the Kootenay Plains, and one at Saskatchewan River Crossing (C. Finlay pers. comm.). The adjacent Rocky Mountains of southern British Columbia mark the previously accepted eastern range limit for this species in Canada.

Habitat:
In other parts of this species' range, breeding habitat is described as sparse woodland, burns, and logged-over areas where snags are retained to provide nest sites and perches (Godfrey 1986). The three nests in west-central Alberta were in nest boxes in open areas (C. Finlay pers. comm.).

Nesting:
Early arrival in spring allows the Western Bluebird to choose the best cavity sites available. The female constructs the nest inside a natural tree cavity, an old woodpecker cavity, or a nest box, 1-9 m above the ground. It is a slight cup consisting of grass with leaves, twigs, rootlets, feathers, and hair mixed in. The female typically lays 4-6 pale blue unmarked eggs, very similiar to the clutch of an Eastern Bluebird. The male assists in the incubation duties, which last 12-16 days (Godfrey 1986). Both adults feed the nestlings a diet that primarily consists of insects. Once the young have fledged, a time not reported in literature, the male often assumes full responsibility for their care, while the female renests and begins her second brood.

Remarks:
The male Western Bluebird has bright blue upperparts that are deeper in color than other bluebird species, duller blue sides of the head and throat, often a chestnut patch on the back, reddish-brown sides, flanks, and breast, and a pale blue-gray abdomen. The female is somewhat similar, but much duller. She has a brownish-gray head and back, a prominent white eye ring, and a grayish throat tinged with blue. It is distinguished from its eastern cousin by its deeper blue coloration, chestnut back patch, when present, and blue (male) or gray (female) throat. The reddish-brown breast quickly distinguishes it from our native Mountain Bluebird. Two white wing bars, a stubby bill, and a smaller overall size distinguish the similar Lazuli Bunting. The male Western Bluebird's song is a short "few few fawee". The call consists of a soft "few" note.
Hawking from a low perch and gleaning foliage are the common methods employed to capture their insect diet. Flycatching is also used, occasionally. In season, grasshoppers, beetles, caterpillars, ants, spiders, and flies are eaten, as are some berries such as raspberry and elderberry. Berries make up the largest portion of their winter diet.
Because there are very few provincial records of this species, precise migration dates are not available.

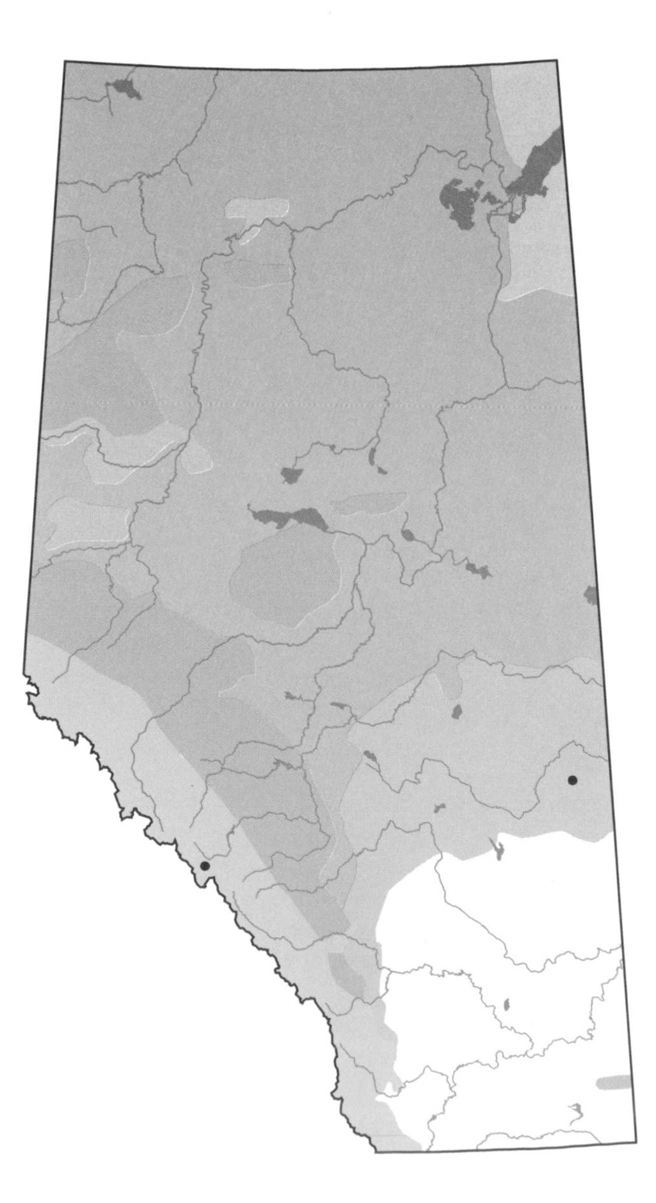

Mountain Bluebird

Sialia currucoides

MOBL photo credit: T. Webb

Status:
Alberta's most common bluebird has rebounded from once low numbers earlier in the century to a healthy status at present. This is a result, largely, of dedicated volunteers who have established and maintained bluebird trails. The first formal Bluebird Trail in Alberta was established near Elk Island Park in 1971 by Joy Finlay and the Edmonton Junior Naturalists.

Distribution:
The Mountain Bluebird nests in suitable habitat in all natural regions of the province. High breeding confirmation in central and southern Alberta, including the mountain parks, is no doubt attributable to the monitoring of numerous nest boxes. Although scattered observations have been made further north, breeding confirmation was reported only in the Peace River district and the Margaret Lake area. Salt and Salt (1976) reported breeders in the Fort McMurray area.

Habitat:
Mountain Bluebirds prefer areas of open woodland with scattered trees for shelter purposes and adjacent open areas for insect foraging. This habitat includes the woodland openings, edges, burns, and farmland of the Rocky Mountain, Foothills, Parkland, Grassland, and southern Boreal Forest natural regions.

Nesting:
Being quite territorial, male bluebirds establish their domain with a variety of flight and behavioral displays, often centred about the nest site. Obligate cavity nesters, Mountain Bluebirds seek out woodpecker holes, cliff crevices, and nest boxes in which to place their loose cup-shaped nests of grass, shredded bark, twigs, hair, and feathers. They are also known to use buildings and holes in earth banks. Although nest-building duties are generally shared, the male collects the materials and the female constructs the nest. The female lays 4-7 pale blue (rarely white) eggs, which she predominantly incubates for 13-14 days. The nestlings are tended by both parents and fledge after 22-23 days. The female may leave before the first young fledge to start a second nest within the same male's territory, if the first brood appears successful. Nest box and pair bond fidelity has also been recorded between years in Alberta (Hoffman and Pletz 1988). Fledglings have been known to collect food for the second brood.

Remarks:
The Mountain Bluebird male is all blue, and the much duller female is brownish gray above and gray below, with blue on the wings, rump, and tail. These birds are distinguished from other bluebirds by their lack of red plumage and by the paler blue of the female and young. They prefer to hover when foraging for insects, rather than attacking from a low perch. They feed mainly on crickets, grasshoppers, and beetles. Worms and wild fruit are also taken.
These birds arrive in numbers in mid-March (Backhouse 1986) to occupy nest boxes. Fall migration occurs in mid-August to mid-Sepember, with some lingering into November (Pinel at al. in prep.).

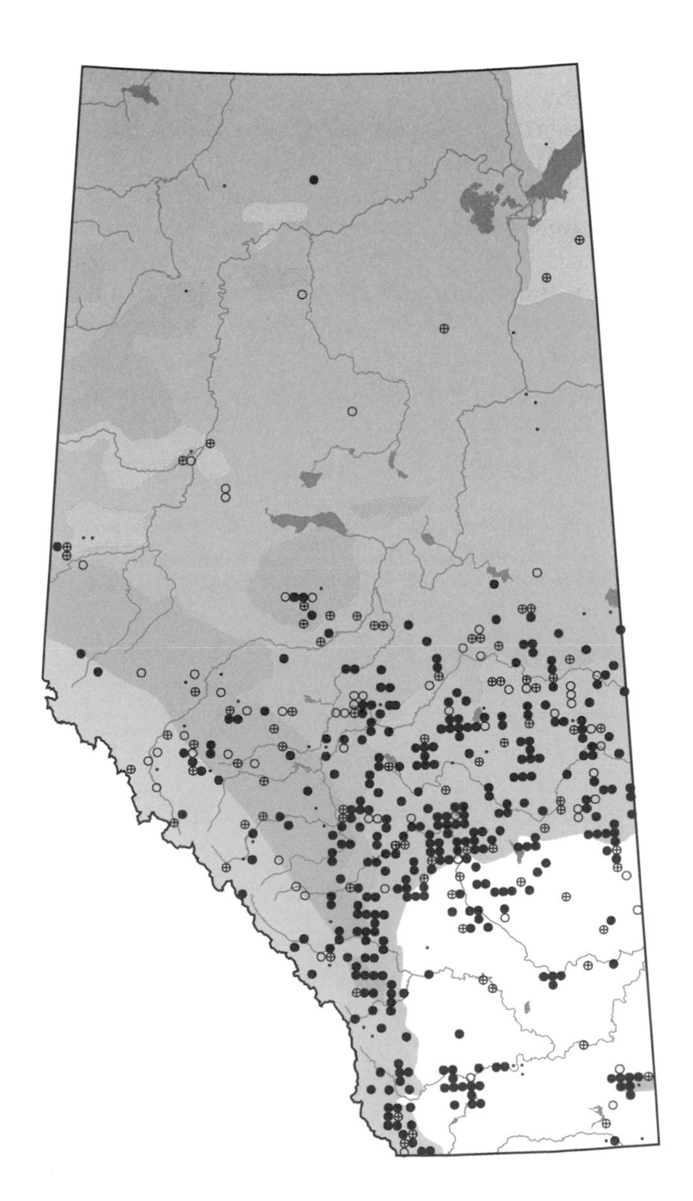

Townsend's Solitaire

Myadestes townsendi

TOSO

Status:
The provincial population of this species is healthy and not at risk (Wildlife Management Branch 1991). In the mountain parks, it is recorded as a common resident in summer, but less common during migration and a rare visitor in winter (Holroyd and Van Tighem 1983).

Distribution:
The Townsend's Solitaire breeds throughout the Rocky Mountain and Foothills natural regions from the American border north to Grande Prairie. It was recorded in 36% of surveyed squares in the Rocky Mountain region and 19% in the Foothills region. Historically, it has been recorded breeding in the Cypress Hills (Salt and Salt 1976) and this was confirmed by Atlas records. It often wanders eastward during migration and it has been observed, at this time, at Clyde, Elk Island National Park, Camrose, Rosebud, and Brooks (Godfrey 1986).

Habitat:
Townsend's Solitaires select open or broken spruce and fir forests, often on steep terrain that provides a few cliffs for nesting. Not birds of the deep woods, they inhabit areas of margins, burns, and timberline, always on slopes and with single standing trees. In the Rocky Mountain region of Alberta, the highest densities of these birds were found in Engelmann spruce/fir/lodgepole pine forests (Holroyd and Van Tighem 1983). When recorded overwintering in the mountain parks, they occur in Douglas fir forests in the montane ecoregion (Holroyd and Van Tighem 1983).

Nesting:
The solitary and retiring lifestyle of the Townsend's Solitaire has resulted in little knowledge of this bird's breeding biology. The untidy, loosely constructed nest of twigs and pine needles is lined with grass and moss and positioned on or near the ground. These nests have been observed in the banks of roads, trails, and river cutbanks, as well as under overhanging tree roots, in tree stumps, and in cavities and crevices. The female typically lays 3-5 dull white eggs, which are spotted and scrawled with browns—quite distinct from other thrushes. The length of incubation and by whom is not known. Two broods may be raised in the south.

Remarks:
The Townsend's Solitaire is a slender brownish-gray bird with a white eye ring, a short broad bill, and a longish tail (Godfrey 1986). In flight, a buffy wing patch and white outer edge of the tail are visible. These birds are easily distinguished from other thrushes, though they are similar in appearance to Northern Mockingbirds. However, these two species have provincial distributions that do not overlap and they can be separated by the white eye ring, buffy wing patches, and pale gray underparts of the solitaire. Their cheerful song, also heard out of the breeding season, is melodious and sustained, similar to that of the Purple Finch. The voice call is a high pitched "peek" and the alarm call a sharp "chuck, chuck".
In spring and early summer, these birds forage in bluebird-like fashion, perching on a low branch near a margin and pouncing on insects and larvae below. It will occasionally flycatch. Later in the season, it switches more to berries, including kinnikinnick and juniper.
The spring migration period is from early April to late May, peaking in early May. Fall migration begins in late August and can extend into October, with numerous overwintering records, mostly from the area of Banff townsite (Salt and Salt 1976; Holroyd and Van Tighem 1983). Migrants travel to the western United States and throughout Mexico, where winter territories are established based on available food supply.

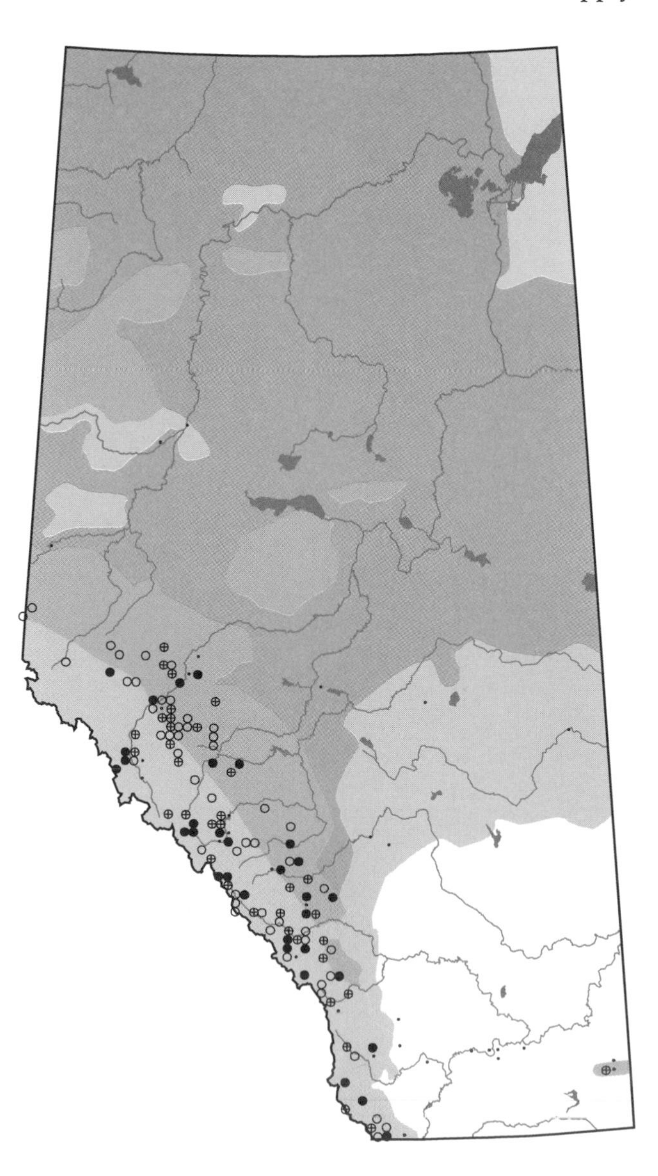

Veery

Catharus fuscescens

VEER

Status:
The Veery is relatively common in the Parkland Natural Region, where approximately 50% of the records for this species occurred. Salt and Salt (1976) reported that early investigators found it in the Peace River district but that it had become scarce or absent. This was confirmed by the Atlas surveys, which documented only one record for this species in that area in the five-year survey period.

Distribution:
This wood thrush breeds in suitable habitat in the Parkland region, where it was recorded in 14% of surveyed squares. It was also recorded in 5% of surveyed squares in the Foothills, Grassland, and Boreal Forest natural regions. Surveys confirmed breeding north to Grande Prairie, Slave Lake, and Cold Lake.
Salt and Salt (1976) reported that the Veery had only been recorded in the Rocky Mountain region as a rare summer visitor. Holroyd and Van Tighem (1983) recorded a maximum of three summer records, all in Banff National Park. Atlas surveys recorded 15 breeding confirmations from Jasper south to Waterton Lakes.

Habitat:
Breeding habitat includes deciduous or deciduous/coniferous woodland dominated by aspen and poplar. Preference is shown for the moist areas created by a thick, darkened understory, as well as the second-growth willow or alder shrubbery along water bodies. Open woodland, with an understory of deciduous shrubbery, may also be used.

Nesting:
The Veery positions its bulky, cup-shaped nest on or near the ground. The nest is usually found in thick shrubbery or heavy timber along the edge of a stream. Placed on a mass of dead leaves, the nest is formed with twigs, grass, and bark strips, and it is lined with surrounding vegetation, including dry leaves, rootlets, and fine grasses. While constructing the nest, the bird may be quite tolerant of observers (Bent 1964f). The female lays a single clutch of 3-5 greenish-blue eggs, which are rarely marked with a few brown spots. A close sitter, she incubates the eggs for 10-13 days. Studies in Ontario showed that the Veery's nest is frequently parasitized by the Brown-headed Cowbird (Peck and James 1983). The monogamous male assists in tending to the young, which are initially fed soft caterpillars and grubs and later receive broken-up insects. Young fledge in 10-12 days.

Remarks:
This aggressive and territorial thrush is generally brown above, light gray on the sides, and white spotted with brown underneath. It differs from other thrushes in being more brown above, having less distinctive spotting on the breast, and having no discernible eye ring. Most often heard at dusk, the flute-like song rolls and descends in scale in a rendition of "veery, veery, vaary, vaary" (Salt and Salt 1976). This elusive, though unwary, bird is most often spotted foraging on the ground for spiders, ants, beetles, and grasshoppers. It may initially be confused with the Fox Sparrow, which forages by jumping up and scratching leaves backwards. However, the Veery, in contrast, exposes food by scratching with its bill. Watching from a low perch, this bird may also swoop to the ground after prey, returning to its perch afterwards. It also occasionally flycatches. Increasing quantities of wild fruits are eaten in the late summer and fall. It has been recorded eating the hairy caterpillar of the gypsy moth.
This typically hardy thrush arrives in mid- to late May to coincide with insect and foliage appearance. Fall migration is in early September, though some birds may remain into October.

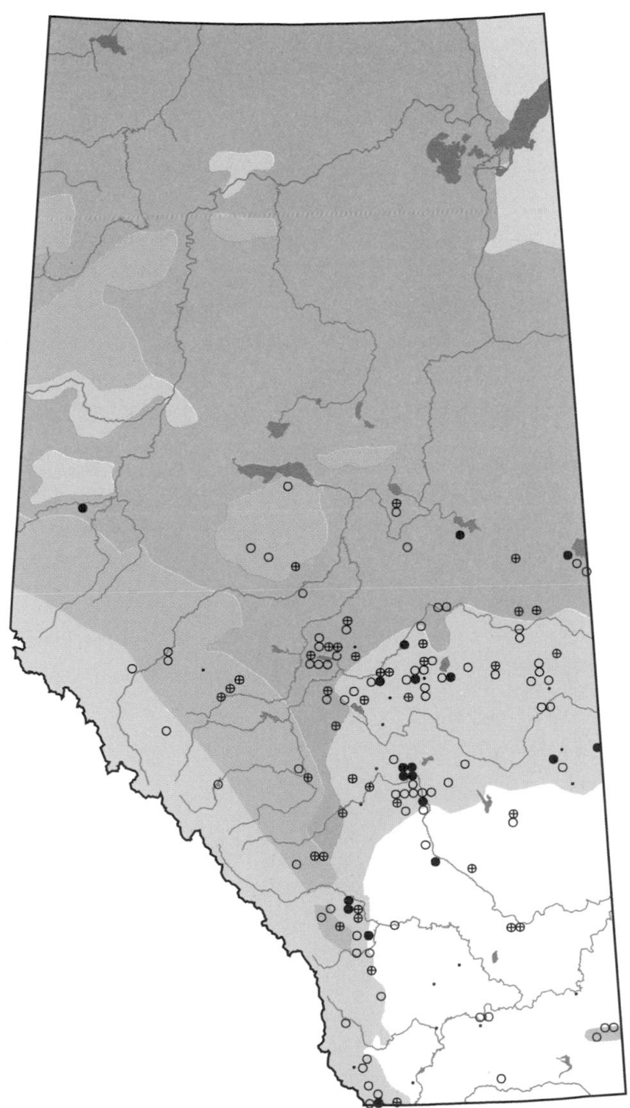

Gray-cheeked Thrush

Catharus minimus

GCTH

Status:

Although uncommon, this species has been seen in all regions of the province as a migrant (Salt and Salt 1976). Pinel et al. (in prep.) further categorized it as a scarce spring migrant in all areas except the Rocky Mountain region, where it is rare. However, it was once considered common and apparently widely distributed near timberline in Banff National Park (Clarke and McTaggart-Cowan 1945: from McNicholl 1985). It is not known what has caused this apparent change in status. The Wildlife Management Branch (1991) considered it to be a sensitive species, though not immediately at risk. However, present status may be affected by significant habitat loss on its wintering grounds.

Distribution:

A breeder of the northern and Atlantic regions of Canada, the Gray-cheeked Thrush's nearest established distribution adjacent to Alberta includes north-central British Columbia, the southern Mackenzie region of the Northwest Territories, and northeastern Saskatchewan. Between 1973 and 1976, several breeding records were obtained in the Caribou Mountains of northern Alberta (Höhn and Marklevitz 1974). Although breeding season records are most numerous for Jasper National Park, especially from 1976-1980, additional singing males have been reported in Banff National Park, Calgary, and Edmonton (Pinel et al. in prep.). No records for this species were obtained during the Atlas Project.

Habitat:

Distributed primarily in the subarctic regions of northern Canada, it uses areas of scrub willow, alders, and dwarf spruce (Salt and Salt 1976). In mountain areas, dense stunted spruce at timberline is also used. In the Caribou Mountains, a displaced area of subarctic habitat, this species was reported breeding in burns that hosted a scattering of young spruce (Höhn and Marklevitz 1974). In Jasper National Park, breeding season observations have been among tree islands of subalpine fir near timberline (Haddow et al. 1978).

Nesting:

The female constructs a relatively large cup-shaped nest of finely woven grass, leaves, and shredded bark, and lines it with grass, rootlets, and decayed leaves. It is usually positioned from ground level to 3.3 m up a bush or tree. The female lays 3-5 pale blue eggs, which are spotted with browns, and she incubates them for 12-14 days. Although only the female broods the nestlings for the first few days, the male does assist in their care, bringing them insects and removing faecal sacs. The young fledge after 11-13 days.

Remarks:

Adults have plain grayish-olive upperparts, tails, faces, and ear coverts that are gray with dusky streaks, white throats with brown streaks on each side, pale buffy breasts that are spotted with dark brown, pale olive sides, white remaining underparts, and indistinct eye rings. Similar to the other thrushes, this species is distinguished from the Hermit Thrush by its lack of a rusty tail, from the Veery by its grayer upperparts and a more heavily spotted breast, and from the Swainson's Thrush, which it most closely resembles, by its gray, not buffy, cheeks and a very light eye ring. The male produces a song and alarm call that resemble those of the Veery. The Gray-cheek's song usually differs by rising near the end.

Foraging on the ground in the open, the Gray-cheeked feeds primarily on insects, switching somewhat to berries, such as blueberries and cranberries, in the fall and winter. In migration, it may be found in loose mixed thrush flocks in a variety of wooded or shrubby habitats, where it ground-feeds on insects and worms.

Arriving in the province anywhere from the second week of May to early June (Pinel et al. in prep.), the Gray-cheeked Thrush departs the area during the month of September.

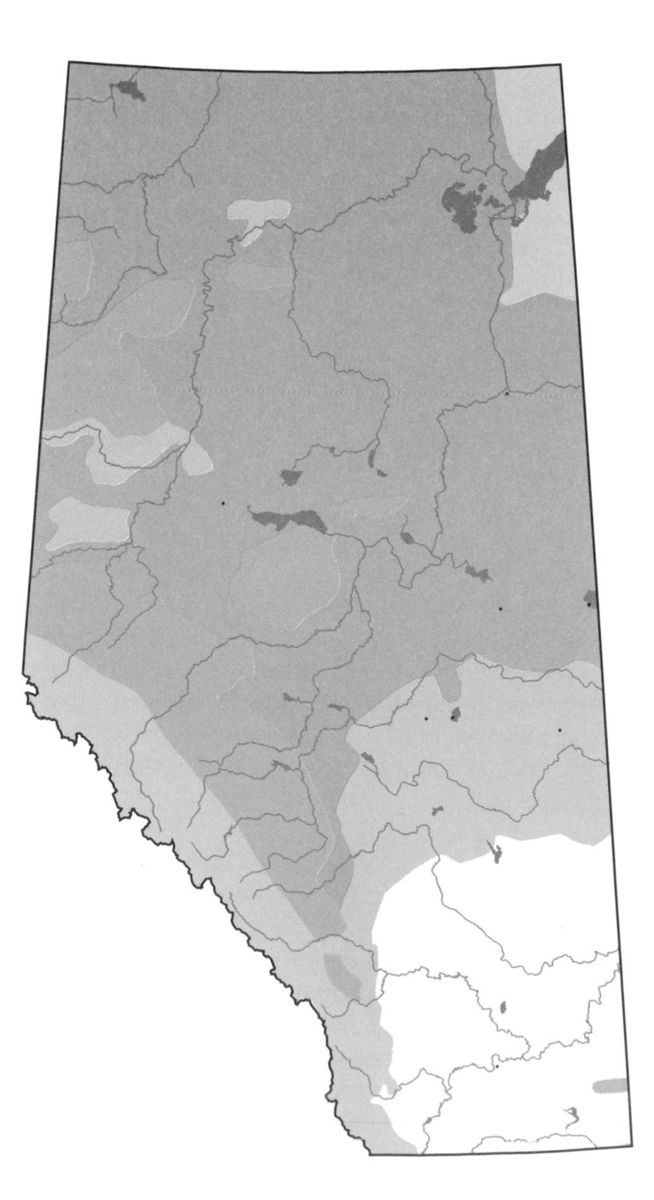

Swainson's Thrush

Catharus ustulatus

SWTH photo credit: T. Webb

Status:
The Wildlife Management Branch (1991) considered this species to be not at risk, experiencing widespread distribution in secure habitats and a healthy population. This bird is very common in the Rocky Mountain Natural Region (Holroyd and Van Tighem 1983).

Distribution:
The Swainson's Thrush breeds in the forested regions of Alberta. It was recorded in 42% of surveyed squares in the Rocky Mountain region, 46% in the Foothills region, over 30% in the Canadian Shield and Boreal Forest regions, and 11% in the Parkland region. There were only four breeding records in the Grassland region, where it is considered a transient.

Habitat:
This species shares similar habitat requirements with the Hermit Thrush, and both may be heard often in the forested areas of all but the Grassland Natural Region. The Swainson's prefers the forest margins, wooded lakeshores, and river banks of damp coniferous woodlands. It can tolerate tracts with either fewer or older coniferous trees than the Hermit Thrush.

Nesting:
The female constructs a nest of twigs, grass, moss, lichens, and weed stems, and lines the interior with dead leaves, grass, and rootlets. This well-camouflaged structure is usually positioned close to the trunk of a low evergreen or bush. Occasionally, it may be in tall deciduous shrubbery. It is usually no more than 2.2 m above the ground. The 3-5 greenish-blue eggs are spotted with light brown. Incubation, by the female, lasts 12-14 days and she broods the clutch for the first few days. Tended to by both adults, the nestlings are fed insects and, perhaps, some fruit. The young fledge after 10-12 days. The number of broods raised in a summer is not known. The Swainson's Thrush is a rare cowbird host.

Remarks:
This cautious and secretive bird is far more often heard than seen, quietly skirting about to investigate an intrusion and usually slipping back into the security of the deep woods without ever a notice. Should an intruder initially succeed in approaching an individual Swainson's, this bird will respond either by going to ground or by turning its back to the observer and remaining motionless. Cryptic plumage assists in either strategy. Adults have dark olive-brown upperparts, buffy eye rings and cheeks, a bright buffy breast with dusky spots extending from the sides of the throat, olive-brown sides, and a white abdomen and undertail coverts. The Swainson's Thrush may be distinguished from the Hermit Thrush by the latter's rusty-brown tail. The Veery has a more distinctly spotted breast than the Swainson's. Singing mostly in the morning, the male's song begins with a low note and continues with notes that roll or spiral upward. The same rendition is repeated following lengthy pauses. When disturbed, he will either continue singing at noticeably greater distances or frequently utter an alarm call, a low "whick", and investigate. The female may also use this call. At other times, a second call note is used, a high-pitched "queep".

The Swainson's Thrush primarily hover gleans and ground gleans in order to forage on ants, beetles, crickets, caterpillars, flies, wasps, and some soft-skinned fruit. It also occasionally flycatches.

Arriving later than the Hermit Thrush, the Swainson's appears locally in the second week of May. Fall migration is largely completed in mid-September, with some lingering into October. Unlike most warblers, post-breeding molt often does not begin until migration has started.

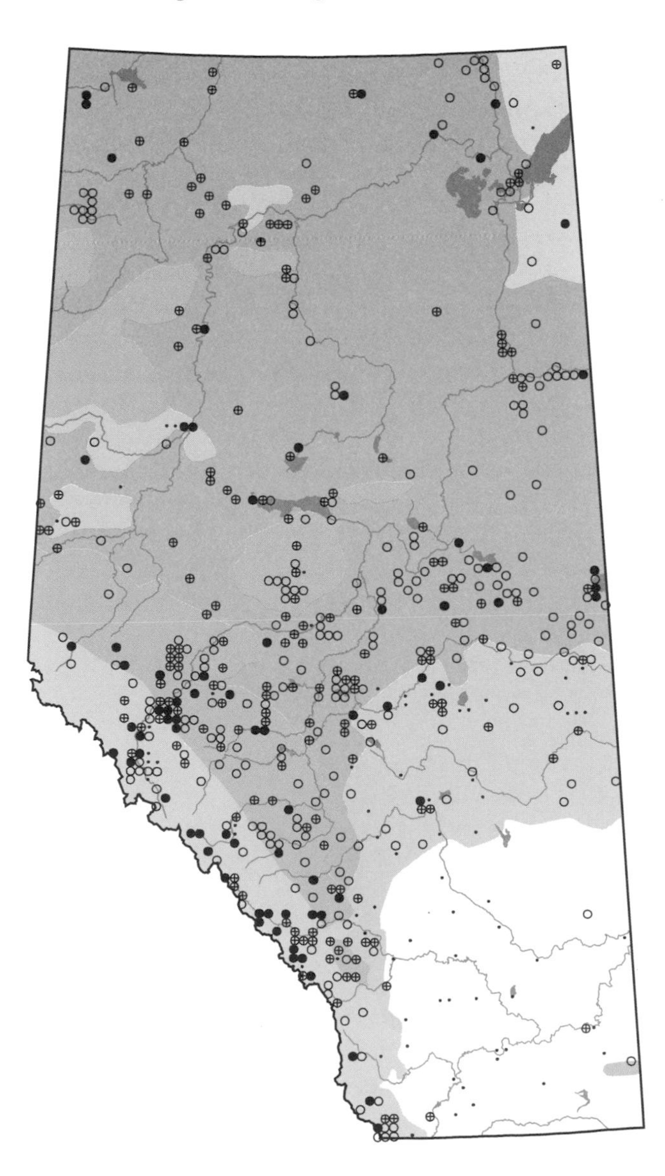

Hermit Thrush

Catharus ustulatus

HETH

Status:

This thrush has a widespread distribution, is considered to have a healthy population and secure habitat, and is common in the southern portion of the Boreal Forest and Foothills regions. In the mountain national parks, it is an uncommon spring and fall migrant and a common summer resident (Holroyd and Van Tighem 1983).

Distribution:

The Hermit Thrush breeds in suitable habitat in all areas of the province, with the exception of the Grassland region. It was recorded in 40% of the surveyed squares in the Rocky Mountain and Foothills regions, 30% in the Canadian Shield and Boreal Forest regions, and 11% in the Parkland region.

Habitat:

More versatile in habitat requirements than the Swainson's Thrush, the Hermit Thrush breeds in fairly heavily wooded forests, particularly those with a good mix of both coniferous and deciduous trees. It can also be found in pure coniferous stands, as well as second growth burns and logged-over clearings. In the mountains, it is found more at the higher elevations of the subalpine than is the Swainson's Thrush.

Nesting:

The nest is usually on or near the ground within a small natural depression in a dense portion of the wood. It is well-hidden, often covered by low plants or branches or, rarely, in a low bush or sapling. This may explain the fact that less than 15% of the breeding records for this species were in the confirmed category. Constructed of twigs, grass, and moss, the cup-shaped nest is lined with plant fibres, fine rootlets, and pine needles. The clutch typically consists of 3-6 eggs, which are light greenish blue and rarely have any spots. Incubation is conducted by the female for 12 days, and the monogamous male often feeds and guards her at this time. Tended by both parents, the young are initially fed mostly insect larvae and later receive small beetles, moths, spiders, and grasshoppers. Like many ground nesters, a low percentage of young live to fledge, 10-12 days in this case. This species is known to renest after nest failure.

Remarks:

Similar in appearance to the Swainson's Thrush and the Veery, this thrush has olive-brown upperparts, a pale eye ring, whitish underparts, wedge-shaped spots on the upper breast, and streaks on the sides of the throat and flanks. It can be distinguished by its reddish rump and tail, which it habitually raises upon alighting. Both sexes are alike. The immature is pictured above.

The male's song, most often heard in the early morning or evening, is an often-praised, flute-like arrangement of three short phrases. Each phrase is higher in pitch than the previous one and is bracketed by definite pauses. The final note is high and thin and does not carry well. A soft chuck call is similar to that of the Swainson's Thrush and is often heard as the bird investigates a disturbance and summarily departs, usually unseen. Even when spotted, this wary bird is easily flushed when approached. Its calculating nature is evident as it flies about and perches motionless for a time, blending in with the vegetation. It then flies off to repeat this. This behavior is exhibited both while in the trees and while foraging for ants, beetles, and spiders on or near the ground. In late summer, wild berries and fruit are largely taken.

The Hermit Thrush returns to the province in mid- to late April, usually before the Swainson's Thrush. Fall migration, carried out at night, occurs in mid-September.

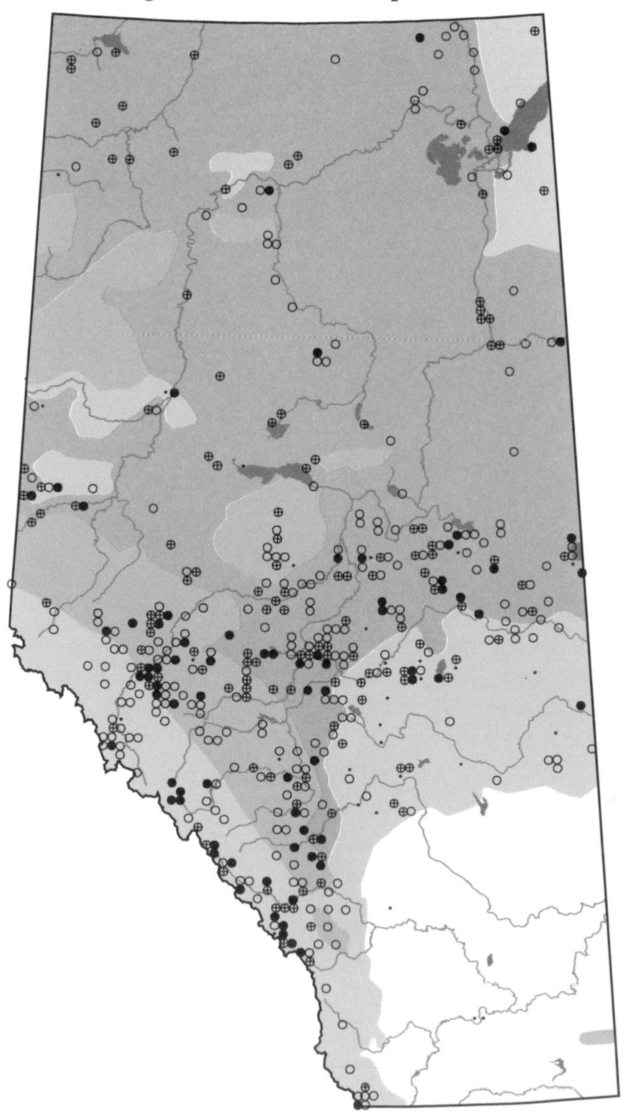

American Robin

Turdus migratorius

AMRO photo credit: T. Webb

Status:

The American Robin is one of a minority of species that have benefitted from the settlement and urbanization of North America. In pre-settlement times, the Robin was a forest species. With the new habitat provided by the settlement of the prairies and the development of the suburban landscape, its range and numbers have expanded significantly. It was one of the most common species encountered during the Atlas surveys, with 1,900 records (57% of which were confirmed breeding records).

Distribution:

This familiar species is found in all natural regions of the province. It was recorded in 30% of surveyed squares in the Grassland region, 61% in the Parkland, 58% in the Boreal Forest, 50% in the Canadian Shield, 70% in the Foothills, and 63% in the Rocky Mountain region.

Habitat:

The American Robin is a bird of both the woodlands and urban settings, although it is now most common in areas of human habitation. Wherever it settles, the robin's primary needs are open, grassy ground for feeding and sturdy trees and shrubbery for nesting. In forested areas, this species inhabits open and broken woodlands, forest edges along rivers, lakes and natural openings, and second growth in burns or cut-over areas. In the mountain parks, Holroyd and Van Tighem (1983) reported that the highest densities of robins in the breeding season occur in open subalpine larch/heath forest and white spruce/prickly rose/horestail forests. In agricultural areas, woodlots or thickets with nearby open fields provide for all the robin's needs. In cities and towns, residential areas with close-cropped lawns, ornamental plantings, and shade trees provide near-optimum American Robin habitat.

Nesting:

The nest of the American Robin is usually found in a tree or shrub, most often in a fork or on a strong branch, and usually 1-5 m above ground. It will also nest on a ledge or in a recess of a building, on a nesting platform under an eave, on a fence post, and, occasionally, on the ground. The nest is a substantial structure made of twigs, coarse grass, stems, and sometimes string, rags, or other man-made materials. It has a smooth inner cup made of mud, with an additional soft lining of fine grasses.

The female lays 3-5 blue eggs in her first clutch and 3-4 in her second. Incubation is 11-14 days by the female, with occasional help from the male. When the nestlings have fledged (after 14-16 days), the female immediately starts building a nest for her second clutch. The male then takes over care of the fledglings.

Remarks:

This is one of the best known birds in the province, easily recognized by its dark gray upper parts and reddish-orange underparts. It is not a true robin, but actually a large thrush, and has no close kinship with the European Robin, after which it is named. The song of the male in March signals the true arrival of spring for many people.

The robin's main foods are worms, insects, berries, and other fruits, with the earthworm being the main prey in settled areas. Even though widely reported to the contrary, experiments have shown that this thrush locates earthworms by sight and not by sound (Ehrlich 1988).

It arrives early in March, often before the winter snow has melted, and moves south by October. There are a few reports of birds attempting to overwinter in Edmonton and Calgary.

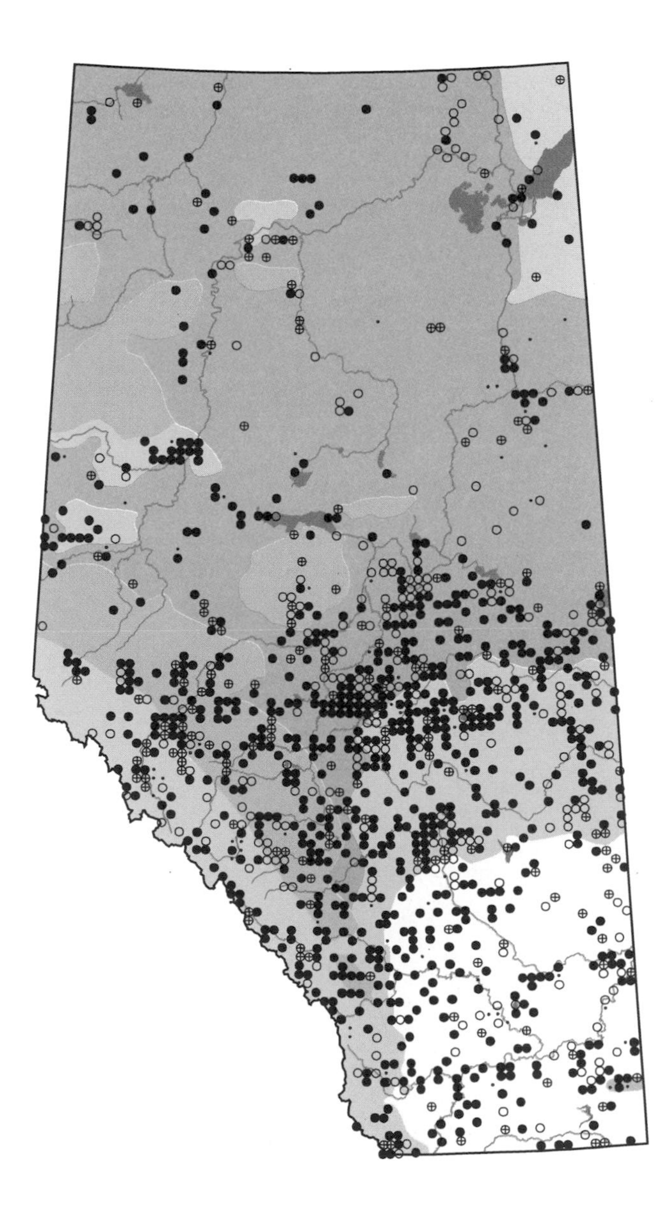

Varied Thrush

Ixoreus naevius

VATH

Status:

Although not considered at risk, the Varied Thrush is considered to be a sensitive species with special concerns (Wildlife Management Branch 1991). Unrestricted forestry practices may limit the distribution of this species, because the appearance of this species may be forest-size dependent. Little provincial research has been conducted into densities, habitat association, and territory size, especially as they relate to present forestry practices (Quinlan et al. 1990). In the mountain parks, the Varied Thrush is most visible in spring as a migrant, is a fairly common summer resident and is an uncommon fall migrant (Holroyd and Van Tighem 1983).

Distribution:

This thrush breeds in Waterton, Banff, and Jasper national parks, as well as the Willmore Wilderness Park. It is also known to breed in the Foothills region north to the Grande Cache area, with additional sightings south of Lesser Slave Lake. It was recorded in 35% of surveyed squares in the Rocky Mountain and Foothills regions, with sporadic sightings in the Boreal Forest region. Fall migrants commonly wander further east.

Habitat:

Breeding predominantly in coniferous stands, this bird uses the dark and damp understory characteristic of relatively old second-growth forests.

Nesting:

An early nester, the Varied Thrush usually builds its nest against the trunk of a small conifer, positioned 1-5 m above ground. The bulky nest is constructed of twigs and mosses, bound with some mud and grass and lined with fine grasses and shredded bark. The female lays 3-4 pale blue eggs, which are usually sparsely spotted with brown. The eggs are somewhat lighter colored than those of the American Robin. Incubation is carried out by the female for 13-14 days. This species is easily disturbed at the nest. If approached during nest-building or early incubation, the female may abandon the nest. Later in the cycle, the female will quickly flush from her nest and be joined by her mate. They will become quite vociferous and demonstrative as they circulate about the area. Nest overcrowding often prompts some young to leave the nest prematurely, while they still have short tails and are quite helpless (Munro 1963). No information is available on the fledging period of this thrush.

Remarks:

This thrush resembles the American Robin in size and general appearance. However, the male has a broad black breast band, an orange-buff line trailing from above the eye, and orange-brown wing bars, patches, and throat. The female is similar in appearance, but duller overall. Young resemble the female, but with more olive-brown above and buff mixing with the orange. The breast is spotted dark like that of the young American Robin. More often heard than seen, the song of the Varied Thrush was well characterized by Godfrey (1986) as "an eerie resonant whistle in various pitches." These prolonged notes are made without inflection and are separated by distinctive pauses.

These birds forage principally on the forest floor. Here they scratch away leaves with both feet, in the same fashion as a White-throated Sparrow. They feed on spiders, ants, beetles, larvae, and worms.

This hardy bird is among the earliest songbirds to make an appearance, arriving in numbers between early April and mid-May (Holroyd and Van Tighem 1983). Fall migration occurs in September, with some remaining into early November. Winter records for this species come from the southern half of the province, mostly in the Calgary area.

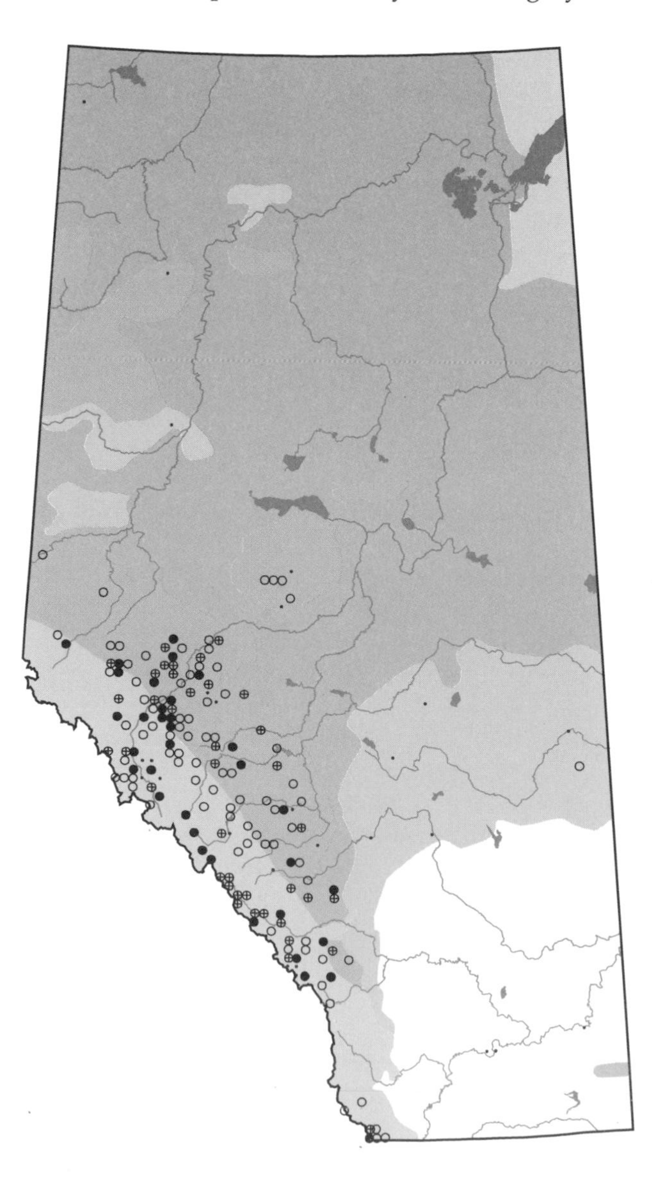

Gray Catbird

Dumetella carolinensis

GRCA photo credit: T. Webb

Status:
In Alberta, the Gray Catbird is fairly common locally in the Grassland and Parkland regions, and the southern portion of the Boreal Forest region. Holroyd and Van Tighem (1983) reported this species to be a very rare summer visitor to the mountain parks and unrecorded for the rest of the year.

Distribution:
This species occurs north to Fort Assiniboine and Lac La Biche, and west to Waterton Lakes National Park, Banff, and Drayton Valley. It was recorded in 20% of surveyed squares in the Parkland region, 11% in the Grassland, and 7% in the Boreal Forest and Rocky Mountain regions.

Habitat:
Gray Catbirds prefer dense shrubby habitat, such as that often found along woodland edges, roadsides, and streams. They also use similar areas adjacent to wetlands and those found in coulees. This species will use coniferous woods, but the frequency of its preferred habitat decreases with deeper penetration into the Boreal Forest, thus accounting for its only partial invasion of this region (Cadman 1987). In the mountain parks, the few records were in the montane area, where suitable habitat occurs in the form of willow shrubbery along the edges of streams and wetlands (Holroyd and Van Tighem 1983).

Nesting:
During courtship, the male sings to announce his presence and pursues the female in flight. He performs a strutting display, with wings dropped and tail erect, and turns to display his chestnut undertail coverts.

This species usually selects nesting sights in low dense vegetation, such as shrubs and saplings, and the nest is usually positioned less than 3 m above ground. It is composed of coarse sticks, twigs, plant stems, grass, and leaves, and is lined with fine plant material. It is usually well concealed. The female, primarily, builds the structure, with some assistance from the male, which brings material to the nest site.

The clutch consists of 3-5 greenish-blue eggs, which are incubated by the female for 12-13 days. Some mate feeding, by the male, occurs during this period (Bent 1964f). The young are tended by both parents and are fed almost entirely on insects. Fledging occurs approximately 10 days after hatching.

This species is double brooded.

Remarks:
The Gray Catbird is aptly named for its distinctive catlike mewing, which it inserts in its rich variety of musical and discordant song. This species is an accomplished mimic and, unlike the Brown Thrasher, does not usually repeat phrases.

The Gray Catbird is a rather plain-looking bird, which is slate gray with a blackish crown and tail. The bird's only bright visual feature is its chestnut-colored undertail coverts.

This species eats insects, spiders, other invertebrates, berries, and, occasionally, over half of its total diet can be cultivated fruit (Ehrlich et al. 1988). The food this species eats is so diversified in nature that the catbird can be considered omnivorous (Bent 1964f).

This species arrives in the province from mid- to late May and has departed again by early September (Salt and Salt 1976), although it is not often observed during migration, because of its tendency to travel at night (Root 1988).

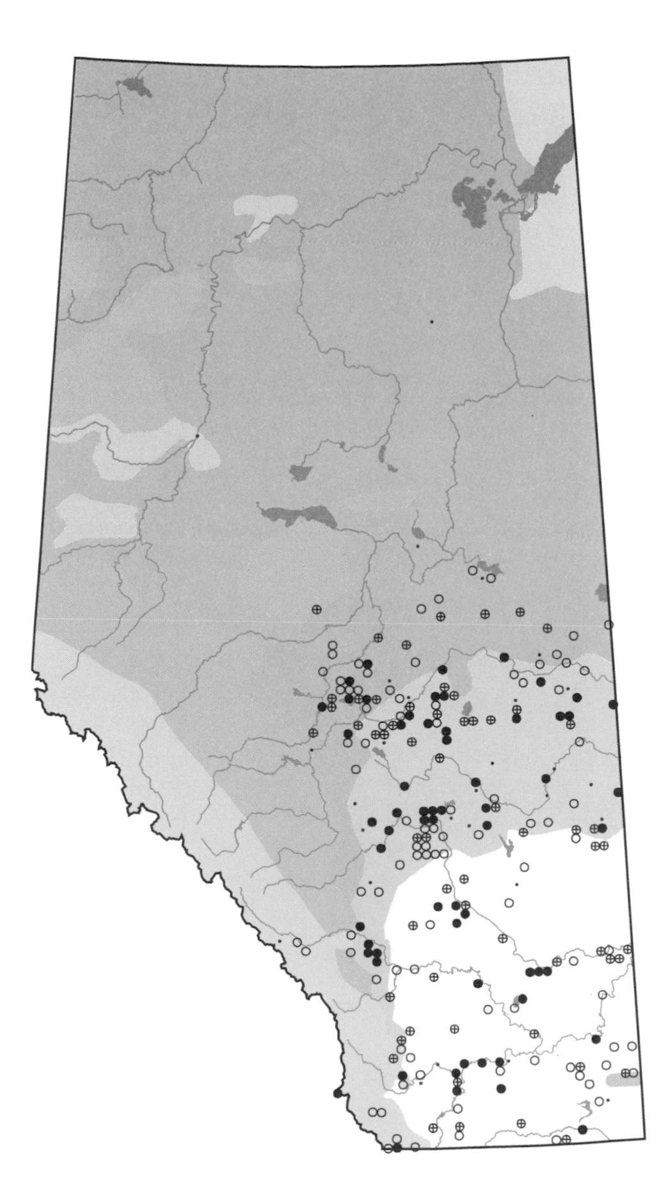

Northern Mockingbird

Mimus polyglottos

NOMO

Status:
The Northern Mockingbird is considered to be very uncommon in the province. There were only nine breeding records for this species in the five year Atlas survey period. Salt and Salt (1976) reported that this species occurs in the province locally and irregularly. Pinel et al. (in prep.) indicate that historical records in Alberta show clearly that the Northern Mockingbird is an irregular vagrant that occasionally nests in the province.

Distribution:
The Northern Mockingbird is at the northern limit of its range in Alberta, normally being found in the United States and Mexico. The first nesting record for the province was at Didsbury in 1928 (Salt and Wilk 1958). Other nesting records have come from Brooks, Cadogan, and Lake Newell. Confirmed nesting records during Atlas surveys were from Suffield, Blindloss, and Sylvan Lake.

Habitat:
This species is highly adaptable; it uses a variety of semi-open habitats, such as rural fields, hedgerows, gardens, and open woods of both deciduous and coniferous varieties. Northern Mockingbirds are often found near human habitations that have ornamental trees and shrubbery.

Nesting:
During mating season, Northern Mockingbirds can be seen "dancing" together, which entails facing each other with tails and heads high, hopping up and down and darting at each other. This display is performed both in courtship and in territorial disputes between males.
The nest is made by both sexes, with the male building a foundation of twigs and the female lining it with grass and rootlets; it is a procedure that takes 4-8 days to complete (Ehrlich et al. 1988). The nest is usually located in a small tree or shrub 1-4 m above ground. It may be placed either near or far from the trunk, but, in either case, it is usually well concealed.
The female lays 3-5 eggs, which are pale greenish blue and are spotted with browns. She incubates the eggs for 12-13 days and broods the young for several days, bringing them insects on the bill. The young leave the nest at 11-14 days and the male assumes responsibility until the young are independent, while the female renests (Ehrlich et al. 1988).
In fall, the male and female may remain together and defend a common feeding territory, or they may separate and defend separate feeding territories.

Remarks:
The Northern Mockingbird is a rather plain looking bird with brownish gray upper parts and blackish tail and wings. Its underparts are light gray and it has a yellow eye. In flight, there is a conspicuous white patch on the wing. This species can be distinguished from the shrikes by its brownish color and lack of black face mask.
As its name implies, the Northern Mockingbird imitates other birds' songs. Its singing usually consists of a mixture of its own songs with those of other birds intermixed; phrases are repeated several times. Their memory of borrowed songs is impressive, and they are able to reproduce songs eight months after exposure (Bent 1948).
This species is highly territorial. There are reports of dogs and cats being chased from breeding territories by this bird.
In summer, this species' diet consists primarily of insects, but it also includes snails, small vertebrates, and berries. In winter, however, vegetable matter makes up a large proportion of the diet.
This species is non-migratory, which is, likely, a significant factor limiting their northward expansion. There is only one wintering record for this species, in the Calgary area (Salt and Salt 1976).

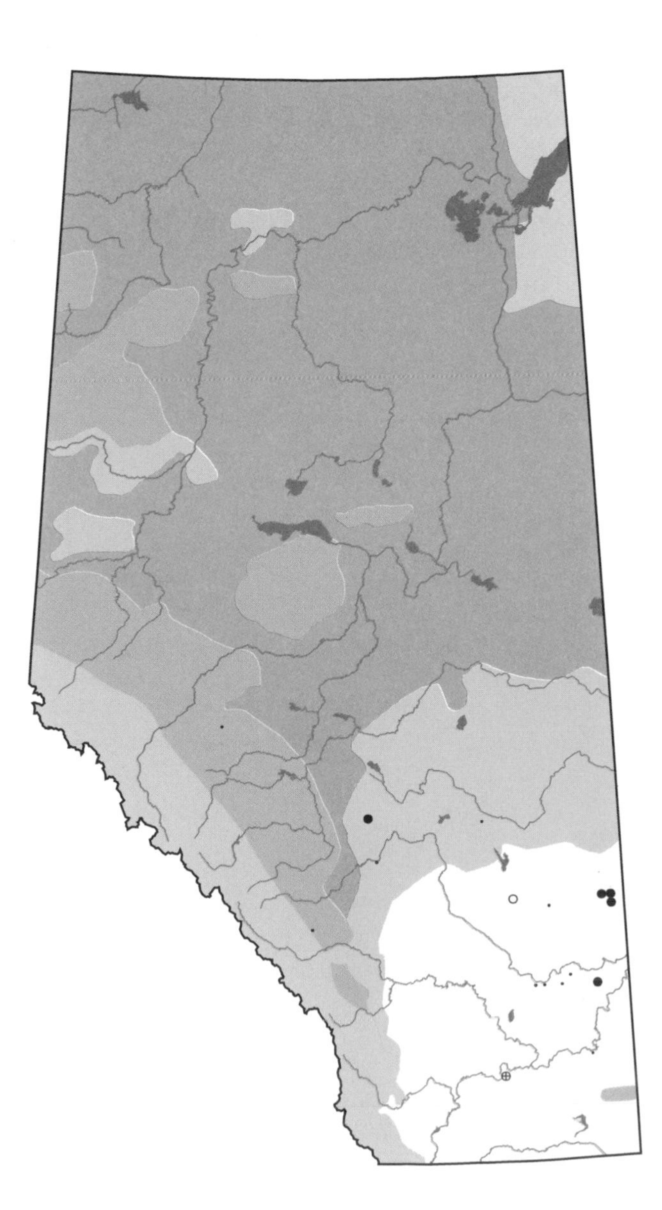

Sage Thrasher

Oreoscoptes montanus

SATH photo credit: B. Randall/VIREO

Status:
Salt and Salt (1976) considered the Sage Thrasher to be a rare vagrant in Alberta. Only four sightings had been documented between 1924 and 1958 and none during the 1960s and 70s. However, during the Atlas Project, there were three instances when breeding was confirmed, as well as a possible breeder at another location.

Distribution:
Before the Atlas Project, very rare sightings were made in Orion, Walsh, Calgary, and Drumheller. In the years between 1987-1991, breeding was confirmed in areas near Manyberries and south of Medicine Hat. A possible breeder was recorded in the Red Deer River valley area west of Empress.

Habitat:
Breeding activity in the province is confined to the semi-arid plains of southeastern Alberta. Although other types of vegetative cover may be present, sagebrush is the dominant species of plant in the area in which the Sage Thrasher breeds. Godfrey (1986) also included the thickets on arid hillsides as acceptable habitat. Such habitat may exist in the coulees and valleys of this region, where this species has been known to feed on seeds and berries (Salt and Salt 1976).

Nesting:
In courtship, the monogamous male will attempt to lure a mate with a two part performance. First, the male carries out a flight display with wings tremoring. Once he alights upon a prominent object, he will hold his wings over his head and flutter briefly.
Usually located low in the fork of a sage or other type of bush, the large bulky nest consists of a mass of twigs with a lining of grasses, wool, and rootlets. It is also often partly arched over. Well concealed in whatever location, it may also, occasionally, be on the ground under a bush. The 4-5 greenish-blue eggs are spotted and blotched with chestnut. Both adults participate in the incubation for 13-17 days, and in the rearing of the young, which fledge after 11-14 days. This species may raise two broods (Ehrlich et al. 1988). The Sage Thrasher reacts to cowbird parasitism by rejecting the foreign eggs.

Remarks:
The best opportunity to get a good look at this species occurs when the male sings from a prominent elevated perch. Otherwise, this highly terrestrial bird prefers to run and hide than flush. Low flight is occasionally used. Smaller than the Brown Thrasher, Sage Thrasher adults have brownish-gray upperparts with faint dusky streaks, darker wings than the Brown's, with two narrow off-white wing bars, longish darker tails with outer feathers tipped in white, buffy white underparts streaked with brown on the sides of the throat, breast, and sides, and yellow eyes. The streaking below is highly reduced by late summer. Resembling another prairie dweller, the Northern Mockingbird, the Sage Thrasher also holds its tail high when running and, when sitting, raises its tail quickly and lowers it slowly.
A prolonged effort, the energetic song of the male Sage Thrasher has been recorded up to 2 1/2 minutes long (Bent 1964f). It is similiar in composition to that of the Brown Thrasher, although it doesn't include well-defined pauses. It is also vaguely similar to the Gray Catbird song, and may be given in flight. Call notes include a "chuck, chuck" and a high "wheeur".
A ground feeder, the Sage Thrasher mostly feeds on weevils, beetles, spiders, caterpillars, crickets, and insect eggs and larvae, as well as such berries as currant and gooseberry. Migration dates are not available for the province.

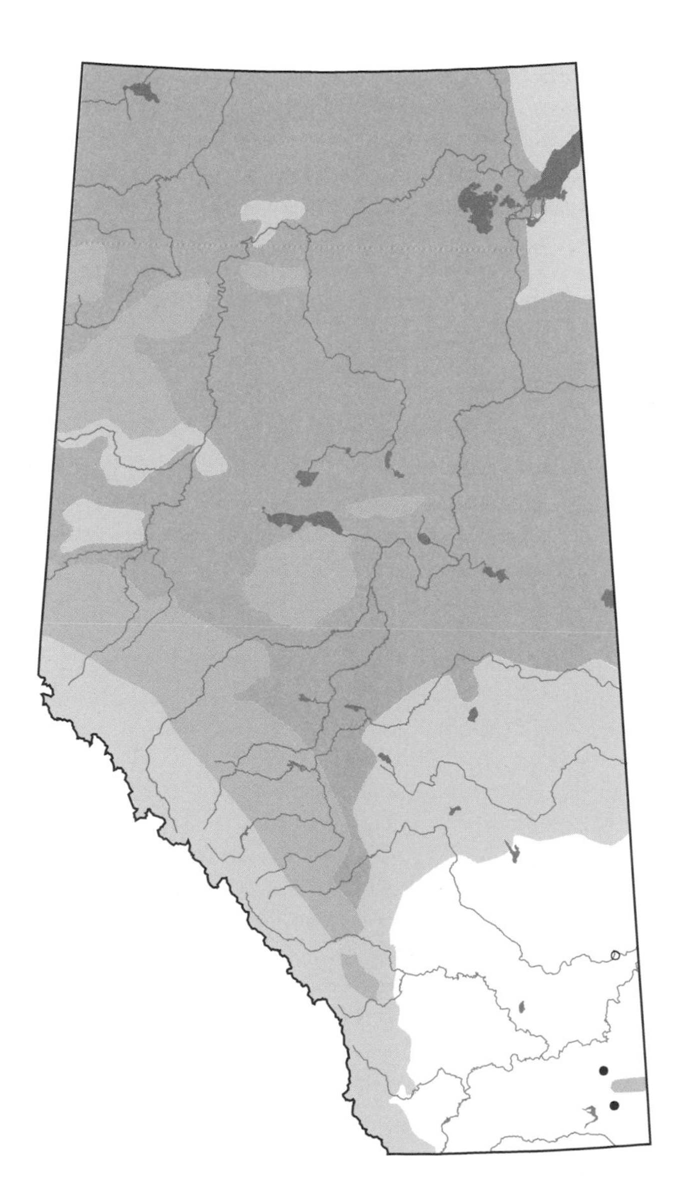

Brown Thrasher

Toxostoma rufum

BRTH photo credit: R. Gehlert

Status:

In Alberta, the Brown Thrasher is relatively common in the Grassland Natural Region. It is widespread and local in the Parkland Natural Region has not been determined.

Distribution:

This species is found from Cold Lake and Edmonton south, east of the Rocky Mountains (Salt and Salt 1976). Alberta is at the extreme northwest edge of the Brown Thrasher's range in North America (Pinel et al. in prep.). Atlas data indicate that breeding does take place in the Rocky Mountain region in the southwest corner of the province.

Habitat:

The Brown Thrasher prefers habitat in the early successional habitats, such as woodland edges, hedges, thickets, and the brush along riversides. This preference often causes localized distributions (Cadman 1987). In Alberta, the preferred habitat is tall shrubbery in the coulees and valleys of grasslands (Pinel et al. in prep.). In the Parkland region, fence rows and dense brush bordering aspen groves are often used (Salt and Salt 1976).

Nesting:

In courtship, the female responds to the male's song by picking up a twig and hopping towards him, fluttering her wings and chirping. This can elicit a male to pick up some dead leaves and hop towards her (Ehrlich et al. 1988).

The pair will build a nest in a low shrub or bush on the ground, or near the base of a tree or shrub. These nests are rarely above 1 m from the ground. Nest materials that may be used are grasses, plant stalks, bark strips, pine needles, mud, and feathers. Nest linings are often made of rootlets but may include hair, leaves, and fine grasses.

The female typically lays 4-5 pale bluish or greenish eggs, which are spotted with browns. These are incubated by both sexes for 10-12 days, and the young are tended, also by both parents, for 9-13 days.

This species may renest in the same nest or may construct a new one for second broods, or in nesting attempts after failed broods.

Remarks:

The Brown Thrasher has reddish upperparts and buffy underparts, and is heavily streaked with dark browns on the breast and sides. It has two white wing bars and a yellow eye.

Although it will sing from a treetop perch, if startled, it will often seek the dense cover present in its select breeding habitat. This ability to disappear into the underbrush becomes a primary behavior when Brown Thrashers have young or eggs, and they are very difficult to find during this period. However, when threatened, they will aggressively defend their young.

The male Brown Thrasher has the largest documented song repertoire in North America, including more than 1100 song types (Ehrlich et al. 1988). This bird may acquire songs of other birds, but thes songs do not compose a major proportion of their song repertoire. Distinct to its singing habitats, as opposed to the Gray Catbird, the Brown Thrasher tends to repeat phrases.

This species feeds primarily on the ground, sifting through the dirt for invertebrates; it also gleans from foliage, eating fruit and seeds. Bent (1964f) also reported Brown Thrashers hawking insects on the wing.

This species arrives in Alberta in mid-May and leaves in late August/early September. In Alberta, there have been recent records of successful overwintering in Turner Valley and Edmonton (Hingston 1991).

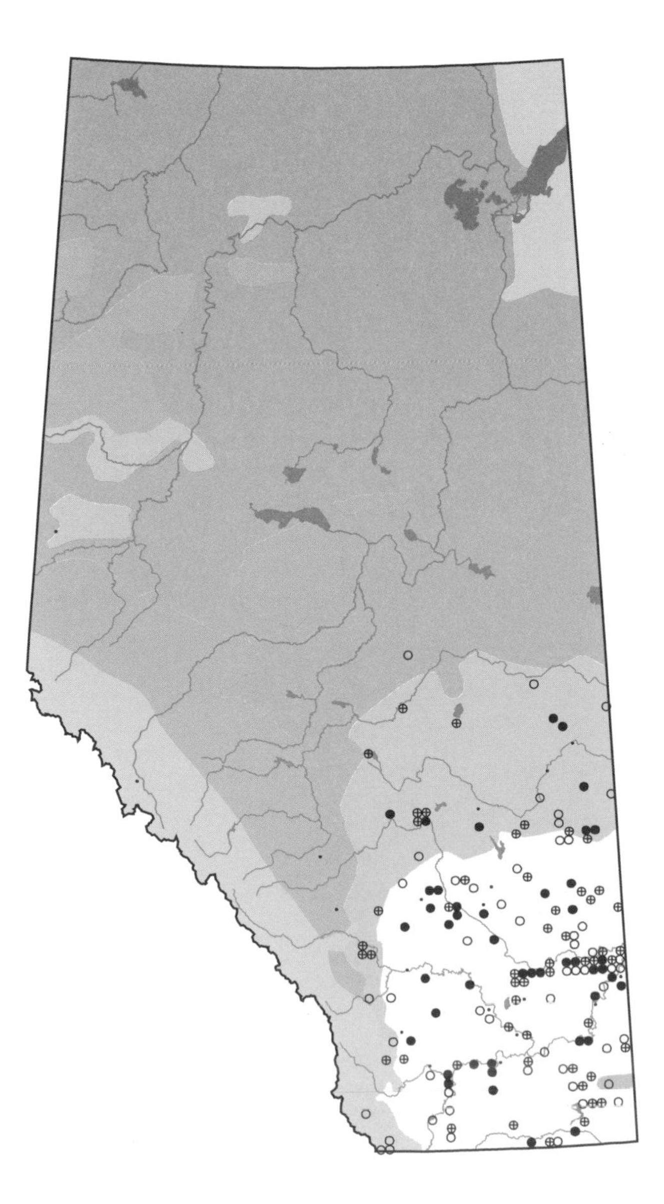

American Pipit

Anthus rubescens

AMPI

Status:

The American Pipit (formerly known as the Water Pipit) is a common spring and fall migrant and a very common summer resident in the mountain parks, but it is not recorded in winter (Holroyd and Van Tighem 1983).

Distribution:

The American Pipit is found across North America in tundra areas and in the western mountain region. In Alberta, this species is found mainly in the Rocky Mountain Natural Region, where it was recorded in 18% of surveyed squares. Of 50 total breeding records during Atlas surveys, only five were outside of this region and they were all in the Foothills region. It is a common transient through all parts of central and southern Alberta.

Habitat:

The American Pipit prefers moister habitats to dry ones (Root 1988). It can be found in alpine regions, generally on open habitats above timberline, often in areas where Horned Larks and White-tailed Ptarmigans are also found (Salt and Salt 1976). It occurs in the upper subalpine ecoregions, except during migration. The American Pipit inhabits a wide variety of open habitats, from wetlands to fairly dry tundra, with highest densities during the breeding season in mountain avens/snow willow tundra (Holroyd and Van Tighem 1983).

Nesting:

The song flight of the male American Pipit is the most conspicuous part of courtship for this species. It will spiral upwards while singing and then drop back down, often many times in succession (Bent 1965b). The male is monogamous.

The female builds the nest over a 4-5 day period. It is a cup of plant stems, moss, and grass constructed in a hollow in a bank or cliff, near a tuft of grass, or among rocks. It is lined with fine twigs, grasses, and feathers. The female lays 4-6 gray eggs, which are marked with brown. The eggs are incubated for 13-14 days by the female alone and she is fed away from the nest by the male during this time. Once hatched, the young are tended by both parents for 14-16 days before leaving the nest and up to an additional 14 days afterwards (Harrison 1978). A single brood is raised.

Remarks:

The American Pipit is a small brownish bird with a thin bill, buffy eye line, white throat, darkly streaked breast, and dark legs. It also has white on the outer tail feathers. Although the Vesper Sparrow and some longspurs show white on the outer tail feathers, the American Pipit's slim bill separates it from those and other sparrows (Godfrey 1986). The different habitat and its black legs differentiate this species from Sprague's Pipit. The above picture shows this bird in its fall plumage.

This bird feeds on the ground, where it walks rather than hops and its tail jerks nervously. It feeds on insects, small molluscs and crustaceans, seeds, and vegetable matter. The vast majority of the American Pipit's diet is insects, although vegetable matter is more significant in spring and fall. It may occasionally feed in shallow water.

This species has an undulating flight pattern and migrates in large loosely associated flocks, feeding in cultivated fields and on beaches.

They arrive in late April and May and leave the province in August with occasional stragglers being recorded as late as mid-November. It winters in a range from the coastal and southern United States south through to Guatemala and El Salvador.

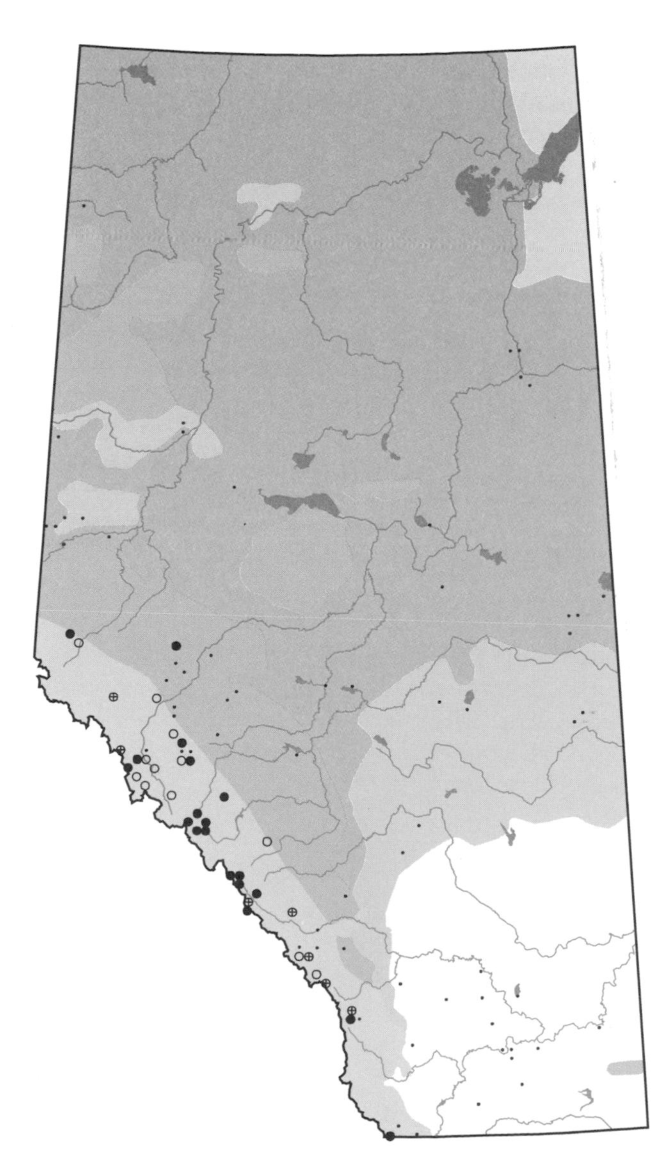

Sprague's Pipit

Anthus spragueii

SPPI photo credit: A. Morris/VIREO

Status:

The status of Sprague's Pipit in Alberta not well known. It is a typical bird of the short grass plains but, unlike the meadowlark, it has not been able to adapt to the effects of grain farming upon its environment (Salt and Salt 1976). Cultivated areas, roadsides, and burns, although superficially similar to natural habitat, appear to be unsuitable for this species (Salt and Salt 1976;Godfrey 1986). This lack of adaptation to cultivated lands has reduced its potential range.

Distribution:

Sprague's Pipit is at the northwest perimeter of its continental breeding range in Alberta. Salt and Salt (1976) and Godfrey (1986) reported the breeding range of this bird to be southern Alberta west to the Foothills region and north through the Parkland region to Athabasca and into the Peace River district. Atlas surveys have only recorded breeding evidence of this species north to Barrhead and Bonnyville. Confirmed breeding was only recorded in the Grassland and southern Parkland regions. There were breeding records in the Rocky Mountain region in Kananaskis Country and Waterton Lakes National Park.

Habitat:

The habitat of Sprague's Pipit is described by Wershler et al. (1991) as bush grassland, including dry lake bottoms, some moderately grazed areas, and some grassy sites in the sandhills. This species is a common associate of the Baird's Sparrow and is intolerant of heavy grazing.

Nesting:

The male spends much time in flight during breeding season, flying very high, singing on the rising actions in flight (Bent 1965b). Like the American Pipit, the male Sprague's Pipit will plummet to the ground after performing its high flying song. Upon his arrival, the female will often fly off of the nest to greet him.

Sprague's Pipit nests on the ground, usually concealing the nest under grassy overhangs. Concealment, and the females tendency to flush while intruders are still a long distance off, make the nest of this bird very difficult to find. This may explain the fact that, of 236 breeding records, only 16 were in the confirmed category.

The nest is a cup of fine dried grasses with some coarser material around the outside. The female lays 3-5 light gray eggs, which are marked with dark grays and browns.

The incubation period for this species is not known, but the young are thought to leave the nest 10-11 days after hatching.

Remarks:

Sprague's Pipit is similar in appearance to the American Pipit except that it has a longer bill,its upper parts are buff and distinctly streaked, and it has flesh-colored legs. It has large dark eyes.

The common name for this species, pipit, comes from the Latin *"to chirp"* (Ehrlich et al. 1988). The song is a series of seven or eight tinkling double notes on a descending scale, usually as the bird circles high in the air (Godfrey 1986).

Unique to this bird is its habit of always landing on the ground and never in trees (Bent 1965b). This is a secretive and solitary bird that does not show itself in open areas when disturbed but hides in grass cover. When flushed, it will only fly for a short distance, low to the ground, and will suddenly drop down out of sight. This bird feeds on insects while walking around on the ground.

They arrive in late April and early May. There is some conjecture about fall migration, which is variously reported, between July, August, and September (Pinel et al. in prep.).

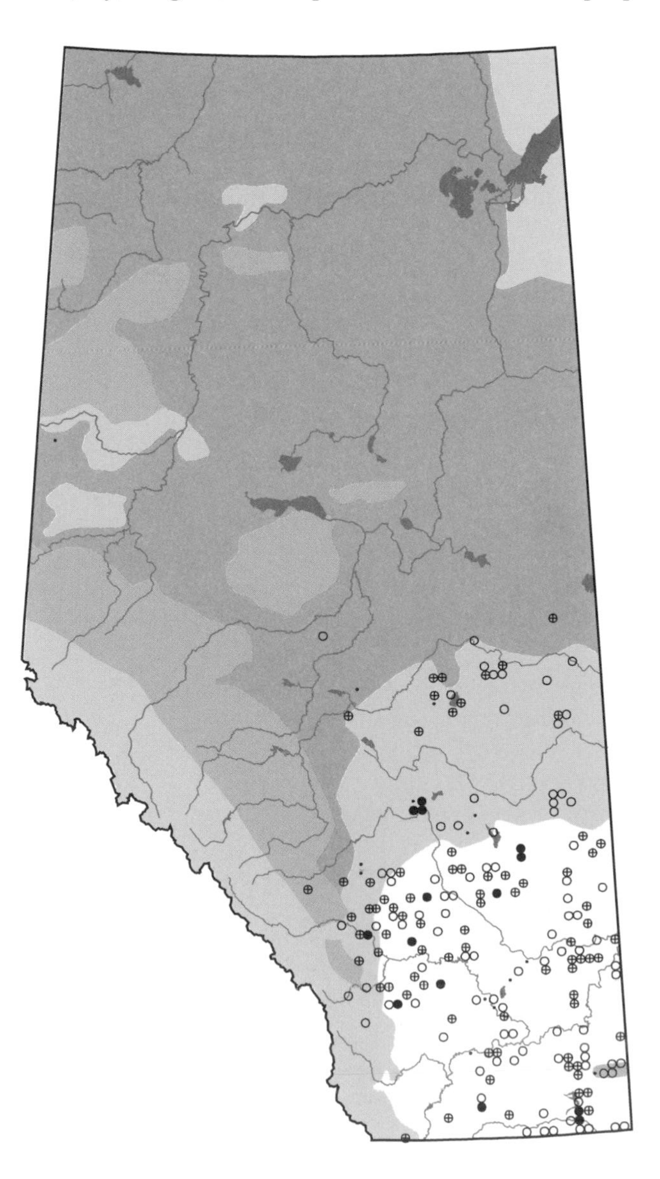

Bohemian Waxwing

Bombycilla garrulus

BOWA photo credit: R. Gehlert

Status:
The Wildlife Management Branch (1991) reports this species to have a healthy population and secure habitat. The Bohemian Waxwing is a common migrant and fairly common resident in the Mountain Parks, wintering in valleys and spending its summers in areas of higher elevation (Holroyd and Van Tighem 1983).

Distribution:
Bohemian Waxwings are found in northern Eurasia and North America, but their range penetrates south in the mountains into the United States. Salt and Salt (1976) reported the breeding distribution of this species to be northern Alberta south through the Peace River district to the Rocky Mountains and their foothills, as far as Banff, and east to the McLeod River. Atlas data indicate that this species is widespread in the Boreal Forest region, where it was recorded in 6% of surveyed squares. It was also recorded in 7% of surveyed squares in the Rocky Mountain and Foothills regions. In the Parkland region (2% of surveyed squares), it was recorded father east than was previously reported, with one probable breeding record east of Red Deer.

Habitat:
This species will use both coniferous and mixed woodlands, preferring edges and openings in the forest. They choose wintering areas which have and an abundancy of fruit producing trees and shrubs.

Nesting:
Bohemian Waxwings often nest near to a lake or stream in conifers, usually 1-6 m above the ground. Their loosely colonial nesting habits make them locally distributed, so they are easily missed if surveys do not coincide with the small areas in which they occur (Cadman et al. 1987).
They construct their nests with twigs and grass and line them with fine grass, moss, hair, down, and other fine plant materials. The female lays 4-6 pale blue eggs, which are marked with dark browns. She incubates them for 13-14 days and is fed, at this time, by the male.
The young are tended by both parents for 15-17 days before leaving the nest (Harrison 1978). The parents feed the young regurgitated insects and berries.

Remarks:
Waxwings were named for the red, drop-shaped, waxlike tips of their secondaries, which reminded people of sealing wax (Ehrlich et al. 1988). The Bohemian Waxwing is deemed attractive by many people, because of its sleek brown plumage, black patch through the eye, and yellow tipped tail. It has a crest on the head and may have red, white and yellow markings on the wing. Its larger size, chestnut undertail coverts, and gray chest distinguishes it from the Cedar Waxwing.
This bird is highly gregarious and is known to pass berries down a row of birds until one of them eats it. They have a definite preference for mountain ash and juniper berries, but also eat saskatoons, choke cherries, rose hips, raspberries, strawberries, and cedar berries. In breeding season, they eat mostly insects, which they hawk on the wing, but they rely, primarily, on berries in winter time.
Bent (1965b) reported that Bohemian Waxwings are so "polite" that, when attacked by Robins, they will simply wait for the Robin to eat its fill before feeding themselves.
Bovey (1988) reported that, in years where there are insufficient berries to support them, they migrate further south. They migrate in spring in April/May and again in fall in September/October.

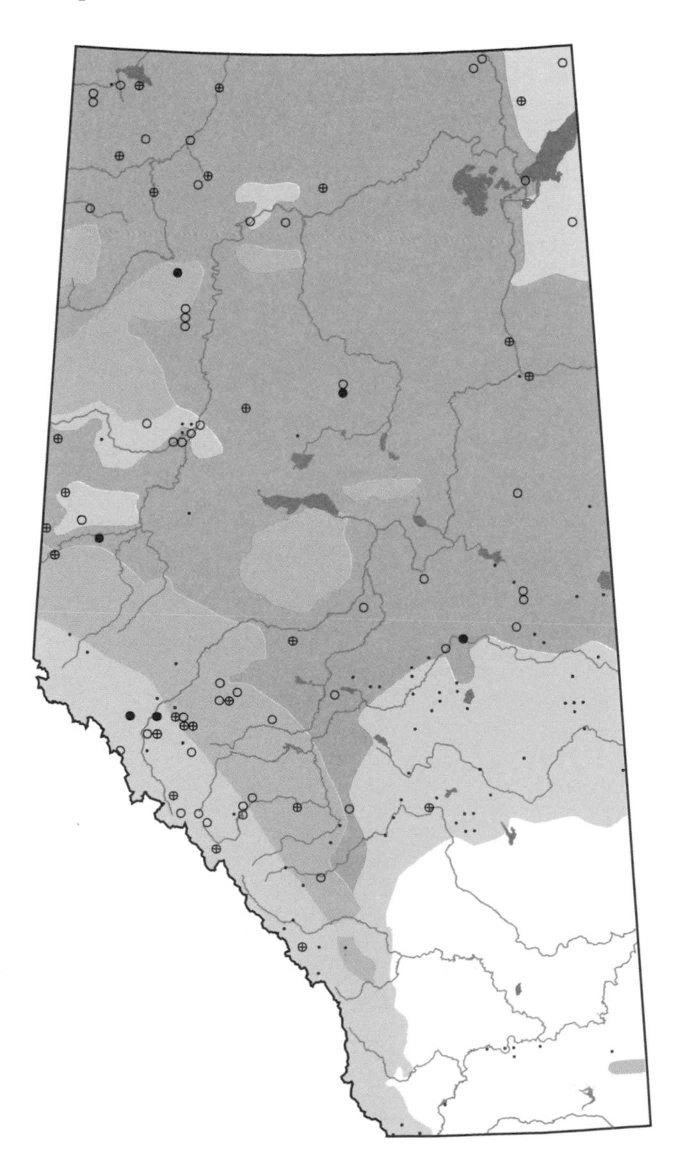

Cedar Waxwing

Bombycilla cedrorum

CEWX photo credit: T. Webb

Status:
The Cedar Waxwing is considered to have secure habitat and a healthy population in Alberta. It is common in the Parkland and southern Boreal Forest regions.

Distribution:
Cedar Waxwings are found from southern Alberta north to Bistcho Lake and Lake Athabasca. This species is uncommon in the mountains in spring through fall, and it has not been recorded there in winter (Holroyd and Van Tighem 1983). It was recorded in 42% of surveyed squares in the Boreal Forest region, 49% in the Parkland, 33% in the Foothills, 26% in the Rocky Mountain, and 13% in the Grassland region.

Habitat:
This waxwing uses a variety of habitats, but it tends to use areas that are open with edge habitat, such as near a stream or lake. It generally prefers deciduous woodlands, but will use areas near human habitation that have ornamental trees and shrubs. In this Grassland region, this species is found in wooded valleys and coulees.

Nesting:
This species' initial courtship display consists of a male hopping towards a female with a berry in his mouth. If she is not receptive, she will either leave, give a threat display, or attack him (Skutch 1976). However, if the female is receptive, she will accept the berry and hop away and then hop back and give back the berry. The male then repeats this action and they track the berry back and forth until the female eats it. Associated with this "dancing" display are fast circular flights of the nest area.
For the maintenance of the bond during the nesting cycle, the female begs from, and is fed by, the male (Putnam 1949).
During the nesting season, Cedar Waxwings are territorial directly around the nest area, but there is no aggressive behavior away from that area.
The nest is constructed of course and fine grass, hair feathers, and other plant matter, and lined with similar finer materials. They usually place the nest higher than 2 m above the ground, in either a coniferous or deciduous tree or tall shrub. The female lays 4-5 pale bluish gray or greenish-blue eggs, which are spotted with dark brown. The eggs are incubated by the female for 12-14 days before hatching. During this time, the female still begs from, and is fed by, the male (Putnam 1949).
Cedar Waxwings are known rejecters of cowbird eggs, and will eject foreign eggs if possible. Since Cedar Waxwings have small bills, it is difficult for them to eject eggs without endangering their own. Rothstein (1976) suggests this as a reason for these birds' higher rejection rates earlier in the nesting period.
This species is double-brooded.

Remarks:
These brownish, crested birds are much endeared to humans because of their gentle and gregarious habits. The tips of their tails are yellow and they have black masks. They are distinguished from Bohemian Waxwings by their yellow abdomens, white undertail coverts, and lack of white or yellow on the wings. Root (1988) reported that the waxlike tips on their wings and tails protect the tips from breaking and reduces the wear from fluttering in dense vegetation.
Cedar Waxwings are one of the latest migrants, arriving in late May/early June (Salt and Salt 1976). This results in coincident hatches and fledges with high berry production in August. They migrate south in August and September. There have been records of wintering in Edmonton, Calgary, Trochu, Vauxhall, and Lethbridge (Pinel et al. in prep.).

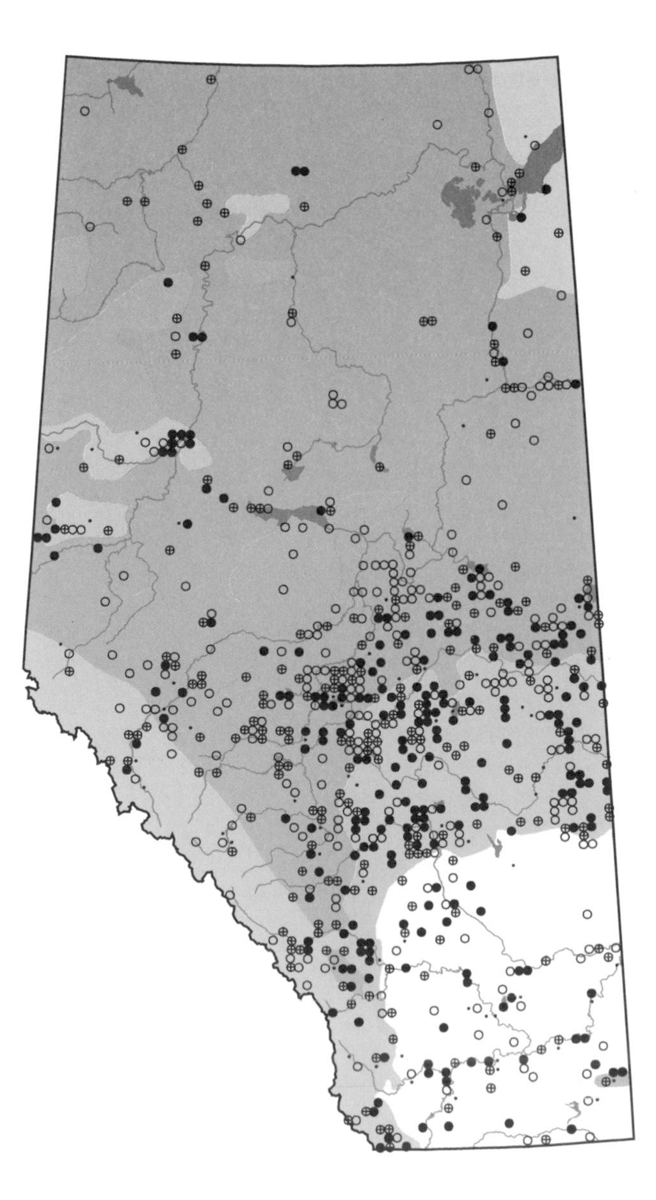

Northern Shrike

Lanius excubitor

NOSK

Status:
The Northern Shrike is considered to have a stable population and secure habitat in Alberta (Wildlife Management Branch 1991). Atlas data indicate that it is uncommon to rare; there were only four breeding records for this species in the five year survey period. These records were all from the southern fringe of the Boreal Forest Natural Region. Salt and Salt (1976) reported that it is a regular transient and rare winter visitor in the southern part of the province. It is an uncommon to very rare migrant in the mountain parks (Holroyd and Van Tighem 1983).

Distribution:
This species breeds across northern Canada from Labrador to the Yukon. Godfrey (1986) described the breeding range of the Northern Shrike as the northeast part of the province, in the Lake Athabasca area. The only confirmed breeding for this species during Atlas surveys was northwest of Rocky Mountain House.

Habitat:
Northern Shrikes can be found in open deciduous and coniferous woodlands, thickets, bogs, and taiga.

Nesting:
This species nests in the branches of bushes or trees, usually 2-4 m above ground. The nest is a bulky, untidy affair made of twigs, grass, moss, wool, hair, and feathers. The centre cup is lined with rootlets, hair, and feathers.
The female lays 5-7 grayish-white eggs, which are marked with spots of gray or brown. These are incubated by both sexes, predominately the female, for 15-16 days. Once hatched, the female broods the young for several days. The male provides food to the female during incubation and brooding. The young fledge at 19-20 days and continue to receive assistance from their parents until around 35 days (Harrison 1978).
This species is monogamous and raises one brood per year.

Remarks:
The Northern Shrike has blue-gray upper parts, black wings and tail, and a black mask across the face. It has light gray underparts and a white wing patch, which can be seen in flight. The mask distinguishes it from the Northern Mockingbird. The Northern Shrike's slightly larger size and narrower mask, which does not cover the forehead, distinguishes it from the Loggerhead Shrike. Also, the Northern has a heavier, longer, more heavily hooked bill (Godfrey 1986). In settled parts of southern Canada, only the Northern Shrike is encountered in winter, while the Loggerhead Shrike is encountered in summer (Godfrey 1986).
This species is an aggressive predator, catching small birds on the wing, pouncing on small mammals and killing both with a blow to the back of the neck with its beak. It will also hawk insects and, characteristically, hunt from a perch. Once prey are caught, they are impaled in a thorny bush or on barbed wire, much like a butcher hanging meat; this trait, and the habit of caching food in this manner, resulted in the common name "butcher bird". Shrikes are lacking the powerful talons of raptors, so they impale their prey to enable them to tear off bite-sized pieces.
This species is a true song bird, which is demonstrated by its lengthy warbling and song. Northerns will occasionally imitate other birds, such as the catbird.
Northern Shrikes arrive in northern Alberta in April and leave in late September and October. They reportedly winter in central Alberta north to Edmonton, and there are reports of wintering in the Grande Prairie area (Pinel et al. in prep.).

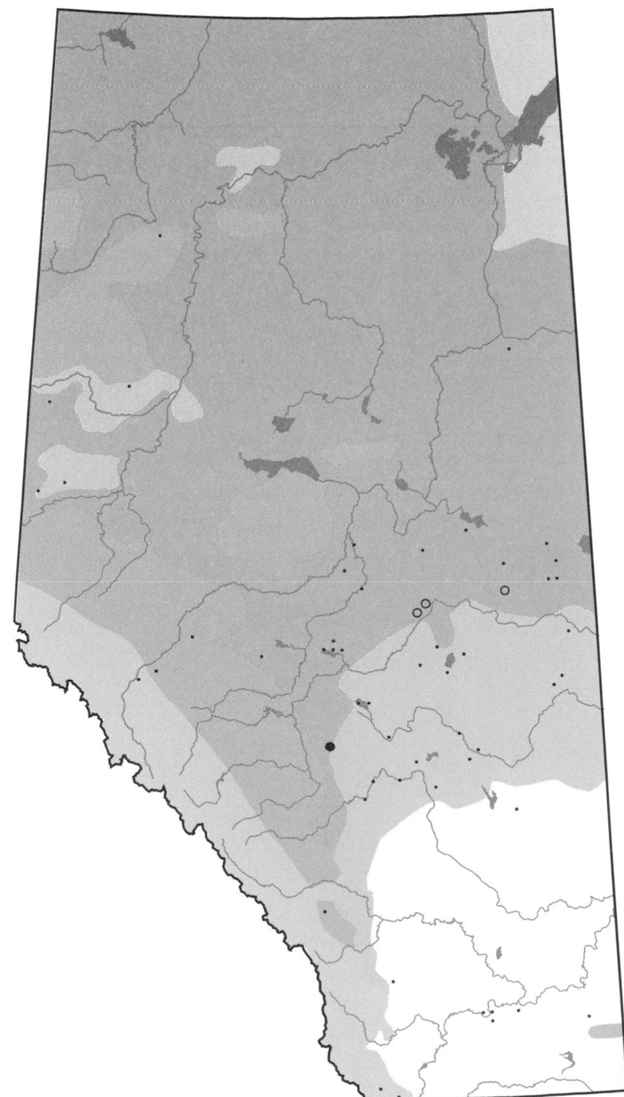

Loggerhead Shrike

Lanius ludovicianus

LOSH

Status:
There was a significant decline in the North American population of the Loggerhead Shrike from 1966 until 1979, which coincided with a documented decline in the Alberta population from 1972 to 1983 (Cadman 1985). Consequently, the Loggerhead Shrike was designated as a threatened species in Canada by the Committee on the Status of Endangered Wildlife in Canada (COSEWIC), in 1987. In Alberta, the population is thought to be declining and to be less than 500 pairs (Wildlife Management Branch 1991). The causes for these declines are not clearly understood.

Distribution:
The Loggerhead Shrike is found across southern Canada from Alberta to Nova Scotia. In Alberta, Atlas data indicate that it breeds predominantly in the Grasslands Natural Region, where it was recorded in 26% of surveyed squares. It was also found sporadically in the Parkland and southern Boreal Forest regions, where it was recorded in less than 5% of surveyed squares. Historical records indicate that the range of this species expanded northward to Fairview in the 1950s and has subsequently retracted (Cadman 1985).

Habitat:
The Loggerhead Shrike is found mainly in lightly wooded river valleys and coulees. In the Red Deer area, this species has been noted as having a preference for thorny buffaloberry bushes found beside railway tracks (Smith 1991).

Nesting:
The courtship of the Loggerhead Shrike involves mate feeding by the male, flight displays, and mock pursuits (Ehrlich et al. 1988).
The pair builds a nest in a tree or shrub, usually 1-4 m above ground. This nest is usually well concealed and is often placed near the trunk. It is a bulky cup of twigs, weed stems, and other plant matter, which is lined with plant down, bark, hair, rootlets, and feathers.
The female lays 4-7 grayish eggs, which are marked with browns, and she incubates for 14-16 days while being fed by the male. Once hatched, the female will brood the young for several days while the male continues to supply food. Fledging is at 17-21 days, but the young are not independent for 2-4 weeks after this (Harrison 1978; Ehrlich et al. 1988). This species often has 2-3 broods per season in the southern United States, but more than one brood has rarely been observed in Canada.

Remarks:
Of the 73 species of shrikes in the world, only the Loggerhead Shrike is exclusive to North America. This species, like the Northern Shrike, is a predatory bird, eating small birds and vertebrates. However, the majority of the Loggerhead Shrike's diet is large insects, mainly grasshoppers, crickets, and beetles. Much like the Northern Shrike, this bird impales its food for storage and to anchor its prey for consumption, to compensate for the talons it lacks. This "butcher bird" will often have a number of caches, to which it may return as late as eight months after the kill (Ehrlich et al. 1988).
The Loggerhead Shrike is very similar in appearance to the Northern Shrike, but is slightly smaller than that species, has a thicker mask which covers part of its forehead, and has less of the faint barring across the chest.
Loggerhead Shrikes arrive in Alberta during the last week of April, only a short time after the Northern Shrikes have left for the north. They usually move southward before the end of August, at least a month before Northern Shrikes return from the north (Salt & Salt 1976). There are few documented records of overwintering, including one from Okotoks (Pinel et al. in prep.).

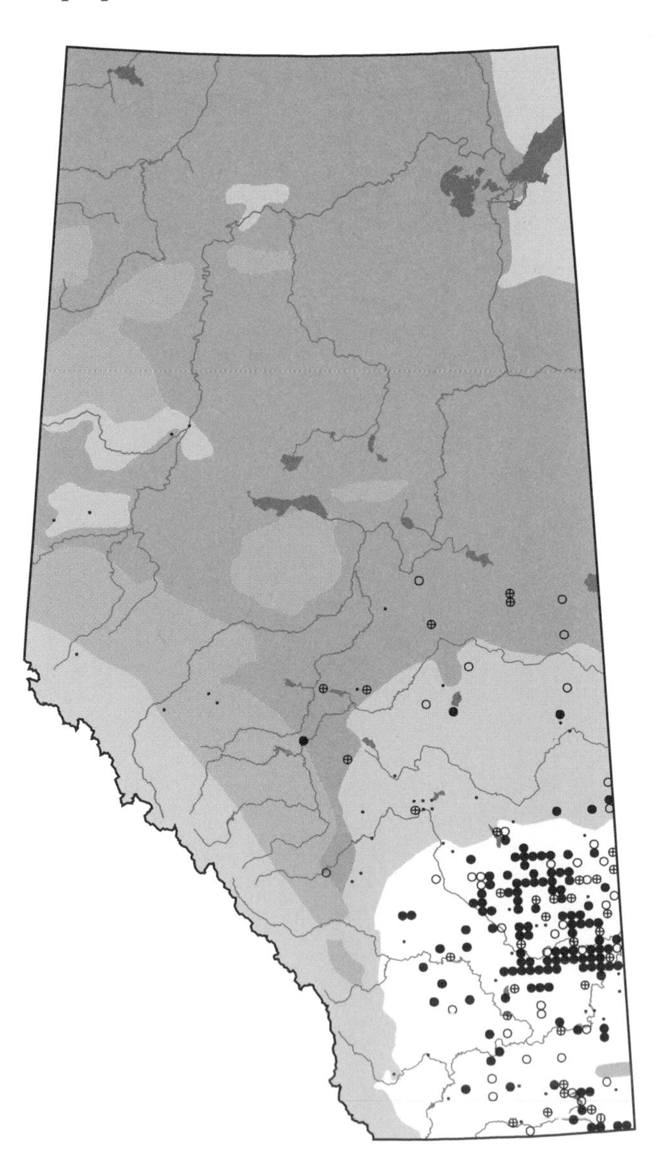

European Starling

Sturnus vulgaris

EUST

Status:
European Starlings were introduced into North America in 1890 in New York City. They have spread across much of the continent and have expanded from the initial 100 birds to an estimated 200 million individuals. Cadman et al. (1987) reported a decline in Ontario from 1967 to 1983 but suggested this was probably a levelling off to a stable population size.

Distribution:
European Starlings first appeared in Alberta at Camrose in late 1934, and they are now found throughout the province. It would appear that they have reached the limits of their distribution, now being found from the Canadian Arctic to Mexico. In Alberta, they were found breeding in all natural regions, with the exception of the Canadian Shield. They were recorded in 34% of surveyed squares in the Grassland region, 46% in the Parkland, 31% in the Boreal Forest, 24% in the Foothills, and 17% in the Rocky Mountain region.

Habitat:
Starlings are highly adaptable, and are found in city, town, and country habitats where there are tree cavities, or some reasonable facsimile, for nesting. Open spaces for feeding are also important (Godfrey 1986).

Nesting:
Starlings nest in holes in trees, buildings, among rocks, or in nest boxes. The nest is a mass of stems, leaves, and other plant materials, with a cup lined with feathers and fine grass.
The male begins nest construction before pairing, and the female completes the nest. The female lays 5-7 pale blue eggs, which are incubated by both sexes for 12-15 days once all of them are laid. Power et al. (1980) reported that the male Starling watches his mate very closely prior to the completion of the clutch to be sure that she does not mate with another male. If she did, he would end up caring for another male's offspring, reducing his own reproductive potential.
The young leave the nest at 20-22 days and spend a few more days with the parents afterward. This species is double-brooded.

Remarks:
The European Starling has a long tapered bill and a short square tail. In breeding plumage, the male is glossy black with stems of green and purple. His bill is yellow and he has a bluish patch near the mandible. The female is similar in appearance, but has a creamy or pinkish patch near the mandible and a yellow iris. In winter plumage, their bills are black instead of yellow and a plain streaked appearance replaces the glossy iridescence of their breeding plumage.
They eat many insects, which would appear to have been the motivation for those who introduced them initially. They also eat some small fruits, and they are able to eat tent caterpillars by squeezing out their insides and spitting out the hairy exterior (Bent 1965b).
One of the many complaints against this bird is its suspected competition for nesting holes with Eastern Bluebirds, Tree Swallows, and Red-headed Woodpeckers. Other complaints against them include noise, the killing of trees in Toronto from profuse droppings, and the crash of a plane in Boston in 1960, killing 62 passengers (Schneider 1990). Several unsuccessful methods have been tried to control the starling population across North America, including bounty systems, fire hoses, detergent to eliminate waterproofing and cause hypothermia, and glue painted to the limbs of roosting trees.
The European Starling arrives in the province in late February/early March, and leaves again in late September/early October (Pinel et al. in prep.).

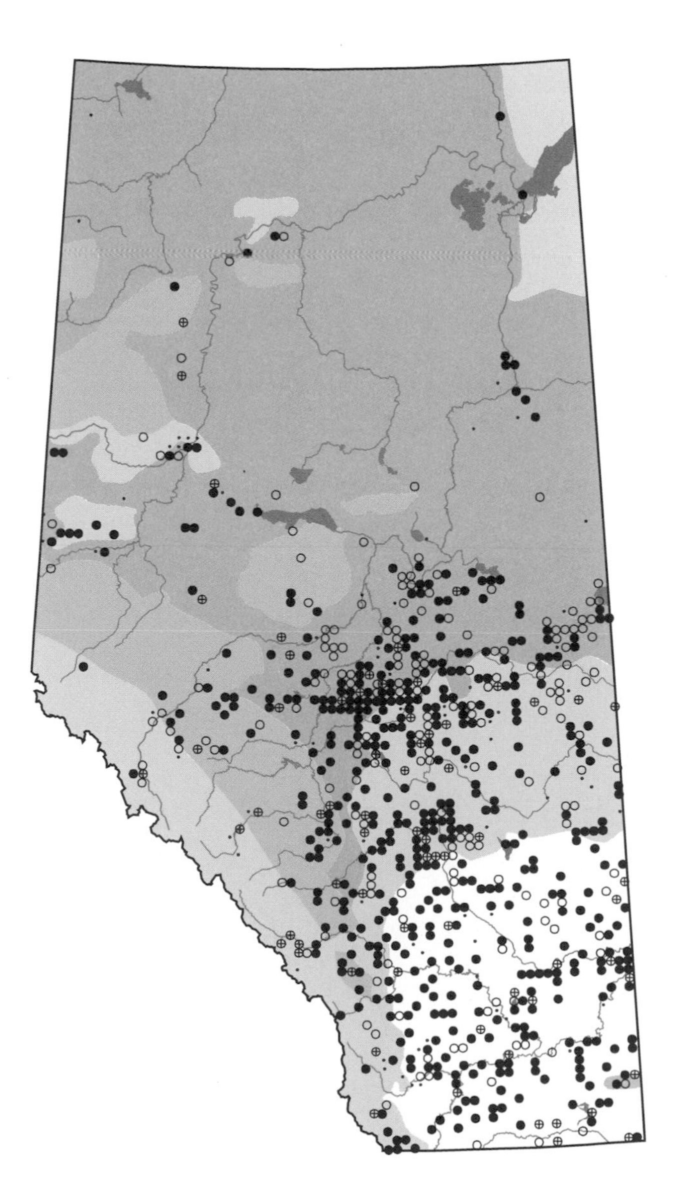

Solitary Vireo

Vireo solitarius

SOVI photo credit: R. Gehlert

Status:
The Solitary Vireo is widely distributed in Alberta but, as its name suggests, it is never found in great abundance. It is the least common vireo in the province (Salt 1973). Godfrey (1986) reported that, of the two subspecies found in Alberta, *V. s. solitarius* breeds throughout Alberta and *V. s. cassinii* is a migrant in the southwestern part of the province but does not breed here. Recent studies (McGillivray and Barlow unpubl.) have demonstrated large genetic and vocal differences between the two subspecies, implying that they might be separate species.

Distribution:
V. s. solitarius is widely distributed through the Boreal Forest region in north and central Alberta. Its range extends south along the eastern slopes of the Rockies to the Clearwater River west of Sundre (B. McGillivray pers. comm.). *V. s. cassinii* breeds in southwestern Alberta north to Kananaskis. An area of vocal intergrades exists between Kananaskis and the Clearwater River. Solitary Vireos are not known to breed in Cypress Hills Provincial Park. They were recorded in 21% of surveyed squares in the Boreal Forest region, 23% in the Foothills, 12% in the Canadian Shield and Rocky Mountain regions, and 5% in the Parkland region.

Habitat:
Alberta's other vireos prefer deciduous woods in breeding season, but the Solitary seeks out stands of jack pine or lodgepole pine or, less commonly, deciduous-coniferous forests for nesting. These pine stands are found on sandy well-drained soils throughout the Boreal Forest and Foothills natural regions. In the mountains, the Solitary Vireo is most prevalent in spruce and pine stands with intermixed deciduous trees and shrubbery (Holroyd and Van Tighem 1983).

Nesting:
The Solitary Vireo nests in a tree or sapling, conifer preferred, generally less than 4.5 m from the ground, but sometimes higher. The nest is a shallow, rounded cup which is usually suspended by the rim at a twig fork near the end of a horizontal branch. It is built by both male and female. Birch bark strips, dry leaves, moss, lichens, plant down, and sometimes paper, foil, or plastic are bound together with insect silk and spider webs to form the outer walls; moss, fine grasses, rootlets, and, occasionally, hair and feathers, are used for a lining. The nest exterior is often decorated with lichens and plant down.
The typical clutch is 3-5 white eggs with brown spots. Incubation, which is done by both sexes, extends for about 14 days. Both parents tend and feed the young, which fledge after 14 days (Ehrlich et al. 1988). One brood, occasionally two, is raised per season.

Remarks:
The plumage of all Solitary Vireos is intermediate between the blue-headed forms of "pure" *solitarius* in eastern Canada, and the pale gray-headed *cassinii* of British Columbia. The yellow wash on the flanks of Alberta *solitarius* is reduced relative to eastern birds. Vocalizations are intermediate as well, with elements of *cassinii* appearing in all Alberta *solitarius*. However, the birds of southwestern Alberta (*cassinii*) are easily distinguished from those in the rest of Alberta (*solitarius*) by their heavily burred (frequency modulated) vocalizations (B. McGillivray pers. comm.).
They are almost entirely insectivorous, though a few fruits are added to the winter diet.
The males arrive in the province in mid-May; the females arrive in late May. The birds depart from late August to early September, with some lingering into October.

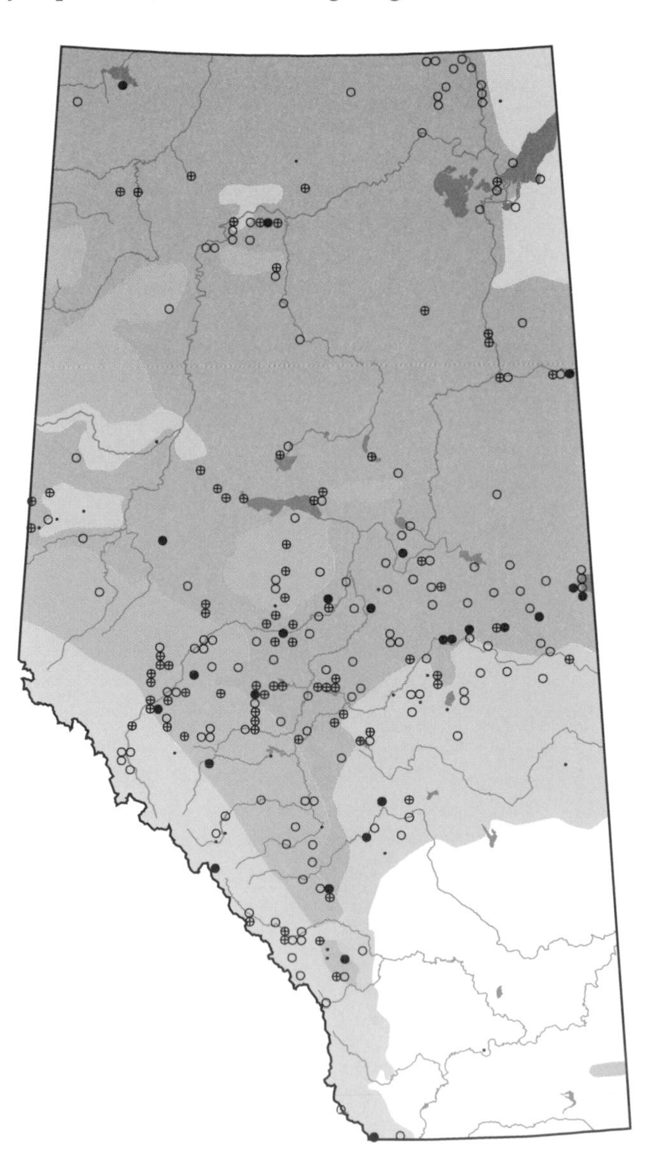

Warbling Vireo

Vireo gilvus

WAVI

Status:
The Warbling Vireo is the most common vireo of the Parkland and Foothills regions. Recent studies of Warbling Vireo morphology, song, and genetics (Murray 1992; McGillivray and Barlow unpubl.) have determined that the two subspecies, *V. g. gilvus* and *V. g. swainsonii*, are disjunct enough to be considered separate species.

Distribution:
V. g. gilvus, the eastern subspecies, is found in the Parkland region from Edmonton south and east to the Milk River. The western edge of *gilvus* distribution is approximately Highway 2 south from Edmonton to Red Deer, then southeast toward Drumheller. *V. g. swainsonii* is widespread in the Boreal Forest region, but is increasingly rare toward the Northwest Territories and Saskatchewan borders. *V. g. swainsonii* is found in the Rocky Mountain and Foothills regions south to Waterton Lakes National Park and east in southern Alberta at least to the town of Milk River. *Swainsonii* is also the breeding form found in the Cypress Hills. The Warbling Vireo was recorded in 26% of surveyed squares in the Boreal Forest, 28% in the Parkland, 47% in the Foothills, 33% in the Rocky Mountain, and 11% in the Grassland region.

Habitat:
The eastern form, *V. g. gilvus*, prefers aspen groves to the edges of aspen forests, generally near sloughs. Like the Red-eyed Vireo, the eastern Warbling Vireo is found in city parks and mature residential areas. Golf courses create excellent habitat for eastern Warbling Vireos. The western form, *V. g. swainsonii*, prefers thick aspen stands in mixed deciduous-coniferous forests, but is also found in aspen groves and shrubby avalanche slopes (Holroyd and Van Tighem 1983). *Swainsonii* can be found in fairly dense forests, but *gilvus* is generally associated with water.

Nesting:
The nest of the Warbling Vireo is built high in a deciduous tree and is suspended by its rim from a fork well out on a horizontal branch. In the prairies, a full bush may be used if no tall trees are available (Salt 1973). The nest site is chosen by the female, but the nest itself is built by both sexes. It is a neat thin-walled cup of grass and bark strips, bound with spider webs and insect cocoons, and lined with fine grass. The female lays a clutch of 3-5 eggs, which are dull white in color and spotted with browns. Both sexes incubate for 12-14 days, brood the young, and feed them. Fledging takes place after about 16 days.

Remarks:
The two subspecies, *gilvus* and *swainsonii*, do not differ substantially in plumage, although *gilvus* is greener on the back and shows more yellow on the flanks. *Gilvus* averages 25% larger than *swainsonii*. Vocally, these two subspecies are again similar, but *swainsonii* has more breaks and peaks in its song, giving it a choppy character (B. McGillivray pers. comm.).
A key area for contact of these two forms is the meeting of Parkland with Boreal Forest northwest and northeast of Edmonton. Modification of these regions, caused by additional clearing and logging of aspen, will change the distribution of these two forms and may modify the dynamics of gene exchange.
Swainsonii returns to southwestern Alberta in early May, whereas *gilvus* appears about two weeks later (McGillivray unpubl.) This suggests differing wintering areas for the two subspecies and provides additional support for the genetic and morphological evidence of specific status for these two taxa. In fall, msot birds have left the province by the end of August (Pinel et al. in prep.).

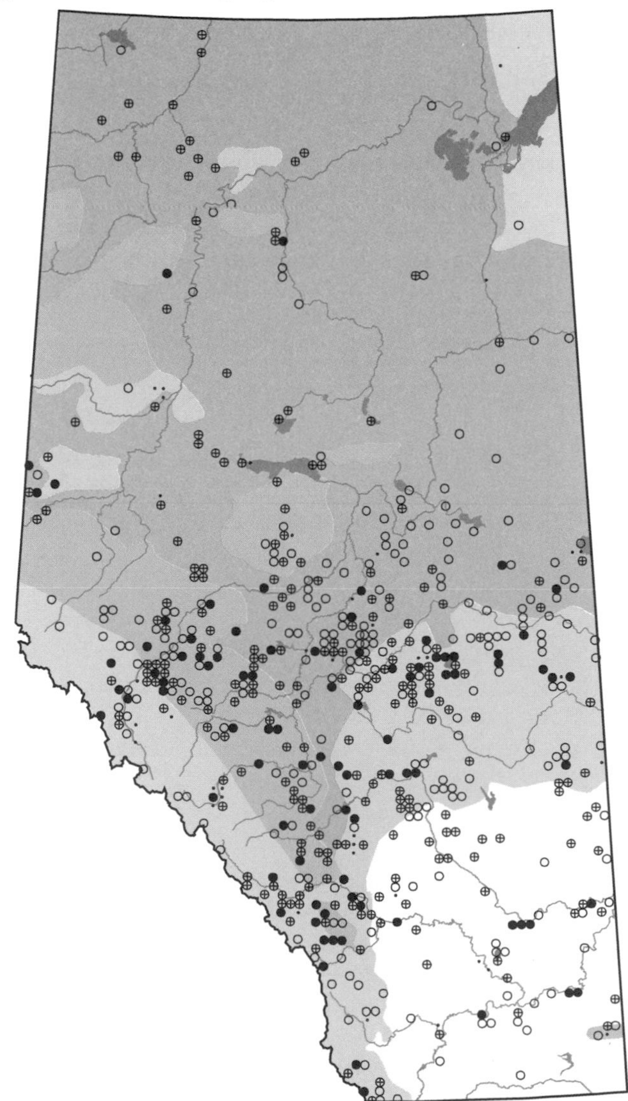

Philadelphia Vireo

Vireo philadelphicus

PHVI

Status:

The Philadelphia Vireo is relatively uncommon throughout its range, but it is a regular and sometimes quite common breeding species in the east and central areas of Alberta's northern forests (Salt 1973). This is the least known of Alberta's vireos and abundance and distributional limits are underestimated because many are assumed to be Red-eyed Vireos, from overlap and song (B. McGillivray pers. comm.).

Distribution:

Salt and Salt (1976) reported that it is found in the northern and east-central parts of Alberta south to Cold Lake and Sundre and west to Breton and Peace River. Atlas data confirm that this vireo prefers the southern Boreal region, but also indicate that breeding may occur in the far northern portion of the province, with records from the Lake Athabasca area in the east and the Hay-Zama lakes area in the west. It was recorded in 15% of surveyed squares in the Boreal Forest region, and in approximately 5% of surveyed squares in the Parkland and Foothills regions. There was only one probable breeding record for the Rocky Mountain region, although Pinel et al. (in prep.) has reported extralimital breeding in Jasper National Park and Bow Valley Provincial Park.

Habitat:

The Philadelphia Vireo prefers open mixed deciduous forests, such as the aspen poplar or birch/poplar forests of central Alberta (Salt and Salt 1976). It also inhabits second growth on burns or logged areas and willow and alder thickets bordering muskegs and streams.

Nesting:

The Philadelphia Vireo builds its nest in a deciduous tree or tall shrub at heights up to about 14 m. It is usually higher than the nests of Red-eyed Vireos, whose habitat the Philadelphia shares. The preferred nesting tree species are aspen, balsam poplar, and birch. Most nests are woven pendant cups suspended, in typical vireo fashion, from the horizontal fork of a lateral branch, sometimes close to the trunk. The female selects the site and does all nest building, but is closely accompanied by the male in these activities. The nest is constructed of bark strips, grasses, plant fibres, plant down, insect and spider webs, and lichen. It is lined with grasses, plant down, pine needles, and sometimes animal hair.

The female typically lays 3-5 white eggs, which are lightly spotted with brown. Both male and female incubate for 11 to 14 days, and both adults feed the nestlings and fledglings. Young birds leave the nest after 12-14 days. Whether this species raises one or two broods is not confirmed.

Remarks:

The Philadelphia Vireo is named for the location in which it was first collected. It is a grayish, olive-green bird with a gray crown, yellow underparts, a white streak over the eye, and a dusky streak through it. The Philadelphia's thick blunt bill distinguishes it from warblers of similar coloration. This vireo may be misidentified as the Red-eyed Vireo; Rice (1977 Ph.D. thesis U. of T.) demonstrated that the Philadelphia Vireo song is close enough to that of the Red-eyed Vireo to be interpreted by Red-eyes as conspecific song. Both species will respond to recorded tape of either song.

This species eats mainly insects, though a few fruits are taken in summer and fall. Insects are also fed to the young. Philadelphias often hang chickadee-like from branches or clusters of leaves when feeding.

Males and females migrate together, arriving in central Alberta in mid-May. They depart again with warblers in the last half of August, and are rarely seen after the beginning of September.

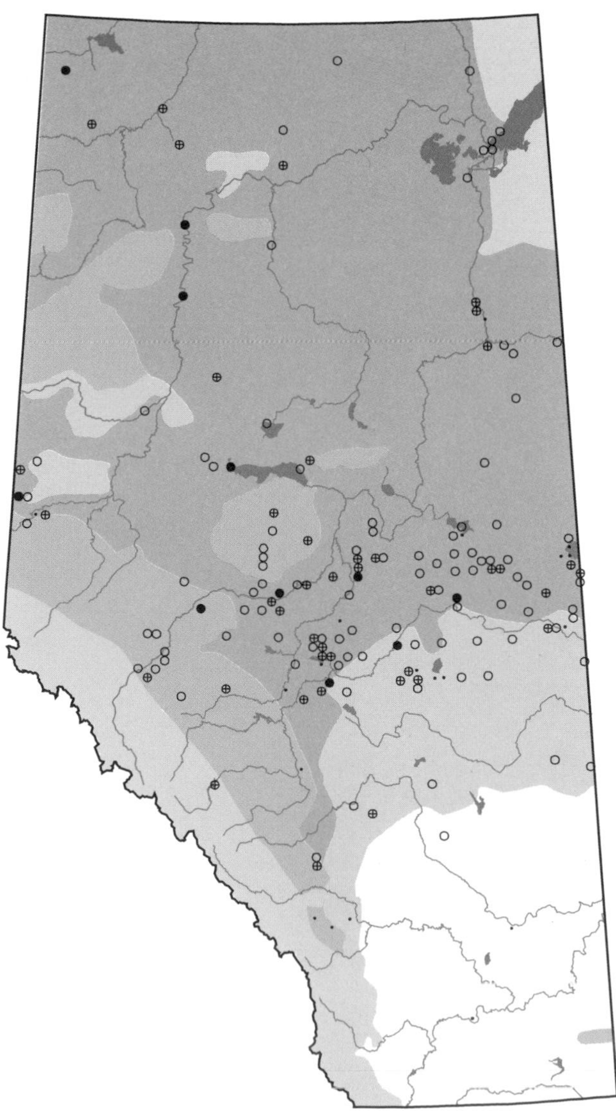

Red-eyed Vireo

Vireo olivaceus

REVI photo credit: T. Webb

Status:
This is the most common vireo found in Alberta, except in the Rocky Mountain and southern Parkland regions, where the Warbling Vireo is the most abundant.
This species' preference for mature growth deciduous woods may force distribution shifts if such habitats are impacted by burns, logging and forest clearing.

Distribution:
The Red-eyed Vireo breeds across southern Canada and into the northern United States. It can be found in every natural region in Alberta. During the Atlas Project, it was found to be most common in the Boreal Forest region (recorded in 48% of surveyed squares) and in the Parkland region (34% of surveyed squares. It was recorded in 26% of surveyed squares in the Foothills, 10% in the Rocky Mountain region, and only 5% in the Grassland region. It was also recorded in the Cypress Hills.

Habitat:
This species inhabits open deciduous woods and tall shrubbery, and less commonly mixed wood forests. It may be found in farm shelterbelts, cottonwoods in southern river valleys and coulees, the aspen groves of the parklands, and montane deciduous forest and shrubbery. It also frequents second growth in burned-over or logged areas, shrubs such as willow or alder along lakes and watercourses, in balsam poplar in northern mixed woods, and in shade trees in residential areas. A minimum 1/2 ha is needed for territory, so a single pair can occupy an isolated patch of trees (Cadman et al. 1987). The species reaches its greatest densities in continuous forest stands, where mature trees support a moderate to dense canopy and understory layers (Brewer et al. 1991).

Nesting:
The Red-eyed Vireo generally chooses a deciduous tree or shrub for a nest site, or occasionally a conifer. The site is very often at the forest edge beside roads, shorelines, or clearings. The nest is usually 1.5 m to 3 m above ground, at eye level or a little above, but some nests may be built at heights up to 17 m. The female selects the nest site and builds the nest, but is accompanied closely by the male. She weaves a pendant cup of bark strips, grass, rootlets, paper from wasp nests, moss, lichen, leaves, and plant fibres, attaching it by the rim to a horizontal branch fork. The nest is lined with grasses, rootlets, pine needles, and sometimes hair. The clutch is 3-4 dull white eggs with brown and blackish markings. Nests are commonly parasitized by cowbirds; occasionally, the vireo will build a floor over the intruding egg. The female alone incubates for 12-14 days, but both sexes feed the young. Fledging takes place after about 12 days.

Remarks:
The Red-eyed Vireo is a dull olive-green above and white below, with a gray crown, a prominent eye stripe, and red eyes. It has neither barred wings nor eye rings. It is a very persistent singer and may deliver over 20,000 songs a day (Lawrence 1953). It sings through most of a hot summer day when other birds are silent.
Insects and other small invertebrates form the bulk of this vireo's diet, and these are also fed to the nestlings. Berries may also be eaten in small amounts.
Male Red-eyed Vireos appear in the spring shortly before the females, arriving in late May. They often travel in groups of 2-3 birds and are less likely to be seen with warblers than the Solitary Vireo. Birds depart, often in family groups, in late August and early September.

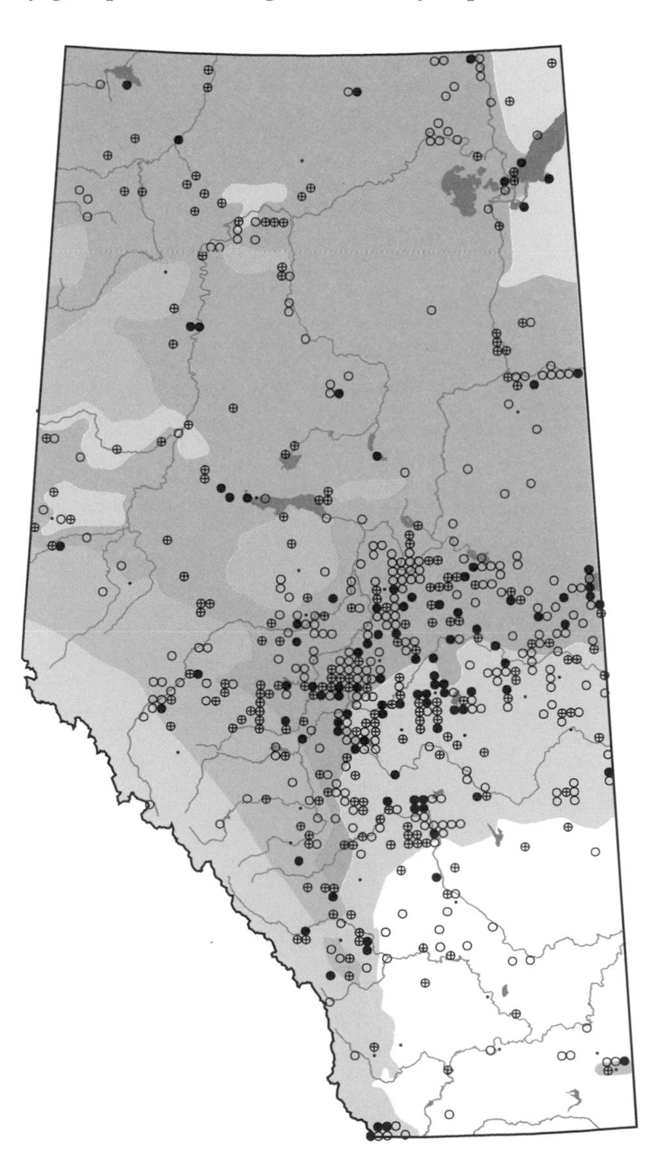

Tennessee Warbler

Vermivora peregrina

TEWA

Status:

Although the Tennessee Warbler was previously considered uncommon in the Rockies (Holroyd and Van Tighem 1983), the Atlas project recorded this species in 17% of surveyed squares in the Rocky Mountain Natural Region. The Wildlife Management Branch (1991) has not determined this species' status in Alberta. Although once considered common in the Boreal Forest and Foothills regions (Salt 1973), a large drop in captured migrants was experienced in the early 1980s (Jones 1986) in central Alberta. Atlas records were highest in these two regions.

Distribution:

There are indications of the Tennessee Warbler breeding in all regions of the province. Of the squares that were surveyed, it was most commonly recorded in the Foothills (39%) and Boreal Forest (36%) natural regions, as well as the lesser surveyed Canadian Shield (43%). However, there is only one possible breeding confirmation in the Grasslands region, and none for the Alberta portion of the Cypress Hills.

Habitat:

Although breeding habitat may be quite variable, deciduous woods of some kind are always nearby. Specifically, it may be open deciduous woods, mixed wood with a poplar dominance, burns, willow and alder thickets around muskeg, or tall aspen along woodland streams. Holroyd and Van Tighem (1983) indicated that it also inhabits spruce forests in Jasper and Banff national parks.

Nesting:

The female constructs a cup-shaped nest of grass and lines it with rootlets, hair, and fine grasses. The nest is open to sunlight and is often built in a moss hummock along the edge of a wet area or burn. Rarely, it is located in a bush above ground. The nest is usually obscured from view by overhanging vegetation. The female lays 4-7 white eggs marked with brown, which is a relatively greater number than most warbler species. Clutch sizes may be larger in response to forest insect outbreaks (Ehrlich et al. 1988). The female incubates her clutch for 11-12 days. Despite its relative abundance, in relation to other local warblers, little else is known of its breeding biology.

Remarks:

The spring male is gray on the top of his head with a whitish line above the eye and a dusky line through it. He is generally olive-green above, white below, including the undertail coverts, has no wing bars, and has a sharper and more slender bill than the vireos. The secretive female is greenish on the top of the head, has a yellowish line over the eye, and is tinged in yellow below. Young and fall adults are olive-green above and have a yellow line over the eye and yellow underparts, excluding a whitish belly and undertail coverts. The absence of yellow undertail coverts and partial eye ring distinguish it from the Orange-crowned Warbler in fall.

The male's song, sung from a tall tree or brush, is clear, strong, and distinctive. The three-phrase song may be described as "ticka-ticka-ticka, swit-swit-swit, chouchou-chouchouchou". The initial phrases are slower and more deliberate than the run-together notes of the final phrase.

While on breeding grounds, Tennessees forage along medium canopy heights. In fall migration, they will commonly appear at all levels, often bickering and chasing. Caterpillars, flies, and aphids are consumed, as are some wild raspberries.

Spring arrival peaks during the third week of May, and the majority leave central Alberta by late August. A few accompany Yellow-rumped Warbler flocks in September.

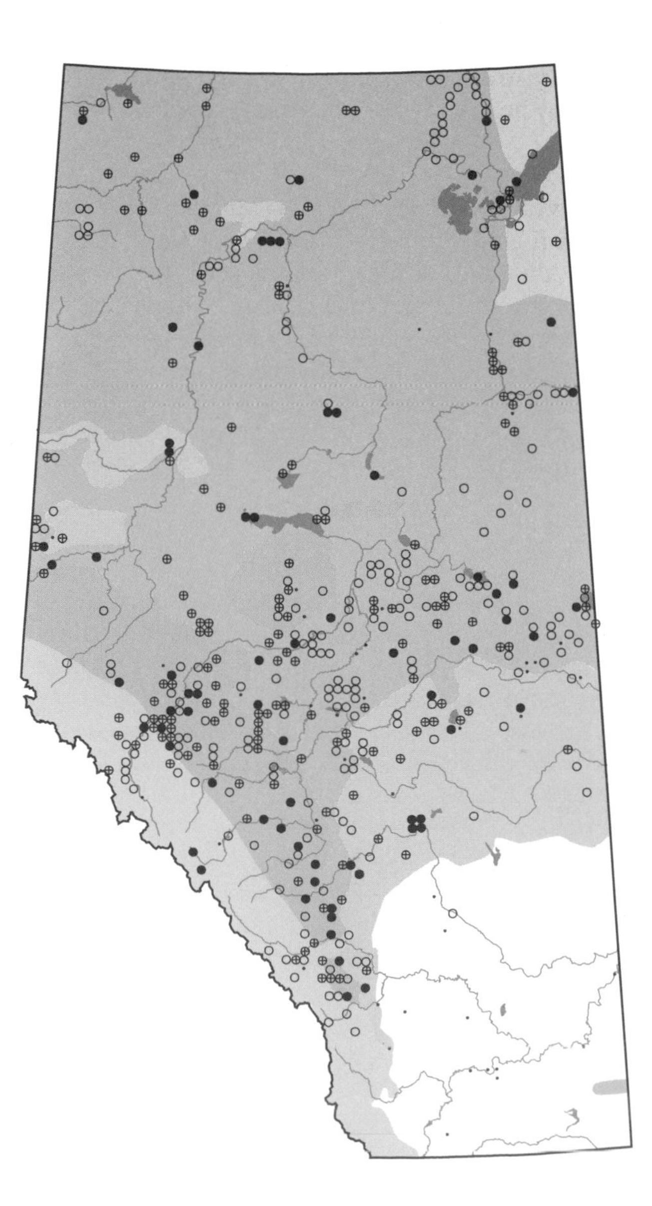

Orange-crowned Warbler

Vermivora celeta

OCWA

Status:
The Wildlife Management Branch (1991) determined the status of this species to be "not at risk", based on a secure habitat, widespread distribution and a healthy population. It is a common summer resident in Banff and Jasper national parks (Holroyd and Van Tighem 1983), as well as fairly common in the adjacent foothills (Salt 1973). Proportionately, it was most frequently recorded in these two regions (Rocky Mountain - 18.5% and Foothills - 36% of surveyed squares). A fairly common migrant in the Grassland and Parkland, few pairs remain to nest. It is more common in northern forests, though locally distributed (Salt 1973).

Distribution:
The Orange-crowned Warbler has very local distribution throughout all the natural regions (Pinel et al. in prep.). This is particularly true in the Grassland Natural Region, where it was recorded in only 1% of the squares surveyed. Atlas records confirm historical records (Salt 1976) for the Cypress Hillsand the Neutral Hills, and the Red Deer and Bow rivers. Although there was a record in Waterton Lakes National Park, breeding confirmation was not established.

Habitat:
Preferring a generally brushy habitat, the Orange-crowned Warbler breeds in the tangled deciduous thickets of remote coulees in the prairies, willow and alder thickets in the mountains, and similar shrubby habitat around the beaver ponds and woodland pools of the northern forests (Salt and Salt 1976). Frequently, these areas include open deciduous stands. Godfrey (1986) also included the second growth of clearings and burns.

Nesting:
Despite their widespread appearance, Orange-crowned Warblers do not nest in large numbers. Information relating to breeding activities is scanty. Whether or not the male assists in building the nest is unknown. It is a cup constructed of grass, shreds of bark, and moss, that is lined with fine grasses, hair, and feathers. This nest, large relative to the bird's size, is situated on the ground or, occasionally, low in a bush. It is well hidden from above by overhanging grass or leaves. The 4-6 white eggs are finely dotted around the larger end with reddish browns and, often, lilac. It is not known whether the male assists in the 12-14 day incubation period. Both adults tend to the young, who go on to fledge after 8-11 days.

Remarks:
The generally drab male has olive-green upperparts that are brightest on the rump. A dull orange crown patch is normally concealed. Pale eyelids appear as a partial eye ring and a pale yellowish line extends over the eye. Gray-yellow underparts become gray on the sides and indistinct dusky streaks line the breast. The female is similar but duller overall, with little or no crown patch.

Its call note is a short "chip", while its song consists of a monotonous trill, weaker than most, dropping in pitch in the middle and off at the end.

It primarily feeds on insects, including those in plant galls, but will also eat some berries and fruits.

An early spring migrant, this species arrives from late April to mid-May. Although some may appear in mixed warbler flocks from mid-August to mid-September, most depart with conspecifics in the last two weeks of September. Small numbers of Yellow-rumped Warblers may accompany them.

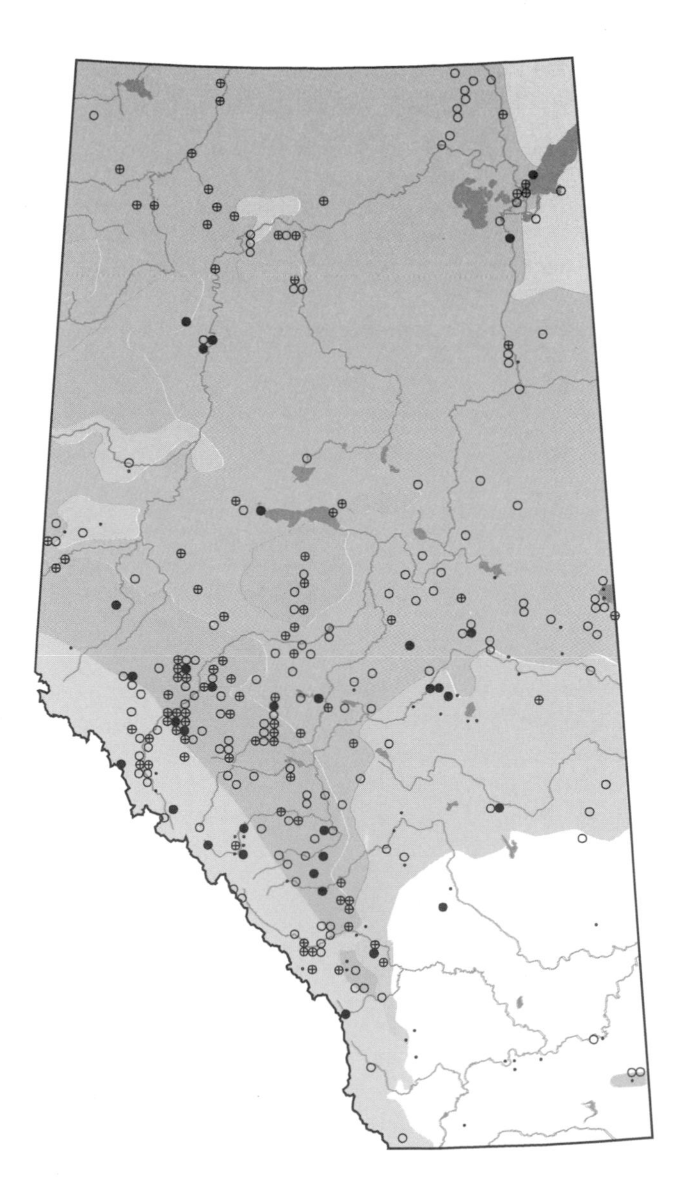

Nashville Warbler

Vermivora ruficapilla

NAWA

Status:
The Nashville Warbler is a very rare breeder in Alberta. Although spring migrants are occasionally found in Jasper and Banff national parks, particularly in the Vermillion Lakes area, there is only one account of a confirmed breeder, at Link Creek in 1980 (Lein and Wagner 1982).

Distribution:
Alberta lies between the breeding ranges of the two Nashville Warbler subspecies in Canada. To the west, *V. r. ridgwayi* ranges along the Rocky Mountains of eastern British Columbia. The westward range limit of *V. r. ruficapilla* extends into Saskatchewan (Godfrey 1986). Because Alberta is on the periphery of both ranges and because field confirmation is lacking, it is not known which, or if both, subspecies wander into the province (Pinel et al. in prep.). There are 23 historical records for the province, most of which are from Banff National Park. However, others have wandered as far as Fort Saskatchewan and the Cypress Hills (Pinel et al. in prep.). There were no breeding records reported in the five year period of the Atlas Project.

Habitat:
Based on known breeding habitat in British Columbia, the Nashville Warbler breeds in fairly open deciduous woods or the deciduous portions of mixed woodlands (Salt and Salt 1976). Godfrey (1986) described appropriate habitat as being sparse immature deciduous or mixed woods, especially aspen and birch, with low bushes on wet or dry ground. Habitat also includes the second growths of old clearings and burns and sparsely treed bogs. Observations in Banff National Park have been confined to the montane area (Holroyd and Van Tighem 1983).

Nesting:
The female constructs a cup-shaped nest of grass, leaves, and moss, and lines the inside with finer grasses. This small ground nest is well concealed by shrubs or a mossy hummock. The 4-5 eggs are white and speckled with browns. They may sometimes be unmarked (Harrison 1978). The female incubates the eggs for 11-12 days, with some assistance from her mate. The female also assumes the majority of the parental care of the nestlings. The young fledge at 11 days.

Remarks:
The male Nashville Warbler has a pale blue-gray nape, face, and top of the head, a white eye ring, a pale line from the base of the bill to the eye, olive-green remaining upperparts, no wing bars, a yellow throat, and underparts that become whitish on the abdomen. The yellow throat distinguishes it from the similar Connecticut Warbler, because the latter species has a gray throat. The female Nashville is similar to her mate, but duller. Easily confused with the Orange-crowned Warbler in fall, the Nashville Warbler can be separated by a head that has become more brownish, a white eye ring, and a yellow throat that is sharply defined from gray cheeks.

The male's song resembles that of the Tennessee Warbler, but is less robust and consists of two, not three, parts. It may be notated as "see it, see it, see it, ti-ti-ti-ti-ti". The second portion is lower in pitch and much faster.

A typically active "warbler-like" forager among birch and poplar branches and bushes, the Nashville Warbler is entirely insectivorous. It will also glean from the ground.

The few available records suggest a spring arrival during the first two weeks in May. Departure dates are not known.

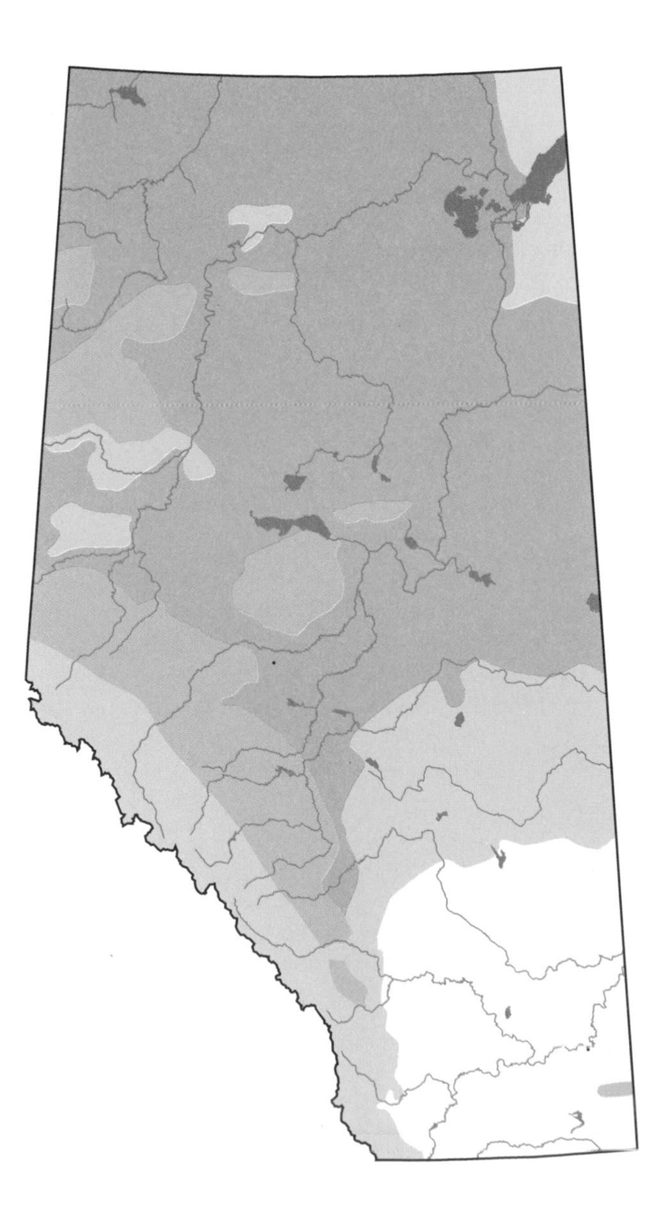

Yellow Warbler

Dendroica petechia

YEWB photo credit: T. Webb

Status:
This species is the most common yellow-colored warbler in the Parkland Natural Region. In the Grassland region, it is often the only local warbler (Salt 1973). It is an uncommon migrant and fairly common summer resident in Jasper and Banff national parks (Holroyd and Van Tighem 1983). Yellow Warbler numbers have rebounded following the reduction in the use of defoliants and insect-killing agents (Salt 1973).

Distribution:
This warbler enjoys wide distribution throughout the province. It was most frequently recorded in the Parkland (54% of surveyed squares) and the Boreal Forest (50%) natural regions. Because it adapts well to disturbance, distribution in the Boreal Forest may increase in the future.

Habitat:
The Yellow Warbler owes its wide distribution to its ability to nest in a variety of areas, usually near water. In the Parkland region, it may be found in alder or willow tangles or, occasionally, in new-growth aspen over a dense undergrowth. In heavily forested areas, nesting is confined to riparian growth along streams and woodland edges. In mountain regions, wet, deciduous shrubby meadows, shrubby lower avalanche slopes, and burns in the lower valleys are used for nesting. This is the only warbler that will use man-made habitats, to the point where it may also be found in ornamental shrubbery and caragana hedges.

Nesting:
Beginning in the last week of May, the female constructs a compact, firm nest of grass and plant fibres and lines it with hair, fine grass, and plant down. This is securely positioned between 1 m and 3 m up a deciduous bush or tree, usually in the fork of a branch. Rarely is it in a conifer. Females lay a total of 4-5 eggs, which are white speckled with brown. Incubation by the female lasts 10-11 days. The young are brooded by the female for the first few days and are tended by both parents. Fledging is at nine days. A second brood may possibly be raised (Salt 1973). Very commonly parasitized by the cowbird, the female will respond most often by building a new nest over the egg(s). However, she may choose to desert the nest or continue to lay and incubate the clutch, the latter inevitably resulting in the high or total mortality of her own chicks. Nesting in the proximity of Red-winged Blackbirds, which do not tolerate intrusion by cowbirds, may reduce parasitism (Bent 1963d).

Remarks:
Misidentified as a "wild canary" by some, the male is generally a brilliant yellow with olive-green wings and reddish breast streaks. Surprisingly well camouflaged, he is more easily seen soon after arrival when trees have yet to leaf out. The female is similarly colored, being more green above, drabber below and, rarely, with faint breast streaking. The Yellow Warbler can only be mistaken for the Wilson's Warbler, whose male has a black cap. Both sexes of Wilson's lack yellow on the tail.
The male Yellow Warbler uses two song types. A clear and distinctive breeding song, "see see see titi see", is used in May and June with little variation. Later, a variable territorial song is sporadically used, easily confused with the songs of neighboring species, including the American Redstart and the Chestnut-sided Warbler.
Insectivorous, this species eats mostly caterpillars, cankerworms, beetles, plant lice, and aphids, but will also feed upon some wild raspberry.
Yellow Warblers arrive in numbers in mid-May. Their southern migration is completed by early September, largely in mixed-warbler flocks.

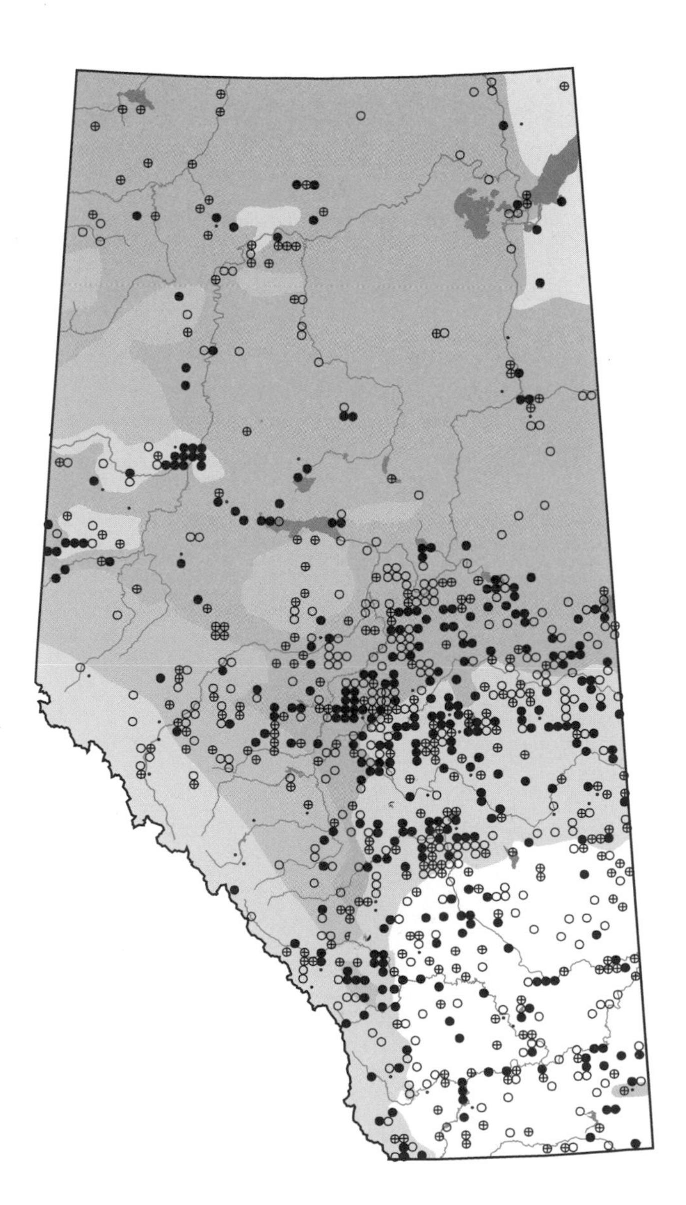

Chestnut-sided Warbler

Dendroica pensylvanica

CSWA photo credit: R. E. Gehlert

Status:
Once considered rare in North America, the Chestnut-sided Warbler benefitted from the clearing of native woodlands by European settlers (Bent 1963d). As forested lands and burns regenerated, and as marginal agricultural lands were abandoned, this warbler became abundant in areas where it was once rare. In Alberta, historical records before 1972 are very scarce. There were 32 reports between 1972 and 1980 (Pinel et al. in prep.), while Atlas records number only 12, one of which is confirmed as breeding.

Distribution:
Both historical and Atlas distributions have been largely confined to the Boreal Forest Natural Region. Of the squares surveyed there, only 1% included records for this species. There were single migrant records for the Parkland and Grassland natural regions. The general lack of familiarity by local birders with this species, and its similarity to the Yellow Warbler in song and, perhaps, in habitat, may serve to underestimate its local distribution.

Habitat:
A bird of second-growth deciduous bushes and thickets in eastern Canada, local habitat usage has been based on only two accounts. In each case, fairly open deciduous woodland was used, one with a loose understory of dogwood, cranberry, alder, and willow (Salt 1973), the other with a dense undergrowth of beaked hazelnut (Pinel et al. in prep.).

Nesting:
The nest is positioned less than a metre up a tree or bush. A loosely constructed cup-shaped nest made of grass and plant fibres, and lined with grass and hair, is built by the female. The 3-5 creamy white eggs are marked with brown and are incubated by the female for 12-13 days. Both adults feed the young. Normally a close sitter, the female, once flushed, may drop to the ground and slowly move away, using a distraction display. Similarly threatened, the male may respond by spreading his tail, vibrating his wings, and flitting about nearby branches. Young fledge after 10-12 days. Known to be a frequent cowbird host in other parts of its range, this species may respond by building a new nest floor over the parasitized clutch.

Remarks:
This active but not particularly wary warbler carries a distinctive plumage. The male has a bright yellow crown bordered in black, a broad black line down the sides of the throat, white around the ear sides of the neck, two whitish wing bars, white underparts, and broad chestnut patches extending from each side of the throat to the flanks. The similarly patterned female has a duller crown and reduced black and chestnut coloration. Fall birds are greenish above and white below, and have white eye rings and yellowish wingbars. Males sometimes show chestnut on their sides.

The male uses variations of two songs. The breeding song, sung frequently earlier in the season, is often mistaken for the clear and robust breeding song of the Yellow Warbler. Notated as "very, very pleased to meetcha", it is sung very quickly and the last phrase is emphasized. A song from a streamside thicket is most likely from a Yellow Warbler, while a song emanating from a shrubby hillside is most likely a Chestnut-sided Warbler (Rising 1973). A variable territorial song is used less regularly later in the season.

Almost entirely insectivorous, these warblers will glean, hawk, hover, and, occasionally, flycatch for their meals. When insects become scarce, some seeds and berries are consumed.

Little is known of migration dates. The earliest records have been for late May. The fall migration period is unknown.

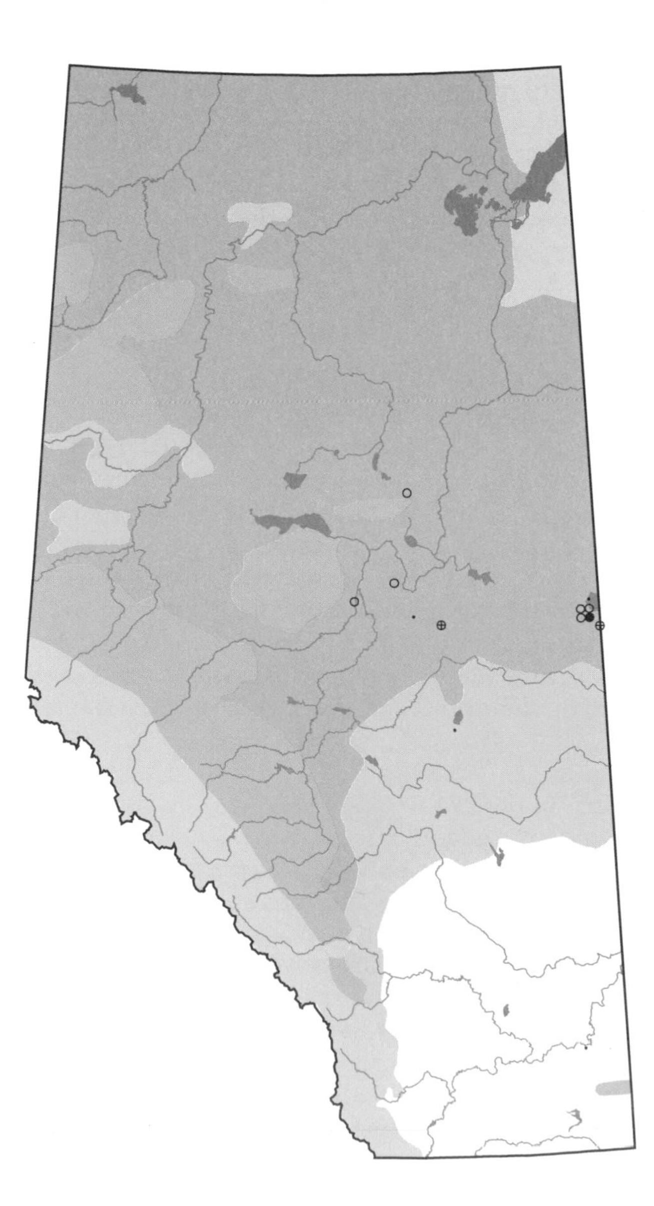

Magnolia Warbler

Dendroica magnolia

MNWA photo credit: T. Webb

Status:
The provincial status of the Magnolia Warbler has not been satisfactorily determined. It has historically been considered uncommon (Salt 1973). Atlas records concur with the findings of Holroyd and Van Tighem (1983), that it is a rare summer resident of Jasper National Park and that it is an uncommon transient in the Grassland region.

Distribution:
During the Atlas Project, this species was recorded most frequently in the Boreal Forest and Canadian Shield natural regions, in 14.6% and 12.5% of surveyed squares, respectively. Although historical distribution (Salt and Salt 1976) was considered to extend from northern Alberta south to the North Saskatchewan River and Banff, Atlas records include the Parkland region south to Red Deer and to Sundre in the Foothills region. No breeding records were found in the Rocky Mountain region south of the Athabasca River valley.

Habitat:
The Magnolia Warbler generally breeds in open coniferous or mixed wood stands, often near water. It is most frequently found among the low or young coniferous saplings and willow of woodland borders and openings. In its few mountain locations, it frequents areas of stunted alder and water birch.

Nesting:
In an area of deciduous bushes or, more often, small evergreens, both sexes construct a fragile cup in a tree crotch or cradled amongst branchlets. The nest is made of small twigs and coarse grass and is lined with fine black rootlets and hair. Although often below 2 m, the nest is well hidden from sight. The 3-5 white eggs are marked with brown and grays, especially at their larger end. The female incubates the eggs for 11-13 days. While both adults attend the young, only the female broods them for the first few days. Both adults will use a distraction display around the nest, in addition to a squeaky call given by the female (Bent 1963d). Once the young have fledged (8-10 days), the family unit abandons its territory and becomes noisy and visible for a short time. A second brood has not been recorded in Alberta. This species is an uncommon cowbird host.

Remarks:
The male has a gray crown, a white brow stripe extending from the eye, a black mask and back, yellow throat, underparts, and rump, and a broad white, incomplete mid-tail band. The female (pictured on this page) is patterned similarly but is duller, with an olive back, duller wing bars and eye-stripe and reduced streaking. Young resemble the female, but have a brownish-olive head, neck, and back and dingier white areas. In fall, adults are duller overall, with the eye-stripe, mask, streaking, and wing bars reduced or absent.
The male's song varies over the length of the breeding season, becoming weaker and more hesitant later (Cadman et al. 1987). A common song, "wisha wisha wisha witsy", rises along its course. Call notes include "chip" and a distinctive, metallic, and lispy "tlep".
Although the male tends to forage higher than his mate, both adults feed on beetles, flies, worms, and spiders, often from the undersurface of the leaves.
Migration is conducted within mixed-warbler flocks arriving in the spring after mid-May and departing during the last week of August and the first week of September.

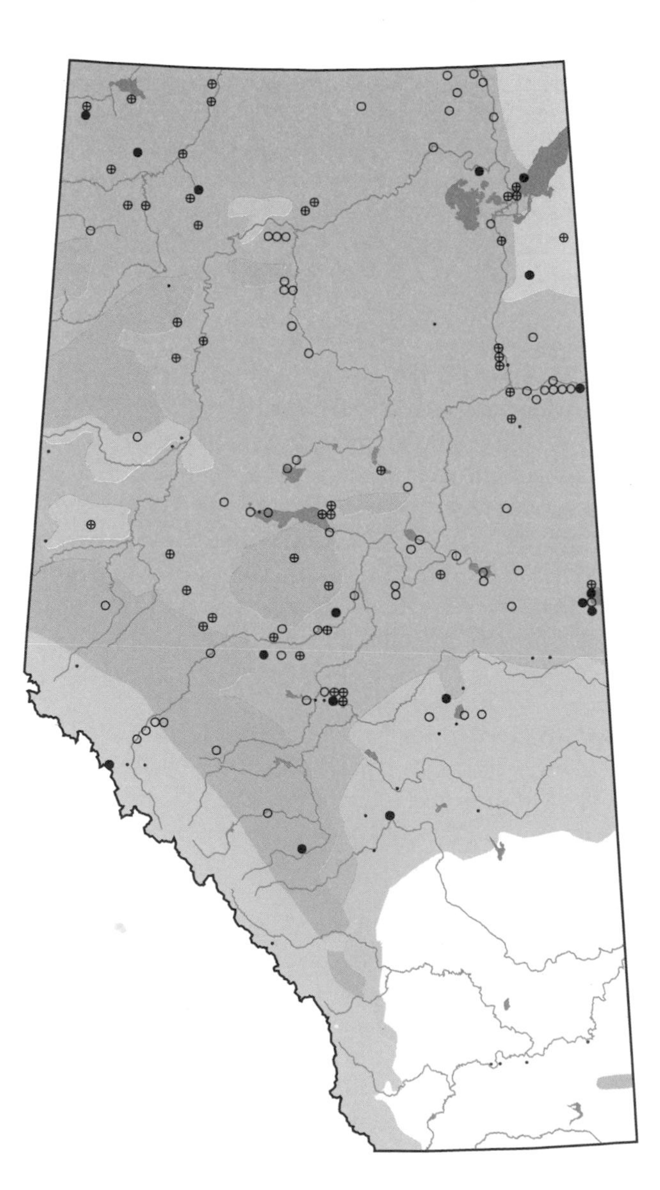

Cape May Warbler

Dendroica tigrina

CMWA

Status:
Atlas records corroborate historical reports, which consider the Cape May Warbler to be an uncommon breeder in northern Alberta (Salt and Salt 1976), an uncommon migrant in the Parkland (Salt 1973), a rare transient in southern Alberta (Salt and Salt 1976), and a very rare visitor to Banff and Jasper national parks (Holroyd and Van Tighem 1983). During the Atlas project, only 2% of the surveyed squares in the Boreal Forest Natural Region contained records. However, this species may be underrepresented, due to effective nest concealment, sparse distribution, limited coverage of the region, and the male's weak song. The Wildlife Management Branch (1991) classified this species as vulnerable and at risk, though the threat may not be immediate and little information is available to accurately determine its status.

Distribution:
Historically (Salt and Salt 1976), this species bred locally in northern Alberta south to Boyle, Lac La Nonne, Glenevis, and Sturgeon Lake. Atlas records extend this range, adding Hinton as the southwest limit, Fort Assiniboine in the south-central area, and Cold Lake to the southeast. However, there are no records for the Canadian Shield or the Parkland region of the Peace River district. There are a few extralimital historical records for Caroline, Priddis, Jasper, Banff, Longview, and Valleyview (Pinel et al. in prep.).

Habitat:
The Cape May Warbler breeds in the dense mature white spruce stands of coniferous and mixed wood forests. It prefers open stands and edges. Some of the spruce must rise above the forest, to be used as singing posts by the male.

Nesting:
Like many other tree-nesting wood warblers of the Boreal Forest, little information has been reported concerning the breeding biology of this species. The monogamous male is very aggressive in defence of his territory. A bulky nest made of moss and grass and lined with grass, feathers, and hair, is constructed within the crown of a tall conifer. Well-concealed from below, the nest is normally built within 2 m of the top of a tree. The female lays from 5-9 white eggs, which are blotched with browns. This is the largest clutch of all wood warblers. The female enters the tree from a distance below the nest and works her way up the trunk. In departing, she dives downward towards the ground. Incubation period, incubator, fledging period, and number of broods have not been reported.

Remarks:
The adult male is black on top of the head with a like-colored stripe through the eye, has chestnut cheeks and line over the eye, yellow on the sides of the neck, olive-green upperparts streaked with black, a yellow rump, largely white wing coverts, and a blackish tail. The individual pictured on this page has begun his late summer moult. Duller overall, the female has very pale yellow or whitish underparts with dusky streaks, an olive ear patch and upperparts streaked with a dusky color, and two distinct white wing bars.
Males sing a weak sibilant song consisting of four or more high-pitched "see" or "seet" notes. Sung mostly during the first few weeks after their arrival, this song resembles the song of the Bay-breasted Warbler.
The Cape May will glean foliage for caterpillars and insects at lower levels and flycatch at upper levels.
This species is not commonly observed on spring migration, arriving from the east in the third week of May in mixed warbler flocks. The majority depart the province in late August or early in September in mixed-species flocks.

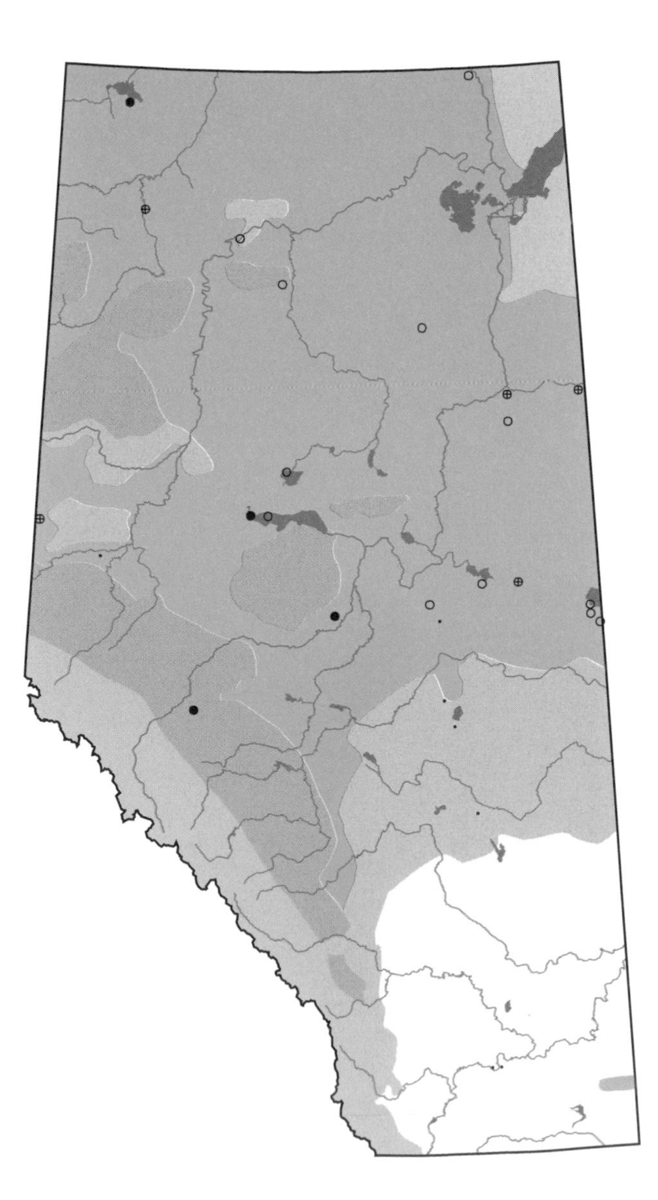

Yellow-rumped Warbler

Dendroica coronata

YRWA photo credit: T. Webb

Status:
Perhaps the most common wood warbler in Canada, the Yellow-rumped is widespread in Alberta and, in some areas, ubiquitous. It is a common migrant in both the Grassland and Parkland regions. With a total of 911 breeding records during the Atlas Project, status apparently remains unchanged.

Distribution:
Historically (Salt 1973), the eastern subspecies, the Myrtle, has bred in northern and central Alberta, the Rocky Mountains, and adjacent foothills south to Banff and Bragg Creek. The western equivalent, the Audubon's, bred in the Cypress Hills, where Myrtles appeared only as migrants, and the southwestern mountains and foothills north to Jasper. Within the range of overlap between Jasper and Banff, the two subspecies commonly interbreed. Although subspecies of the Yellow-rumped Warbler were not always recorded during the Atlas Project, Atlas distribution data concur with past findings. As in the past, small numbers were found to be breeding in the appropriate remaining habitat of the wooded river valleys of the northern margin between the Parkland and the Boreal Forest regions.

Habitat:
The Yellow-rumped Warbler breeds in coniferous woods of all kinds. It prefers open mature woods that have a few deciduous trees intermixed or a small stand nearby. It will also nest in evergreen-dominated mixed wood stands. In areas of muskeg, it nests in stands of black spruce and birch.

Nesting:
The female constructs a rather loose and bulky nest of twigs, grass, strips of bark, and moss. The interior is lined with hair, plant fibres, and, characteristically, feathers. The nest is usually positioned on the branch of an evergreen, usually from 3 m to 5 m above the ground. Nests are rarely located on the ground. The 4-5 eggs are spotted and blotched with browns and blue-gray. Incubation by the female lasts 12-13 days and, while both adults tend the young, only the female broods them. They are easily disturbed around the nest and will readily perform a distraction display. Young fledge after 12-14 days. Within this species' North American range, two broods are often raised. The Audubon's subspecies is an uncommon cowbird host, but the Myrtle is commonly parasitized. It will counteract with the same alternatives as the Yellow Warbler. Whatever tact is chosen, a large percentage of adults raise Brown-headed Cowbirds.

Remarks:
The male Yellow-rumped has a yellow cap, flanks, and rump. It is blue-gray above with black streaks on the back, breast, and sides, white underparts and wing bars, and a white stripe over the eye. A white throat denotes a Myrtle Warbler; a yellow, an Audubon's. Females are similar, but duller.

The male's distinctive song, a quick repetition of 6-8 echoing "che" notes, may either rise or fall at the end. A "chip" call note is also distinctive.

This species will hawk, hover, or glean in searching out beetles, plant lice, and flies. It switches to mainly berries and seeds when early snowfalls arrive.

Because Audubons migrate into Alberta along the mountains and coast west of the province, few are seen locally as transients, especially in the fall. Usually the first warblers to arrive locally, some males begin arriving in flocks in late April and early May. The remaining males and the females arrive in mixed flocks by mid-May. Fall migration begins in mixed warbler flocks in mid-August, and the majority have departed by mid-September.

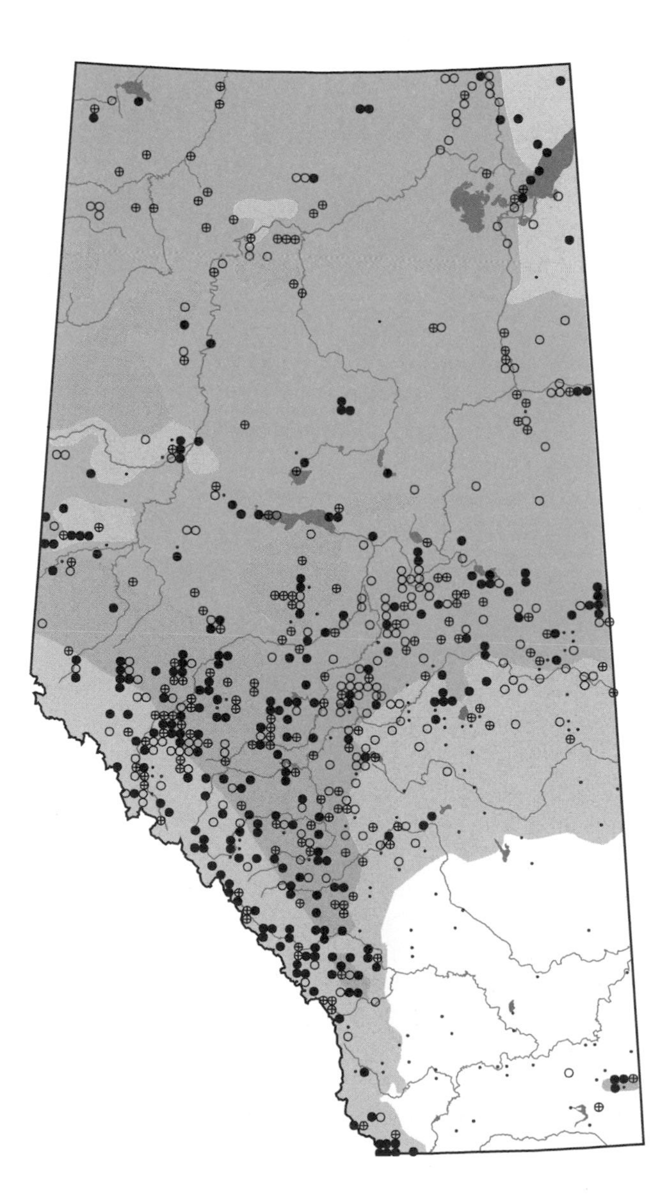

Townsend's Warbler

Dendroica townsendi

TOWA

Status:

This species is generally an uncommon summer resident of the Rocky Mountains of southwestern Alberta. Holroyd and Van Tighem (1983) listed this species as very common in the valleys of the main ranges. It is fairly common to common in the Banff and Kananaskis area, but uncommon north and south of these two park areas (Salt 1973; Salt and Salt 1976).

Distribution:

Fifty-four breeding records were gathered in the Rocky Mountain Natural Region, from Waterton Lakes National Park in the south to the Athabasca River Valley in Jasper National Park to the north. Eleven of these were confirmed breeders. This included one in the latter area on the margin bordering with the Foothills Natural Region. There is a single historical breeding record for the Neutral Hills (Pinel et al. in prep.). Although Godfrey (1986) stated that this species rarely wandered east of the Rockies, Pinel et al. (in prep.) counter that it is a regular fall migrant in Calgary. It has also bred farther east, in the Saskatchewan portion of the Cypress Hills (Cullop 1979).

Habitat:

Townsend's Warblers breed in dense stands of mature coniferous forests. Common characteristics of this habitat are nearby water and a thick canopy. The degree of undergrowth does not appear to be important. In Alberta, breeding is conducted in the subalpine coniferous forests of mountain slopes and valleys (Salt 1973). It most frequently appears in stands of white spruce or Engelmann spruce and subalpine fir (Holroyd and Van Tighem 1983).

Nesting:

Many aspects of the Townsend's Warbler's breeding biology are unknown or require confirmation. This may be a result of the remoteness of its habitat and its life history among tall conifers. Breeding activity begins in late June with the territory acquisition, singing, and mating. The nest is a rather large mass of shredded bark and moss. The relatively shallow interior is lined with moss and hair. Positioned far out on a horizontal limb, the nest is usually 2-5 m above ground, but has been recorded as high as 30 m. The 3-5 eggs are white and speckled with brown. Incubation is thought to be 12 days. The female is a rather close sitter and, when flushed, will drop to the ground and disappear.

Remarks:

The male is detailed with a black crown, cheeks, and eye stripes, a broad yellow line above the eyes and under the cheek, a green back streaked with black, a black tail with a lot of white on the outer feathers, a black throat forming streaks on the upper breast and sides, a yellow breast, and a white abdomen, undertail coverts, and wing bars. Similar to the Black-throated Green Warbler, it is distinguished by a black crown and cheeks. The female is similar to the male, with a more olive crown and cheeks and a yellow throat. She is pictured above.

The male's breeding song, although quite variable, has two main renditions: "zwee, zwee, zwee, zwee, zweezit", with the last two notes higher, and "zweesy, zweesy, zweesy, zeezee". The songs are similar to that of the Black-throated Green, but are more wheezy and lispy.

Predominantly insectivorous, this species has been observed eating weevils, beetles, caterpillars, spiders, plant galls, and some seed.

Most Townsend's Warblers arrive in the Banff area from the west and southwest by the last week of May. They travel in mixed species flocks, including chickadees and nuthatches. Peaking in mid-August, fall migration is complete by early September.

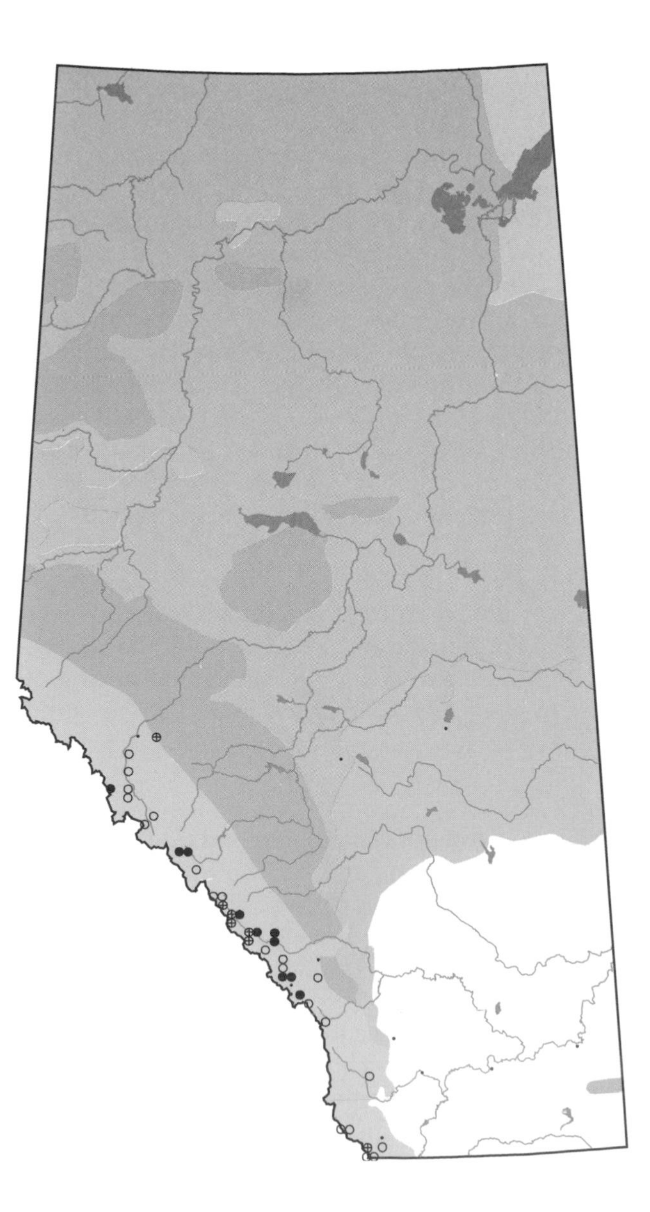

Black-throated Green Warbler

Dendroica virens

BGNW

Status:
Atlas records would seem to confirm historical reports (Salt 1973; Salt and Salt 1976; Pinel et al. in prep.) that the Black-throated Green Warbler is sparsely and locally distributed throughout its breeding range in Alberta. It is uncommon even in apparently appropriate habitat, including only 6% of the squares surveyed in the Boreal Forest region. It is a scarce transient south of its breeding range.

Distribution:
The historical breeding range of this species is relatively unchanged. It breeds in the Boreal Forest region from Wood Buffalo National Park in the north, up the Peace River to the British Columbia border, south to the Athabasca River and east to Cold Lake. Salt (1973) stated that it formerly bred in the Glenevis, Edmonton, and Battle Lake areas, until its habitat was removed. Of the 65 records from the Atlas Project, eight were in the Foothills Natural Region. Pinel et al. (in prep.) also states that there was increased activity in the latter region in the 1970s. This may constitute a local expansion of its range.

Habitat:
Breeding habitat of the Black-throated Green Warbler in Alberta has been reported as mature coniferous or mixed wood forest with large stands of white spruce (Salt and Salt 1976). However, Francis and Lumbis (1979) found two territories in riparian balsam poplar/aspen poplar, with a few scattered tall white spruce.

Nesting:
The nest is built on the horizontal limb of an evergreen, and positioned from 6-11 m up a small or large tree. A difficult nest to find, it may also be near the top of a tree. The female builds a deep, compact nest of fine twigs and shredded bark, and lines it with fine grasses and rootlets. She lays 4-5 creamy white eggs, which are spotted and blotched with browns. Incubation occurs over 12-13 days, and it is unclear whether the male assists in this duty (Ehrlich et al. 1988). The female tends to the nestlings, with some assistance from the male. Fledging after 8-10 days, the young continue to be cared for beyond this time. Although generally considered an uncommon cowbird host, one series of reports revealed a parasitism rate of 34% (Peck and James 1983).

Remarks:
The contrast-colored male has yellowish upperparts, a dusky line through the eye, black wings, throat, upper breast, and sides, and white wing bars and remaining underparts. The Black-throated Green shares a similar song, habitat, and, in part, Foothills range, with the Townsend's Warbler, but may be distinguished from that species by its green crown, mainly yellow face without any black patch, and mostly white and black underparts. Resembling her mate, the female is duller overall, with a yellow throat and reduced black on the sides and breast. She is pictured above.
Frequent singers long into the summer, the males use two songs, one at the territory boundary and one around the female at the nest (Ehrlich et al. 1988). The primary song is a distinctive, unhurried and drowsy "zee zee zee zoo zee". Salt and Salt (1976) described the other song as a "series of hurried sibilant notes, `seeseeseeseeseeseeseesee - sweesee'."
Probably feeding entirely on insects during nesting, this species hawks, hover gleans, and gleans bark to obtain food. Like many warblers, the male tends to forage higher up than the female.
Arriving in mixed warbler flocks from the east in the last two weeks of May, they forage on insects attracted to the catkins of willows and poplars. More commonly seen as a migrant in the fall, particularly in the southern third of the province, departure through southern Alberta occurs between mid-August and mid-September (Pinel et al. in prep.).

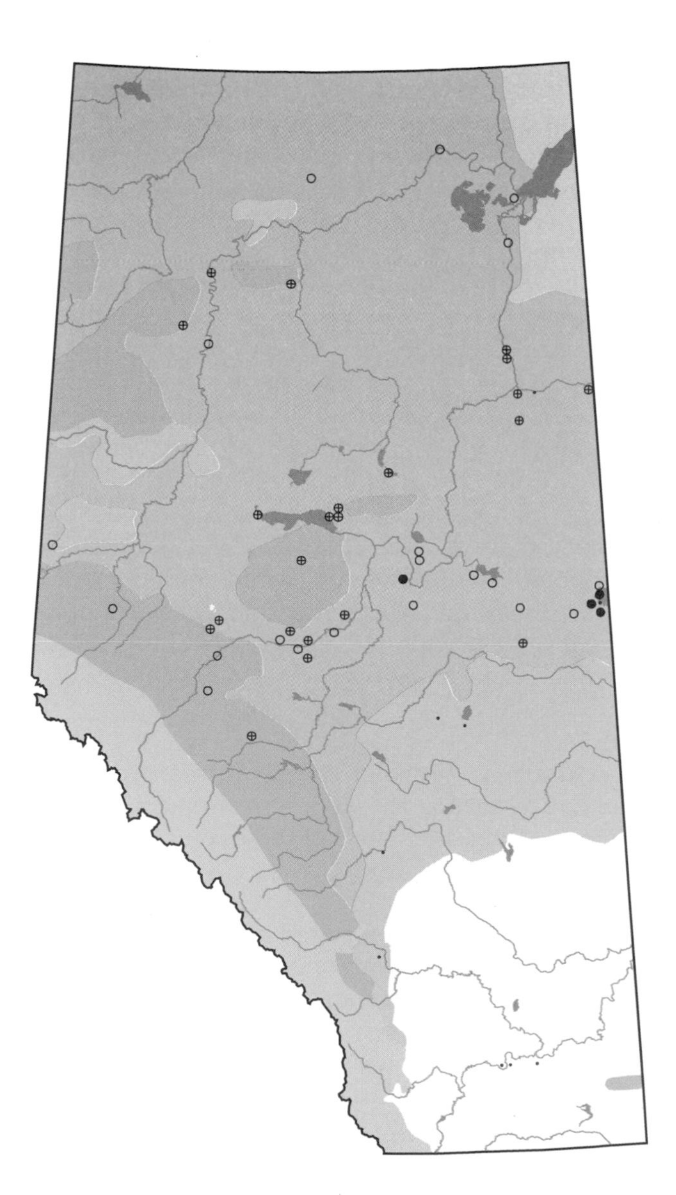

Blackburnian Warbler

Dendroica fusca

BLWA photo credit: R. Gehlert

Status:
Sporadic appearances in the province mark the western limit of the North American range of the Blackburnian Warbler. Historical (Salt 1973; Salt and Salt 1976; Pinel et al. in prep.) and Atlas records are scattered and scarce. Only 1.6% of the squares surveyed in the Boreal Forest Natural Region furnished any level of breeding. The Wildlife Management Branch (1991) considered this species to be at risk and vulnerable to disturbance, although limited local information exists to allow an accurate determination. Intolerant of disturbance to the forest structure (Cadman et al. 1987), the Blackburnian Warbler may be adversely affected by further clearing in the Boreal Forest region.

Distribution:
Historically, the Blackburnian Warbler has bred sporadically in east-central Alberta where appropriate habitat was available. This included the Grande Centre, Skeleton Lake, High River, Glenevis, and Edmonton areas, disappearing from the last two areas as habitat was cleared (Salt 1973). Pinel et al. (in prep.) suggests an area bounded by Bonnyville, Cold Lake, Lac La Biche, and Boyle. However, this report also notes two unusual reports from the Gregoire Lake and Banff areas. The Atlas Project had a single record in the Foothills Natural Region, while, in the Boreal Forest region, distribution was from the Cold Lake area northwest through Slave Lake and up to an area near the confluence of the Peace and Chinchaga rivers.

Habitat:
The Blackburnian Warbler breeds in mature balsam fir and/or white spruce forests or in mixed wood forests incorporating extensive stands of these conifers.

Nesting:
Not a great deal has been noted regarding the breeding biology of the Blackburnian Warbler. The female constructs a cup-shaped nest of twigs and lichens in about four days. She lines the nest with grass, fine rootlets, and hair, and most often positions it on a horizontal branch of a conifer. Occasionally, the nest will be placed in a fork or crotch. Nest height ranges from 2-26 m above ground. The female conducts the incubation for 11-13 days. She is not easily flushed from the nest at this time. Both adults tend to the nestlings, but only the female broods the clutch. There is no information on time to fledging. Although reported by others as an uncommon cowbird host (Ehrlich et al. 1988), an Ontario study reported one-third of nests were parasitized (Cadman et al. 1987).

Remarks:
The contrasting colors of the male Blackburnian Warbler are distinctive. The female pictured here, is similar to the male, only with yellow replacing orange, an olive cheek patch, and two white wing bars. Although ranges are not expected to overlap, females and juveniles could be confused with female Townsend's Warblers.
The male's weak song, notated as: "zip, zip, zip, zip, ze e e eeeee", does not carry well. Godfrey (1986) described a second song as a rolling, high-pitched "chickety chickety chickety chick".
Primarily insectivorous, Blackburnians hover glean, bark glean, and hawk most of their food among tree tops. Some berries will also be eaten in fall. The male tends to forage higher up than his mate.
Few observations have been made during migration locally. These birds arrive in central Alberta from the east after the middle of May. Observations during the fall migration period have been very scanty. There have been two reports from Calgary in late August (Pinel et al. in prep.).

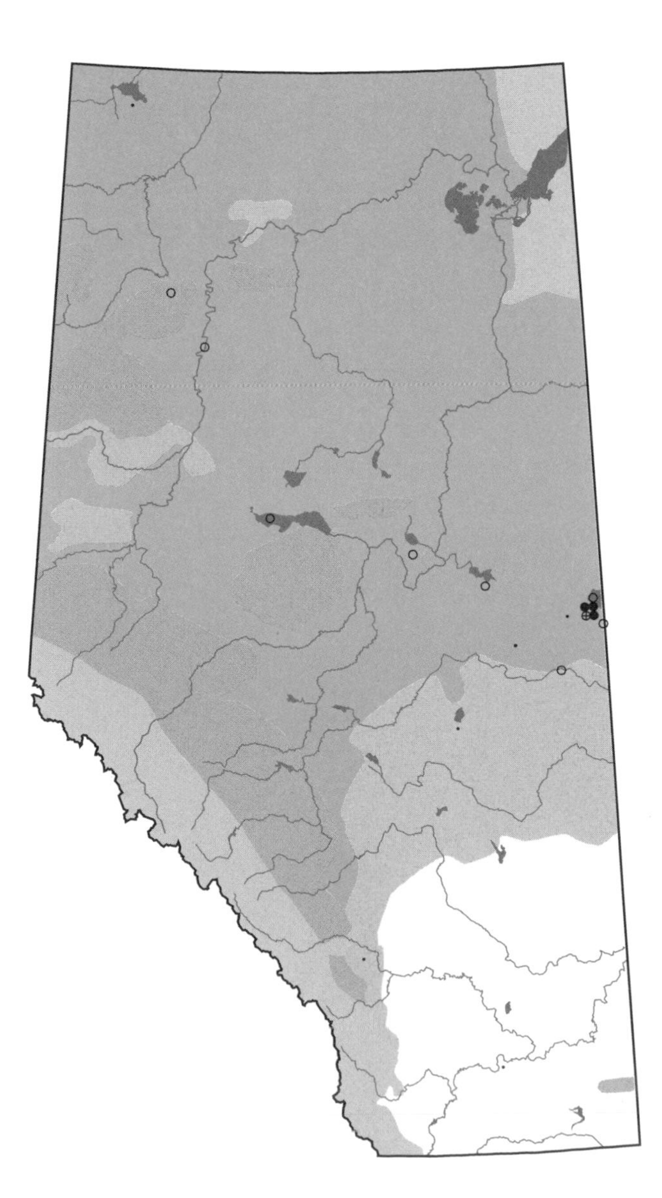

Palm Warbler

Dendroica palmarum

PAWA photo credit: R. Gehlert

Status:
The status of the Palm Warbler is undetermined, because insufficient local information has been collected to determine an accurate status. However, the drainage of bogs will reduce suitable nesting locations.

Distribution:
Breeding in northern and central Alberta, the Palm Warbler's generally accepted range has been south to Grande Prairie, Edson, and Edmonton, and east to Cold Lake. Salt and Salt (1976) considered Rocky Mountain House to be another possible locality. In the 1970s, breeding season records were obtained for Alder Flats and for Sunwapta Falls in Jasper National Park (Pinel et al. in prep.). Atlas records for Vimy and Ranfurly deviated slightly from the previously accepted range. Although this species was, perhaps, a regular fall migrant in the south (Pinel et al. in prep.), Atlas observations support the statement by Salt and Salt (1976), that it is an uncommon transient in the region.

Habitat:
The Palm Warbler breeds in muskeg areas hosting sparse stands of black spruce, tamarack, birch, alder, willow, and cranberry. These receding muskeg areas may be dry or wet and include woodland bogs. A treed margin is characteristic and some shrubbery may be present. To the east, in Saskatchewan, upland coniferous second growth is used (A. Smith pers. comm.).

Nesting:
The cup-shaped nest, built by the female, is positioned on or near the ground. Often this is on a dry mossy hummock, occasionally shaded by a low shrub or the lowest branch of a conifer seedling. Less commonly, it may be in a low conifer sapling. The nest is constructed of grass, moss, weed stalks, and shreds of bark, and is lined with fine grass, feathers, and rootlets. The typical clutch is 4-5 white eggs, speckled with brown. The nest may be parasitized by the Brown-headed Cowbird. However, Palm Warblers may bury the intruding eggs in the bottom of the nest under their own. It is not known if both sexes contribute to the 12-day incubation period, although the female is known to be a close sitter. Both adults tend to the chicks, even occasionally feeding young that are able to fly and forage on their own. Although the nestlings leave the nest at 12 days, they are unable to fly until several days later. Cover is taken under herbage.

Remarks:
The similarly colored adults have chestnut crowns, olive-brown upperparts, yellowish-olive rumps, a stripe over the eye, bright yellow throats and undertail coverts, white abdomens and buffy wing bars. A habit of flipping the tail while at rest is characteristic.
The male's song, sung from the top of a nearby black spruce, pine, or tamarack, may be confused with the Chipping Sparrow, Dark-eyed Junco or Swamp Sparrow. His sporadic trill-like song is a simple note rapidly and monotonously repeated, "see, see, see, see, see", 5-7 times. Distinct swells may be noticeable between notes.
The male spends most of his time feeding on the ground or in low shrubs. This species also forages by hover gleaning. Almost wholly insectivorous, Palm Warblers mainly consume beetles, ants, spiders, caterpillars, and grasshoppers. They have also been recorded eating berries (Bent 1963d).
These warblers arrive in central Alberta around the first week of May. Fall migration takes place during the last two weeks of September. Late migrants will appear until mid-October, before leaving south to winter in the southeastern United States and the West Indies. Rarely seen around palm trees at any time of the year, their name is a misnomer.

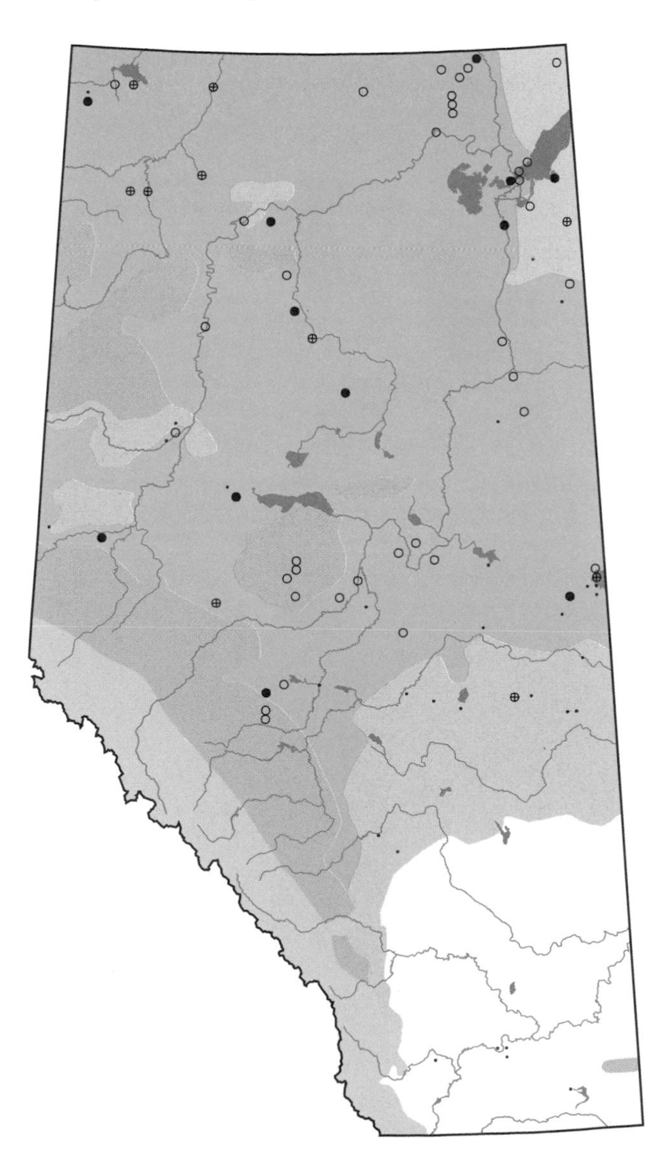

Bay-breasted Warbler

Dendroica castanea

BBWA

Status:

Atlas data would appear to confirm Salt's (1973) contention that the Bay-breasted Warbler is uncommon in Alberta. It is a very rare visitor to Jasper and Banff national parks (Holroyd and Van Tighem 1983). Although this status has not been clearly defined, viable population levels may be at risk as further clearing activities remove dense spruce stands. The cyclic population levels experienced in Ontario (Cadman et al. 1987) and in Saskatchewan (A. Smith pers. comm.), corresponding to spruce budworm outbreaks have not been studied in Alberta.

Distribution:

This species is known to breed in the Boreal Forest Natural Region, although breeding was not recorded in northwestern Alberta. Only 1.5% of surveyed squares in this region contained records of this species. It has only been recorded as a rare migrant in the Grassland and Parkland regions.

Habitat:

Extensive stands of spruce provide preferred breeding habitat, while a mixed stand of spruce, pine, and tamarack may also be used. Secondarily, only mixed wood stands with a heavy predominance of coniferous trees will be selected for nesting. Godfrey (1986) stated that wood edges and openings and second growth stands may be used within its Canadian range.

Nesting:

The Bay-breasted Warbler constructs a bulky structure of twigs and grass, lining it with rootlets and hair and positioning it on the horizontal branch of a coniferous tree. The nest is placed 1-6 m above ground level, distinguishing it from lower nesting, neighboring Blackpoll Warblers. The typical clutch consists of 4-6 creamy white eggs, which are marked with brown. The close-sitting female incubates the eggs for 12 days. Although both adults tend to the young, only the female has been observed standing over the nest with her wings extended, to protect the nestlings from heavy rain or extreme heat. The female will only reluctantly flush when almost stepped upon and will then endeavor to lure away the intruder by feigning injury. If this is not successful, the bird will flit about and scold the intruder. The male will courtship-feed his mate during incubation and breeding. The young fledge at 11 days of age.

Remarks:

The spring male has a chestnut crown, throat, upper breast, and sides, as well as a buff neck, black face, and two white wing bars. The underparts are white tinged with buff. The spring female has the same general markings, although more patchy and fainter. This distinctly colored species becomes obscure in the fall. It is similar to the Blackpoll Warbler at that time, and can sometimes be distinguished by its chestnut sides and buffy undertail coverts. Yellowish-green underparts distinguish it from the white of the Chestnut-sided Warbler. The Bay-breasted male has a very high and sibilant song, which is phrased as "seee - seese - seese - seee". This may drop in pitch towards the end and is sung from mid-canopy elevations, separating it from the Cape May Warbler. It is also sung slower, less regularly, and at a lower pitch than the more common, similar sounding Blackpoll Warbler. Singing frequency has been positively correlated with population densities (Cadman 1987). It has been suggested that the female may sing weakly from the nest (Griscom and Sprunt 1979), although this has yet to be confirmed.

This species returns to Alberta by the third week of May, later than most migrants. In fall, it migrates within mixed warbler flocks from mid-August to mid-September, foraging indiscriminately at intermediate levels.

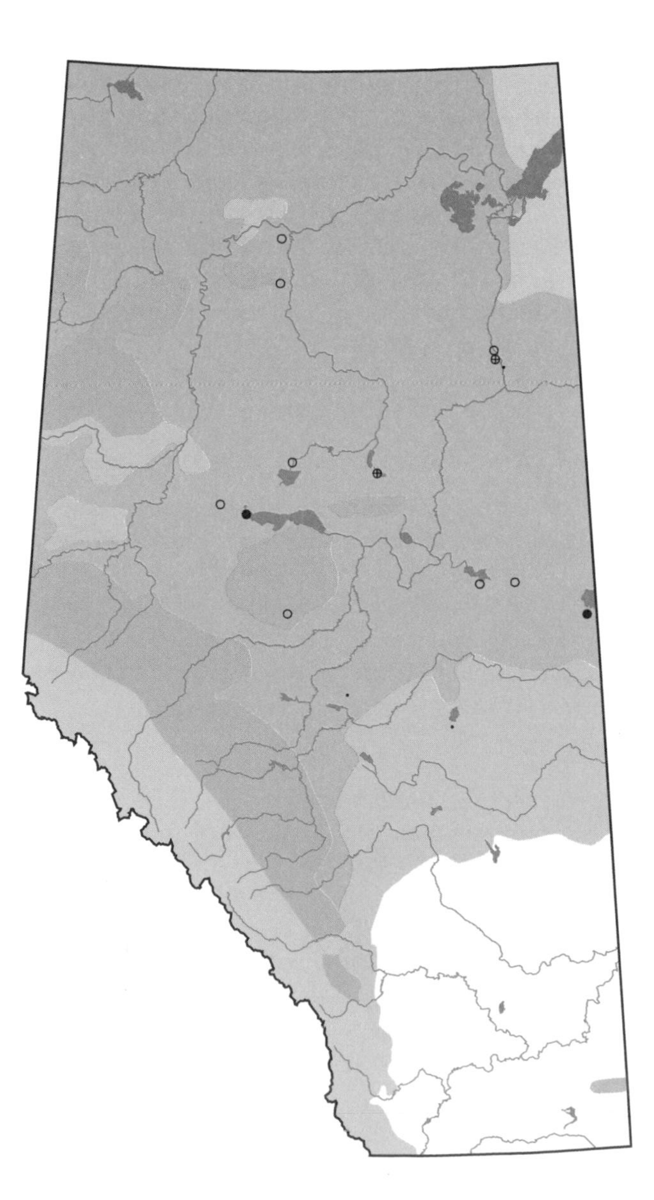

Blackpoll Warbler

Dendroica striata

BPWA photo credit: R. Gehlert

Status:
The current status of the Blackpoll Warbler has not been determined by the Wildlife Management Branch (1991). Salt (1973) considered it a fairly common and regular migrant of the Grassland and Parkland natural regions. Although an uncommon summer resident in Banff and Jasper national parks, this species was considered a fairly common summer resident in the Boreal Forest region of northern Alberta.

Distribution:
The Blackpoll Warbler's local distribution is restricted to fairly extensive tracts of mature spruce (Salt 1973). During the Atlas Project, the Blackpoll was distributed predominantly in northern and southern Alberta, south to Elk Island National Park, Caroline, and Banff Townsite. It was recorded in 4% of surveyed squares in the Foothills region, 5% in the Boreal Forest, and 6% in the Canadian Shield region. Although there were proportionately more records for the Rocky Mountain region (8%), this may be a result of more extensive coverage in this region. All five records for the Parkland region were located on the fringe between that region and the more typical habitat of the Boreal Forest.

Habitat:
The Blackpoll Warbler breeds in coniferous forests. Preferring mature spruce, including stunted trees, this species may also be found in mixed woodlands hosting a spruce dominance. It may also breed along mixed wood edges and logged or burned-over areas where a new growth of spruce, aspen or alder exists. On migration, this warbler may be found in either deciduous or coniferous stands.

Nesting:
The female constructs a nest of grass, lichens, twigs, and moss, and lines it with fine grasses, feathers, and hair. This is situated against the trunk of a spruce or alder (under 2 m) or occasionally on the ground. This somewhat bulky, cup-shaped structure is similar to the nest and location used by the Magnolia Warbler. The 4-5 creamy white eggs are speckled with browns and lilac. The female, a close sitter, incubates the eggs for 11-12 days, and is occasionally fed by the male during the incubation period. The chicks are initially fed larvae and aphids by both adults, switching to spiders and larger insects by day six (Bent 1963d). After 10-12 days, the young fledge from the nest, and the adult male probably leaves the nesting grounds (Salt 1973). Ehrlich et al. (1988) suggested that two broods may be raised.

Remarks:
The adult male Blackpoll in spring is similar in appearance to the Black-and-white Warbler. The former can best be distinguished by its solid black cap, white cheeks, and pale brown legs. The female, having no black cap, is quite different. Her greenish-gray back is streaked with black, her whitish underparts are tinged with yellow and faintly streaked in black, and she has a pale line over the eye. Losing most of its distinguishing plumage in fall, the Blackpoll can be separated from most warblers by its light wing bars, from the Chestnut-sided Warbler by its faint dusky streaks on yellowish underparts, and from the Bay-breasted Warbler by its white undertail coverts and yellow feet. Otherwise, it has olive-green upperparts with dusky streaks, a pale line over the eye, and, in fall, no black cap.
The male's song, a weak, two-second long, rapidly repeated "tititi. . .", is similar to that of the Bay-breasted Warbler. It also rises in volume in the middle before tapering off.
Being insectivorous, the Blackpoll will glean trees and spider webs and will flycatch. Few seeds are consumed.
First arriving in south and central Alberta in mid-May, these birds arrive singly or in twos or threes. The reverse migration in mixed flocks lasts several weeks and peaks throughout central Alberta in late August. Most Blackpolls have left the province by the end of September.

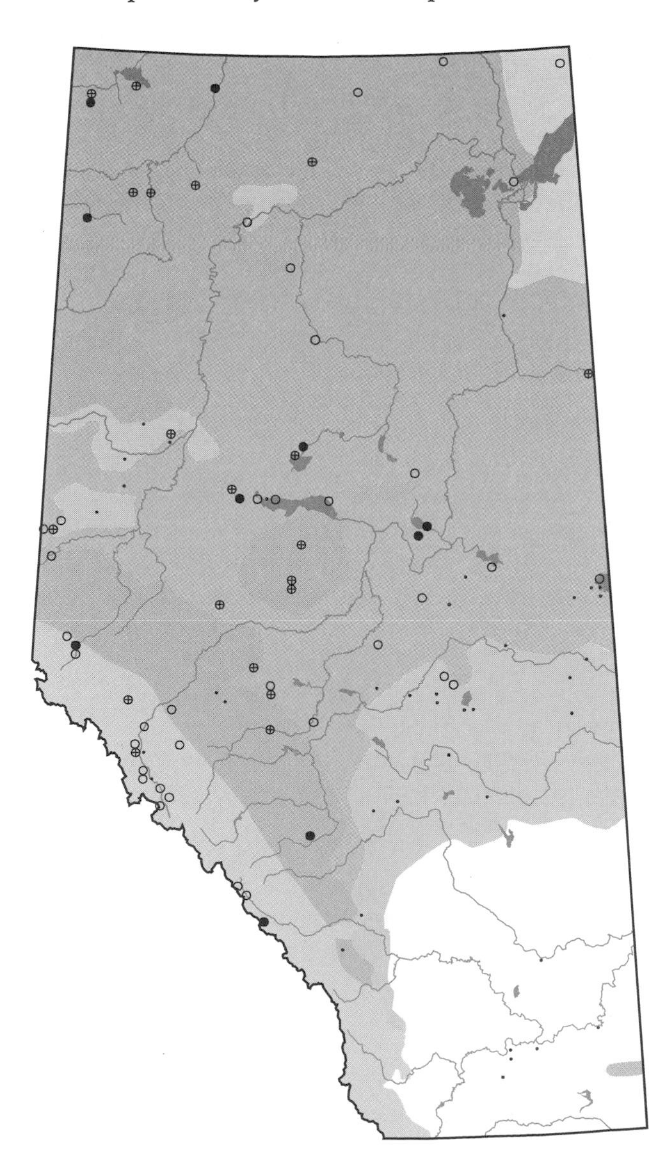

Black-and-white Warbler

Mniotilta varia

BAWW

Status:

Salt (1973) reported this species as an uncommon migrant of the southern prairie and a regular migrant in the Parkland Natural Region, where it was once more common. It was also suggested to be widely distributed and sometimes fairly common in the Boreal Forest region, despite reductions caused by land drainage and clearing. This species is very sensitive to the fragmentation of its forested breeding habitat (Ehrlich et al. 1988).

Distribution:

Historical breeding distribution has been in northern and central Alberta south to Battle Lake, the Battle River, and Cadogan. During the Atlas project, 88% of all records were in the Boreal Forest Natural Region. There was no confirmation of breeding within the Parkland Natural Region or the Cypress Hills. A lone confirmation was made southwest of Drayton Valley in the Foothills Natural Region. It has been seen only rarely in Jasper National Park (Salt 1984) and in Banff (Salt and Salt 1976).

Habitat:

Preferring moist areas, often near water, the Black-and-white Warbler breeds in the deciduous or mixed wood stands of the Boreal Forest. It may also be found in the thick growth of alder and willow that borders muskegs and woodland pools.

Nesting:

The female constructs a nest on a slight depression made in the ground; it is made of bark strips, grass, moss, and rootlets, and is lined with hair. This stout cup is usually sheltered, or partly so, by some near or overhanging structure, such as a mossy bank, a fallen tree, a stump, a bush, or within the exposed roots of a blowdown. The location is well concealed by surrounding leaf litter. The 4-5 white eggs are spotted all over with brown and lavender. Nests are often parasitized by cowbirds. The female alone incubates the eggs for 10-13 days. If flushed from the nest, the female may perform a distraction display. Both adults will tend to the young, which fledge after 8-12 days. Becoming self-reliant a short time later, the young will leave to wander widely before migration. It is not known if a second brood is raised.

Remarks:

Well indicated by its name, the Black-and-white Warbler is boldly striped black-and-white above and below and has a white belly. A broad white central stripe down the black crown and a thick white eyebrow stripe distinguish it from a similar Blackpoll Warbler. The female resembles the male, but has a white throat, fewer black streaks below, and its white coloring is duller.

The male's song, a high-pitched, squeaky and rolling "weesy-weesy-weesy-weesy-weesy", does not carry a long distance. The call note, a sharp "pit", is similar to the "chip" of the Blackpoll Warbler.

Not a particularly shy bird, this species can be readily identified by its distinctive creeping habit. Long claws enable it to move along tree branches and trunks in nuthatch fashion, examining bark cracks and crevices and leaf clusters for wood-boring and other insects. Unlike Brown Creepers and woodpeckers, the tail feathers are not used to lend support. It will also occasionally flycatch.

Black-and-white Warblers arrive in central Alberta in mid-May with earlier mixed-warbler flocks. The majority of these birds leave the province in the last two weeks of August. This species is rarely seen beyond the first week of September.

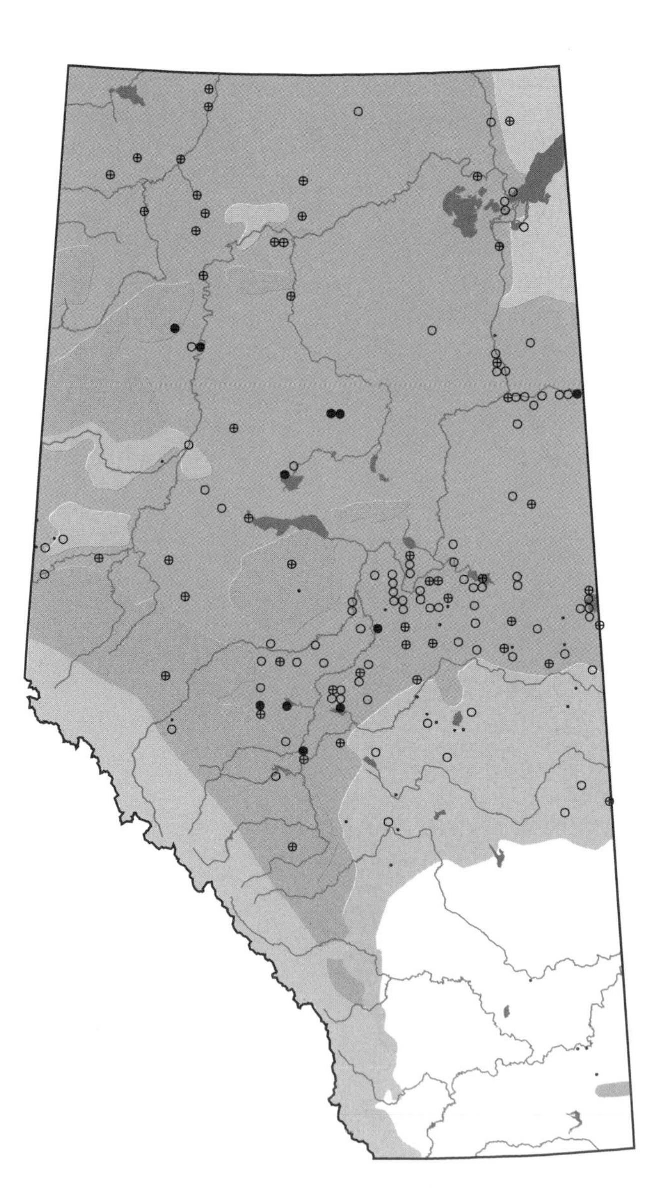

American Redstart

Setophaga ruticilla

AMRE photo credit: T. Webb

Status:

Data from spring migration counts in the northeast United States indicate a significant decline in American Redstart numbers over the past 40 to 50 years (Hill and Hagan 1991). The loss of tropical forest habitat in the winter range of this warbler is a possible cause of this decline. Although not considered, by the Wildlife Management Branch (1991), to be at risk at present, little information has been recorded to accurately determine its status.

Distribution:

The American Redstart breeds in northern and central Alberta south to about Stettler and Provost. In the Rocky Mountain and Foothills regions, it breeds south to the Waterton Lakes and Police Outpost area. It also nests in the Cypress Hills. Of those squares surveyed in each natural region, the American Redstart was recorded at the following levels: Boreal Forest 23%, Canadian Shield 13%, Foothills 11%, Rocky Mountain 9%, Parkland 7%, and Grassland 1%. As a transient, the American Redstart is fairly common throughout Alberta.

Habitat:

Choice breeding habitat for the American Redstart is deciduous woodland or the deciduous parts of mixed wood forests, particularly those areas with dense undergrowth near water. Tangled willows along streams or willow-alder thickets bordering muskeg pools are ideal. The Redstart is most common along the western and northern borders of the Parkland region where the Parkland merges with the Boreal Forest, in the valleys of the Rocky Mountain and Foothills regions, and in the northern deciduous woodlands.

Nesting:

This wood warbler nests in bushes or small trees, often in a crotch or group of upright branches 2-8 m above ground. It never nests in conifers. The nest, constructed by the female, is a firm compact cup of grass and shredded bark. It is lined with plant fibres, hair, and, sometimes, spider webs.

The female lays 3-5 eggs, usually four. The eggs are white and are spotted with browns and a little purplish gray. The female incubates for 12 days; young are tended by both parents. The nestlings fledge at 8-10 days. Pairs probably produce one brood per season. They are frequent cowbird hosts.

Remarks:

The Redstart is one of our best known warblers, mainly because of the brilliant colors of the male and the bird's energetic personality. The male has a characteristic habit of dropping its wings and fanning its tail, thus displaying the bright orange patches located there.

This warbler has a highly varied insect diet; it will forage for insects and spiders among the trees and bushes, but will also gracefully hawk insects from mid-air.

The Redstart arrives after mid-May with the main wave of migrant warblers and leaves the province towards the end of August or early September, also in the company of other migrating warblers and vireos.

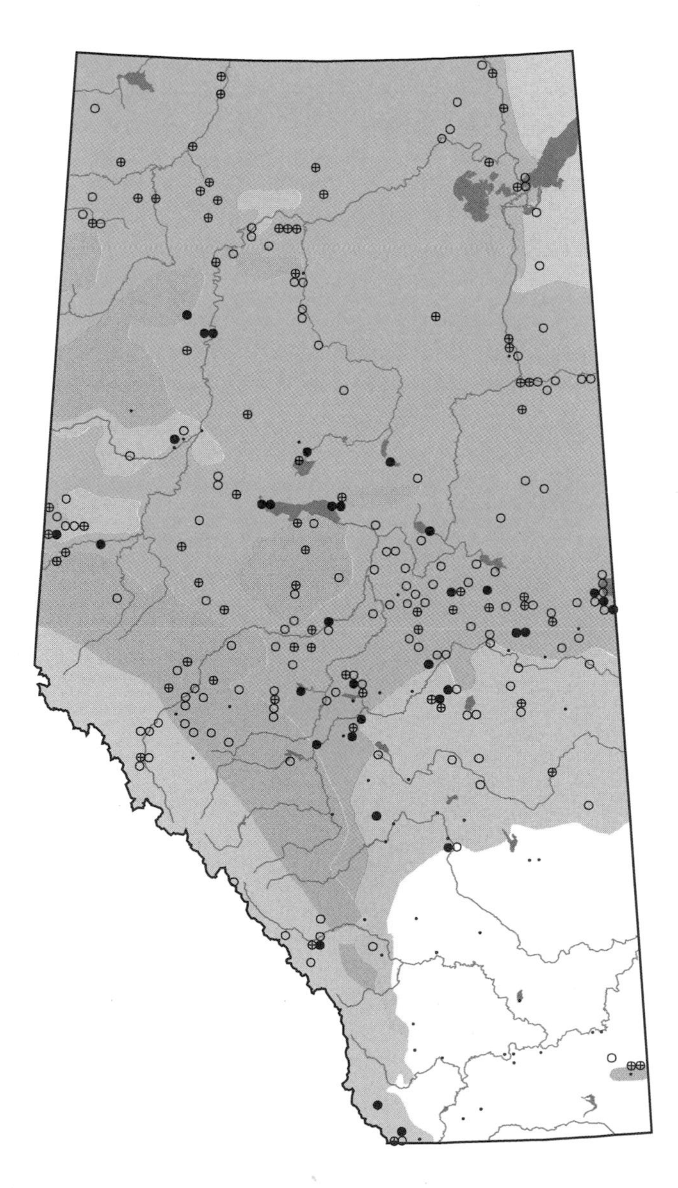

Ovenbird

Seiurus aurocapillus

OVEN

Status:
Though not at risk at this time, the status of the Ovenbird is unknown, although probably stable. Increased commercial harvest of mature deciduous forest within its habitat is a cause of concern (Wildlife Management Branch 1991). Less common in the Parkland region than in the Boreal Forest, where it is very common, it is often restricted to the fairly extensive tracts of deciduous woods that may remain in broad river valleys (Salt 1973). Local presence is sensitive to logging (Quinlan et al. 1990) and other clearing activities.

Distribution:
During the Atlas Project, records were obtained for the Boreal Forest, Parkland, and Foothills natural regions. Possible records were found in the Cypress Hills and in Waterton Lakes National Park. Although distribution was limited, probable nesters were found around the area of the Athabasca River's exit from the Rocky Mountain Natural Region.

Habitat:
The Ovenbird breeds in deciduous woods that host a sparse bushy undergrowth. It will tolerate mixed wood stands if these include extensive aspen or poplar. The highest breeding densities found in Alberta occur in about 30-year old aspen stands (Quinlan et al. 1990). These closed canopies block out extensive shrubby growth and provide good leaf litter and herbaceous growth. A forest interior species, the Ovenbird is intolerant of clearcuts (Quinlan et al. 1990).

Nesting:
Adults endeavor to return to their previous breeding territory each year. As the male stands guard, the female constructs a domed nest of leaves, grass, and moss, and lines it with rootlets and hair. A side entrance provides access to this well-concealed structure. This species' name was originally derived from the appearance of the nest, which was thought to resemble a dutch oven. It is situated on the ground in a relatively open area within its woodland habitat. Interference during nest-building or early in the egg-laying may result in nest abandonment. The 3-6 white eggs are speckled with brown and gray. Some females are polyandrous. Incubation by the close-sitting female is carried out for 11-14 days. Once flushed, the female will feign injury (Griscom and Sprunt 1979). The female is rarely fed by the male while she is on the nest. The hatchlings are tended to by both parents and are coaxed from the nest after 8-10 days by the offering of food. Once fledged, the brood is divided between the adults. One study (Hann 1937) reported heavy cowbird parasitism.

Remarks:
These shy and elusive warblers have pale olive-green upperparts, two blackish crown stripes bordering a stripe of brownish orange, a pale eye ring, whitish underparts behind a black-lined throat, and streaked breast and sides. Although thrush-like in overall appearance, they have warbler-type bills and streaks instead of spots. This species is also similar to the Northern Waterthrush, but has no pale line over its eye, does not habitually flit its tail, and has distinctive crown markings. It does not hop while on the ground like most warblers do, but rather walks with a bounce.

The male's distinctive song is a series of 5-12 identical phrases of "teacher", rising in volume and stronger than neighboring species. A sharp call note is used to scold intruders near the nest site. The Ovenbird also has a musical flight song, which is most often given at dusk (Griscom and Sprunt 1979).

This insectivore spends most of its time on the ground searching the leaf litter for spiders and snails.

The Ovenbird arrives in Alberta by the third week of May, and departs in the last two weeks of August. Stragglers leave by mid-September.

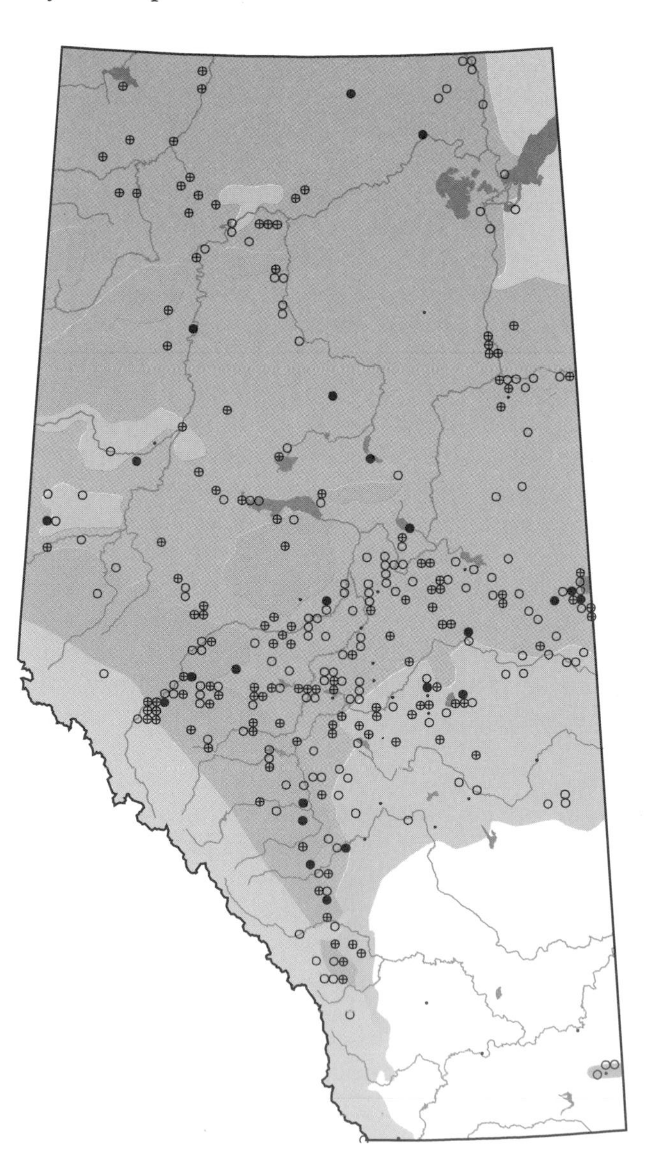

Northern Waterthrush

Seiurus noveboracensis

NOWA

Status:
Salt (1973) considered the Northern Waterthrush to be common in the Boreal Forest region of northern Alberta, less common in the Rocky Mountains, Foothills, and southern extension of the Boreal Forest, a regular transient in the Grassland, and rare in the Cypress Hills.

Distribution:
Data collected during the Atlas Project generally concur with historical reports (Salt and Salt 1976; Godfrey 1986). It is fairly common in the Boreal Forest and Rocky Mountain natural regions, with records in 14% of squares surveyed in each region. It was consistently found in the Canadian Shield (12.5%) and in the Foothills (10%) region. In addition, there were 11 records from the Parkland, including the Peace River District, six of which were confirmed breeding records.

Habitat:
A bird of the heavily forested parts of Alberta, the Northern Waterthrush breeds in areas which include extensive shrubbery and nearby standing water that meets the woods. These mixed wood stands, more often than not, have no preponderance of spruce, aspen, or birch. This includes the wooded margins of lake and stream shores with a thick understory of willow and alder. Flooded woodland is also acceptable.

Nesting:
Like other northern warblers, little has been reported concerning the breeding biology of this species. The cup-shaped nest may be built in a variety of locations, almost all of which are on the ground. These locations include the following: sunken into the moss or ground, on a bank, at the foot of a stump, or, frequently, in the roots of an upturned tree, often over a resulting pool of water. The nest is constructed of moss, leaves, and twigs, and it is lined with fine grass and hair. The typical clutch is 4-5 creamy white eggs which are spotted with browns. Incubation is thought to be 12-13 days and fledging is in 10 days.

Remarks:
Despite its name and general appearance, the Northern Waterthrush is not a thrush, but a wood warbler. However, it does exhibit thrush-like behavior, and lives near water. Sharing the same plumage, adults have dark olive brown upperparts, a pale buff eye line, and white underparts that are sometimes tinged with pale yellow. Its smaller size, warbler-type bill, and line over the eye readily distinguish it from thrushes.
Walking causes a characteristic "teetering" motion, similar to the Spotted Sandpiper. This incessant motion, created by the head bobbing and the tail tipping, separates it from all other woodland birds.
The male's song consists of three introductory notes, moves on to three faster notes, and ends with 4-6 wire-like "chew" notes, which run together. Sung frequently, it has an almost staccato quality and may carry a fair distance, especially when sung from upper levels. A "clink" call note is also distinctive.
The Northern Waterthrush feeds close to or on the ground, gleaning fallen logs and under leaf litter and wading in shallow water and pools. It consumes terrestrial and aquatic insects, worms, small crustaceans, and molluscs, and, occasionally, minnows.
Sometimes travelling in mixed warbler flocks, but more often in small groups of waterthrushes, males will sing on migration, usually from within the treed margins of ponds. These birds arrive in the province in the second and third week of May and most depart by the end of August. There are occasionally stragglers into early September (Pinel et al. in prep.).

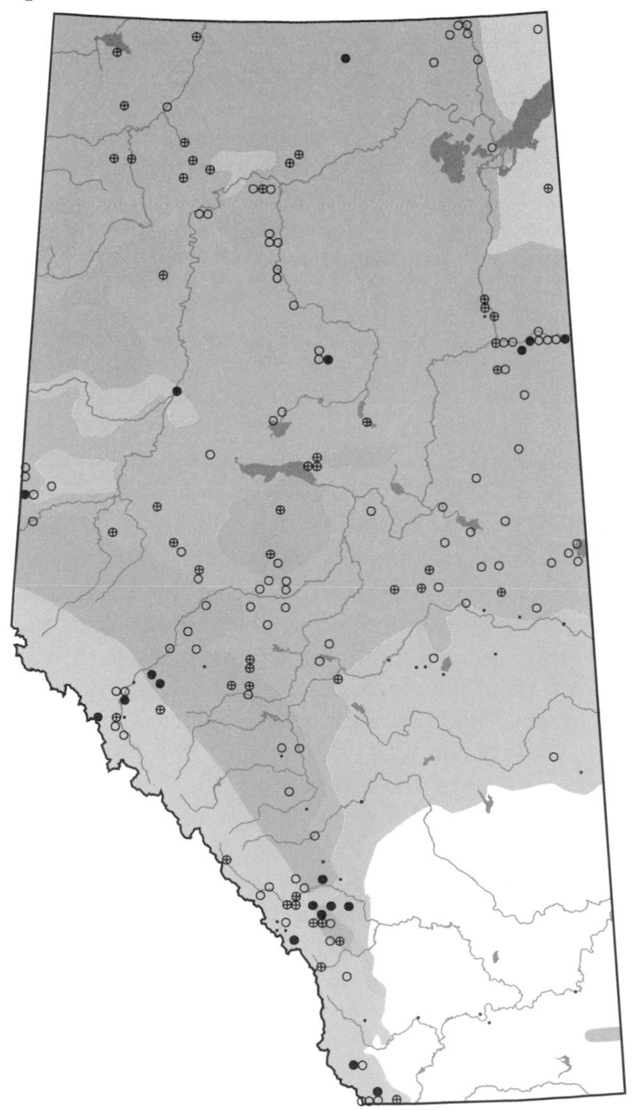

Connecticut Warbler

Oporornis agilis

COWA

Status:

Throughout its North American breeding range, little information has been collected to determine the status of this species. Although Atlas efforts were largely concentrated on the breeding season, as opposed to the migration periods, data collected would seem to confirm that the Connecticut Warbler is a scarce migrant in southern Alberta, as suggested earlier by Salt (1973). He also suggested that it was a fairly common breeder within its range in northern Alberta, though only locally distributed. Atlas findings generally agree with this observation.

Distribution:

Historical distribution of the Connecticut Warbler has been from the Peace River district east to Cold Lake and south roughly to the North Saskatchewan River. In central Alberta, this species' distribution ranged southward through the Battle Lake area to Sundre (Salt and Salt 1976). Atlas records agree with other reports since 1976 (Pinel et al. in prep.), that the breeding range extends to Hinton and Bottrel in west-central Alberta. The majority of records are for the Boreal Forest and Foothills natural regions. As testament to its elusive nature, only seven of the total 114 records confirmed breeding.

Habitat:

Over its entire Canadian breeding range, the Connecticut Warbler nests in a variety of habitats, although it generally selects relatively open tree stands without dense or tall undergrowth. It seems to tolerate a large range of moisture conditions, from stands of relatively mature aspen to dry, sandy ridges and knolls with open poplar woods or sapling thickets, or, less commonly, to spruce and tamarack bogs. Mixed wood stands that are predominantly deciduous may also be used.

Nesting:

Consistent with other aspects of the life history of this species, very little is known about its breeding biology. The nest, a deep, compact cup of grass and plant fibres, is lined with finer grasses. It is situated on the ground, or very close to it, and often at the base of a small sapling or weed. In appropriate habitat the nest may be built in a mossy hummock. The 4-5 eggs are creamy white and speckled with brown. The female does not land or take off at the nest. Instead, she approaches stealthfully and departs the nest by walking to and from an area 10 -15 m away.

Remarks:

The male is bluish-gray on the head, throat, and upper breast, olive-green above, yellow below, and has a complete white eye ring. Resembling the Mourning and MacGillivray's Warblers, it is distinguished by the lack of any black on the head or breast and by a complete eye ring. It may also be confused with the smaller Nashville Warbler, although the latter has a bright yellow throat that contrasts with gray cheeks. The breeding range of these two species do not overlap in Alberta. The female Connecticut is similar to the male, only duller. Young have brownish-olive upper parts with a pale buffy eye ring, olive-brown throat, and breast, olive sides and the remaining underparts yellow.

The male sings from treetops near the nest and occasionally from the undergrowth while feeding. Two renditions can be described as: "chip-chuppety-chuppety-chuppety" or "chippychuppy, chippychuppy, chippychuppy."

Feeding along the ground, the Connecticut is thought to consume spiders, small insects, and larvae. Seeds and berries are also occasionally included in their diet.

This species passes through Alberta largely unnoticed on migration. One of the later migrants, Connecticut Warblers do not arrive in Alberta until late May. Departure for their wintering grounds in north and central South America is in late August or early September.

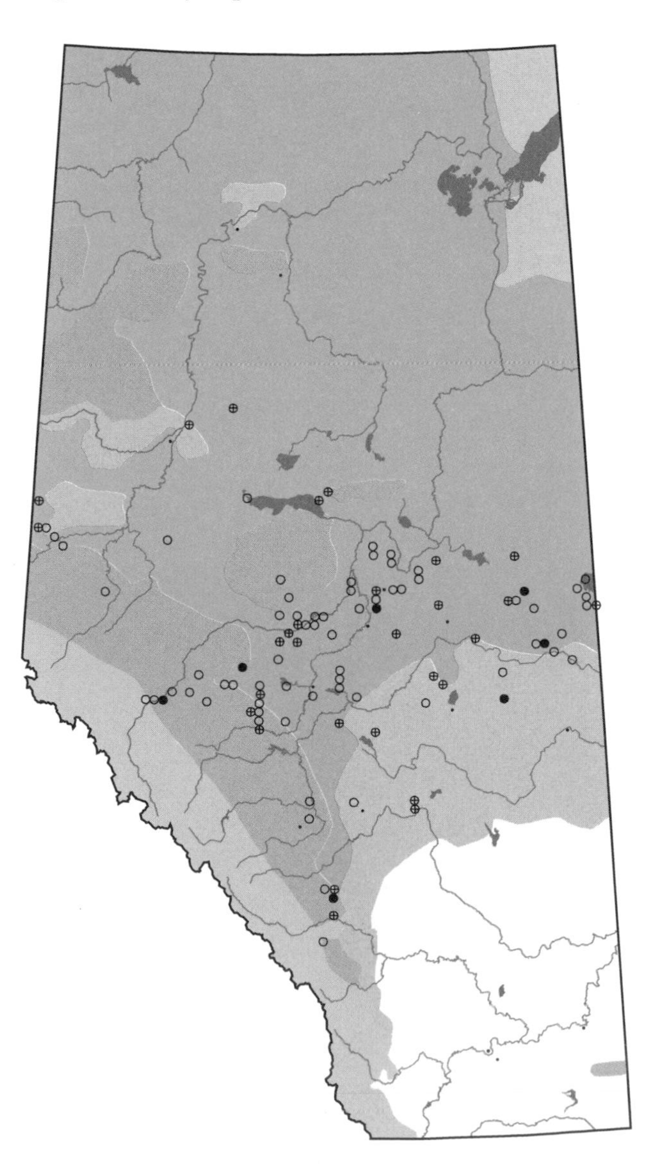

Mourning Warbler

Oporornis philadelphia

MOWA

Status:
Insufficient information has been collected to determine the provincial status of the Mourning Warbler (Wildlife Management Branch 1991).

Distribution:
Atlas records generally agree with the historical distribution. However, more records were found along some river valleys of the Foothills region, as well as inside the margin separating the Foothills and Boreal Forest natural regions. It is a very rare transient in the Grassland region. Although it was most commonly recorded in the Boreal Forest (16.3% of squares surveyed), there are still no observations for the province's northwestern corner. It was recorded in 5% of surveyed squares in the Parkland Natural Region, mainly in the northern portion. In the Foothills Natural Region, the range limit of the closely related MacGillivray's Warbler, it was recorded in 9% of the surveyed squares.

Habitat:
The Mourning Warbler breeds in deciduous woods that have a partially open canopy. This may exist in areas of high windfall or along woodland edges and openings. A dense lower canopy is provided by a thick undercover of tall shrubs. Some mixed wood may be tolerated.

Nesting:
The female constructs a nest on or near the ground. It is a bulky mass of dead leaves, weeds, and grass, which is lined with rootlets, hair, and fine grasses. The nest is often located in a cranberry, choke cherry, raspberry, or rose bush. The eggs, creamy with brown speckles, are incubated by the female for 12-13 days. The male will courtship-feed his mate during the incubation period. When approached on the nest, the female will either sit close or quietly slip off the nest and flush 6-8 m away. While both adults tend to the young, only the female broods them. The adults are easily agitated at this time. Leaving the nest after 7-9 days, the young have undeveloped tail feathers and are initially unable to fly. Their parents will use a distraction display during this vulnerable period. Starting to fly by their second week, they will not leave their parents until 2-3 weeks after fledging.

Remarks:
The male has a bluish-gray head and neck, black in front of the eye and on the throat, a bluish-gray upper breast tipped in black, olive-green upperparts, bright yellow underparts, sides tinged with olive-green, and no eye ring. The Mourning Warbler is difficult to distinguish from Connecticut and MacGillivray's Warblers, especially in fall. Some hybridization has been reported with the MacGillivray's Warbler in the area between Kananaskis and Rocky Mountain House (Cox 1973). The female is generally similar to the male, but the head, neck, and upper breast are slate-gray and a faint and/or incomplete eye ring exists in the fall. Young are dark olive-green above and yellow below, with pale gray throats, buffy breasts, and, often, faint eye rings.
The male sings a clear, strong, and quite recognizable song of seven evenly spaced notes: "chee-chee-chee-chee-chee-choo-choo." There may be some variation from this. The male does not sing near the nest nor very often during this period, but will continue to sing well after the young have fledged. A strong call note, "tchip", is pronounced in two syllables.
It feeds close to the ground, probably on insects and spiders. Late spring migrants, Mourning Warblers do not arrive until late May or early June, when the foliage has emerged. Very quiet during fall migration, most of these birds have departed by early September.

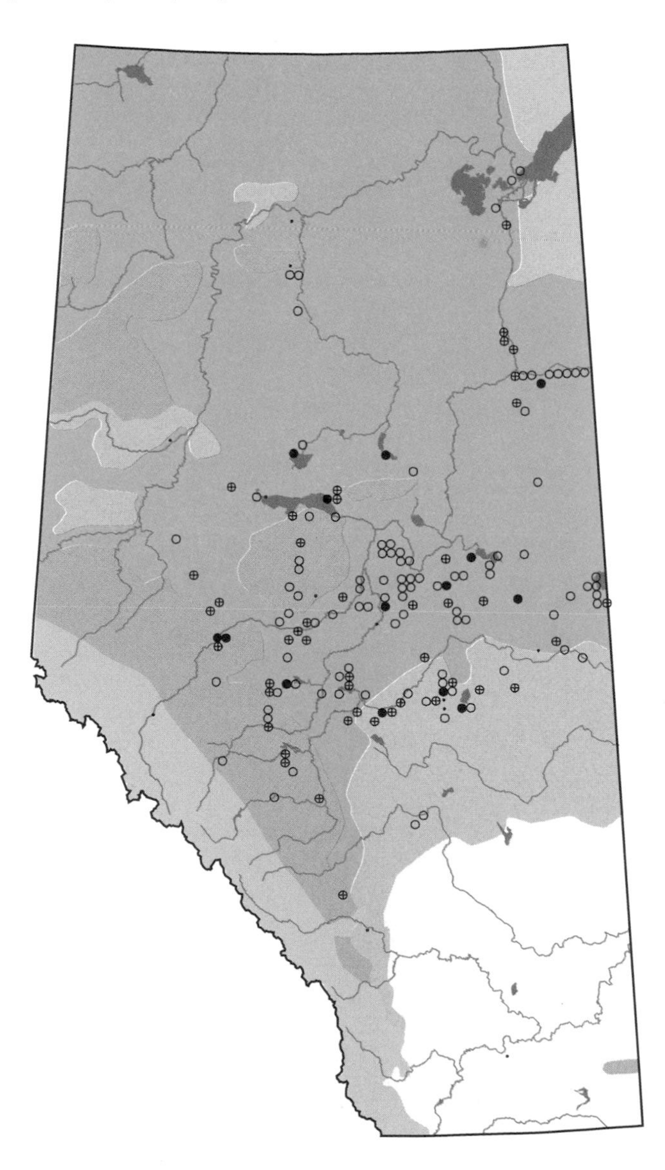

MacGillivray's Warbler

Oporornis tolmiei

MGWA

Status:
Although not considered at risk, insufficient local information is available to determine the provincial status of the MacGillivray's Warbler. Salt (1973) suggested this species to be a fairly common summer resident in mountain and foothills areas, whereas Holroyd and Van Tighem (1983) considered it to be uncommon in Banff and Jasper national parks. It was recorded in 16% of the surveyed squares in the Rocky Mountain Natural Region.

Distribution:
Records in the Rocky Mountain and Foothills natural regions generally mark the eastern range limit of this species. Historical (Salt and Salt 1976) and Atlas records also exist for the Cypress Hills and the Red Deer, Bow, and Highwood river valleys. Hybridization with its eastern cousin, the Mourning Warbler, has been recorded in the past (Salt 1973) along a north/south line west of Edmonton and Red Deer, where the two range limits approach and occasionally overlap. It was most recorded in the Rocky Mountain Natural Region at 16% of the surveyed squares. Adjacent to this area, the MacGillivray's Warbler was recorded in 3% of Foothills region squares. It was rarely encountered outside these two regions.

Habitat:
Wherever this warbler breeds, it seems to prefer areas thick in dense brush with water nearby. In montane habitat, this includes deciduous stands along mountain slopes, forest edges and openings, and burns. In prairie coulees, tangles of saskatoon, choke cherry, willow, and alder are utilized, with an undergrowth of gooseberry, currant, and rose. In either region, the dense shrub canopy promotes moist ground conditions.

Nesting:
Many specifics of this species' breeding biology are not known. The nest, situated in bushes and usually no higher than 6 m above the ground, is a loosely woven compact cup of grass and bark shreds. It is lined with fine grasses, rootlets, and hair. The 3-5 creamy white eggs, speckled with brown, are incubated 11-13 days by the female. Tended to by each adult, the young fledge after 8-9 days. The MacGillivray's Warbler is an uncommon cowbird host.

Remarks:
The breeding male has a slate-gray head, neck, and upper breast, with a few black feathers on the throat and breast. It has olive-green upperparts, yellow undertail coverts, black in front of the eye, and white eyelids appearing as an incomplete eye ring. It is most easily confused with the male Connecticut Warbler, which has a paler hood, paler underparts, and a complete eye ring, and the male Mourning Warbler, which has no eye ring at all. In addition, whereas the Connecticut Warbler walks, this species hops and lacks any dark apron. Female MacGillivray's Warblers have paler heads and no black markings. Young have brownish-olive upperparts and sides, grayish-olive throats and upper breasts, yellow underparts, and pale eyelids.

Usually singing from a 5-7 m high perch, the male's clear notes are similar to those of the Mourning Warbler, only quicker. Salt (1973) described it as "chee-chee-chee-chee-chi-chi-chi-chooey-choey", with the final two notes dropping in pitch and slurring. A harsh call note, "tchek", is also used.

It forages on caterpillars, insects, and larvae under the cover of the underbrush. Juveniles have been seen taking sap from Red-naped Sapsucker wells excavated in willows, but this has not been recorded in Alberta.

This species arrives in Alberta during the last week of May and departure for wintering grounds is completed by mid-September.

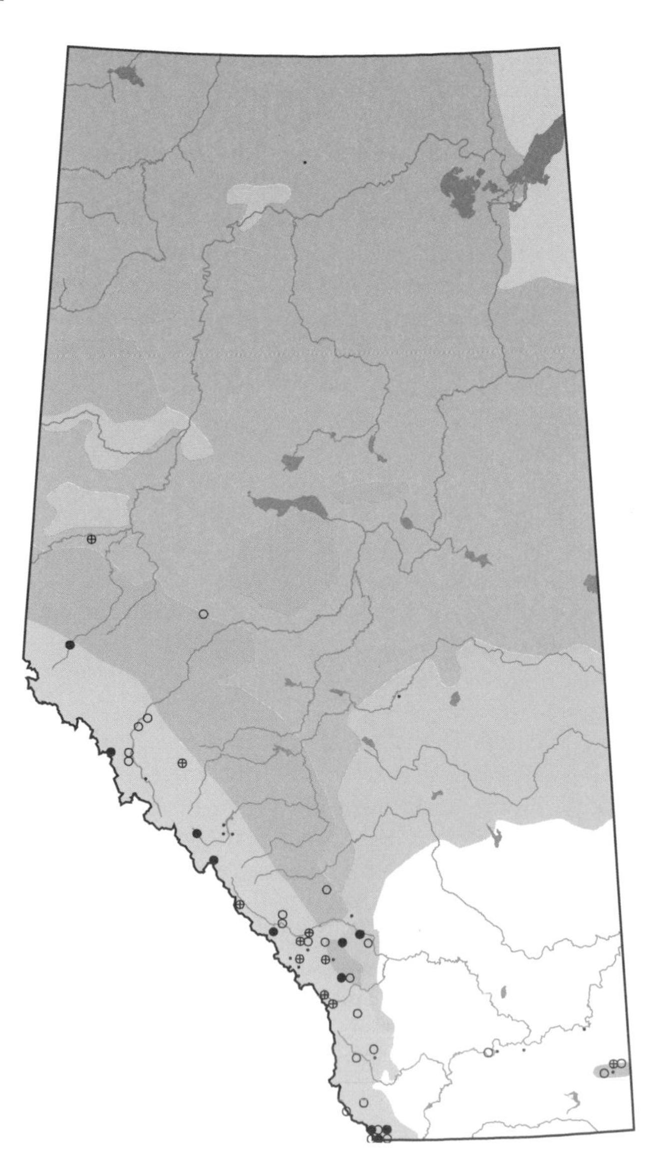

Common Yellowthroat

Geothlypis trichas

COYE photo credit: T. Webb

Status:
As indicated by its name, the Common Yellowthroat is common, given appropriate local habitat, throughout Alberta. However, it is scarce on the prairie and in the higher elevations of the mountains (Salt 1973). Further clearing of forested lands may increase habitat and abundance.

Distribution:
This species breeds in all natural regions of Alberta, including the mountain national parks. During the Atlas Project, it was most frequently encountered in the Foothills and Boreal Forest (39% and 38% of surveyed squares, respectively). Grassland distribution is restricted to the brushy margins of water bodies and coulees (Salt 1973).

Habitat:
Not a bird of the deep woods, the Yellowthroat generally breeds in more open, damp, and brushy areas. This includes the dense willow shrubbery of open marshy areas and the low buckbrush, willow-alder tangles, scrub alders, cattails, or poplar saplings along woodland edges. In northern areas, habitat may be bogs, muskeg and wet meadows, along riverbanks and lakeshores, and around beaver ponds.

Nesting:
The female constructs a bulky nest on the ground or in thick herbage or a bush, most often on or near still water. The nest consists of a loose mass of grasses, herbs, leaves, and strips of rushes where available. It is lined with fine grasses, rootlets, and hair. The 3-6 creamy white eggs found inside the nest are speckled with black, brown and gray. Incubation is by the female for 12 days. If approached while on the nest, the female will quietly slip off and creep away before flushing. No distraction display is used. Although only the female broods the nestlings, both adults tend to the young. Despite fledging at 9-10 days, the young are tended for two weeks longer than most warblers. Two broods are raised (Ehrlich et al. 1988). One of the most frequently parasitized nests, the Yellowthroat's eggs are often moved and replaced or punctured by the larger Brown-headed Cowbird. The female Yellowthroat may build a new nest on top of an earlier parasitized nest.

Remarks:
The crown and upper surfaces of wings, and tail of the male are olive-green, while the breast and undertail coverts are yellow. The abdomen is whitish. He is distinguished by a broad black mask over the eyes, forehead, and face, which is margined above and behind by light gray, giving him the appearance of a bandit. The female is similar, but with grayish olive above and paler yellow below. The black mask is replaced with a pale yellow line over and around the eye. A yellow throat and breast separate the female in fall from similar-appearing Mourning, MacGillivray's and Connecticut Warblers.
Easily agitated, the male's loud, clear, and distinctive song, "witchety - witchety - witchety", is often the only clue to its movements. The first syllable is emphasized, although the song's length and structure may vary, even by the same individual. It also uses a husky "tsick" call while raising its tail. The Yellowthroat may occasionally mimic neighboring species, including flight songs.
They eat spiders and a few seeds, usually from lower tree levels or bushes. It occasionally forages on the ground and will feed on taller bushes.
Migrating in mixed-warbler flocks at night, the Yellowthroat experiences heavy losses to its overall population as a result of collisions with erected structures. They arrive in Alberta in the last two weeks of May and the majority depart the province by mid-September.

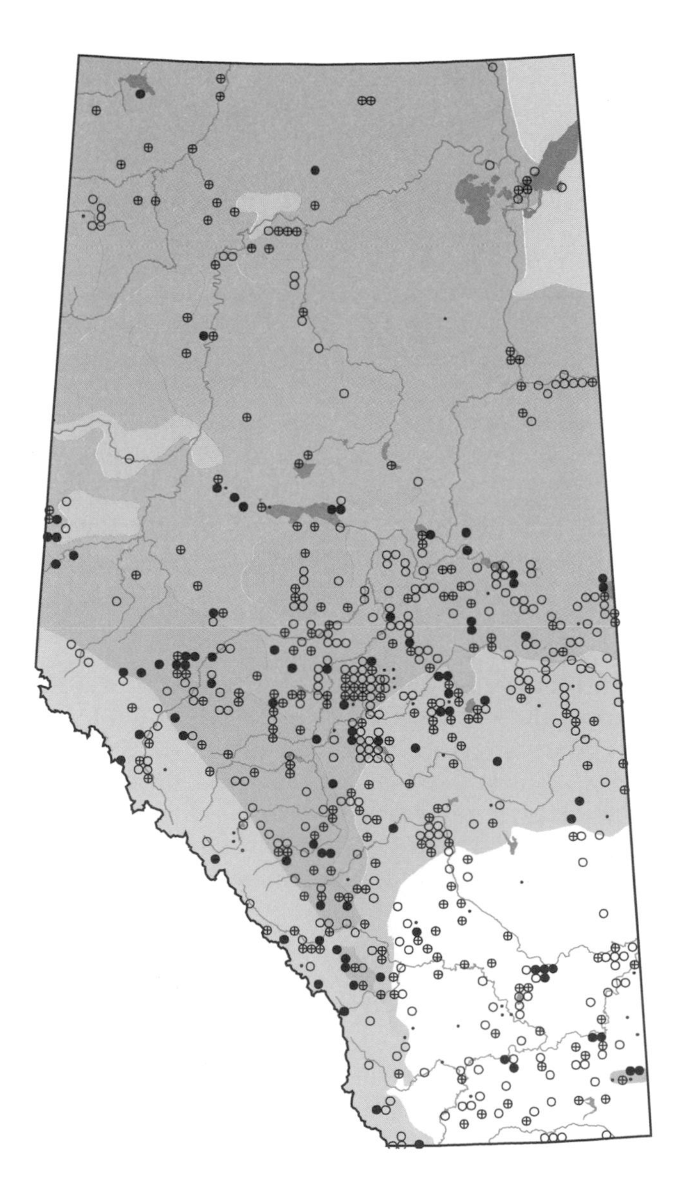

Wilson's Warbler

Wilsonia pusilla

WIWA photo credit: R. Gehlert

Status:
Previous reports classified the Wilson's Warbler as an uncommon summer resident in northern Alberta (Salt 1973) and a common summer resident in Banff and Jasper national parks (Holroyd and Van Tighem 1983). Atlas results substantiate this classification. As a migrant, it is a regular visitor to the Grassland region (Salt 1973) and uncommon in Banff and Jasper national parks (Holroyd and Van Tighem 1983).

Distribution:
Salt and Salt (1976) reported provincial distribution to include northeast Alberta south to Lac La Biche and Boyle. In addition, the range extended in the northwest only along the northern border. In the south, the only range reported was the Rocky Mountains and adjacent foothills. It once bred in central Alberta (Salt 1973). Atlas results include the above range, as well as all of northwest Alberta south to Fort Assiniboine, Edson and Hinton. This species was encountered most often in the Rocky Mountain Natural Region (in 39% of surveyed squares) and less so in the Boreal Forest region (7%). However, it was recorded in 31% of those squares surveyed in the adjacent Canadian Shield region. It was found in the Foothills (20%) and, less often, in the Parkland region (2%).

Habitat:
More indicative of the habitat available beyond the mountains, the Wilson's Warbler breeds in willow alder thickets on river flats and beaver meadows or in similar vegetation along lakeshores and woodland streams.
In the Rockies, it resides in the dense alder and shrubby thickets of mountain meadows, avalanche slopes (preferred), subalpine fir krummholz at timberline, and burns.

Nesting:
Arriving earlier in the spring, males establish territories in preparation for mating activities. Over a five-day period, the female builds a bulky, cup-shaped nest of leaves, shreds of bark, grass, and moss, and lines the inside with fine grass. This well-concealed nest is usually below 1 m and inside thick shrubbery, on the ground at the base of a small tree, or in a dry hummock. Inside this structure, the female lays her single clutch of 4-6 white eggs, each speckled with browns. She incubates this clutch for 10-13 days. Both adults participate in feeding the nestlings and removing their faecal sacs. The young fledge after 10-11 days.

Remarks:
The adult male has a black crown patch with a yellow forehead, face, throat, and underparts. His ear coverts and remaining upperparts are olive-green. Similar to her mate, the female's crown patch is usually obscured by olive.. Only possibly confused with the Yellow Warbler, the Wilson's Warbler is distinguished by its black crown patch, when visible, and a slate gray rather than a yellow and olive tail. It also has a nervous habit of twitching its tail.
The male's weak song, a series of rapid, almost staccato, chattering notes, drops in pitch at the end (Godfrey 1986). Varying in pitch, volume, and tempo, it sometimes resembles the song of the Orange-crowned Warbler. However, the Wilson's song is generally higher pitched.
Their diet consists of insects and occasional berries, and foraging methods include hover gleaning, bark gleaning, and occasional flycatching. In fact, this species has a broad-based bill and prominent bristles typical of flycatching birds.
Arriving in the province in the middle of May within mixed warbler flocks, this species' migration peaks in late May in Jasper and Banff national parks. With the males departing first, a fall migration may occasionally linger into early October, although the majority leave during the last week of August and first week of September.

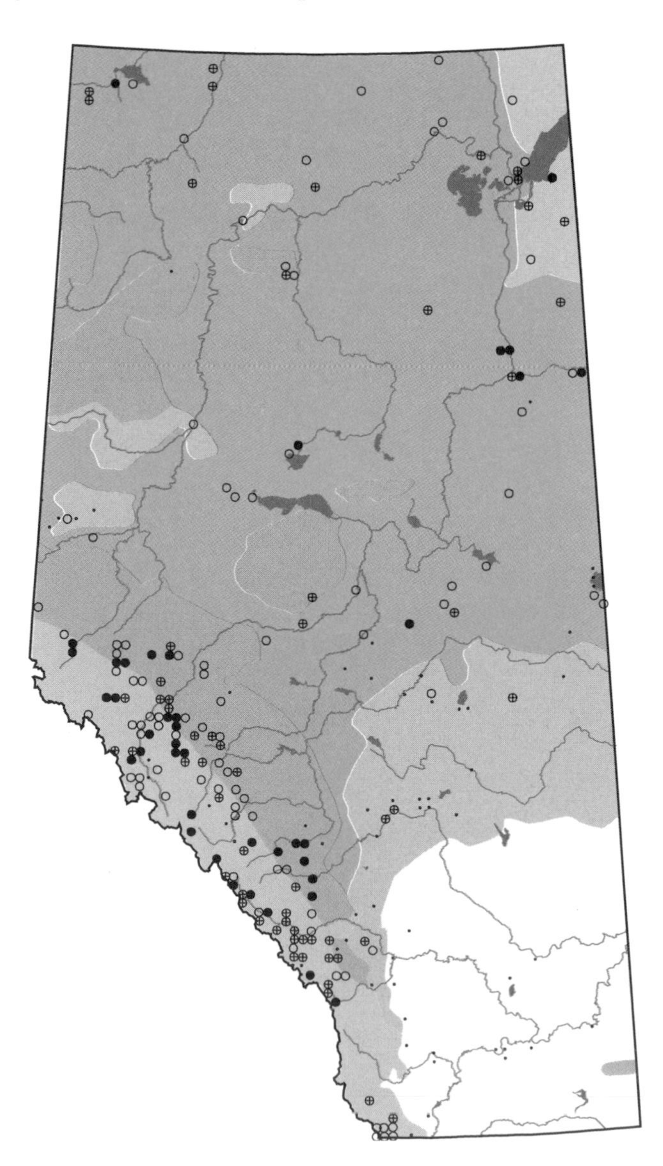

Canada Warbler

Wilsonia canadensis

CAWA photo credit: R. Gehlert

Status:
Historically (Salt 1973), the Canada Warbler has been considered scarce in Alberta, which marks the western limit of its North American range. In particular, it has been scarce in the Boreal Forest Natural Region, scarce and irregular in the Parkland, a rare transient in the south, and a very rare visitor in the Rocky Mountain region.

Distribution:
Salt and Salt (1976) considered the breeding range of this species to be northeastern and central Alberta, particularly those areas drained by the Peace and Athabasca rivers, south to Glenevis and Cold Lake. Observations had been made at Sturgeon Lake and Grande Prairie in the Peace River District. Godfrey (1986) included northwestern Alberta to the British Columbia border.
Atlas distribution includes all of northern Alberta south to the Peace River District, east to Lesser Slave Lake, south over the Athabasca River to the Rocky Mountain House area, and east through the Red Deer region to the Saskatchewan border.

Habitat:
Breeding habitat in Alberta includes thick stands of willow and alder along streams, and dense shrubs and bushes in swamps near the forest edge. In general, tall sunlit thickets in or near quiet water appear to be essential (Salt and Salt 1976). Other common features include the moist low-lying areas and glades of deciduous or mixed woodland. Francis and Lumbis (1979) specifically described appropriate habitat in the Fort MacKay area as mesic deciduous or mixed wood, taller than 10 m, with a substantial deciduous undergrowth taller than 1.5 m and a ground slope equal to or greater than 15 degrees (Pinel et al. in prep.).

Nesting:
Like many other warblers that tend to breed in remote forests, little is known of the breeding biology of this species. The female builds a cup-shaped nest on or near the ground in a variety of localities: under a sapling, in a mossy hummock, in a decaying log or stump, in the roots of an overturned tree, or, rarely, in a cavity in a bank. It is a bulky mass of leaves and grass, which is lined with plant fibres and fine rootlets. The 3-5 white eggs are spotted with browns. Incubation is carried out by the female, possibly with some assistance (Ehrlich et al. 1988). Incubation time has not been recorded. The male often displays anticipatory feeding behavior; food is brought to the nest before hatching. When flushed from the nest, the female may perform a distraction display. Both adults will tend to the young, even after fledging. The Canada Warbler is a common cowbird host.

Remarks:
The male has a black forehead and crown, bluish-gray upperparts with no wing bars or tail patches, a yellow line extending from the base of the bill back to a yellow eye ring, a black area beneath the eye extending down the side of the throat and forming a streaked necklace across the upper breast, bright yellow underparts, and white undertail coverts. The duller female has a gray forehead and face and a dusky necklace.
Used more frequently before mating, the male's song is roughly notated as "chip - chupety, swee, ditchety". There is a definite pause following the "chip".
This species eats mainly insects and spiders and catches them in three different ways. It may hover and glean, foliage glean, or ground glean. It will also flycatch along the margins of its shrubby habitat.
Canada Warblers arrive in the province from the east in late May or early June. Fall migration is conducted during the last two weeks of August. It is problably the last warbler to arrive in spring and the first to leave in fall.

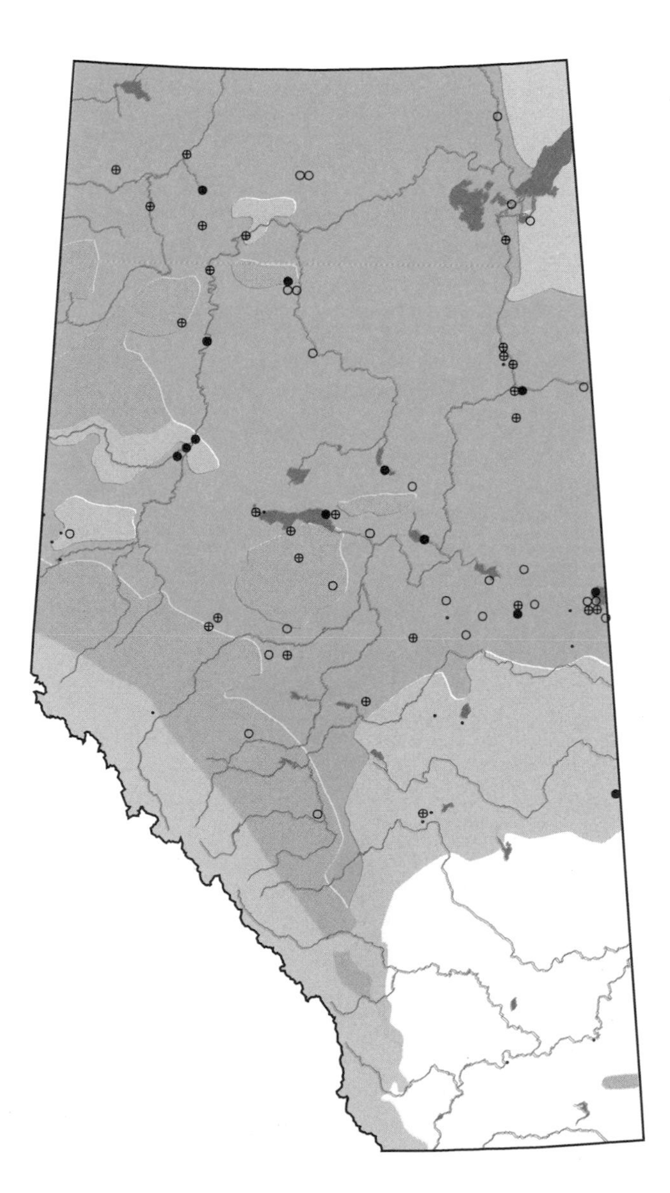

Yellow-breasted Chat

Icteria virens

YBCH

Status:

First observed in 1941, the Yellow-breasted Chat has expanded its range since that time (Salt and Salt 1976). More recently, Wallis (1977) found it to be fairly common in appropriate habitat along the lower Red Deer River.

Distribution:

Salt and Salt (1976) outlined provincial distribution as southeastern Alberta north to Empress and Trochu and west to Lethbridge and Beiseker. More specifically, it was predominantly distributed in the Milk, South Saskatchewan, and Red Deer river valleys and emanating coulees. The slopes of the Cypress Hills were also included. From observations during the 1970s, Pinel et al. (in prep.) included the areas south of the Cypress Hills and around Manyberries. During the Atlas Project, distribution agreed with these earlier observations, with the inclusion of one record from Waterton Lakes National Park.

Habitat:

This unique warbler of the Grassland region nests among the dense willow, birch, saskatoon, and rose undergrowth of the ancient cottonwood stands of prairie rivers and creeks. In particular, it seems to prefer thickets of buffaloberry, hawthorn, and rose growing along shady banks (Salt and Salt 1976). Further away in coulees and gulleys, it may breed in shrubbery with few or no tall trees present (Salt 1973).

Nesting:

The female constructs a large, bulky cup-shaped nest of coarse grass and leaves and lines the interior with fine grasses. This well-concealed structure is positioned within a bush or in a small tree, between 1 m and 2 m above the ground. Given optimal habitat, this species may nest in loose colonies, although individual territories are defended.

The female lays 3-5 white eggs, which are spotted with brown. Incubation, conducted by the female, is at least 11 days. The female broods the nestlings. They are tended to by both adults, which feed them by regurgitation for the first few days and switch to beetles, grasshoppers, butterflies, and hairless caterpillars later. Young require at least eight days before fledging. Because this species is a late nester, fledging does not occur until mid-July. This species is known to raise two broods in other parts of its range. The Yellow-breasted Chat is frequently parasitized by cowbirds.

Remarks:

The largest of the North American warblers (17.8 cm), the adult male Yellow-breasted Chat is grayish green above with conspicuous white lines bordering the black eye patch. The underparts consist of a bright yellow throat and breast and a white abdomen. A long tail and heavy bill are distinguishing features. The female, similar in appearance, has paler white lines and is paler yellow below. Juveniles resemble the adults, but are greener above.

Solitary and shy, the Chat is most often detected by the male's distinctive song or chatter. Usually sung from a low perch encased by thick shrubbery, it is a series of squeals, squawks, chuckles, and cackles, all mixed with occasional loud clear whistles. However, the first-time observer should be careful, for his song resembles those of the Gray Catbird, Northern Mockingbird, and Brown Thrasher.

Although both are foliage gleaners, females tend to forage lower in the shrubbery and on the ground. This species eats weevils, beetles, ants, moths, and caterpillars. Berries make up a large portion of their diet in late summer.

Because of sparse habitation in their breeding areas and their proximity to the border, few migration data are available. The chat is thought to arrive in late May to early June. Records are too few to make an accurate determination of fall migration. There have been a few sightings west of its range in August and one in September (Pinel et al. in prep.).

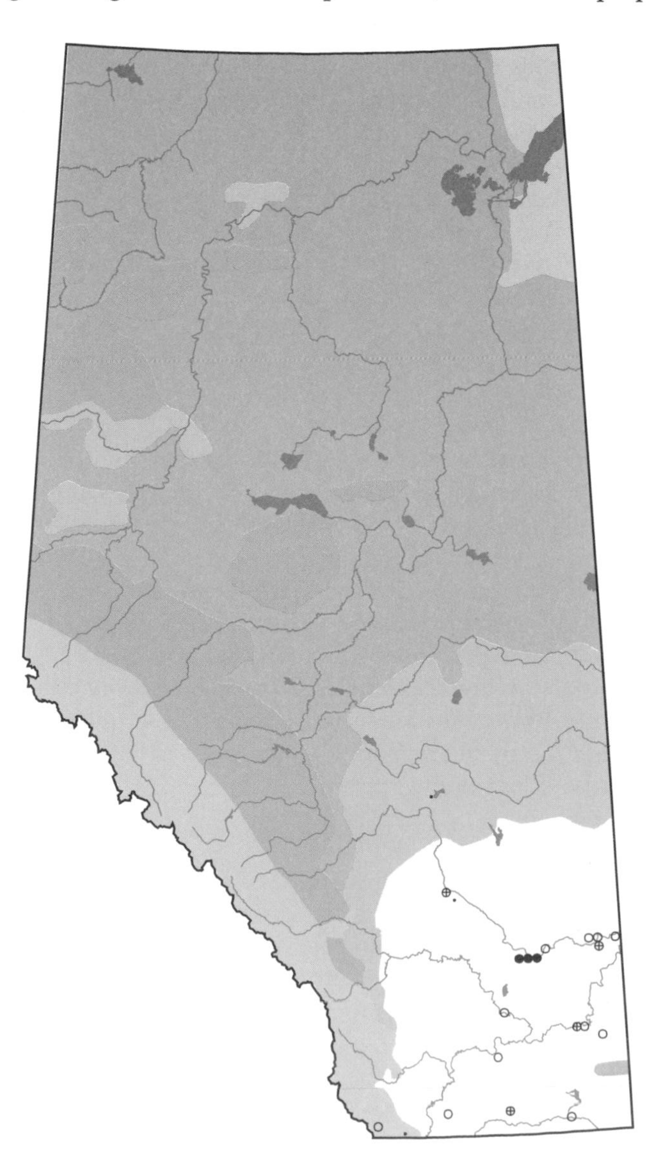

Western Tanager

Piranga ludoviciana

WETA

Status:
The Western Tanager is common in the southern portion of the Boreal Forest region, and is widely distributed in the northern parts of the province. Holroyd and Van Tighem (1983) reported this species as an uncommon summer resident in the mountain parks. Atlas surveys recorded confirmed breeding in all the mountain parks.

Distribution:
The Western Tanager breeds in western North America from Alaska, the southern Mackenzie, and central Saskatchewan, south to Baja California and Texas. Salt and Salt (1976) reported the distribution of this species as northern Alberta, south to about the North Saskatchewan River, and in the mountains and foothills south to the Waterton Lake National Park. Alberta Bird Atlas data show that the Western Tanageris also found in the Parkland Natural Region between the North Saskatchewan and Red Deer rivers, where there were 46 breeding records. It was recorded in 44% of surveyed squares in the Canadian Shield, 24% in the Boreal Forest, 20% in the Rocky Mountain, 14% in the Foothills, and 8% in the Parkland region. It is a scarce transient in most of southeastern Alberta (Salt and Salt 1976). It also nests sparsely in the Cypress Hills (Godfrey 1986), with one probable breeding record from Atlas surveys.

Habitat:
Boreal and montane forests are the main habitat of the Western Tanager in nesting season. The species prefers open coniferous and mixed wood forests, though it is occasionally seen in deciduous woods with few or no conifers and avoids dense coniferous forests (Salt and Salt 1976). In the mountains, highest densities are found in Douglas fir and aspen/lodgepole pine forests (Holroyd and Van Tighem 1983). During migration, the Tanager frequents a wider variety of habitats, but is always found in forests and usually among the higher branches of trees (Salt and Salt 1976).

Nesting:
The nest of this species is a flat, loosely constructed cup of twigs and grasses, lined with grass, fine rootlets, and hair. It is built in a conifer (occasionally in a deciduous tree) at various heights up to 15 m, usually in a fork well out on a branch (Harrison 1978).
The female lays 3-5 eggs, which are pale blue with brown speckles. She alone incubates for 13 days, and is not easily flushed from the nest. Both sexes tend and feed the nestlings. The young leave the nest at about 13-15 days (Ehrlich et al. 1988). The number of broods is not reported.

Remarks:
The male Western Tanager is a striking bird with a bright red head, yellow-green bill, and canary yellow neck, rump, and underparts. His back, wings, and tail are black, and he has two conspicuous pale yellow or whitish wing bars. In the fall, his head is olive green and is tinged with red.
The female is green olive above and yellow below, with grayish brown wings and tail and pale wing bars. She resembles a female oriole, but has a heavier bill, greener plumage, and a calmer, more sluggish nature.
The male's song resembles that of a robin, but is hoarser and lower (Godfrey 1986).
This species eats insects and fruits, feeding its young mainly insects and larvae. It feeds among trees and bushes, and sometimes hawks insects out of the air.
Western Tanagers arrive in southern Alberta in early to mid-May and later in the month in the mountains. They depart again in the second week of August, with a few stragglers remaining into mid-September. No overwintering is reported.

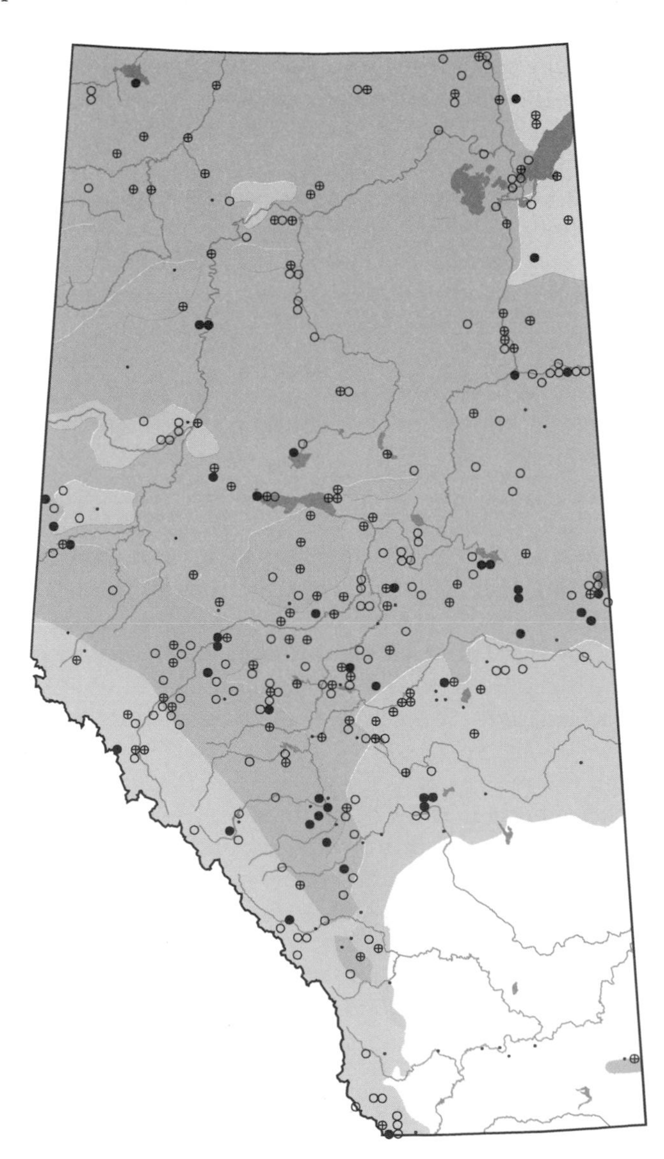

Rose-breasted Grosbeak

Pheucticus ludovicianus

RBGR photo credit: T. Webb

Status:
Although the precise status in Alberta is not known, the Rose-breasted Grosbeak is a fairly common breeding bird where suitable habitat exists.

Distribution:
This colorful grosbeak breeds in central and eastern North America from northeastern British Columbia and the south Mackenzie District, east across southern Canada and the northeastern United States. Atlas data confirm the northern breeding range of this species as reported by Salt and Salt (1976) and Godfrey (1986). However, they reported the southern limits to be Rocky Mountain House and Red Deer, with Salt and Salt (1976) stating that it probably breeds further southwest to Bottrel and the Porcupine Hills. Atlas surveys have produced confirmed breeding records to south of Calgary. It was recorded breeding in 26% of surveyed squares in the Boreal Forest region, 18% in the Parkland region, 18% in the Foothills regions, and less than 3% in the other natural regions.

Habitat:
Rose-breasted Grosbeaks nest in deciduous and mixed woods, favoring areas where there are mature trees close to tall shrubbery that is not too dense. They also use areas of second growth. This grosbeak often nests at shrubby edges and clearings, along roads or shorelines or in overgrown fields and pastures. It is active at all levels of the canopy.

Nesting:
This species nests in trees or tall bushes, usually deciduous, generally 1.5 m to 4.5 m from the ground. Smaller sapling trees are preferred. Male and female may both participate in building the nest. It is a shallow flimsy cup of twigs, bark strips, rootlets, fine grasses, and, sometimes, conifer needles, with a lining of fine twigs, rootlets, and hair (Harrison 1978). The nest is sometimes so loosely woven that the eggs can be seen through the bottom.
The typical clutch is 3-5 light blue to greenish-blue eggs, which are spotted with browns and purple. Both sexes incubate for 12-13 days, during which time the male may sing on the nest. The male also helps brood and feed the young, and takes over first brooding if the female renests. Fledging is at 9-12 days, and fledglings depend on their parents for a further three weeks. It is reported that the male teaches the fledglings how to open seeds (Bent 1968a; Rising 1982). The species is usually single-brooded, but double broods have been reported.

Remarks:
The male Rose-breasted Grosbeak is a black-and-white bird with a rosy red breast and wing linings. His song is robin-like, but is more energetic, richer, and more rapidly delivered (Godfrey 1986). The female is a brownish-streaked bird with yellow wing linings. She is sometimes confused with the female Purple Finch, but is much larger. She is also similar to the female Black-headed Grosbeak, but is whiter underneath and her breast is more heavily streaked than that species. The Rose-breasted Grosbeak interbreeds with the Black-headed Grosbeak where their ranges overlap in the western plains (Rising 1982).
These grosbeaks eat insects, seeds, and wild fruits, in almost equal amounts. In spring, buds and blossoms are added to the diet. The young are fed both insects and seeds.
Rose-breasted Grosbeaks arrive in central Alberta in the third week of May, travelling alone or in pairs. The birds leave again in the latter half of August, with a few laggards staying into early September (Salt and Salt 1976). No overwintering has been recorded.

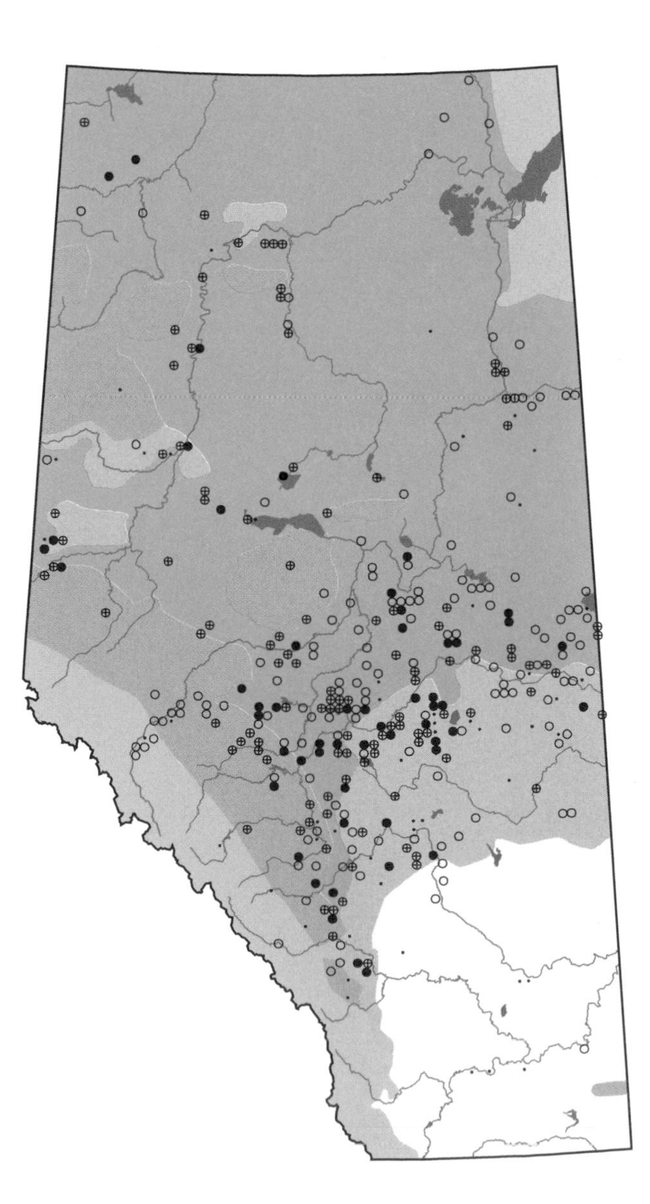

Black-headed Grosbeak

Pheuticus melanocephalus

BHGR

Status:

The range of the Black-headed Grosbeak in the province is extremely limited, and the species as a whole may be in natural decline as a result of interbreeding with the more common Rose-breasted Grosbeak (Wildlife Management Branch 1991). It is an uncommon breeder; there were only 17 records for this species over the five-year survey period of the Atlas and only two of these were confirmed breeding records (in the same square).

There are two subspecies of the Black-headed Grosbeak, with *P. m. melanocephalus* breeding in southern Alberta and the interior of southern British Columbia. *P. m. maculatus*, slightly smaller in size, breeds in coastal southwestern British Columbia (Godfrey 1986).

Distribution:

The range of the Black-headed Grosbeak extends from southern British Columbia, southern Alberta, and central Nebraska south to Mexico. Atlas data indicate that it breeds in the very southern portion of the province, with the majority of records being in Waterton Lakes National Park and surrounding area. Individual records farther east and north agree with Salt and Salt (1976), that this species nests at High River and the Red Deer River valley. In Alberta, the ranges of the Black-headed Grosbeak and Rose-breasted Grosbeak overlap in the Kananaskis area.

Habitat:

This species inhabits open deciduous and mixed woodlands, particularly where there is an understory of tall shrubbery. This includes shorelines, roads, woodland meadows, and other edges. In general, the Black-headed Grosbeak occupies habitat that, if further north, would be used by the Rose-breasted Grosbeak (Salt and Salt 1976).

Nesting:

Black-headed Grosbeaks nest in trees or tall bushes, usually deciduous, often near water. Nests are built in a crotch or fork, generally 1-6 m from the ground. The female builds the nest and is accompanied by the male on flights for nest material. She weaves a loose bulky structure of twigs, stems of forbs, dead leaves, rootlets, and grass. It is lined with fine grasses and rootlets (Harrison 1978). As with the Rose-breasted Grosbeak nest, the eggs can sometimes be seen through the nest bottom.

The typical clutch is 3-4 eggs, which are pale green or blue and spotted with brown. Both sexes incubate (12-13 days) and may sing on the nest. They are both close sitters. Male and female also cooperate in tending the nest, brooding, and feeding the young. Fledging takes place after 12 days, after which the male leaves the nesting grounds and the fledged young are cared for by the female. One brood is raised per season.

Remarks:

The male Black-headed Grosbeak is conspicuous with his black head, black wings with white wing bars, and cinnamon-brown breast, sides, and rump. The brownish-streaked female is very similar to the female Rose-breasted Grosbeak, but is more buffy and less white, and has an almost unstreaked breast.

This species eats insects and spiders and some vegetable matter, including fruits, buds, flowers, and seeds. Nestlings are first fed a soft mash, then soft animal matter such as caterpillars, and later, a variety of insects and vegetation (Weston 1947). The birds forage at any level in trees and bushes. They may hawk for insects or they may feed on the ground.

Both male and female Black-headed Grosbeaks arrive singly in the spring, the males a few days before the females. The birds first appear in late May or early June. Fall departure is probably in late August, with a few stragglers lingering into September (Pinel et al. in prep.).

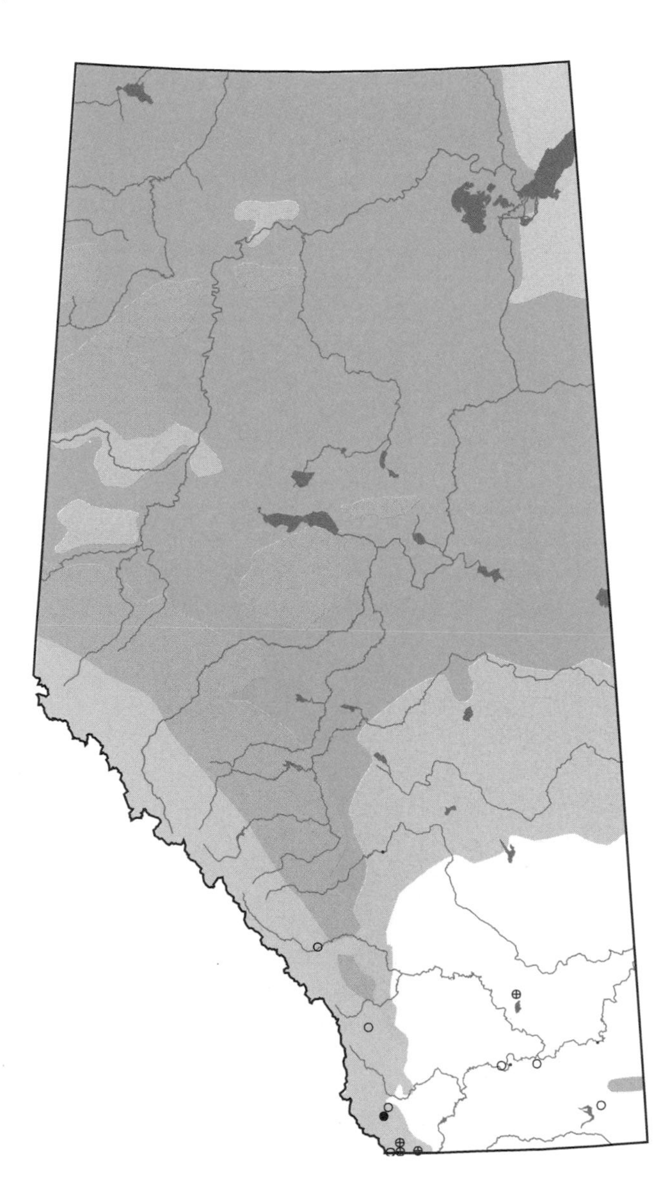

Lazuli Bunting

Passerina amoena

LZBU

Status:
The numbers of Lazuli Buntings have increased in this century with the creation of early successional stages following logging, and the growth of riparian thickets formed by agricultural irrigation systems in arid regions. However, the species also seems to have disappeared from former areas of its range as a result of encroaching urbanization (Ehrlich et al. 1988). In general, where suitable habitat exists, this species is relatively common; in Alberta, it is at the northern periphery of its range and may be seen less regularly than elsewhere.

Distribution:
The Lazuli Bunting breeds from southern British Columbia, Alberta, and Saskatchewan south to Baja California and Oklahoma. Salt and Salt (1976) reported that it nests in southwestern Alberta in the Rockies and their foothills from Jasper south to Waterton Lakes, and east to Castor, Beynon, Lethbridge, and Brooks, and that it occurs casually in the Cypress Hills. Atlas data agreed with this distribution with the exception that no breeding evidence was reported in Jasper National Park. Holroyd and Van Tighem (1983) reported that there are no recent records of Lazuli Bunting in Jasper National Park. Abrahamson (1980) reported it as occurring casually at Sylvan Lake, which was also confirmed by Atlas records. Males that are unsuccessful in attracting mates leave the area, which results in geographic and numerical discrepancies in distribution from year to year (Pinel et al. in prep.).

Habitat:
This species breeds in areas with thickets or other shrubby growth, particularly along water courses or on hillsides, also in burns and clearings. In the mountains, Lazuli Buntings are found in the lower valleys in either deciduous woods with willow and alder thickets or aspen groves with dense undergrowth (Salt and Salt 1976; Holroyd and Van Tighem 1983). In the prairies, dense shrubbery in coulees or along creeks is favored.

Nesting:
Lazuli Buntings nest low in thick tangles of bushes, saplings, or coarse forbs, usually not more than 3 m off the ground. The nest, built by the female in a crotch or fork, is a construction of grass, leaves, rootlets, plant fibres, and forb stems. It is lined with fine grasses and hair.
The female lays 3-4 bluish-white eggs, which she incubates for 12 days. She also broods and feeds the nestlings, though some males assist with the feeding. The young leave the nest after 10-12 days, and may then be fed by the female alone, by both parents, or by the male if the female renests. Up to three broods are reported for this species.

Remarks:
The pattern and color of the male Lazuli Bunting resembles a bluebird, but the lighter blue coloring, white wing bars, stubby bill, and sparrow-sized body help distinguish him. The female Lazuli can be distinguished from all sparrows by her plain, unstreaked, brownish-blue back and rump.
This species tends to be inconspicuous except when the male sings his lively trill from a high open perch in a tree, telephone wires, or the top of the tallest shrub in his territory. He is a persistent singer and often sings in the heat of the day when other birds are silent. Lazuli Buntings feed in trees, bushes, and weeds, or on the ground, where they search for insects and small seeds. The young are fed mainly larvae and insects, particularly grasshoppers.
Male Lazuli Buntings migrate separately from females, arriving on the breeding grounds in Alberta in late May to mid-June. They depart again in late August.

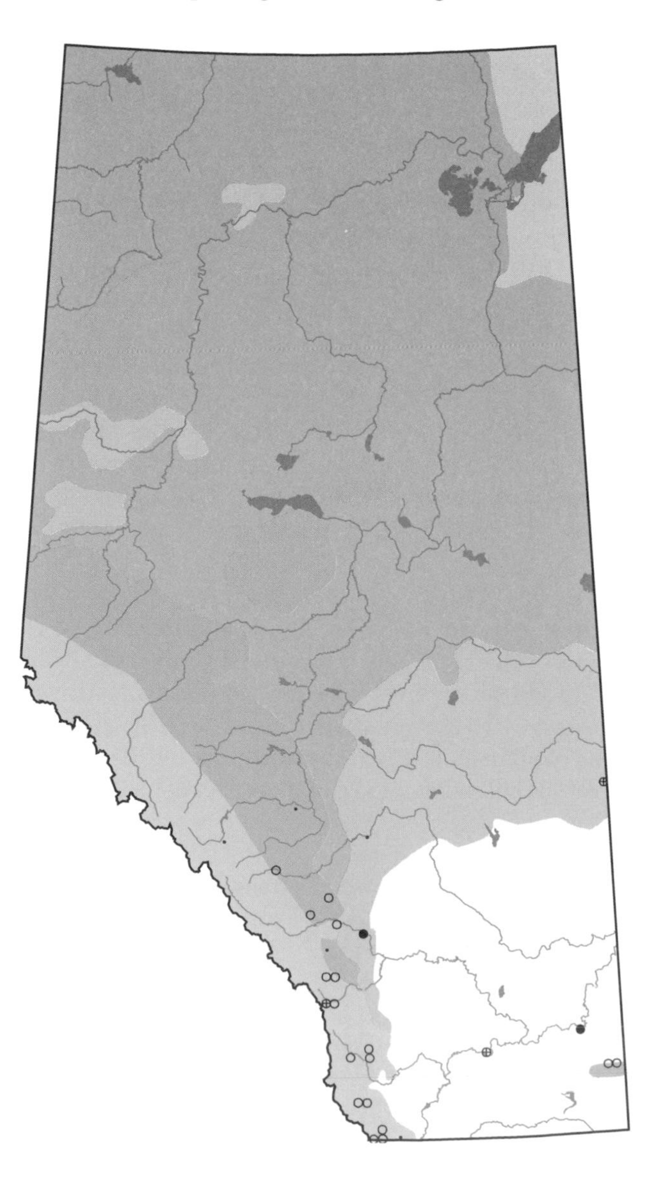

Rufous-sided Towhee

Pipilo erythrophthalmus

RSTO

Status:
The status of this species in Alberta has not been determined, although it appears to be a common resident in the Grassland Natural Region, specifically in the badlands of the Red Deer and Milk rivers. It is less common in the southern portion of the Parkland Natural Region. However, Breeding Bird Surveys and other observations in eastern Canada indicate population declines starting in 1976 (Silieff and Finney 1981; Cadman et al. 1987). Of four subspecies of Rufous-sided Towhee found in Canada, only one, *P. e. arcticus*, breeds in Alberta (Godfrey 1986). There is a possibility of *P. e. oregonus* occuring in the mountains of southwestern Alberta (Pinel et al. in prep.).

Distribution:
The Rufous-sided Towhee is found primarily in the Grassland and southern Parkland natural regions. This catbird-sized sparrow breeds in southern Alberta north to Big Valley (irregularly to Coronation, Camrose and Wainright) and west to Fort MacLeod and Calgary; it has also been reported in the Waterton Lakes area (Salt and Salt 1976).

Habitat:
It is usually seen on or near the ground in dense, brushy cover and commonly frequents the shrubbery along prairie coulees, streams, rivers, the tangles at forest edges and the undergrowth of open woodland. This species avoids areas of intensive agriculture and dense forests.

Nesting:
The nest site selection and building are performed by the female alone. The nest is built on the ground or a little above, in dense, low shrubby growth. A cup of grass, leaves, twigs, weed stalks, and bark strips is constructed, and is then lined with fine grasses and rootlets.
The clutch is typically 3-5 eggs, pale greenish white and spotted with reddish brown and lavender. The female incubates 12-13 days, and she alone broods the nestlings. At this time she is very secretive about the nest. Both parents feed the young by regurgitation, and clean the nest. Fledging is at 10-12 days. In Alberta, the Rufous-sided Towhee is single-brooded. It is frequently parasitized by Brown-headed Cowbirds, and readily accepts the cowbird eggs.

Remarks:
Distinguishing characteristics of this species are black, white and chestnut underparts, a sparrow-like bill, a long tail, and a red iris. Western birds (once called Spotted Towhees) have more white on the back and wing coverts than eastern birds, and western females have much darker brown (nearly black) heads and upper parts (Godfrey 1986).
These birds tend to be inconspicuous, spending much of their time feeding on the ground under thickets and shrubbery. They scratch with both feet at once, revealing seeds and insects, which are their major fare. Spiders, caterpillars, snails, and berries are also eaten.
The male is more visible in summer, when he can be found singing his distinctive call. This call includes a whining nasal alarm note and a song that ends in a buzzy trill (Godfrey 1986).
Spring migrants have been observed early in April, but most arrive in the province in the first two weeks of May. Most birds move south in early September, with some birds lingering into October. Overwintering is rare, with sightings reported from Edmonton and Red Deer.

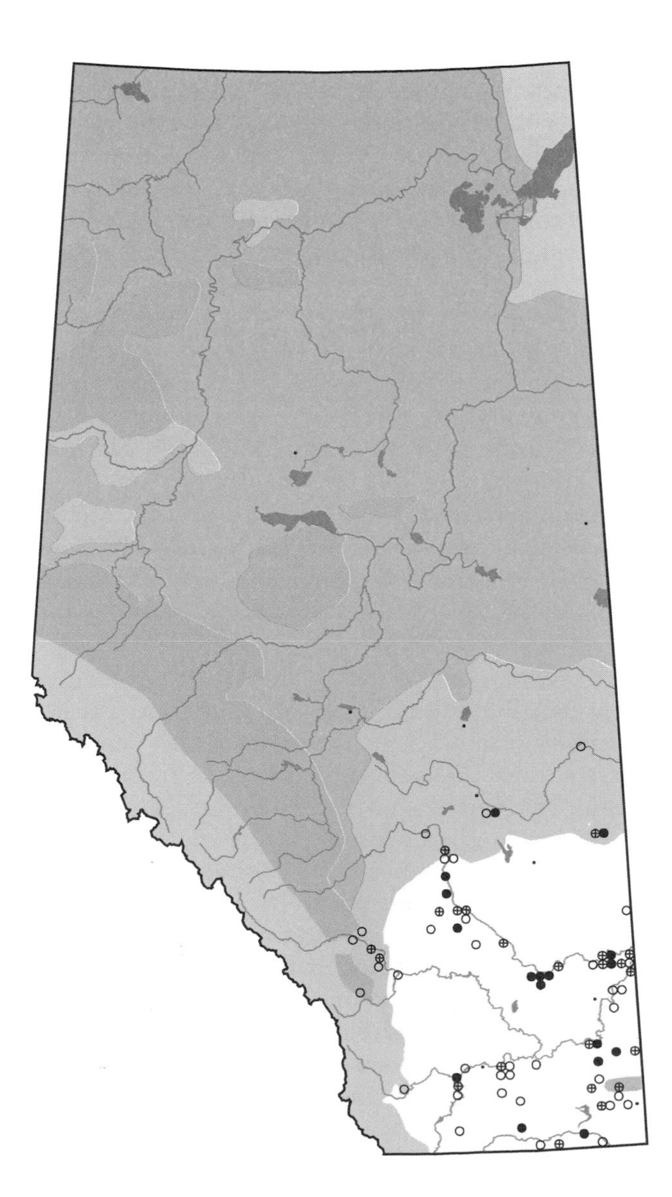

American Tree Sparrow

Spizella arborea

ATSP

Status:

As a nesting species, this bird is very rare in Alberta, with one previous confirmed record from the Cameron Hills (Wershler 1983) and only three confirmed records from Atlas data.

Distribution:

The American Tree Sparrow breeds across the subarctic of North America. In Alberta, the species has been found breeding in the Cameron Hills, a subarctic environment in northwest Alberta. It is also a common spring and fall transient throughout the province, although there are very few observations from the Rocky Mountain region.

Habitat:

In breeding season, this sparrow prefers areas of tundra with low shrubs such as willow, dwarf birch and alder, and scrub conifers along streams and bogs. The birds nesting in the Cameron Hills were in an area of low open black spruce with an understory of Labrador tea, peat moss, and reindeer lichens (Wershler 1983).

Habitat in winter and during migration is usually weedy or fallow fields, fencerows, gardens, and similar areas where birds have access to weed seeds (Godfrey 1986).

Nesting:

Nests are usually built on the ground, often in a tussock of grass, mossy hummock or depression and concealed by grasses or shrubs. These birds occasionally nest in bushes or low trees. The female constructs the nest. The outer layer consists of heavy grass, weed stems, rootlets, moss, lichens, and bark shreds; the inner layer consists of fine dry grass and the lining is of feathers or hair (Bent 1968a).

The female lays 3-6 eggs which are a pale, green-blue with brown speckles. She alone incubates for 12-13 days, but both parents tend and feed the young. Young birds leave the nest at 9-10 days, but are unable to fly for another 5-6 days (Harrison 1978). The parents feed the fledglings for another two weeks after fledging. One brood is raised per season.

Remarks:

A chestnut cap, single dark spot in the middle of a pale breast, and two bold white wing bars are distinguishing features of the American Tree Sparrow. It is most similar to the Chipping Sparrow, but the latter has a black line through the eye and no breast spot.

The American Tree Sparrow was named by early settlers after the European Tree Sparrow because both have rufous caps. However, the name of this sparrow is somewhat misleading, since it is usually found on the ground or in shrubbery and undergrowth rather than in trees (Root 1988).

The species is not particularly difficult to locate on its breeding grounds; a territorial male perches conspicuously on a low tree or shrub and sings frequently (Cadman et al. 1987). The song, two or more higher notes ending in a warble, is distinctive and quite far-carrying. In breeding season, the birds forage in the thick tangle of shrubs for seeds and insects. Young birds up to 3-4 weeks of age are fed only animal matter. During migration and in winter, the sparrows are mainly seed eaters, consuming vast amounts of weed seeds, an aid to prairie farmers.

This species is among the earliest of the spring migrants. Early arrivals appear in late March and numbers peak in mid-April. Most leave Alberta by late April to early May and return in mid-September. They leave the province again by the end of October. Small numbers sometimes overwinter in southern Alberta, with reports as far north as Edmonton, Fort Saskatchewan, and the Peace River/Grimshaw area.

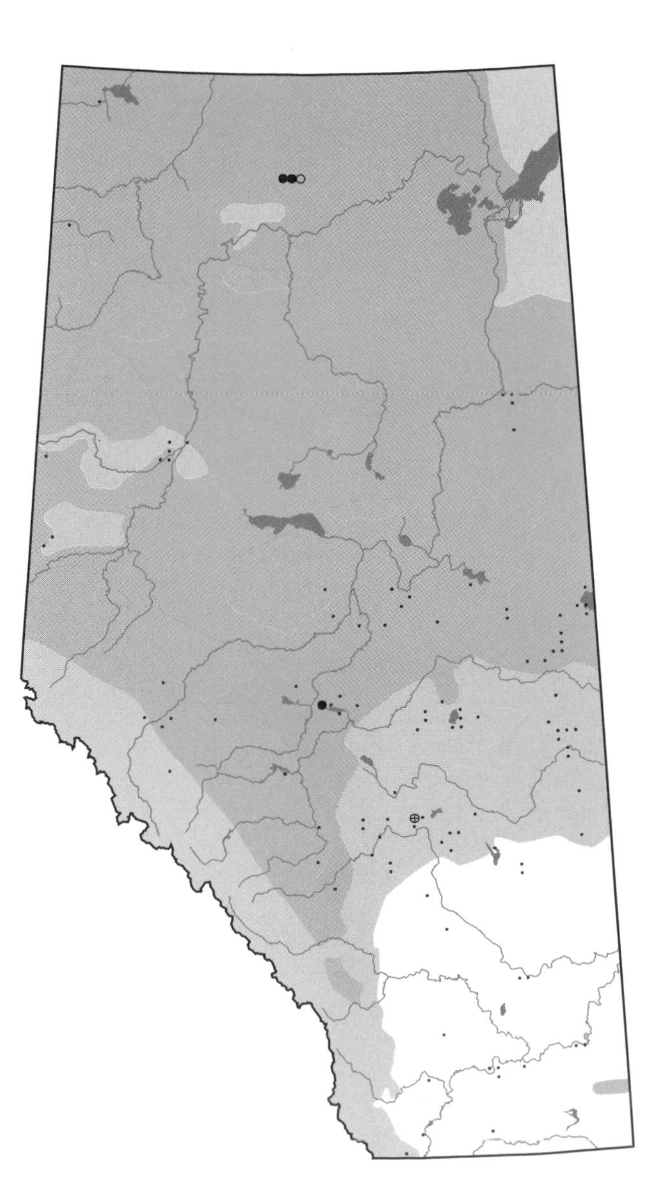

Chipping Sparrow

Spizella passerina

CHSP

Status:
The Chipping Sparrow population dropped dramatically around the turn of the century, probably due to the insurgence of the House Sparrow (Summers-Smith 1963). However, the species is now common and abundant throughout most of its range, where suitable habitat exists. Its adaptation to man-made environments has undoubtedly been a major part of its present success (Cadman et al. 1987). Of the two subspecies present in Canada, one, *S. p. boreophila*, is found in Alberta (Godfrey 1986).

Distribution:
This species breeds across Canada from the northern limit of the boreal forest south to the border. It is found throughout Alberta, but is scarce in the Grassland Natural Region, where breeding habitat is restricted to wooded river valleys.

Habitat:
In breeding season, the Chipping Sparrow frequents open deciduous, mixed wood, and coniferous forests, and the openings and edges of woodlands. It is a common summer inhabitant of Alberta farmsteads and residential areas with gardens, lawns, and shrubbery. It nests in shrubs and trees and feeds in adjacent meadows, lawns, or other open grassy areas. In the mountain parks, it is most abundant in dry montane coniferous forest, especially more open stands (Holroyd and Van Tighem 1983). In migration, birds are often found in a variety of disturbed habitats such as fields, and along roads and railways.

Nesting:
Chipping Sparrows nest in bushes, vines, and small trees, and very rarely on the ground. They prefer a conifer and the nest is generally less than 3 m off the ground. It is usually situated close to an open, short grass area. The nest is a compact cup of dead grass, weed stems, and rootlets. It has a lining of animal hair and sometimes fine vegetable material. In most accounts the female is reported to build the nest, but Reynolds and Knapton (1984) report both sexes involved in gathering material and construction. The typical clutch is 3-5 eggs, blue and lightly spotted with browns and sometimes lavender. Chipping Sparrow nests are often parasitized by cowbirds. The female alone incubates 11-13 days. She will sometimes be fed by the male during this period. He will also occasionally take a second mate at this time. The female broods the nestlings for most of their first 4-5 days, with both parents feeding the young and guarding the nest. The young leave the nest at 9-12 days and fly at 14 days. Birds then wander in family groups prior to migration.

Remarks:
This small common sparrow is easily distinguished by its rufous crown, unstreaked gray breast and black eye stripe with white eye line above it. It is sometimes confused with the American Tree Sparrow, but the latter has a dusky breast spot and very different eye markings. The song of the Chipping Sparrow is a rapid monotone "trill" and its call is a high, hard "chip", from which it gets its name.
Birds spend much of their time feeding inconspicuously on the ground in short grasses, hedgerows, or low shrubs. In breeding season, the birds focus on insect prey, along with a few seeds; in winter seeds are the main source of food.
The arrival of the first spring migrants seems variable, beginning as early as mid-April, but peak numbers are recorded in southern Alberta in mid-May (Pinel et al. in prep.). Fall migration begins in late July and early August and peaks in the first half of August. The last Chipping Sparrows have usually left the province by the third week of September, although some have been recorded in early October.

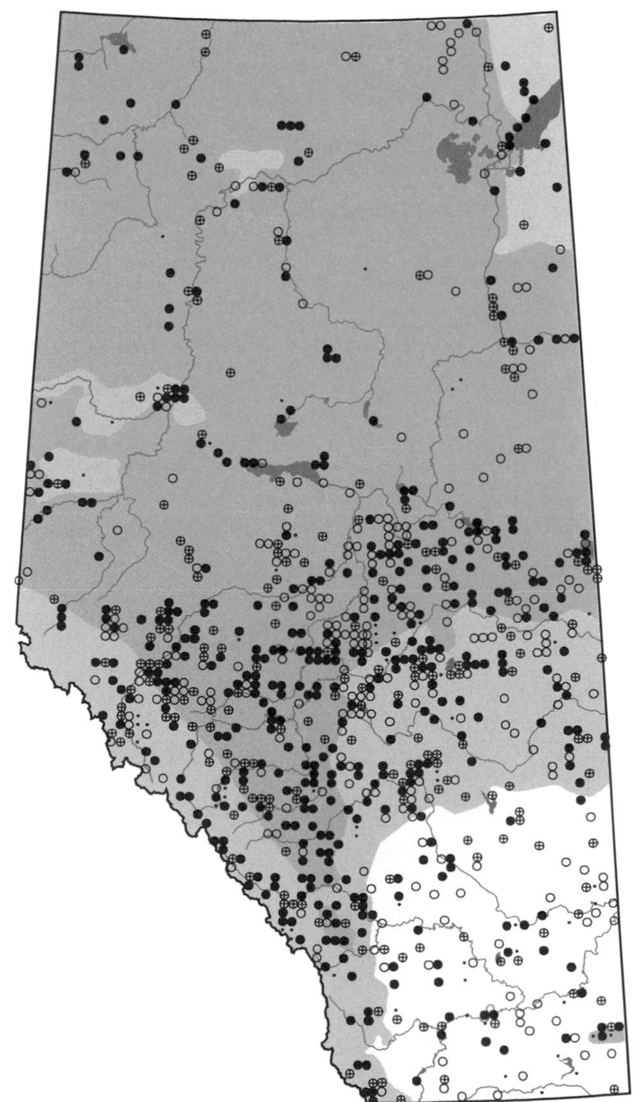

Clay-colored Sparrow

Spizella pallida

CCSP photo credit: T. Webb

Status:
The Clay-colored Sparrow population within Alberta is healthy and stable (Wildlife Management Branch 1991).

Distribution:
This species breeds in the Canadian interior from the south Mackenzie to the Quebec border (Cadman et al. 1987) and south to the American border. It is found in suitable habitat throughout Alberta. It is most common in the Parkland and Grassland regions, where it was recorded in 52% and 35% (respectively) of squares surveyed. In the Rocky Mountain Natural Region it is usually restricted to river valleys.

Habitat:
This is a bird of field and brush. It is found in shrubby uncultivated fields and pastures, tall shrubbery in meadows, brushy openings, edges and burns, and thickets along streams, ponds, lakes, muskegs, and swamps. In the mountains, this species inhabits willow and birch shrubbery in meadows and forest/grassland interfaces (Holroyd and Van Tighem 1983). Birds are sometimes found in suitable habitat in urban situations (McNicholl 1977; Salt 1964).

Nesting:
The nest is built close to the ground, usually no higher than 2 m, in a dense grass clump at the base of a shrub or in the low branches of a bush or small tree. It is usually well-concealed in entanglement or by branches. The female builds the nest, but is accompanied by the male on trips to collect material. She weaves a cup of grasses, weed stems, and rootlets, and provides a lining of fine grasses, rootlets, and, occasionally, hair.
Clutches are typically 3-5 eggs. Eggs are green-blue and are spotted with browns and blacks, usually around the larger end. Nests are often parasitized by cowbirds, but the sparrows tend to desert infiltrated nests (Salt 1964). Both sexes incubate for 10-11 days. The male frequently carries food to the female during this period. He is a vigorous defender of the nest. Hatched young are tended by both parents, and both sexes brood. The young leave the nest at 7-9 days and hide in the bushes. They can fly at 14-15 days. This species may occasionally raise two broods. Males generally return to previous nesting territory; females have a low rate of return.

Remarks:
The Clay-colored Sparrow is a small bird with a plain white breast and streaked upperparts. It has a brown crown with black streaks and a whitish central stripe and also a brown ear patch bordered above and below by a thin, dark brown line. A conspicuous pale stripe separates the ear patch and dark whisker stripe. This species' song, a short series of insect-like buzzes, is distinctive. Males are not shy and retiring and will sing for long periods from their singing posts.
These sparrows forage for both seeds and insects during breeding season, also eating catkins and buds in the spring. Adults collect all food for their young away from the breeding territory, an unusual behavior for an open country sparrow (Quinlan et al. 1990).
Spring migrants arrive in late April or early May, and generally leave again in August and into the last two weeks of September. In the fall, some migrate with Chipping Sparrows and Brewer's Sparrows. Unlike many sparrows, these species migrate during the day (Root 1988).

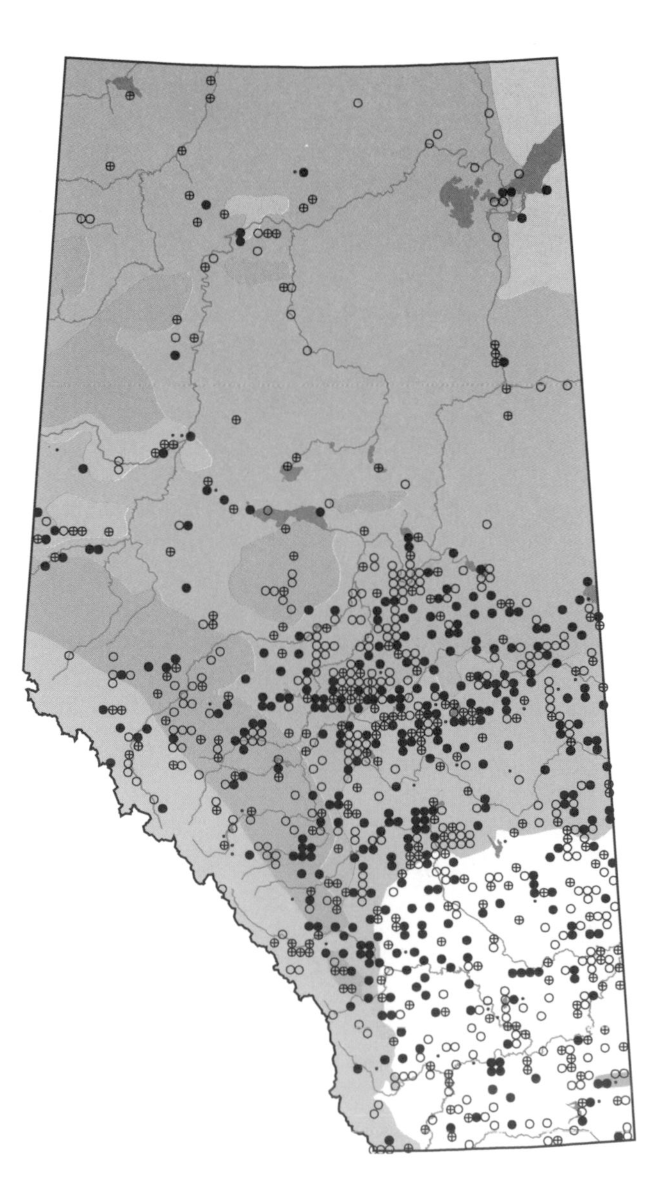

Brewer's Sparrow

Spizella breweri

BWSP

Status:
The status of the Brewer's Sparrow in Alberta is uncertain, but population decline is suspected. The prairie population of this species may be in jeopardy if natural sagebrush habitat, upon which it depends, continues to be lost (Wildlife Management Branch 1991).

Distribution:
There are two subspecies of Brewer's Sparrow in Alberta, separated by geography and habitat preference as well as biological and behavioral characteristics. *S. b. breweri* breeds in the extreme south and southeast of the province, west to Brooks, Taber and Milk River and north to the Blindloss-Empress area. Calgary and Drumheller are extralimital ranges for this subspecies. Populations within this breeding range are localized and the species is not common (Pinel et al. in prep.). *S. b. taverneri*, once known as the Timberline Sparrow, is found in the Rocky Mountains from Jasper National Park south to the Waterton Lakes.

Habitat:
S. b. breweri, the prairie subspecies, inhabits the semi-arid plains, where short grass, cacti, and low bushy shrubbery, mainly sagebrush and silverberry, are found.
S. b. taverneri prefers mountain meadows at the timberline, particularly where there are thickets of willow and birch/willow or subalpine fir krummholz (Holroyd and Van Tighem 1983).

Nesting:
In both the prairies and mountains, this species nests low in small shrubs. The nest is a compact cup of weed stems, grass and rootlets lined with fine grasses and hair.
The clutch is usually 3-5 eggs, greenish blue and speckled with reddish brown. The female incubates 11-13 days; it is unknown whether or not the male participates. He may bring the female food during courtship and while she is incubating (Nordin et al. 1988). Brewer's Sparrow nests are sometimes parasitized by cowbirds, but the nest is usually then abandoned and a new nest constructed (Biermann et al. 1987). Young birds fledge at 8-9 days. Little information is available on parental care.

Remarks:
Brewer's Sparrow was named in 1850 in honor of Thomas Mayo Brewer, a prominent Boston physician and naturalist (Bent 1968). It is a small slender sparrow with streaking above, plain breast, brown crown with fine black streaks, white eye ring, and pale brown ear patch. It is most similar to the Clay-colored Sparrow, but lacks the white crown stripe, the gray hind neck and the distinct face markings of that species. Brewer's Sparrows are very wary; when flushed they quickly fly to cover or retreat on the ground.
The song of this sparrow is surprisingly varied and sustained, with many abrupt changes in pitch and tempo. It combines buzzes and canary-like trills with various nervous lisps and chips (Godfrey 1986).
In the breeding season the diet of this species includes insects, spiders, and seeds. The young are fed primarily on soft-bodied insects. In winter the diet shifts almost exclusively to seeds.
Birds arrive in southern Alberta in early May, and towards the end of the month in the Rockies. They move south with other small sparrows in August and September.

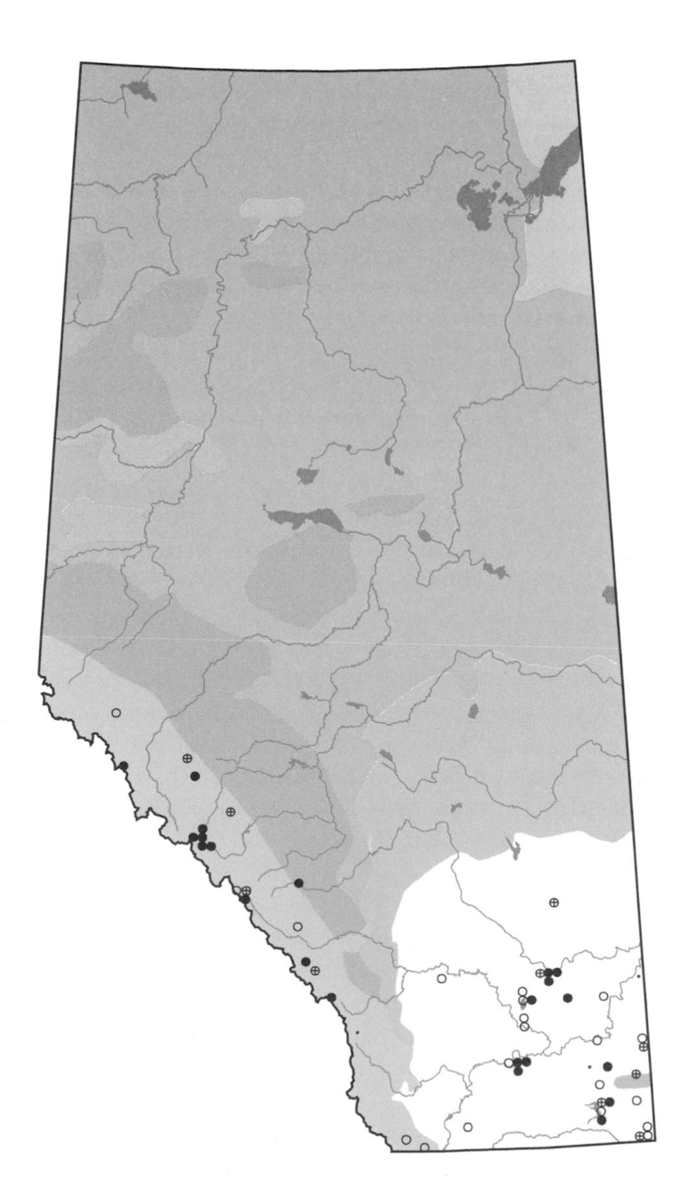

Vesper Sparrow

Pooecetes gramineus

VESP

Status:

Of the two subspecies found in Canada, the paler, grayer race, *P. g. confinis*, nests in Alberta (Godfrey 1986).

The status of the Vesper Sparrow in Alberta has not been fully determined. Atlas data confirm Salt and Salt (1976), that it is one of the most common and best known species in the Grassland Natural Region. It is also common in the Parkland and southern Boreal Forest regions. Reports from Ontario and Michigan indicate that, although it is a common species in those areas, there appears to be a decline in abundance (Cadman et al. 1987; Brewer et al. 1991). In both areas, changes in farming practices, use of chemicals, and large-scale tillage have been implicated in this decline.

Distribution:

This species breeds throughout the province, and was recorded in 44% of surveyed squares in the Grassland Natural Region, 39% in the Parkland region, and approximately 20% in the Boreal Forest and Rocky Mountain regions. It is scarce in the extreme north and data from the Rocky Mountain Natural Region show that it is generally restricted to the valleys.

Habitat:

This is a bird of open, weedy, fairly dry situations which is attracted to grassy margins along roads, railways, fields, fencelines, grassy weedy fields and pastures, meadows, recent burns, and grassy coulee slopes. In the mountains, it inhabits dry, sparse grassland where there are a few scattered pines or spruces (Holroyd and Van Tighem 1983).

Nesting:

The nest is built on the ground, often in a slight depression or hollow. It is usually well-hidden in grass or weeds, or under small trees and shrubs. It is occasionally left exposed. It is a loosely woven cup of grasses, rootlets, and, occasionally, weed stems. It is lined with fine grasses and hair.

There are typically 3-5 eggs in a clutch, whitish and speckled with browns. Incubation is 12-13 days by the female, possibly with help from the male. The female, if flushed, will generally give a broken wing display, or fan her tail to show her white tail feathers. (Cadman et al. 1987). The nest may sometimes be parasitized by cowbirds. Only the female broods but both parents feed the young. This species is commonly double-brooded and the male cares for the first brood when the female renests. The young leave the nest at 9-13 days, unable to fly, and are tended by their parents another 20-22 days.

Remarks:

The Vesper is a streaked, pale grayish sparrow, distinguishable by its white outer tail feathers (junco-like), chestnut shoulders and white eye ring. It has an attractive song which it often sings in twilight after sunset, when other birds have stopped singing, hence the name "vesper" (Root 1988). This species was originally known as the Bay-winged Bunting (Bent 1968).

Most of this bird's time is spent on or close to the ground, where it hunts for insects such as grasshoppers, beetles, cutworms, and moths of destructive species. This species also includes weed seeds in its diet (Bent 1968a). Unlike the American Tree Sparrow, which is known for its habit of water bathing, the Vesper Sparrow has a propensity for dust bathing (Ehrlich et al. 1988).

Vesper Sparrows arrive in the province from mid-April to early May and begin to depart again in August. Most birds having left by mid-September.

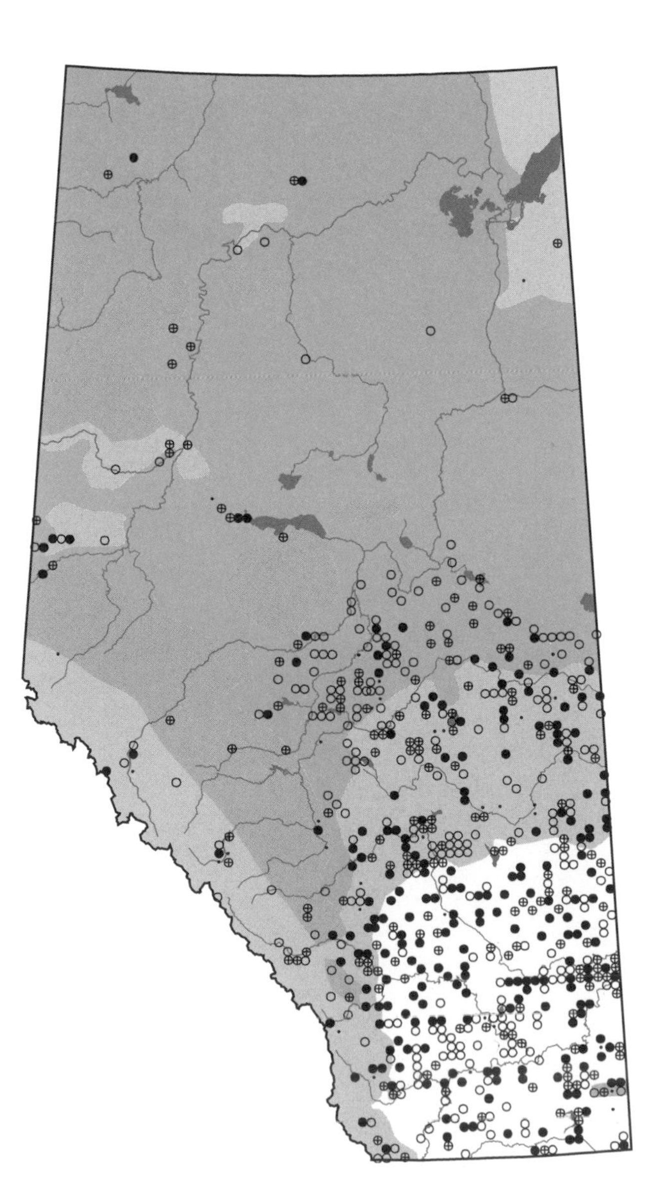

Lark Sparrow

Chondestes grammacus

LASP photo credit: R. Gehlert

Status:

The status of this species is Alberta has not been clarified (Wildlife Management Branch 1991). Of the two subspecies of this sparrow in Canada, *C. g. strigatus*, with paler upper parts and narrower streaks, inhabits Alberta (Godfrey 1986).

Distribution:

The Lark Sparrow occurs in North America primarily west of the Mississippi River, breeding in southern British Columbia, the southern portions of the prairie provinces and southern Ontario. It breeds in Alberta in the eastern portion of the Grassland region, west to Lethbridge and Calgary, and north to Big Valley, Cadogan and Czar. It is local in ditribution, favoring river valleys and associated coulees, with populations along the lower Red Deer, South Saskatchewan and Milk rivers and in the Middle Sandhills (Pinel et al. in prep.). Atlas data indicate that it nests north into the Parkland region in the Wainwright area.

Habitat:

In Alberta, this species is associated with semi-open areas in river valleys and coulees, as well as sandhills. It prefers grassland-badland ecotones with sagebrush, sandhills with scattered cottonwoods, river bottoms in sandy areas, and sandy parkland (Wershler et al. 1991). It is occasionally found in well-grazed pasture, abandoned fields, or residential areas with shade trees and shrubs. In general, the habitat contains a mixture of brush, shrubbery, grass, and bare earth (Salt and Salt 1976).

Nesting:

This sparrow nests on the ground, usually under weeds, a grass tuft or other low vegetation, or sometimes in a bush. Nests on the ground are a depression lined with grass, rootlets and hair; nests off the ground are constructed of grasses and weed stems, with similar lining. Both male and female search for the nest site, with the female making final selection and building the nest. This species may re-use its own nest and occasionally uses those of other species (Ehrlich et al. 1988).

The average clutch is 3-6 eggs, whitish and scribbled with browns and blacks which usually wreath the larger end. Cowbirds may parasitize the nest, but eggs are tolerated only late in incubation or late in the season. The female incubates 11-13 days. She is a tight sitter and can be closely approached. If flushed, she may run along the ground, tail spread, feigning injury. Both sexes feed and tend the young, which leave the nest at 9-10 days, barely able to fly. After fledging, birds tend to leave the nesting area and form flocks (Bent 1968). This species is single-brooded, with a prolonged nesting period (Harrison 1978).

Remarks:

One of North America's largest sparrows, the Lark Sparrow is easily recognized by its chestnut, black and white head markings, dark breast spot and white-tipped tail. Its sweet, melodious song, often sung at night, and also in flight, is also distinctive. The male bird has a habit of dropping his wings and spreading his tail during courtship. He is also quite pugnacious during this period; mid-air combats and chases with other males are common. This sparrow spends much of its time on the ground in the open, feeding on insects and weed seeds. Grasshoppers and crickets are particular favorites.

Birds arrive in southern Alberta in mid- to late May and depart in late July to early August. There is one overwintering record from Red Deer in 1982 (Young 1983).

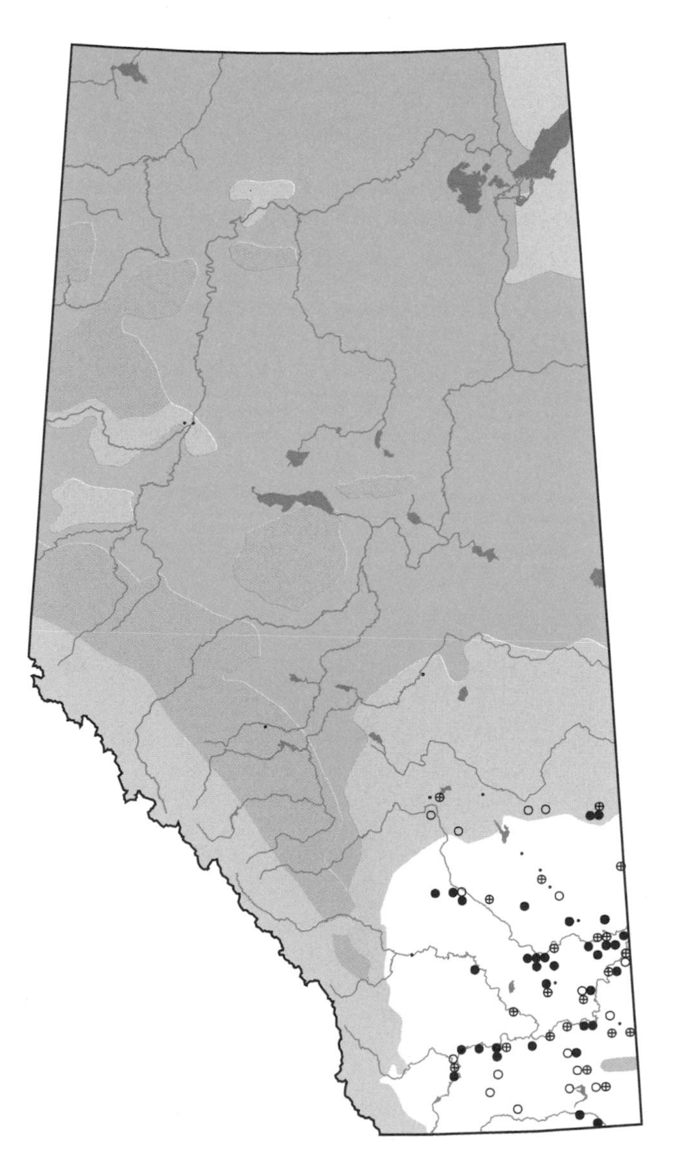

Lark Bunting

Calamospiza melanocorys

LKBU

Status:
The status of the Lark Bunting in Alberta is not precisely known (Wildlife Management Branch 1991). However, destruction of native prairie in the mid-western United States has resulted in the disappearance of this species from the eastern and northeastern portions of its range (Ehrlich et al. 1988).

Distribution:
Atlas data confirm that this species breeds in the southern parts of the prairie provinces and nests locally in southern Alberta north to Sibbald, Castor and Youngstown, and west to Calgary and Fort MacLeod, as reported by Salt and Salt (1976). All confirmed breeding records during the Atlas Project occurred in the Grassland Natural Region, with only seven of 154 breeding records occurring outside of this region. It has occurred casually as far north as Dewberry, Beaverhill Lake, and Vegreville, with one sighting at Fort McMurray (Kennedy 1980). It has also occurred west into the Rockies near Banff and Waterton.

Habitat:
Dry, grassy plains and prairies are the preferred habitat of this sparrow, particularly those areas where sagebrush and cactus abound.

Nesting:
The nest is a cup of dry grasses and weed stems lined with fine grasses, plant down, rootlets, and hair. It is built in a small depression on the ground, and is usually sheltered under a grass tuft, weed clump or small bush (Harrison 1978).
The clutch is 3-7 eggs, typically 4-5, which are blue and otherwise unmarked. The eggs are incubated mainly, if not solely, by the female for 12 days. Males stay in close proximity to the nest. Nestlings are tended by both parents and fledge at 8-9 days. This species may sometimes be double-brooded.

Remarks:
The male Lark Bunting is black with bold white wing patches. He resembles the Bobolink somewhat, but the latter has a whitish rump and buffy hind neck, as well as white on the wings (Godfrey 1986). The female Lark Bunting is a brown and white streaked sparrow with a pale line over the eye. She resembles a heavily built Vesper Sparrow, but is browner and always shows a white wing patch (Salt and Salt 1976).
The rich and varied song of the male is often sung in flight. After a rapid ascent to 7-10 m, he floats down, butterfly-like, singing as he goes (Bent 1968).
This species feeds mainly on the ground, gleaning insects and seeds from the prairie grasses. The young are fed mostly insects, with grasshoppers forming a major portion of the diet. Later in the summer the diet shifts to grass seeds, weed seeds, and grain. Lark Buntings do not establish or maintain strong territories and, once nesting is over, form large flocks.
The Lark Bunting begins to arrive in late May and seems to be erratic in occurrence. It appears early in some years and later in others. In some years (1974-76) there were many sightings of the Lark Bunting, but there were no reports in 1978 and few in 1979 (Pinel et al. in prep.). Fall migration begins in late July and early August. A few stragglers may remain into early September.

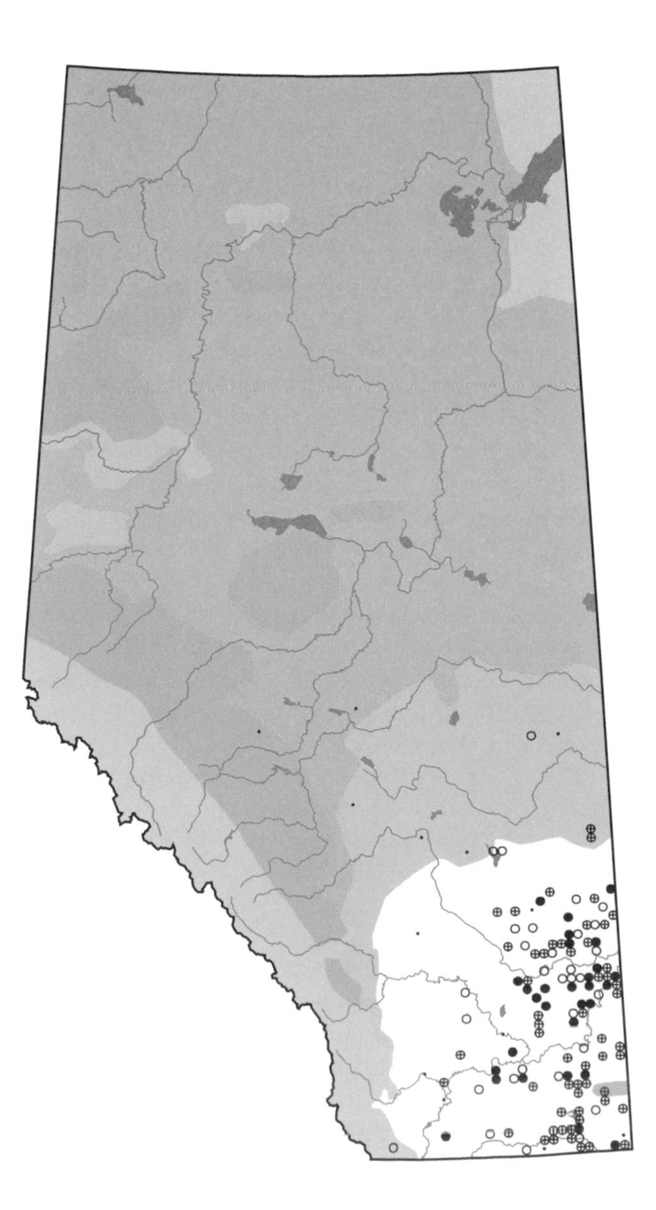

Savannah Sparrow

Passerculus sandwichensis

SVSP

Status:
The Savannah Sparrow population in Alberta is healthy, stable and widespread (Wildlife Management Branch 1991). Of the nine subspecies of this sparrow found in Canada, only one, *P. s. nevadensis*, breeds in Alberta (Godfrey 1986).

Distribution:
This species breeds throughout Canada, north to the Arctic coast; it is found in all parts of Alberta. It was found in over 50% of squares surveyed in the Parkland region and in about 30% of other regions except the Canadian Shield, where it was recorded in 12% of surveyed squares.

Habitat:
A bird of open habitats, the Savannah Sparrow is the most widely distributed of Alberta sparrows during the breeding season (Salt and Salt 1976). In northern areas it favours sedge meadows, marshes, bogs, burns, and clearcuts. In the mountains it is most abundant in moist meadows (Holroyd and Van Tighem 1983). In central and southern Alberta, preference for habitat is given to edges of prairie sloughs, marshes, moist grasslands, hayfields, overgrown meadows and fields, and any damp, low-lying area with dense vegetation. The Savannah Sparrow avoids areas of short grass, but is almost always found where vegetation is less than 1 m tall (Quinlan et al. 1990).

Nesting:
The nest is constructed on the ground in a shallow, scratched out hollow. It is built of coarse grass, plant stems and occasionally some moss, and is lined with fine grasses and hair. It is usually well-concealed by over-hanging grass, weeds, sedges or other vegetation.
The female usually lays 4-5 eggs, though second clutches are smaller. Eggs are whitish and blotched with browns and purple. Incubation averages 12 days, by the female, possibly with some assistance from the male. Later in incubation the female sits tight to the nest and, if disturbed, may run from the nest a short distance before flushing (Cadman et al. 1987). She may also perform a rodent-run distraction display to confuse the intruder. Both parents tend the young, which fledge after 7-10 days. Two broods are raised per season.

Remarks:
This small, streaked sparrow was named after Savannah, Georgia, where it was first collected (Bent 1968). It has a relatively short, slightly forked tail, yellowish crown stripe and distinctive yellow line over the eye. The Song Sparrow, which it somewhat resembles, has a rounded tail and less pink in the legs. The Baird's Sparrow, also similar, has buffy coloring in the face, instead of yellow. The Savannah Sparrow can be distinguished from the Vesper Sparrow by its lack of white tail borders.
Savannah Sparrows spend most of their time on the ground, where they glean large numbers of insects and seeds, with the latter becoming more important later in the year. These birds do not scratch debris out of the way, preferring to forage where the ground is bare. When approached, they often run to cover or flutter a few inches above the ground, drop to the ground, then continue their retreat. When undisturbed, males will sing heartily from a tall weed or a bit of bush or fencepost, but rarely higher (Salt and Salt 1976).
This species first arrives in Alberta in late April, with migration peaking in early May. The birds depart again in September, with a few birds lingering into October. Savannah Sparrows feed in the day and migrate at night. Overwintering has been reported in Edmonton (Slater 1987).

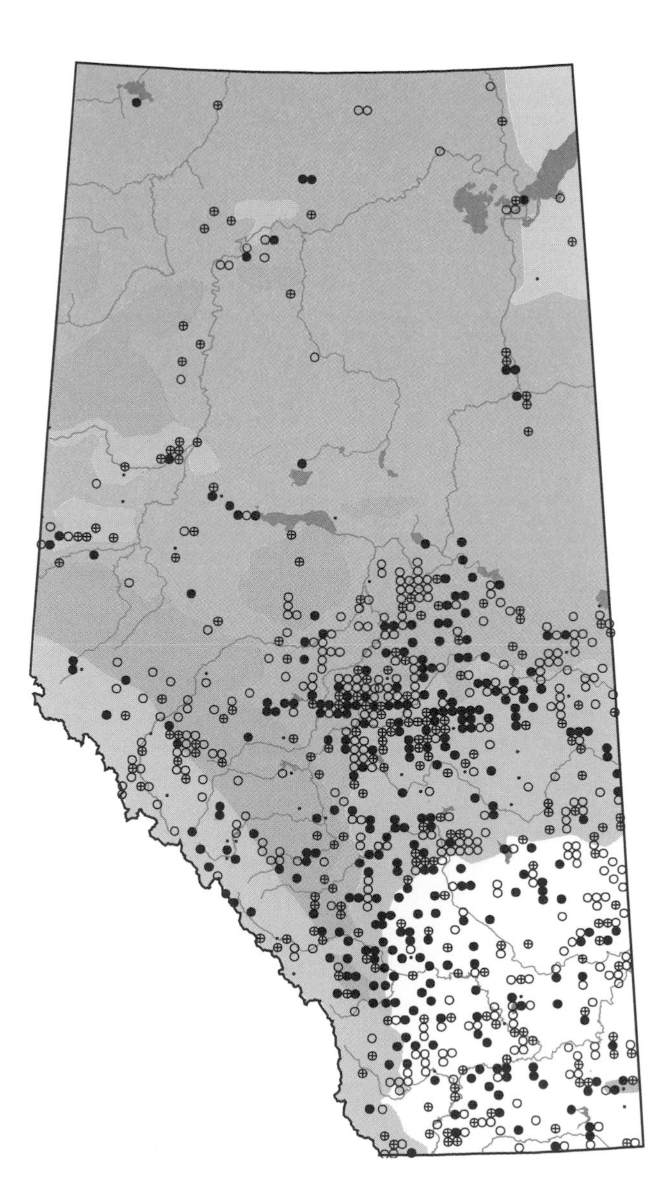

Baird's Sparrow

Ammodramus bairdii

BDSP

Status:

Baird's Sparrow is a species of the undisturbed grassland of the Great Plains. Unfortunately, little of this grassland is left in Alberta, primarily due to agricultural practices. The Baird's Sparrow's restricted and generally unadaptable habitat preferences have made it highly susceptible to this habitat loss. It is considered to be a "threatened" species by the Committee on the Status of Endangered Wildlife in Canada (De Smet and Miller 1988), but has not been designated as "endangered" under the Alberta Wildlife Act (Wildlife Management Branch 1991).

Distribution:

The major breeding range of this species is south of Stettler and east of the Red Deer River, with a western extension to the Calgary area. Farther south, it occurs mainly east of Township 23, west of the fourth meridian. There are scattered records north and west of this range. There is a nesting record for Beaverhill Lake and the bird has been observed as far north as Lake Isle, Mayerthorpe and Elk Island Park. To the west, Baird's Sparrow has been reported in the Banff area.

Habitat:

In Alberta, Baird's Sparrow is found in the Prairie Natural Region. Its major habitat is ungrazed or lightly grazed native grassland, especially fescue grassland. This grassland typically has extensive areas of tall, thick tussocky grasses with patches of snowberry, buck brush or rose. Occasionally, Baird's Sparrow is seen in cultivated areas in heavy stands of hay, or where heavier growth of grass exists in grazed pastures. Within upland grassland habitats, Baird's Sparrow is found in dry shallow ponds, depressions, and drainages. These moister sites provide the best grass cover following periods of grazing, mowing, or drought. In general, Baird's Sparrow will seldom nest unless there is a great tangle of grasses at the ground level. Disturbance to habitat created by cropping, mowing, or heavy grazing always results in Baird's Sparrow abandoning the area.

Nesting:

Baird's Sparrow breeds in open grassland, preferring areas of tall grass. It nests on the ground, often in a scrape, within a tangle of grasses. The female selects the site. The nest may be concealed under a grass tuft or small shrub, or it may be exposed from above. The nest is made of dead grass and is lined with fine grass, hair, or the setae of mosses.

Four to six eggs are laid which are grayish white and blotched and lined with browns. Incubation is 11-12 days, by the female alone. The female broods and tends the young on her own for the first few days, then is assisted by the male. The young fledge at 8-10 days. This species is normally single brooded (De Smet 1991) but has been reported as being double and even treble-brooded (Harrison 1978).

Remarks:

This rarely observed sparrow is short-tailed and grayish with a buffy tinge on the head and across the breast, and a necklace of short black streaks on the breast. It is very shy and wary, and will run off into dense grass or fly low into cover if disturbed. Its song is comprised of two, three or more zips followed by a lower-pitched musical trill.

During nesting season small insects are the main food of this sparrow, with grasshoppers and spiders the primary food of nestlings. In the wintering grounds seeds form the bulk of its diet.

Baird's Sparrow arrives on the breeding grounds in the third week of May. Fall migration dates are unknown; once the males stop singing, they are virtually impossible to find (Pinel et al. in prep.).

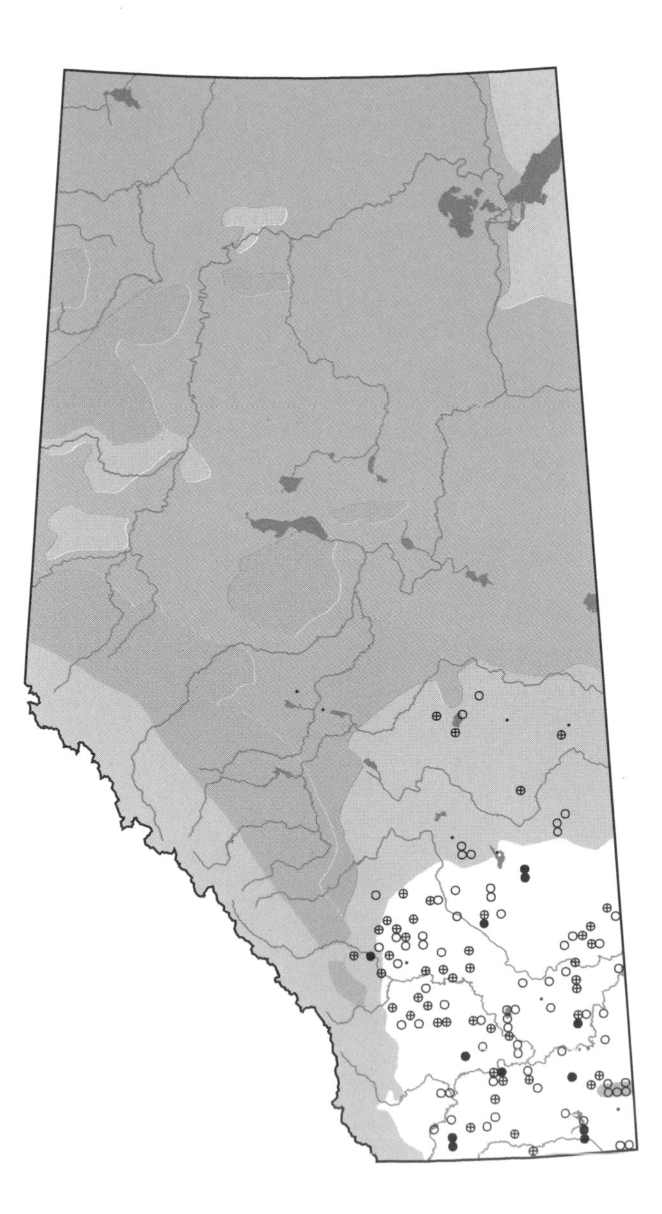

Grasshopper Sparrow

Ammodramus savannarum

GRSP

Status:

The Grasshopper Sparrow was first sighted in Alberta in 1945 (Salt and Salt 1976). However, its status is the province has not been precisely determined (Wildlife Management Branch 1991). The species was on the Blue List from 1974 to 1986, with declines noted in many parts of its range (Tate 1986).

There are two subspecies of this sparrow found in Canada. *A. s. perpallidus*, a paler bird with longer wing and tail, and smaller bill than other races, is found in Alberta (Godfrey 1986).

Distribution:

This sparrow breeds from southern Canada to the southern United States. First reported in Alberta in the Lost River area, it is found locally in the province's southeast, particularly in parts of the Milk River drainage system, the Middle Sandhills, and the Cypress Hills. It was recorded as far west as the Rockies in the south and Lake Wabamun in the north.

Habitat:

Grasshopper Sparrows are found primarily in the Prairie Natural Region, rarely in parkland. They favor a mixture of lush grasses and low, relatively open shrubbery, and, in Alberta, are most typically found in sandhills (Wershler et al. 1991). Characteristic habitat for this species includes overgrown pastures, hayfields, dry short grass plains, and, in the Cypress Hills, seasonally wet meadows. The situation is generally drier than that chosen by the Savannah Sparrow, which is also found in grassland areas (Godfrey 1986).

Nesting:

The nest of this species is sunk in a small depression in the ground, and consists of a tightly woven cup of grasses lined with finer grasses and hair. It is usually well-hidden and partially domed over by grasses or other vegetation. Nesting is sometimes semi-colonial, with a small number of pairs nesting in a limited area.

The clutch is usually 3-6 white eggs, which are spotted with reddish brown. Incubation is 11-12 days by the female, with the male helping to guard the nest. The incubating female is a close sitter and, if disturbed, she will run first before flying. She may attempt a broken wing display to lure away the intruder. She also returns to her nest on foot. Both parents feed and tend the nestlings. The young birds leave the nest at nine days, not yet able to fly. If danger appears, they run mouse-like through the grass (Bent 1968). Two broods may be raised in a season.

Remarks:

The adult Grasshopper Sparrow is a small, short-tailed bird with a buffy, unstreaked breast and unstreaked sides, and an unusually flat forehead. Other similar sparrows, such as the Baird's and Savannah, have streaking on the sides. The male sings a characteristic buzzy, grasshopper-like song from the highest perch in his territory, generally a bush, weed stalk, or fence post.

This sparrow is quite inconspicuous and even secretive in breeding season. It spends most of its time on the ground, where it hunts for insects (particularly grasshoppers and beetles), spiders, snails, and other small invertebrates, as well as seeds. Flushed birds may run through the grass, or may take to the air with a distinctive zig-zag flight before diving back into cover (Bent 1968).

These birds generally arrive at their nesting grounds in the last half of May. There is no data on the timing of fall migration.

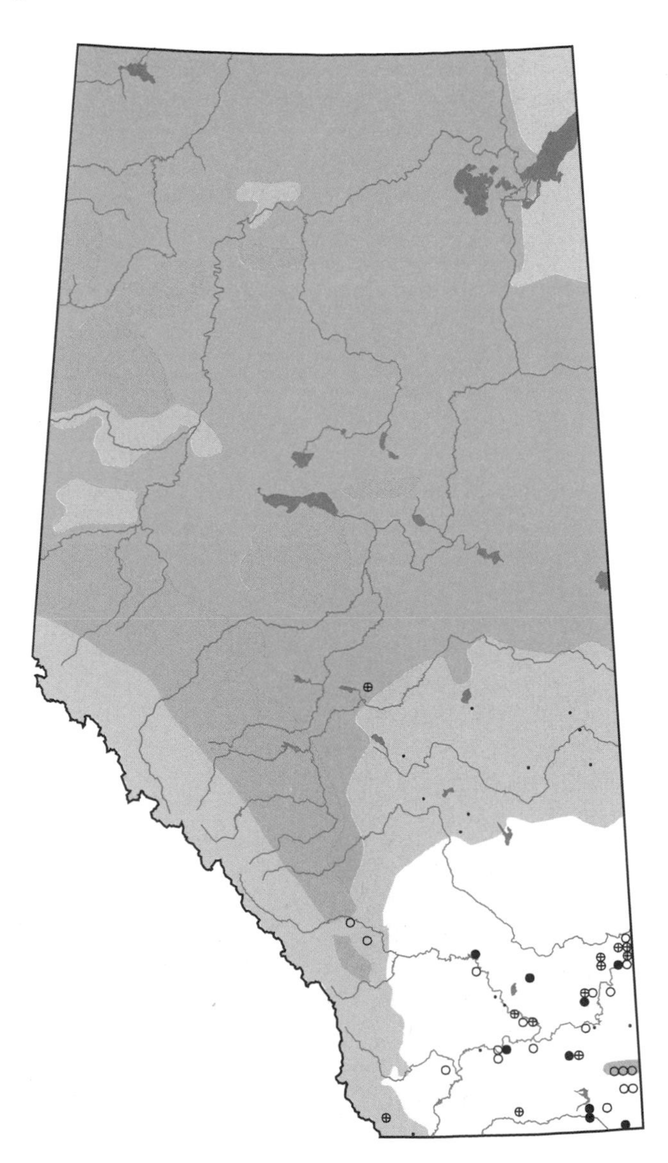

Le Conte's Sparrow

Ammodramus leconteii

LCSP

Status:
The status of the Le Conte's Sparrow in Alberta is not well known, although it appears to be common in the southern Boreal forest and Parkland natural regions. It is reported as an uncommon resident in the Rocky Mountain region and appears to breed locally in Jasper (Holroyd and Van Tighem 1983).

Distribution:
This species breeds from the Northwest Territories south into Montana. Atlas data confirm Godfrey's (1986) report that this sparrow is found in suitable habitat throughout northern and central Alberta, south to about Brooks and Lethbridge, and west to Jasper. It was recorded in 33% of surveyed squares in the Boreal Forest Natural Region, 18% in the Canadian Shield region, 28% in the Parkland region, and 21% in the Foothills region. It was also recorded in approximately 5% of surveyed squares in the Rocky Mountain and Grassland regions.

Habitat:
In northern and central parts of the province, the Le Conte's favors sedge meadows, the thick grass, shrub tangles at the edges of marshes and bogs, and low, damp parts of cultivated hayfields. Further south, bushes along creeks or tall grasses beside sloughs are typical habitat. In the mountains, flooded grass and sedge meadows or wet grass and willow tangles are characteristic (Holroyd and Van Tighem 1983).

Nesting:
The Le Conte's Sparrow usually nests in a wettish area on or close to the ground in clumps of dead grass or at the base of a bush. The nest is woven of coarse grasses and lined with fine grasses and hair. It is usually extremely well-concealed in thick tangles of vegetation. This species sometimes nests semi-colonially, but pairs maintain separate territories during the breeding season (Cadman et al. 1987).
Clutches are typically 4-5 eggs, greenish white and speckled with browns. Nests are occasionally parasitized by cowbirds. The female incubates for 11-13 days and tends the young alone at first, though the male may assist later. Little information is available on the fledging period or number of broods of this species.

Remarks:
This small, buffy-orange sparrow is most similar to the Savannah and Sharp-tailed Sparrows. Its buffy-orange underparts distinguish it from the Savannah and its light buff rather than dark gray median crown stripe separate it from the Sharp-tailed (Salt and Salt 1976). Its black side streaks and gray ear patch help differentiate it from the Grasshopper Sparrow, which is also somewhat similar.
The Le Conte's is one of the most elusive and secretive of Alberta's sparrows. The birds spend most of their time on the ground or low in dense grass or bushes, and tend to run on the ground, rather than fly, when disturbed. Though the male may perch atop a dead rush or reed stem, he often sings from the concealment of vegetation, and is rarely seen in flight. His song, easily unnoticed, is a high, weak, insect-like buzz, lasting only a second (Godfrey 1986).
The main foods of this sparrow are insects, spiders and seeds; the young are fed exclusively on insects.
The birds arrive in late April and early May and move south again in early September.

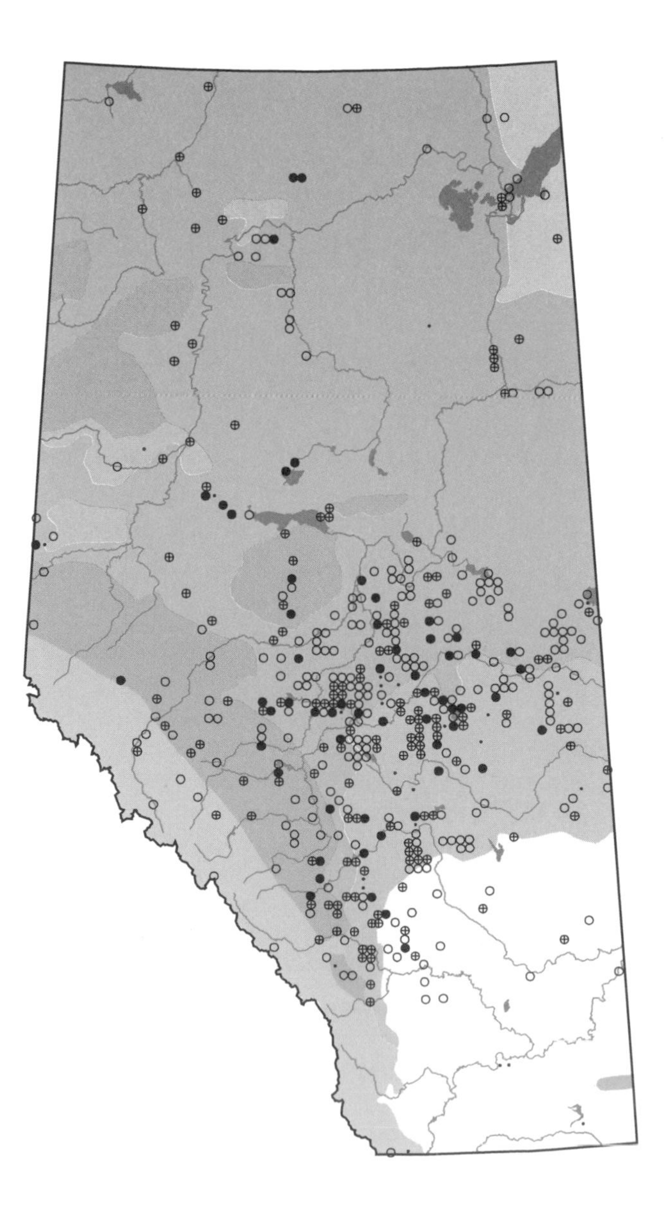

Sharp-tailed Sparrow

Ammodramus caudacutus

STSP

Status:
The Sharp-tailed Sparrow breeds in three disjunct areas of North America (Murray 1969). Of the three Canadian subspecies, a darker variety, *A. c. nelsoni,* is found in Alberta. Little detailed information is available on the status of this sparrow in Alberta (Wildlife Management Branch 1991).

Distribution:
Salt and Salt (1976) and Godfrey (1986) both reported that the Sharp-tailed Sparrow bred locally across the northern half of the province, in the Boreal Forest and Parkland natural regions, southward to about Red Deer. Atlas data indicate that this species breeds farther south and west to Bragg Creek and Standard. As in previous reports, no breeding was confirmed in the Rocky Mountain Natural Region.

Habitat:
This species' typical habitat is sedge marshes with scattered willows. It favors margins of ponds, lakes, and marshes in woodland areas; particularly where there are sedges, cattails or bulrushes in shallow water, or wet grassy meadows. Its habitat is wetter than that preferred by Le Conte's Sparrow.

Nesting:
The nest of the Sharp-tailed Sparrow is built on the ground, or a few centimetres above it, usually in dense, tall grasses or sedges in moist areas bordering water. The nest is very difficult to locate (Bent 1968a; Peck and James 1987), and this may contribute to the fact that breeding was onnly confirmed in two of the squares surveyed during the Atlas Project.
The female constructs the nest, weaving a loose cup of grasses and sedges, then adding a lining of finer grasses. The nest is usually well concealed by over-hanging vegetation. This species nests in small loose colonies and males exhibit no territorial behavior. They criss-cross one another's paths, use common singing perches and mate indiscriminately (Murray 1969).
The clutch is typically 4-6 eggs, pale-greenish and strongly marked with browns. The female alone incubates the eggs for 11 days and feeds the young. Nestlings fledge at 10 days but depend on the female another 20 days after leaving the nest. The nest of this species is occasionally parasitized by the Brown-headed Cowbird and, in Alberta, this species is likely single-brooded in Alberta.

Remarks:
The Sharp-tailed Sparrow is an inconspicuous, furtive little sparrow, recognizable by its buffy face, grayish earpatch, dark gray median crownline, gray hindneck, and black and white stripes on the back. It is distinguished from Le Conte's Sparrow by its dark gray crown stripe, dark hind neck, and lightly streaked breast. The male's song is a high, wheezy nasal buzz which he sings on the top of a grass tuft or hidden in the bushes; he may remain silent much of the day, only to sing at night.
Birds hunt for insects and seeds on the ground. They brush stems aside with their bills as they walk through dense grass and, in a characteristic behavior, may stretch their heads and necks upward to survey the surroundings (Bent 1968). They traverse open areas by running and, if flushed, they may take to the air for a short time, then quickly drop to the ground and conceal themselves in vegetation.
This species arrives on the nesting grounds in late May and early June. Fall migration data is lacking, but it appears that most Sharp-tailed Sparrows leave the province by the second week of September (Pinel et al. in prep.).

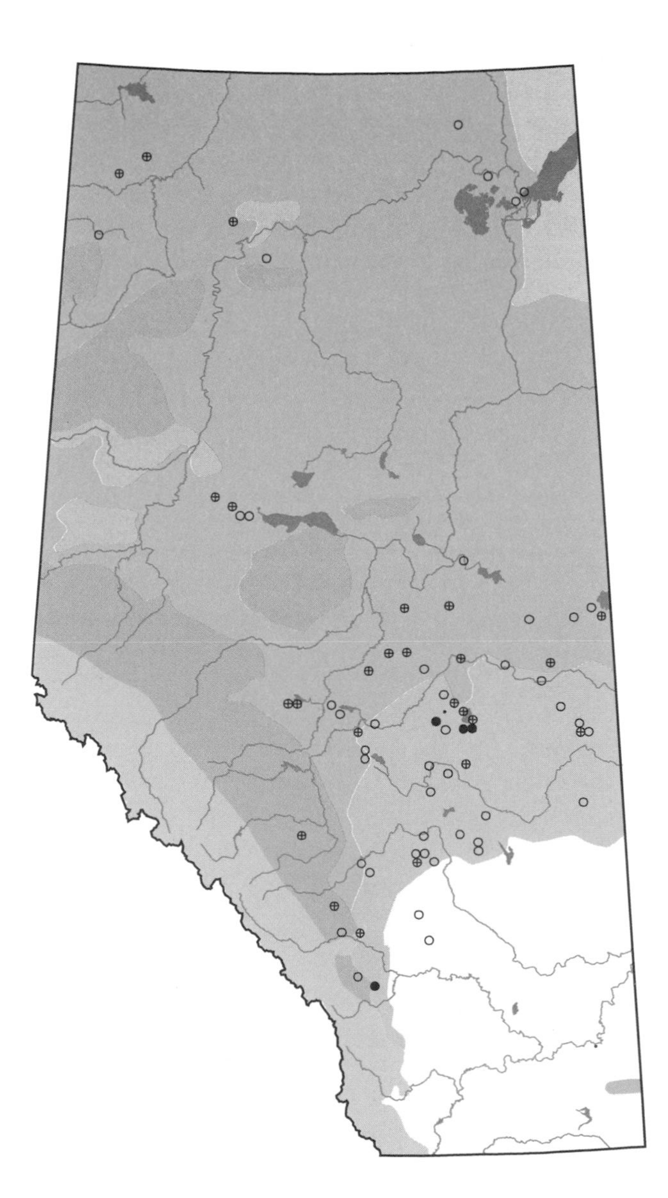

Fox Sparrow

Passerella iliaca

FOSP

Status:
Three of six subspecies nesting in Canada are found in Alberta. *P. i. zaboria,* breeding in north and central Alberta, is rusty brown but is darker and grayish; *P. i. altivagons,* found in Jasper and south to Banff, is similar to *zaboria,* but the browns are less rufescent and the upper parts are more vaguely streaked. *P. i. shistacea,* located in the Waterton Lakes area, has a dark gray head and back and is lightly streaked (Godfrey 1986). Holroyd and Van Tighem (1983) reported that the Fox Sparrow is a fairly common resident of the Rocky Mountain Natural Region of Alberta. Atlas data also confirm Salt and Salt (1976) who reported that, in northern Alberta, it is local in distribution and common nowhere. It was observed as a transient in the Parkland and Grassland natural regions.

Distribution:
The Fox Sparrow breeds across northern Canada from the treeline south to northern parts of the provinces and in the western mountains to the border. The Boreal Forest, Canadian Shield, Rocky Mountain, and Foothills natural regions are the primary habitat of Alberta's Fox Sparrows. The distribution of breeding records from Atlas surveys coincides with the distribution of this species as reported by Salt and Salt (1976) and Godfrey (1986).

Habitat:
This species favors dense woodland thickets and brushy edges, tangled willows and alders along streams and lakeshores, scrubby woods, burnt lands, and cutover land. In the mountains, these birds are most commonly found in Engelmann spruce - subalpine fir open forest and edges at the treeline or in scrub on avalanche slopes (Holroyd and Van Tighem 1983).

Nesting:
Fox Sparrows nest on the ground under a bush or tree, usually in a thicket or low in a shrub or tree. The nest, generally well-concealed, is a bulky, deep cup constructed of grass, twigs, moss, and rootlets; and lined with grass, hair, and feathers.
The female lays 3-5 greenish-white eggs which are speckled with browns. Incubation is 12-14 days, primarily by the female, who also broods the young. Adults may affect a broken wing display if the nest or young are threatened. Both parents feed the young, which fledge at 9-11 days. This species is thought to be double-brooded (Bent 1968a).

Remarks:
The coloration of this robustly built sparrow varies greatly geographically. East of the Rockies, it is distinguished by its fox reds, especially in the tail and rump, and its underparts, which are heavily marked with triangular spots. Birds with gray heads and backs are found in Alberta's southwest mountain areas. The Fox Sparrow male has a lively song with a few clear whistles followed by varied short buzzy trills.
These birds spend most of their time on the ground in dense thickets, where they forage by hopping forward and back with both feet at once. While their foraging may be noisy, they are generally quite shy and wary, so there is still much to be learned of their habits and behavior. They eat insects, spiders, millipedes, and other small terrestrial invertebrates, as well as seeds, berries, and other small fruits.
The Fox Sparrow arrives in Alberta just after mid-April. The last birds leave the mountain areas by late August to early September. Most leave southern Alberta by early October, but some linger into late October. One record of overwintering is reported from Cardston in 1986 (Slater 1987).

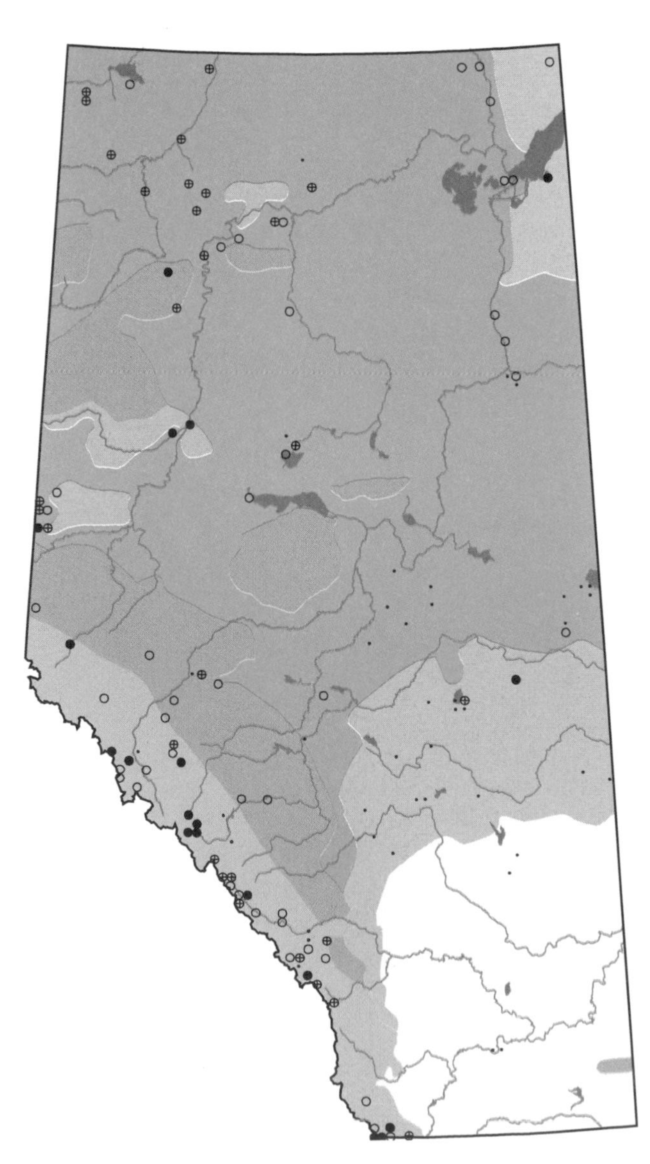

Song Sparrow

Melospiza melodia

SOSP photo credit: T. Webb

Status:
This sparrow is one of the most common and widely distributed in Canada. Atlas data confirm that this is also true for Alberta.
Of the 31 subspecies of Song Sparrow found in North America (eight in Canada), two breed in Alberta: *M. m. merrilli*, a very dark race found in the southwestern mountains, and *M. m. juddi*, throughout the rest of the province (Godfrey 1986).

Distribution:
The Song Sparrow breeds in Canada from the treeline south to the border. It nests in suitable habitat in all of Alberta's natural regions, but is less common and local in the Grassland region, where it is confined to more heavily wooded coulees (Salt and Salt 1976). It was only found in 13% of the squares surveyed in this region.

Habitat:
This species nests in all of the province's natural regions. Favored habitat includes low shrubby growth along the margins of ponds, lakes, and streams; brushy woodland openings, the forest edge, farmland thickets, hedgerows, bushy pasture, scattered aspen groves, and shrubbery around buildings.

Nesting:
Nesting is either on the ground or in a bush or small tree. Early nests, often built before leaves are out, are usually on the ground. The female chooses the nest site and builds the nest. She uses grasses, weeds, bark, and leaves, and lines it with finer grasses, roots, and hair. The nests are usually well concealed under clumps of grass, other vegetation, logs, roots, and branches.
The clutch varies from 3-6 eggs, pale bluish to grayish green and blotched with browns. Song Sparrows are frequently parasitized by cowbirds. The female incubates 12-13 (occasionally 14-15) days and also broods the young. She sits very tight to the nest and may sit until almost stepped on. She is likely to run in the undergrowth before flushing. Both parents feed and tend the nestlings. Young birds leave the nest at 10 days, fly well at 17 days, and depend on adults for another 18-20 days. Once they fledge, young birds may be split into two single-parent families, until the female re-nests, at which point the male takes over care of the early brood. Two, or sometimes three, broods are raised per season. Males, female and young tend to return to previous territories. Pairs often mate over successive years.

Remarks:
Song Sparrow color is highly variable, but its brown back is larger and more strongly streaked than the Lincoln's Sparrow. It has heavily streaked underparts, a dark central breast spot and dark lines bordering the sides of the throat. This bird also lacks the Lincoln's buffy breast band. The call note and song of the Song Sparrow are also easily recognized and distinctive. The male may sing as many as 6-8 songs a minute. He sings early in the spring and continues late into the summer when other birds are silent.
This sparrow eats a variety of insects and seeds, particularly weed seeds, scratching for them in the leaf litter of the woodland floor, or out in open ground. Small and inconspicuous, except for its song, this bird stays close to ground and flies only short distances, mainly from bush to bush, pumping its tail up and down as it goes.
Birds arrive in central Alberta in mid-April and leave in September, and early October, with some stragglers persisting into November. Though migrating Song Sparrows may be abundant, they do not flock. Overwintering has been recorded in the Banff and Canmore area, Jasper, Edmonton, Calgary, Lethbridge, Medicine Hat, Strathmore, and Red Deer.

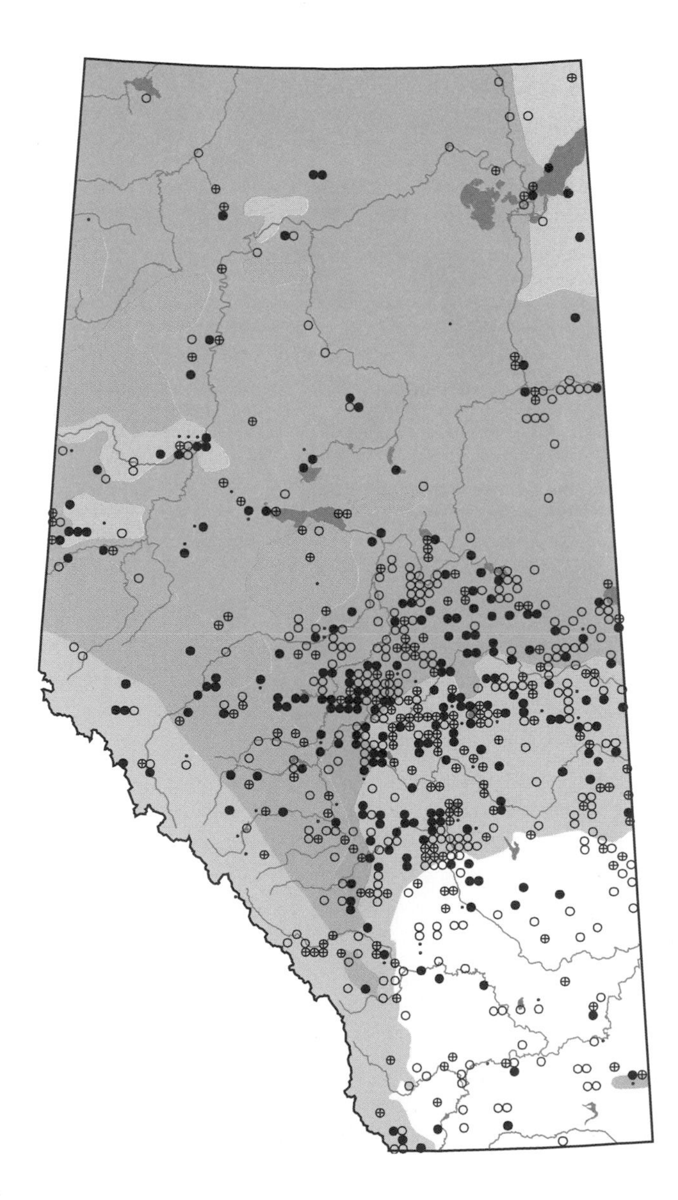

Lincoln's Sparrow

Melospiza lincolnii

LISP photo credit: T. Webb

Status:

Of the three Canadian subspecies, *M. l. alticola* is the only one that breeds in Alberta (Godfrey 1986). Atlas data confirm previous reports that it is a fairly common summer resident in northern Alberta (Salt and Wilk 1958) and a very common resident in the mountain parks (Holroyd and Van Tighem 1983).

Distribution:

Lincoln's Sparrow is found in the northern half of the province and also in the mountains and foothills, south to the American border and east to Calgary. It was recorded in 38% of squares surveyed in the Boreal Forest Natural Region, 31% in the Canadian Shield, 50% in the Foothills, 34% in the Rocky Mountains, and 15% in the Parkland region.

Habitat:

During the nesting season in northern areas, this species is found in bogs, wet meadows, the willow and alder thickets bordering muskeg, marshes, and woodland streams, and similar wet and brushy places. In the mountain parks, it is most common in the montane and lower subalpine areas, where it inhabits wetlands with emergent sedge vegetation and clumps of willow shrubbery, as well as forest edges bordering lakes, bogs, and slow streams (Holroyd and Van Tighem 1983). In migration, it prefers thickets, roadsides, brush piles, and hedges, and also brushy forest edges and the grassy or weedy open places near them.

Nesting:

The Lincoln's Sparrow nest is generally on the ground in marsh or muskeg, often sunk in a grass or moss hummock at the base of a bush or small tree. It is usually well-hidden by grasses and sedges. A deep cup of coarse grasses, sometimes sedges, is constructed and lined with finer grasses, rootlets, and hair.

The typical clutch is 4-5 eggs which are greenish white and blotched with browns. The female alone incubates for 13 days; the male is inconspicuous during this period. Unlike many other ground sparrows, the Lincoln's is rarely parasitized by cowbirds. The young are tended by both parents and leave the nest at 9-12 days. This species may, at times, be double-brooded.

Remarks:

This shy and elusive sparrow was named by Audubon after Thomas Lincoln, who accompanied him to Labrador and first collected the species there (Root 1988). The broad, buffy, black streaked band across its breast and sides distinguishes it from other sparrows. The male has a quick, bubbling song which is somewhat reminiscent of a Purple Finch. He sings from a perch in the bushes or from a low branch in a tree. If disturbed he will cease, steal away into the shrubbery, and begin to sing again a safe distance away (Salt and Wilk 1958).

The birds forage on the ground and are rarely seen far from dense cover. They scratch with both feet in search of insects, spiders, other small invertebrates, and seeds.

The Lincoln's Sparrow appears in Alberta in late April to early May. Fall migration from the province is in late September and early October. It is particularly elusive in migration when, in silence, it often passes through unseen. No overwintering is reported.

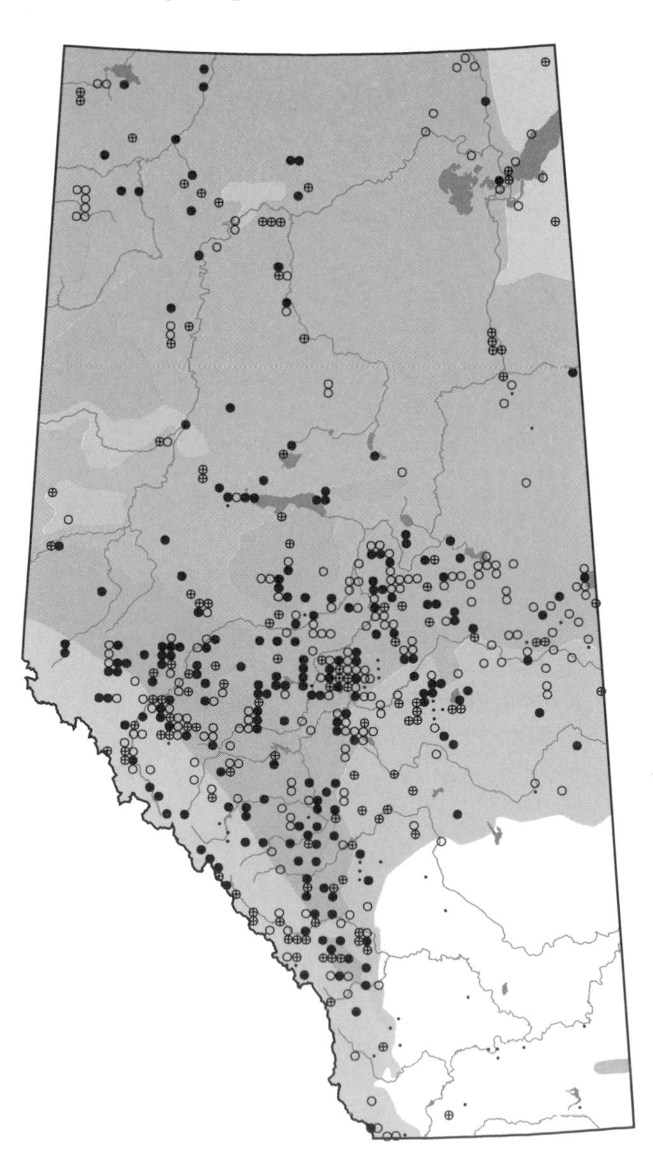

Swamp Sparrow

Melospiza georgiana

SWSP

Status:
There are two subspecies of Swamp Sparrow breeding in Canada, but only one, *M. g. ericrypta*, nests in Alberta (Godfrey 1986). The status of this species in Alberta has not been fully determined, but it is common in the Boreal Forest Natural Region, where approximately 65% of the confirmed breeding records originated.

Distribution:
A bird of eastern North America, the Swamp Sparrow is found east of the mountains south of the tree line. It breeds locally in northern and central Alberta south to Olds and Vermillion. It was recorded in 43% of surveyed squares in the Canadian Shield Natural Region, 25% in the Boreal Forest region, 11% in the Foothills region, and 5% in the Parkland region. It was not recorded breeding in the Grassland region.

Habitat:
In nesting season, this sparrow favors margins of ponds, lakes, streams with tall emergent vegetation such as cattails or willow and alder thickets, swamps and bogs with shrubs and small conifers, marshes with tangles of vegetation, and sedge meadows. In migration, it also prefers these habitats, but is also found in weedy fields near water or sometimes in dry thickets at forest edges and openings (Godfrey 1986).

Nesting:
Nests are on or just above ground in a grass or sedge tussock, or suspended in reeds over water, or in small shrubs. Most nests are completely hidden from above by vegetation. The nest is a bulky weave of coarse, dry grasses, cattails, and sedges; the inner cup is lined with finer grasses and sedges. The female probably builds the nest.
Clutches are typically 3-5 eggs, which are variable in color, but most often are bluish green and marked with browns. Incubation is 13 days by the female. Males may be polygamous at times, but assist the female in feeding the young. He may also feed the female on the nest while she broods. Young birds leave the nest at 9-10 days. This species is thought to be double-brooded.

Remarks:
This wetland sparrow is distinguished by its chestnut crown and wings, whitish throat and plain gray chest (sometimes with a vague dusky breast spot). The male's song is a clear, slow trill which he sings from a cattail or bush.
The Swamp Sparrow is a rather wary and secretive bird. It rarely flies more than a few metres at a time, except in migration. Instead, it keeps close to the ground in bushes or reeds, climbing nimbly up and down cattails or running quickly through the grass.
Its diet, which includes both insects and seeds, has a higher proportion of animal matter than other *Melospiza* species and most other sparrows (Root 1988). Because of this difference in diet, its skull, jaw muscles and bill are smaller than most others in its family. While little is known of the Swamp Sparrow's feeding habits, it has been observed wading in shallow water, sandpiper-style, picking up insects and seeds from the surface (Bent 1968).
Birds arrive in central Alberta in mid-May and leave again in late September and early October. A Swamp Sparrow was observed overwintering in Calgary during the 1990 Christmas Bird Count (Slater 1991).

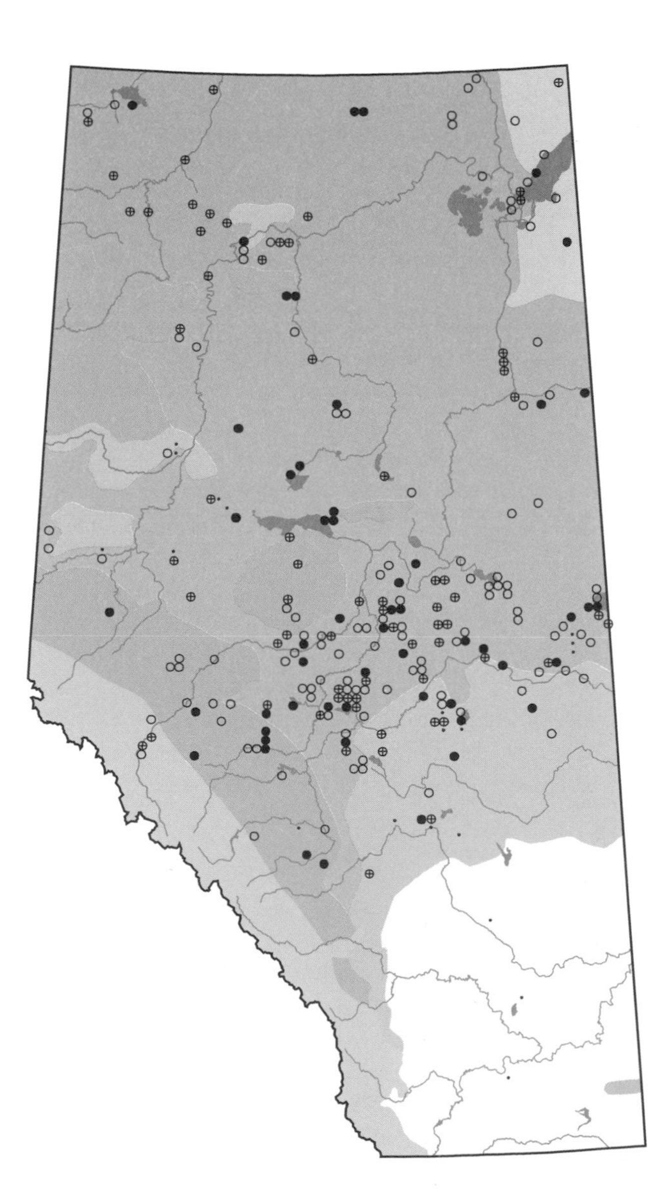

White-throated Sparrow

Zonotrichia albicollis

WTSP

Status:
This northern species is a relatively common summer resident within its range.

Distribution:
The White-throated Sparrow breeds in forested areas of Canada from the Yukon to Newfoundland, mainly east of the Rockies, and south to the northern Appalachians. It nests mainly in northern and central Alberta, where it was recorded in 52% of squares surveyed in the Boreal Forestregion, 52% of the Foothills region, 44% of the Canadian Shield region, and 38% of the Parkland region. Although it was previously reported as less common in the Rocky Mountain region, with nesting at Brule, Jasper and west of Calgary (Salt and Salt 1976), it was recorded in 13% of squares surveyed in this region.

Habitat:
This large, long-tailed sparrow breeds in coniferous, deciduous and mixed woodlands, preferring the forest edge, brushy openings and second growth to mature dense stands. It is seen along wooded lakeshores and rivers, muskegs and bogs with patches of conifers, logged areas and old burns with deadfall, the brushy edges of woodland fields and roadsides, and clearings around cottages. In the Rocky Mountain region, it occurs where boreal elements reach into the region. In this region, it frequents willow thickets and edges of balsam poplar - white spruce forests on montane river floodplains and dunes (Holroyd and Van Tighem 1983).

Nesting:
This species generally nests on the ground or near it, in forest openings and edges, often in shrubbery. Ground nests are usually well-concealed beneath low shrubs, in grass tufts, moss with grass and sedges or weed clumps and ferns (Peck and Peck 1983). Elevated nests are low in thick bushes, or in the lower branches of trees. The nest is built by the female. She uses grass, twigs, woodchips, mosses, and conifer needles for the exterior, and finer grasses, rootlets, and hair for an inner lining.
The clutch is typically 3-5 eggs, bluish or greenish white and blotched with several shades of brown. The nest of the White-throated Sparrow is sometimes parasitized by Brown-headed Cowbirds. The female alone incubates, 11-14 days, and broods the nestlings. Both parents bring food and tend the young. Nestlings leave the nest at 7-12 days and fly 2-3 days later. The species can be double-brooded. Males return to previous nesting grounds in successive years,and females do so to a lesser extent.

Remarks:
The White-throated Sparrow is recognized by its striped head, white throat patch which sharply contrasts with its grayish breast and sides of the head, and yellow spot in front of the eye. The bill is always dark, never pink as in White-crowned Sparrows. Its clear song, "Ah, sweet Canada, Canada, Canada", is typical of the northern woods. This species has two distinct breeding plumages; some birds of both sexes have white head stripes, others tan. The white-striped birds mate with tan-striped birds of the opposite sex (Lowther 1961; Rising 1982).
The birds scratch for insects and seeds on the ground and also take seeds directly from plants.
White-throated Sparrows arrive in central Alberta in late April and early May and depart again late in September, with some birds lingering into mid-October. Overwintering is reported from Grande Prairie, Goose Lake, Edmonton, Calgary, Jasper, Banff, Canmore, Crowsnest Pass, and Waterton Lakes.

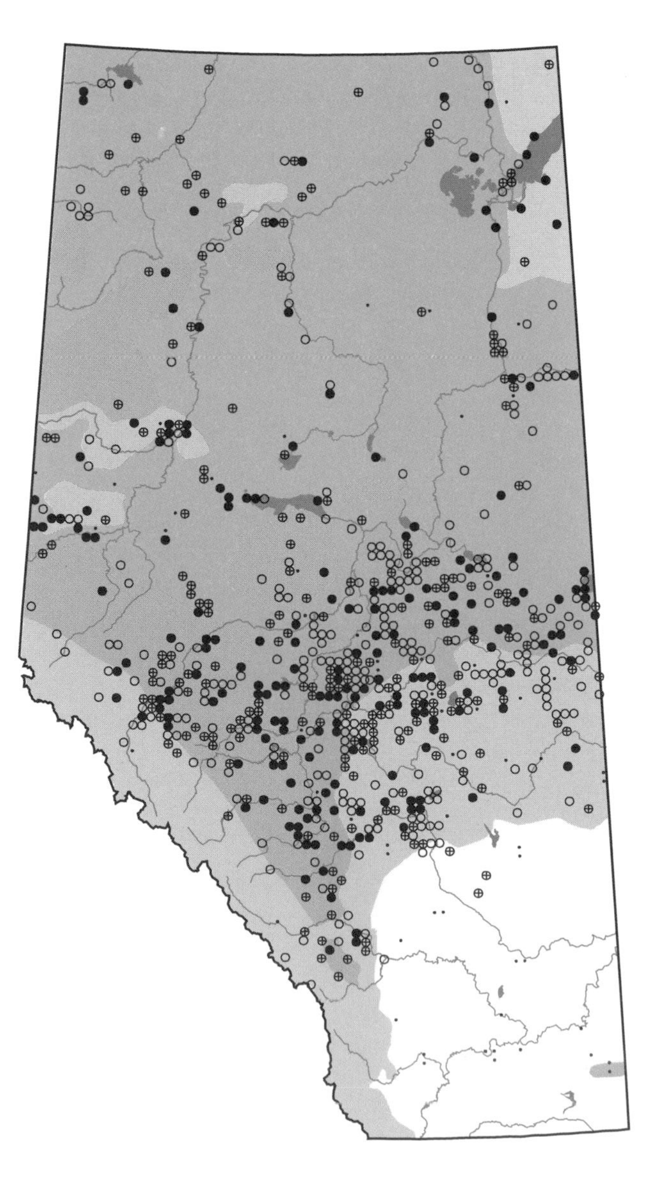

Golden-crowned Sparrow

Zonotrichia atricapilla

GCSP

Status:
The exact status of this species in the province has not been determined. However, they are fairly common in the main ranges of Jasper National Park and the northern part of Banff National Park (Holroyd and Van Tighem 1983).

Distribution:
The Golden-crowned Sparrow is a bird of extreme western North America, breeding in Canada from south central Yukon to southern B.C. and southwestern Alberta. Its distribution in Alberta is primarily in the Rocky Mountain Natural Region. Atlas data confirm Salt and Salt (1976) who reported that this sparrow breeds in Jasper and Banff national parks, but is not recorded in Waterton Lakes National Park. There were Atlas records (of observation level only) near Fawcett, Killam, Red Deer, Alix, and Gleichen.

Habitat:
In breeding season, it is found in montane thickets or low conifer growth near the timberline and often along scree slopes or in high meadows. The Golden-crowned Sparrow breeds in the upper subalpine and the lower edges of alpine, and in meadows where there are scattered clumps of stunted subalpine fir and Engelmann spruce (Holroyd and Van Tighem 1983).

Nesting:
Nests are usually built on the ground at the base of a small shrub or in a bush under overhanging plants and, rarely, on a low branch of a shrub within thick cover (Harrison 1978). The nest is a dense cup of grass, leaves, bark, and small twigs lined with fine grasses and feathers; it is usually well hidden. The female lays 4-5 eggs, pale green or buffy and mottled with browns. Little else is known of this species' breeding biology and nesting habits. Both parents feed the young and probably raise one brood per season.

Remarks:
This large, grayish, long-tailed sparrow is similar to the White-crowned Sparrow, but has a broad yellow median crown stripe with black borders rather than black and white stripes. These two species share similar habits, though the Golden-crowned is more wary, and they sometimes flock and forage together.

The Golden-crowned feeds on the ground near cover. It will hop out into the open, pick up seeds and insects, then dash back into the bushes, regardless of danger (Salt and Salt 1976). Sprouting annuals and buds are included in the spring diet, and nestlings are fed mainly insects.

This bird is most conspicuous when singing. The male's song consists of three to four clear, whistled, descending notes, sometimes transcribed as "Oh, dear me". This species is sometimes called the "rainbird" because of its habit of repeating its song over and over on dark days preceding rain (Bent 1968). The male usually sings from the top of a bush or at its periphery.

These sparrows generally arrive in southwestern Alberta in mid-May. They begin their fall migration in late July and early August, and have been recorded in the province as late as October in Calgary (Pinel and Robinson 1973) and Red Deer (Cole 1941). No overwintering is reported.

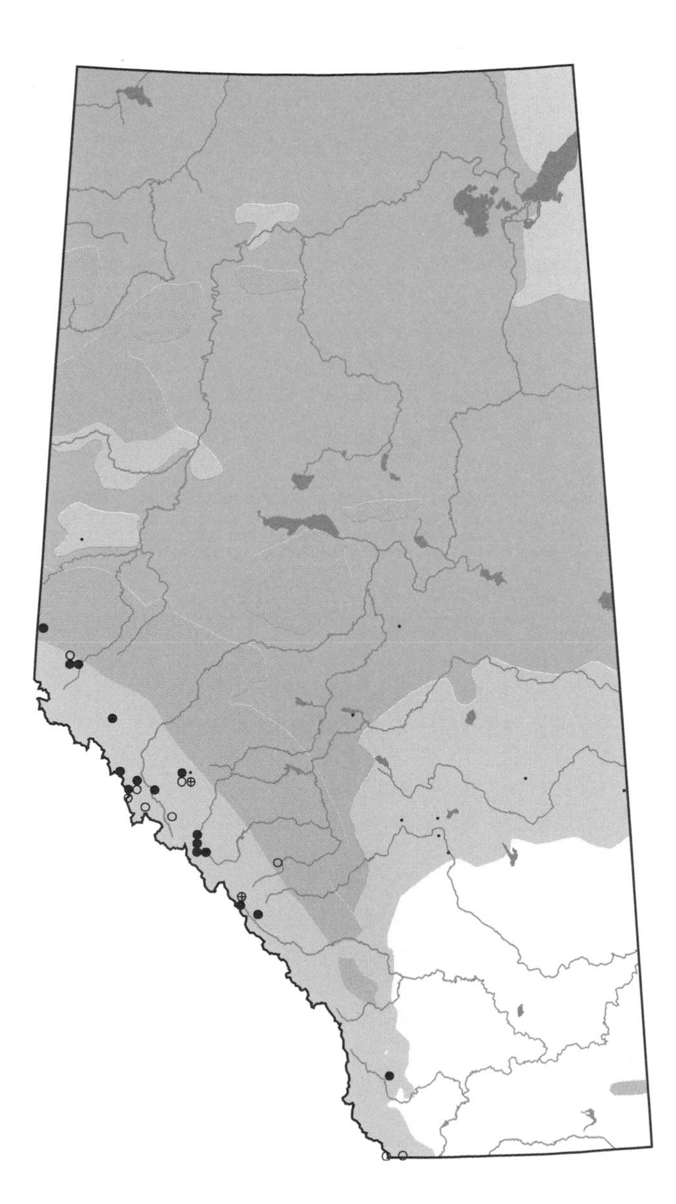

White-crowned Sparrow

Zonotrichia leucophrys

WCSP

Status:
The White-crowned Sparrow is one of the most common sparrows in western North America. Two of the four subspecies, *Z. l. gambelii* and *Z. l. oriantha* breed in Alberta. The population of this sparrow in the province appears to be stable.

Distribution:
This sparrow breeds in Canada from the Yukon across northern Canada to Newfoundland, and south in the western mountains to the border. In Alberta, the species nests in the northern part of the province, south through the Peace River District into the mountains and foothills, and also in the Cypress Hills. It is the most common, by far, in the Rocky Mountain region, where it was recorded in 45% of the squares surveyed. It was also recorded in 22% of surveyed squares in the Foothills region and only 3% of Boreal and Parkland squares. It is common as a transient in the south and southeast parts of the province.

Habitat:
The preferred habitat of this species is woody shrubbery in open situations. This provides open patches of grass and ground for foraging, with nearby shrubbery for escape and nesting. It is commonly found in thickets along woodland edges, old burns or cutovers, birch and willow patches near streams, bogs and the tundra edge, and open stunted woodland. In the mountains, it is most abundant in shrubby meadows, sedge meadows, and shrubby spruce/fir forests, habitats found on avalanche slopes, burns, active flood plains, subalpine valley floors, and timberline areas (Holroyd and Van Tighem 1983).

Nesting:
The White-crowned Sparrow nest is built on the ground, often at the base of a bush or tree, or low in bushes. Ground nests are often concealed in mossy hummocks or grass clumps. The female chooses the site and also builds the nest. It is a bulky construction of twigs, coarse grass, bark shreds, moss, and lichens, and lined with fine grass, rootlets and hair.
The clutch is usually 4-6 eggs, whitish and speckled with browns. The female incubates the eggs for 12-14 days, and may feed the young the first few days after hatching. Thereafter both parents tend the young. The male may feed the female while she is brooding. If disturbed, the female may use a rodent-run or broken wing display to lure away the intruder. The young leave the nest after 9-11 days and depend on adults another 25-30 days. If the female renests the male may feed the fledged young. While this species may fledge four broods in some parts of its range, usually one brood (occasionally two) is raised in northern areas (Ehrlich 1988).

Remarks:
The black and white stripes on the crown of this sparrow distinguish it from all other sparrows except the White-throated. However, the White-crowned is grayer than the White-throated, has a pinkish or yellowish rather than a dark bill, lacks a yellow spot in front of the eye and has no well-defined throat patch. The male White-crowned song is three to four clear, whistled notes followed by wheezy trills. He often sings at night and during spring migration.
These sparrows eat insects, spiders, seeds, and occasionally, fruit, moss capsules, blossoms, and leaves (Ehrlich et al. 1988). Seeds are their primary winter food.
They begin to arrive in southern Alberta from mid- to late April and peak in early May. They leave the province in September, with some lingering into late October. Overwintering is reported from Banff, Medicine Hat, Hinton, and Snakeshead.

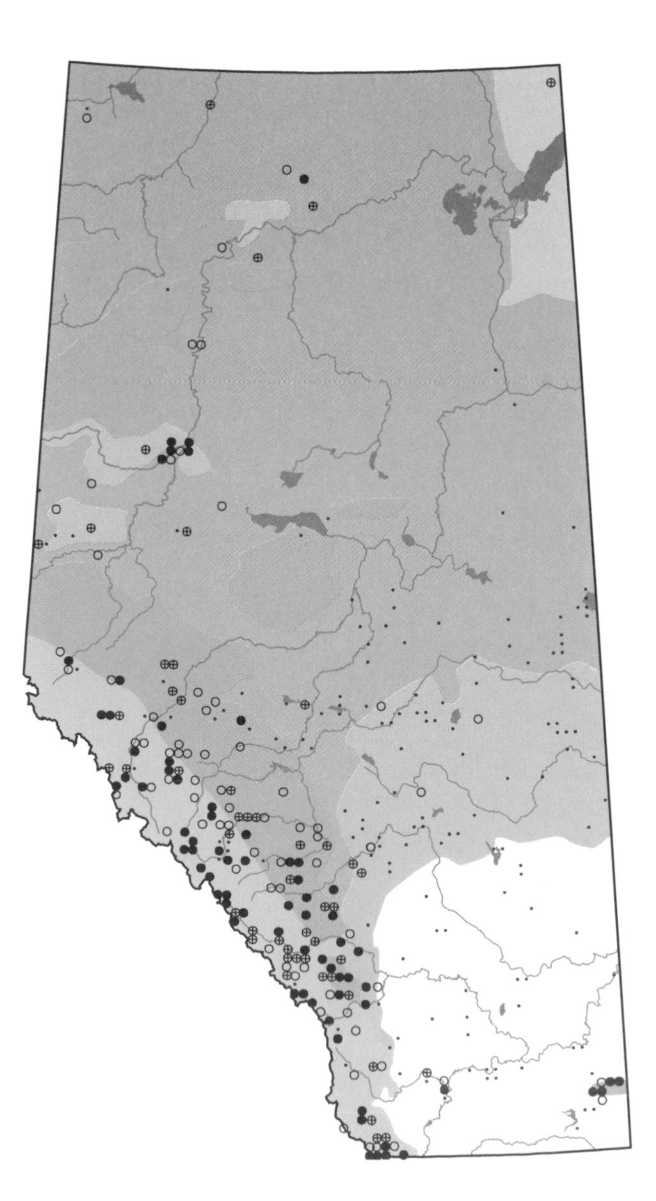

Dark-eyed Junco

Junco hyemalis

DEJU

Status:
Although its status in Alberta has not been examined in detail, the Dark-eyed Junco is a commonly observed nesting species.

Distribution:
Atlas data show that the Dark-eyed Junco breeds in all natural regions of the province, but are least common in the Grassland Natural Region. This species is a complex of populations formerly recognized as separate species. *J. h. cismontanus* nests in Jasper National Park; *J. h. montanus*, elsewhere in southwestern Alberta; *J. h. mearnsi*, in Cypress Hills; *J. h. hyemalis*, elsewhere in the province (Godfrey 1986). There is interbreeding of the different races where there ranges meet and overlap.

Habitat:
In nesting season, juncos are found in coniferous and mixed woodlands, particularly in openings, clearings, edges, burned and cutover areas, occasionally wooded park and cottage areas, and residential gardens. In winter and during migration, they frequent various weedy and brushy places such as fields, roadsides, fencerows, gardens, and the edges of woodland, grainfields, and meadows.

Nesting:
The junco nest is usually built on the ground and well-concealed under vegetation. It is often built beside or under tree roots, stumps, logs or rocks; sometimes it may be built in a moss or lichen hummock, and frequently in a cavity or crevice on a steep slope, embankment, or vertical rock face. On rare occasions birds will nest in low shrubs or trees, or on ledges or in niches in buildings. The nest, generally placed in a depression, is constructed of grass, rootlets, moss and hair. The female builds the nest and the male sometimes assists by carrying nesting material.
The typical clutch is 3-6 eggs, pale bluish white and speckled with reddish browns. Nests are occasionally parasitized by cowbirds. The female incubates for 11-13 days and broods the young. Both parents tend and feed the nestlings, which leave the nest at 9-13 days unable to fly. They depend on adults for another three weeks after leaving the nest. Two broods are reported, with one brood more common at higher elevations and latitudes.

Remarks:
Various populations once considered separate species (the Slate-colored, Oregon and Gray-headed Junco) are now regarded as one species, the Dark-eyed Junco (American Ornithologists' Union 1983). This sparrow, with its slaty head and breast, white abdomen, white outer tail feathers, and pink bill is probably the most distinctive and easy to recognize of Alberta's sparrows. Its song, a ringing metallic trill, is similar to the Chipping Sparrow, but more musical (Godfrey 1986).
Dark-eyed Juncos prefer to forage on the ground, where they scratch for seeds, insects and old berries. They hop back and forth in the debris, but not as vigorously as the Fox Sparrow. Like other sparrows, they eat more insects in summer, and feed their young primarily on animal fare.
The first arrivals are usually recorded in the province in the last two weeks of March, and peak in April. They begin fall migration in late August and early September and fall migration peaks in late September and the first two weeks of October. Most are gone by the end of October, though small numbers reguarly winter in southern Alberta. This species has been observed overwintering in Edmonton, Devon, Penhold, Red Deer, Caroline, Jasper, Canmore, Banff, Calgary, Lethbridge, and Waterton.

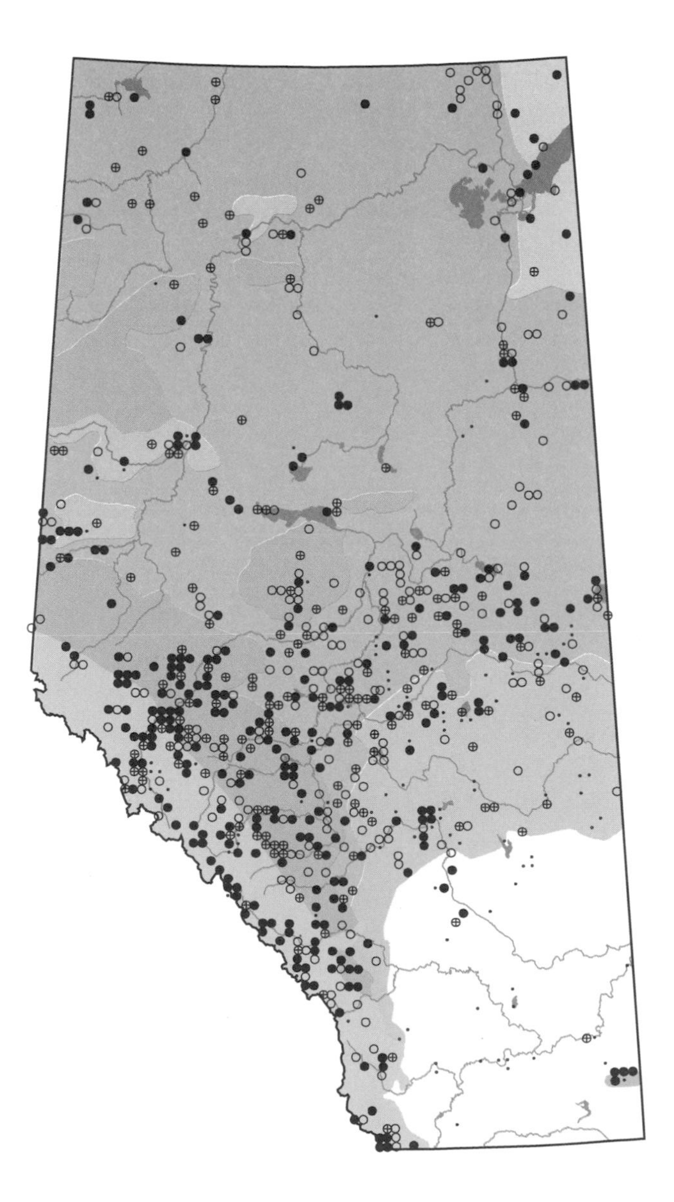

McCown's Longspur

Calcarius mccownii

MCLO

Status:

While it was formerly abundant on native grasslands, this species' breeding range and, to some extent, wintering range, have contracted in this century (Ehrlich et al. 1988). Although the status of this species in Alberta is incompletely known, it is locally common in the southeast portion of mixed grassland within the Grassland Natural Region (Wershler et al. 1991).

Distribution:

McCown's Longspur breeds in the southern prairies of Canada. In Alberta, it nests locally in the Grassland Natural Region, north to Hanna and Youngstown and west to Lethbridge, Vulcan and Drumheller. It has been sighted north to beaverhill Lake, Athabasca, and Lesser Slave Lake (Salt and Salt 1976). Its major range is south of the Cypress Hills, especially in the Milk River/Lost River area west to Pakowki Lake, with another population located north and west of Medicine Hat (Wershler et al. 1991).

Habitat:

This is a bird of short grass plains which requires moderately to heavily grazed sites in native mixed grasslands, showing a preference for drier, sandier sites (Wershler et al. 1991). It occasionally nests in cultivated fields and stubble.

Nesting:

The McCown's Longspur nest is in a small depression in the ground, often at the base of a grass tuft, weed clump, or bush. The female collects the nest material within her territory, and builds the nest, a cup of coarse grass lined with finer grasses and, sometimes, hair, plant down or wool (Harrison 1978). This species is gregarious and retains something of its colonial tendencies in breeding season. Within loosely-formed flocks, a pair will maintain possession of a territory, but abandon it when young fledge (Mickey 1943).

Typical clutches are 3-4 eggs, white to pale olive and streaked with browns and, at times, purplish gray. The female incubates for 12 days, during which time she may be fed by the male. She is a close sitter and will remain on the nest until practically stepped on. Both male and female brood and tend the young and remove excrement from the nest. The young leave the nest at about 10 days and fly at approximately 14 days. This species may, on occasion, be double-brooded.

Remarks:

The male McCown's has a black head and gray on the nape and back of the neck. The back is light brown and streaked. The underparts are white whith a black crescent on the breast. The female is a brown bird similar to the female Chestnut-collared Longspur, but lacking the streaked breast. In both sexes, the tail pattern in flight shows a black inverted T. The male has a melodious flight song, a series of warbles and twitters, which he sings as he floats to earth with tail spread and wings held motionless at an angle above his back (Salt and Salt 1976). He is a vigorous defender of his territory and often engages in aerial combats to keep out intruding males.

This species forages on the ground, walking about in search of seeds and insects; it feeds its young exclusively on insects, never regurgitated food (Mickey 1943).

Spring migrants *

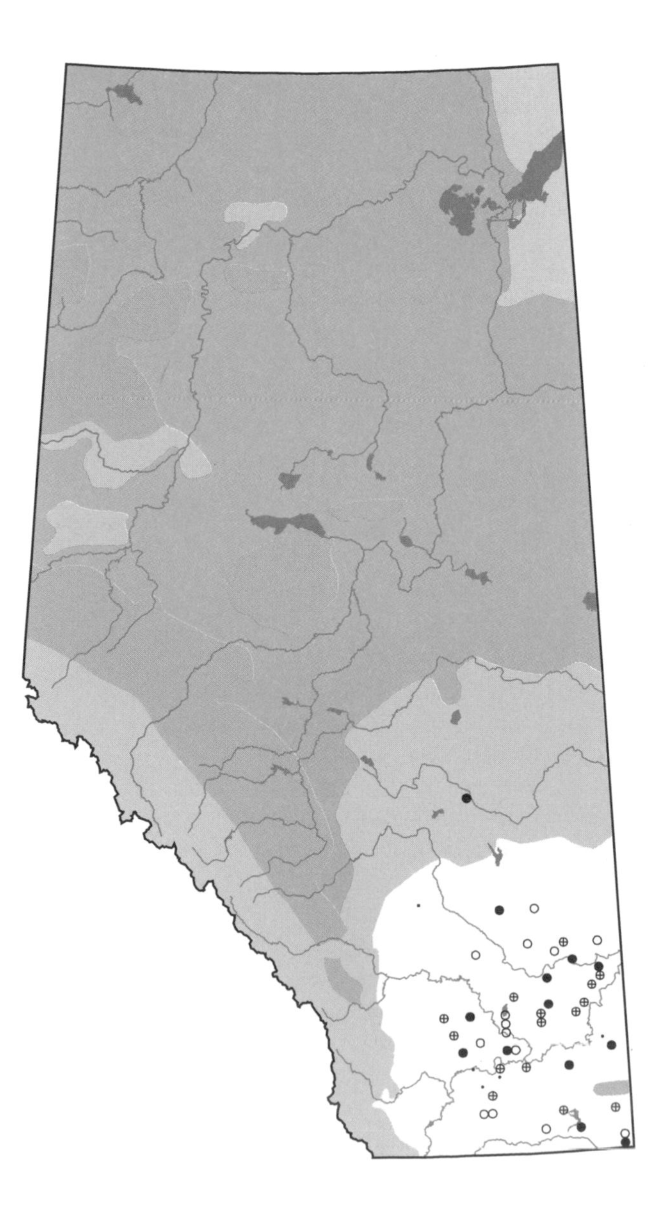

Chestnut-collared Longspur

Calcarius ornatus

CCLO

Status:
The abundance of the Chestnut-collared Longspur has decreased with the destruction of prairie grasslands from both over-grazing and farming. In Alberta, the species is found locally in good numbers, but its reliance on native grasslands, which continue to be at risk, require that it be kept under observation (Wildlife Management Branch 1991).

Distribution:
This species nests locally in southern Alberta, ranging a bit farther west and north than the McCown's Longspur, with which it often associates. The Chestnut-collared is found west to about Lethbridge and Calgary and north to the southern fringe of the Parkland Natural Region. It has been observed as a vagrant west to Banff and north to Fort McMurray and Fort MacKay.

Habitat:
The Chestnut-collared Longspur breeds in the Prairie Natural Region, most commonly in mixed grassland and locally in the northern fescue Grasslands (Wershler et al. 1991). It frequents moderate to heavily grazed grassland areas and poor, uncultivated fields with sparse growth. It prefers moister sites with somewhat more dense vegetation than those chosen by McCown's Longspur. Unlike the McCown's, it will not nest in cultivated fields, but will sometimes colonize recently mowed sites.

Nesting:
The nest of this longspur is a scrape on the ground in low and slightly moist situations with light to moderately thick grass. It is occasionally well-concealed, but is often open and exposed in sparse grass. The nest will frequently be placed beside dry horse or cattle droppings, or some other landmark. The female builds the nest, using grasses and weed stems and lining it with fine grasses and sometimes hair. Birds will sometimes nest in loose colonies with McCown's Longspurs.
The clutch is 3-6 eggs, whitish or buffy and speckled with browns and lavender. The female incubates 10-12 days and, relying on protective coloration, sits very close to the nest. If flushed, she may engage in distraction display. Both male and female brood and tend the young, which leave the nest at 9-11 days. They are able to fly a few days later and are independent 24 days after hatching. This species is commonly double-brooded and the male takes over care of the first brood when the female renests.

Remarks:
His chestnut collar, black underparts, and white or buffy face separate the breeding Chestnut-collared Longspur from other longspurs. The brown female (pictured above) is similar to McCown's, but has a faintly streaked breast. Females and fall birds are best identified while in flight; they show a large white triangle on each side of the tail at this time. The male, quite pugnacious on the breeding territory, also has a conspicuous flight song. He rises from the ground on rapidly beating wings, circles and undulates at the peak of his ascent, then glides down, singing, with wings fluttering and tail spread (Ehrlich et al. 1988). His song is spirited and melodious.
These longspurs forage on the ground picking up insects, other small invertebrates, and seeds in nesting season, and mainly seeds in the winter. Nestlings are fed insects, particularly grasshoppers.
Birds arrive in southern Alberta in mid- or late April. By August flocks have reformed and the flight south begins. By September only a few stragglers remain, but Pinel et al. (in prep.) report that a few flocks remain until late September in some years.

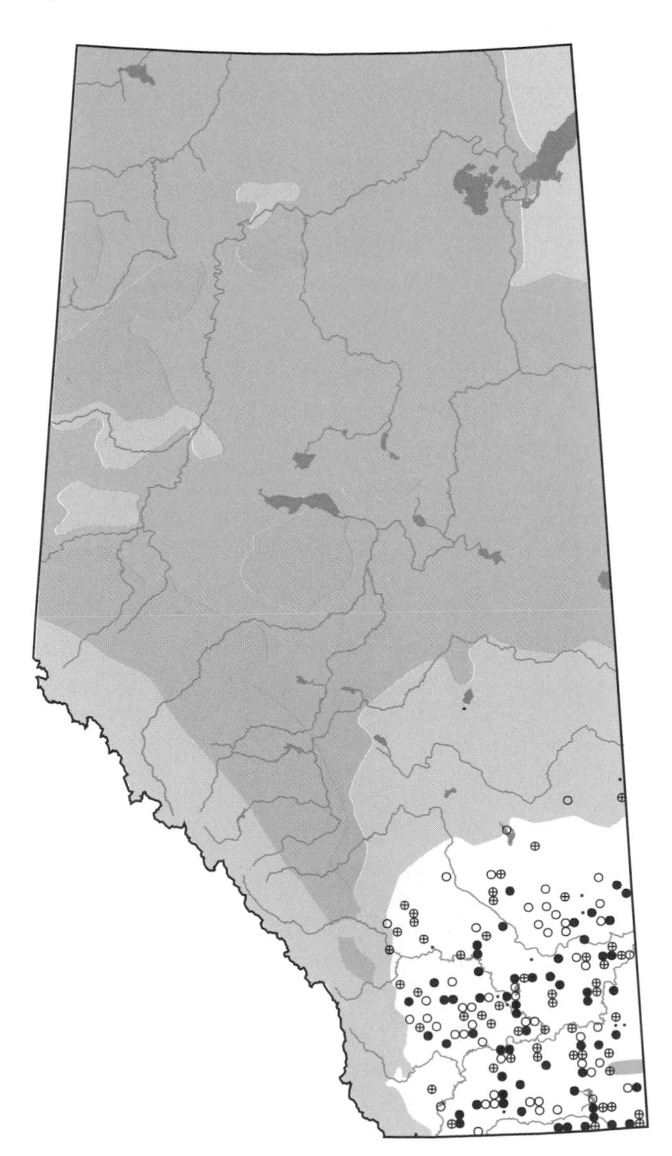

Bobolink

Dolichonyx oryzivorus

BOBO

Status:

The creation of pasture lands and hayfields by early settlers enabled the Bobolink to spread both eastward and westward from its North American pre-colonial range. However, changes in agricultural practices, loss of habitat, extensive shooting around the turn of the century, and capture for use as a cage species resulted in a general disappearance of the species from the eastern and northeastern parts of this expanded range (Read 1991).

The Bobolink is uncommon in Alberta, though northern and western expansion of its range is thought to be still occurring (Salt and Salt 1976). This may be attributed to sporadic occurrence at the edge of its range (P. Stepney pers. comm.).

Distribution:

This species breeds in southern Canada west to southeastern British Columbia and in the northern United States. It is most numerous in the upper mid-western states and prairie provinces (Read 1991). Salt and Salt (1976) reported it breeding irregularly in central and southern Alberta, north to Wainwright, Athabasca, and Whitecourt and west to Waterton Lakes, Turner Valley, Calgary, and Pigeon Lake. Atlas data confirms this and shows that the distribution is sporadic and inconsistent from one year to the next. Typically, only one to a few pairs are encountered at a given location. It is possible that "over-shooting" during migration is the main determinant of their occurrence in Alberta; sightings are often of solitary males (P.Stepney pers. comm.).

Habitat:

This is a bird of open meadow and pasture land, preferring moist areas of tall grass and hayfields; it is not attracted to native prairie. In autumn, prior to migration, Bobolinks congregate in marshlands, farmland, and other open places.

Nesting:

The Bobolink nest is found in the tall grass of a damp meadow or hayfield, usually in a depression in the ground, but sometimes elevated and woven into surrounding grass and plant stems. Several pairs may nest together in a loose colony, and males may acquire up to four mates. The female builds the nest, which is a loosely woven, shallow cup of grasses, plant stalks, and rootlets, and is lined with fine grasses.

Clutches are 4-7 eggs, which are whitish and strongly marked with browns and lilac. The female incubates 10-13 days. She is highly secretive about the nest and, if disturbed, will run some distance through the grass before flushing. She never flies directly to the nest, but lands several feet away, and walks to it through the grass. Both parents feed the young, but if a male has more than one mate, he will concentrate his attention on the first brood. The young fledge at 10-14 days, then hide in the grass until they can fly several days later.

Remarks:

The male Bobolink has black underparts, whitish scapulars, a grayish white rump, and a large buffy patch on the hind neck. He is the only North American songbird that is all black below and pale above. In contrast, the female Bobolink has the drab look of a sparrow, except that she is a little larger and is more yellow.

The male may sing his buoyant, bubbling melody from a perch, while in flight, or, if he is courting, as he slowly descends from 5 m up. The species' characteristic call note, "pink", helps distinguish females and juveniles from other sparrows in the fall.

On the nesting grounds, Bobolinks feed on insects, and seeds. The young are fed mainly insects. After nesting, flocks wander, living on weed seeds and waste grain.

This species arrives in Alberta in late May to late June and leaves the province by early August.

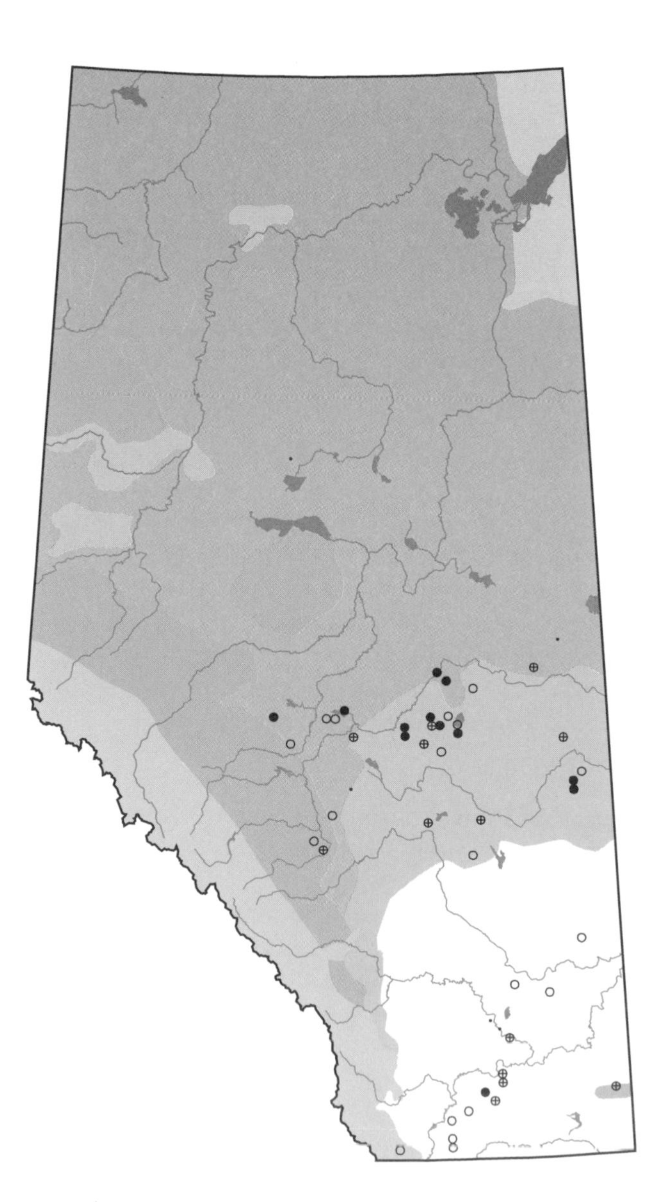

Red-winged Blackbird

Agelaius phoeniceus

RWBL photo credit: T. Webb

Status:

The Red-winged Blackbird is one of the most abundant and widespread birds on the continent, with a fall population once estimated at 400 million (Savage 1985). The species' fondness for certain agricultural crops, particularly oats, corn, and sunflower seeds, has made it the target of population reduction efforts, most of which have been unsuccessful.

Of the four Canadian subspecies, *A. p. arctolegus* is found in Alberta, where it is abundant and common.

Distribution:

This species breeds across Canada from the southern Yukon and British Columbia to Newfoundland. Atlas data indicate that it nests in suitable habitat throughout Alberta, with records ranging from 61% of surveyed squares in the Parkland to 23% in the Rocky Mountain region.

Habitat:

The Red-winged Blackbird prefers freshwater marshes for breeding and common Alberta habitat includes reed beds in sloughs, cattails and bulrushes along the margins of lakes, and willow-alder thickets beside streams and ponds (Salt and Salt 1976). It is rarely seen far from water during breeding season, but will, rarely, nest in drier habitats such as upland fields. The birds forage in wetland areas, stubble fields, ploughed land, and various open places. After the nesting season, these birds form flocks and forage over the countryside, returning to marshes to roost at night.

Nesting:

This species usually nests in thick reeds, sedges or bushes over or near water, occasionally in weeds in a field, or, rarely, in a tree. The female builds the nest, typically a loosely woven deep cup of coarse grass and marsh vegetation lined with finer grasses. In wet situations, the nest is woven into and supported by vegetation at the water's edge. Though highly territorial, nesting birds are often grouped together by habitat distribution. A dominant male may often have several females. The female maintains a sub-territory within the territory of her mate, and excludes intruding females from it.

The average clutch is four eggs, pale bluish green and scrawled with browns, black, and purple. Nests are frequently parasitized by cowbirds. The female incubates for 10-12 days and may or receive assistance from the male in feeding the young. The male defends his nests vigorously and intervenes aggressively in quarrels between females. The young fledge in 10-14 days and then leave the nesting territory with the female, who feeds them another two weeks. The male usually stays on his territory with the remaining females. Occasionally two or three broods may be raised per season.

Remarks:

The male is a striking bird with completely black plumage except for his red and buffy wing epaulets. The female (pictured above) is a heavily streaked, brownish black bird with a light streak over the crown and above the eye. The male's song, an exuberant "con keree", is distinctive.

In the spring, these blackbirds eat both waste grains and weed seeds, and are often seen at home feeders. Insects become more important during nesting. The young are fed mostly insects and, for a brief period after they have fledged, Red-wings are often encountered foraging in the tree canopy. After nesting, the birds switch again to seeds and grains.

Male Red-wings are among the earliest spring migrants, arriving in the province before mid-April. The females follow two weeks later. Most birds have left the province by the end of October. Overwintering is rare, but has been reported at all three mountain parks, Rocky Mountain House, Camrose, Calgary, and Lethbridge.

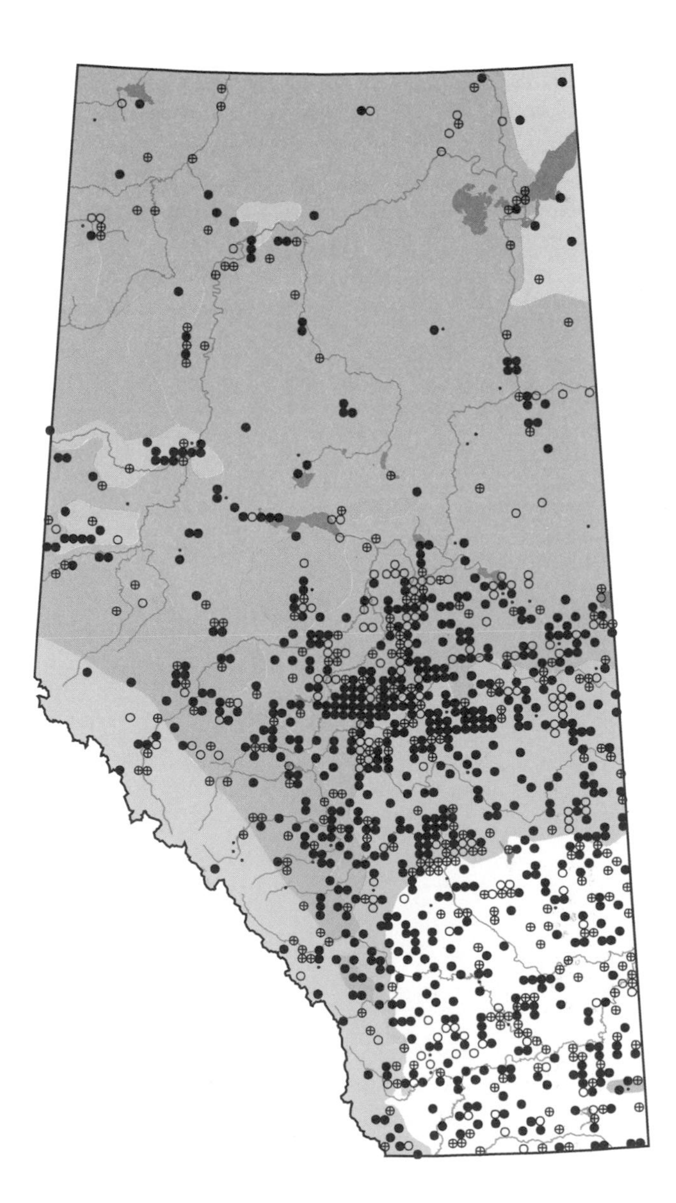

Western Meadowlark

Sturnella neglecta

WEME photo credit: T. Webb

Status:

The status of this species in Alberta has not been fully determined. However, the species appears to be much less common in the northern portion of its range, particularly the Peace River Parklands, than it was thirty years ago (P. Stepney pers. comm.).

Distribution:

The Western Meadowlark breeds in southern Canada from central British Columbia east to southeastern Ontario. It followed settlers north and westward and reached Red Deer in 1892 and Edmonton in 1897 (Salt and Salt 1976). In Alberta, the northwest edge of its North American range, it nests in the central and southern parts of the province north to the Athabasca River and Peace River District. In the Rocky Mountain Natural Region, there were no breeding records north of Saskatchewan River Crossing. This species is found primarily in the Grassland (49% of surveyed squares) and, to a lesser extent, in the Parkland (28% of surveyed squares) natural regions. It was found less frequently in the foothills (7%) and Rocky Mountain (11%) regions. Pinel et al. (in prep.) reports extralimital breeding in the Fort McMurray area on more than one occasion between 1976 and 1980. There were no breeding records for the Western Meadowlark in northeastern Alberta during Atlas surveys from 1987 to 1991.

Habitat:

The Western Meadowlark inhabits grassy plains and river valleys, pasture land, uncultivated fields, and the grass verges in country roads, breeding wherever there is a thick growth of weeds and grasses. When found in the mountains, it favors montane grasslands, roadsides, and shrubby alluvial meadows (Holroyd and Van Tighem 1983). In the fall, flocks of meadowlarks will be seen on stubble fields, roadsides, and cultivated areas.

Nesting:

This species nests on the ground in a natural or scraped depression, always in open grassy areas. The nest is a well-concealed domed canopy of grass, bark, and plant stems woven into the surrounding vegetation. Its entrance is on the side and a tunnelled grass runway often leads to it. The female builds the nest, and lays 3-7 white eggs, which are speckled with browns. She incubates these alone for 13-15 days; the young are tended by both parents. The young birds leave the nest at about 12 days, before they are able to fly, and are fed for a futher 14 days (Harrison 1978). A male Western Meadowlark may acquire up to three mates which nest separately within his territory. They are double-brooded over most of their range.

Remarks:

The Western Meadowlark is a stocky, robin-sized bird with a short tail, a strong pointed bill, a grayish brown back, and a yellow breast with a black V on the bib.

The male's rich, melodious song, which he delivers from fence posts and telephone poles, can be heard from nearly 1 km away. He defends his territory with various displays, including chase-like aerial duets and "jump flights".

The species spends most of its time in the grass foraging for insects, other small invertebrates, and seeds. Young are fed mainly insects. In the fall, waste grain and weed seeds form a more substantial part of the diet. If approached on the ground, a Western Meadowlark will walk away, flicking its tail spasmodically to reveal white outer tail feathers, and giving its alarm call, until forced to fly (Salt and Salt 1976).

Birds arrive in southern Alberta at the end of March and early April, and most depart again by the end of October. Overwintering has been reported from Edmonton, Tofield, Vegreville, Rimbey, Snake's Head, Stettler, Calgary, Lethbridge, and Claresholm (Pinel et al. in prep).

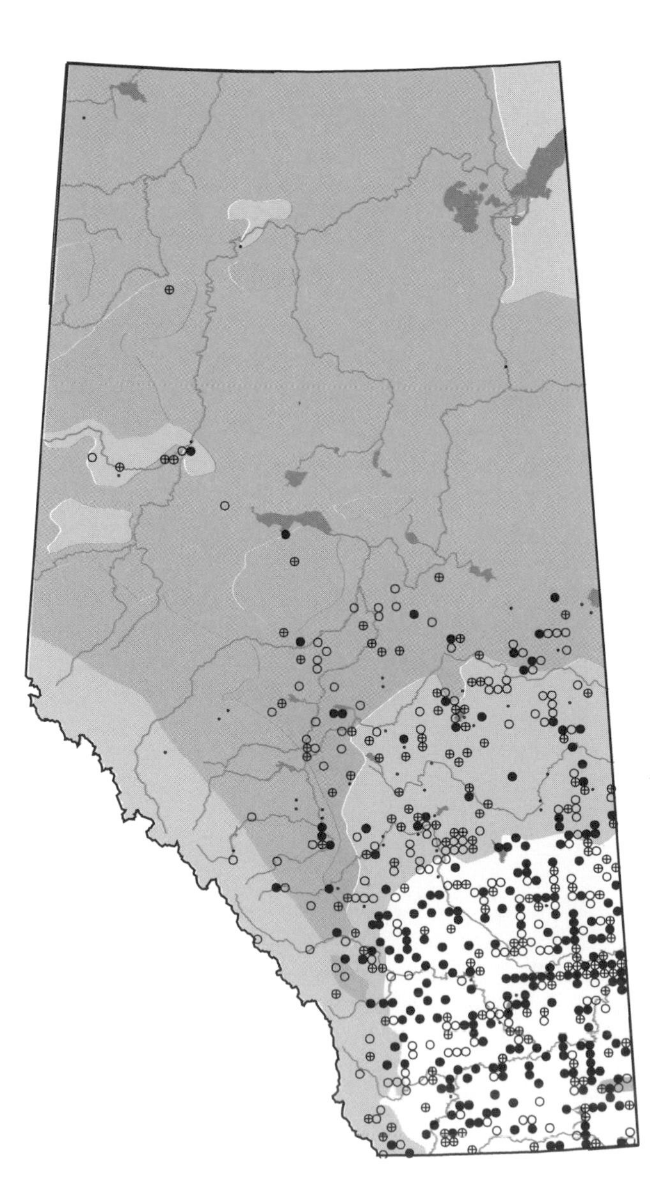

Yellow-headed Blackbird

Xanthocephalus xanthocephalus

YHBL photo credit: T. Webb

Status:

Populations of the Yellow-headed Blackbird in Alberta are quite strongly influenced by water conditions on the prairies. In dry years, many more are found in central and northern Alberta than are found in wet years (P. Stepney pers. comm.). The numbers actually vary considerably from one year to the next in many localities.

Distribution:

This showy blackbird breeds in western Canada from central British Columbia east to western Ontario and south into the western United States. In Alberta, it breeds north to La Crete and Lake Athabasca and it is found most commonly in the Parkland (45% of surveyed squares) and Grassland (23% of surveyed squares) natural regions. It is seldom encountered in the northern and western portions of the province. Pinel et al. (in prep.) states that this species has become more abundant in southern Alberta in the last 25 years, perhaps as a result of the development of permanent wetland areas.

Habitat:

This blackbird tends to frequent deeper sloughs and marshes, and river and lake edges with dense emergent vegetation. However, in drier years, smaller sloughs are also used. It forages in marshlands, grain fields, freshly ploughed ground, barnyards, and similar locations (Godfrey 1986), and also stubble fields and disturbed sites on migration (Holroyd and Van Tighem 1983).

Nesting:

Unlike other blackbirds, this species never nests in bushes or trees on dry land. Instead, the nest is found in dense emergent vegetation, almost always over water. Built by the female, it is a deep-bowled, compact cup woven from dead, water-soaked grasses and reeds and fastened to cattails or reeds. It has an inner lining of fine leaf strips and grasses, and a partial canopy is occasionally constructed above the rim. Yellow-headed Blackbirds nest in small colonies, but are highly territorial. Nesting areas frequently are returned to annually.

The typical clutch is 3-4 eggs, which are whitish and suffused with spots of browns and grays. The female incubates 11-14 days and also broods the young. A male may be polygamous, with up to five mates, and he may or may not assist the female in feeding the young. A polygamous male who helps with feeding is usually most attentive to the first nest on his territory. All males are vigorous defenders of the nest sites. Young birds depart the nest at 9-12 days, but are unable to fly until about 20 days (Harrison 1978). This species is usually single-brooded, but may be double-brooded on occasion.

Remarks:

The male Yellow-headed Blackbird is easily recognized by his rich yellow head and neck, black eye patch, and patch of white on the wings. The female is brown and mottled, but pale yellow coloration about the head and breast allows her to be distinguished from other blackbird females.

Insects, including beetles, grasshoppers, and caterpillars, are the main food of this species. They also eat grain, grass seeds, and weed seeds. The young are fed mostly insects, by regurgitation at first.

This species begins to arrive in Alberta in mid-April (Pinel et al. in prep.), the males arriving a week or two ahead of the females. In the fall, Yellow-headed Blackbirds often flock together with grackles, Red-winged Blackbirds, and cowbirds, and they leave the province by the end of September. Overwintering is extremely rare, with one report from Lethbridge in 1980.

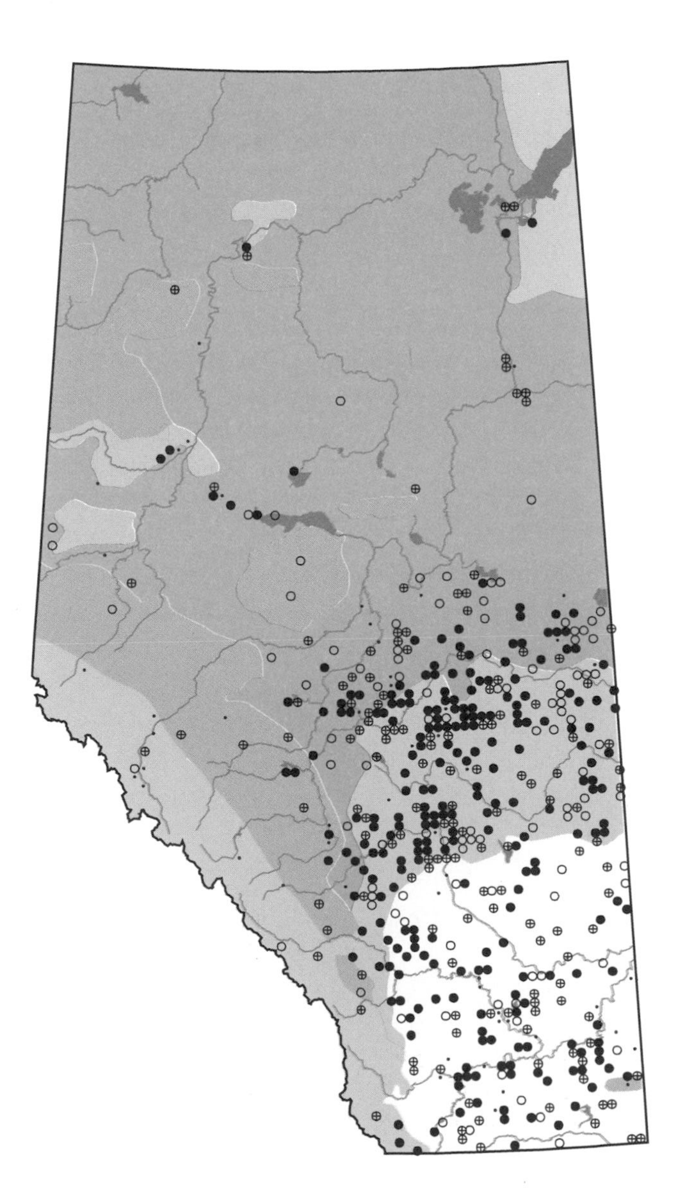

Rusty Blackbird

Euphagus carolinus

RUBL

Status:
The status of the Rusty Blackbird in Alberta is undetermined. It is the least conspicuous of the blackbirds, and is typically a solitary nester in Boreal Forest wetlands. It is often confused with the Brewer's Blackbird and, as a result, very little is known about this species overall (P. Stepney pers. comm.).

Distribution:
This species breeds in Alaska and across Canada from the treeline south to the central portions of the western provinces, northern Ontario, Quebec, and the maritime provinces. It breeds regularly in northern Alberta in suitable habitat, in low numbers, and less commonly in central and southwestern parts of the province. Birds nest south to Pigeon Lake, Red Deer, Sundre, and Okotoks, and into the mountains at Jasper (Salt and Salt 1976). Records in central and southern portions of the province should be confirmed with a specimen or photograph, because this bird is often confused with the Brewer's Blackbird (P. Stepney pers. comm.). It is only a rare summer visitor to Banff and, in southern Alberta, it is only seen during migration.

Habitat:
The Rusty Blackbird frequents alder-willow bogs, muskegs, beaver dams, and other openings in wet coniferous and mixed forest, particularly swampy shores beside lakes, streams, and rivers.

Nesting:
A small conifer or bush, usually near or over water, is the preferred nesting site of this species. The nest is usually positioned against the trunk on a horizontal branch, or in a fork, at a height of less than 3 m. Built by the female, it is a bulky, well-constructed cup with a rough, untidy exterior of twigs, grass, and moss, a middle layer of rotting vegetation which dries hard, and a lining of fine grasses. Unlike many other blackbirds, the Rusty is not colonial; seldom can more than one or two pairs found in a given locality.
The female lays 4-5 pale blue-green eggs, which are strongly marked with brown and a little gray. She alone incubates for 14 days. The male brings food to the female at this time, calling her off the nest to collect his offering. He is a vigorous defender of the nest. The nestlings are tended by both parents, and leave the nest at 13 days.

Remarks:
The male Rusty Blackbird in breeding plumage can be difficult to distinguish from the male Brewer's Blackbird. However, the Rusty is much less glossy, is more evenly black all over, and lacks the Brewer's strong purple iridescence on the head. The songs of the two species are also quite different; the Rusty's song is short and ends on a high note that sounds like a rusty hinge (Godfrey 1986). The female Rusty is a dull slate color like the Brewer's female, but has yellow rather than brown eyes. In fall, both sexes are distinctively rusty above and buffy below.
While not shy or wary, this species is more quiet and inconspicuous than other blackbirds, at least in breeding season. It also eats more animal matter than other blackbirds, feeding on insects, spiders, crustaceans, snails, and sometimes salamanders and small fish. Vegetable matter, more important in the fall and winter, includes grains, weed seeds, and fruit. Most foraging is done on the ground.
In spring, the birds arrive on the nesting grounds in small groups, generally by mid-April, with the main influx coming from the east (Pinel et al. in prep.). They migrate eastwardly in large flocks in early October, though some stragglers linger into November. Overwintering is reported from several locations in western parts of central and southern Alberta, including St. Paul, Wabamun Lake, Drayton Valley, Calgary, Medicine Hat, Snake's Head, Camrose, and Banff.

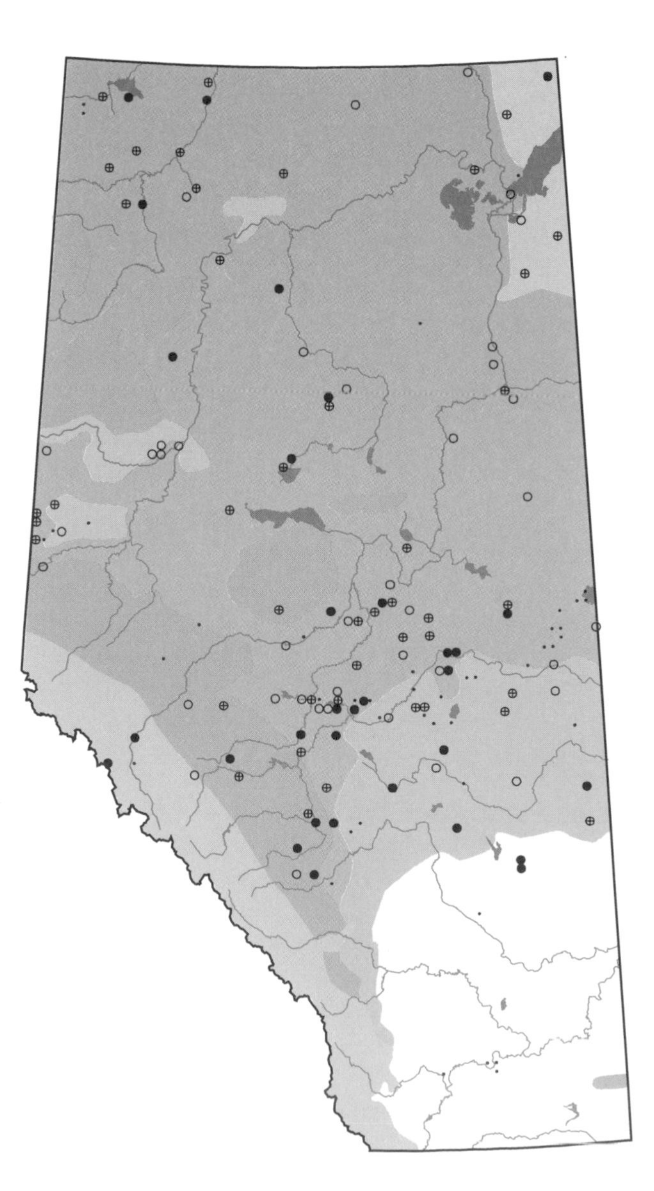

Brewer's Blackbird

Euphagus cyanocephalus

BRBL photo credit: T. Webb

Status:
Populations of the Brewer's Blackbird have increased and its range has extended as a result of settlement and the spread of agriculture (Ehrlich et al. 1988). In Alberta, populations are healthy and stable. Their range has expanded considerably as they colonize modified areas and roadways in Northern Alberta (P. Stepney pers. comm.)

Distribution:
This familiar blackbird breeds in Canada from central British Columbia to western Ontario. In this century, the Brewer's has expanded its range significantly to the north and northeast (Stepney and Power 1973; Stepney 1975).
In Alberta, this species breeds in the southern, central, and northwestern parts of the province. It was found nesting at Hay River and Fort Simpson in 1977, making it very likely that it nested along much of the highway to Hay River (Stepney 1979).
The Brewer's is the common blackbird of the Grassland and Parkland natural regions, having been recorded in over 40% of surveyed squares. They were found in 37% of surveyed squares in the Boreal Forest and in 15% of Rocky Mountain and Foothills region squares.

Habitat:
This blackbird breeds in a variety of habitats, including grassy prairies, pastures, other open areas with patches of trees and shrubs, brushy growth along water courses, irrigation canals, roads, railway rights-of-way, hedgerows, aspen groves, marshy edges, and cutover areas or burns. In the mountains, the Brewer's is found on willow and birch wetlands and meadows (Holroyd and Van Tighem 1983). They roost in reeds and trees.

Nesting:
Nesting may be on the ground, in low bushes, or in low trees, often close to water or marshy areas. Ground nests are generally in depressions or sunk in ground vegetation such as moss. They are situated near clumps of grass or at the base of a shrub or tree. These birds are loosely colonial, maintaining distinct territories within the colony.
The female chooses the nest site, defending her choice aggressively, and builds the nest. It is a stout cup of twigs and grasses, often mixed with mud. It is lined with fine rootlets, hair, and fine grasses.
The typical clutch is 5-6 pale greenish-gray eggs, with a heavy blotching of browns. Brewer's nests are often parasitized by cowbirds. The female incubates 12-14 days and may be fed on the nest by the male at this time. Brewer's males are normally monogamous, but may be polygamous and have up to four mates. The young fledge in 11-13 days, afterwhich they are moved by their parents out of the nesting area (Cadman et al. 1987) and fed for another 12-13 days. Brewer's are sometimes double brooded.

Remarks:
This species can be confused with the Common Grackle, Brown-headed Cowbird, and, in spring, the Rusty Blackbird. It is most like the Rusty, but the male has a purplish sheen to its head which is absent in that species, and the female Brewer's has brown rather than yellow eyes (characteristic of the female Rusty).
The common fare of the Brewer's is insects, seeds, and fruit, and also crustaceans, snails, and waste grain. The young are fed mostly insects and spiders.
In spring, these birds arrive in mid-April. In the fall, they form large flocks, often with other species of blackbirds and grackles, and depart in September and October. Overwintering has been reported from Medicine Hat, Erskine, Lacombe, Red Deer, Ponoka, Banff, Canmore, Edmonton, and Fort McMurray.

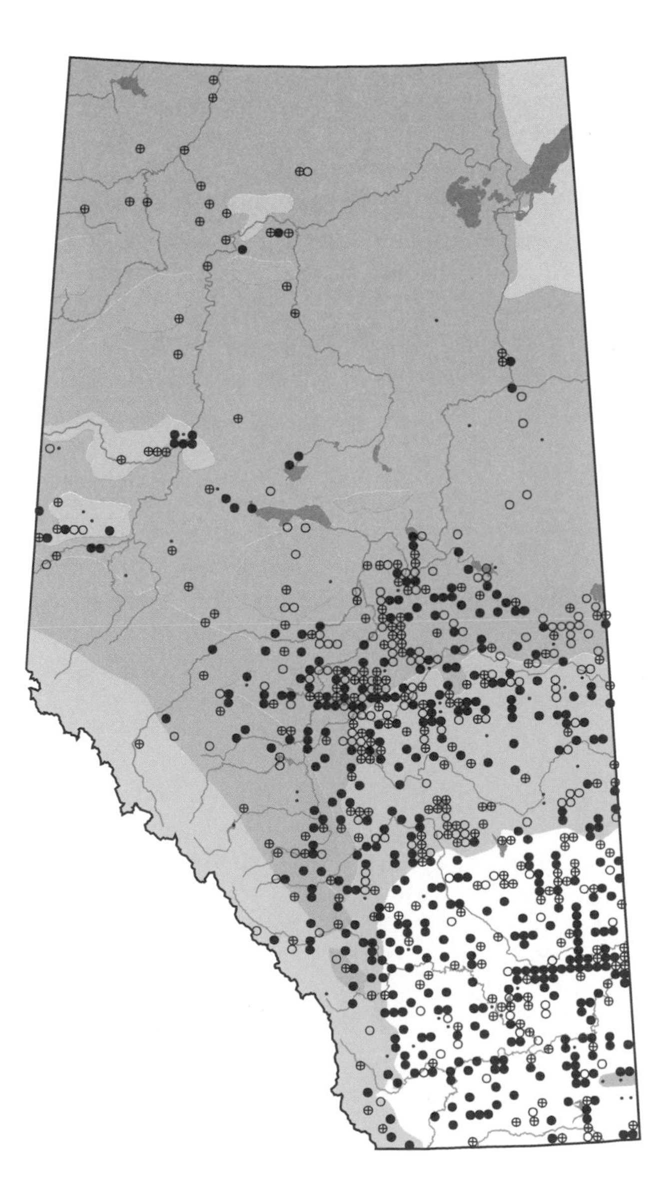

Common Grackle

Quiscalus quiscula

COGK

Status:
The Common Grackle is Canada's largest blackbird of regular occurrence (Godfrey 1986). Its status in Alberta has not been fully determined, but it appears to be common in the Parkland and southern Boreal Forest regions.

Distribution:
This species breeds across North America south of the tundra and east of the Rockies. It is found throughout much of Alberta, but is rare in the Rocky Mountain and Foothills natural regions, where it was recorded in only 1% of surveyed squares. It is most often found in the Parkland, where it was recorded in 22% of surveyed squares, and southern Boreal Forest (18% of surveyed squares) regions. It is found in suitable habitat in the Grasslands region (8% of surveyed squares), and is widespread throughout the northern portions of the province, including the Canadian Shield, though its density is low and its occurrence is sporatic.

Habitat:
It breeds in a variety of sites, but favors damp, open woodlands, the shores of lakes and streams, and wet meadows with scattered patches of brush (Salt and Salt 1976). It forages on the ground and is commonly seen in open places such as shorelines, marshes, fields, pastures, farmyards, lawns, parks, gardens, and golf courses. The grackle is often associated with human habitation.

Nesting:
This noisy and cocky blackbird's nest can be found in trees, preferrably conifers, but sometimes in shrubs, vines, marsh vegetation, stumps, or logs, or man-made structures such as abandoned buildings and bridges. In Alberta, grackle nests are usually found in cattails or dead willows over water (P. Stepney pers. comm.). In the southern and central parts of the province, several pairs will form a loose colony; in the northern areas, isolated pairs or two or three loosely aggregated pairs, are the rule (P. Stepney pers. comm).
The nest is a large bulky cup of sticks, weeds, grass, reeds, and string. It is usually plastered together with mud and lined with grasses and rootlets. The female builds the nest. The typical clutch is 4-6 pale greenish or bluish eggs, which are heavily marked with black and browns. Incubation is 12-14 days by the female, which also broods the young. Both parents help feed the nestlings, which leave the nest at 12-15 days. The young stay in the vicinity only 2-3 days after fledging (Harrison 1978). This species is occasionally double-brooded.

Remarks:
Up close, the Common Grackle is a handsome bird, with purple iridescence on its head, neck, and upper breast, and a bronze-green sheen elsewhere (Godfrey 1986). From a distance, the bird appears to be uniformly black. Its bright yellow eyes and long, wedge-shaped, keeled tail are also distinctive. The male proclaims himself with a voice that sounds like "an un-oiled wheelbarrow" and, though not musical, it is not, altogether, an unpleasant song (Godfrey 1986).
The grackle's main summer food consists of insects, other small invertebrates, and, occasionally, the eggs and nestlings of other birds. Outside of breeding season, it also eats waste grain, seeds, fruit, and garbage. When not breeding, the species is quite gregarious and is often seen in fields with other blackbirds and starlings.
These birds arrive in the province in April and depart again late in September, with a few stragglers lingering into October. Overwintering has been noted in Stettler, Calgary, Ponoka, Devon, Calmar, Ardrossan, and Edmonton.

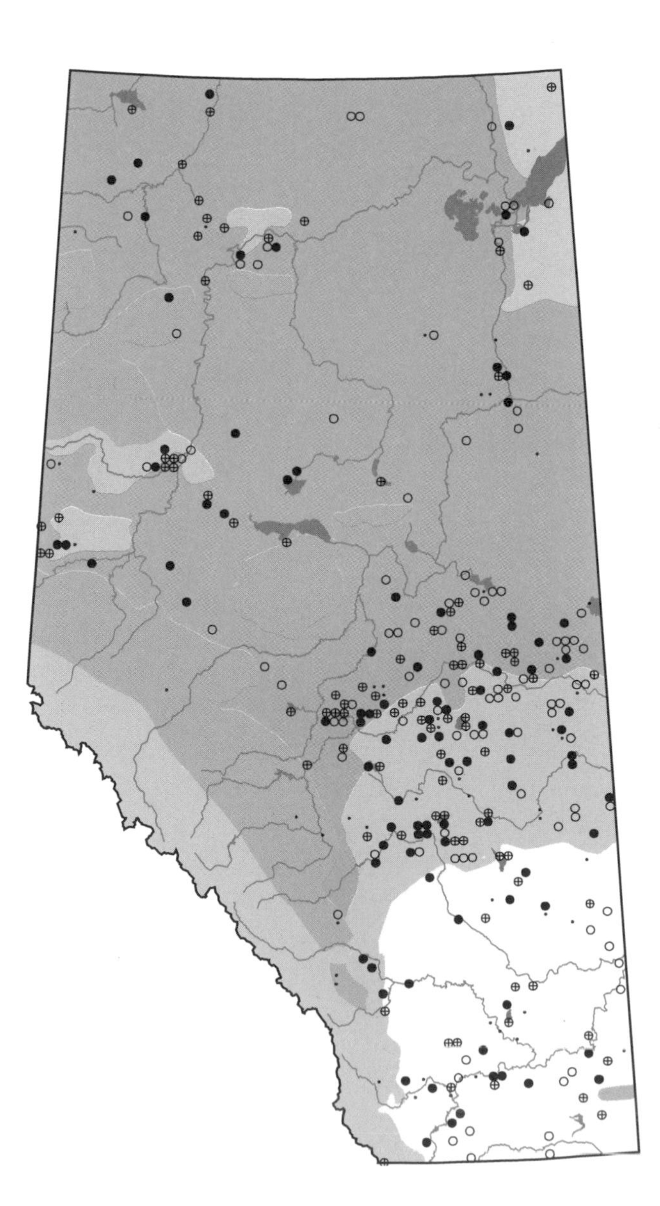

Brown-headed Cowbird

Molothrus ater

BHCO

Status:
The Brown-headed Cowbird is common throughout Alberta. In the past two hundred years, it has gone through one of the greatest range expansions of any North American bird species (Graham 1989).

Distribution:
This familiar blackbird was recorded breeding in all regions. At one time, it followed the buffalo herds and its range was limited to the prairies. However, as European settlers cleared the forests for farmland and introduced cattle, its range expanded. With this expansion it came into contact with many bird species not previously exposed to the brood parasitism it practices, and it now poses a threat to some species (Graham 1989).

Habitat:
The cowbird forages on the ground in open areas such as pastures, cultivated fields, residential lawns, and forest and roadside clearings. It seeks out livestock, horses, and other large mammals—mainly because of insects they stir up and attract (Godfrey 1986). The species parasitizes nests in a wide range of habitats, although nests in farmland, overgrown fields, fence rows, residential areas, and woodland edges are often preferred (Peck and James 1983).

Nesting:
This species lays its eggs in the nests of over 200 North American species. Warblers, sparrows, finches, blackbirds, flycatchers, and thrushes are the most frequent hosts.
The female lays one whitish egg, which is blotched with browns, and will sometimes remove an egg of the host species. Usually, an egg is laid after the host finishes laying, in order to ensure synchrony of feeding and rearing (Jasper 1988). More than one cowbird female may lay in the same nest. Females lay an average of 25-50 eggs each season. This great fertility balances the low fledging rate of the species (Graham 1989).
Some host species are intolerant of a cowbird egg; they may throw it out, build a new nest floor over it, or abandon the nest. However, many do raise the young bird, usually at the expense of their own nestlings. Cowbirds hatch after 11-12 days of incubation, most often a little sooner than the eggs of the host. The young cowbird is generally larger than the host nestlings and has a larger mouth gape, so it receives most of the food. Young fledge after 10 days, and continue to beg persistently until independent.

Remarks:
The male of the species has a glossy black body and may be confused with other blackbirds at a distance. However, its brown head and short, conical-shaped bill are distinctive. The brownish-gray female is similar to female Rusty and Brewer's Blackbirds but, again, the bill is diagnostic.
Cowbirds commonly eat insects, and also spiders, snails, grain, grass, and forb seeds. They are gregarious and, after breeding, may form large flocks with other blackbirds and starlings for foraging and roosting.
The species arrives in central Alberta towards the end of April and departs again by late September, with numbers peaking in early August.

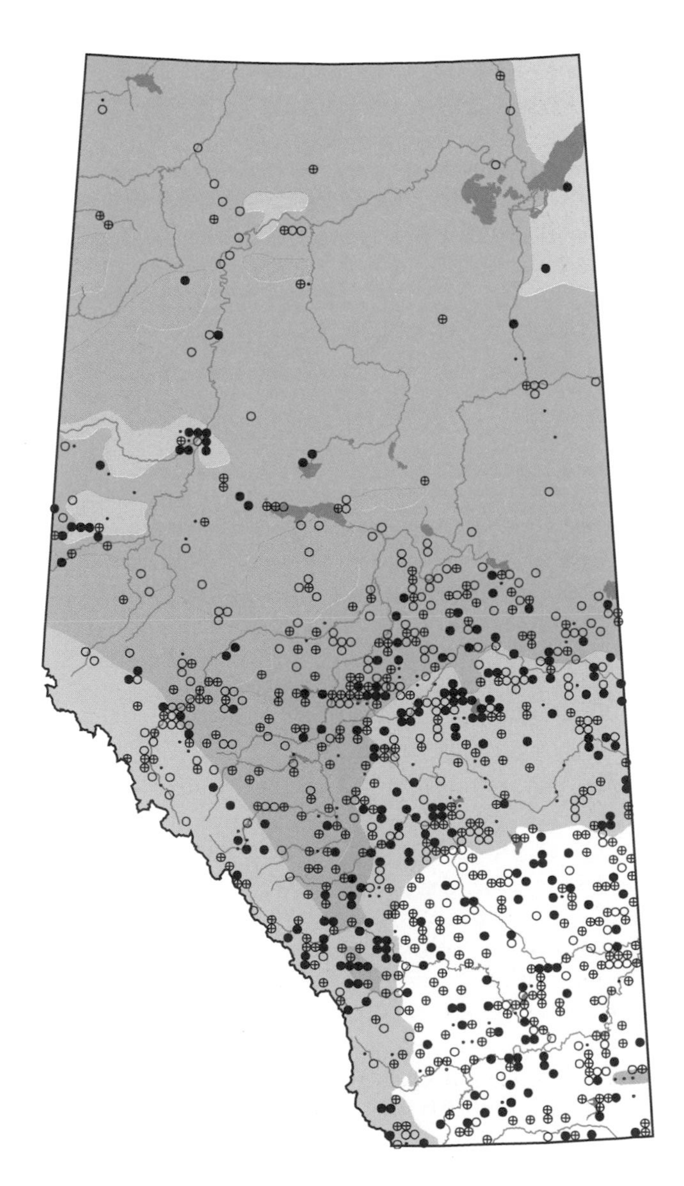

Northern Oriole

Icterus galbula

NOOR photo credit: T. Webb

Status:
The Baltimore Oriole and Bullock's Oriole were once classified as distinct species, but are now considered to be conspecific and, as one species, have been re-named the Northern Oriole (American Ornithologists' Union 1983). This oriole is common in central Alberta east of the Rockies.

Distribution:
The two subspecies of Northern Oriole occupy different ranges in the province. The former Bullock's Oriole (*I. g. bullocki*) breeds in southern Alberta in the lower reaches of the South Saskatchewan and Red Deer rivers, and in the Milk River valley. The Baltimore Oriole (*I. g. galbula*) nests throughout the rest of the species' Alberta range. The two races interbreed along the eastern parts of the Milk River, and many birds in this area show atypical markings (Salt and Salt 1976).
Atlas data indicate that the Northern Oriole is found mainly in the Parkland region (52% of surveyed squares), Boreal Forest (31%), Grassland (21%), and Foothills (15%) regions.

Habitat:
The Northern Oriole is a species which prefers the tall trees of open deciduous woods, often in edge habitat. Characteristic habitat includes cottonwoods along prairie creeks and rivers, or aspen, poplar, and birch along lakeshores or other watercourses and country roads (Salt and Salt 1976). It will also be attracted to a grove of trees in a field or mature shade trees in an urban park or garden.

Nesting:
This species nests in deciduous trees, and, occasionally, tall deciduous shrubs. The nest is a deep bag-like pouch suspended from the end of a horizontal branch generally at a height of 7-10 m. It is built by both the male and female, and is woven of plant fibres, grass, shredded bark, hair, and string. It is suspended by the rim by long shreds of similar material. The nest is lined with hair, plant down, fine grasses, and moss. Its entrance is usually at the top.
The typical clutch is 4-5 grayish-white eggs with black and brown scrawlings. The female incubates alone for 12-15 days. Both parents tend the young, which fledge at 12-14 days. Parents and young leave the nesting area soon after fledging; the female and young form flocks, and the male remains solitary.

Remarks:
The male of this species is a striking bird, easily recognizable by his orange and black markings. He reaches sexual maturity in his first year, but does not wear his bright plumage until his second year. The male Baltimore has an all black head and back, black wings with orange shoulder patch, orange underparts and rump. The male Bullock's has a black crown, orange stripe through the eye, orange cheeks and sides of neck, large white wing patch, and yellower rump and underparts. Northern's have a loud, melodious song — "Peter, here Peter, Peter" — but the male sings very little once the birds are paired (Cadman et al. 1987).
The females of both races are yellowish olive with white wing bars. The Baltimore female has more yellowish underparts, and the Bullock's female is grayer above.
In nesting season, this oriole feeds on insects high in trees. Its young are fed soft-bodied insects and larvae. The Northern Oriole is one of the few species tolerant of hairy larvae such as tent caterpillars. Later in the season, fruits such as raspberries, Saskatoons, and pincherries have a prominent place in the diet (Salt and Salt 1976).
The Northern Oriole arrives in Alberta in mid- to late May and departs again by late August, with a few stragglers staying on into September. No overwintering is reported.

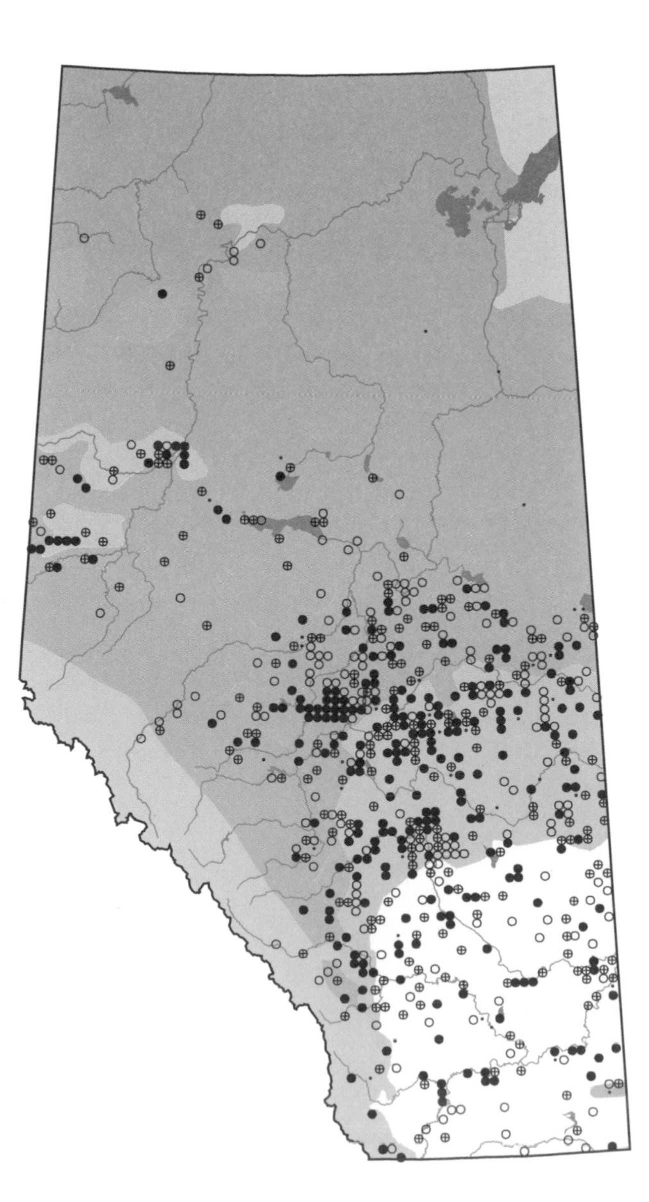

Rosy Finch

Leucosticte arctoa

ROFI photo credit: K. Morck

Status:
The three races of Rosy Finch were classified as a single species in 1983 (American Ornithologists' Union 1983). The Gray-crowned Rosy Finch (*L. a. tephrocotis*) is the subspecies breeding in Alberta. The exact status of this finch in the province is unclear, but Holroyd and Van Tighem (1983) describe it as a "very common resident" during the summer in the mountain parks.

Distribution:
The Rosy Finch breeds in the mountains of western North America from the Bering Sea south to California. In Alberta, it is found in the Rocky Mountain region, breeding from Jasper south to the Waterton Lakes. Its distribution is somewhat local, but it may be expected throughout the area (Salt and Salt 1976).

Habitat:
During breeding season, this species inhabits mountain rocks and cliffs among the glaciers and snows above the timberline. It is locally common in tundra and heath meadows at and above the timberline, and is also found on grassy avalanche slopes and talus (Holroyd and Van Tighem 1983). In the fall, the birds gather in flocks and retreat to the lower mountain valleys and foothills. At that time, they may be seen in open areas such as river channels, fields, farmyards, roadsides, railroad clearings, and at feeders. In the winter, birds roost in sheltered crevices and nooks in cliffs and rocks, cave entrances, and sometimes man-made structures. In Alberta, they have also been observed roosting in the nests of Cliff Swallows and, once, in the nest of an American Dipper (Nordstrom and Butler 1976).

Nesting:
This species nests in crevices or under rocks at high altitudes in the mountains. The nest site is chosen by the female. She also builds the nest, a bulky mass of grass and rootlets lined with fine grass, rootlets, plant down, moss, and feathers. Nests are often used over successive years. The Rosy Finch is semi-colonial and only weakly territorial, but males are often observed fighting in breeding season. Pairs are monogamous.
The female incubates 4-5 white eggs for 12 to 14 days. She is often fed by her mate during incubation. The young are tended by both parents, who develop buccal pouches (openings in the floor of the mouth) to carry food to the nestlings. The young birds leave the nest at 18-20 days and are then fed by their parents for a further two weeks. The family unit remains intact until the fall.

Remarks:
The Rosy Finch is brown with a pinkish rump, wings, and belly. The gray-crowned form has a black cap, gray head, and brown cheeks and ear region. The song of the male is a rich, goldfinch-like warble (Godfrey 1986) and his call is a sharp chirp, reminiscent of the House Sparrow.
A gregarious species, the Rosy Finch is also quite tame. It spends much of its time on the ground where it gleans insects, spiders, seeds, and berries, depending more on seeds and grains in the winter. The young are fed mostly insects, at first by regurgitation.
Spring migration into the foothills and mountains occurs from mid-March to early May, with peak numbers in April (Pinel et al. in prep.). In fall, these finches descend from the alpine heights to lower altitudes and, in southern Alberta, out into the Grassland region. The greater distances that the birds move away from the mountains appear to be related to lower levels of precipitation and, hence, lower seed production in the previous summer (Hendricks and Swenson 1983). Fall migration begins in late August, peaking in late September and early October. Christmas bird counts have recorded this finch in Medicine Hat, Lethbridge, Calgary, Crowsnest Pass, Jasper, Edson, and Edmonton.

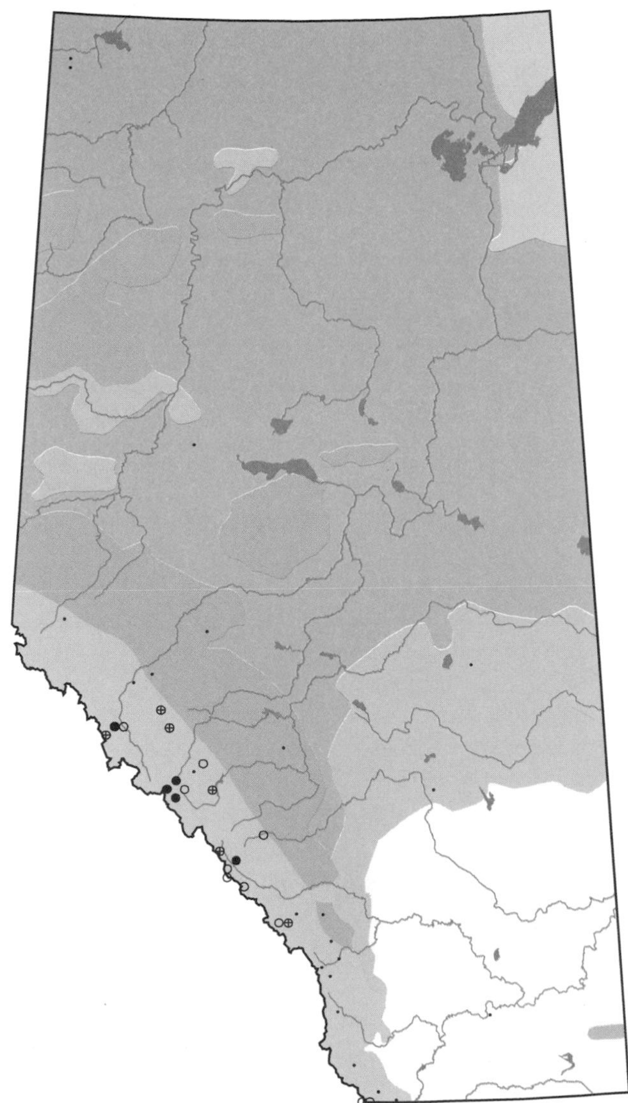

Pine Grosbeak

Pinicola enucleator

PIGR photo credit: R. Fairweather

Status:
The exact status of this species in Alberta is not known (Wildlife Management Branch 1991). It is rarely seen in the province during the nesting season, but is quite commonly observed in the winter.

Distribution:
A holarctic species, the Pine Grosbeak breeds in Canada in coniferous forests north to the treeline, south to the American border in the mountains, and in Newfoundland in the east. In Alberta, it breeds in the Rockies in Banff and Jasper national parks, where it is a permanent resident, and probably at Waterton Lakes. It breeds locally in northern Alberta, where it has been observed in nesting season at Salt Prairie and in Wood Buffalo National Park (Salt and Salt 1976), and was recorded as a probable breeder in Wood Buffalo National Park by the Atlas project.

Habitat:
The Pine Grosbeak breeds in coniferous and mixed wood forests, and shows a decided preference for evergreens at all times of the year (Salt and Salt 1976). It often frequents towns and cities during migration and in winter, and may be seen in deciduous trees including crabapples, ashes, and other ornamentals, and also in shrubbery and at feeders. It occasionally bathes in soft snow during the winter.

Nesting:
This species nests in conifers, usually 3-4.5 m from the ground. The nest, built by the female, is a rather loose construction of twigs, grasses, rootlets, and moss lined with fine grasses, rootlets, and hair. The secretive nature of these birds during the nesting season makes them difficult to observe. In other areas, it has been recorded that the male seldom sings at this time (Cadman et al. 1987). However, in Alberta, this bird frequently sings in late winter and spring (D. Prescott pers. comm.).
The female lays and incubates 4-5 eggs, which are a pale green-blue spotted with browns and purplish gray. During the 13-14 days of incubation, the male feeds the female, calling her off the nest to pass her the food. Both sexes feed the young and develop sacs in their cheeks for the transportation of food. The nestlings fledge in 13-20 days. The species is thought to be single-brooded.

Remarks:
Pine Grosbeaks are robin-sized birds with heavy, dark bills. The adult male has a dull red head, neck, back, and breast, and gray wings and tail; the female is plain gray with a yellowish or orange crown and rump. Both have conspicuous white wing bars. Male Purple and Cassin's Finches are also reddish, but are smaller and lack white wing bars.
This finch is primarily a seed eater, with preference for seeds from the cones of conifers. It also eats the seeds of deciduous trees and bushes, weed and grass seeds, berries, buds, mast, and a few insects in the summer.
Because of its retiring nature during the nesting period, Albertans are most likely to see this species in winter, when it forms small flocks of up to two dozen birds. It wanders irregularly and erratically during winter, but is commonly observed in central and southern parts of the province. It usually arrives in these areas in October and is usually gone by late March.

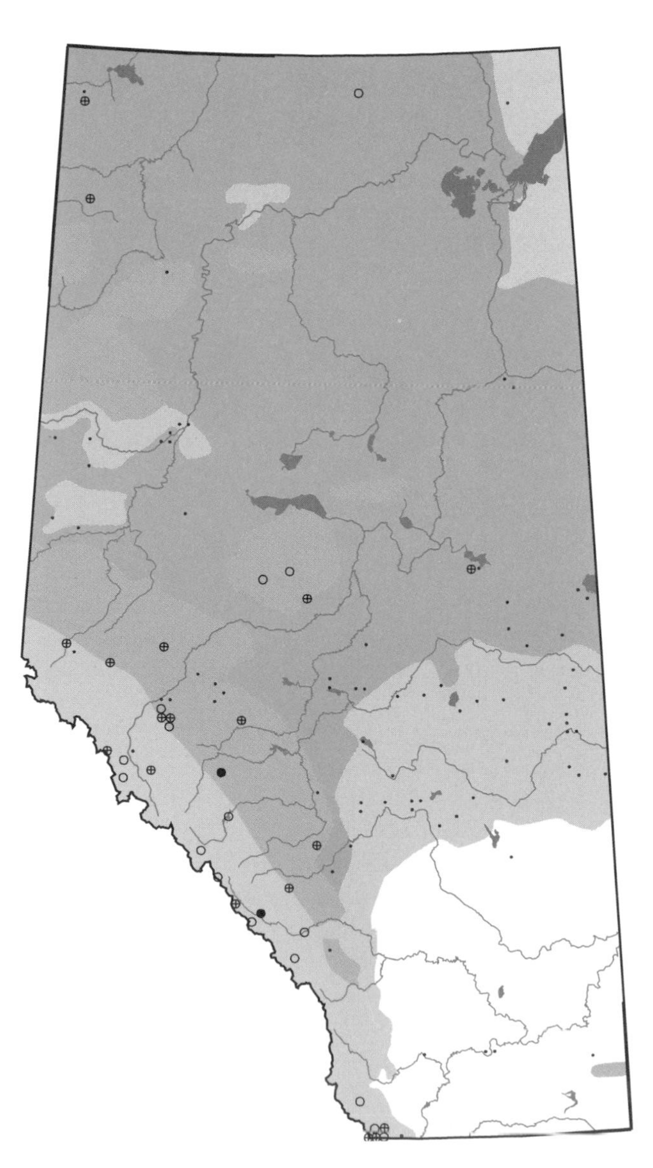

Purple Finch

Carpodacus purpureus

PUFI photo credit: T. Webb

Status:
The exact status of this species in Alberta is unclear, but it is commonly observed in the province during the nesting season and migration. There are three subspecies of Purple Finch in Canada, of which a pale race, *C. p. taverneri*, breeds in Alberta (Godfrey 1986).

Distribution:
The Purple Finch breeds across the forested parts of Canada and south to Baja California and New York. It nests locally in northern and central Alberta, where it was recorded in approximately 20% of the squares surveyed in the Boreal Forest, Parkland, and Foothills natural regions. In southwestern Alberta, it has been found at Bottrel, in the Porcupine Hills, and in the Banff-Jasper area of the Rockies (Salt and Salt 1976). Atlas data indicate that it is a rare visitor in the Grasslands region, but that there is possible nesting in the Waterton Lakes area. Birds observed in southeastern Alberta are transients.

Habitat:
This species prefers open mixed and coniferous woodlands, with conifers being favored for nesting, and deciduous species often chosen for feeding. In central Alberta, these finches commonly occur in cities and towns, where they feed in ornamental fruit and shade trees (Salt and Salt 1976). In the mountains, they are usually observed in pine and spruce forests with aspen or balsam poplar (Holroyd and Van Tighem 1983).

Nesting:
The nest of the Purple Finch is usually built in a conifer (occasionally in a deciduous tree or hedge) at heights up to 18 m. It is most often positioned on a horizontal branch some distance from the trunk, well-hidden in foliage. Both male and female build the nest, which is a small, neat cup of twigs, grasses, and rootlets, and is lined with fine grasses, rootlets and hair.
A typical clutch is 4-5 eggs, pale blue and spotted with dark brown and black. The female incubates for 13 days and is fed on the nest by the male at this time. Both parents tend the young. They leave the nest after 14 days and are fed by their parents for another three weeks. This species is single-brooded or, occasionally, double-brooded.

Remarks:
Purple Finches are sparrow-sized birds. The mature male is a rosy-red and the female an olive-gray. They are most easily confused with Cassin's Finch, to which they are closely related. However, the Purple Finch male is more uniformly red than the Cassin's male, and the Purple female has broader streaking below than her counterpart. The Purple Finch also has white unstreaked undertail coverts. The Purple male's call note, a metallic "pink", is distinctive. The male Purple Finch has an energetic and colorful courtship display, and both sexes like to sunbathe.
Like other finches, the Purple Finch eats primarily seeds, mainly from trees, weeds, and grasses. Buds, catkins, berries, some insects, and the centres of blossoms are also important parts of the diet. The young are fed mainly regurgitated seeds.
Spring migrants arrive late in April (Sadler 1976). In the fall, birds tend to flock and move south and east with sparrows, usually in late September. A few stragglers remain into October. Banding recoveries suggest that Purple Finches breeding in Alberta appear to move to the upper midwestern states for the winter (Collister 1989).
Christmas Bird Count information indicates that this species does not commonly overwinter in Alberta, but it has been reported in Edmonton, Elk Island, Rocky Mountain House, Ponoka and Stettler.

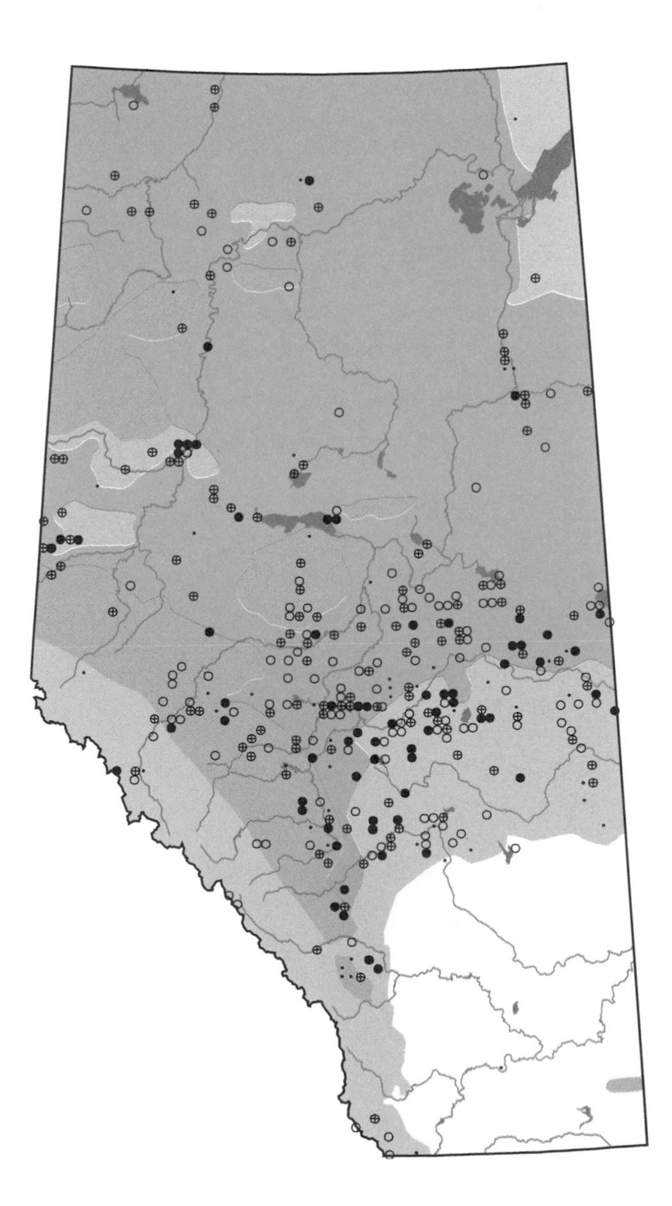

Cassin's Finch

Carpodacus cassinii

CAFI

Status:
In Alberta, this species is at the northeastern limits of its normal range and is considered to be an uncommon breeding species in the province. It is very closely related to the much more widely distributed Purple Finch (Godfrey 1986).

Distribution:
The Cassin's Finch breeds in western North America from southern British Columbia and southwestern Alberta south to California and New Mexico. In Alberta, it was recorded breeding in Waterton Lakes National Park and north to about Canmore and the Bow Valley (Godfrey 1986), although there is one probable breeding record from the Atlas surveys in the Hinton area. The two confirmed breeding records were in the Crowsnest Pass and Waterton Lakes National Park. Salt and Salt (1976) reported that a pair was observed in May near Pyramid Lake in Jasper National Park.

Habitat:
This finch is a bird of high, open montane coniferous forests and, sometimes, mixed wood forests. Salt and Salt (1976) reported that it tends to keep close to conifers, but, in spring, is sometimes seen feeding in adjacent aspens and poplars. The species forages both high in trees and on the ground in clearings or at the forest edge, or rarely in bushes. It sometimes frequents feeders (Brunton et al. 1978).

Nesting:
The nest of the Cassin's Finch will be found in a large conifer, usually well out along a branch, 3-25 m from the ground. The nest is a loose cup constructed of twigs, grasses, rootlets, weed stems, and lichens, and lined with fine grasses, rootlets, and, occasionally, shredded bark, plant fibres, and hair. The species is semi-colonial and the male defends only a small zone around the nest and female. The birds tend to be nomadic and nest in a different location each year (Samson 1976).
The typical clutch is 4-5 eggs which are greenish blue and spotted with browns and lavender. The female incubates for 12-14 days and broods the young, during which time she may be fed on the nest by the male. Nestlings are tended by both parents. Very little information is available on the nestling period; Ehrlich et al. (1988) suggest 14 days. Both young and adults leave the nesting area once the young have fledged. The species may occasionally be double-brooded.

Remarks:
This species is very similar to the Purple Finch in habit, song and coloration. However, the underparts of the Cassin Finch male are paler red, the back is browner and the crimson crown contrasts sharply with brownish upper parts. The female Cassin's Finch is more gray and less olive than the Purple Finch female, and she has narrower, sharper streaks on the underparts. The Cassin's call note, a "kee up" is quite distinctive from the Purple's metallic "pink", and is a good field indicator.
The Cassin's Finch is gregarious and birds flock together except in breeding season. The birds eat mainly buds, catkins, berries, and seeds (particularly conifer seeds), as well as some insects in nesting season. The young are fed by regurgitation Like other finches, the Cassin's Finch irrupts southward when the northern cone crop is small (Bock and Lepthien 1976).
Very little is known of the movements of this bird in Alberta during the spring and autumn, although it appears to arrive in the province in late May or early June (Pinel et al. in prep.). It has been recorded overwintering in Edmonton (Anonymous 1984).

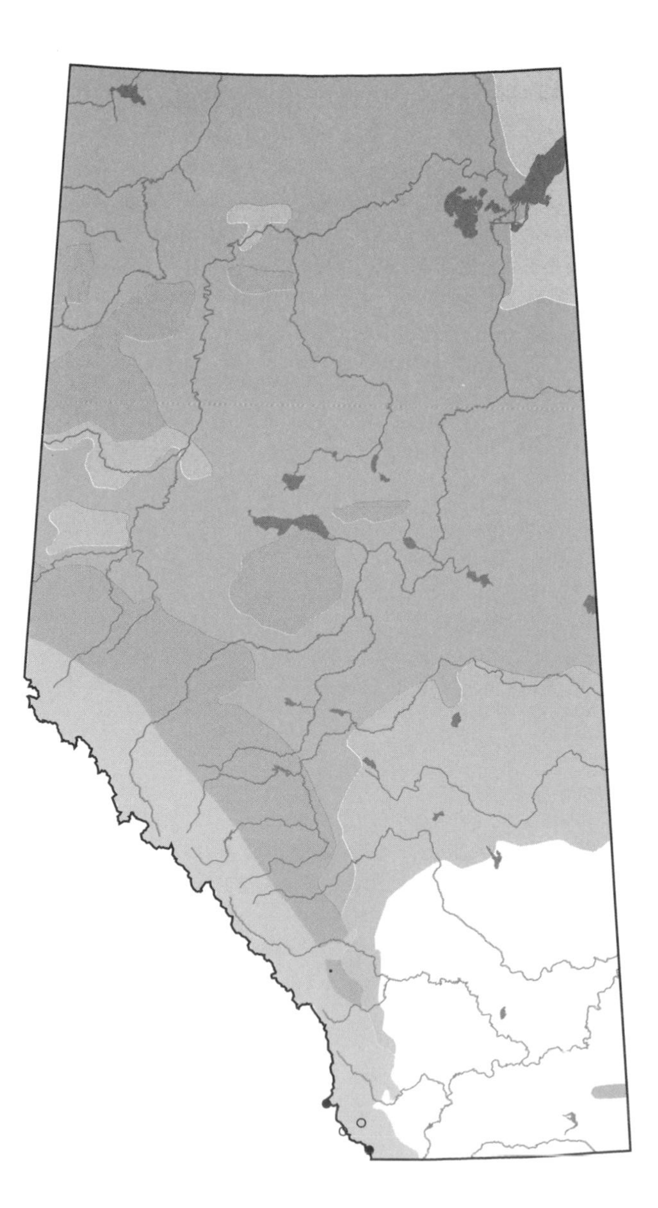

House Finch

Carpodacus mexicanus

HOFI

Status:
Not as yet recorded as a breeding species in any of the prairie provinces, this species' range has profoundly expanded in this century. One of the most abundant species in the western United States, it has also become abundant within its range in British Columbia. Range expansion will probably continue, with one study (Robbins et al. 1986 from Cadman et al. 1987) suggesting that there is an average 21% increase in the eastern population annually. Individuals and pairs have been reported (32 reports up to 1990) for Alberta over all seasons, although no conclusive evidence of nesting exists.

Distribution:
Originally limited to the southeastern United States and Mexico, it has expanded its range north and eastward, first breeding in British Columbia in the Penticton area in 1935. It has since moved along major river valleys to include the inhabited areas of southwestern and south-central British Columbia, including the Columbia valley just west of the Alberta border. The eastern population was introduced into New York in 1940. It quickly expanded with North Dakota presently being the eastern breeding community nearest to Alberta (Bancroft and Parsons 1991). Although the western population is nearest Alberta, and is probably the source of our local sightings, it has traditionally been less expansive than its eastern counterpart (Bancroft and Parsons 1991). At least one pair of House Finches spent the summer of 1981 in Canmore (Holroyd and Van Tighem 1983).

Habitat:
This species breeds in open and quite dry areas, although some water must be available. This includes cultivated areas with limited shrubbery, trees, orchards, and buildings. It is most abundant in urban areas, unlike the Purple and Cassin's finches.

Nesting:
Highly adaptive, the female may construct a nest in a variety of locations, including trees, bushes, building ledges, nest boxes, tree cavities, and even the old nests of other species. Likewise, she will use a range of nest materials, dependent on their availability: straw, grass, paper, string, rags, etc. The clutch is typically 4-5 eggs, which are bluish and spotted with brown and black. The same nest may be used for another brood or in a subsequent year. Although only the female incubates the clutch, for 12-14 days, she is occasionally fed by the male. Both adults tend to the chicks, feeding them, by regurgitation, a diet of mostly seeds. Fledging occurs after 11-19 days.

Remarks:
The adult male House Finch has a rosy rump, forehead, and stripe over the eyes. The remaining upperparts are brownish gray with a reddish wash. The tail and wings are dusky, and the wings have two faint wing bars. The upper breast and throat are reddish, while the remainder of the underparts are whitish with heavy dusky brown streaks. The female is brown above with gray-brown streaks extending onto her white upper breast and throat. The lack of a definite face pattern separates her from the Cassin's and Purple finches. Generally similar to these two species, the House Finch is distinguished by its smaller size, more downcurved and stubbier bill, heavy streaking below, and, in the male, less extensive but brighter red, coloration. This species' diet consists of seeds, fruit, buds, and tree sap.
Highly gregarious after the breeding season, House Finches migrate only short distances. The winter range includes most of the western United States, excluding high mountain regions, as well as extreme southern British Columbia. In the east, it is confined to a portion of the eastern seaboard.

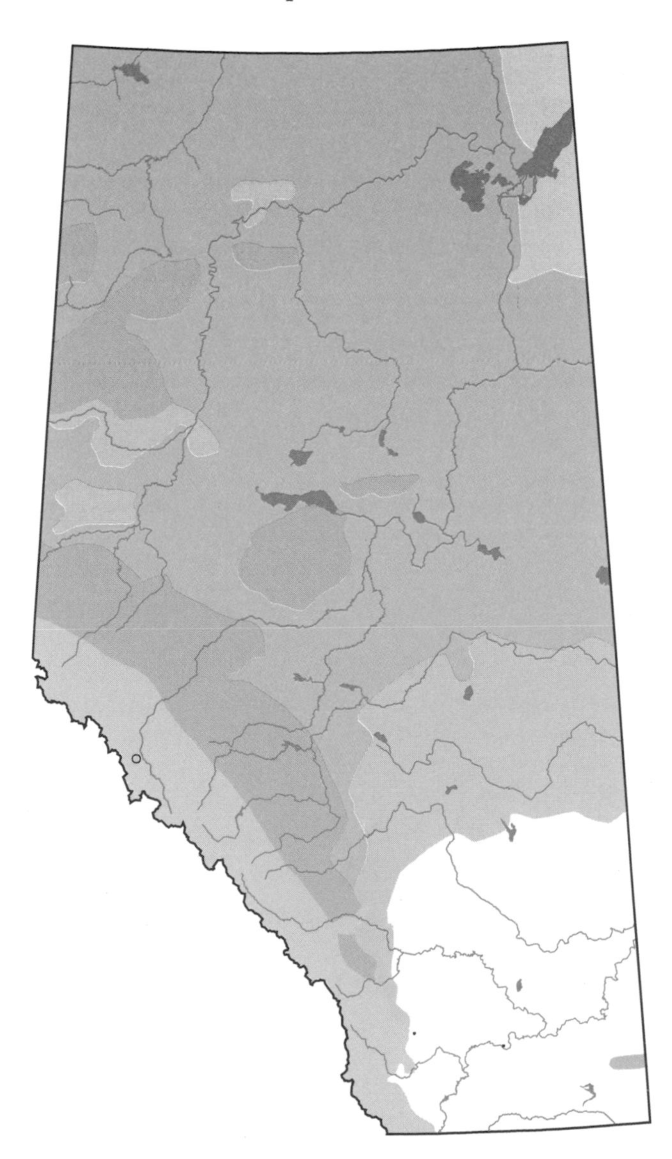

Red Crossbill

Loxia Curvivostra

RECR

Status:
The status of the Red Crossbill in Alberta is undetermined, even though it is a regular and fairly common resident in the Rockies, Foothills and Cypress Hills. Its appearance is more variable in more northerly areas (Salt and Salt 1976).

Distribution:
This species breeds in coniferous forests across North America. It wanders erratically in response to the abundance or scarcity of conifer cones, the seeds being a staple part of the diet. Population irruptions can occur when cone crops fail, forcing birds far from their normal range (Bock and Lepthien 1976). This nomadic behavior, and the variation in the species' nesting season, create difficulty in defining the bird's breeding range.
Atlas data indicate that the Red Crossbill breeds in northern and southwestern parts of the province, south and east to the edge of the Parkland region, and also in the Cypress Hills. It was recorded in 17% of the squares surveyed in the Rocky Mountain region, 5% of surveyed Foothills region squares, and 1% in the Boreal Forest region.

Habitat:
The Red Crossbill is found in coniferous and mixed forests containing mature conifers. Occasionally, birds may be seen feeding in deciduous trees or foraging on the ground for dropped seeds (Salt 1952).

Nesting:
Red Crossbills may nest at any time of the year, depending on the abundance of cone seed, but breeding most often occurs in late winter to early summer and in the late summer to fall (Griscom 1937; Cadman et al. 1987).
The nest is usually built in a conifer at various heights up to 18 m. It is placed along a branch and is well-concealed in foliage. The female constructs the nest, a cup made from twigs, bark shreds, weed stems, and grass lined with moss, plant down, feathers, and fur.
The female lays 3-5 pale, bluish green eggs with spots of lavender and brown. She incubates these for 12-15 days, and also broods the young. She is fed by the male on the nest during incubation and brooding. Both parents feed the young, which leave the nest after 17-22 days. The young depend on adults for at least another 3-4 weeks after fledging. The mandibles of the young birds do not cross over until a few weeks after fledging. The species is usually single-brooded, but has been reported as double-brooded in some areas (Ehrlich et al. 1988).

Remarks:
All crossbills have elongated bills that cross over at the tips, an adaptation for removing seeds from conifer cones. The birds insert their mandibles into the cone to hold apart the scales, and lift out the seeds with their tongues (Benkman 1987). The male Red Crossbill is brick red, the female is yellowish gray with a yellow rump. The species is very similar to the White-winged Crossbill in behavior and plumage, but lacks the white wing-bars of the latter species.
Red Crossbills are highly gregarious and may be seen in small flocks at any time of the year. They generally feed high in conifers and use both bill and feet to climb, and often hang upside down. While feeding, the birds are quiet and not easily seen, but, during breeding periods, the males will fly around the tops of trees singing, calling and chasing (Bent 1968).
Conifer seeds are the Crossbill's main bill of fare, but seeds, buds, fruits of deciduous trees, spruce buds, berries, and a few insects and larvae are also eaten. Young birds are fed a regurgitant of softened seed pulp (Newton 1972).
On Christmas Bird Counts, this species has been noted at a variety of locations, including Bow Summit, Crowsnest Pass, Banff, Canmore, Hinton, Waterton Lakes, Calgary, Fort Smith, Snakeshead, Edmonton, and Elk Island National Park.

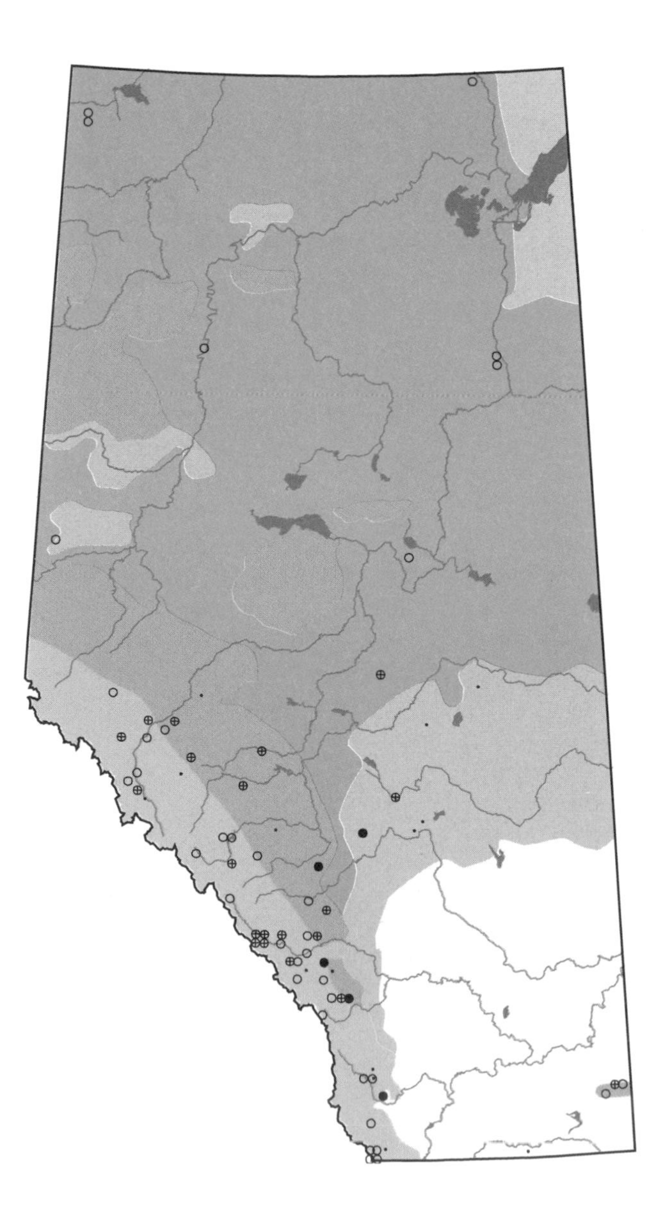

White-winged Crossbill

Loxia leucoptera

WWCR

Status:
The White-winged Crossbill is a fairly common resident of the Rocky Mountain Natural Region (Holroyd and Van Tighem 1983). It is fairly common in the Boreal Forest Natural Region.

Distribution:
This crossbill breeds in boreal coniferous forests across North America. Its range overlaps with that of the Red Crossbill, though the White-winged Crossbill is generally found further north. It is resident throughout much of its range, but may move southward in winter.

In Alberta, the White-winged Crossbill was found in 6% of squares surveyed in the Canadian Shield Natural Region, 10% in the Boreal Forest region and 22% in both the Foothills and Rocky Mountain regions. It is a permanent resident in Banff and Jasper national parks (Holroyd and Van Tighem 1983). It may also nest in the Cypress Hills (Godfrey 1986). It is an irregular visitor in the Grassland and Parkland regions, and is most often seen in the winter in these regions.

Like the Red Crossbill, the White-winged is an erratic wanderer with fluctuations occurring from year to year within its normal breeding range. It is also subject to major population irrruptions when a cone crop fails (Bock and Lepthien 1976).

Habitat:
White-winged Crossbills frequent coniferous and mixed woodlands, generally where spruce predominate (Salt and Salt 1976). In the mountains, they occur in all types of coniferous forest, preferring closed forests with mature trees (Holroyd and Van Tighem 1983). They are often seen at openings and edges, and sometimes feed on the ground. They are less likely to feed in deciduous trees and shrubs than the Red Crossbill (Godfrey 1986).

Nesting:
This species will nest at almost anytime of the year, but spring and summer nesting is most common in Alberta (Salt and Salt 1976). A pair may leave a flock to nest, or several pairs may nest semi-colonially. The nest is usually built in a conifer 1-20 m above the ground. It is placed well out on a branch and is hidden in foliage. The female constructs the nest, using twigs, lichens, rootlets, and bark shreds for the exterior, and grasses, moss, hair and feathers for the lining.

The typical clutch is 3-4 eggs, which are pale bluish green and spotted with brown and lavender. The female incubates, probably for 12-14 days (Ehrlich et al. 1988). While she incubates, the male feeds her on the nest and also helps feed the young.

Remarks:
The male White-winged Crossbill is a bright, rosy red and the female is olive-green or grayish with a yellow rump. Conspicuous white wing bars help distinguish this crossbill from the Red Crossbill and Purple Finch. The White-winged Crossbill is much smaller than the Pine Grosbeak and neither the Purple Finch nor the Pine Grosbeak has a crossed bill.

This species may be seen in small flocks at any time of the year, sometimes in the company of Pine Siskins, Red Crossbills and redpolls.

Its favorite food is cone seeds. When a cone is closed, the bird wrenches it from the branch, clamps it down with one foot, and picks out the seeds, letting the scales fall to the ground. This bird also eats other plant seeds, berries, and buds, plus a few insects and larvae.

On Christmas Bird Counts, this species is observed only on an irregular basis. Some locations it has been noted in include the Banff and Canmore area, Bow Summit, Edson, Rocky Mountain House, Crowsnest Pass, Red Deer, Calgary, Snakeshead, Horseshoe Canyon, Erskine, Elk Island National Park, Wabamun Lake, Edmonton, and Fort McMurray.

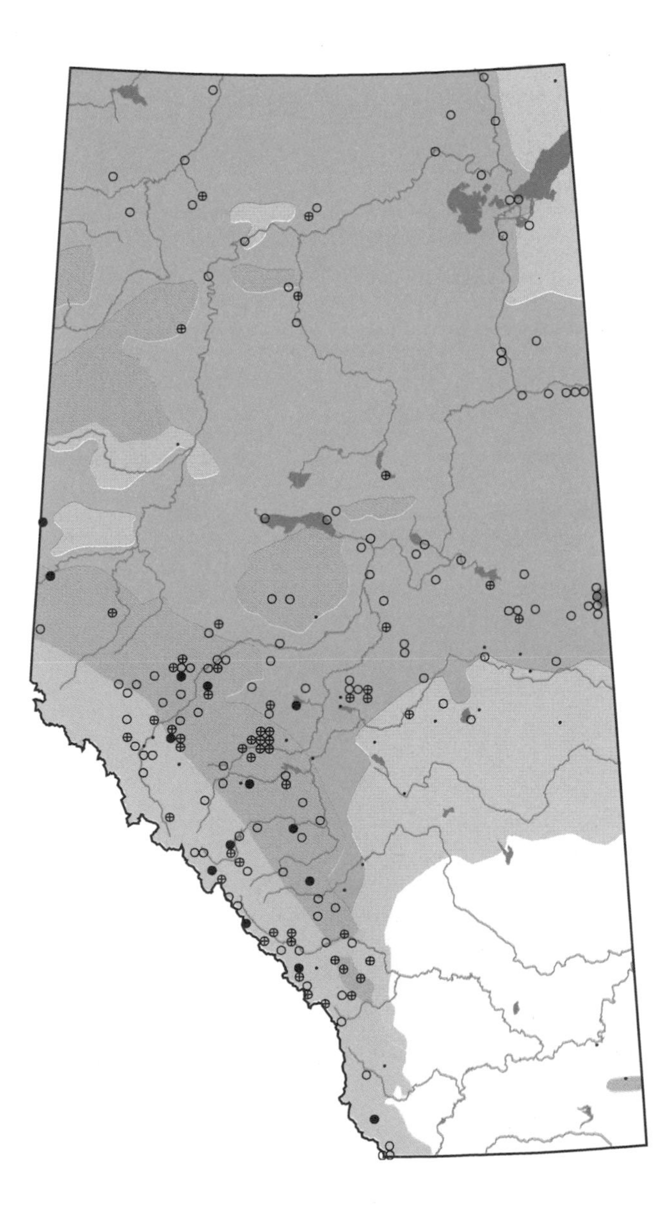

Common Redpoll

Carduelis flammea

CORE

Status:
The Common Redpoll is not a common breeder in Alberta; its usual breeding range is north of the province. However, scattered records show breeding in central parts of the province.

Distribution:
This species breeds in the arctic and subarctic regions of North America. In Alberta, it is normally a winter visitor that appears, sometimes in large numbers, in all parts of the province (Salt and Salt 1976). However, Atlas data show that it does breed sporadically and locally in central Alberta. Other records show birds nesting in Edmonton and Devon (Lister 1975; Godfrey 1976), Camrose (Godfrey 1976), and Red Deer (Snell 1924; McNicholl 1978). Godfrey (1986) also suggested that this species may breed in the province's north east corner. Confirmed breeding was recorded in three squares in the five year Atlas surveying period. Two of these were in the Boreal Forest Natural Region and one was in the Foothills Natural Region.

Habitat:
The breeding records of the Common Redpoll in Alberta indicate an attraction to deciduous trees and shrubs.
In winter, flocks of redpolls may be found in open woodland, shrubs, bushy hedgerows, meadows, fields, weed patches, roadsides, and other disturbed areas. They are particularly fond of birch and alder, and are also attracted to residential feeders. At night, birds retreat into thickets. In the Rocky Mountain Natural Region, it occurs commonly in alluvial areas, roadsides, and disturbed areas in the montane and lower subalpine, where it prefers meadows, alder thickets, tall shrubbery, and forest edges (Holroyd and Van Tighem 1983).

Nesting:
In northern areas, the Common Redpoll nests in dwarf trees and shrubs, on sedge or grass tussocks, or in rock crevices (Godfrey 1986). Alberta reports indicate nesting 1-3 m up in deciduous trees (e.g., maples, choke cherry) and shrubs. The nest is an untidy, thick-walled cup of grass, plant stems and small twigs lined with plant down, hair or other soft materials. The birds apparently exhibit little territorial behavior, and often nest near one another in loose associations.
Common Redpoll eggs are blue and spotted with reddish brown. The female lays 4-6 eggs and incubates for 10-11 days. While incubating, she is fed by the male. After the young hatch, the male initially continues to bring food to the female, who transfers it to the young. Later, both parents feed the nestlings. The young birds fledge in about 12 days. One or two broods are raised in a season (Bent 1968).

Remarks:
The Common Redpoll is chickadee-sized and has a stubby bill, red cap, black chin, and dark streaks on the sides. It often has a pinkish tinge on the breast, the male's being brighter than the female's. This species closely resemble the Hoary Redpoll, which has not been recorded as breeding in Alberta, but has been recorded overwintering. These two species are sometimes seen together in winter flocks. The Hoary has a white rump, while the Common Redpoll's is streaked.
Redpolls eat the seeds of coniferous and deciduous trees and shrubs, and also weed seeds and spilled grain. Insects are taken when they are abundant, particularly during the nesting season. The birds are relatively quiet while feeding, and may be seen hanging sideways or upside down on branches or weed stalks. When on the move, a the birds of a flock are constantly twittering. This species is quite unwary of humans and is easily approached.
These birds are most commonly observed in the province between October and April. Accounts of breeding in Alberta have mostly been in May.

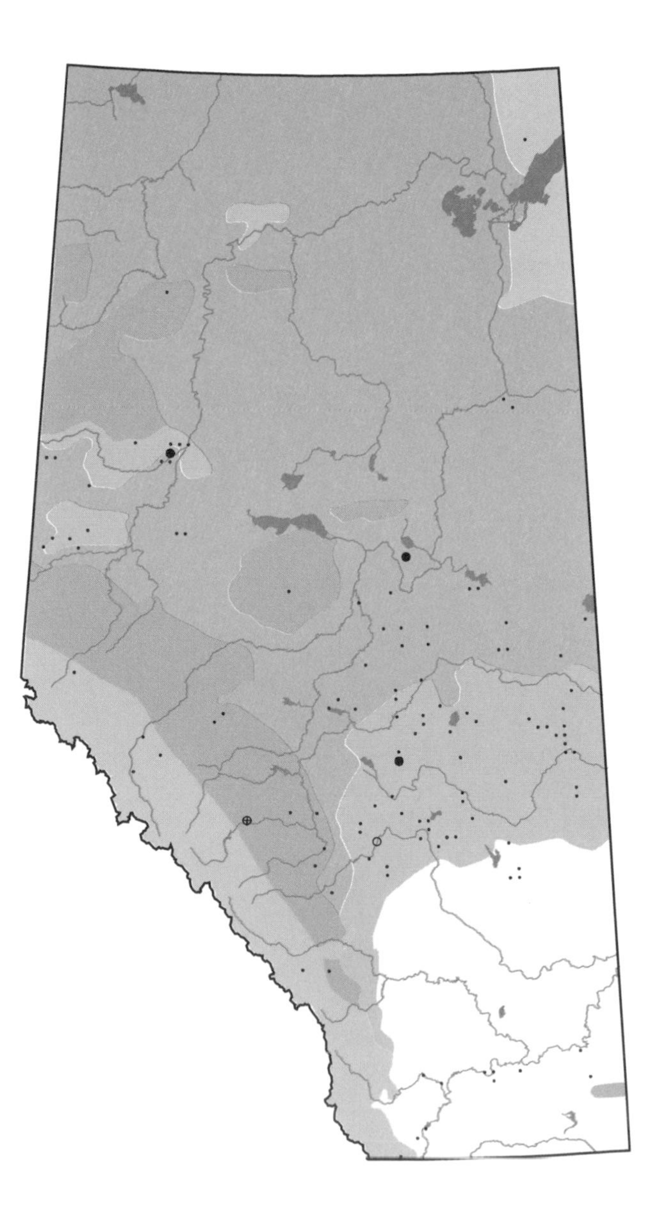

Pine Siskin

Carduelis pinus

PISI

Status:
Pine Siskins are quite abundant in parts of Alberta; they are the most common bird species seen during the summer months in Banff and Jasper national parks (Holroyd and Van Tighem 1983).

Distribution:
As with other nomadic and irruptive finches, the exact breeding range of the Pine Siskin is not easy to define. It may be common and breeding in an area one year, and gone the next.
The Pine Siskin nests throughout Alberta, though with very few records in the Grassland Natural Region. It is most abundant in the Rocky Mountain and Foothills regions, where it was recorded in over 50% of surveyed squares. This species was also recorded in 27% of Boreal Forest region squares surveyed and in 13% of Parkland squares.

Habitat:
They inhabit coniferous and mixed wood forests, and sometimes ornamental groves or shade trees in residential areas (Godfrey 1986). In the mountains, lodgepole pine, Engelmann spruce, subalpine fir, subalpine larch, heath, and aspen woodlands are preferred habitat (Holroyd and Van Tighem 1983). In spring, aspens and poplars are favored.

Nesting:
Pine Siskins nest either singly or semi-colonially. The female chooses the nest site and builds the nest, usually at medium height (2-12 m) in a conifer, although deciduous trees are sometimes used. The nest is built on a lateral limb well away from the trunk, often in a branch fork. It is usually well concealed in branchlets. It is a fairly large, untidy affair of twigs, shredded bark, lichens, grass, and rootlets lined with plant down, fine rootlets, hair, or moss.
The typical clutch contains 3-6 eggs, pale blue and spotted with browns and purple. The female incubates for 13 days and, while on the nest, takes food provided by the male. She broods the young for 7-9 days, taking food from the male feeding the nestlings. After brooding, both parents feed the young, which leave the nest after 14-15 days. They depend on their parents another three weeks after fledging. The species is single-brooded, with double-broods being a possibility (Harrison 1978).

Remarks:
The adult Pine Siskin is heavily streaked with yellowish patches on the wing and sides of tail, which are most conspicuous in flight. It is the same size and has the general shape of a redpoll, but has a sharper bill and lacks the red cap and black chin of the redpoll.
Siskins are very tame and are usually seen in small flocks, sometimes with other finch species, especially in the fall. They often hang upside down while feeding on the seeds of conifers and some deciduous species, such as birch and alder. They also feed on the seeds of weed species, particularly thistles and dandelions, leaves, buds, blossoms, insects, and larvae. Siskins are quite aggressive around feeding stations.
This species arrives over an extended period in late April and May and generally departs the province from mid-August to early November, with numbers peaking early in September (Pinel et al. in prep.). Siskins have been observed in winter in a number of locations in southern Alberta north to Banff, Wabamun Lake, Edmonton, and Fort Saskatchewan.

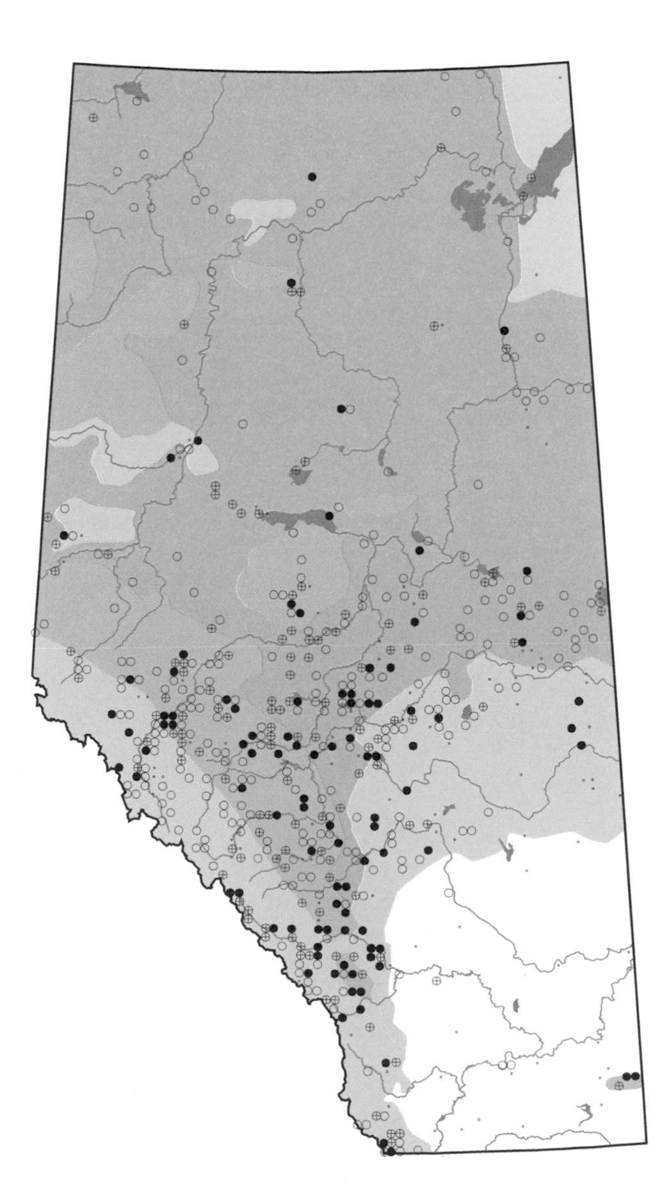

American Goldfinch

Carduelis tristis

AMGO photo credit: T. Webb

Status:
Breeding Bird Survey data indicate that the American Goldfinch is declining in some parts of its range, but increasing in others (Robbins et al. 1986). In Alberta, the goldfinch is relatively common where suitable habitat exists.

Distribution:
The American Goldfinch breeds mainly in central and southern Alberta, with some Atlas records in the Boreal Forest region. In the Grasslands region, it is restricted to wooded valleys and coulees (Salt and Salt 1976).

Habitat:
This species avoids mature forest and is usually seen in overgrown or cultivated fields, pastures, fencerows, roadsides, or along shorelines. It may also frequent residential gardens, farmyards, and cemeteries. Areas with thistles, dandelions, sunflowers, or other plants of the composite family are particularly favored, as they are used for food and nesting material (Holroyd and Van Tighem 1983).

Nesting:
The American Goldfinch nests in late June or early July, well after most other songbirds, likely an adaptation to ensure that there is an abundance of seed when the young hatch. It is highly gregarious and several pairs tend to nest in a small area and socialize on the feeding grounds (Salt and Salt 1976). In many passerine species, the male shows fidelity to a nesting territory; with the American Goldfinch it is the female that returns to a previous nesting area (Middleton 1979).
The nest of this species is usually built in a tree or shrub, generally less than 6 m from the ground. It is most often placed in an upright crotch or fork away from the trunk. The female constructs a compact bowl of plant stems, bark strips, and grasses, binds it with spider webs and cocoon silk, and adds a lining of plant down. The birds may disappear for a short time from the area once the nest is built, possibly to protect the site from predators, but will return later to lay the eggs (Stokes 1986).
The clutch is 4-6 pale blue eggs which the female incubates for 12-14 days. She is fed by the male, which circles above the nest until she calls him down. The female broods the young for a week after hatching, then both sexes assist in tending and feeding. The nestlings leave the nest after 11-17 days and remain with the parents for another month. Double-broods are occasionally reported; the male cares for the first brood while the female renests. This often occurs if the female is polyandrous (Middleton 1988).

Remarks:
The breeding male is a conspicuous bright yellow bird with black cap, wings, and tail. The female is greenish with pale yellow beneath and dark wings.
A distinctive field mark of this species is its rollercoaster undulating flight; at the top of the rise the male calls "per-chick-o-ree". His cheerful song and colorful plumage have earned this species the name "wild canary".
Goldfinches are primarily seed eaters with a preference for seeds of the Compositae, but other plant seeds and, rarely, insects are also taken.
It is among the last of the spring migrants to arrive in the province, appearing in late May or early June. The birds usually depart by mid-September. Overwintering is very rare, with the first record in Alberta occurring in 1979-80 when this species staged a "winter invasion" of the Prairie Provinces (Harris 1980b; Pinel et al. in prep.). Winter records have also been reported from Edmonton (Slater 1985) and Red Deer (Slater 1986).

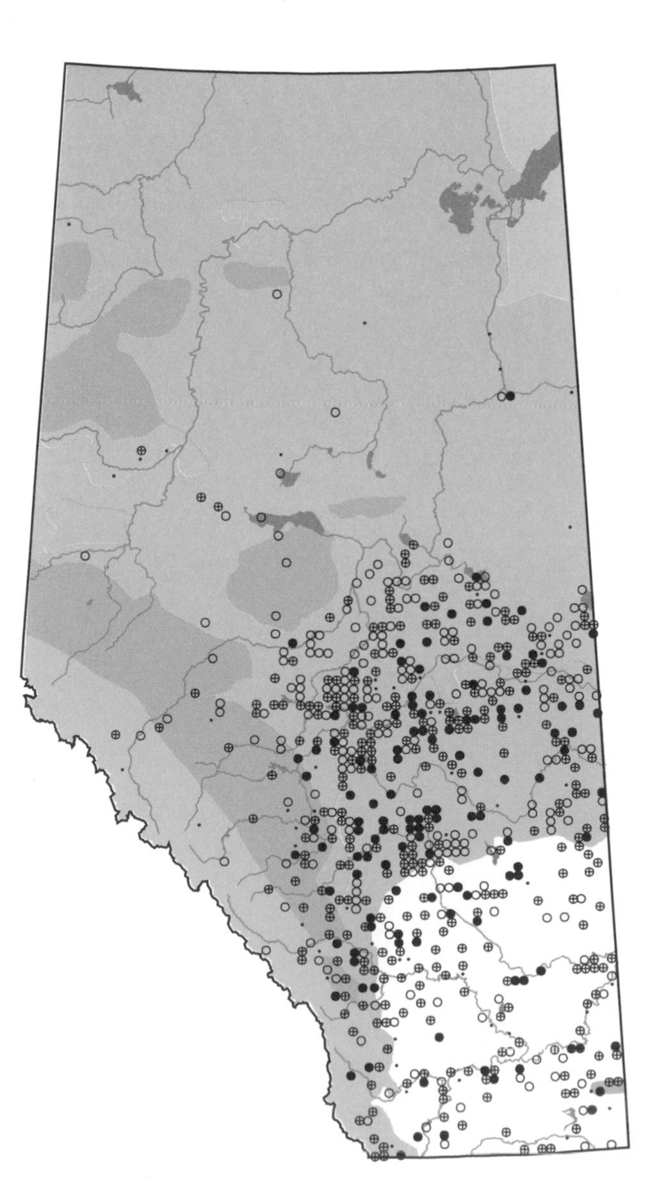

Evening Grosbeak

Coccothraustes vespertinus

EVGR

Status:

The Evening Grosbeak is relatively common throughout its breeding range in Alberta but, because of its secretive nesting habits, and tendency to move into urban areas in winter, it is less likely to be seen in summer than in winter.

Distribution:

This finch nests in northern, central and southwestern Alberta from the Athabasca Delta south to about Edmonton, and through the Rocky Mountain and Foothills natural regions. It is a permanent resident throughout much of central and southern Alberta. Although there have been overwintering records as far north as Peace River and Fort McMurray, they often move to the southern parts of the province in late September or early October.

Habitat:

In breeding season, the Evening Grosbeak inhabits mixed and coniferous forests, including stands of second growth, and occasionally deciduous woods. In the mountains, it appears in open coniferous forests and forest edges (Holroyd and Van Tighem 1983).

In the winter, this species frequents a variety of forests and wooded stands, and is often seen in parks and shade trees in towns and cities. It is particularly attracted to Manitoba maple and ashes, both of which retain their seeds all winter. It is a common visitor to feeders, especially those stocked with sunflower seeds.

Nesting:

The nest is built by the female in a tree, often near the top, at heights up to 18 m. It may be built in a crotch close to the trunk or towards the end of the branch, but it is always well-concealed in foliage. It is a loose, shallow structure made of twigs, grass, moss, rootlets, and bark shreds and lined with rootlets and fine twigs. Very few nests have been located in Alberta; those found have been high in poplars (Hoffmann and Pletz 1988).

The typical clutch size is 3-5 eggs, which are a bluish green with brown and gray markings. The female incubates for 12-14 days and is fed by the male during this time. Nestlings are fed and tended by both parents, and they leave the nest after 13-14 days.

Remarks:

The Evening Grosbeak is a heavy-set bird with a large yellowish conical bill. The male is yellow and has black wings with white patches; the female is grayish yellow, but also has black and white wings, which are most conspicuous in flight.

This species is usually seen in flocks, feeding on the ground or in trees. The birds often sit in trees for long periods of time, chirping continuously. They eat the seeds from a vast array of native and cultivated plants, but have a particular penchant for Manitoba maple. When these maples were planted extensively as shade trees in the east, the Evening Grosbeaks followed, substantially expanding their eastern range (Taverner 1921). The birds also eat buds, cones, crabapples, and other fruits, including berries. During the nesting season, they also eat insects. The young are fed both masticated insects and larvae, and crushed seeds. The birds are gregarious and sociable except at feeding stations, where they belligerently vie for position.

In winter, the Evening Grosbeak ranges widely and is sporadic in occurrence. Moreover, the number of birds overwintering varies considerably, because this finch is also an irruptive migrant (Prescott 1991). In some years, most individuals probably leave the province (Pinel et al. in prep.). Because of its erratic overwintering, its spring migration also varies, occurring as early as mid-March in some years, and not until late May in others.

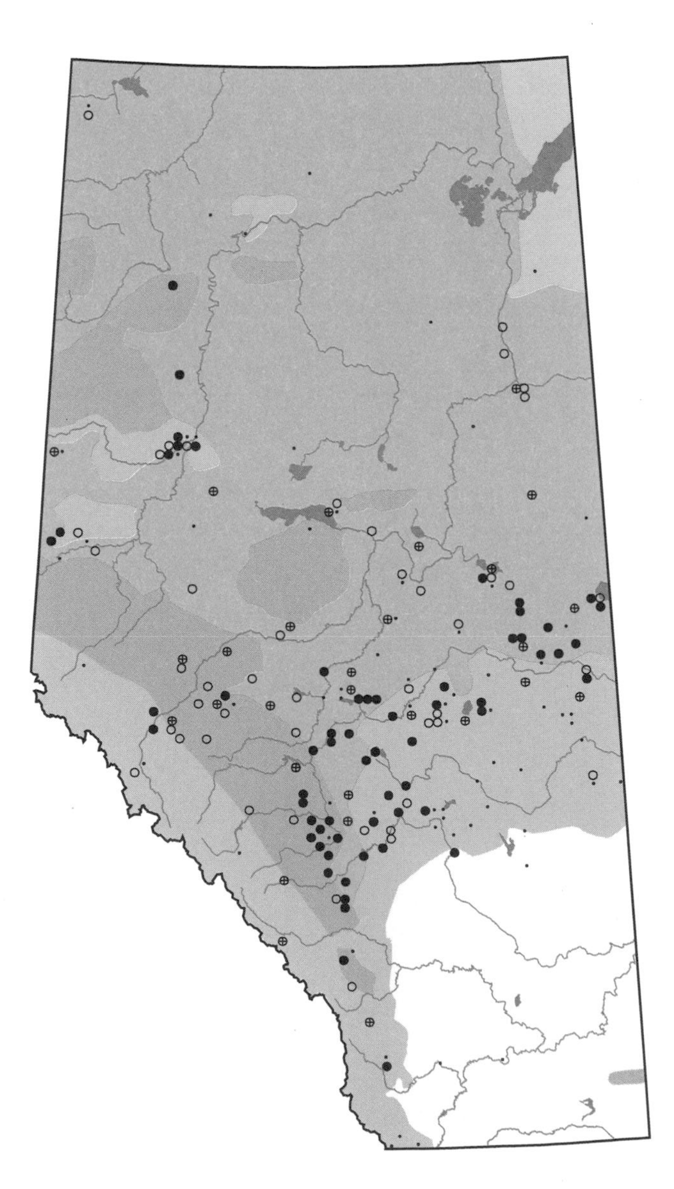

House Sparrow

Passer domesticus

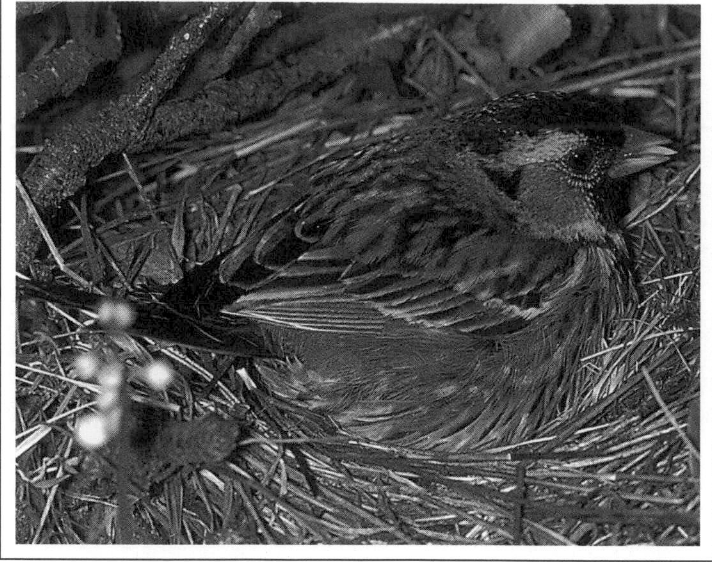

HOSP photo credit: W. Carlyle

Status:
The House Sparrow was introduced into North America in the 1850s, partly because European immigrants longed for familiar birds and partly because they believed this bird would aid in controlling insect pests (Robbins 1973). One of the earliest records for this species in Alberta was in Red Deer in 1898 (Houston 1978). Highly adaptable and opportunistic, it spread rapidly, reaching its present coast to coast range on the continent in 1940. It is flourishing in settled parts of Alberta.

Distribution:
Native to Eurasia and North Africa, the House Sparrow has been introduced into North and South America, the Caribbean, Australia, New Zealand, and southern Africa. It breeds throughout Alberta wherever there is human settlement. The province's previous northerly record of this Old World sparrow was from High Level (McGillivray 1988); however, during the Atlas Project, it was recorded on the Hay River east of Bistcho Lake.

Habitat:
This species is usually associated with human habitation. It is commonly found in cities and towns, particularly residential areas, parks, roadsides, wastegrounds, farms, and cultivated fields.

Nesting:
The House Sparrow nests in holes and crevices in man-made structures, dense vines on buildings, natural and woodpecker cavities in trees, and sometimes tree branches. It will drive Cliff Swallows from their mud nests and use the cavities excavated by Bank Swallows (Godfrey 1986). This sparrow also takes over nest boxes set out for other species. It is loosely colonial and often nests in "clumps" in trees or other suitable sites.
The male or female chooses the nest site, and the nest is chiefly built by the male. In trees, the nest is a neat domed structure with a side entrance. In cavities, the hole will be completely and untidily filled with nesting material and domed over, though the roof may be omitted in small spaces. The nest is built of straw, grass, plant stems, and rubbish (paper, string, cloth), and is lined with feathers.
The female lays 3-7 eggs, most often five. They are variable in color, but are generally white to pale greenish or bluish with fine spots of grays and browns. Incubation is 11-14 days, mainly by the female, with some help from the male. Both parents feed the young, which fledge at 12-18 days. After fledging, the juveniles group together where there is a suitable food supply. One to four broods may be raised in a season (McGillivray 1983). House Sparrows generally mate for life and remain faithful to a breeding territory.

Remarks:
The common name, sparrow, is a misnomer for this species; it is actually a weaver finch. This stocky, noisy little bird is familiar to most Albertans. The male has a black throat and bib, and a chestnut patch back from the eye. His song is a series of loud monotonous non-musical chirps. The female is dull brown above and dull gray below, with an unstreaked breast.
Since the 1880s, many have regarded this species to be an agricultural and aesthetic pest (Cadman et al. 1987). It is well-known for its belligerent behavior towards other bird species, and its tendency to usurp their nesting cavities.
The House Sparrow is primarily a seed-eater, specializing in seeds of cultivated grain crops. It also eats the seeds of grasses and other annual herbs, insects, household scraps, and spillage in parks and near garbage. The young are fed mainly insects.
In Alberta, this species is sedentary. Wintering birds have been noted as far north as Fort McMurray, Wembley, and Grande Prairie.

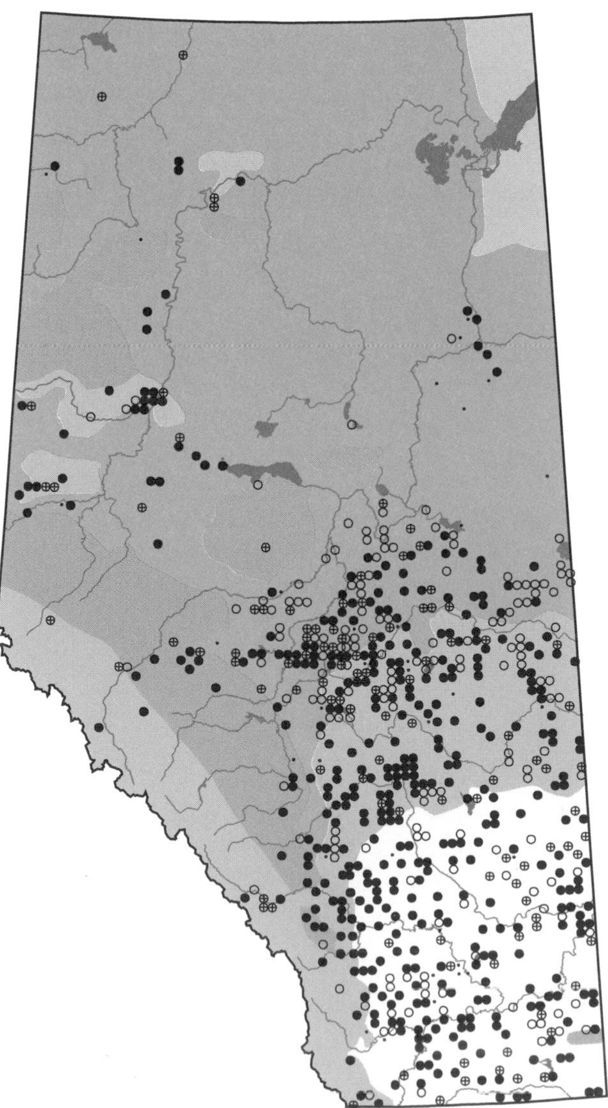

Appendices

The appendices include various reference sections which expand upon and complement the text. The first appendix describes the migrant species and winter visitors. The second appendix is the Birds of Alberta Species List, which outlines the status of each species that has occurred in Alberta and classifies them in one of the following groups: nesting, migrant, migrant/winter visitor, regular winter visitor, vagrant or wanderer, vagrant or possible migrant, or extinct or extirpated. Also included is a list of hypothetical species for which substantiation is not available. This is followed by a reprint of the field data card that was used for record keeping by the individual atlassers. It lists the species which were thought to breed or had once bred in the province. The next reference section is a generalized drawing of a passerine bird identifying the various parts that are often used in detailing bird descriptions. A glossary is included to define those terms particular to the study of ornithology. The next section, entitled Future Records, describes the procedure for handling future records to continue the gatheirng of information to update the database. Following this is a list of individual advisors with whom a personal communication is cited in text. The Bibliography provided in this book is an extensive list of references which were reviewed in the production of the Atlas, including some which were not cited. The appendices section concludes with three indexes of Bird Names, arranged alphabetically: English, French, and Latin.

Migrant Species of Alberta

The migrant species included in the following appendix represent those which may be observed locally en route to their breeding grounds outside of the province, most often in Arctic and subarctic Canada and Alaska. The majority of these species are either waterfowl or shorebirds that often pass through the province, stopping at historical staging areas to rest and feed. Conservation of these staging areas in the province is vital in maintaining the status of the migratory species. The distances covered by migrants passing through Alberta range right up to the demanding Arctic to Antarctic and return migration of the Arctic Tern.

This section includes four species which also breed in northern Canada, although their winter ranges span all or part of Alberta. Their presence is often influenced by available food supply in the local area or in other parts of its winter range. The scarcity of such resources may cause some to wander, often making their local presence unpredictable.

A map of Canada, which includes the names of the northern districts and major Arctic islands, is provided to help the reader to better relate to references made to these areas in this appendix.

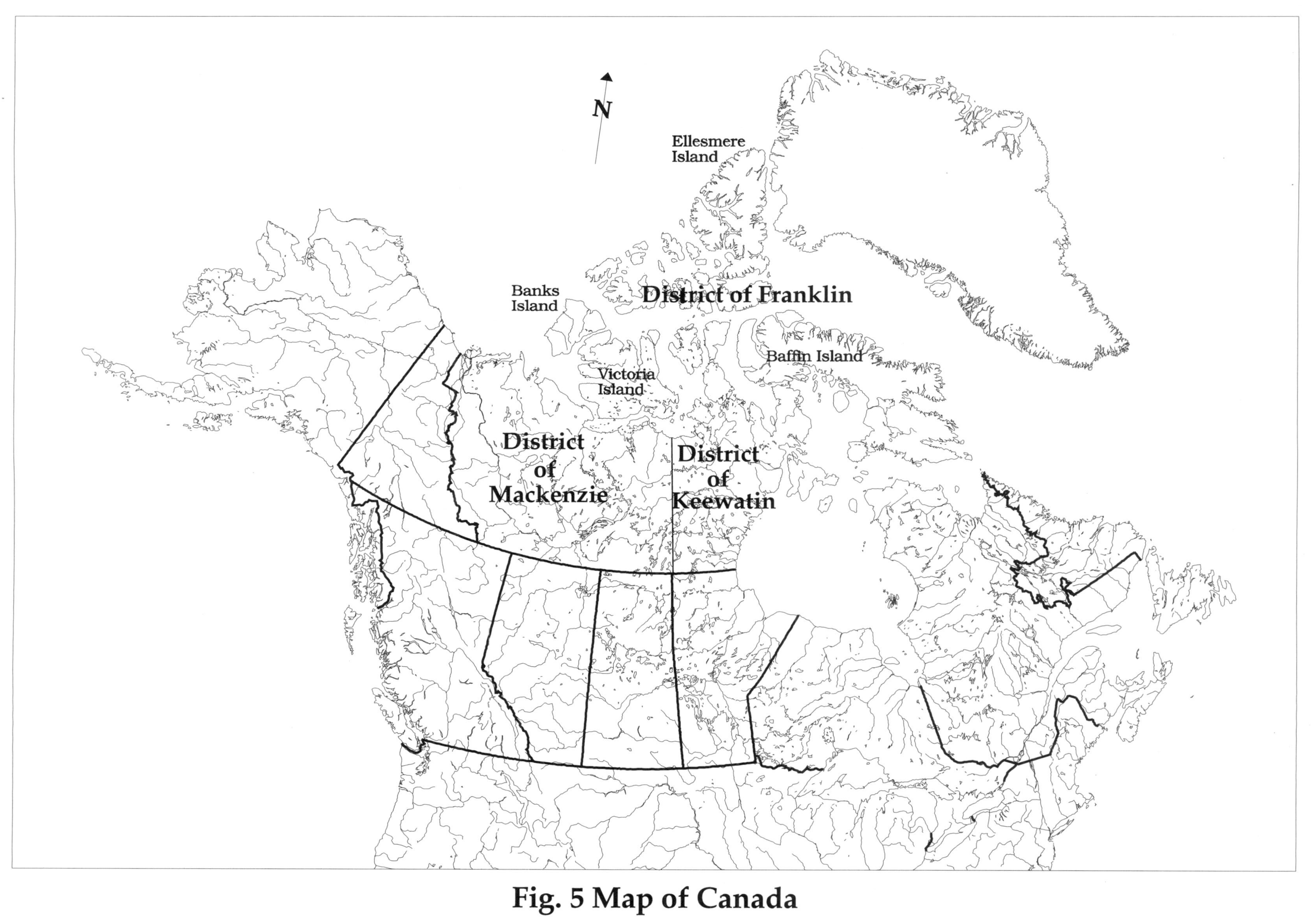

Fig. 5 Map of Canada

Tundra Swan

Cygnus columbianus

TUSW photo credit: L. Bennett

Occurring in the province in the spring from mid-March until the third week of May, the Tundra Swan (formerly the Whistling Swan) is white overall with some rust about the head and neck and black feet, legs, and bill. Smaller than the similar Trumpeter Swan, it is also separated by a yellow or orange spot, if present, on the black area before the eye. Following the snow melt northward, it uses large lakes as staging areas and stubble fields for feeding. On its breeding grounds in Alaska, and the Canadian low Arctic, it prefers to nest amongst the marshy lakes and ponds of the low tundra. On fall migration, these swans pass through the province from early October until mid-November, peaking in late October. Winters are spent along the Pacific and Atlantic coasts from Washington to California and Maryland to North Carolina. There are breeding season records for the Grassland region and winter records for Banff and Jasper national parks.

Greater White-fronted Goose

Anser albifrons

GWFG photo credit: G. Horne

This circumpolar migrant passes through Alberta in late March or early April en route to its breeding grounds in the western and central Arctic. Spring migration peaks in late April or early May. Fall migration is more sporadic, with geese seen in British Columbia across to Illinois, south to Mexico, as well as North Africa, India, China and Japan.
In Alberta, Beaverhill Lake is a major staging area. This grayish-brown goose has no black on its head or neck, and has distinctive yellow legs.

Snow Goose

Chen caerulescens

SNGO

Leaving its winter home, along the Pacific and Atlantic coasts south to Mexico, this migrant reaches Alberta by late March or early April. Its arrival at selected staging lakes in eastern Alberta, where it will often linger for several days, can be predicted locally almost to the day. Its journey ends in the Arctic regions from northeast Asia to across Baffin Island and south along the west coast of Hudson Bay. Once the young have fledged, adults moult their flight feathers, becoming vulnerable to predation. The peak of fall migration occurs over Alberta in October.
The Snow Goose is white with black wing tips. Its pink bill is lined with a black "grin" and its legs are purplish red. The blue phase has a predominately slate-colored body with a white head and neck.

Ross' Goose

Chen rossii

ROGO

Identical in coloration to the Snow Goose, the Ross' Goose is smaller, and more shrill-voiced than its cousin. In addition, it lacks the Snow Goose's "grin," and is less plentiful, with 32,000 estimated in the wild (Godfrey 1986). It arrives in Alberta later in the spring (late April or early May), and earlier in the fall (second half of September) than the Snow Goose. The main migration route extends along eastern Alberta and western Saskatchewan, north to localized breeding areas on the Arctic coast, and south to wintering grounds in California, Texas and Louisiana.

Once a commonly hunted gamebird, Ross' Goose is now protected in some areas by delayed hunting seasons on its preferred staging areas.

Eurasian Wigeon

Anas penelope

EUWI photo credit: D. Wechsler/VIREO

Spring and fall observations are scarce for this species in Alberta, with a reported spring period from mid-March to the end of April. Although similarly patterned to the local nesting American Wigeon, the Eurasian Wigeon has a rusty, not green, head and neck and a creamy buff, not white, forehead and crown. Seen locally only as males or in pairs, they mainly occur in the Grassland region. Increases in local occurrence may be due to an increase in the species' abundance (Pinel et al. 1991). Established breeding areas include Iceland, northern Europe and Asia. Although wintering grounds are in northern Africa and southern Asia, a few regularly overwinter on the west coast of British Columbia.

Greater Scaup

Aythya marila

GRSC photo credit: G. Wood

Very similar in appearance to our more common Lesser Scaup, the Greater Scaup is distinguished by a green, not purple, gloss over its black head and neck. It is an uncommon migrant over the province, with observations from late March to late May and again in September and October. Although large lakes are used locally, they tend to be smaller than those selected by the Lesser Scaup. The Greater Scaup breeds on lakeshores and islands on lakes along a band that includes the Yukon Territories, the southern half of the Northwest Territories, and the south half of Hudson Bay. Breeding status in Alberta is uncertain, with possible breeding activity occurring in the Caribou Mountains. Winters are spent in Lake Ontario, Lake Erie, the Atlantic coast south to the Gulf states, and from Alaska to California along the Pacific Ocean.

Oldsquaw

Clangula hyemalis

OLDS

The Oldsquaw is a regular migrant in northern Alberta, with some additional reports from the mountains and a few from the south. It has been recorded most often in the spring, especially in May, though only in small numbers. Fall records are less plentiful and all are from southern Alberta. The Oldsquaw does not remain long in the province before proceeding to its Arctic breeding grounds along the north coast from Alaska to Labrador, the Keewatin district and the Arctic islands. It winters along both coasts, from the Aleutians to Washington on the Pacific, and from Greenland to South Carolina on the Atlantic. This diving duck, which has a peculiar yodel-like call, can be recognized by a long "pin" tail in the male. Its wings are plain brown with no pattern. During migratory periods, a large amount of white on its head and neck becomes distinctive.

Rough-legged Hawk

Buteo lagopus

RLHA photo credit: D. Wood

This panboreal hawk's North American nesting range extends from Alaska to Newfoundland. It winters between California and Virginia and, occasionally, further south. Beginning migration through Alberta in mid-September, numbers peak locally in mid- or late October. Winter records in Alberta are common, occurring as far north as Donnelly. At this time of year, the bird prefers open fields and marshes.

Spring migrants pass through the province beginning in mid-March, with numbers peaking in late March or early April. It will return to the same nest in successive years, using escarpments, steep banks, or low trees for nesting.

This hawk derives its name from the fact that its legs are completely feathered down to the base of the toes.

Black-bellied Plover

Pluvialis squatarola

BBPL photo credit: W. Burgess

This circumpolar shorebird is a regular spring and fall migrant in Alberta, except in the mountains where it rarely occurs. It breeds on the dry gravel ridges and wet tundra of the southern Arctic islands and parts of the north coastline. Always in the vicinity of fresh or salt water, it winters along the Pacific coast from southwestern British Columbia to Chile and along the Atlantic from New Jersey to Brazil. Adults pass through Alberta in May and, in worn post-breeding condition, transit southwards in late July. At the latter time, males arrive about a week before females. Arriving by the end of August, juveniles may linger into November. Similar in appearance to the Lesser Golden-Plover, this species has grayer upperparts and white undertail coverts. Its call note, a "tee-yu-ee," drops in the middle syllable.

Lesser Golden-Plover

Pluvialis dominica

LGPL photo credit: G. Beyersbergen

Similar to the boldly patterned Black-bellied Plover, the Lesser Golden-Plover has a darker crown, golden color above, a dark rump and tail, and black undertail coverts. Arriving in flocks of no more than 50 between early and mid-May, they frequent areas of short grass, stubble fields, pastures, plowed land, and sometimes beaches. A few Black-bellied Plovers may be included in the flock. Although they are often seen in the Edmonton area, they migrate through all parts of the province, with only a few reports from the mountains. Breeding in the southern Arctic islands, in most of the Yukon, and in the northern half of the Northwest Territories east through Keewatin, they select dry uplands, ridges, or tundra knolls on which to nest. Fall migration, consisting of only juveniles, passes through between the second week of September and the third week of October. Their final destination is the plains of central South America.

Whimbrel

Numenius phaeopus

WHIM photo credit: C. Gordon

Similar to the Long-billed Curlew, the Whimbrel is separated from that species by a shorter down-curved bill, a dark stripe through the eye, a light medial crown stripe, grayish (less pinkish-cinnamon) plumage, and smaller overall size. Spring sightings are usually made from May 5-31 in the Grassland or Parkland regions. At a distance from their main migration routes, individuals and small groups can be seen locally in stubble and short grassy fields, lakeshores, beaches, and the sparsely vegetated portions of marshes. This species breeds in two disjunct areas: Alaska, northern Yukon Territories, and northwestern Mackenzie; and the eastern Keewatin, northeastern Manitoba, and the south shore of Hudson Bay. Various wet or dry tundra conditions are used for nesting. This species is quiet on migration, so there are very few fall migrant observations. Winter range is from the southern coast of the United States to the Galapagos, southern Chile, and Brazil.

Hudsonian Godwit

Limosa haemastica

HUGO photo credit: G. Beyersbergen

This intrepid and unassuming shorebird, once on the verge of extinction, is now making a slow recovery. It arrives from its wintering grounds in southern South America during late April to early June, with peak numbers in late April or early May. It is usually seen in flocks of up to 20 birds, usually in the eastern half of the province. This godwit breeds in two disjunct areas, in the delta of the Mackenzie River and along the southwest coasts of James and Hudson bays. It probably makes the 5,000 km trip from its staging area in James Bay to South America in one non-stop flight (Godfrey 1986).

The Hudsonian Godwit is a large shorebird with a long, slender, upturned bill. It can be seen close to water, where it scavenges through mud and debris in search of small invertebrates.

Ruddy Turnstone
Arenaria interpres

RUTU photo credit: W. Burgess

Arriving singly or in small groups with other shorebirds in May, the Ruddy Turnstone may be distinguished by its stocky build, short neck, unique white, black, and russet coloration, and its woodpecker-like pointed bill, adapted to turn over stones in search of worms and crustaceans. A regular transient of east-central Alberta, frequently occurring at Beaverhill Lake, it stops to feed at sand spits and pebbly beaches. It has wandered as far west as Banff National Park. A circumpolar breeder, in Canada it nests among various Arctic islands. During the breeding period, it may use a variety of habitats, from sandy islands to well-vegetated tundra. Fewer birds make the reverse journey through the province from mid-August to mid-October. Their winter range is from the southern United States to Chile and southern Brazil.

Red Knot
Calidris canutus

REKN photo credit: M. Preston

This regular, though uncommon, spring migrant passes through eastern Alberta, usually in small flocks, between May 10 and June 3, with greatest numbers being seen at Beaverhill Lake. It usually stays for only a week or so before proceeding to northern climes. The majority of Red Knots migrate along the west side of Quebec and along the James and Hudson bays. They nest locally along Arctic islands and coasts of the Old and New Worlds, where they seek out higher stony ridges in moist tundra. The Red Knot winters along both coasts, from California to South America and as far north as Massachusetts on the Atlantic side. More rarely seen in Alberta in the fall, the Red Knot is recorded alone or in small flocks after mid-August. There are no migration records for the mountains. The Red Knot's breeding plumage is an unmistakable robin-red throat and breast. It has a short, straight bill. It prefers broad sandy beaches where it will pick over food items among the debris.

Sanderling
Calidris alba

SAND

This rather chunky sandpiper nests throughout the Arctic islands of North America. En route to its wintering grounds along the coasts from southern British Columbia and Nova Scotia to South America, the Sanderling begins to arrive locally in July. In some years, birds will linger in Alberta until late October. Spring arrivals are from late April to early June, with the peak in late May. Renewing its fat reserves for further migration, it may be seen on the sandy, windswept shores of large lakes, where it follows receding waves, feeding on stranded food particles. As the wave returns the bird moves ahead of it and begins the cycle again. Unlike other sandpipers, this species lacks a hind toe.

Semipalmated Sandpiper

Calidris pusilla

SESA

Beginning in the last half of April, this migrant arrives from its wintering grounds, which extend from the southern United States to northern South America. Numbers peak around the third week of May and most have departed by the end of May.

It nests in Arctic areas from Alaska to Labrador and south to southeast Keewatin district, continuing along the coast of Hudson Bay. Its nest, a small depression in a mound of wet tundra or sand near water, is lined with grass or willow leaves.

Southbound birds reach the province again in July. Numbers peak in August. Most have passed through by mid-September, although some may occasionally linger into mid-October.

The most common "peep" in Alberta in the spring and fall, it is also one of the smallest, about the size of a large sparrow (15 cm). It can be distinguished from other sandpipers by its "clerical collar" (white throat band), black legs, and short stout bill.

Western Sandpiper

Calidris mauri

WESA

This diminutive sandpiper, common only on the British Columbia coastline, was first authentically documented in Alberta in 1972. Since then, there have been dozens of records, almost always in the fall, especially in August.

It breeds along the north and west coasts of Alaska, in sandy places or sedge-grass or heath tundra. It winters in Venezuela and Peru. During migration, almost any source of water will suit it, including salt and fresh. Sparrow-sized, the Western Sandpiper is very similar to the Semipalmated, but breeding plumage of the Western is rustier and more heavily streaked. In autumn, the two are very similar.

Least Sandpiper

Calidris minutilla

LESA

A regular spring migrant travelling in larger flocks, the Least Sandpiper passes through the province from mid-April to the end of May. During this time, it frequents the muddy margins of water bodies, mud flats, wet fields and, at times, beaches and dry fields. This smallest of sandpipers is separated from Semipalmated and Western Sandpipers, and other "peeps", by a rather slender, straight bill, greenish-yellowish legs, webbed toes, and a buffy, not gray, breast band. Breeding in southern Alaska and across subarctic Canada, it selects sedge grass, and mossy bogs and hummocky marshes. In Alberta in 1971, it may have nested at Lake Athabasca and Lillabo Lake (Pinel et al. 1991). A protracted fall migration occurs from late June to mid-October, and ends when these birds eventually depart for southern United States, Central America, and northern South America.

White-rumped Sandpiper

Calidris fuscicollis

WRSA

This "peep" arrives as a rare migrant in east-central Alberta around late May or early June, typically as one or two individuals among other "peeps". Pinel et al. (1991) reported it to be a regular migrant in the Provost-Chauvin area, where it frequents sandy shores of saline lakes. In June of 1978, 1,100 White-rumped Sandpipers were observed at Beaverhill Lake. This sandpiper nests in mounds of wet tundra along the coast and near islands of the Canadian Arctic, and rarely, in Alaska. It winters in southern South America. It is the only small, short-legged sandpiper with a white rump.

Baird's Sandpiper

Calidris bairdii

BASA

This migrant "peep" passes through the province, heading north, in the second half of April, and is usually gone by early June. It breeds on Arctic coasts and islands from Greenland to Siberia. Although it appears throughout the province, it rarely occurs in the mountains. Fall migration takes place in late June and July, and is concluded by mid-September. Stragglers may occasionally linger as late as November.

This bird prefers the drier parts of wet areas, frequenting the beaches and mud flats of larger lakes, as well as the margins of sloughs. An important staging area exists in the Provost-Chauvin area, where there are shallow saline ponds in sandy soil.

The Baird's is a small sandpiper with buff underparts, a "scaly" pattern on its back, and black legs.

Pectoral Sandpiper

Calidris melanotos

PESA photo credit: G. Beyersbergen

This sandpiper is a regular spring and fall transient over most of the province, except in the mountains. It arrives locally from its wintering grounds in northern South America in early May, on its way to localized breeding areas along the Arctic coast and southerly islands, from east Siberia to northern James Bay.

Males leave soon after mating, arriving back in Alberta late in July or early in August, the females following a week or so later. Young birds pass through later yet, usually in September or October. Preferring vegetation-covered, moist, rolling tundra for nesting, the Pectoral Sandpiper typically frequents grassy marshes or wet fields during migration. This medium-sized sandpiper has a straight bill and a well-defined breast band.

Dunlin

Calidris alpina

DUNL photo credit: T. Thormin

Usually travelling singly or in small mixed flocks, this rare spring migrant has been observed from the third week of April to the first week of June. Distinguishing features that separate it from the White-rumped Sandpiper and Sanderling include a rather long, slightly down-curved blackish bill, a reddish back, dark olive legs, and a black abdominal patch. Locally, it uses the muddy margins of water bodies, mud flats, flooded grassland, and sandy beaches and bars. This holarctic species usually breeds in moist to wet grassy or hummocky tundra and salt marshes. Canadian breeding range includes two areas: northern Yukon to northwestern MacKenzie, and eastern Keewatin to the west coast of Hudson Bay and some parts of the bay's south shore. Even less common in the fall, these late migrants pass through from late September to late October, proceeding to winter along the Pacific coast from Alaska to California and along the east coast from Massachusetts to the Gulf of Mexico.

Buff-breasted Sandpiper

Tryngites subruficollis

BBSA photo credit: R. Gehlert

The Buff-breasted Sandpiper's principal migration route takes it through the interior of North America and over the Prairie provinces. It arrives in flocks in Alberta around the middle of May, travelling by night. It frequents fields, where it runs about searching for insects and spiders. It can be quite tame. The largest numbers of migrants occur at Beaverhill Lake, but they are also found irregularly at lakes around Hanna, Calgary and Strathmore, and in Wood Buffalo National Park. There is also one mountain record. Its breeding range is not well known, but it is thought to nest locally in northern Alaska, and in the western Canadian Arctic, along the coast and in the high islands. Its courtship display involves the males chasing each other in displays of bravado. Passing through the province again in August or early September, individually or in small groups, it then proceeds to overwinter in Argentina or Uruguay. Its buffy coloration, dove-like head and yellowish legs make it easy to identify.

Long-billed Dowitcher

Limnodromus scolopaceus

LBDO photo credit: M. Preston

The Long-billed Dowitcher is a common visitor throughout the province. Spring migrants pass through Alberta in the last half of April or early in May, their numbers peaking in mid-May.

It nests in northeast Siberia, northwest Alaska, and probably in the northwest Mackenzie district and northern Yukon area (Godfrey 1986). Southbound fall migrants pass through Alberta starting in July, and the last birds are seen in late October. They winter in the southern United States (California to Florida) and farther south throughout western Mexico and into Guatemala.

The Long-billed Dowitcher tends to stay in tight flocks whether on ground or in the air. It feeds in marshes and damp meadows or along lakeshores, where it wades in the shallow water and probes the mud with the sensitive tip of its long bill, picking out invertebrates. It is a medium-sized sandpiper with a long, straight bill.

Parasitic Jaeger

Stercorarius parasiticus

PAJA

The Parasitic Jaeger is seen in Alberta mainly as an occasional fall transient from late August to early October. There have been summer records from northern Alberta, and this species has been recorded as a regular summer visitor on eastern Lake Athabasca. It is typically a migrant along the open ocean and coasts, sometimes seen on large lakes in the interior.
The Parasitic Jaeger breeds on low and grassy or mossy tundra, or in stony country throughout the Arctic and subarctic of the northern hemisphere. Over Canada, its breeding range extends from the northern Yukon, southeast to the north shore of Hudson Bay in Ontario, along the northern coasts of Quebec and Labrador, and north, probably to southern Ellesmere Island. It winters at sea, in the New World from Maine to Argentina and from Baja to Chile and Australasia.

Smaller than the Pomarine Jaeger, the Parasitic Jaeger is a brown-and-white gull-like bird with longer, pointed middle tail feathers.

Thayer's Gull

Larus thayeri

THGU

There were only 13 Alberta records of this migrant, the distribution extending south to Calgary and Kananaskis. It nests along ledges of cliffs on or near the coasts of the high Arctic islands of Canada and northwestern Greenland. Primarily a saltwater bird, fall migrants do visit lakes and rivers in the Yukon and northern Mackenzie. It winters along the Pacific coast from British Columbia south to California.
Similar to the Herring Gull, the Thayer's Gull's wing tips, however, are paler, and it is smaller overall. Its eyes are brownish with a reddish eye ring. Thayer's Gull was formerly treated as a subspecies of either the Iceland or Herring Gull; it breeds sympatrically with the latter.

Glaucous Gull

Larus hyperboreus

GLGU

This circumpolar migrant breeds along Arctic coasts and islands from Alaska to Greenland, wintering along the Pacific and Atlantic coast south to southern California and New York, as well as on the Great Lakes. According to Salt and Salt (1976), it is probably only the territorially unattached immature birds that venture as far inland as Alberta before migrating south and west. However, there were extralimital occurrences of these birds during the breeding season at Fort MacKay in 1975 (Pinel et al. 1991).
The adult Glaucous Gull is a very large, pure white gull, with a pale slate mantle and wings, a yellow bill with a red spot, and a yellow eye.

Sabine's Gull

Xema sabini

SAGU

This long-voyaging coastal migrant is mostly seen in Alberta in the fall, especially during the third week of September. Spring records, mostly in the first half of June, are less frequent. Typically, however, it is a rare migrant inland, preferring instead to follow the Pacific coast of British Columbia.

Sabine's Gull breeds in Arctic regions of North America and Eurasia. In Canada, this is along the north shore of the mainland and in the northern islands. For nesting, it seeks the low, wet tundra along lakes, coasts and islands. Although occasionally a solitary nester, it usually nests in small colonies, often in association with Arctic Terns. Distinctive features include the following: a forked tail; black legs, wing tips, and bill, with a yellow tip; and a slate triangular pattern on the upper side of spread wings.

Arctic Tern

Sterna paradisaea

ARTE

The amazing trajectory of the Arctic Tern's migration has been determined as a result of banding programs. Annually, it travels from the Arctic regions of North America to the extreme south of South America and back. This is a distance of 30,000 km. Eastern immatures cross the North Atlantic, descend the coast of Africa, and then cross the South Atlantic. After a few weeks they begin the return trip, following the sun. Equally amazing is this hardy species' longevity. One Norwegian fledgling was found in the same colony 27 years later.

A marine bird except when nesting, the Arctic Tern is a rare transient in Alberta, although it is often observed near the east end of Lake Athabasca. The Arctic Tern has a black crown, white face and neck, a slate mantle, and red legs and bill.

Harris' Sparrow

Zonotrichia querula

HASP

A large, long-tailed sparrow, this species is distinguished by black on the nape, top of head, front of the face, chin, throat and upper middle breast, gray sides of the head, a small black ear patch, and two white wing bars. A rare spring migrant, they are found in various thickets, woodland edges, and hedgerows of, mostly, eastern Alberta between May 8 - 31. Breeding along a band from the northwest Mackenzie through southern Keewatin to northern Manitoba, they select the wooded shrubbery and stunted trees along the margin between tree line and tundra. Our latest fall sparrow migrant, the Harris' Sparrow is more frequently observed than in the spring, passing through from early September to early November. It overwinters in southern British Columbia and in Iowa, south to California and Texas. There have been winter records for Banff, Strathmore and Lethbridge.

Lapland Longspur

Calcarius lapponicus

LALO

The Lapland Longspur is a common migrant throughout Alberta, beyond the mountain areas. It also winters frequently in the province, as far north as Vimy, sweeping stubble-fields in large flocks, for seeds and grains, often alongside Snow Buntings. It leaves Alberta in late March or early April for its breeding grounds above the tree line, where it chooses spots in the tundra with grass or shrub cover for nesting. Beginning in late August, it departs for wintering grounds as far south as Texas and Virginia, with fall migrant numbers peaking over Alberta in the first half of October.

The male in breeding plumage has a black crown, face and neck, and a chestnut hindneck. His melodious breeding song is heard by few other than indigenous people.

Smith's Longspur

Calcarius pictus

SMLO photo credit: T. Thormin

Smith's Longspur is primarily a local and uncommon migrant in southern Alberta, arriving in the province in peak numbers in mid-May, by which time Lapland Longspurs have already passed through (Salt and Salt 1976). It returns again in the first half of September.

Smith's Longspur is far more common in the spring than in the fall, and more likely to occur in eastern than in western Alberta, suggesting a more easterly fall migration route. Beaverhill Lake is probably an important spring stopover area (Pinel et al. 1991).

Smith's Longspur breeds in grassy, hummocky tundra, from northern Alaska, through the low Arctic and subarctic of northwest continental Canada and east to northern Ontario. It winters in the central and south-central United States.

Like all longspurs, the Smith's has a very long hind toenail. It is buffier below than the Lapland, and males have distinctive head markings in the spring.

Gyrfalcon

Falco rusticolus

GYRF photo credit: R. Fyfe

The Gyrfalcon is a holarctic breeder, in Canada nesting sparsely across most of the Yukon and NWT, north to Ellesmere, except for a large inverted triangle bisected approximately by the southern Mackenzie River.

Young birds wander widely in the south, occasionally into the northern United States. When wintering, they prefer snow-covered prairie or windswept mountain areas. These birds return north to establish their first breeding territory in the spring.

In Arctic regions, the Gyrfalcon's staple fare is the Willow Ptarmigan, which it stoops on from high overhead, and kills with a sudden blow of its feet. In Alberta, the Gyrfalcon is mostly seen in late October or November, especially in the Calgary area, where it avails itself of large Mallard populations along the Bow River. At such times, it is usually seen in its gray phase.

The Gyrfalcon may be almost white, with dark bars, it may be ashy brown with white streaks, or any intermediate coloration. It can be distinguished from other falcons by its large size, pointed wing tips, and rapid wing strikes. The largest falcon, highly prized by feudal, and, more recently, Arab princes, the Gyrfalcon is known as the "King of the falcons."

Snowy Owl

Nyctea scandiaca

SNOW photo credit: R. Fyfe

This circumpolar species breeds and generally resides year-round along Arctic coasts and islands from Alaska to Labrador. Movements southward vary significantly, according to the availability of its prey in Arctic regions, particularly the lemming. Migrants arrive in the province around mid-November and leave again by late March.

This bird will return to the same area in consecutive years, if game continues to be plentiful. It prefers open fields and will use favourite perching spots day after day. Hunting is mostly nocturnal, with considerable activity taking place at dawn and dusk.

The male is mostly pure white, usually with some dark bars outside of the face and throat. Females and, especially, young have more barring overall. This owl has no ear tufts.

Breeding season records exist for Ft. McMurray, Lake Louise, and Lesser Slave Lake (Pinel et al. 1991). One study (Boxall and Lein 1989) indicated the leading causes of death among Snowy Owls are collision (63%), starvation (14%), and gunshot (13%).

Snow Bunting

Plectrophenax nivalis

SNBU

The circumpolar Snow Bunting winters regularly throughout the province, as far north as Ft. McMurray. It frequents open fields, descending on stubble fields in flocks ranging from two to 60 birds, eating seeds and grain. When wintering, it will stay on the ground even at night. It leaves for its breeding grounds, along Arctic coasts and islands from Alaska to Greenland, in late March or early April. When breeding, it prefers rocky slopes or mossy tundra. It returns south in October, to winter as far south as California or Georgia.

Hoary Redpoll

Carduelis hornemanni

HORE

Although Alberta records for the Hoary Redpoll are scarce, this species may be under-represented because of its similarity with the Common Redpoll. The Hoary Redpoll breeds in Arctic regions of the Old and New Worlds. In Canada, this is locally in Ellesmere and northern Baffin islands and along the Arctic coast from Mackenzie Bay, throughout the Keewatin district, to Churchill, Manitoba. Most Alberta records are from the period between December and April, making it a winter visitor. The Hoary winters in much of its breeding range, as well as into most of southern Canada. Formally grouped taxonomically with the Common Redpoll, it is distinguished from its more plentiful cousin by less distinct streaking on its sides and undertail coverts, and a paler overall coloration. These species often flock together.

BIRDS OF ALBERTA SPECIES LIST

David M. Ealey

The following checklist of bird species for Alberta has been revised and updated (to July 30, 1992) from Ealey and McNicholl (1991). The Latin name is adjacent the English name for each species. The list follows the taxonomic order and nomenclature according to the American Ornithologists' Union (1983a, 1983b, 1985, 1989). The status of each species is indicated immediately beofre the English name as either nesting (* - 270 species), migrant (M - 32), migrant/winter visitor (M/WV - 1), regular winter visitor (WV - 3), vagrant or wanderer (V - 60), vagrant or possible migrant (V/M - 1), migrant and possible vagrant (M/V - 1), or extinct or extirpated (E - 4), for a total of 372 species.

Some species reported to nest are more commonly considered as vagrants or migrants, so the symbol V or M, respectively, follows the *. The majority of the nesters are year-round residents or summer residents. The seasonal occurrence of bird species can vary from one region of Alberta to another, so the typical status designated represents that for the majority of each species' range in the province. Criteria that help determine whether a species should be considered a regular part of the provincial fauna as opposed to a vagrant are consecutive years' breeding records, occurrence in at least a few different sites, and at least a fair likelihood of detection if appropriate habitat and region are visited, during the appropriate season.

Species whose status in Alberta is hypothetical (29) are listed following the main species list. Hypothetical species are those for which no substantiation exists, i.e., there is no specimen, identifiable body part, recording of a distinctly identifiable vocalization, or identifiable photograph. However, there must be acceptable, written documentation of the record for the sighting to be deemed adequate.

Physical evidence of occurrence is required for scientific acceptability of first records, but the time has long gone when it is considered acceptable to rigorously collect specimens merely to establish occurrence. Indeed, a great body of unsubstantiated, but properly documented sight records now forms the basis of much of our modern knowledge of distribution and status of birds. Therefore, placement of a species on a hypothetical list should not be considered a slur on the documentation of a species or on the observer, but should be looked upon as encouragement to substantiate the species' occurrence. Photographic or tape-recorded evidence is strongly recommended wherever possible.

Some of the hypothetical species at the end of the following list are poorly documented, but have been retained from earlier publications (e.g., Salt and Salt 1976) in the interest of a conservative approach. Review of many of these records is ongoing, so the removal of some species from the hypothetical list may occur if the documentation is deemed inadequate.

Class: AVES

Order: GAVIIFORMES (Loons)

Family: GAVIIDAE

Latin name	Status	English name	
Gavia stellata	* - M	Red-throated Loon	☐
Gavia pacifica	* - M	Pacific Loon	☐
Gavia immer	*	Common Loon	☐
Gavia adamsii	V	Yellow-billed Loon	☐

Order: PODICIPEDIFORMES (Grebes)

Family: PODICIPEDIDAE

Latin name	Status	English name	
Podilymbus podiceps	*	Pied-billed Grebe	☐
Podiceps auritus	*	Horned Grebe	☐
Podiceps grisegena	*	Red-necked Grebe	☐
Podiceps nigricollis	*	Eared Grebe	☐
Aechmophorus occidentalis	*	Western Grebe	☐
Aechmophorus clarkii	*	Clark's Grebe	☐

Order: PELECANIFORMES (Pelicans and Cormorants)

Family: PELECANIDAE

Latin name	Status	English name	
Pelecanus erythrorhynchos	*	American White Pelican	☐

Family: PHALACROCORACIDAE

Latin name	Status	English name	
Phalacrocorax auritus	*	Double-crested Cormorant	☐

Order: CICONIIFORMES (Herons, Egrets and Ibises)

Family: ARDEIDAE

Latin name	Status	English name	
Botaurus lentiginosus	*	American Bittern	☐
Ardea herodias	*	Great Blue Heron	☐
Casmerodius albus	V	Great Egret	☐
Egretta thula	V	Snowy Egret	☐
Egretta caerulea	V	Little Blue Heron	☐
Egretta tricolor	V	Tricolored Heron	☐
Bubulcus ibis	V	Cattle Egret	☐
Butorides striatus	V	Green-backed Heron	☐
Nycticorax nycticorax	*	Black-crowned Night-Heron	☐

Family: THRESKIORNITHIDAE

Latin name	Status	English name	
Plegadis chihi	*	White-faced Ibis	☐

Order: ANSERIFORMES (Ducks, Geese and Swans)

Family: ANATIDAE

Latin name	Status	English name	
Tribe: Cygnini			
Cygnus columbianus	M	Tundra Swan	☐
Cygnus buccinator	*	Trumpeter Swan	☐
Tribe: Anserini			
Anser albifrons	M	Greater White-fronted Goose	☐
Chen caerulescens	* - M	Snow Goose	☐
Chen rossii	M	Ross' Goose	☐
Branta bernicla	V	Brant	☐
Branta canadensis	*	Canada Goose	☐
Tribe: Cairinini			
Aix sponsa	*	Wood Duck	☐
Tribe: Anatini			
Anas crecca	*	Green-winged Teal	☐
Anas rubripes	*	American Black Duck	☐
Anas platyrhynchos	*	Mallard	☐
Anas acuta	*	Northern Pintail	☐
Anas querquedula	V	Garganey	☐
Anas discors	*	Blue-winged Teal	☐
Anas cyanoptera	*	Cinnamon Teal	☐
Anas clypeata	*	Northern Shoveler	☐
Anas strepera	*	Gadwall	☐
Anas penelope	M	Eurasian Wigeon	☐
Anas americana	*	American Wigeon	☐
Tribe: Aythyini			
Aythya valisineria	*	Canvasback	☐
Aythya americana	*	Redhead	☐
Aythya collaris	*	Ring-necked Duck	☐
Aythya fuligula	V	Tufted Duck	☐
Aythya marila	M	Greater Scaup	☐
Aythya affinis	*	Lesser Scaup	☐

Tribe: Mergini			
Somateria spectabilis	V	King Eider	☐
Histrionicus histrionicus	*	Harlequin Duck	☐
Clangula hyemalis	M	Oldsquaw	☐
Melanitta nigra	V	Black Scoter	☐
Melanitta perspicillata	* - M	Surf Scoter	☐
Melanitta fusca	*	White-winged Scoter	☐
Bucephala clangula	*	Common Goldeneye	☐
Bucephala islandica	*	Barrow's Goldeneye	☐
Bucephala albeola	*	Bufflehead	☐
Lophodytes cucullatus	*	Hooded Merganser	☐
Mergus merganser	*	Common Merganser	☐
Mergus serrator	*	Red-breasted Merganser	☐

Tribe: Oxyurini			
Oxyura jamaicensis	*	Ruddy Duck	☐

Order: FALCONIFORMES (Vultures, Hawks, Eagles and Falcons)

<u>Family: CATHARTIDAE</u>			
Cathartes aura	*	Turkey Vulture	☐

<u>Family: ACCIPITRIDAE</u>			
Pandion haliaetus	*	Osprey	☐
Haliaeetus leucocephalus	*	Bald Eagle	☐
Circus cyaneus	*	Northern Harrier	☐
Accipiter striatus	*	Sharp-shinned Hawk	☐
Accipiter cooperii	*	Cooper's Hawk	☐
Accipiter gentilis	*	Northern Goshawk	☐
Buteo platypterus	*	Broad-winged Hawk	☐
Buteo swainsoni	*	Swainson's Hawk	☐
Buteo jamaicensis	*	Red-tailed Hawk	☐
Buteo regalis	*	Ferruginous Hawk	☐
Buteo lagopus	M	Rough-legged Hawk	☐
Aquila chrysaetos	*	Golden Eagle	☐

<u>Family: FALCONIDAE</u>			
Falco sparverius	*	American Kestrel	☐
Falco columbarius	*	Merlin	☐
Falco peregrinus	*	Peregrine Falcon	☐
Falco rusticolus	M/WV	Gyrfalcon	☐
Falco mexicanus	*	Prairie Falcon	☐

Order: GALLIFORMES (Pheasants, Grouse, Ptarmigan and Turkey)

<u>Family: PHASIANIDAE</u>			
Perdix perdix	*	Gray Partridge	☐
Alectoris chukar	E	Chukar	☐
Phasianus colchicus	*	Ring-necked Pheasant	☐
Dendragapus canadensis	*	Spruce Grouse	☐
Dendragapus obscurus	*	Blue Grouse	☐
Lagopus lagopus	*	Willow Ptarmigan	☐
Lagopus leucurus	*	White-tailed Ptarmigan	☐
Bonasa umbellus	*	Ruffed Grouse	☐
Centrocercus urophasianus	*	Sage Grouse	☐
Tympanuchus cupido	E	Greater Prairie-Chicken	☐
Tympanuchus phasianellus	*	Sharp-tailed Grouse	☐
Meleagris gallopavo	*	Wild Turkey	☐
Colinus virginianus	E	Northern Bobwhite	☐

Order: GRUIFORMES (Rails, Coots and Cranes)

<u>Family: RALLIDAE</u>			
Coturnicops noveboracensis	*	Yellow Rail	☐
Rallus limicola	*	Virginia Rail	☐
Porzana carolina	*	Sora	☐
Fulica americana	*	American Coot	☐

<u>Family: GRUIDAE</u>			
Grus canadensis	*	Sandhill Crane	☐
Grus grus	V	Common Crane	☐
Grus americana	*	Whooping Crane	☐

Order: CHARADRIIFORMES (Plovers, Sandpipers, Phalaropes, Jaegers, Gulls and Terns)

<u>Family: CHARADRIIDAE</u>			
Pluvialis squatarola	M	Black-bellied Plover	☐
Pluvialis dominica	M	Lesser Golden-Plover	☐
Charadrius mongolus	V	Mongolian Plover	☐
Charadrius alexandrinus	V	Snowy Plover	☐
Charadrius semipalmatus	* - M	Semipalmated Plover	☐
Charadrius melodus	*	Piping Plover	☐
Charadrius vociferus	*	Killdeer	☐
Charadrius montanus	*	Mountain Plover	☐

<u>Family: RECURVIROSTRIDAE</u>			
Himantopus mexicanus	*	Black-necked Stilt	☐
Recurvirostra americana	*	American Avocet	☐

<u>Family: SCOLOPACIDAE</u>			
Tringa melanoleuca	*	Greater Yellowlegs	☐
Tringa flavipes	*	Lesser Yellowlegs	☐
Tringa solitaria	*	Solitary Sandpiper	☐
Catoptrophorus semipalmatus	*	Willet	☐
Heteroscelus incanus	V	Wandering Tattler	☐
Actitis macularia	*	Spotted Sandpiper	☐
Bartramia longicauda	*	Upland Sandpiper	☐
Numenius borealis	M - E	Eskimo Curlew	☐
Numenius phaeopus	M	Whimbrel	☐
Numenius americanus	*	Long-billed Curlew	☐
Limosa haemastica	M	Hudsonian Godwit	☐
Limosa fedoa	*	Marbled Godwit	☐
Arenaria interpres	M	Ruddy Turnstone	☐
Aphriza virgata	V	Surfbird	☐
Calidris canutus	M	Red Knot	☐
Calidris alba	M	Sanderling	☐
Calidris pusilla	M	Semipalmated Sandpiper	☐
Calidris mauri	M	Western Sandpiper	☐
Calidris minutilla	M	Least Sandpiper	☐
Calidris fuscicollis	M	White-rumped Sandpiper	☐
Calidris bairdii	M	Baird's Sandpiper	☐
Calidris melanotos	M	Pectoral Sandpiper	☐
Calidris acuminata	V/M	Sharp-tailed Sandpiper	☐
Calidris alpina	M	Dunlin	☐
Calidris ferruginea	V	Curlew Sandpiper	☐
Calidris himantopus	M	Stilt Sandpiper	☐
Tryngites subruficollis	M	Buff-breasted Sandpiper	☐
Philomachus pugnax	V	Ruff	☐
Limnodromus griseus	*	Short-billed Dowitcher	☐
Limnodromus scolopaceus	M	Long-billed Dowitcher	☐
Gallinago gallinago	*	Common Snipe	☐
Phalaropus tricolor	*	Wilson's Phalarope	☐
Phalaropus lobatus	* - M	Red-necked Phalarope	☐
Phalaropus fulicaria	V	Red Phalarope	☐

<u>Family: LARIDAE</u>			
Stercorarius parasiticus	M/V	Parasitic Jaeger	☐
Stercorarius longicaudus	V	Long-tailed Jaeger	☐
Larus pipixcan	*	Franklin's Gull	☐
Larus minutus	V	Little Gull	☐
Larus philadelphia	*	Bonaparte's Gull	☐
Larus canus	*	Mew Gull	☐
Larus delawarensis	*	Ring-billed Gull	☐
Larus californicus	*	California Gull	☐
Larus argentatus	*	Herring Gull	☐
Larus thayeri	M	Thayer's Gull	☐
Larus glaucoides	V	Iceland Gull	☐
Larus fuscus	V	Lesser Black-backed Gull	☐
Larus glaucescens	V	Glaucous-winged Gull	☐
Larus hyperboreus	M	Glaucous Gull	☐
Larus marinus	V	Great Black-backed Gull	☐
Rissa tridactyla	V	Black-legged Kittiwake	☐
Xema sabini	M	Sabine's Gull	☐
Sterna caspia	*	Caspian Tern	☐
Sterna hirundo	*	Common Tern	☐
Sterna paradisaea	M	Arctic Tern	☐
Sterna forsteri	*	Forster's Tern	☐
Chlidonias niger	*	Black Tern	☐

<u>Family: ALCIDAE</u>			
Cepphus grylle	V	Black Guillemot	☐
Synthliboramphus antiquus	V	Ancient Murrelet	☐

Order: COLUMBIFORMES (Doves and Pigeons)

<u>Family: COLUMBIDAE</u>			
Columba livia	*	Rock Dove	☐
Columba fasciata	V	Band-tailed Pigeon	☐
Zenaida macroura	*	Mourning Dove	☐
Ectopistes migratorius	E	Passenger Pigeon	☐

Order: CUCULIFORMES (Cuckoos)

<u>Family: CUCULIDAE</u>			
Coccyzus erythropthalmus	*	Black-billed Cuckoo	☐
Coccyzus americanus	V	Yellow-billed Cuckoo	☐

Order: STRIGIFORMES (Owls)

Family: STRIGIDAE

Otus asio	V	Eastern Screech-Owl	☐
Otus kennicottii	V	Western Screech-Owl	☐
Bubo virginianus	*	Great Horned Owl	☐
Nyctea scandiaca	WV	Snowy Owl	☐
Surnia ulula	*	Northern Hawk Owl	☐
Glaucidium gnoma	*	Northern Pygmy-Owl	☐
Athene cunicularia	*	Burrowing Owl	☐
Strix varia	*	Barred Owl	☐
Strix nebulosa	*	Great Gray Owl	☐
Asio otus	*	Long-eared Owl	☐
Asio flammeus	*	Short-eared Owl	☐
Aegolius funereus	*	Boreal Owl	☐
Aegolius acadicus	*	Northern Saw-whet Owl	☐

Order: CAPRIMULGIFORMES (Nightjars)

Family: CAPRIMULGIDAE

Chordeiles minor	*	Common Nighthawk	☐
Phalaenoptilus nuttallii	V	Common Poorwill	☐

Order: APODIFORMES (Swifts and Hummingbirds)

Family: APODIDAE

Cypseloides niger	*	Black Swift	☐

Family: TROCHILIDAE

Archilochus colubris	*	Ruby-throated Hummingbird	☐
Archilochus alexandri	V	Black-chinned Hummingbird	☐
Calypte anna	V	Anna's Hummingbird	☐
Calypte costae	V	Costa's Hummingbird	☐
Stellula calliope	*	Calliope Hummingbird	☐
Selasphorus rufus	*	Rufous Hummingbird	☐

Order: CORACIIFORMES (Kingfishers)

Family: ALCEDINIDAE

Ceryle alcyon	*	Belted Kingfisher	☐

Order: PICIFORMES (Woodpeckers)

Family: PICIDAE

Melanerpes lewis	* - V	Lewis' Woodpecker	☐
Melanerpes erythrocephalus	V	Red-headed Woodpecker	☐
Sphyrapicus varius	*	Yellow-bellied Sapsucker	☐
Sphyrapicus nuchalis	*	Red-naped Sapsucker	☐
Sphyrapicus ruber	V	Red-breasted Sapsucker	☐
Sphyrapicus thyroideus	V	Williamson's Sapsucker	☐
Picoides pubescens	*	Downy Woodpecker	☐
Picoides villosus	*	Hairy Woodpecker	☐
Picoides tridactylus	*	Three-toed Woodpecker	☐
Picoides arcticus	*	Black-backed Woodpecker	☐
Colaptes auratus	*	Northern Flicker	☐
Dryocopus pileatus	*	Pileated Woodpecker	☐

Order: PASSERIFORMES (Perching Birds)

Family: TYRANNIDAE - Flycatchers

Contopus borealis	*	Olive-sided Flycatcher	☐
Contopus sordidulus	*	Western Wood-Pewee	☐
Empidonax flaviventris	*	Yellow-bellied Flycatcher	☐
Empidonax alnorum	*	Alder Flycatcher	☐
Empidonax traillii	*	Willow Flycatcher	☐
Empidonax minimus	*	Least Flycatcher	☐
Empidonax hammondii	*	Hammond's Flycatcher	☐
Empidonax oberholseri	*	Dusky Flycatcher	☐
Empidonax occidentalis	*	Cordilleran Flycatcher	☐
Sayornis phoebe	*	Eastern Phoebe	☐
Sayornis saya	*	Say's Phoebe	☐
Myiarchus crinitus	*	Great Crested Flycatcher	☐
Tyrannus verticalis	*	Western Kingbird	☐
Tyrannus tyrannus	*	Eastern Kingbird	☐
Tyrannus forficatus	V	Scissor-tailed Flycatcher	☐

Family: ALAUDIDAE - Larks

Eremophila alpestris	*	Horned Lark	☐

Family: HIRUNDINIDAE - Swallows

Progne subis	*	Purple Martin	☐
Tachycineta bicolor	*	Tree Swallow	☐
Tachycineta thalassina	*	Violet-green Swallow	☐
Stelgidopteryx serripennis	*	Northern Rough-winged Swallow	☐
Riparia riparia	*	Bank Swallow	☐
Hirundo pyrrhonota	*	Cliff Swallow	☐
Hirundo rustica	*	Barn Swallow	☐

Family: CORVIDAE - Jays, Crows and Allies

Perisoreus canadensis	*	Gray Jay	☐
Cyanocitta stelleri	*	Steller's Jay	☐
Cyanocitta cristata	*	Blue Jay	☐
Nucifraga columbiana	*	Clark's Nutcracker	☐
Pica pica	*	Black-billed Magpie	☐
Corvus brachyrhynchos	*	American Crow	☐
Corvus corax	*	Common Raven	☐

Family: PARIDAE - Chickadees

Parus atricapillus	*	Black-capped Chickadee	☐
Parus gambeli	*	Mountain Chickadee	☐
Parus hudsonicus	*	Boreal Chickadee	☐
Parus rufescens	V	Chestnut-backed Chickadee	☐

Family: SITTIDAE - Nuthatches

Sitta canadensis	*	Red-breasted Nuthatch	☐
Sitta carolinensis	*	White-breasted Nuthatch	☐

Family: CERTHIIDAE - Creepers

Certhia americana	*	Brown Creeper	☐

Family: TROGLODYTIDAE - Wrens

Salpinctes obsoletus	*	Rock Wren	☐
Thryothorus ludovicianus	V	Carolina Wren	☐
Troglodytes aedon	*	House Wren	☐
Troglodytes troglodytes	*	Winter Wren	☐
Cistothorus platensis	*	Sedge Wren	☐
Cistothorus palustris	*	Marsh Wren	☐

Family: CINCLIDAE - Dippers

Cinclus mexicanus	*	American Dipper	☐

Family: MUSCICAPIDAE - Kinglets, Bluebirds and Thrushes

Regulus satrapa	*	Golden-crowned Kinglet	☐
Regulus calendula	*	Ruby-crowned Kinglet	☐
Polioptila caerulea	V	Blue-gray Gnatcatcher	☐
Oenanthe oenanthe	V	Northern Wheatear	☐
Sialia sialis	* - V	Eastern Bluebird	☐
Sialia mexicana	* - V	Western Bluebird	☐
Sialia currucoides	*	Mountain Bluebird	☐
Myadestes townsendi	*	Townsend's Solitaire	☐
Catharus fuscescens	*	Veery	☐
Catharus minimus	*	Gray-cheeked Thrush	☐
Catharus ustulatus	*	Swainson's Thrush	☐
Catharus guttatus	*	Hermit Thrush	☐
Hylocichla mustelina	V	Wood Thrush	☐
Turdus migratorius	*	American Robin	☐
Ixoreus naevius	*	Varied Thrush	☐

Family: MIMIDAE - Catbirds and Thrashers

Dumetella carolinensis	*	Gray Catbird	☐
Mimus polyglottos	*	Northern Mockingbird	☐
Oreoscoptes montanus	*	Sage Thrasher	☐
Toxostoma rufum	*	Brown Thrasher	☐
Toxostoma bendirei	V	Bendire's Thrasher	☐

Family: MOTACILLIDAE - Pipits

Anthus rubescens	*	American Pipit	☐
Anthus spragueii	*	Sprague's Pipit	☐

Family: BOMBYCILLIDAE - Waxwings

Bombycilla garrulus	*	Bohemian Waxwing	☐
Bombycilla cedrorum	*	Cedar Waxwing	☐

Family: LANIIDAE - Shrikes

Lanius excubitor	*	Northern Shrike	☐
Lanius ludovicianus	*	Loggerhead Shrike	☐

Family: STURNIDAE - Starlings

Sturnus vulgaris	*	European Starling	☐

Family: VIREONIDAE - Vireos

Vireo solitarius	*	Solitary Vireo	☐
Vireo gilvus	*	Warbling Vireo	☐
Vireo philadelphicus	*	Philadelphia Vireo	☐
Vireo olivaceus	*	Red-eyed Vireo	☐

Family: EMBERIZIDAE - Wood-Warblers, Buntings, Sparrows, Blackbirds,and Allies

Subfamily: Parulinae

Vermivora peregrina	*	Tennessee Warbler	☐
Vermivora celata	*	Orange-crowned Warbler	☐
Vermivora ruficapilla	* - M	Nashville Warbler	☐
Parula americana	V	Northern Parula	☐
Dendroica petechia	*	Yellow Warbler	☐
Dendroica pensylvanica	*	Chestnut-sided Warbler	☐
Dendroica magnolia	*	Magnolia Warbler	☐
Dendroica tigrina	*	Cape May Warbler	☐
Dendroica caerulescens	V	Black-throated Blue Warbler	☐
Dendroica coronata	*	Yellow-rumped Warbler	☐
Dendroica nigrescens	V	Black-throated Gray Warbler	☐
Dendroica townsendi	*	Townsend's Warbler	☐
Dendroica virens	*	Black-throated Green Warbler	☐
Dendroica fusca	*	Blackburnian Warbler	☐
Dendroica pinus	V	Pine Warbler	☐
Dendroica palmarum	*	Palm Warbler	☐
Dendroica castanea	*	Bay-breasted Warbler	☐
Dendroica striata	*	Blackpoll Warbler	☐
Mniotilta varia	*	Black-and-white Warbler	☐
Setophaga ruticilla	*	American Redstart	☐
Seiurus aurocapillus	*	Ovenbird	☐
Seiurus noveboracensis	*	Northern Waterthrush	☐
Oporornis formosus	V	Kentucky Warbler	☐
Oporornis agilis	*	Connecticut Warbler	☐
Oporornis philadelphia	*	Mourning Warbler	☐
Oporornis tolmiei	*	MacGillivray's Warbler	☐
Geothlypis trichas	*	Common Yellowthroat	☐
Wilsonia citrina	V	Hooded Warbler	☐
Wilsonia pusilla	*	Wilson's Warbler	☐
Wilsonia canadensis	*	Canada Warbler	☐
Icteria virens	*	Yellow-breasted Chat	☐

Subfamily: Thraupinae

Piranga olivacea	V	Scarlet Tanager	☐
Piranga ludoviciana	*	Western Tanager	☐

Subfamily: Cardinalinae

Cardinalis cardinalis	V	Northern Cardinal	☐
Pheucticus ludovicianus	*	Rose-breasted Grosbeak	☐
Pheucticus melanocephalus	*	Black-headed Grosbeak	☐
Passerina amoena	*	Lazuli Bunting	☐
Passerina cyanea	V	Indigo Bunting	☐
Spiza americana	V	Dickcissel	☐

Subfamily: Emberizinae

Pipilo erythrophthalmus	*	Rufous-sided Towhee	☐
Aimophila cassinii	V	Cassin's Sparrow	☐
Spizella arborea	*	American Tree Sparrow	☐
Spizella passerina	*	Chipping Sparrow	☐
Spizella pallida	*	Clay-colored Sparrow	☐
Spizella breweri	*	Brewer's Sparrow	☐
Pooecetes gramineus	*	Vesper Sparrow	☐
Chondestes grammacus	*	Lark Sparrow	☐
Calamospiza melanocorys	*	Lark Bunting	☐
Passerculus sandwichensis	*	Savannah Sparrow	☐
Ammodramus bairdii	*	Baird's Sparrow	☐
Ammodramus savannarum	*	Grasshopper Sparrow	☐
Ammodramus leconteii	*	Le Conte's Sparrow	☐
Ammodramus caudacutus	*	Sharp-tailed Sparrow	☐
Passerella iliaca	*	Fox Sparrow	☐
Melospiza melodia	*	Song Sparrow	☐
Melospiza lincolnii	*	Lincoln's Sparrow	☐
Melospiza georgiana	*	Swamp Sparrow	☐
Zonotrichia albicollis	*	White-throated Sparrow	☐
Zonotrichia atricapilla	*	Golden-crowned Sparrow	☐
Zonotrichia leucophrys	*	White-crowned Sparrow	☐
Zonotrichia querula	M	Harris' Sparrow	☐
Junco hyemalis	*	Dark-eyed Junco	☐
Calcarius mccownii	*	McCown's Longspur	☐
Calcarius lapponicus	M	Lapland Longspur	☐
Calcarius pictus	M	Smith's Longspur	☐
Calcarius ornatus	*	Chestnut-collared Longspur	☐
Plectrophenax nivalis	WV	Snow Bunting	☐

Subfamily: Icterinae

Dolichonyx oryzivorus	*	Bobolink	☐
Agelaius phoeniceus	*	Red-winged Blackbird	☐
Sturnella neglecta	*	Western Meadowlark	☐
Xanthocephalus xanthocephalus	*	Yellow-headed Blackbird	☐
Euphagus carolinus	*	Rusty Blackbird	☐
Euphagus cyanocephalus	*	Brewer's Blackbird	☐
Quiscalus quiscula	*	Common Grackle	☐
Molothrus ater	*	Brown-headed Cowbird	☐
Icterus galbula	*	Northern Oriole	☐

Family: FRINGILLIDAE (Finches and Grosbeaks)

Fringilla montifringilla	V	Brambling	☐
Leucosticte arctoa	*	Rosy Finch	☐
Pinicola enucleator	*	Pine Grosbeak	☐
Carpodacus purpureus	*	Purple Finch	☐
Carpodacus cassinii	*	Cassin's Finch	☐
Carpodacus mexicanus	V	House Finch	☐
Loxia curvirostra	*	Red Crossbill	☐
Loxia leucoptera	*	White-winged Crossbill	☐
Carduelis flammea	* - WV	Common Redpoll	☐
Carduelis hornemanni	WV	Hoary Redpoll	☐
Carduelis pinus	*	Pine Siskin	☐
Carduelis tristis	*	American Goldfinch	☐
Coccothraustes vespertinus	*	Evening Grosbeak	☐

Family: PASSERIDAE (Weaver Finches)

Passer domesticus	*	House Sparrow	☐

SPECIES WHOSE STATUS IS HYPOTHETICAL

Dendrocygna bicolor	Fulvous Whistling-Duck
Chen canagica	Emperor Goose
Elanoides forfiatus	American Swallow-tailed Kite
Buteo lineatus	Red-shouldered Hawk
Lagopus mutus	Rock Ptarmigan
Tringa erythropus	Spotted Redshank
Arenaria melanocephala	Black Turnstone
Calidris ruficollis	Rufous-necked Stint
Eurynorhynchus pygmeus	Spoonbill Sandpiper
Scolopax minor	American Woodcock
Stercorarius pomarinus	Pomarine Jaeger
Rhodostethia rosea	Ross' Gull
Pagophila eburnea	Ivory Gull
Tyto alba	Barn Owl
Caprimulgus vociferus	Whip-poor-will
Chaetura pelagica	Chimney Swift
Chaetura vauxi	Vaux's Swift
Aeronautes saxatilis	White-throated Swift
Picoides albolarvatus	White-headed Woodpecker
Tyrannus savana	Fork-tailed Flycatcher
Gymnorhinus cyanocephalus	Pinyon Jay
Pica nuttalli	Yellow-billed Magpie
Sitta pygmaea	Pygmy Nuthatch
Catherpes mexicanus	Canyon Wren
Motacilla alba/M. lugens	White Wagtail/Black-backed Wagtail
Vireo flavifrons	Yellow-throated Vireo
Vermivora chrysoptera	Golden-winged Warbler
Vermivora virginiae	Virginia's Warbler
Amphispiza bilineata	Black-throated Sparrow

LITERATURE CITED

American Ornithologists' Union. 1983a. Check-list of North American Birds. 6th ed. American Ornithologists Union, Lawrence, Kansas. 877 pp.

American Ornithologists' Union. 1983b. Thirty-fourth supplement to the American Ornithologists' Union check-list of North American birds. Auk 99(suppl.):1CC-16CC.

American Ornithologists' Union. 1985. Thirty-fifth supplement to the American Ornithologists' Union check-list of North American birds. Auk 102(3):680-686.

American Ornithologists' Union. 1989. Thirty-seventh supplement to the American Ornithologists' Union check-list of North American birds. Auk 106:532-538.

Ealey, D.M., and M.K. McNicholl. 1991. A bibliography of Alberta ornithology. Natural History Occasional Paper No. 16, Provincial Museum of Alberta, Edmonton. 751 pp.

Salt, W.R., and J.R. Salt. 1976. The birds of Alberta. Hurtig Publishers, Edmonton. 498 pp.

Bird Atlas Field Data Card

Species	Code	Breeding Evidence				Abun
		O	PO	PR	CO	
Red-throated Loon *	RTLO					
Pacific Loon *	PALO					
Common Loon	COLO					
Yellow-billed Loon *	YBLO					
Pied-billed Grebe	PBGR					
Horned Grebe	HOGR					
Red-necked Grebe	RNGR					
Eared Grebe	EAGR					
Western Grebe	WEGR					
Clark's Grebe *	CLGR					
American White Pelican	AWPE					
Double-cr. Cormorant	DCCO					
American Bittern	AMBI					
Great Blue Heron	GBLH					
Blk-crwnd Night-heron	BCNH					
White-faced Ibis *	WFIB					
Tundra Swan	TUSW					
Trumpeter Swan *	TPSW					
Gr. White-frntd Goose	GWFG					
Snow Goose	SNGO					
Ross' Goose *	ROGO					
Brant *	BRAN					
Canada Goose	CAGO					
Wood Duck *	WODU					
Green-winged Teal	GWTE					
Mallard	MALL					
Northern Pintail	NOPI					

Species	Code	Breeding Evidence				Abun
		O	PO	PR	CO	
Blue-Winged Teal	BWTE					
Cinnamon Teal †	CITE					
Northern Shoveler	NOSV					
Gadwall	GADW					
Eurasian Wigeon *	EUWI					
American Wigeon	AMWI					
Canvasback	CANV					
Redhead	REDH					
Ring-necked Duck	RNDU					
Greater Scaup †	GRSC					
Lesser Scaup †	LESC					
Harlequin Duck	HADU					
Oldsquaw	OLDS					
Black Scoter	BLSC					
Surf Scoter *	SUSC					
White-winged Scoter	WWSC					
Common Goldeneye	COGO					
Barrow's Goldeneye	BOGO					
Bufflehead	BUFF					
Hooded Merganser *	HOME					
Common Merganser	COME					
Red-breasted Merganser	RBME					
Ruddy Duck	RUDU					
Turkey Vulture *	TUVU					
Osprey	OSPR					
Bald Eagle	BAEA					
Northern Harrier	NOHA					

Species	Code	Breeding Evidence				Abun
		O	PO	PR	CO	
Sharp-shinned Hawk †	SSHA					
Cooper's Hawk †	COHA					
Northern Goshawk †	NOGO					
Broad-winged Hawk	BWHA					
Swainson's Hawk	SWHA					
Red-tailed Hawk	RTHA					
Ferruginous Hawk *	FEHA					
Rough-legged Hawk	RLHA					
Golden Eagle	GOEA					
American Kestrel	AMKE					
Merlin	MERL					
Peregrine Falcon *	PEFA					
Gyrfalcon *	GYRF					
Prairie Falcon	PRFA					
Gray Partridge	GRPA					
Ring-necked Pheasant	RGNP					
Spruce Grouse	SPGR					
Blue Grouse	BLGS					
Willow Ptarmigan †	WIPT					
White-tailed Ptarmigan †	WTPT					
Ruffed Grouse	RUGR					
Sage Grouse *	SAGR					
Greater Prairie-Chicken *	GPCH					
Sharp-tailed Grouse	STGR					
Turkey *	TURK					
Yellow Rail *	YERA					
Virginia Rail	VIRA					

Species	Code	Breeding Evidence				Abun
		O	PO	PR	CO	
Sora	SORA					
American Coot	AMCO					
Sandhill Crane	SACR					
Whooping Crane *	WOCR					
Black-bellied Plover	BBPL					
Lesser Golden-Plover	LGPL					
Semipalmated Plover	SEPL					
Piping Plover *	PIPL					
Kildeer	KILL					
Mountain Plover *	MTPL					
Black-necked Stilt *	BNST					
American Avocet	AMAV					
Greater Yellowlegs †	GRYE					
Lesser Yellowlegs †	LEYE					
Solitary Sandpiper	SOSA					
Willet	WILL					
Wandering Tattler *	WATA					
Spotted Sandpiper	SDSA					
Upland Sandpiper	UPSA					
Eskimo Curlew *	ESCU					
Whimbrel *	WHIM					
Long-billed Curlew	LBCU					
Hudsonian Godwit	HUGO					
Marbled Godwit	MAGO					
Ruddy Turnstone	RUTU					
Red Knot	REKN					
Sanderling	SAND					

Comments/Notes:

Species	Code	Breeding Evidence				Abun
		O	PO	PR	CO	
Semipalm. Sandpiper	SESA					
Western Sandpiper †	WESA					
Least Sandpiper	LESA					
White-rmped Sandpiper	WRSA					
Baird's Sandpiper †	BASA					
Pectoral Sandpiper	PESA					
Sharp-tailed Sandpiper *	SHSA					
Dunlin	DUNL					
Stilt Sandpiper	SLSA					
Buff-brsted Sandpiper *	BBSA					
Short-billed Dowitcher †	SBDO					
Long-billed Dowitcwher *	LBDO					
Common Snipe	COSN					
Wilson's Phalarope	WIPH					
Red-necked Phalarope *	RDNP					
Pomarine Jaeger *	POJA					
Parasitic Jaeger *	PAJA					
Long-tailed Jaeger *	LTJA					
Franklin's Gull	FRGU					
Bonaparte's Gull	BOGU					
Mew Gull *	MEGU					
Ring-billed Gull	RBGU					
California Gull	CAGU					
Herring Gull	HGGU					
Glaucous Gull *	GLGU					
Caspian Tern *	CATE					
Common Tern	COTE					

Species	Code	Breeding Evidence				Abun
		O	PO	PR	CO	
Arctic Tern *	ARTE					
Forster's Tern †	FOTE					
Black Tern	BLTE					
Rock Dove	RODO					
Mourning Dove	MODO					
Black-billed Cuckoo	BBCU					
Eastern Screech-Owl *	ESOW					
Western Screech-Owl *	WSOW					
Great Horned Owl	GHOW					
Snowy Owl	SNOW					
Northern Hawk-Owl	NHOW					
Northern Pygmy-Owl	NPOW					
Burrowing Owl *	BUOW					
Barred Owl	BAOW					
Great Gray Owl	GGOW					
Long-eared Owl	LEOW					
Short-eared Owl	SEOW					
Boreal Owl	BOOW					
Northern Saw-whet Owl	NSWO					
Common Nighthawk	CONI					
Common Poorwill *	COPW					
Black Swift *	BLSW					
Chimney Swift *	CHSW					
Vaux's Swift *	VASW					
Ruby-thrtd Hummingbird	RTHU					
Calliope Hummingbird	CAHU					
Rufous Hummingbird	RUHU					

Species	Code	Breeding Evidence				Abun
		O	PO	PR	CO	
Belted Kingfisher	BEKI					
Lewis' Woodpecker *	LEWO					
Rd-hded Woodpecker *	RHWO					
Yellow-bell. Sapsucker	YBSA					
Red-naped Sapsucker †	RNSA					
Downy Woodpecker	DOWO					
Hairy Woodpecker	HAWO					
Three-toed Woodpecker	TTWO					
Black-bked Woodpecker	BBWO					
Northern Flicker	NOFL					
Pileated Woodpecker	PIWO					
Olive-sided Flycatcher	OSFL					
Western Wood-Pewee	WWPE					
Yellow-bell. Flycatcher †	YBFL					
Alder Flycatcher †	ALFL					
Willow Flycatcher †	WIFL					
Least Flycatcher †	LEFL					
Hammond's Flycatcher †	HAFL					
Dusky Flycatcher †	DUFL					
Western Flycatcher †	WEFL					
Eastern Phoebe	EAPH					
Say's Phoebe	SAPH					
Gr.-crested Flycatcher *	GCFL					
Western Kingbird	WEKI					
Eastern Kingbird	EAKI					
Horned Lark	HOLA					
Purple Martin	PUMA					

Species	Code	Breeding Evidence				Abun
		O	PO	PR	CO	
Tree Swallow	TESW					
Violet-green Swallow	VGSW					
R.Rough-winged Swallow	NRWS					
Bank Swallow	BKSW					
Cliff Swallow	CLSW					
Barn Swallow	BRSW					
Gray Jay	GRJA					
Steller's Jay	STJA					
Blue Jay	BLJA					
Clark's Nutcracker	CLNU					
Black-billed Magpie	BBMA					
American Crow	AMCR					
Common Raven	CORA					
Blk-capped Chickadee	BCCH					
Mountain Chickadee	MOCH					
Boreal Chickadee	BOCH					
Red-breasted Nuthatch	RBNU					
White-breasted Nuthatch	WBNU					
Brown Creeper	BRCR					
Rock Wren	ROWR					
House Wren	HOWR					
Winter Wren †	WIWR					
Sedge Wren *	SEWR					
Marsh Wren	MAWR					
American Dipper	AMDI					
Golden-crowned Kinglet	GCKI					
Ruby-crowned Kinglet	RCKI					

Species	Code	Breeding Evidence				Abun
		O	PO	PR	CO	
Eastern Bluebird*	EABL					
Western Bluebird*	WEBL					
Mountain Bluebird	MOBL					
Townsend's Solitaire	TOSO					
Veery	VEER					
Gray-cheeked Thrush†	GCTH					
Swainson's Thrush	SWTH					
Hermit Thrush	HETH					
American Robin	AMRO					
Varied Thrush	VATH					
Gray Catbird	GRCA					
Northern Mockingbird*	NOMO					
Brown Thrasher	BRTH					
Water Pipit	WAPI					
Sprague's Pipit	SPPI					
Bohemian Waxwing	BOWA					
Cedar Waxwing	CEWX					
Northern Shrike*	NOSK					
Loggerhead Shrike	LOSH					
European Starling	EUST					
Solitary Vireo	SOVI					
Warbling Vireo	WAVI					
Philadelphia Vireo†	PHVI					
Red-eyed Vireo	REVI					
Tennessee Warbler	TEWA					
Orng.-crwnd Warbler	OCWA					
Nashville Warbler*	NAWA					

Species	Code	Breeding Evidence				Abun
		O	PO	PR	CO	
Northern Parula*	NOPA					
Yellow Warbler	YEWB					
Chestnut-sided Warbler	CSWA					
Magnolia Warbler	MNWA					
Cape May Warbler	CMWA					
Bl.-thrtd Blue Warbler*	BTBW					
Yellow-rumped Warbler	YRWA					
Bl.-thrtd Gray Warbler*	BGYW					
Townsend's Warbler	TOWA					
Bl.-thrtd Green Warbler	BGNW					
Blackburnian Warbler*	BLWA					
Pine Warbler*	PIWA					
Palm Warbler	PAWA					
Bay-breasted Warbler	BBWA					
Blackpoll Warbler	BPWA					
Black-and-White Warbler	BAWW					
American Redstart	AMRE					
Ovenbird	OVEN					
Northern Waterthrush	NOWA					
Connecticut Warbler	COWA					
Mourning Warbler†	MOWA					
MacGillivray's Warbler†	MGWA					
Common Yellowthroat	COYE					
Wilsons' Warbler	WIWA					
Canada Warbler	CAWA					
Yellow-breasted Chat	YBCH					
Western Tanager	WETA					

Species	Code	Breeding Evidence				Abun
		O	PO	PR	CO	
Rose-brsted Grosbeak	RBGR					
Blk-headed Grosbeak*	BHGR					
Lazuli Bunting	LZBU					
Indigo Bunting*	INBU					
Rufous-sided Towhee	RSTO					
American Tree Sparrow	ATSP					
Chipping Sparrow	CHSP					
Clay-colored Sparrow	CCSP					
Brewer's Sparrow†	BWSP					
Vesper Sparrow	VESP					
Lark Sparrow	LASP					
Lark Bunting	LKBU					
Savannah Sparrow	SVSP					
Baird's Sparrow*	BDSP					
Grasshopper Sparrow	GRSP					
Le Conte's Sparrow	LCSP					
Sharp-tailed Sparrow†	STSP					
Fox Sparrow	FOSP					
Song Sparrow	SOSP					
Lincoln's Sparrow	LISP					
Swamp Sparrow†	SWSP					
White-throated Sparrow	WTSP					
Golden-crwnd Sparrow	GCSP					
White-crwnd Sparrow	WCSP					
Harris' Sparrow	HASP					
Dark-eyed Junco	DEJU					
McCown's Longspur	MCLO					

Species	Code	Breeding Evidence				Abun
		O	PO	PR	CO	
Lapland Longspur	LALO					
Smith's Longspur*	SMLO					
Chestnt-collrd Longspur	CCLO					
Snow Bunting	SNBU					
Bobolink	BOBO					
Red-winged Blackbird	RWBL					
Western Meadowlark	WEME					
Yellow-hded Blackbird	YHBL					
Rusty Blackbird†	RUBL					
Brewer's Blackbird	BRBL					
Common Grackle	COGK					
Brown-headed Cowbird	BHCO					
Northern Oriole	NOOR					
Rosy Finch	ROFI					
Pine Grosbeak	PIGR					
Purple Finch	PUFI					
Cassin's Finch*	CAFI					
House Finch*	HOFI					
Red Crossbill	RECR					
White-winged Crossbill	WWCR					
Common Redpoll	CORE					
Pine Siskin	PISI					
American Goldfinch	AMGO					
Evening Grosbeak	EVGR					
House Sparrow	HOSP					

Breeding Codes

Observed
X – species identified, but no indication of breeding.

Possible
H – species observed, or breeding calls heard, in suitable nesting HABITAT.

Probable
P – PAIR observed in suitable nesting habitat.
T – TERRITORY presumed through ***territorial behaviour*** in the same location ***on at least 2 occasions a week or more apart.***
C – COURTSHIP behaviour between a male and a female.
V – VISITING probable nest-site, but no further evidence obtained.
N – NEST-BUILDING or excavation of nest-hole by wrens and woodpeckers.

Confirmed
NB – NEST-BUILDING or adult carrying nesting material; use for all species except wrens and woodpeckers.
DD – DISTRACTION DISPLAY or injury feigning.
UN – USED NEST or eggshells found.
FL – recently FLEDGED young or downy young.
ON – OCCUPIED NEST indicated by adult entering or leaving nestsite, or adult seen incubating.
CF – CARRYING FOOD; adult seen carrying food or faecal sac for young.
NE – NEST with EGGS.
NY – NEST with YOUNG.

Adbundance Codes

Code		Code	
0 –	1	5 –	251-500
1 –	2-10	6 –	501-1000
2 –	11-50	7 –	1001-5000
3 –	51-100	8 –	5001-10,000
4 –	101-250	9 –	more than 10,000

Additional Species

Species	Code	O	PO	PR	CO	Ab	UTM 1 km or 100 m	Doc Form (✓)

* Verification of breeding status (PO, PR, CO) of these species requires that a rare bird report form be completed.

† Take special care in dentifying these species for they are easily confused with others.

Parts of a Bird

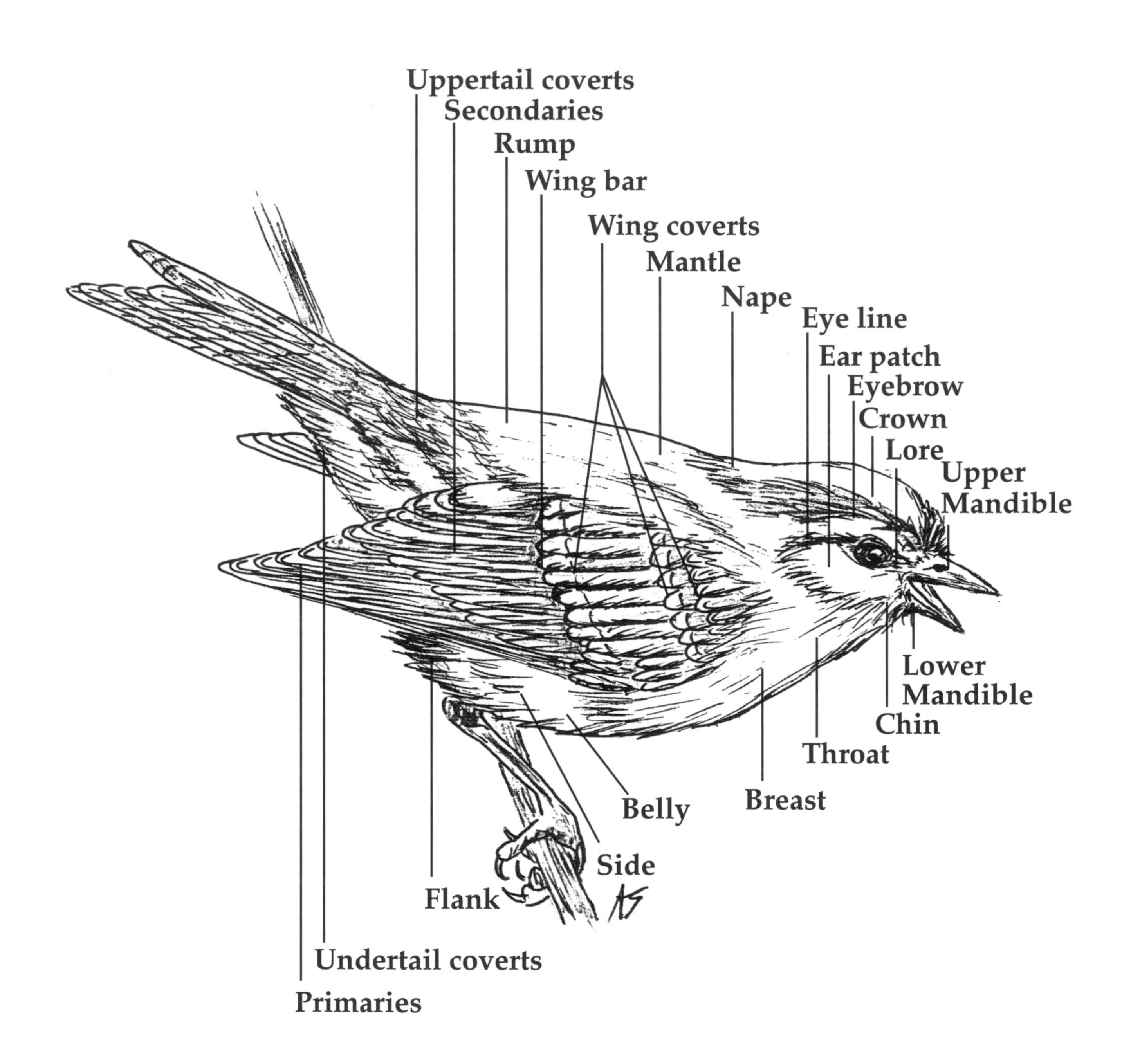

GLOSSARY

Accipiters: broad-winged raptors of the Order Falconiformes that use their strong feet and long talons to seize and kill their prey.

Altricial: newly hatched young that are helpless and hence confined to the nest for some time.

Anting: behavior of passerines in which the bird squats on the ground and ants are allowed and encouraged to move freely through the plumage and over the skin. The reason for this behavior is not known, but it may help maintain healthy feathers.

Arboreal: frequently conducting various or all activities in and around trees.

Atlas: in ornithology, a collection of maps showing bird distribution within a defined region based on dot distribution within a defined regular grid, over a defined limited time period in which fieldwork was carried out.

Avian: referring to birds.

Blowdown: a tree that has been blown down by wind, often with root mass exposed.

Blue List: The National Audubon Society's list to provide early warning of those North American species undergoing population or range reductions.

Bog: wetland ecosystem characterized by an accumulation of peat, acid conditions, and dominance of sphagnum moss.

Brood Parasitism: the habit of laying eggs in the nests of other birds, often of a different species, effectively forcing the host parents to raise the chick from the introduced egg. The term usually refers to the regular use of another host species to raise young.

Brooding: the act of one adult sitting over its nestlings during their first few days of life.

Cambium: a layer of soft tissue between the bark and wood of a tree.

Carnivorous: having a diet that is composed entirely of animal matter.

Cirque: a glacially carved scoop in a mountainside.

Clutch: a total group of eggs laid by one female at a single nesting.

Cob: special term for a male swan.

Colony: a group of birds of the same species nesting in very close proximity.

Composite Plants or Compositae: family of plants including the aster, dandelion, and daisy, in which the florets are borne in a close head resembling a single flower.

Coniferous: bearing cones.

Conspecific: of two or more species or other forms, meaning that they belong to the same species or should be so regarded, although considered by others to be specifically distinct.

COSEWIC: The Committee on the Status of Endangered Wildlife in Canada.

Coulee: dry, steep-sided valley common in southern Alberta.

Coverts: small feathers that cover the base of larger feathers of the wing and tail or cover a specific area.

Courtship Feeding: feeding of one member of a breeding pair by its mate, as with a male bringing food to a female, in the days before egg laying.

Crepuscular: being active in the dim light of dusk and dawn rather than in full night or day.

Crest: larger feathers on the top of the head that may be erectile or may extend beyond the back of the head.

Crop: a swelling in the oesophagus (throat), used by some species to store food before digestion.

Crop Milk: in pigeons and doves, a secretion of the crop, resembling mammalian milk in composition.

Cupnest: a nest built in the shape of a cup.

Cygnet: special term for a young swan.

Deciduous: shedding leaves annually at end of growing season.

Decurved: curved downward.

Dihedral: pertaining to the upward or, rarely, the downward slant of a wing.

Disjunct: disjoined or separated.

Diurnal: being active during the light of day.

Drake: a male duck.

Eclipse Plumage: dull, inconspicuous plumage that precedes and follows the brighter breeding plumage.

Ecotone: the transition zone between two structurally different communities; an edge.

Egg Dumping: intraspecific brood parasitism in which birds lay or "dump" eggs in the nests of fellow members of their own species.

Emergent vegetation: aquatic vegetation with a portion that rises freely above the water's surface.

Endangered species: any indigenous species of fauna or flora whose existence in Canada is threatened with immediate extinction through all or a significant portion of its range.

Extirpated species: any indigenous species of fauna or flora no longer existing in the wild in Canada but existing elsewhere in the world.

Extinct species: any species of fauna or flora formerly indigenous to Canada but no longer existing anywhere on Earth.

Extralimital: observation or record existing beyond the species' previously accepted range.

Eye Stripe: a stripe running horizontally from the base of the bill through the eye.

Eyrie: term (also "eyre", "aerie", "aery") for the nest of a bird-of-prey, often an eagle *Aquila* species, etc.

Faeces: intestinal waste materials.

Faecal sac: an ejected sac of intestinal waste materials.

Fen: wetland dominated by sedge.

Feral: a domestic bird or species which has escaped captivity and become wild.

Fledging: term usually applied to the acquisition by a young bird of its first true feathers.

Fledging Period: the interval between hatching and being able to fly; in flightless species, the equivalent interval is between hatching and independence from the parents.

Flight feathers: the long, vaned feathers used to generate lift or manoeuvre in flight.

Flycatch: to feed by hawking flying insects in mid-air or by diving from a perch onto insects on the ground.

Generalist: a species that forages for a variety of foods.

Glean: to pick off food from a surface.

Gregarious: grouping together in flocks.

Gular: pertaining to the throat.

Habitat: place where a plant or animal lives.

Hatching Asynchrony: eggs in a clutch hatching with a significant interval between first and last. If food is in short supply, the youngest—and therefore smallest—chicks will die, but some of the brood will survive. In this way, the size of the brood is decreased without starving and weakening all the chicks.

Hatching Synchrony: the hatching of different eggs in a clutch at the same time. This can be achieved by deliberately leaving the early eggs in the clutch without incubating them until the last but one egg has been laid. The eggs then commence their embryonic development at the same time. In many precocial species, the eggs hatch together, within as little as an hour of each other, as a result of communication among the developing eggs; those laid later must accelerate their development and early ones must slow down to achieve synchronous hatching.

Hawking: used to describe a bird of any kind flying in search or pursuit of prey; to attack by swooping or striking.

Holarctic Region: the Palaearctic and Nearctic realms combined. Though each of these realms has its own distinct species, a large number of species are common to both.

Hoverglean: to remove food from an elevated surface while in flight; directional movement temporarily ceases.

Hummock: a wooded ridge of land adjacent to a marsh or swamp.

Indigenous: native to a given area or region.

Insectivorous: having a diet composed of insects.

Interspecific: between individuals of different species.

Intraspecific: between individuals of same species.

Irruption: a form of migration common to several bird species, in which, forced by changing conditions to leave their usual territory, they invade another in search of food.

Krumholtz: stunted trees characteristic of the transition zone between alpine tundra and subalpine coniferous.

Leks: displays in which male birds take up stances and posture in small areas, generally called courts, usually on arenas or display grounds, and mate with as many females as possible.

Marsh: wetland dominated by rushes, cattails, and sedges.

Mixed wood: forest including a mixed of both coniferous and deciduous trees.

Mobbing: a collective attack on a predator, often a bird of another species, by a group of birds, sometimes of more than one species. Usually there is no physical contact, just harassment.

Monogamy: term used to cover a wide range of mating relationships, generally referring to a breeding pattern in which a male and female remain together for at least one breeding cycle, having a more or less exclusive mating relationship, and work together to rear their young.

Molt or Moult: the shedding and replacement of plumage feathers.

Montane: pertaining to the mountain region, from valley floor to subalpine.

Muskeg: an environment, consisting of grassy bog, characteristic of much of northern Canada.

Nape: the back part of the neck.

Nocturnal: being active in full night conditions.

Obligate Cavity Nester: a species which will only nest in a cavity.

Omnivorous: having a varied and unspecialized diet.

Ornithology: the scientific study of birds.

Papillae: unsheathed feathers and supporting gland.

Passerine: term commonly used to refer to birds in the order Passeriformes, the "perching" birds. Birds of other orders are referred to as "non-passerines".

Peep: one of five small, gray sandpipers that may be seen, at some time, in the province: Semipalmated, Least, Baird's, Western, and White-rumped.

Pellet: a compact mass composed of those undigested portions of a bird's food that have been retained in the stomach by a mechanical barrier before being regurgitated and ejected through the mouth, rather than evacuated as faeces.

Pen: special term for a male swan.

Pigeon Milk: See **Crop Milk**.

Pileated: crested or capped; referring to the shape of the crown feathers.

Polyandry: mating system, much rarer than polygamy, in which the female forms a bond and mates with several males.

Polygamy: a general term for breeding patterns other than monogamy. More specifically, polygamy can be divided into polygyny and polyandry. Promiscuity is a mixture of polygyny and polyandry and usually means there is no lasting pair bond.

Polygyny: Mating system in which a male mates with several females, either by defending a large territory and attempting to attract as many females as possible, or competing with other males for status at a lek and mating there with as many females as possible.

Precocial: active immediately after hatching.

Preen: to groom feathers with the bill.

Race: the same as subspecies.

Raft: an assembled, swimming flock of waterbirds.

Rare species: any indigenous species of fauna or flora that, because of its biological characteristics, or because it occurs at the fringe of its range, or for some other reason, exists in low numbers or in very restricted areas in Canada but is not a threatened species.

Raptor: bird of prey.

Recurved: curved upward.

Regurgitate: to bring from raw to completely digested food up from the crop to feed young.

Resident: non-migratory, remaining in the same area year round.

Riparian: along the banks of rivers and streams.

Rookery: a place used for breeding and nesting by a colony of birds.

Rufous: a flat red shade with undertones of brown.

Scrape: a shallow depression excavated in the ground in which to nest.

Secondaries: or "secondary feathers", any one of the flight feathers borne on the forearm (ulna), as contrasted with the "primaries" borne on the manus. The secondaries are customarily numbered inwards from the carpal joint.

Secondary cavity nester: a species which secondarily chooses cavities as nest sites.

Snag: an upright, or nearly upright, dead tree in rotted condition.

Specialist: a species which specializes in the foraging and consumption of a single food source.

Species: a group of birds that interbreed and produce fertile young.

Speculum: a distinctly colored area on the wing of a bird.

Subaqueous: functioning under water.

Subspecies: a distinct population of a species that is slightly different from other populations of the same species.

Sympatric: occurring in the same geographical area—contrasted with Allochronic.

Taiga: Boreal Forest.

Tail fanning: spreading out the feathers of the tail laterally.

Terrestrial: conducting activities on the ground.

Territorial fidelity: showing preference for the same nest territory each year.

Territory: that area which is defended by a bird or pair of birds for nesting or feeding.

Threatened species: any indigenous species of fauna or flora that is likely to become endangered in Canada if the factors affecting its vulnerability do not become reversed.

Torpor: temporary condition of respiratory slowdown enabling an individual to adapt to an environmental stress.

Transient: a species that appears in an area on migration but neither breeds or overwinters there.

Vagrant: a wanderer outside the normal migration range of the species.

Vulnerable species: any indigenous species of fauna or flora that is particularly at risk because of low or declining numbers, occurrence at the fringe of its range or in restricted areas, or for some other reason, but is not a threatened species.

Zygodactylous: having two toes directly forward and two back.

FUTURE RECORDS

The Atlas of Breeding Birds of Alberta and the contributing database are largely the result of the efforts of over a thousand volunteers from across the province. Because of the limited availability of field and office resources, volunteer effort was critical in the successful completion of the Atlas Project. Volunteers were required for both field and office duties, to record, check, and compile data, to fundraise, to complete office work and to carry out public relations. In many cases, atlassers had to endure pouring rain, swarms of mosquitoes, cold temperatures, or scorching sun to complete their assigned squares. The publication which grew out of this commitment, the Atlas, is an accurate, yet generalized account intended to appeal to a broad spectrum of users. This ranges from the complete novice wanting to learn more about birding and ornithology to the professional biologist who desires a single, comprehensive source on local data.

However, as appealing as the Atlas may be, perhaps the most important result of the Atlas Project was the creation of the database. It is intended that this inventory be available for use by resource managers, scientists, researchers, conservation groups, planners, naturalists, and the general public.

The continuing legacy of the Atlas Project, the database will remain open to include future records and to answer further requests. The Federation of Alberta Naturalists (FAN) has been entrusted by the participating agencies to manage this database. In response to this, FAN has established the Database Management Committee to oversee the proper storage, update, and retrieval of records. Future records are to be submitted to them, at the address below:

Federation of Alberta Naturalists
P.O. Box 1472
Edmonton, Alberta
T5J 2N5

Safeguards will be in place to ensure not only that the correct data is entered, but also that the data is entered correctly. This increased record handling will require further funding. Therefore, a system of charges will be established to follow up on future data requests.

Beyond the term of the Atlas Project, records will be accepted in a manner similar to that used during the Project. Field Data Cards will be available from the above address. Once again, the level of breeding evidence should be recorded, as well as your name, address, record date, and area. Although abundance data was not faithfully recorded during the Atlas Project, this kind of information should be included in future records, because it may be important to some future study. Even the more common species, such as the robin or starling, require reporting, to determine population trends and changes in distribution. As occurred during the Atlas Project, the database will only be as accurate as the time and care taken in submitting records. Because of insufficient resources, a continued emphasis is placed on the volunteer, a resource that is increasingly counted upon to assist in scientific and conservation endeavors. Records need not be restricted to specific surveying activities, but can also be compiled while carrying out other activities, such as camping, hiking, travelling through an area, or any outdoor activity. Special emphasis should be placed on remote areas where possible. As always, survey records from professional groups and agencies are equally accepted.

In order to effectively use the existing database, data must be taken over a protracted period, thus enabling comparisons to be made and trends to be determined. *The Atlas of Breeding Birds of Alberta* is a first step in that direction, but continued volunteer support will be required to determine the long term effectiveness of such efforts. Equipped with new or honed skills in surveying, individuals have the skills and the opportunity to participate in other worthy surveying methods and projects, all contributing to establishing better conservation strategies. Interested parties may contact FAN, their local natural history group, or one of the participating agencies listed on page two to register in these other activities.

It is anticipated that the concerted Atlas effort shown during this Prohect will be repeated in the next 10 to 15 years.

Personal Communications

The following people contributed to the species accounts, in their capacity as Technical Advisors, by offering personal communications with Atlas staff. Their personal experience with these specific species helped to make the accounts more interesting and accurate. We are indebted to them for their assistance.

Harry Armbruster - Osprey and Falcons
Wildlife Technician, Canadian Wildlife Service - Edmonton, Alberta

Steve Brechtel - Pelicans
Non-game Status Biologist, Fish and Wildlife Division - Edmonton, Alberta

D.L. (Lynne) Dickson - Loons
Population aand Assessment Biologist, Canadian Wildlife Service - Edmonton, Alberta

H.L. (Loney) Dickson - Shorebirds
Non-game Program Leader, Canadian Wildlife Service - Edmonton, Alberta

Gary Erickson - Wild Turkey
Regional Head of Wildlife Management, Fish and Wildlife Division - Lethbridge, Alberta

Cam Finlay - Swallows and Thrushes
Nature Columnist, The Edmonton Journal - Edmonton, Alberta

Richard Fyfe - Falcons
Consulting Biologist - Fort Saskatchewan

Geoff Holroyd - Falcons
Research Scientist, Canadian Wildlife Service - Edmonton, Alberta

Bruce McGillivray - Woodpeckers, Kinglets, and Vireos
Assistant Director Natural History and Collections Administration, Provincial Museum of Alberta - Edmonton, Alberta

Dave Moore - Herons
Wildlife Biologist, Fish and Wildlife Division - St. Paul, Alberta

David Prescott - Finches
Professor, Department of Biological Sciences, University of Calgary - Calgary, Alberta

Blair Rippin - Grebes
Regional Head of Wildlife Management, Fish and Wildlife Division - St. Paul, Alberta

Al Smith - Warblers
Wildlife Technician, Canadian Wildlife Service - Saskatoon, Saskatchewan

Phil Stepney - Blackbirds
Director, Provincial Museum of Alberta - Edmonton, Alberta

C. Stoneman - Turkey Vulture
Naturalist

Bruce Turner - Ducks
Wildlife Biologist, Canadian Wildlife Service - Edmonton, Alberta

The individuals listed above may have acted as Technical Advisors on accounts other than the ones listed, as did several other persons who are listed in the Acknowledgments section at the beginning of the book. The listings on this page refer to those comments that added specific personal experiences that would otherwise not be found in general literature.

BIBLIOGRAPHY

The bibliography section is divided into three main categories - Historical References, General References, and References by Family. Literature used in writing the History of Ornithology is included in the Historical References section. Literature used throughout the text occurs in the General References section. Literature used specifically for a particular family will be found in the References by Family section. The latin family names are denoted in bold face, and the common family name is in all capitals. Not all references in this bibliography were necessarily cited in the text.

Historical References

Ainley, M. 1987. William Rowan: Canada's first avian biologist. Picoides 1:6-8.

Bartram, W. 1791. Travels through North and South Carolina.

Bent, A.C. 1907-1908. Summer birds of southwestern Saskatchewan. Auk 24:407-430; 25:25-35.

Blakiston, T. 1861-1862. On birds collected and observed in the interior of British North America. Ibis 3:314-320; 4:3-10, 5:39-87 and 121-155.

Bonaparte, C.L. 1825-1833. American Ornithology. 4 volumes.

Brisson, M.-J. 1760. Ornithologie.

de Buffon, G.-L. 1770-1783. Histoire naturelle des oiseaux.

Burpee, L.J. 1907. The journal of Anthony Henday, 1754-55. Trans. Roy. Soc. Can. Ser. 3, Sect. 11:307-364.

Chamberlain, M. 1887. A catalogue of Canadian Birds. J. and A. McMillan, St. John, N.B.

Chapman, F.M. 1908. Camps and Cruises of an Ornithologist. D. Appleton and Co., New York.

Clarke, C.H.D., and I. McTaggart-Cowan. 1945. Birds of Banff National Park, Alberta. Can. Field Nat. 59:83-103.

Coues, E. 1878. Field notes on birds observed in Dakota and Montana along the 49th parallel during the seasons of 1873 and 1874. Bull. U.S. Geol. and Geog. Survey Terr. 4:545-661.

Coues, E. 1903. Key to North American Birds. Fifth ed. Two Volumes, Dana Estes and Co., Boston, Mass.

Dawson, G.M. 1875. Report on the geology and resources of the region in the vicinity of the 49th parallel, from the Lake of the Woods to the Rocky Mountains. Dawson Brothers, Montreal.

Ealey, D.M., and M.K. McNicholl. 1991. A bibliography of Alberta ornithology (2nd ed.). Natural History Occasional Paper No. 16. Prov. Museum of Alberta, Edmonton.

Farley, F.L. 1922. Summer birds of the Lac La Biche and Fort McMurray region. Can. Field-Nat. 36:72-75.

Farley, F.L. 1925. Changes in the status of certain animals and birds during the past fifty years in central Alberta. Can. Field-Nat. 39:200-202.

Farley, F.L. 1932. Birds of the Battle River Region. Inst. of Applied Art Ltd., Edmonton 1-83.

Fearnhough, N.J. 1940. Is the Long-billed Curlew doomed? Oologist 57:93-94.

Glover, R. 1958. Introduction **In:** Hearne, S. 1958. A journey from Prince of Wales's Fort in Hudson Bay to the Northern Ocean. MacMillan Co. [of Canada Ltd.] Toronto, [Ont.]

Godfrey, W.E. 1952. Birds of the Lesser Slave Lake-Peace River areas, Alberta. Nat'l.Mus. Can. Bull. 126:142-175.

Grant, G.M. 1873. Ocean to Ocean: Sandford Fleming's Expedition through Canada in 1872. James Campbell and Son, Toronto.

Harper, F. 1915. The Athabasca-Great Slave lake expedition, 1914. Geol. Surv. Can. Dept. Mines, Summary rept. 1914. Vol. 50, No. 22, Sessional paper No. 26:159-163.

Hearne, S. 1958. A journey from Prince of Wales's Fort in Hudson Bay to the Northern Ocean. MacMillan Co. [of Canada Ltd.] Toronto, [Ont.]

Henderson, A.D. 1923(a). Cycles of abundance and scarcity in certain mammals and birds. J. Mammal. 4:264-265.

Henderson, A.D. 1923(b). The return of the Magpie. Oologist 40:142.

Henderson, A.D. 1926. Bonaparte's Gull nesting in Northern Alberta. Auk 43:288-294.

Henderson, A.D. 1941. The breeding waders of the Belvedere district, Alberta, Canada. Oologist 58:14-19.

Höhn, E. O. 1962. The names of economically important or conspicuous mammals and birds in the Indian languages of the District of Mackenzie, N.W.T. and in Sarcee. Arctic 15:299-316.

Höhn, E.O. 1973. Mammal and bird names in the Indian languages of the Lake Athabasca area. Arctic 26:163-171.

Höhn, E.O. 1981. The history of the Edmonton Bird Club. Alberta Naturalist Special Issue No. 1. 48-51.

Hopwood, V.G. (ed.). 1971. David Thompson: Travels in western North America. MacMillan Co. [of Canada Ltd.] Toronto, [Ont.]

Horsbrugh, C.B. 1915. Ornithological notes from the Alix and Buffalo Lake District, Province of Alberta, Canada. Ibis 10:670-689.

Houston, C.S., and M.J. Bechard. 1982. Oology on the Northern Great Plains: an historical review. Blue Jay 40:154-157.

Houston, C.S., and M.J. Bechard. 1990. A.D. Henderson, Alberta's foremost oologist, 1878 - 1963. Blue Jay 48:85-96.

Houston, C.S., and M.I. Houston. 1987. Samuel Hearne, naturalist. Beaver 67:23-27.

Houston, C.S., and M.I. Houston. 1990. Andrew Graham, Fur Trader and Naturalist. Picoides 4(2):4-6.

Houston, C.S., and M.I. Houston. 1991. Thomas Hutchins 1742(?)-1790. Picoides 5(1):9-11.

Lindsey, D. (ed.). 1991. The modern beginnings of subarctic ornithology. Manitoba Record Society, Winnipeg, Man.

Lister, R. 1979. The birds and birders of Beaverhill Lake. Edmonton Bird Club, Edmonton.

Mackenzie, A. 1802. Voyages from Montreal on the river St. Laurence through the continent of North Anerica to the frozen and Pacific oceans in the years 1789 and 1793. T. Caddell, Jr. and W.B. Strand. London, England.

Macoun, J. 1883. Manitoba and the Great North-West. Thomas C. Jack, London, England.

Macoun, J. 1900. Catalogue of Canadian Birds. Part I. (Water Birds, Gallinaceous Brids, and Pigeons). Ottawa: S.E. Dawson, Queen's Printer. 218 pp.

Macoun, J. 1903. Catalogue of Canadian Birds. Part II. (Birds of Prey, Woodpeckers, Fly-catchers, Crows, Jays and Blackbirds). Ottawa: S.E. Dawson, Queen's Printer. 199 pp.

Macoun, J. 1904. Catalogue of Canadian Birds. Part III. (Sparrows, Swallows, Vireos, Warblers, Wrens, Titmice and Thrushes). Ottawa: S.E. Dawson, Queen's Printer. 345 pp.

Macoun, J. 1922. Autobiography of John Macoun: Canadian explorer and naturalist, 1831-1920. Ottawa Field-Naturalists' Club, Ottawa.

Macoun, J., and J.M. Macoun. 1909. Catalogue of Canadian Birds. Government Printing Bureau. Ottawa.

McGillivray, W.B., and R.I. Hastings. (eds.). 1988. Natural History of the Bistcho Lake Region, Northwest Alberta. Natural History Occasional Paper No. 10. Prov. Mus. Alberta, Edmonton.

McGillivray, W.B., and R.I. Hastings. (eds.). 1990. Natural History of the Andrew Lake Region, Northeastern Alberta. Natural History Occasional Paper No. 12. Prov. Mus. Alberta, Edmonton.

McGillivray, W.B., and R.I. Hastings. (eds.). 1992. Natural History of the Winnifred Lake Region, eastern Alberta. Natural History Occasional Paper No. 18. Prov. Mus. Alberta, Edmonton.

McNicholl, M.K., P.H.R. Stepney, P.C. Boxall and David A.E.Spalding. 1981. A bibliography of Alberta Ornithology. Nat. Hist. Occ. Pap. No. 3. PMA, Edmonton.

Milton, V., and W.B. Cheadle. 1865. The North-West Passage by Land. Cassell, Petter and Galpin, London, England.

Myres, M.T., and D.J. Stiles. 1981. History of the Calgary Field-Naturalists Society. Alberta Naturalist Spec. Publ. No.1. 40-48.

Pennant, T. 1784. Arctic Zoology Vol II, Class 2, Birds. H. Hughs, London, England.

Petitot, E. 1884. On the Athabasca district of the Canadian North-West Territory. Canadian Record of Science 1:27-54.

Pielou, E.C. 1991. After the Ice-age: The return of life to Glaciated North America. University of Chicago Press, Chicago, Ill.

Preble, E.C. 1908. A biological investigation of the Athabaska-Mackenzie Region. North American Fauna No. 27. U.S. Dept. of Agriculture, Washington.

Raine, W. 1892. Bird-nesting in north-west Canada. Hunter, Rose and Co., Toronto.

Raine, W. 1904. Discovery of the eggs of the Solitary Sandpiper. Can. Field-Nat. 18:135-138.

Rand, A.L. 1948. Birds of southern Alberta. Nat'l. Mus. Can. Bull. 111:1-105.

Randall, T.E. 1933. A list of the breeding birds of the Athabasca District, Alberta. Can. Field-Nat. 47:1-6.

Riley, J.H. 1913. Birds observed or collected on the expedition of the Alpine Club of Canada to Jasper Park, Yellowhead Pass, and Mount Robson region. Can. Alpine Journal Special Issue pp. 47-75.

Ross, B.R. 1861. List of species of mammals and birds collected in Mackenzie's river district during 1860-1861. Canadian Naturalist and Geologist. 6:441-444.

Ross, B.R. 1862. List of mammals, birds, and eggs, observed in the Mackenzie's river district, with notices. Canadian Naturalist and Geologist 7:137-155.

Salt, W.R. 1948. Certain aspects of the male reproductive system of some passerine birds. M.Sc. Thesis, University of Alberta, Edmonton. 48pp.

Salt, W.R., and A.L. Wilk. 1958. The Birds of Alberta. Queen's Printer, Edmonton.

Selwyn, A.R.C. 1874. Notes on a journey through the North-West Territory from Manitoba to Rocky Mountain House. Canadian Naturalist and Quarterly Journal of Science 7:193-215.

Soper, J.D. 1940. Preliminary faunal report on Elk Island National Park, Alberta, Canada. Canadian Wildlife Service, Unpubl. report.

Soper, J.D. 1942. The birds of Wood Buffalo Park and vicinity, Northern Alberta and District of Mackenzie, N.W.T. Canada. Trans. Roy. Can. Inst. 24 no. 51:19-97.

Soper, J.D. 1947. Observations on mammals and birds in the Rocky Mountains of Alberta. Can. Field-Nat. 61:143-173.

Soper. J.D. 1949(a). Birds observed in the Grande Prairie-Peace River Region of northwestern Alberta, Canada. Auk 66:233-257.

Soper, J.D. 1949(b). Notes on the fauna of the former Nemiskam National Park and vicinity, Alberta. Can. Field-Nat. 63(5):167-182.

Spalding, D.A.E. 1981. Naturalists in Alberta: A brief history. Alberta Naturalist Spec. Issue No. 1:1-15.

Spalding D.A.E. 1988. The early history of dinosaur discovery in Alberta and Canada. Alberta 1:17-26.

Spry, I. 1963. The Palliser Expedition, The MacMillan Co. of Canada Ltd., Toronto, Ont.

Swainson, W., and J. Richardson. 1832. Fauna Boreali-Americana; or the Zoology of the northern parts of British America. Part Second, The Birds. John Murray, London, England.

Taverner, P.A. 1919. The birds of the Red Deer River, Alberta. Auk 36:1-21, 248-265.

General References

American Ornithologists' Union. 1973. Thirty-second supplement to the American Ornithologists' Union checklist of North American birds. Auk 90:411-419.

American Ornithologists' Union. 1983. Checklist of North American birds, 6th ed. A. O. U., Washington

American Ornithologists' Union. 1985. Thirty-fifth supplement to the American Ornithologists' Union checklist of North American birds. Auk 102:680-686.

American Ornithologists' Union. 1987. Thirty-sixth supplement to the American Ornithologists' Union checklist of North American birds. Auk 104:591-596.

American Ornithologists' Union. 1989. Thirty-seventh supplement to the American Ornithologists' Union checklist of North American birds. Auk 106:532-538.

American Ornithologists' Union. 1991. Thirty-eighth supplement to the American Ornithologists' Union checklist of North American birds. Auk 108:750-754.

Anonymous. 1984. The birds of the Calgary area, 1982. Calgary Field Naturalists Society.

Anonymous. 1986. Checklist of Canadian birds. The Royal Ontario Museum, Toronto.

Anonymous. 1977. Birds protected in Canada under the Migratory Birds Convention Act. Canadian Wildlife Service, Occasional Paper. 18 pp.

Arbib, R. 1977. The blue list for 1978. Amer. Birds 31(6):1087-1096.

Arbib, R. 1978. The blue list for 1979. Amer. Birds 32(6):1106-1113.

Arbib, R. 1980. The blue list for 1980. Amer. Birds 33(6):830-835.

Audubon, J.J. 1840. The birds of America. J.B. Chevalier, Philadelphia.

Baker, B.W., and L.H. Walkinshaw. 1946. Bird notes from Fawcett, Alberta. Can. Field-Nat. 60(1):5-10.

Bendire, C., and B. Major. 1895. Life histories of North American birds, from the parrots to the grackles, with special reference to their breeding habits and eggs. Smithsonian Contributions to knowledge No. 985. Smithsonian Institute Washington, D.C.

Bent, A.C. 1961(a). Life histories of North American birds of prey. Part I. Reprinted from Smithsonian Institution United States National Museum Bulletin 167, 1937. Dover Publications, Inc., New York.

Bent, A.C. 1961(b). Life histories of North American birds of prey. Part II. Reprinted from Smithsonian Institution United States National Museum Bulletin 170, 1938. Dover Publications, Inc., New York.

Bent, A.C. 1962(a). Life histories of North American wildfowl. Part I. Reprinted from Smithsonian Institution United States National Museum Bulletin. Bulletin 126, 1923. Dover Publications, Inc., New York.

Bent, A.C. 1962(b). Life histories of North American wildfowl. Part II. Reprinted from Smithsonian Institution United States National Museum Bulletin 130, 1925. Dover Publications, Inc., New York.

Bent, A.C. 1962(c). Life histories of North American shorebirds. Part I. Reprinted from Smithsonian Institution United States National Museum Bulletin 142, 1927. Dover Publications, Inc., New York.

Bent, A.C. 1962(d). Life histories of North American shorebirds. Part II. Reprinted from Smithsonian Institution United States National Museum Bulletin 146, 1929. Dover Publications, Inc., New York.

Bent, A.C. 1962(e). Life histories of North American gallinaceous birds. Reprinted from Smithsonian Institution United States National Museum Bulletin 162, 1932. Dover Publications, Inc., New York.

Bent, A.C. 1963(a). Life histories of North American diving birds. Reprinted from Smithsonian Institution United States National Museum Bulletin 107, 1919. Dover Publications, Inc., New York.

Bent, A.C. 1963(b). Life histories of North American gulls and terns. Reprinted from Smithsonian Institution United States National Museum Bulletin 113, 1921. Dover Publications, Inc., New York.

Bent, A.C. 1963(c). Life histories of North American marsh birds. Reprinted from Smithsonian Institution United States National Museum Bulletin 135, 1926. Dover Publications, Inc., New York.

Bent, A.C. 1963(d). Life histories of North American wood warblers. Part I and Part II. Reprinted from Smithsonian Institution United States National Museum Bulletin 203, 1953. Dover Publications, Inc., New York.

Bent, A.C. 1964(a). Life histories of North American petrels and pelicans and their allies. Reprinted from Smithsonian Institution United States National Museum Bulletin 121, 1922. Dover Publications, Inc., New York.

Bent, A.C. 1964(b). Life histories of North American woodpeckers. Reprinted from Smithsonian Institution United States National Museum Bulletin 174, 1939. Dover Publications, Inc., New York.

Bent, A.C. 1964(c). Life histories of North American cuckoos, goatsuckers, hummingbirds and their allies. Reprinted from Smithsonian Institution United States National Museum Bulletin 176, 1940. Dover Publications, Inc., New York.

Bent, A.C. 1964(d). Life histories of North American jays, crows and titmice. Part I and II. Reprinted from Smithsonian Institution United States National Museum Bulletin 191, 1946. Dover Publications, Inc., New York.

Bent, A.C. 1964(e). Life histories of North American nuthatches, wrens, thrashers and their allies. Reprinted from Smithsonian Institution United States National Museum Bulletin 195, 1948. Dover Publications, Inc., New York.

Bent, A.C. 1964(f). Life histories of North American thrushes, kinglets and their allies. Reprinted from Smithsonian Institution United States National Museum Bulletin 196, 1949. Dover Publications, Inc., New York.

Bent, A.C. 1965(a). Life histories of North American flycatchers, larks, swallows and their allies. Reprinted from Smithsonian Institution United States National Museum Bulletin 179, 1942. Dover Publications, Inc., New York.

Bent, A.C. 1965(b). Life histories of North American wagtails, shrikes, vireos and their allies. Reprinted from Smithsonian Institution United States National Museum Bulletin 197, 1950. Dover Publications, Inc., New York.

Bent, A.C. 1965(c). Life histories of North American blackbirds, orioles, tanagers and their allies. Reprinted from Smithsonian Institution United States National Museum Bulletin 211, 1958. Dover Publications, Inc., New York.

Bent, A.C. 1968(a). Life histories of North American cardinals, grosbeaks, buntings, towhees, finches, sparrows and their allies. Part I, II and III. Reprinted from Smithsonian Institution United States National Museum Bulletin 237, 1968. Dover Publications, Inc., New York.

Blokpoel, H., and F.L. Waite. 1970. Birds observed in the area around Cold Lake, Alberta. Unpublished manuscript on file, Provincial Museum of Alberta, Edmonton, Alberta.

Bovey, R. 1988. Birds of Edmonton. Lone Pine Publishing, Edmonton, Alberta.

Brewer, R., G.A. McPeek and J.R. Adams Jr. (eds.) 1991. The atlas of breeding birds of Michigan. Michigan State University Press, East Lansing, Michigan.

Brooke, M., and T. Birkhead. 1991. The Cambridge Encyclopedia ofOrnithology. Cambridge University Press, Cambridge, England. 362 pp.

Burleigh, T.D. 1972. Birds of Idaho. The Caxton Printers, Ltd., Caldwell, Idaho.

Cadman, M.D., P.F.J. Eagles and F.M. Helleiner. 1987. Atlas of the breeding birds of Ontario. University of Waterloo Press, Waterloo, Ontario.

Campbell, B., and E. Lack (eds.). 1985. A Dictonary of Birds. British Ornithologists' Union, Buteo Books, South Dakota. 670 pp.

Campbell, R.W., N.K. Dawe, I. McTaggart-Cowan, J.M. Cooper, G.W. Kaiser and M.C.E. McNall. 1990. The birds of British Columbia, Vol. 1. Non-passerines. Introduction, loons through waterfowl. Royal British Columbia Museum in association with Canadian Wildlife Service, Environment Canada, Victoria B.C.

Campbell, R.W., N.K. Dawe, I. McTaggart-Cowan, J.M. Cooper, G.W. Kaiser and M.C.E. McNall. 1990. The birds of British Columbia, Vol. 2. Non-passerines. Diurnal birds of prey through woodpeckers. Royal British Columbia Museum in association with the Canadian Wildlife Service, Environment Canada, Victoria B.C.

Canadian Wildlife Service. 1980. Birds protected in Canada under the Migratory Birds Convention Act. Occasional Paper No. 1, 1980 ed. Environment Canada.

Clarke, C.H.D., and I. McTaggart-Cowan. 1945. Birds of Banff National Park, Alberta. Can. Field-Nat. 59(3):83-103.

Collias, N.E. 1984. Nest building and bird behavior. Princeton University Press, Princeton, N.J.

Cook, F.R., and D. Muir. 1984. The Committee on the Status of Endangered Wildlife in Canada (COSEWIC): history and progress. Can. Field-Nat. 98(1):63-70.

Cottonwood Consultants Ltd. 1983. A biophysical system overview for ecological reserves planning in Alberta. Vol. 1. Regional overviews. Prepared for Alberta Recreation and Parks, Edmonton, Alberta.

Dekker, D. 1991. Prairie water. Watchable wildlife at Beaverhill Lake, Alberta. BST Publications, Edmonton, Alberta. 144 pp.

Dunn, E.H. 1991. Population trends in Canadian songbirds. Bird Trends 1 (summer):2-11.

Ebel, G.R.A., and A. Johns. 1977. Edmonton birds: an annual report for 1975. Manuscript on file, Provincial Museum of Alberta, Edmonton, Alberta.

Ehrlich, P.R., D.S. Dobkin and D. Wheye. 1988. The birders handbook: a field guide to the natural history of North American birds. Simon and Schuster, Inc., N.Y.

Erskine, A.J. 1977. Birds in boreal Canada: communities, densities and adaptations. Canadian Wildlife Service Report Series No. 41.

Erskine, A.J. 1978. The first ten years of the co-operative breeding bird survey in Canada. Canadian Wildlife Service Report Series No. 42.

Farley, F.L. 1932. Birds of the Battle River Region. Institute of Applied Art Ltd., Edmonton, Alberta.

Farrand, J. Jr. (ed.) 1983. The Audubon Society master guide to birding, Vol. 2. Random House Ltd., Toronto.

Fish and Wildlife Division. 1984. Status of the fish and wildlife resource in Alberta. Alberta Energy and Natural Resources, Edmonton, Alberta.

Flack, J.A.D. 1976. Bird populations of aspen forests in Western North America. Ornithol. Monogr. No. 19, 1976.

Francis, J., and K. Lumbis. 1979. Habitat relationships and management of terrestrial birds in northeastern Alberta. Prepared for AOSERP by Canadian Wildlife Service, Edmonton, Alberta. AOSERP Report 78. 365 pp.

Geissler, P.H., and B.R. Noan. 1981. Estimates of avian population trends from the North American breeding bird survey. Studies in Avian Biology 6:42-51.

Gentry, T.G. 1882. Nests and eggs of birds of the United States. J.A. Wagenseller, Philadelphia.

Gilroy, D. 1967. An album of prairie birds. Modern Press, Prairie Books Service, Saskatoon, Saskatchewan.

Gilroy, D. 1976. Prairie birds in color. Western Producer Prairie Books, Saskatoon, Saskatchewan.

Godfrey, W.E. 1950. Birds of the Cypress Hills and Flotten Lake Regions, Saskatchewan. Natl. Museum Canada, Bull. No. 120, Biological Series No. 40, Ottawa, Ontario.

Godfrey, W.E. 1986. The birds of Canada. Rev. ed. National Museum of Natural Sciences, National Museums of Canada. 595 pp.

Gollop, B. 1989. The nesting season - June 1 - July 31, 1989: Prairie Provinces Region. Amer. Birds 43(5):1330-1332.

Goodfellow, P. 1977. Birds as builders. Arco Publishing Co., Inc., New York.

Goossen, J.P. 1987. A contribution to the Alberta Bird Atlas: bird counts from northern Alberta (1986). Unpublished Canadian Wildlife Service report, Edmonton, Alberta.

Hancock, D.A., and J. Woodford. 1973. Birds of Alberta, Saskatchewan and Manitoba. General Publishing Co., Don Mills, Ontario.

Hardy, W.G. 1975. Alberta. A natural history. M.G. Hurtig, Publishers. (4th ed), Edmonton, Alberta

Harris, W.C. 1985. The nesting season. Prairie Provinces Region. Amer. Birds 39:927-928.

Harrison, C. 1978. A field guide to nests, eggs and nestlings of North American birds. Collins, New York.

Harrison, H.H. 1975. A field guide to birds' nests. Houghton Mifflin Co., Boston.

Hayman, P., J. Marchant and T. Prater. 1986. Shorebirds. An identification guide. Houghton Mifflin Co., Boston.

Headstrom, R. 1970. A complete field guide to nests in the United States. Iwes Washburn, New York.

Hill, N.P., and J.M. Hagan III. 1991. Population trends of some northeastern North American landbirds: a half century of data. Wilson Bull. 103(2):165-182.

Höhn, E.O. 1972. The birds of the Lake Athabasca area, Mackenzie District, NWT, Alberta and Saskatchewan. Privately Printed.

Höhn, E.O. 1973. The birds of the Peace-Athabasca Delta and of the Lake Athabasca region. Canadian Wildlife Service. 32 pp.

Höhn, E.O. 1976. Additional bird and mammal observations in the Caribou Mountains, Alberta. Blue Jay 34:176-177.

Höhn, E.O. 1976. Addendum - Caribou Mountains. Blue Jay 34:194.

Holroyd, G.L., and H. Coneybeare. 1990. The compact guide to birds of the Rockies. Lone Pine Publishing, Edmonton, Alberta.

Holroyd, G.L., G. Burns and H.C. Smith. (eds.) 1991. Proceedings of the Second Endangered Species and Prairie Conservation Workshop. Provincial Museum of Alberta, Natural History Occasional Paper No. 15. Edmonton, Alberta.

Holroyd, G.L., and K.J. Van Tighem. 1983. The Ecological (Biophysical) Land Classification of Banff and Jasper National Parks, Vol. 3, The Wildlife Inventory. Canadian Wildlife Service, Edmonton, Alberta. 691 pp.

Johnsgard, P.A. 1979. Birds of the great plains: breeding species and their distribution. University of Nebraska Press. 539 pp.

Kondla, N.G., H.W. Pinel, C.A. Wallis and C.R. Wershler. 1973. Avifauna of the Drumheller area, Alberta. Can. Field-Nat. 87:377-393.

Koonz, W.H., and P.W. Rakowski. 1985. Status of colonial waterbirds nesting in southern Manitoba. Can. Field-Nat. 99(1):19-29.

Livingston, J.A. 1966. Birds of the northern forest. McClelland and Stewart Ltd., Toronto/Montreal.

Long, J.L. 1981. Introduced birds of the world. Universe Books, New York.

MacCallum, B., and G.R.A. Ebel. 1985. Ecological considerations for the wildlife on the Champion forest management area, Alberta. 275 pp.

Mackenzie, J.P.S. 1977. Birds in peril. McGraw-Hill Ryerson Ltd., Toronto, Montreal, Halifax, Vancouver.

Macoun, J.J., and J.J. Macoun. 1909. Catalogue of Canadian birds. Geological Survey of Canada, Ottawa. 761 pp.

McElroy, T.P. Jr. 1974. The habitat guide to birding. A guide to birding east of the Rockies. Nick Lyons Books, New York.

McNicholl, M.K., P.H.R. Stepney, P.C. Boxall and D.A.E. Spalding. 1981. A bibliography of Alberta ornithology. Provincial Museum of Alberta Natural History Occasional Paper No. 3, Edmonton, Alberta.

McTaggart-Cowan, I. 1955. Birds of Jasper National Park, Alberta, Canada. Wildlife Management Bulletin, Series 2, No. 8, Canadian Wildlife Service.

Munro, J.A. 1963. Birds of Canada's mountain parks. National ParksBranch. 75 pp.

Munson, B., D. Ealey, R. Beaver, K. Bishoff and R. Fyfe. 1980. Inventory of selected raptor, colonial and sensitive species in the Athabasca oil sands area of Alberta. Prep. for the Alberta Oil Sands Environmental Research Program (AOSERP) by CWS, Project LS22.3.2. 66 pp.

National Parks Service. 1971. Checklist of the birds of Banff National Park. Canada Dept. of Indian Affairs and Northern Development, National Parks Service, Banff, Alberta.

National Parks of Canada Branch, Dept. Interior, Canada. 1927. Official Canadian record of bird-banding returns. Can. Field-Nat. 41(7):157-171.

Nero, R.W. 1963. Birds of the Lake Athabasca region, Saskatchewan. Special publication No. 5, Saskatchewan Natural History Society, Regina. 143 pp.

Newton, I. 1979. Population ecology of raptors. Buteo Books, Vermillion, S. Dakota.

Owens, R.A., and M.T. Myres. 1973. Effects of agriculture upon populations of native passerine birds of an Alberta fescue grassland. Can. J. Zool. 51:697-713.

Paca, L.G. 1953. Introduction to western birds. d'Angelo Publishing Co., Carmel-by-the-Sea, California.

Palmer, R.S. (ed.). 1962. Handbook of North American birds. Vol. 1. Loons through flamingos. Yale University Press, New Haven.

Palmer, R.S. (ed.). 1976. Handbook of North American birds. Vol. 2. Waterfowl (first part), whistling ducks, swans, geese, sheld-ducks, dabbling ducks. Yale University Press, New Haven. 521 pp.

Palmer, R.S. (ed.). 1976. Handbook of North American birds. Vol. 3. Waterfowl (concluded), eiders, woodducks, diving ducks, mergansers, stifftails. Yale University Press, New Haven. 560 pp.

Palmer, R.S. (ed.). 1988. Handbook of North American birds. Vol. 4. Family Cathartidae, New World condors and vultures - Family Accipitridae (first part), Osprey, kites, Bald Eagle and allies, accipiters, Harrier, buteo allies. Yale University Press, New Haven. 433 pp.

Palmer, R.S. (ed.). 1988. Handbook of North American birds. Vol. 5. Family Accipitridae (concluded), buteos, Golden Eagle, Family Falconidae, Crested Caracara, falcons. Yale University Press, New Haven. 465 pp.

Park, J.L. 1988. The breeding bird survey in Alberta, 1968-87. Alta. Nat. 18(2):53-58.

Park, J.L. 1990. Breeding bird survey report, 1988-89. Alta. Nat. 20(3):89-92.

Peck, G.K., and R.D. James. 1983. Breeding birds of Ontario: nidiology and distribution, Vol. 1. Non-passerines. Life Sci. Misc. Publ., Royal Ontario Museum, Toronto.

Peck, G.K., and R.D. James. 1983. Breeding birds of Ontario: nidiology and distribution, Vol. 2. Passerines. Life Sci. Misc. Publ., Royal Ontario Museum, Toronto.

Peterson, R.T. 1961. Field guide to western birds. Houghton Mifflin Co.,Boston.

Pinel, H.W., and R.J. Butot. 1978. Highlights of bird observations in the Calgary area, Alberta 1971-1977. Blue Jay 36:159-163.

Pinel, H.W., W.W. Smith and C.R. Wershler. 1991. Alberta Birds, 1971-1980. Vol 1. Non-passerines. Provincial Museum of Alberta Natural History Occasional Paper No. 13, Edmonton, Alberta.

Pinel, H.W., W.W. Smith and C.R. Wershler. In Prep. Alberta Birds, 1971 - 1980. Vol. 2. Passerines. Provincial Museum of Alberta Natural History Occasional Paper. Edmonton, Alberta.

Poston, B., D.M. Ealey, P. Taylor and G.B. McKeating. 1990. Priority migratory bird habitats of Canada's prairie provinces. Habitat Conservation Section, Canadian Wildlife Service, Environment Canada, Edmonton, Alberta.

Powell, J.M., T.S. Sadler and M. Powell. 1975. Birds of the Kananaskis Forest Experimental Station and surrounding area: an annotated checklist. Information Report NOR-X-133 Northern Forest Research Centre, Edmonton, Alberta.

Quinlan, R.W., W.A. Hunt, K. Wilson and J. Kerr. 1990. Habitat requirements of selected wildlife species in the Weldwood Forest Management Agreement Area. A final report submitted to the Weldwood Forest Management Agreement Area Intergrated Resource Management Steering Committee.

Rand, A.L. 1948. Birds of southern Alberta. National Museum of Canada Bulletin No. 111, Biological Series No. 37, Ottawa, Ontario.

Reed, C.A. 1965. North American birds' eggs. Dover Publications, Inc. New York (Revised ed).

Reily, E.M. Jr. 1968. The Audubon illustrated handbook of American birds. McGraw Hill Book Co., Toronto.

Robbins, C.S., D. Bystrak and P.H. Geissler. 1986. The breeding bird survey: its first fifteen years, 1965-1979. U.S. Fish and Wildlife Service Resource Publication 157, Washington.

Robbins, C.S., J.R. Sauer, R.S. Greenberg and S. Droege. 1989. Population declines in North American birds that migrate to the neotropics. Proceedings of the National Academy of Science 86:7658-7662.

Rogers, T.H. 1974. The nesting season, June 1 - July 31, 1974. Northern Rocky Mountain - Intermountain region. Amer. Birds 28(5):925-929.

Rogers, T.H. 1974. The fall migration, August 1 - November 30, 1973. Northern Rocky Mountian - Intermountain region. Amer. Birds 28(1):78-83.

Rogers, T.H. 1976. The fall migration, August 1 - November 30, 1975. Northern Rocky Mountain - Intermountain Region. Amer. Birds 30(1):97-101.

Root, T. 1988. Atlas of wintering North American birds. An analysis of Christmas bird count data. University of Chicago Press, Chicago and London.

Rowan, W., and E.O. Höhn. 1950. A provincial list of the birds of Alberta. Occasional Paper No. 1. (rev. ed.) Edmonton Bird Club, Edmonton, Alberta.

Sadler, T.S., and M.T. Myres. 1976. Alberta birds, 1961-1970 with particular reference to migration. Provincial Museum of Alberta Natural History Occasional Paper No. 1., Edmonton, Alberta.

Salt, W.R. 1961. Recent additions to the avifauna of Alberta. Auk 78:427-428.

Salt, W.R., and A.L. Wilk. 1966. The birds of Alberta. Dept. of Industry and Development, Government of Alberta, Edmonton, Alberta, 2nd ed. Revised.

Salt, W.R., and J.R. Salt. 1976. The birds of Alberta with their ranges in Saskatchewan and Manitoba. Hurtig Publishers, Edmonton, Alberta.

Savage, C. 1985. The wonder of Canadian birds. Western Producer Prairie Books, Saskatoon, Saskatchewan.

Scott, S.L. 1983. Natural Geographic Society field guide to the birds of North America. Kingsport Press, Kingsport, Tennessee.

Scotter, G.W., T.J. Ulrich and E.T. Jones. 1990. Birds of the Canadian Rockies. Western Producer Prairie Books, Saskatoon, Saskatchewan. 170 pp.

Seel, K.E. 1969. An annotated list of the avifauna of Waterton Lakes National Park, Alberta. Manuscript on file, Provincial Museum of Alberta, Edmonton, Alberta.

Shortt, R. 1989. (Results of the 1987 May species count - birds). Alta. Nat. 19(2):52-54.

Shortt, R. 1989. (Results of the 1988 May species count - birds). Alta. Nat. 19(2):54-62.

Shortt, R. 1990. (Results of the 1989 May species count - birds). Alta. Nat. 20(3):114-120.

Shortt, T.M. 1977. Wild birds of Canada and the Americas. Paqurian Press Ltd., Toronto.

Silieff, E. 1980. The co-operative breeding bird survey in Canada, 1980. Canadian Wildlife Service, Environment Canada.

Skutch, A.F. 1976. Parent birds and their young. University of Texas Press, Austin and London.

Slater, A. 1985. Alberta Christmas bird counts, 1984. Alta. Nat. 15(3):89-94.

Slater, A. 1985. A compilation of Alberta Christmas bird counts from 1973 to 1984. Alta. Nat. 15(4):117-119.

Slater, A. 1986. Alberta Christmas bird counts, 1985. Alta. Nat. 16(4):126-134.

Slater, A. 1988. Alberta Christmas bird counts, 1987. Alta. Nat. 18(3):81-90.

Slater, A. 1989. Alberta Christmas bird counts, 1988. Alta. Nat. 19(4):143-156.

Slater, A. 1990. Alberta Christmas bird counts, 1989. Alta. Nat. 20(2):68-81.

Slater, A.C. 1991. Alberta Christmas bird counts, 1990. Alta. Nat. 21(2):58-71.

Smith, W.W., and C.A. Wallis. 1976. Preliminary investigation of the birds of Pakowki Lake, Alberta. Blue Jay 34:168-171.

Soper, J.D. 1942. The birds of Wood Buffalo Park and vicinity, northern Alberta and district of Mackenzie, N.W.T., Canada. Trans. Royal Canadian Institute No. 51, Vol. 24. Part 1:19-98.

Soper, J.D. 1947. Observations on mammals and birds in the Rocky Mountains of Alberta. Can. Field-Nat. 61(5):143-173.

Soper, J.D. 1949. Birds observed in Grande Prairie - Peace River region of northwestern Alberta, Canada. Auk 66:233-257.

Soper, J.D. 1960. An account of the mammals and birds collected and otherwise observed in Jasper National Park, Alberta, during the season of 1960. Canadian Wildlife Service (Unpubl.), Edmonton. 29 pp.

Spalding, D.A.E. (ed.). 1980. A nature guide to Alberta. Provincial Museum of Alberta Publication No. 5. Hurtig Publishers in conjunction with Alberta Culture.

Storms, R.W. 1986. The three newest splits that affect Alberta birders. Alta. Bird Record 4(2):46-47.

Tate, J. 1981. The blue list for 1981. Amer. Birds 35:3-10.

Tate, J. 1986. The blue list for 1986. Amer. Birds 40:227-236.

Tate, J., and D.J. Tate. 1982. The blue list for 1982. Amer. Birds 36:126-135.

Terborgh, J. 1989. Where have all the birds gone? Princeton University Press, Princeton, N.J.

Terres, J.K. 1980. The Audubon Society Encyclopedia of North American birds. Alfred A. Knopf, Inc., N.Y. 1109 pp.

Tuck, G., and H. Heinzel. 1978. A field guide to seabirds of Britain and the world. William Collins Sons and Co. Ltd., Glasgow. 292 pp.

Tull, C.E. 1975. Ground surveys of terrestrial breeding bird populations along the proposed gas pipeline, Alberta, Saskatchewan and British Columbia, May and June, 1974. Arctic Gas Biol. Report Ser. 31(1):1-116.

Wetmore, A. 1964. Song and garden birds of North America. National Geographic Society, Washington, D.C.

Wetmore, A. 1965. Water, prey and game birds of North America. National Geographic Society, Washington, D.C.

Wildlife Management Branch. 1991. The status of Alberta wildlife. Fish and Wildlife Division, Alberta Forestry, Lands and Wildlife. 49 pp.

Young, V.L. 1982. Alberta Christmas bird counts - 1981. Alta. Nat. 12:1-13.

Young, V.L. 1984. Alberta Christmas bird counts, 1983. Alta. Nat. 14(3):91-96.

References by Family

Order Gaviiformes

Family Gaviidae: LOONS

Alvo, R. 1987. The acid test. Living Bird 6(2):25-30.

Anderson, D.W., H.G. Lumsden and J.J. Hickey. 1970. Geographical variation in eggshells of Common Loons. Can. Field-Nat. 84(4):351-356.

Decker, D. [sic - Dekker]. 1980. Notes from Beaverhill: fall 1979. Edmonton Nat. 8(1):17-19.

Henderson, A.D. 1924. The Common Loon in Alberta. Condor 26(4):143-145.

Henderson, A.D. 1924. With canoe and camera on some Alberta Lakes. Oologist 41(2):14-18.

Höhn, E.O. 1972. Arctic Loon breeding in Alberta. Can. Field-Nat. 86(4):372

Höhn, E.O. 1976. Additional bird and mammal observations in the Caribou Mountains, Alberta. Blue Jay 34:176-177.

Höhn, E.O., and P. Marklevitz. 1974. Noteworthy summer observations of birds in the Caribou Mountains, Alberta. Can. Field-Nat. 88(1):77-78.

Johnsgard, P.A. 1987. Diving birds of North America. University of Nebraska Press, Lincoln.

Kennedy, A.J. 1981. Interspecific aggressive display by a Common Loon. Murrelet 62(1):20-21.

Kuyt, E., and J.P. Goossen. 1986. Arctic Loon nesting in Alberta. Alta. Nat. 16(2):61-64.

McNicholl, M.K. 1988. Common Loon distribution and conservation problems in Canada. pp. 196-214. **In:** Strong, P.I.V., ed., Papers from the 1987 conference on loon research and management. North American Loon Fund, Meredith, N.H.

Olson, S.T., and W.H. Marshall. 1952. The Common Loon in Minnesota. Minnesota Museum of Natural History Occasional Paper No. 5. 77 pp.

Petersen, M.R. 1989. Nesting biology of Pacific Loons, *Gavia pacifica*, on the Yukon-Kuskokwim delta, Alaska. Can. Field-Nat. 103(2):265-269.

Raffan, J. 1984. The mysterious life of the Common Loon. Nature Canada 13(3):28-33.

Spalding, D.A.E. 1974. Loons in Alberta. Alta. Nat. 4(2):37-40.

Sperry, M.L. 1987. Common Loon attacks on waterfowl. J. Field Ornith. 58(2):201-205.

Vermeer, K. 1973. Some aspects of the breeding and mortality of Common Loons in east-central Alberta. Can. Field-Nat. 87(4):403-408.

Vermeer, K. 1973. Some aspects of the nesting requirements of Common Loons in Alberta. Wilson Bull. 85:429-435.

White, M. 1978. The diver. Edmonton Nat. 6(1):18

Order Podicipediformes

Family Podicipedidae: GREBES

Cool, N. 1982. Red-necked Grebe survey on Astotin Lake. Unpublished report. Warden Service, Elk Island National Park. 11 pp + App.

De Smet, K.D. 1982. Status report on the Red-necked Grebe (*Podiceps grisegena*) in Canada. Committee on the Status of Endangered Wildlife in Canada, Ottawa.

Deusing, M. 1939. Nesting habits of the Pied-billed Grebe. Auk 56:367-373.

Fitzharris, T. 1980. The grebes of Canada. Beaver 311(1):36-42.

Fitzharris, T. 1987. Priceless antiques. Living Bird 6(1):12-19.

Forbes, L.S. 1984. The nesting ecology of the Western Grebe in British Columbia. Canadian Wildlife Service Report, Delta, B.C. 20 pp.

Goodwin, C.E. 1984. Getting to know the grebes. Nature Canada 13(2):24-29.

Hicklin, H.S. 1984. Housing problems. Pica 5(2):33.

Höhn, E.0. 1988. Antagonistic behavior of Horned Grebes. Blue Jay 46:92.

Johnsgard, P.A. 1987. Diving birds of North America. University of Nebraska Press, Lincoln.

Kevan, C.L. 1970. An ecological study of Red-necked Grebes on Astotin Lake, Alberta. M.Sc. Thesis, University of Alberta. 80 pp.

Kondla, N.G. 1974. Comments on Horsbrugh's Buffalo Lake grebe observations. Alta. Nat. 4(2):46-47.

Kristensen, J., and W.R. Nordstrom. 1979. Western Grebe colony, Cold Lake. Unpubl. report, prep. for Provincial Parks Div. Alberta Recreation, Parks and Wildlife and Esso Resources Canada Ltd.

Lynch, W. 1977. Coots disturb Eared Grebe nests. Blue Jay 35(3):173.

McAllister, N.M. 1958. Courtship, hostile behavior, nest establishment and egglaying in the Eared Grebe (*Podiceps caspicus*). Auk 75:290-311.

Munro, J.A. 1941. The grebes: studies of waterfowl in British Columbia. British Columbia Provincial Museum Occasional Paper No. 3, Victoria. 71 pp.

Nuechterlein, G.L. 1981. Courtship behavior and reproductive isolation between Western Grebe color morphs. Auk 98:335-349.

Nuechterlein, G.L., and R.W. Storer. 1989. Mate feeding by Western and Clark's Grebes. Condor 91:37-42.

Packham, R., and B. Rippin. 1980. Human disturbances on a Western Grebe colony on Cold Lake and the status of Western Grebes in part of N.E. Alberta. Fish and Wildlife Div., Alberta Energy and Natural Resources.

Ratti, J.T. 1979. Reproductive separation and isolating mechanisms between sympatric dark-phase and light-phase Western Grebes. Auk 96:573-586.

Riske, M.E. 1976. Environmental and human impact upon breeding grebes in central Alberta. Ph.D. Thesis, University of Calgary.

Storer, R.W. 1965. The color phases of the Western Grebe. Living Bird 4:59-63.

Storer, R.W. 1969. The behavior of the Horned Grebe in spring. Condor 71(2):180-205.

Storer, R.W., and G.L. Nuechterlein. 1985. An analysis of plumage and morphological characters of the two color forms of the Western Grebe (*Aechmophorus*). Auk 102(1):102-119.

Valadka, A. 1988. Eared Grebe, Horned Grebe and American Coot interactions. Blue Jay 46(4):199-200.

Wetmore, A. 1924. Food and economic relations of North American grebes. U.S.D.A. Dept. Bulletin No. 1196.

Order Pelecaniformes

Family Pelecanidae: PELICANS

Beaver, R.D. 1980. Breeding behavior of White Pelicans in the Birch Mountains, northeastern Alberta. M.Sc. Thesis, University of Alberta, Edmonton. 103 pp.

Brechtel, S.H. 1978. Legislative protection for the White Pelican and Double-crested Cormorant in Alberta. Alta. Nat. 8(1):78-80.

Brechtel, S.H. 1981. A status report, management proposal and selected bibliography for the White Pelican, Double-crested Cormorant and Great Blue Heron in Alberta - 1980. Fish and Wildlife Division, Alberta Energy and Natural Resources (Unpubl. report). 113 pp.

Brechtel, S.H. 1987. The White Pelican, pp.145-149. **In:** Holroyd, G.L., W.B. McGillivray, P.H.R. Stepney, D.M. Ealey, G.C. Trottier and K.E. Eberhart, eds. Proceedings of the Workshop on Endangered Species in the Prairie Provinces. Provincial Museum of Alberta, Natural History, Occasional Paper No. 9. 367 pp.

Ealey, D. 1979. The distribution, foraging behaviour, and allied activities of the White Pelican in the Athabasca Oil Sands area. Report prep. for the Alberta Oil Sands Environmental Research Program, prep. by Environment Canada, Canadian Wildlife Service. AOSERP Report 83. 70 pp.

Halkett, A. 1896. An Ottawa naturalists journey westward. Ottawa Naturalist 10(6):113-118.

Koonz, W.H. 1987. Update status report of the American White Pelican *Pelecanus erythrorhynchos*. Committee on the Status of Endangered Wildlife in Canada (COSEWIC), Ottawa i + 16 pp.

Markham, B.J. 1978. Status report on White Pelican *Pelecanus erythrorhynchos* in Canada. Status report, Committee on the Status of Endangered Wildlife in Canada (COSEWIC), Ottawa. 19 pp. + tables, figures.

Markham, B.J., and S.H. Brechtel. 1978. Status and management of three colonial waterbird species in Alberta. Proc. Colonial Waterbird Group Conf. 1978:55-64.

Nordstrom, W. 1978. Lake Newell: Colonial birds in jeopardy. Alta. Nat. 8:107-111.

Stepney, P.H.R. 1987. Cormorants and pelican surveys by the Provincial Museum. **In:** Beaverhill Bird Observatory 1985 Annual Report. Holroyd, G.L., ed. Edmonton Nat. 15(2):5-14.

Stepney, P.H.R. 1987. Management considerations for the American White Pelican in Alberta. Pp. 155-171. **In:** Holroyd, G.L., W.B. McGillivray, P.H.R. Stepney, D.M. Ealey, G.C. Trottier and K.E. Eberhart, eds. Proceedings of the Workshop on Endangered Species in the Prairie Provinces. Provincial Museum of Alberta, Natural History, Occasional Paper No. 9. 367 pp.

Vermeer, K. 1967. Colonies of Double-crested Cormorants and White Pelicans in Alberta. Unpubl. report, Canadian Wildlife Service, Edmonton, Alberta. 9 pp.

Vermeer, K. 1969. Colonies of Double-crested Cormorants and White Pelicans in Alberta. Can. Field-Nat. 83:36-39.

Vermeer, K. 1970. Distribution and size of colonies of White Pelicans *Pelecanus erythrorhynchos* in Canada. Can. J. Zool. 48(5):1029-1032.

Family Phalacrocoracidae: CORMORANTS

Brechtel, S.H. 1978. Legislative protection for the White Pelican and Double-crested Cormorant in Alberta. Alta. Nat. 8(1):78-80.

Brechtel, S.H. 1981. A status report, management proposal and selected bibliography for the White Pelican, Double-crested Cormorant and Great Blue Heron in Alberta - 1980. Unpubl. report, Fish and Wildlife Division, Alberta Energy and Natural Resources. 113 pp.

Brechtel, S.H. 1983. The reproductive ecology of Double-crested Cormorants in southern Alberta. M.Sc. Thesis, University of Alberta. 119 pp.

Lewin, V., and B. Van Scheik. 1983. The breeding ecology of the Double-crested Cormorant (*Phalacrocorax auritus*) in Alberta with special reference to habitat modification, thermal tolerances and egg parameters. Research Progress Report, prep. for Alberta Fish and Wildlife Division, prep. by University of Alberta. 31 pp.

Markham, B.J. 1979. Status report on Double-crested Cormorant *Phalacrocorax auritus* in Canada. Status report, Comm. on the Status of Endangered Wildlife in Canada (COSEWIC), Ottawa. 18 pp. + figures, tables.

Stepney, P.H.R. 1987. Cormorant and pelican surveys by the Provincial Museum. **In:** Beaverhill Bird Observatory 1985 Annual Report. Holroyd, G.L., ed. Edmonton Nat. 15(2):5-14.

Van Scheick, W.J. 1985. Thermal aspects of the reproductive ecology of the Double-crested Cormorant (*Phalacrocorax auritus*) in southern Alberta. Ph.D. Thesis, University of Alberta, Edmonton. 120 pp.

Van Scheick, W.J., and V. Lewin. 1986. Double-crested Cormorant, *Phalacrocorax auritus*, egg ejection. Can. Field-Nat. 100(4):561-562.

Vermeer, K. 1967. Colonies of Double-crested Cormorants and White Pelicans in Alberta. Unpubl. report, Canadian Wildlife Service, Edmonton. 9 pp.

Vermeer, K. 1969. Colonies of Double-crested Cormorants and White Pelicans in Alberta. Can. Field-Nat. 83:36-39.

Vermeer, K. 1970. Arrival and clutch initiations of Double-crested Cormorants at Lake Newell, Alberta. Blue Jay 28(3):124-125.

Vermeer, K. 1973. Great Blue Heron and Double-crested Cormorant colonies in the prairie provinces. Can. Field-Nat. 87:427-432.

Weseloh, D.V., D. Dekker, S. Brechtel and R. Burns. 1975. Notes on the Double-crested Cormorants, White Pelicans and Great Blue Herons of Beaverhill Lake, summer 1975. Alta. Nat. 5:132-137.

Weseloh, D.V., S. Brechtel, L. Bogaert, R. Burns and J. Keizer. 1977. A survey of the colonial nesting waterbirds of Beaverhill Lake, 1976. Edmonton Nat. 5(6):183-192.

Order Ciconiiformes

Family Ardeidae: HERONS AND BITTERNS

Allen, R.P., and F.P. Mangels. 1940. Studies of the nesting behavior of the Black-crowned Night Heron. Proc. of Linnean Society of New York 50-51:1-28.

Anderson, R. 1980. Great Blue Heron report, Edmonton region 1980. Unpubl. report, Fish and Wildlife Division, Alberta Energy and Natural Resources.

Armstrong, D., M. Glossop and S. Lamden. 1984. Fish Creek Provincial Park resource. Management plan. Unpubl. report. Alberta Recreation and Parks. 97 pp. + tables.

Autotte, C. 1988. Encounter with a pumper. Living Bird 7(4):28-31.

Barger, N.R. Sr. 1963. American Bittern *Botaurus lentiginosus*. Wisconsin Conservation Bulletin 28(6):28-29.

Brechtel, S.H. 1981. A status report, management proposal and selected bibliography for the White Pelican, Double-crested Cormorant and Great Blue Heron in Alberta - 1980. Unpubl. report, Fish and Wildlife Division, Alberta Energy and Natural Resources. 113 pp.

Cottrille, W.P., and B.D. Cottrille. 1958. Great Blue Heron: behavior at the nest. Misc. Publ. No. 102, Museum of Zoology, University of Michigan. 15 pp.

Desgranges, J.L., and P. Laporte. 1979. Second tour of inspection of Quebec heronies, 1978. Canadian Wildlife Service Progress Note No. 105. 12 pp.

Henny, C.J., and M.R. Bethers. 1971. Population ecology of the Great Blue Heron with special reference to western Oregon. Can. Field-Nat. 85(3):205-209.

Kristensen, J. 1981. Great Blue Heron (*Ardea herodius*) colony in the Peace-Athabasca Delta, Alberta. Can. Field-Nat. 95:95-96.

L'Arrivee, L.P., and H. Blockpoel. 1990. Seasonal distribution and site tenacity of Black-crowned Night-Herons, (*Nycticorax nycticorax*) banded in Canada. Can. Field-Nat. 104(4):534-539.

Lang, B. 1974. Census of five Great Blue Heron colonies in southern Alberta, with a discussion of methods. Alta. Nat. 4:16-22.

McFetridge, R.J., and W.M. Glasgow. 1977. Great Blue Heron survey 1977 Edmonton region. Unpubl. report, Fish and Wildlife Division, Alberta Recreation, Parks and Wildlife. 10 pp. + App.

Nette, T. 1983. Brazeau Dam Osprey and heron survey, June 7, 1983. Unpubl. report, Alberta Fish and Wildlife Division. 4 pp.

Noble, G.K., and M. Wurm. 1942. Further analysis of the social behavior of the Black-crowned Night-Heron. Auk 59:205-224.

Paulsen, A.C. 1982. Great Blue Heron colonies in Central Region Wildlife Division May 1982. Unpubl. report, Fish and Wildlife Division, Alberta Energy and Natural Resources. Unnumbered.

Provost, M.W. 1947. Nesting of birds in the marshes of northwest Iowa. Amer. Midl. Nat. 38(2):485-503.

Salt, W.R. 1961. Recent additions to the avifauana of Alberta. Auk 78:427-428.

Saunders, A.A. 1949. A bittern "pumps" from a perch in a tree. Auk 66(2):196.

Smith, W.W., and C.A. Wallis. 1976. Preliminary investigation of the birds of Pakowki Lakes, Alberta. Blue Jay 34:168-171.

Vermeer, K. 1969. Great Blue Heron colonies in Alberta. Can. Field-Nat. 83(3):237-242.

Vermeer, K. 1973. Great Blue Heron and Double-crested Cormorant colonies in the prairie provinces. Can. Field-Nat. 87:427-423.

Weseloh, D.V., D. Dekker, S. Brechtel and R. Burns. 1975. Notes on the Double-crested Cormorants, White Pelicans and Great Blue Herons of Beaverhill Lake, summer 1975. Alta. Nat. 5:132-137.

Weseloh, D.V., S. Brechtel, L. Bogaert, R.D. Burns and J Keizer. 1977. A survey of the colonial nesting waterbirds of Beaverhill Lakes, 1976. Edmonton Nat. 5:183-192.

Williams, J. 1983. The Great Blue Herons at Fish Creek Provincial Park: monitoring study 1982. Alta. Nat. 13(1):1-4.

Wolford, J.W. 1966. An ecological study of the Black-crowned Night-Heron in southern Alberta. M.Sc. Thesis, University of Alberta. 55 pp.

Wolford, J.W., and D.A. Boag. 1971. Distribution and biology of Black-crowned Night-Herons in Alberta. Can. Field-Nat. 85(1):13-19.

Family Threskiornithidae: IBISES AND SPOONBILLS

Bray, M.P., and D.A. Klebenow. 1988. Feeding ecology of White-faced Ibises in a Great Basin Valley, USA. Colonial Waterbirds 11(1):24-31.

Burger, J., and L.M. Miller. 1977. Colony and nest site selection in White-faced and Glossy Ibises. Auk 94:664-676.

Callin, E.M. 1978. White-faced Ibis at Last Mountain Lake, Saskatchewan. Blue Jay 36:123.

Capen, D.E. 1977. The impact of pesticides on the White-faced Ibis. Ph.D. dissertation, Utah State Univ., Logan, Utah.

Gollop, J.B. 1978. White-faced Ibis and Cattle Egret at Little Quill Lake. Blue Jay 36:122.

Goossen, J.P., H. Judge, D.M. Ealey and D.C. Duncan. In Prep. Disribution and breeding status of the White-faced Ibis *Plegadis chihi* in Canada. Can. Field Nat.

Ivey, G.L., M.A. Stern and C.G. Carey. 1988. An increasing White-faced Ibis population in Oregon. Western Birds 19:105-108.

Krebs, J.R. 1974. Colonial nesting and social feeding as strategies for exploiting food resources in the Great Blue Heron (*Ardea herodias*). Behavior 51:99-134.

Kushlan, J.A. 1979. Feeding ecology and prey selection in White-faced Ibis. Condor 81:376-389.

Lahrman, F.W. 1976. White-faced Ibis in Saskatchewan. Blue Jay 34:238.

Lister, R. 1964. Northern Great Plains. Audubon Field Notes 18:461.

Ryder, R.A. 1967. Distribution, migration and mortality of the White-faced Ibis (*Plegadis chihi*). North American Bird-Banding 38(4):257-277.

Schuler, H.C.E. 1977. The White-faced Ibis in southern Alberta. Alta. Nat. 7(2):146-148.

Smith, W.W., and C.A. Wallis. 1976. Preliminary investigation of the birds of Pakowki Lake, Alberta. Blue Jay 34:168-171.

Taylor, D.M., C.H. Trost and B. Jamison. 1989. The biology of the White-faced Ibis in Idaho. Western Birds 20:125-133.

United States Fish and Wildlife Service. 1985. White-faced Ibis management guidelines, Great Basin population. Portland, Oregon. 31 pp.

Order Anseriformes

Family Anatidae: SWANS, GEESE, AND DUCKS

Allen, A.W. 1986. Habitat suitability index models: Lesser Scaup (breeding). U.S. Fish and Wildlife Service Biological Report No. 82(10.117). 16 pp.

Anderson, D.R., and C.J. Henny. 1972. Population ecology of the Mallard. I. A review of previous studies and the distribution and migration from breeding areas. U.S. Dept. of Interior, Fish and Wildlife Service, Resource Publ. No. 105. 166 pp.

Banko, W.E. 1960. The Trumpeter Swan: its history, habits and population in the United States. North American Fauna 64 Bureau of Sport Fisheries and Wildlife, Washington, D.C. 214 pp.

Bartonek, J.C. 1972. Summer foods of American Wigeon, Mallards and Green-winged Teal near Great Slave Lake, NWT. Can. Field-Nat. 86:373-376.

Bartonek, J.C., K.E. Gamble, R.J. Blohm, H.W. Miller, R.K. Brace, R.S. Pospahala, F.D. Caswell and M.M. Smith. 1984. Status and needs of the Mallard. Trans. N. Amer. Wildl. Nat. Res. Conf. 49:501-518.

Beacham, E.D. 1957. A breeding record for the Wood Duck in Alberta. Can. Field-Nat. 71(1):35.

Bellrose, F.C. 1980. Ducks, geese and swans of North America. Stockpole Books, Harrisburg, PA (3rd ed.).

Bennett, L.J. 1938. The Blue-winged Teal, its Ecology and Management. Collegiate Press, Inc., Ames, Iowa.

Beswick, B. 1977. Observations of the Harlequin Duck (*Histrionicus histrionicus* L.) on Maligne Lake, Jasper National Park, summer 1977. Unpubl. report, Warden Service, Parks Canada, Jasper. 8 pp.

Bortner, J.B., F.A. Johnson, G.W. Smith and R.E. Trost. 1991. 1991 status of waterfowl and fall flight forecast. U.S. Dept. Interior, Fish and Wildlife Service, Office of Migratory Bird Management, Laurel, Maryland.

Bouvier, J.M. 1974. Breeding biology of the Hooded Merganser in southwestern Quebec, including interaction with Common Goldeneyes and Wood Ducks. Can. Field-Nat. 88:323-330.

Boyd, H. 1981. Prairie dabbling ducks, 1941-1990. Canadian Wildlife Service Progress Notes No. 119. 9 pp.

Breault, A.M., and J.P.L. Savard. 1991. Status report on the distribution and ecology of Harlequin Ducks in British Columbia. Technical Report Series No. 110, Canadian Wildlife Service, Pacific and Yukon Region, B.C.

Brechtel, S.H. 1977. A management plan for Trumpeter Swans in Alberta. Alberta Fish and Wildlife Division.

Brown, P.W., and M.A. Brown. 1981. Nesting biology of the White-winged Scoter. J. Wildl. Manage. 45(1):38-45.

Butot, R. 1974. Menage a trois, or the end of a species? Calgary Field Nat. 5(11):281-282.

Butot, R. 1975. Trumpeter Swan winters in Calgary. Calgary Field Nat. 6:303.

Canadian Wildlife Service. 1973. Mallard. Information Canada, Ministry of the Environment, Hinterland who's who No. 14.

Canadian Wildlife Service. 1973. Black Duck. Information Canada, Ministry of the Environment, Hinterland who's who No. 17.

Canadian Wildlife Service. 1974. Redhead. Information Canada, Ministry of the Environment, Hinterland who's who No. 41.

Canadian Wildlife Service. 1974. Bufflehead. Information Canada, Ministry of the Environment, Hinterland who's who No. 40.

Canadian Wildlife Service. 1975. Canvasback. Information Canada, Ministry of Environment, Hinterland who's who No. 48.

Canadian Wildife Service. 1975. Wood Duck. Information Canada, Ministry of the Environment, Hinterland who's who No. 50.

Canadian Wildlife Service. 1984. Canada Goose. Information Canada, Ministry of the Environment, Hinterland who's who No. 6.

Carmel, R. 1986. Mallard attacks Ring-billed Gull. Alta. Nat. 16(4):135.

Cartwright, B.W. 1941. Recent western records of Black Duck (*Anas rubripes*). Can. Field-Nat. 55(5):78.

Chasko, G.C., and M.R. Canover. 1988. Too much of a good thing. Living Bird 7(2):8-13.

Cole, D. 1973. Distribution, numbers and productivity of large Canada Geese in southern Alberta. Unpubl. report, Fish and Wildlife Division, Alberta Lands and Forestry. 35 pp.

Cottam, C. 1939. Food habits of North American diving ducks. United States Department of Agriculture, Technical Bulletin No. 643, Washington, D.C.

Cross, D.H., ed. 1988. Waterfowl management handbook. United States Dept. of Interior, Fish and Wildlife Service Leaflet 13, Washington, D.C.

Davison, D.W. 1925. Nesting of the Canada Goose in a tree. Can. Field-Nat. 39(9):197-198.

Delacour, J., and E. Mayr. 1945. The family Anatidae. Wilson Bull. 57:3-55.

Dow, H., and S. Frodga. 1984. Factors affecting reproductive output of the Goldeneye Duck (*Bucephala clangula*). J. Animal Ecol. 53:679-692.

Drewien, R.C., and P.F. Springer. 1969. Ecological relationships of breeding Blue-winged Teal to prairie potholes. **In:** Saskatoon wetlands seminar, Canadian Wildlife Service Report Series, No. 6 pp. 102-105.

Duncan, D.C. 1987. Nest-site distribution and overland brood movements of Northern Pintails in Alberta. J. Wildl. Manage. 51(4):716-723.

Duncan, D.C. 1987. Nesting of Northern Pintails in Alberta; laying date, clutch size and renesting. Can. J. Zool. 65:234-246.

Dwernychuk, L.W. 1968. Some aspects of the ecology of island-nesting waterfowl at Miquelon Lake, Alberta. M.Sc. Thesis, University of Alberta, Edmonton. 155 pp.

Erskine, A.J. 1960. Further notes on interspecific competition among hole-nesting ducks. Can. Field-Nat. 74(3):161-162.

Erskine, A.J. 1972. Populations, movements and seasonal distribution of mergansers. Canadian Wildlife Service Report Series No. 17, Environment Canada.

Erskine, J.A. 1972. Buffleheads. Canadian Wildlife Service Report Series Mono. Ser. No. 4.

Ewaschuk, E., and D.A. Boag. 1972. Factors affecting hatching success of densely nesting Canada Geese. J. Wildl. Manage. 36(4):1097-1106.

Ewaschuk, E., and D.J. Neave. 1971. The status of the large Canada Goose in Alberta. Alberta Waterfowl Technical Committee, Special Report - 1. 40 pp + App.

Fisher, B.M. 1975. Aggressive behavior of a female Baldpate. Edmonton Nat. 3(2):3-4.

Girard, G.L. 1939. Notes on the life history of the shoveler. Trans. N. Amer. Wildl. Conf. 4:364-371.

Giroux, J.-F. 1981. Interspecific nest parasitism by Redheads on islands in southeastern Alberta. Can. J. Zool. 59(11):2053-2057.

Giroux, J.-F. 1981. Use of artificial islands by nesting waterfowl in southeastern Alberta. J. Wildl. Manage. 45(3):669-679.

Goodman, A.S. 1974. The influence of habitat on Lesser Scaup, Ring-necked Duck and Blue-winged Teal populations in the Saskatchewan River Delta. M.Sc. Thesis, University of Alberta, Edmonton.

Gray, B.J. 1970. Reproduction, energetics and social structure of the Ruddy Duck. Ph.D. Thesis, University of California, Davis.

Greenwood, R.J., A.B. Sargeant, D.H. Johnson, L.M. Cowardin and T.L. Shaffer. 1987. Mallard nest success and recruitment in prairie Canada. Trans. 52nd N. Amer. Wildl. Nat. Res. Conf.:298-309.

Grieb, J.R. 1970. The shortgrass prairie Canada Goose population. Wildl. Monog. No. 22. 49 pp.

Hanson, H.C. 1965. The Giant Canada Goose. Southern Illinois University Press.

Harwood, M. 1991. What's going on here? Audubon 93(2):99.

Henderson, A.D. 1924. Nest of the Canvas-back duck, Lake Magean. Oologist 41(10):125.

Henderson, A.D. 1925. An unusual nest of the Mallard. Can. Field-Nat. 39(2):44.

Hennan, E., and B. Munson. 1979. Species distribution and habitat relationships of waterfowl in Northeastern Alberta. Alberta Oil Sands Environmental Research Program. Project LS 22.1.2 AOSERP Report No. 81.

Hester, F.E., and J. Dermid. 1973. The world of the Wood Duck. J.B. Lippincott Co., Philadelphia and New York. 160 pp.

Hochbaum, H.A. 1959. The Canvasback on a prairie marsh. Stackpole, Harrisburg, PA and Wildlife Management Institute, Washington, D.C. (2nd ed.).

Holton, G.R. 1982. Habitat use by Trumpeter Swans in the Grande Prairie region of Alberta. M.Sc. Thesis, University of Calgary, Calgary, Alberta.

Johnsgard, P.A. 1975. Waterfowl of North America. Indiana University Press, Bloomington.

Johnson, D.H., and T.L. Shaffer. 1987. Are Mallards declining in North America? Wildl. Soc. Bull. 15(3):340-345.

Keith, L.B. 1961. Three further records of parasitic egg laying by ducks. Auk 78(1):93.

Kemper, J.B. 1973. A summary of biological data pertaining to two proposed thermal generating plants at Dodds and Langdon, Alberta. Unpubl. report prep. for Montreal Engineering by Canadian Wildlife Service. 25 pp. + maps.

Kirby, R.E. 1988. American Black Duck breeding habitat enhancement in the northeastern United States: a review and synthesis. U.S. Dept. of Interior, Fish and Wildlife Service, Washington, D.C.

Kitchen, D.W., and G.S. Hunt. 1969. Brood habitat of the Hooded Merganser. J. Wildl. Manage. 33(3):605-609.

Klett, A., H.F. Duebbert, C.A. Faanes and K.F. Higgins. 1986. Techniques for studying nest success of ducks in upland habitats in the prairie pothole region. U.S. Dept. of Interior, Fish and Wildlife Service, Resource Publication No. 158, Washington, D.C.

Kondla, N.G. 1973. Canada Goose goslings leaving cliff nest. Auk 90(4):890.

Kortright, F.H. 1953. The ducks, geese, and swans of North America. Stackpole Co., Harrisburg, PA. 474 pp.

Krauss, H. 1960. Regional reports: nesting season June 1 to August 15, 1960. Northern Great Plains region. Audubon Field Notes 14(5):456-459.

Lang, B., V. Lang, and D.V. Weseloh. 1972. Comments on the status and distribution of Cinnamon Teal in the Calgary area. Calgary Field Nat. 4:55-57.

Lang, V. 1973. Winter record of Harlequin Duck within Calgary. Calgary Field Nat. 4:146.

Leitch, W.G. 1964. Black Duck breeding record for Alberta. Can. Field-Nat. 78(3):199.

Loewen, A.C. 1970. Waterfowl study, Pocahontas District. 1970 supplement. Unpubl. report, Warden Service, Parks Canada, Jasper. 12 pp.

Loewen, A.C. 1970. Waterfowl study, Pocahontas District. Unpubl. report, Warden Service, Parks Canada, Jasper. 13 pp.

Lokemoen, J.T., H.F. Duebbert and D.E. Sharp. 1990. Homing and reproductive habits of Mallards, Gadwalls and Blue-winged Teal. Wildl. Monog. No. 106. 28 pp.

Long, R.J. 1970. A study of nest-site selection by island-nesting anatids in Central Alberta. M.Sc. Thesis, University of Alberta, Edmonton. 123 pp.

MacKay, R.H. 1978. Status of endangered species in Canada: Trumpeter Swan. Committee on the Status of Endangered Wildlife in Canada, Ottawa.

Mansell, W.C. 1978. Sawbill: An affectionate look at the Common Merganser. Nature Canada 7(4):56-61.

McKelvey, R.W., K.J. McCormick and L.J. Shandruk. 1988. The status of Trumpeter Swans (*Cygnus buccinator*) in western Canada, 1985. Can. Field-Nat. 102(3):495-499.

McNicholl, M.K. 1980. Two decades of Mallards on Alberta Christmas bird counts. Alta. Nat. 10(Suppl. 1):1-4.

Morse, T.E., J.L. Jakabosky and V.P. McCrow. 1969. Some aspects of the breeding biology of the Hooded Merganser. J. Wildl. Manage. 33:596-604.

Munro, J.A. 1942. Studies of waterfowl in British Columbia. Bufflehead. Can. J. Res. 200:133-160.

Munro, J.A., and W.A. Clemens. 1939. The food and feeding habits of the Red-breasted Mergansers in British Columbia. J. Wildl. Manage. 3(1):46-53.

Myres, M.T. 1969. Ring-necked Ducks breeding in Banff National Park. Calgary Field Nat. 1(2):5.

Nichols, J.D., R.S. Pospahala and J.E. Hines. 1982. Breeding-ground habitat conditions and the survival of Mallards. J. Wildl. Manage. 46:80-87.

Nieman, D.J., and R.J. Isbister. 1974. Population status and management of trumpeters in Saskatchewan. Blue Jay 32:97-101.

Nieman, D.J., E.M. Wright and R.J. Isbister. 1974. Investigations into the status of the western prairie Canada Goose population. Unpubl. report, Canadian Wildlife Service, Saskatoon. 27 pp.

Nordstrom, W. 1984. Trumpeter Swans northeast of Edson. Alta. Nat. 14:45-47.

Palmer, R.S. (ed.). 1976. Handbook of North American birds volume 2. Waterfowl (first part). Whistling ducks, swans, geese, sheld ducks, dabbling ducks. Yale University Press. New Haven and London.

Palmer, R.S. (ed.). 1976. Handbook of North American birds volume 3. Waterfowl (concluded). Eiders, wood ducks, diving ducks, mergansers, stifftails. Yale University Press. New Haven and London.

Pinel, H.W., and R.J. Butot. 1978. Highlights of bird observations in the Calgary area, Alberta 1971-1977. Blue Jay 36:159-163.

Poston, H.J. 1969. Home range and breeding biology of the shoveler. M.Sc. Thesis, Logan, Utah.

Poston, H.J. 1974. Home range and breeding biology of the shoveler. Canadian Wildlife Service Report Series No. 25.

Poysa, H. 1983. Resource utilization pattern and guild structure in a waterfowl community. Oikos 40:295-307.

Randall, T.E. 1984. Notes on the Gadwall at Kazen Lake, Saskatchewan. Blue Jay 42:197-198.

Randall, T.E. 1946. Cinnamon Teal (*Querquedula cyanoptera*) (Viellot) breeding in Alberta. Can. Field-Nat. 60:136.

Rogers, J.P. 1962. The ecological effects of drought on reproduction of the Lesser Scaup, *Aythya affinis* (Eyton). Ph.D. Thesis, University of Missouri.

Rusch, D.H., C.D. Ankney, H.B. Boyd, J.R. Longcore, F. Montalbano III, J.K. Ringelman and V.D. Stotts. 1989. Population ecology and harvest of the American Black Duck: A review. Wildl. Soc. Bull. 17(4):379-406.

Savard, J.P.L. 1987. Status report on Barrow's Goldeneye. Technical Report Series No. 23, CWS, Pacific and Yukon region, British Columbia.

Savard, J.P.L. 1991. Waterfowl in the aspen parkland of central British Columbia. Technical Report Series. No. 132. CWS, Pacific and Yukon Region, British Columbia.

Savard, J.P.L., G.E.J. Smith and J.N.M. Smith. 1991. Duckling mortality in Barrow's Goldeneye and Bufflehead broods. Auk 108(3):568-577.

Schmutz, J.K., W.D. Wishart, J. Allen, R. Bjorge and D.A. Moore. 1988. Dual use of nest platforms by hawks and Canada Geese. Wildl. Soc. Bull. 16(2):141-145.

Scott, P., and the Wildlife Trust. 1972. The Swans. Houghton Mifflin Company, Boston.

Shandruk, L. 1987. A review of habitat requirements and management priorities for the Canadian breeding population of Trumpeter Swans. **In:** Proceedings of the Workshop on Endangered Species of the Prairie Provinces. Holroyd G.L., W.B. McGillivray, P.H.R. Stepney, D.M. Ealey, G.C. Trottier, K.E. Eberhart, eds. Provincial Museum of Alberta, Natural History, Occasional Paper No. 9, Edmonton, Alberta.

Shandruk, L. 1991. The interior Canadian subpopulation of Trumpeter Swans. **In:** Proceedings of the Second Endangered Species and Prairie Conservation Workshop. Holroyd G.L., G. Burns and H.C. Smith, eds. Provincial Museum of Alberta, Natural History Occasional Paper No. 15, Edmonton, Alberta.

Shandruk, L.J. 1991. A survey of Trumpeter Swans in Alberta, Saskatchewan and Northwest Territories. Technical Report Series No. 119. Canadian Wildlife Service, Western and Northern Region, Alberta.

Shandruk, L.J., and R.W. McKelvey. 1990. Status of the Trumpeter Swan, interior Canada subpopulation, 1990. Unpubl. report, Canadian Wildlife Service, Environment Canada, Edmonton, Alberta.

Shields, G.F., and A.C. Wilson. 1987. Subspecies of the Canada Goose (*Branta canadensis*) have distinct mitochondrial DNA's. Evolution 41:662-666.

Siegfried, W.R. 1982. Segregation in feeding behaviour of four diving ducks in southern Manitoba. **In:** Waterfowl Ecology and Management: Selected Readings. Ratti, J.T., C.D. Flate and W.A. Wentz, eds. Wildlife Soc. Publ.

Siferd, T.D. 1982. Mink, *Mustela vison*, attacks Trumpeter Swan, *Cygnus buccinator*, cygnet. Can. Field-Nat. 96:357-358.

Smith, A.G. 1971. Ecological factors affecting waterfowl production in the Alberta parklands. U.S. Dept. Interior. Fish and Wildlife Service, Resource Publication 98. 49 pp.

Smith, R.I. 1970. Response of Pintail breeding populations to drought. J. Wildl. Manage. 34(4):943-946.

Sousa, P.J. 1985. Habitat suitability index models: Blue-winged Teal (breeding). U.S. Fish and Wildlife Service Biological Report No. 82(10,114). 36 pp.

Sousa, P.J. 1985. Habitat suitability index models: Gadwall (breeding). U.S. Fish and Wildlife Service Biological Report No. 82(10.100). 35 pp.

Stephen, W.J.D. 1979. Toward population goals for Canvasback. Canadian Wildlife Service Progress Notes, No. 96. 9 pp.

Stokes, D., and L. Stokes. 1986. Mallards. Living Bird 5(4):22-23.

Stokes, D., and L. Stokes. 1987. Canada Geese. Living Bird 6(2):22-23.

Stoudt, J.H. 1982. Habitat use and productivity of Canvasbacks in southwestern Manitoba. 1961-72. United States Dept. of Interior, Fish and Wildlife Service, No. 248.

Strong, R.M. 1912. Some observations on the life history of the Red-breasted Merganser, (*Mergus merganser*) Linn. Auk 29:479-488.

Suchy, W.J., and S.H. Anderson. 1987. Habitat suitability index models: Northern Pintail. U.S. Dept. of Interior, Fish and Wildlife Service, Research and Development, Washington, D.C., Biological Report No. 82 (10.145). 23 pp.

Sugden, L.G. 1960. Barrow's Goldeneye using crow nests. Condor 65(4):330.

Sugden, L.G. 1973. Feeding ecology of Pintail, Gadwall, American Widgeon and Lesser Scaup ducklings in southern Alberta. CWS Report Series No. 24. 45 pp.

Tome, M. 1991. Diurnal activity budget of female Ruddy Ducks breeding in Manitoba. Wilson Bull. 103(2):183-189.

Turner, B., and R. McKelvey. 1983. Proposed guidelines for transplanting Trumpeter Swans in Canada. Unpubl. Admin. Rept., Canadian Wildlife Service. Edmonton, Alberta.

U.S. Fish and Wildlife Service. 1991. Waterfowl production survey southern Alberta: July 1991. 9 pp.

U.S. Fish and Wildlife Service. 1991. Waterfowl production survey southern Alberta: May 1991. 19 pp.

Van Wagner, C.E., and A.J. Baker. 1986. Genetic differentiation in populations of Canada Geese (*Branta canadensis*). Can. J. Zool. 64:940-947.

Vermeer, K. 1969. Some aspects of the breeding biology of the White-winged Scoter at Miquelon Lake, Alberta. Blue Jay 27(2):72-73.

Vermeer, K. 1970. A study of Canada Geese, *Branta canadensis*, nesting on islands in southeastern Alberta. Can. J. Zool. 48(2):235-240.

Weller, M.W. 1964. Distribution and migration of the Redhead. J. Wildl. Manage. 28(1):64-103.

Weseloh, D.V., and L.M. Weseloh. 1975. Male Wood Duck courting female Mallard. Alta. Nat. 5(3):85-87.

Wilby, D. 1954. Goldeneye Duck. Blue Jay 12(1):5.

Winkler, T. 1991. The Elk Island Park Trumpeter Swan reintroduction project. **In:** Proceedings of the Second Endangered Species and Prairie Conservation Workshop. Holroyd G.L., G. Burns and H.C. Smith, eds. Provincial Museum of Alberta, Natural History Occasional Paper No. 15, Edmonton, Alberta.

Wright, E.M. 1968. A comparative study of farm dugouts and natural ponds and their utilization by waterfowl. M.Sc. Thesis, University of Alberta, Edmonton, Alberta.

Zicus, M.C. 1990. Renesting by a Common Goldeneye. J. Field Ornith. 61:245-248.

Zicus, M.C. 1990. Nesting biology of Hooded Mergansers using nest boxes. J. Wildl. Manage. 54(4):637-643.

Order Falconiformes

Family Cathartidae: AMERICAN VULTURES

Davis, D. 1983. Breeding biology of Turkey Vultures. pp. 271-286, **In:** Vulture biology and management. Wilbur, S.R. and J.A. Jackson, eds. University of California Press, Los Angeles.

Ellenwood, M. 1989. Cinnamon Teal near Bonnyville. Alberta Naturalists. 19(2):69.

Farley, F.L. 1911. Turkey Buzzard. Ottawa Naturalist 25(5):88.

Horning, J.E. 1923. Turkey Vultures in Alberta. Auk 40(2):324-325.

Oeming, A.F. 1957. Turkey Vultures nesting in Northern Alberta. Can. Field-Nat. 71:152.

Stewart, P.A. 1985. Need for new direction in research on Black and Turkey Vultures in the U.S.A. Vulture News 13:8-12.

Taverner, P.A. 1927. Some recent Canadian records. Auk 44(2):217-228.

Wilbur, S.R., and J.A. Jackson (eds.). 1983. Vulture biology and management. University of California Press, Berkeley, CA. 550 pp.

Family Accipitridae: OSPREYS, KITES, EAGLES, HAWKS, AND ALLIES

Adamcik, R.S., A.W. Todd and L.B. Keith. 1979. Demographic and dietary responses of Red-tailed Hawks during a Snowshoe Hare fluctuation. Can. Field-Nat. 93(1):16-27.

Angell, T. 1969. A study of the Ferruginous Hawk: Adult and brood behavior. Living Bird 8:225-241.

Anonymous. 1956. Alberta controversy re: protection of birds of prey. Blue Jay 14(1):15-16.

Anonymous. n.d. The status of the Osprey (*Pandion haliaetus*) in Canada. 10 pp., COSEWIC file at CWS.

Armstrong, D., M. Glossop and S. Lamden. 1984. Fish Creek Provincial Park resource management plan. Unpubl. report, Alberta Recreation and Parks. 97 pp. + tables.

Banasch, U. 1991. Ferruginous Hawk nesting densities on Class I and II habitat in Saskatchewan. **In:** Proceedings of the Second Endangered Species and Prairie Conservation Workshop. Holroyd, G.L., G. Burns and H.C. Smith, eds. Provincial Museum of Alberta Natural History Occasional Paper No. 15, Edmonton, Alberta.

Baresco, D. 1989. Plains "high". Alta. Nat. 19(3):104.

Beebe, F.L. 1974. Field studies of the Falconiformes (vultures, eagles, hawks and falcons) of British Columbia. B.C. Provincial Museum Occasional Paper No. 17, Victoria B.C. 163 pp.

Beebe, F.L. 1976. Hawks, falcons and falconry. Hancock House Publishers Ltd., Saanichton, B.C.

Boag, D.A. 1955. Golden Eagle Study. Alberta Biol. Stn. Rept. 6:33-37.

Boag, D.A. 1977. Summer food habits of Golden Eagles in southwestern Alberta. Can. Field-Nat. 91(3):296-298.

Bock, C.E., and L.W. Lepthien. 1976. Geographical ecology of the common species of buteo and parabuteo wintering in North America. Condor 78(4):554-557.

Bortolotti, G.R. 1984. Trap and poison mortality of Golden Eagles (*Aquila chrysaetos*) and Bald Eagles (*Haliaeetus leucocephalus*). J. Wildl. Manage. 48:1173-1179.

Boyd, M.G. 1972. Bald Eagles of the Peace Athabasca Delta (northern Alberta) - summer 1971. Unpubl. report, University of Alberta. 19 pp + App.

Breckenridge, W.J. 1935. An ecological study of some Minnesota Marsh Hawks. Condor 37:268-276.

Brown, L.H., and D. Amadon. 1968. Eagles, hawks and falcons of the world. McGraw-Hill Book Co. Ltd, Toronto, Ont. 945 pp.

Brownell, V.R., and M.J. Oldham. 1983. Status report of the Bald Eagle (*Haliaeetus leucocephalus*) in Canada. Prepared for Ontario Ministry of Natural Resources, Toronto, Ontario.

Bruns, E.H. 1970. Winter predation of Golden Eagles and Coyotes on Pronghorn Antelopes. Can. Field-Nat. 84(3):301-304.

Cade, J.J. 1982. Falcons of the world. William Collins Sons, London, England.

Call, M.W. 1978. Nesting habits and surveying techniques for common western raptors. United States Department of the Interior. Bureau of Land Management technical note TN-316. Washington, D.C., 115 pp.

Cash, K.J. 1989. Three adult Swainson's Hawks tending a nest. Condor 91(3):727-728.

Clark, W.S., and B.K. Wheeler. 1987. A field guide to hawks: of North America. Houghton Mifflin Company, Boston, Massachusetts. (Peterson field guide series 35.

Dekker, D. 1983. The Bald Eagle: hunter or scavenger? Alta. Nat. 13(2):43-45.

Dekker, D. 1984. Migrations and foraging habits of Bald Eagles in east-central Alberta 1964-1983. Blue Jay 42:199-205.

Dekker, D. 1984. Occurrence of Golden Eagles at Beaverhill Lake, Alberta. Alta. Nat. 14:54-55.

Dekker, D. 1985. Hunting behavior of Golden Eagles, (*Aquila chrysaetos*) migrating in southwestern Alberta. Can. Field-Nat. 99:383-385.

DeSmet, K.D. 1987. A status report on the Golden Eagle (*Aquila chrysaetos*) in Canada. Committee on the Status of Endangered Wildlife in Canada, Ottawa.

DeSmet, K.D., and M.P. Conrad. 1991. Status, habitat requirements, and adaptations of Ferruginous Hawks in Manitoba. **In:** Proceedings of the Second Endangered Species and Prairie Conservation Workshop. Holroyd, G.L., G. Burns and H.C. Smith, eds. Provincial Museum of Alberta Natural History Occasional Paper No. 15, Edmonton, Alberta.

Duncan, S. 1980. An analysis of the stomach contents of some Sharp-shinned Hawks (*Accipiter striatus*). J. Field Ornith. 5(2):178.

Fitch, H.S., F. Swenson and D.F. Tillotson. 1946. Behavior and food habits of the Red-tailed Hawk. Condor 48:205-237.

Fitzner, R.E. 1978. Behavioral ecology of the Swainson's Hawk (*Buteo swainsoni*) in southeastern Washington. Ph.D. Thesis, Washington State University, Pullman, WA. 194 pp.

Flood, N.J., and G.R. Bortolotti. 1984. Status of the Sharp-shinned Hawk (*Accipiter striatus*) in Canada. Status report, Committee on the Status of Endangered Wildlife in Canada, Ottawa. 86 pp.

Fyfe, R.W. 1968. The correlation of eggshell thickness and DDE residue levels in prairie raptors, Western Region Pesticide Section progress report No. 4. Canadian Wildlife Service, Edmonton. 2 pp. + figures.

Fyfe, R.W. 1976. Status of Canadian raptor populations. Can. Field-Nat. 90:370-375.

Gerrard, J.M. 1975. The Bald Eagles in Canada's northern forests. Nature Canada 2(3):10-15.

Gerrard, J.M. 1983. A review of the current status of Bald Eagles in North America. **In:** Bird, D.M., ed., Biology and Management of Bald Eagles and Ospreys, Proceedings of First International Symposium on Bald Eagles and Ospreys, Montreal, 28-29 October 1981. Harpell Press, Ste. Anne de Bellevue, Quebec.

Gerrard, J.M., and T.M. Ingram, eds. 1985. The Bald Eagle in Canada. Whitehorse Plains Publ., Headingley.

Gerrard, J.M., P. Gerrard, W.J. Maher and D.W.A. Whitfield. 1975. Factors influencing nest site selection of Bald Eagles in northern Saskatchewan and Manitoba. Blue Jay 33(3):169-176.

Gilmer, D.S., and R.E. Stewart. 1983. Ferruginous Hawk populations and habitat use in North Dakota. J. Wildl. Manage. 47:146-157.

Godfrey, W.E. 1970. Canada's endangered birds. Can. Field-Nat. 84(1):24-26.

Grant, J. 1957. A breeding record of Cooper's Hawk in Alberta. Can. Field-Nat. 71:82.

Greene, E. 1988. Birds of a feather flock together. Nature Canada 17(1):7-8.

Henderson, A.D. 1923. Nesting habits of the Broad-winged Hawk. Oologist 40(11):182.

Henderson, A.D. 1924. Nesting habitats of the American Goshawk. Can. Field-Nat. 38(1):8-9.

Herrick, F.H. 1924. Nests and nesting habits of the american eagle. Auk 41(2):213-231.

Herrick, F.H. 1932. Daily life of the american eagle: early phase. Auk 49:307-323.

Hodson, K. 1968. Porcupine in Ferruginous Hawk's nest. Blue Jay 26(4):180-181.

Hoffman, W. 1988. Cooper's Hawk uses an artificial nest platform. Alta. Nat. 18(1):24-26.

Höhn, E.O. 1983. Female Cooper's Hawk breeding in brown plumage, apparently for four years. Blue Jay 41(4):208-210.

Höhn, E.O. 1986. Some observations on breeding Broad-winged Hawks. Blue Jay 44(1):44-46.

Holroyd, J.C. 1965. Golden Eagle, Bighorn Valley, Banff National Park, 1961. Unpubl. report in Parks Canada files, Banff, Alberta. 1 pp.

Hopkins, D. 1990. Bald Eagle at Coyote Lake. Edmonton Nat. 18(3):16.

Houston, C.S. 1990. Saskatchewan's Swainson's Hawks. Amer. Birds 44(2):215-220.

Houston, C.S. 1991. Ferruginous Hawk nesting success: a 19-year study. **In:** Proceedings of the Second Endangered Species and Prairie Conservation Workshop. Holroyd, G.L., G. Burns and H.C. Smith, eds. Provincial Museum of Alberta Natural History Occasional Paper No. 15, Edmonton, Alberta.

Houston, C.S., and M.J. Bechard. 1984. Decline of the Ferruginous Hawk in Saskatchewan. Amer. Birds 38:166-170.

Kuyt, E. 1990. Noteworthy raptor records from the NWT-Alberta border area. Alta. Nat. 20(4):143-145.

Landals, A. 1983. Goshawk vs Hare. Edmonton Nat. 11(1):12.

Lincer, J.F., W.S. Clark and M.N. LeFrane, Jr. 1979. Working bibliography of the Bald Eagle. Nat. Wildl. Fed. Scientific and Technical Series No. 2, Washington, D.C. 219 pp.

Luttich, S.N., D.H. Rusch, E.C. Meslow and L.B. Keith. 1970. Ecology of Red-tailed Hawk predation in Alberta. Ecology 51(2):190-203.

Luttich, S.N., L.B. Keith and J.D. Stephensen. 1971. Population dynamics of the Red-tailed Hawk (*Buteo jamaicensis*) at Rochester, Alberta. Auk 88(1):75-87.

Matray, P.F. 1974. Broad-winged Hawk nesting and ecology. Auk 91(2):307-323.

McGahan, J. 1968. Ecology of the Golden Eagle. Auk 85:1-12.

McInvaille, W.B. Jr., and L.B. Keith. 1974. Predator-prey relations and breeding biology of the Great Horned Owl and Red-tailed Hawk in central Alberta. Can. Field-Nat. 88:1-20.

Meng, H. 1959. Food habits of nesting Cooper's Hawks and Goshawks in New York State and Pennsylvania. Wilson Bull. 71:169-174.

Meng, H.K. 1951. The Cooper's Hawk. Ph.D. Thesis, Cornell University, Ithaca, New York. 216 pp.

Mitchell, T. 1987. An aerial duel: Merlin and Cooper's Hawk. Pica 7(1):37-38.

Moore D. 1987. The Ferruginous Hawk in Alberta. **In:** Proceedings of the Workshop on Endangered Species in the Prairie Provinces. Holroyd, G.L., P.H.R. Stepney, G.C. Trottier, W.B. McGillivray, D.M. Ealey and K.E. Eberhart, eds. Provincial Museum of Alberta Natural History Occasional Paper No. 9, Edmonton, Alberta.

Mueller, H.C., D.D. Berger and G. Allez. 1977. The periodic invasion of Goshawks. Auk 94(4):652-663.

Munson, B., D. Ealey, R. Beaver, K. Bishoff and R. Fyfe. 1980. Inventory of selected raptor, colonial and sensitive bird species in the Athabasca Oil Sands area of Alberta. Report for Alberta Oil Sands Environment Research Program by Canadian Wildlife Services. AOSERP Project LS 22.3.2 xvii + 65 pp.

Nero, R.W. 1987. Additional records of Bald Eagle predation of waterfowl. Blue Jay 45(1):28-29.

Newton, I. 1979. Population ecology of raptors. Buteo Books, Vermillion, SD.

Olendorff, R.R. 1976. Food habits of North American Golden Eagles. Amer. Midl. Nat. 95:231-236.

Park, J.L. 1987. Red Squirrel harasses Sharp-shinned Hawk. Alta. Nat. 17(1):16.

Paul, P.J. 1976. Migrating eagles - Cadomin, Alberta. Blue Jay 34(2):102.

Penak, B.L. 1983. The status of the Cooper's Hawk (*Accipiter cooperii*) in Ontario, with an overview of the status in Canada. Committee on the Status of Endangered Wildlife in Canada, Ottawa.

Pinel, H.W., and C.A. Wallis. 1972. Unusual nesting record of Red-tailed Hawk in southern Alberta. Blue Jay 30:30-31.

Ratcliff, B.D. 1987. Ferruginous Hawk, report for Manitoba. **In:** Proceedings of the Workshop on Endangered Species in the Prairie Provinces. Holroyd, G.L., P.H.R. Stepney, G.C. Trottier, W.B. McGillivray, D.M. Ealey and K.E. Eberhart, eds. Provincial Museum of Alberta Natural History Occasional Paper No. 9, Edmonton, Alberta.

Rosenfeld, R.N. 1984. Nesting biology of Broad-winged Hawks in Wisconsin. Raptor Res. 18(1):6-9.

Rothfels, M. 1981. Coexistence of the Red-tailed and Swainson's Hawks in southern Alberta. M.Sc. Thesis, University of Calgary, Calgary. 156 pp.

Rusch, D.H., and P.D. Doerr. 1972. Broad-winged Hawk nesting and food habits. Auk 89(1):139-145.

Salter, R. 1974. Bald Eagle surveys in southern Mackenzie District and northern Alberta, May, July, August, 1973. **In:** Bird distribution and populations ascertained through aerial survey techniques W.W.H. Gunn and J.A. Livingston, eds., Arctic Gas Biological Report Series, Vol. II, prepared by L.G.L. Ltd. 170 pp.

Saunders, M. 1987. Breeding biology of the Northern Harrier. Pica 7(2):15-17.

Savage, C. 1987. Eagles of North America. Western Producer Prairie Books, Saskatoon.

Savage, C. 1988. Circling to the Sun. Nature Canada 17(1):34-41.

Schmutz, J.K. 1977. Relationships between three species of the genus *Buteo* (Aves) coexisting in the prairie-parkland ecotone of southeastern Alberta. M.Sc. Thesis, Department of Zoology, University of Alberta, Edmonton, Alberta.

Schmutz, J.K. 1984(a). Artificial nests for Ferruginous and Swainson's Hawks. J. Wildl. Manage. 48:1009-1013.

Schmutz, J.K. 1984(b). Ferruginous and Swainson's Hawk abundance and distribution in relation to land use in southeastern Alberta. J. Wildl. Manage. 48(4):1180-1187.

Schmutz, J.K. 1987(a). The effect of agriculture on Ferruginous and Swainson's hawks. J. Range Manage. 40(5):438-440.

Schmutz, J.K. 1987(b). Factors limiting the size of the breeding population. **In:** Proceedings of the Workshop on Endangered Species in the Prairie Provinces. Holroyd, G.L., P.H.R. Stepney, G.C. Trottier, W.B. McGillivray, D.M. Ealey and K.E. Eberhart, eds. Provincial Museum of Alberta Natural History Occasional Paper No. 9, Edmonton, Alberta.

Schmutz, J.K. 1989. Hawk occupancy of disturbed grassland in relation to models of habitat selection. Condor 91:362-371.

Schmutz, J.K. 1991. Population dynamics of Ferruginous Hawks in Alberta. **In:** Proceedings of the Second Endangered Species and Prairie Conservation Workshop. Holroyd, G.L., G. Burns and H.C. Smith, eds. Provincial Museum of Alberta Natural History Occasional Paper No. 15, Edmonton, Alberta.

Schmutz, J.K., and R.W. Fyfe. 1987. Migration and mortality of Alberta Ferruginous Hawks. Condor 89(1):169-174.

Schmutz, J.K., and D.J. Hungle. 1989. Populations of Ferruginous and Swainson's Hawks flucuate in synchrony with ground squirrels. Can. J. Zool. 67:2596-2601.

Schmutz, J.K., and S.M. Schmutz. 1979. rev. 1980. Status of the Ferruginous Hawk (*Buteo regalis*). Dept. of Biology, Queen's University, Ont. Prepared for Committee on the Status of Endangered Wildlife in Canada.

Schmutz, J.K., S.M. Schmutz and D.A. Boag. 1980. Coexistence of three species of hawks (*Buteo* spp.) in the prairie-parkland ecotone. Can. J. Zool. 58:1075-1089.

Sealy, S.G. 1967. Notes on the breeding biology of the Marsh Hawk in Alberta and Saskatchewan. Blue Jay 25(2):63-69.

Snow, C. 1973. Habitat management series for unique or endangered species: the Golden Eagle. U.S. Dept. Inter. Bur. Land Manage. Rep. No. 7. 52 pp.

Stalmaster, M.V., and J.R. Newman. 1978. Behavioral responses of wintering Bald Eagles to human activity. J. Wildl. Manage. 42:506-513.

Stokes, D., and L. Stokes. 1988. Broad-winged Hawks. Living bird 7(4):35.

Storer, R.W. 1966. Sexual dimorphism and food habits in three North American accipiters. Auk 83:423-436.

Struzik, E. 1990. Grassland stalker. Equinox 51:58-67.

Torrance, D.J. 1984. Reproductive success of Swainson's Hawks in two habitats in southwestern Alberta. M.Sc. Thesis, University of Calgary, Calgary. 116 pp.

Wakely, J.S. 1978. Hunting methods and factors affecting their use by Ferruginous Hawks. Condor 80:327-333.

Watson, D. 1977. The hen Harrier. T. and A.D. Poyser Ltd. Berkhansted, Hertfordshire, England. 307 pp.

Weseloh, D.V., and L.M. Weseloh. 1976. Bald Eagle feeding on Richardson's Ground Squirrel. Calgary Field Nat. 7(7):195-197.

Wilkie, G. 1986. Goshawk attacks Mallards. Alta. Nat. 16(3):95

Family Falconidae: CARACARAS AND FALCONS

Beebe, F.L. 1974. Field studies of the Falconiformes of British Columbia. British Columbia Provincial Museum Occasional Paper No. 17, Victoria, B.C. 163 pp.

Beebe, F.L. 1976. Hawks, falcons and falconry. Hancock House Publishers Ltd. Saanichton, B.C.

Bond, R.M. 1943. Variation in western Sparrow Hawks. Condor 45(5):168-185.

Bryson, C. 1989. The turning point. Nature Canada 18(1):32-38.

Buchanan, J.B. 1988. North American Merlin populations: an analysis using Christmas bird count data. Amer. Birds 42:1178-1180.

Cade, T.J., and D.M. Bird. 1990. Peregrine Falcons, (*Falco peregrinus*), nesting in an urban environment: a review. Can. Field-Nat. 104(2):209-218.

DeSmet, K.D. 1985. Status report on the Merlin in North America (*Falco columbarius*). Committee on the Status of Endangered Wildlife in Canada, Ottawa.

Dekker, D. 1979. Characteristics of Peregrine Falcons migrating through central Alberta. Can. Field-Nat. 93:296-302.

Dekker, D. 1980. Hunting success rates, foraging habits and prey selection of Peregrine Falcons migrating through central Alberta. Can. Field-Nat. 94:371-382.

Dekker, D. 1984. Falcon sightings in the Rocky Mountains of Alberta. Alta. Nat. 14(2):48-49.

Dekker, D. 1987. Peregrine Falcon predation on ducks in Alberta and British Columbia. J. Wildl. Manage. 51:156-159.

Dekker, D. 1988. Peregrine Falcon and Merlin predation on small shorebirds and passerines in Alberta. Can. J. Zool. 66:925-928.

Dickson, R. 1988. Prairie Falcon reintroduction to Fish Creek. Alta. Bird Record 6(2):38-39.

Edwards, B.F. 1968. A study of the Prairie Falcon in southern Alberta. Blue Jay 26:32-37.

Edwards, R., G.L. Holroyd, H. Reynolds and B. Johns. 1991. Recovery teams and recovery plans. **In:** Proceedings of the Second Endangered Species and Prairie Conservation Workshop. Holroyd, G.L., G. Burns and H. Smith, eds. Provincial Museum of Alberta Natural History Occasional Paper No. 15, Edmonton, Alberta.

Enderson, J.H. 1964. A study of the Prairie Falcon in the central Rocky Mountain region. Auk 81:332-352.

Erickson, G.L., and H.J. Armbruster. 1974. Goose and raptor survey Athabasca River. Unpubl. report, Fish and Wildlife Div., Alberta Dept. Lands and Forests.

Erickson, G.L., R. Fyfe, R. Bromley, G.L. Holroyd, D. Mossop, B. Munro, R. Nero, C. Shank and T. Wiens. 1988. *Anatum* Peregrine Falcon recovery plan. Canadian Wildlife Service, Environment Canada. 52 pp.

Fox, G.A. 1971. Organochlorides and mercury in Merlin eggs. Can. Field-Nat. 85(4):335.

Fyfe, R. 1987. The Peregrine Falcon. **In:** Proceedings of the Workshop on Endangered Species in the Prairie Provinces. Holroyd, G.L., P.H.R. Stepney, G.C. Trottier, W.B. McGillivray, D.M. Ealey and K.E. Eberhart, eds. Provincial Museum of Alberta Natural History Occasional Paper No. 9, Edmonton, Alberta.

Fyfe, R.W., H. Armbruster, U. Banasch and L. Johnston-Beaver. 1978. Fostering and cross-fostering of birds of prey. pp. 183-193. **In:** Temple, S.A., ed. Endangered birds/management techniques for preserving threatened species. University of Wisconsin Press, Madison.

Fyfe, R.W., J. Campbell, B. Hayson and K. Hodson. 1969. Regional population declines and organochlorine insecticides in Canadian Prairie Falcons. Can. Field-Nat. 83:191-200.

Fyfe, R.W., R.W. Risebrough and W. Walker II. 1976. Pollutant effects on the reproduction of the Prairie Falcons and Merlins of the Canadian Prairies. Can. Field-Nat. 90:346-355.

Fyfe, R.W., R.W. Risebrough and W. Walker II. 1976. Pollutant effects on the reproduction of the Prairie Falcons and Merlins of the Canadian Prairies. Can. Field-Nat. 90:346-355.

Guay, R. 1990. Peregrines like city life. International Wildlife 20(1):29.

Hayden-McLean, J. 1987. Peregrine Falcons - things to know. Canadian Parks and Wilderness Society 4(3):13.

Henny, C.J. 1972. An analysis of the population dynamics of selected avian species. U.S. Fish and Wildlife Service, Wildlife Research Report No. 1. 99 pp.

Hickey, J.J. 1969. Peregrine Falcon populations: their biology and decline. University of Wisconsin Press, Madison, Wisconsin.

Hodson, K. 1976. The ecology of Richardson's Merlin (*Falco columbarius richardsonii*) on the Canadian Prairies. M.Sc. Thesis, University of B.C. 83 pp.

Hodson, K. 1978. Prey utilized by Merlins nesting in shortgrass prairies of southern Alberta. Can. Field-Nat. 92:76-77.

Höhn, E.O. 1986. Roosting habits of an urban Merlin. Blue Jay 44(3):194-196.

Holroyd, G.L., and U. Banasch. 1990. The reintroduction of the Peregrine Falcon, (*Falco peregrinus anatum*), into southern Canada. Can. Field-Nat. 104(2):203-208.

James, P.C., and A.R. Smith. 1987. Food habits of urban-nesting Merlins (*Falco columbarius*) in Edmonton and Fort Saskatchewan, Alberta. Can. Field-Nat. 101(4):592-594.

James, P.C., A.R. Smith, L.W. Oliphant and I.G. Warkentin. 1987. Northward expansion of the wintering range of Richardson's Merlin. J. Field Ornith. 58(2):112-117.

Martin, M. 1978. Status report on the Peregrine Falcon *Falco peregrinus*. Committee on the Status of Endangered Wildlife in Canada, Ottawa. 45 pp.

Monk, J.G., W. Risebrough, D.W. Anderson, R.W. Fyfe and L.F. Kiff. 1988. Within-clutch variation of eggshell thickness in three species of falcons. Chapter 36, pp. 377-383. **In:** Cade, T.J., J.H. Enderson, C.G. Thelander and C.M. White, eds. Peregrine Falcon populations/their management and recovery. The Peregrine Fund, Boise, Idaho.

Murphy, J.E. 1990. The 1985-1986 Canadian Peregrine Falcon, (*Falco peregrinus*), survey. Can. Field-Nat. 104(2):182-192.

Nagy, A.C. 1977. Population trend indices based on 40 years of autumn counts at Hawk Mountain Sanctuary in north-eastern Pennsylvania. **In:** Chancellor, R.D., ed., World Conference on Birds of Prey (1975). Report of proceedings of the International Council for Bird Preservation, Cambridge, England. pp. 243-253.

Nelson, R.W. 1987. Where are the Alberta-released Peregrines? Alta. Nat. 17(1):4-9.

Oliphant, L.W. 1985. North American Merlin breeding survey. Raptor Res. 19:37-41.

Oliphant, L.W. 1991. Falcons and prairie conservation. **In:** Proceedings of the Second Endangered Species and Prairie Conservation Workshop. Holroyd, G.L., G. Burns and H.C. Smith, eds. Provincial Museum of Alberta Natural History Occasional Paper No. 15, Edmonton, Alberta.

Oliphant, L.W., and S. McTaggart. 1977. Prey utilized by urban merlins. Can. Field-Nat. 91:190-192.

Oliphant, L.W., and W.J.P. Thompson. 1979. Recent breeding success of Richardson's Merlin in Saskatchewan. Raptor Res. 12:35-39.

Oliphant, L.W., and E. Haug. 1985. Productivity, population density and rate of increase of an expanding Merlin population. Raptor Res. 19:56-59.

Oliphant, L.W., W.J.P. Thompson and T. Donald. 1976. Present status of the Prairie Falcon in Saskatchewan. Can. Field-Nat. 90:365-368.

Peakall, D.B. 1990. Prospects for the Peregrine Falcon, (*Falco peregrinus*), in the nineties. Can. Field-Nat. 104(2):168-173.

Sherman, A.R. 1913. Nest life of the Sparrow Hawk. Auk 30:406-418.

Smith, A.R. 1978. The Merlins of Edmonton. Alta. Nat. 8:188-191.

Snow, C. 1974. Prairie Falcon. Report No. 8, Habitat management series for unique or endangered species. Bureau of Land Management, U.S. Dept. of Interior.

Van Tighem, K. 1967. Destruction of the Prairie Falcon at Calgary, Alberta. Blue Jay 25(3):108.

Weseloh, D.V., L. Bogaert, and J. Keizer. 1977. Peregrine falcons hunting at Beaverhill Lake. Edmonton Nat. 5:8-9.

White, C.M., R.W. Fyfe and D.B. Lemon. 1990. The 1980 North American Peregrine Falcon, (*Falco Peregrinus*), survey. Can. Field-Nat. 104(2):174-181.

Wisely, A.N. 1981. The 1980 Kananaskis Country Peregrine reintroduction project. Alta. Nat. Special Issue No. 2:62-65.

Woodsworth, G., and K. Freemark. 1981. Status report on the Prairie Falcon *Falco mexicanus* in Canada - 1981. Prepared by Canadian Wildlife Service for Committee on the Status of Endangered Wildlife in Canada, Ottawa. 28 pp.

Order Galliformes

Family Phasianidae: PARTRIDGES, PHEASANTS, GROUSE, TURKEYS, AND QUAIL

Anonymous. 1975. Historical notes. Introduction of pheasants and partridges into Alberta. Calgary Field Nat. 7(2):47-48.

Boag, D.A. 1958. The biology of the Blue Grouse of the Sheep River area. M.Sc. Thesis, University of Alberta, Edmonton.

Boag, D.A. 1963. Significance of location, year, sex and age to the autumn diet of Blue Grouse. J. Wildl. Manage. 27(4):555-562.

Boag, D.A. 1976. The effect of shrub removal on occupancy of Ruffed Grouse drumming sites. J. Wildl. Manage. 40(1):105-110.

Boag, D.A., and M.A. Harris. 1966. Territoriality in Blue Grouse. R.B. Miller Biol. Stn. Rept. 17:31-38.

Boag, D.A., G. Glova and G. Cormie. 1967. Changes in the canopy coverage of the overstory on Blue Grouse summer range. R.B. Miller Biol. Stn. Rept. 18:21-26.

Boag, D.A., S.G. Reebs and M.A. Schroeder. 1983. Egg loss among Spruce Grouse inhabiting lodgepole pine forests. Can. J. Zool. 62(6):1034-1037.

Clark, J. 1983. Provincial Sage Grouse population trend counts, May 3-5, 1983. Unpubl. report, Fish and Wildlife Division, Alberta Energy and Natural Resources. 24 pp.

Clark, J., and L.A. Dube. 1982. Sharp-tailed Grouse dancing ground investigations (Southern Region-1982). Unpubl. report, Fish and Wildlife Division, Alberta Energy and Natural Resources. 15 pp. + App.

Doerr, P.D., L.B. Keith, D.H. Rusch and C.A. Fischer. 1974. Characteristics of winter feeding aggregations of Ruffed Grouse in Alberta. J. Wildl. Manage. 38:601-615.

Fish and Wildlife Branch, Saskatchewan Department of Tourism and Natural Resources. 1978. Greater Prairie Chicken *Tympanuchus cupido pinnatus*. Committee on the Status of Endangered Wildlife in Canada, Ottawa.

Friesen, V.C. 1990. Ruffed Grouse: Crazy-flight conclusions. Blue Jay 48(1):33-34.

Herzog, P.W. 1977. Summer habitat use by the White-tailed Ptarmigan in southwestern Alberta. Can. Field-Nat. 91(4):367-371.

Herzog, P.W. 1978. Food selection by female Spruce Grouse during incubation. J. Wildl. Mange. 42(3):632-636.

Herzog, P.W. 1980. Winter habitat use by the White-tailed Ptarmigan in southwestern Alberta. Can. Field-Nat. 94(2):159-162.

Herzog, P.W., and D.M. Keppie. 1980. Migration in a local population of Spruce Grouse. Condor 82(4):366-372.

Hjorth, I. 1970. Reproductive behavior in Tetraonidae with special reference to males. Vitrevy 7(4):183-590.

Hofman, D.E. 1975. A summary of the current status of Wild Turkeys in Alberta. Unpubl. report, Fish and Wildlife Division, Dept. of Recreation, Parks and Wildlife, Lethbridge, Alberta.

Höhn, E.O. 1984. Willow Ptarmigan in the boreal forest of the prairie provinces. Blue Jay 42:83-88.

IEC Beak Consultants Ltd. 1984. The feasibility of re-introducing a viable population of Greater Prairie Chickens (*Tympanuchus cupido pinnatus*) to Alberta. Report prep. for Fish and Wildlife Division, Alberta Energy and Natural Resources, prep. by IEC Beak Consultants Ltd. Calgary. 64 pp.

Johnsgard, P.A. 1973. Grouse and quails of North America. University of Nebraska, Lincoln and London.

Johnsgard, P.A. 1983. The grouse of the world. University of Nebraska Press, Lincoln.

Johnsgard, P.A. 1986. Pheasants of the world. Oxford University Press, Oxford.

Johnson, G.D., and M.S. Boyce. 1990. Feeding trials with insects in the diet of Sage Grouse chicks. J. Wildl. Manage. 54(1):89-91.

Johnston, A., and S. Smoliak. 1976. Settlements of the grasslands and the Greater Prairie Chicken. Blue Jay 34(3):153-156.

Keppie, D.M. 1975. Clutch size of the Spruce Grouse, *Canachites canadensis franklinii*, in southwest Alberta. Condor 77:91-92.

Keppie, D.M. 1977. Snow cover and the use of trees by Spruce Grouse in autumn. Condor 79:382-384.

Keppie, D.M. 1982. A difference in production and associated events in two races of Spruce Grouse. Can. J. Zool. 60(9):2116-2123.

Keppie, D.M., and P.W. Herzog. 1978. Nest site characteristics and nest success of Spruce Grouse. J. Wildl. Manage. 42:628-632.

Klassen, M. 1991. No chickens left to count. In the case of the Greater Prairie Chicken, should we even be trying? Nature Canada 20(3):17-21.

Klott, J.H., and F.G. Lindzey. 1990. Brood habitats of sympatric Sage Grouse and Columbian Sharp-tailed Grouse in Wyoming. J. Wildl. Manage. 54(1):84-88.

MacDonald, S.D. 1965. Report on studies of nuptial behaviour of Blue Grouse. R.B. Miller Biol. Stn. Rept. 16:26-31.

Manry, D.E. 1989. Species profile: White-tailed Ptarmigan. Wildbird 3(5):14-20.

Minish, B.R. 1990. Report on the status of the Greater Prairie-Chicken *Tympanuchus cupido pinnatus* in Canada. Committee on the Status of Endangered Wildlife in Canada.

Moyles, D. 1987. The Greater Prairie Chicken in Alberta. **In:** Proceedings of the Workshop on Endangered Species in the Prairie Provinces. Holroyd G.L., W.B. McGillivray, P.H.R. Stepney, D.M. Ealey, G.C. Trottier and K.E. Eberhart, eds. Provincial Museum of Alberta Natural History Occasional Paper No. 9, Edmonton, Alberta.

Moyles, D.J. 1981. Seasonal and daily use of plant communities by Sharp-tailed Grouse (*Pedioecetes phasianellus*) in the parklands of Alberta. Can. Field-Nat. 95:287-291.

Mussehl, T.W. 1963. Blue Grouse brood cover selection and land-use implications. J. Wildl. Manage. 27(4):547-555.

Redmond, G.W., D.M. Keppie and P.W. Herzog. 1982. Vegetative structure, concealment and success at nests of two races of Spruce Grouse. Can. J. Zool. 60(4):670-675.

Rogers, G.E. 1963. Blue Grouse census and harvest in United States and Canada. J. Wildl. Manage. 27(4):579-585.

Salt, J.R. 1984. Some notes on White-tailed Ptarmigan in the Alberta Rockies. Alta. Nat. 14(4):121-125.

Shantz, B.R. 1966. Greater Prairie Chicken in southern Alberta. Blue Jay 24(2):78.

Stokes, D., and L. Stokes. 1988. Pheasants. Alberta Bird Atlas project newsletter. 2(2):2. Reprinted in part from the Colorado Bird Atlas Newsletter 1988 (3):6.

Williams, L.E. Jr. 1981. The book of the Wild Turkey. Winchester Press, Oklahoma.

Zahm, G.R. 1987. Prairie boomers. Living Bird 6(2):20.

Order Gruiformes

Family Rallidae: GALLINULES, COOTS

Begin, M.T., and P. Hanford. 1987. Comparative study of retinal oil droplets in grebes and coots. Can. J. Zool. 65(8):2105-2110.

Dickson, R. 1988. Duck numbers dive as potholes turn to dustbowls. Alta. Bird Record 6(3):65-70.

Lang, A. 1990. Status report on the American Coot (*Fulica americana*). Committee on the Status of Endangered Wildlife in Canada, Report.

Lowther, J.K. 1977. Nesting biology of the Sora at Vermilion, Alberta. Can. Field-Nat. 91(1):63-67.

Mousley, H. 1937. A study of a Virginia Rail and Sora Rail at their nests. Wilson Bull. 49:80-84.

Randall, T.E. 1946. Virginia Rail nesting in Alberta. Can. Field-Nat. 60:135.

Ripley, S.D. 1977. Rails of the world: A monograph of the family Rallidae. M.F. Fenely Publ. Ltd., Toronto.

Walkinshaw, L.H. 1940. Summer life of the Sora Rail. Auk 57(2):153-168.

Family Gruidae: CRANES

Allen, R.P. 1952. The Whooping Crane. Research report No. 3, National Audubon Society, New York, NY.

Anonymous. 1972. Sighting of the Whooping Cranes. Calgary Field Nat. 4(1):6-7.

Cooch, F.G., W. Dolan, J.P. Goossen, G.L. Holroyd, B.W. Johns, E. Kuyt and G.J. Townsend. 1988. Canadian Whooping Crane recovery plan. Canadian Wildlife Service, Environment Canada.

Edwards, R., G.L. Holroyd, H. Reynolds and B. Johns. 1991. Recovery teams and recovery plans. **In:** Proceedings of the Second Endangered Species and Prairie Conservation Workshop. Holroyd G.L., G. Burns and H.C. Smith, eds. Provincial Museum of Alberta Natural History Occasional Paper No. 15, Edmonton, Alberta.

Gainer, B. 1986. Former sightings of Whooping Cranes in the Fort Vermilion area. Alta. Nat. 16(3):97.

Gollop, M.A. 1979. Status report on Whooping Crane *Grus americana* in Canada. Status report, Committe on the Status of Endangered Wildlife in Canada, Ottawa, 18 pp. + table, figure.

Henderson, P.A. 1923. Crane migration at Battle Prairie, Peace River district. Oologist 40(3):47.

Johns, B.W. 1986. 1986 spring Whooping Crane migration - prairie provinces. Blue Jay 44(3):174-176.

Johns, B.W. 1987. Whooping Crane sightings in the prairie provinces, 1979-1985. Canadian Wildlife Service Progress Notes No. 169.

Johns, B.W. 1989. Whooping Crane migration report. Blue Jay News 82:10.

Johnsgard, P.A. 1983. Cranes of the world. Indiana University Press, Bloomington.

Koonz, W.H. 1990. Unusual concentrations of Sandhill Cranes during the breeding season. Blue Jay 48(3):157.

Kuyt, E. 1976. Recent clutch size data for Whooping Cranes, including a three-egg clutch. Blue Jay 34(2):82-84.

Kuyt, E. 1978. A modern nesting record for Whooping Cranes in Alberta. Blue Jay 36(3):147-149.

Kuyt, E. 1981(a). Population status, nest site fidelity and breeding habitat of Whooping Crane. **In:** Crane research around the world: Proceedings International Crane Symposium, Sapporo, Japan. Lewis, J.C. and H. Masatomi, eds. pp. 119-125.

Kuyt, E. 1981(b). Clutch size, hatching success and survival of Whooping Crane chicks, Wood Buffalo Park, Canada. **In:** Crane research around the world: Proceedings International Crane Symposium, Sapporo, Japan. Lewis, J.C. and H. Masatomi, eds. pp. 126-129.

Kuyt, E. 1982. Whooping Crane. Hinterland who's who, Canadian Wildlife Service, Environment Canada. 6pp.

Kuyt, E. 1987. Whooping Crane. **In:** Proceedings of the Workshop on Endangered Species in the Prairie Provinces. Holroyd G.L., W.B. McGillivray, P.H.R. Stepney, D.M. Ealey, G.C. Trottier and K.E. Eberhart, eds. Provincial Museum Natural History Occasional Paper No. 9, Edmonton, Alberta.

Kuyt, E. 1988. Whooping Cranes in 1987, another year of progress. Blue Jay 46(3):136-139.

Kuyt, E. 1989. Use of a Whooping Crane nest by a Sandhill Crane. Blue Jay 47(1):33-38.

Kuyt, E. 1990. Whooping Crane numbers in 1989 recover from 1988 setback. Alta. Nat. 20(2):49-52.

Kuyt, E., and J.P. Goossen. 1985. Summary of 1984 Whooping Crane studies. Alta. Nat. 15(1):9-10.

McMillen, J.L. 1988. Conservation of North American cranes. Amer. Birds 42(5):1212-1221.

Melvin, S.M., W.J.D. Stephen and S.A. Temple. 1990. Population estimates, nesting biology and habitat preferences of Interlake, Manitoba, Sandhill Cranes, *Grus canadensis*. Can. Field-Nat. 104(3):354-361.

Smith, L.S., D.R. Blankenship, R.C. Drewien, R.A. Lock, S.R. Derrickson, F.G. Cooch and B.C. Thompson. 1986. Whooping Crane Recovery Plan 1986. U.S. Fish and Wildlife Service.

Tacha, T.C., P.A. Vohs and G.C. Iverson. 1984. Migration routes of Sandhill Cranes from mid-continental North America. J. Wildl. Manage. 48(3):1028-1033.

Wishart, W.M. 1990. One for anyone's life list! Edmonton Nat. 18(3):17-19.

Order Charadriiformes

Family Charadriidae: PLOVERS

Adam, C.I.G. 1984. Piping Plover, *Charadrius melodus*, at Lake Athabasca, Saskatchewan: A significant northward extension. Can. Field-Nat. 98(1):59-60.

Atlantic and the Prairie Piping Plover Recovery Teams. 1991. Canadian Piping Plover recovery plan. Unpublished report, Canadian Wildlife Federation, Ottawa.

Bell, F.H. 1978. Status of the Piping Plover in Canada. Committee on the Status of Endangered Wildlife inCanada, Ottawa.

Cairns, W. 1980. On the north shore. Nature Canada 9(2):47-53.

Cairns, W.E. 1982. Biology and behavior of breeding Piping Plovers. Wilson Bull. 94:531-545.

Coues, E. 1878. Field-notes on birds observed in Dakota and Montana along the Forty-ninth parallel during the seasons of 1873 and 1874. Bull. U.S. Geol. and Geogr. Survey, Survey of the Territories, Vol. IV:545-661.

Dryer, P. 1991. Piping Plover habitat protection through the North Dakota Natural Areas Registry Program. **In:** Proceedings of the Second Endangered Species and Prairie Conservation Workshop. Holroyd G.L., G. Burns and H.C. Smith, eds. Provincial Museum of Alberta Natural History Occasional Paper No. 15, Edmonton, Alberta.

Goossen, J.P. 1989. Piping Plover. Canadian Wildlife Service, Hinterland who's who series.

Goossen, J.P. 1990(a). Piping Plover research and conservation in Canada. Blue Jay 48(3):139-153.

Goosen, J.P. 1990(b). Prairie Piping Plover Conservation: Second annual report (1989). Can. Wildl. Serv. Edmonton, Alberta.

Goossen, J.P. 1991. Action plan for conserving Piping Plover in Prairie Canada. **In:** Proceedings of the Second Endangered Species and Prairie Conservation Workshop. Holroyd G.L., G. Burns and H.C. Smith, eds. Provincial Museum of Alberta Natural History Occasional Paper No. 15, Edmonton, Alberta.

Goossen, J.P., and B. Johnson. 1992. A plover's plight. Birds of the Wild Vol. 1 (2).

Graul, W.D. 1973. Adaptive aspects of the Mountain Plover social system. Living Bird 12:69-94.

Graul, W.D. 1975. Breeding biology of the Mountain Plover. Wilson Bull. 87:6-31.

Graul, W.D. 1980. Grassland management practices and bird communities. **In:** Management of western forest and grasslands for non-game birds. USDA Forest Service technical report INT-86:38-47.

Haig, S. 1985. The status of the Piping Plover in Canada. Committee on the Status of Endangered Wildlife inCanada, Ottawa.

Haig, S.M., and L.W. Oring. 1985. Distribution and status of the Piping Plover throughout the annual cycle. J. Field Ornith. 56(4):334-345.

Haig, S.M., and L.W. Oring. 1988(a). Mate, site and territorial fidelity in Piping Plovers. Auk 105:268-277.

Haig, S.M., and L.W. Oring. 1988(b). Distribution and dispersal in the Piping Plover. Auk 105:630-638.

Johnsgard, P.A. 1981. The plovers, sandpipers and snipes of the world. University of Nebraska Press, Lincoln, Nebraska. 493 pp.

Kershaw, W. 1977. The migration and occurrence of three species of shorebirds in the Edmonton area. Edmonton Nat. 5:177-182.

Knopf, F.L. 1988. Conservation of steppe birds in North America. ICBP Technical Publication 7:27-41.

Kuyt, E. 1982. Semipalmated Plover breeding in Alberta. Alta. Nat. 12:6-9.

Leachman, B., and B. Osmundson. 1990. Status of the Mountain Plover. A literature review. U.S. Fish and Wildlife Service, Fish and Wildlife Enhancement, Golden, Colorado.

Lewis, H. 1988. Piping Plover bibliography. Unpublished draft, Canadian Parks Service, Atlantic Region. 23 pp.

Mayer, P.M. 1991. Conservation management of Piping Plovers in North Dakota. **In:** Proceedings of the Second Endangered Species and Prairie Conservation Workshop. Holroyd G.L., G. Burns and H.C. Smith, eds. Provincial Museum of Alberta Natural History Occasional Paper No. 15, Edmonton, Alberta.

McKeating, G. 1987. A national perspective and management stategy on the Piping Plover. **In:** Proceedings of the Workshop on Endangered Species in the Prairie Provinces. Holroyd G.L., W.B. McGillivray, P.H.R. Stepney, D.M. Ealey, G.C. Trottier and K.E. Eberhart, eds. Provincial Museum of Alberta Natural History Occasional Paper No. 9, Edmonton, Alberta.

McNicholl, M.K. 1985. Profiles in risk status of Canadian birds: 2. Piping Plover. Alta. Nat. 15:135-138.

Recce, S.E. 1984. Endangered and threatened wildlife and plants; Piping Plover proposed as an endangered and threatened species. Federal Register 49:44712-44715.

Rowan, W. 1926(a). Notes on Alberta waders included in the British list. Part 1. Semipalmated and Killdeer Plovers. British Birds 20(1):1-10.

Rowan, W. 1926(b). Notes of Alberta waders included in the British list. Part 2. Golden and Grey Plovers. British Birds 20(2):34-42.

Sidle, J.G. 1990. To list or not to list. Living Bird 9(3):16-23.

Soper, J.D. 1941. The Mountain Plover in western Canada. Can. Field-Nat. 55:137.

Soper, J.D. 1942. The birds of Wood Buffalo Park and vicinity, Northern Alberta and district of Mackenzie, N.W.T., Canada. Trans. Royal Can. Inst. 24:19-97.

Stokes, D., and L. Stokes. 1987. Killdeers. Living bird 6(3):28-29.

Sutton, G.M., and D.F. Parmelee. 1955. Breeding of the Semipalmated Plover on Baffin Island. Bird-Banding 26:137-147.

United States Fish and Wildlife Service. 1985. Determination of endangered and threatened status for the Piping Plover. Federal Register 50(238):50720-34.

Wake, W. 1990. Species in danger. Waterfront property is costly for Piping Plovers. Nature Canada 19(1):53.

Wallis, C. 1987. Mountian Plover. **In:** Proceedings of the Workshop on Endangered Species in the Prairie Provinces. Holroyd, G.L., W.B. McGillivray, P.H.R. Stepney, D.M. Ealey, G.C. Trottier and K.E. Eberhart, eds. Provincial Museum of Alberta Natural History Occasional Paper No. 9, Edmonton, Alberta.

Wallis, C.A., and V. Loewen. 1980. First nesting records of the Mountain Plover in Canada. Alta. Nat. 10(2):63-64.

Wallis, C.A., and C.R. Wershler. 1981. Status and breeding of Mountain Plovers (*Charadrius montanus*) in Canada. Can. Field-Nat. 95:133-136.

Wershler, C.R. 1986. Lost River: Mountain Plover Study. Sweetgrass Consultants Ltd., Calgary, Alberta.

Wershler, C.R. 1987(a). The Mountain Plover in Canada. **In:** Proceedings of the Workshop on Endangered Species in the Prairie Provinces. Holroyd, G.L., W.B. McGillivray, P.H.R. Stepney, D.M. Ealey, C. Trottier and K.E. Eberhart, eds. Provincial Museum of Alberta Natural History Occasional Paper No. 9, Edmonton, Alberta.

Wershler, C.R. 1987(b). The Piping Plover in Alberta. **In:** Proceedings of the Workshop on Endangered Species in the Prairie Provinces. Holroyd G.L., W.B. McGillivray, P.H.R. Stepney, D.M. Ealey, G.C. Trottier and K.E. Eberhart, eds. Provincial Museum of Alberta Natural History Occasional Paper No. 9, Edmonton, Alberta.

Wershler, C.R. 1991(a). A management strategy for Mountain Plovers in Alberta. **In:** Proceedings of the Second Endangered Species and Prairie Conservation Workshop. Holroyd, G.L., G. Burns, H.C. Smith, eds. Provincial Museum of Alberta Natural History Occasional Paper No. 9, Edmonton, Alberta.

Wershler, C.R., and C. Wallis. 1986(a). Status report on the Mountain Plover (*Charadrius monantus*) in Canada. Committee on the status of Endangered Wildlife in Canada, Ottawa.

Wershler, C.R., and C. Wallis. 1986(b). Status of the Piping Plover in Alberta. Prep. by Sweetgrass Consultants for World Wildlife Fund, Canadian Wildlife Service and Alberta Fish and Wildlife.

Weseloh, D.V.C., and L.M. Weseloh. 1983. Numbers and nest site characteristics of the Piping Plover in central Alberta, 1974-1977. Blue Jay 41:155-161.

Family Recurvirostridae: AVOCETS AND STILTS

Bogaert, L. 1979. Black-necked Stilts nesting in Alberta. Alta. Nat. 9(2):86-89.

Chapman, B.A., J.P. Goossen and J. Ohanjanian. 1985. Occurrences of Black-necked Stilts, *Himantopus mexicanus*, in western Canada. Can. Field-Nat. 99(2):254-257.

Dekker, D. 1977. Avocets and habitat. Edmonton Nat. 5(6):144.

Dekker, D., R. Lister, T.W. Thormin, D.V. Weseloh and L.M. Weseloh. 1979. Black-necked Stilts nesting near Edmonton, Alberta. Can. Field-Nat. 93:68-69.

Dickson, R. 1989. Black-necked Stilts nest near Calgary. Pica 9(3):19-23.

Elliot, D.G. 1895. North American shorebirds. Suckling and Galloway, London.

Gibson, F. 1971. The breeding biology of the American Avocet (*Recurvirostra americana*) in central Oregon. Condor 73:444-454.

Giroux, J.-F. 1985. Nest sites and superclutches of American Avocets on artificial islands. Can. J. Zool. 63(6):1302-1305.

Hamilton, R.B. 1969. The comparative behavior of the American Avocet and the Black-necked Stilt (Recurvirostridae). Ornithol. Monogr. No. 17. pp. 1-98.

Hayman, P., J. Marchant and T. Prater. 1986. Shorebirds. An identification guide to the waders of the world. Houghton-Mifflin Co., Boston.

Kondla, N.G. 1977. An unusual American Avocet nest. Blue Jay 35(2):94-95.

Kondla, N.G., and H.W. Pinel. 1973. Known nesting sites of the American Avocet in Alberta. Calgary Field Nat. 5(1):15-16.

Kondla, N.G., and H.W. Pinel. 1978. Clutch size of the American Avocet in the prairie provinces. Blue Jay 36(3):150-153.

Kuyt, E. 1989. A recent record of American Avocet in the Northwest Territories. Alta. Nat. 19(1):27-29.

Soothill, E., and R. Soothill. 1982. Wading birds of the world. Blandford Press, Poole, Dorset.

Family Scolopacidae: SANDPIPERS, PHALAROPES AND ALLIES

Bishop, L.B. 1910. Two new subspecies of North American birds. Auk 27(1):59-63.

Blomquist, S. 1983. Bibliography of the genus *Phalaropus*. Ottenby Bird Observatory Degerhamn, Sweden. 27 pp.

Dekker, D. 1976. Muskrat, Marbled Godwit and Willet feeding on sticklebacks at Beaverhill Lake. Alta. Nat. 6(3):184-185.

Dekker, D. 1979. Long-tailed Jaeger preys on Lesser Yellowlegs. Blue Jay 37(4):221-222.

Erskine, A.J. 1977. Birds in Boreal Canada: Communities, densities and adaptations. Can. Wildl. Serv. Report Series No. 41

Farley, F.L. 1913. Is Bartram's Sandpiper disappearing from the prairies? Ottawa Naturalist 27(5-6):63.

Fearnhough, N.V. 1940. Is the Curlew doomed? Oologist LVII(8).

Goater, C.P., and A.D. Bush. 1986. Nestling birds as prey of breeding Long-billed Curlews (*Numenius americanus*). Can. Field-Nat. 100(2):263-264.

Hall, H.M. 1955. The Greater Yellowlegs. Audubon 57:154-155.

Henderson, A.D. 1924. A nest of the Solitary Sandpiper. Oologist 41(11):132-133 [reprinted in Alta. Nat. 5:129-130].

Henderson, A.D. 1927. Nesting habits of the Lesser Yellowlegs (*Totanus flavipes*). Oologists' Rec. 7(1):13-15.

Henderson, A.D. 1931. The Long-billed Curlew at Belvedere, Alberta. Auk 48(3):418.

Higgins, K.F., and L.M. Kirsch. 1975. Some aspects of the breeding biology of the Upland Sandpiper in North Dakota. Wilson Bull. 87(1):97-101.

Higgins, K.F., L.M. Kirsch, M.R. Ryan and R.B. Renken. 1979. Some Ecological Aspects of Marbled Godwits and Willets in North Dakota. Prairie Nat. 11(4):115-118.

Höhn, E.O. 1965. Phalarope mysteries and some solutions. Beaver 295(1):51-54.

Höhn, E.O. 1967. Observations of the breeding biology of Wilson's Phalarope (*Steganopus tricolor*) in central Alberta. Auk 84:220-224.

Höhn, E.O. 1969. The phalarope. Sci. Amer. 220(6):104-111.

Höhn, E.O. 1981. Northern Phalarope flocks at Miquelon Lake, Alberta. Blue Jay 39(1):41-43.

Höhn, E.O., and P. Marklevitz. 1974. Noteworthy summer observations of birds in the Caribou Mountains, Alberta. Can. Field-Nat. 88(1):77-78.

Höhn, E.O., and D.J. Mussel. 1980. Northern Phalarope breeding in Alberta. Can. Field-Nat. 94:189-190.

Jenni, D.A., R.L. Redmond and T.K. Bicak. 1982. Behavioral ecology and habitat relationships of Long-billed Curlews in western Idaho. Bureau of Land Management, Boise, Idaho. 234 pp.

Johns, J.E. 1969. Field studies of Wilson's Phalarope. Auk 86:660-670.

Johnsgard, P.A. 1981. The plovers, sandpipers and snipes of the world. University of Nebraska Press, Lincoln.

Kuyt, E. 1974. Wilson's Phalarope in breeding plumage near Fort Smith, N.W.T. Blue Jay 32:177-178.

Lang, V., and B. Lang. 1973. Some comments on the distribution and status of the Upland Plover in the Calgary area. Calgary Field-Nat. 4(9):175-177.

McNicholl, M.K. 1981. Egg-teeth of Spotted Sandpipers. N. Am. Bird Bander 6(2):44-45.

Miller, E.H., W.W.H. Gunn, J.P. Myers and B.N. Veprintsev. 1984. Species-distinctiveness of Long-billed Dowitcher Song (Aves: Scolopacidae). Proc. Biol. Soc. Washington 97(4):804-811.

Morrison, R.I.G. 1989. Shorebirds. Canadian Wildlife Service, Hinterland who's who series.

Mouseley, H. 1939. Nesting behavior of Wilson's Snipe and Spotted Sandpiper. Auk 56:129-133.

Murray, B.G. Jr. 1983. Notes on the breeding biology of Wilson's Phalarope. Wilson Bull. 95:472-475.

Oring, L.W. 1973. Solitary Sandpiper early reproductive behavior. Auk 90(3):652-663.

Pankratz, H., and E. Kuyt. 1986. Occurrence in northern Alberta of Wilson's Phalarope and first breeding record for the N.W.T. Alta. Nat. 16(1):9-10.

Randall, T.E. 1927. Nesting of the Marbled Godwit (*Limosa fedoa*). Oologists' Rec. 7(1):4.

Randall, T.E. 1927. Nesting of Wilson's Phalarope (*Steganopus tricolor*). Oologist's Rec. 7(1):3-4.

Redmond, R.L., and D.A. Jenni. 1986. Population ecology of the Long-billed Curlew (*Numenius americanus*) in western Idaho. Auk 103:755-767.

Renaud, W.E. 1980. The Long-billed Curlew in Saskatchewan: status and distribution. Blue Jay 38(4):221-237.

Robbins, C.S., D. Bystrak and P.H. Geissler. 1986. The Breeding Bird Survey: Its First Fifteen Years, 1965-1979. U.S. Dept. of the Interior, Fish and Wildl. Serv. Resource Pub. 157. Washington, D.C.

Rowan, W. 1926. Notes on Alberta waders included in the British list. Part III. Turnstone, Bartram's Sandpiper, Sanderling, Knot and Dunlin. British Birds 20(4):82-90.

Rowan, W. 1926. Notes on Alberta waders included in the British list. Part IV. Sandpipers. British Birds 20(6):138-145.

Rowan, W. 1926. Notes on Alberta waders included in the British list. Part V. Buff-breasted Sandpiper, (*Tryngites subruficollis*). British Birds 20(8):186-192.

Rowan, W. 1927. Notes on Alberta waders included in the British list. Part VI. Dowitcher and Spotted Sandpiper. British Birds 20(9):210-222.

Rowan, W. 1929. Notes on Alberta waders included in the British list. Part VII. British Birds 23(1):2-17.

Sadler, D.A.R., and W.J. Maher. 1976. Notes on the Long-billed Curlew in Saskatchewan. Auk 93:383-384.

Stenzel, L.E., H.R. Huber and G.W. Page. 1976. Feeding behavior and diet of the Long-billed Curlew and Willet. Wilson Bull. 88:314-332.

Stirling, D. 1962. Marbled Godwit range extension in Alberta. Blue Jay 20(4):154.

Street, J.F. 1923. On the nesting grounds of the Solitary Sandpiper and Lesser Yellow-legs. Auk 40:577-583.

Taverner, P.A. 1940. The distribution of the western Solitary Sandpiper. Condor 42(4):215-217.

Timkin, R.L. 1969. Notes on the Long-billed Curlew. Auk 86:750-751.

Tuck, L.M. 1972. The snipes. Canadian Wildlife Service, Monograph Series No. 5, Environment Canada.

Vermeer, K. 1971. Spotted Sandpipers as possible indicators of mercury contamination of rivers. Blue Jay 29(2):59-60.

Family Laridae: SKUAS, GULLS, TERNS, AND SKIMMERS

Anonymous. 1960. Bird Notes. Calgary Bird Club Bull. No. 6.

Asher, C.R. 1974. Sabine's Gull sighting - Bassano, Alta. Sept. 8/73. Calgary Field Nat. 6(6):193.

Bailey, P.F. 1977. The breeding biology of the Black Tern (*Chlidonias niger surinamensis* Gmelin). M.Sc. Thesis, University of Wisconsin, Oshkosh, Wisconsin.

Barth, E.K. 1955. Egg-laying, incubation and hatching of the Common Gull (*Larus canus*). Ibis 97:222-239.

Berezay, K., A. Quinton, T. Sheen, K. Powers, C. Olsen and T. Berry. 1975. Ring-billed and California Gulls on Ross Lake. Alta. Nat. 5(3): 75-76.

Bergman, A.D., P. Swain and M.W. Weller. 1970. A comparative study of nesting Forster's and Black Terns. Wilson Bull. 82:435-444.

Beswick, B., S. Draper, D. Parkes and L. Broadhead. 1975. Ring-billed and California Gulls on St. Mary's Reservoir. Alta. Nat. 5(3):76.

Brock, J. 1969. Bonaparte's Gull. **In:** Fisher, J., R.T. Peterson, S. Cramp, E. Hosking, J. Warham, V. Serventy, J.G. Williams, and B. Coleman. Birds of the World. Vol. 4(1) No. 37. IPL Magazines Ltd., London.

Burger, J. 1988. Foraging behavior in gulls: differences in method, prey and habitat. Colonial Waterbirds 11:9-23.

Burton, J.F. 1969. Franklin's Gull. **In:** Fisher, J., R.T. Peterson, S. Cramp, E. Hosking, J. Warham, V. Serventy, J.G. Williams, and B. Coleman. Birds of the World Vol. 4(1) No. 37 pp.1019-1020. IPL Magazines Ltd., London.

Campbell, R.W., and R.G. Fottit. 1976. The Franklin's Gull in B.C. Syesis 5:99-106.

Carmel, R. 1986. Mallard attacks Ring-billed Gull. Alta. Nat. 16(4):135.

Chapman, M.B. 1986. Factors influencing reproductive success and nesting strategies in Black Terns. Ph.D. Thesis, Simon Fraser University, British Columbia.

Conover, M.R. 1983. Recent changes in Ring-billed and California Gull populations in the western United States. Wilson Bull. 95:362-383.

Cuthbert, N.L. 1954. A nesting study of the Black Tern in Michigan. Auk 71:36-63.

Drury, W.H. 1973. Herring Gull. Hinterlands who's who series, No. 10, Environment Canada, Canadian Wildlife Service.

DuMont, P.A. 1940. Relations of Franklin's Gull colonies to agriculture on the Great Plains. Trans. N. Amer. Wildl. Conf. 5:183-189.

Dwight, J. 1925. The Gulls (Laridae) of the world: their plumages, moults, relationships and distribution. Bull. Am. Mus. Nat. Hist. 52:63-408.

Erskine, A.J. 1968. Birds observed in north-central Alberta, summer 1964. Blue Jay 26:24-31.

Evans, R.M., D.B. Krindle and M.E. Mattson. 1970. Caspian Terns nesting near Spruce Island, Lake Winnipegosis, Manitoba. Blue Jay 28:38-71.

Farley, F.L. 1931. Nesting of Bonaparte's Gull (*Larus philadelphia*) in central Alberta. Can. Field-Nat. 45(6):138-139.

Fox, G.A. 1976. Eggshell quality: its ecological and physiological significance in a DDE-contaminated Common Tern population. Wilson Bull. 88:459-477.

Gerrard, J.M., and D.W.A. Whitfield. 1971. Breeding distribution of Forster's Tern in the prairie provinces. Blue Jay 29(1):19-22.

Godfrey, W.E. 1952. Birds of the Lesser Slave Lake -Peace River areas, Alberta. Natl. Museum Canada, Bull. 126:142-175.

Guay, J.W. 1968. The breeding biology of Franklin's Gull (*Larus pipixcan*). Ph.D. Thesis, University of Alberta, Edmonton. 129 pp.

Hatch, D.R.M. 1973. Hatched egg-shells covering Common Tern eggs. Blue Jay 31(2):91.

Holroyd, G. 1986. Great grebe tour - a great success. Edmonton Nat. 14(2):16.

Holroyd, G.L., and E. Beaubien. 1983. Ring-billed Gull, (*Larus delawarensis*), predation on bat, *Myotis*, injured by an American Kestrel, (*Falco sparverius*). Can. Field-Nat. 97(4):452.

Johnson, J.D. 1983. The Wagner Natural Area - A remarkable success story. Alta. Nat. 13(4):146-149.

Lange, J. 1984. Edmonton Christmas bird count 1983. Edmonton Nat. 12(1):12.

MacPherson, A.H. 1961. Observations on Canadian Arctic *Larus* gulls, and the taxonomy of *L. Thayeri* Brooks. Arctic Inst. North Amer., Tech, Papers, No. 7:1-40.

Martin, M. 1979. Status report on Caspian Tern (*Sterna caspia*) in Canada. Status report, Committee on the Status of Endangered Wildlife in Canada, Ottawa. 43 pp.

McNicholl, M.K. 1971. The breeding biology and ecology of Forster's Tern (*Sterna forsteri*) at Delta, Manitoba. M.Sc. Thesis, University of Manitoba, Winnipeg.

McNicholl, M.K. 1978. Franklin's Gulls at hamburger stands. Alta. Nat. 8(4):197-198.

McNicholl, M.K. 1982. Factors affecting reproductive success of Forster's Terns at Delta Marsh, Manitoba. Colonial Waterbirds 5:32-38.

Meyerriecks, R. 1990. Magnificent Scavenger (Herring Gull). Wildbird 4(6):22-29.

Morris, R.D., and R.A. Hunter. 1976. Factors influencing desertion of colony sites by Common Terns (*Sterna hirundo*). Can. Field-Nat. 90(2):137-143.

Munro, J.A. 1922. Notes on the waterbirds of Lake Newell, Alberta. Can. Field-Nat. 36(5):89-91.

Munro, J.A. 1927. Gull colonies on Lake Newell, Alberta. Can. Field-Nat. 41(3):61.

Munro, J.A. 1936. A study of the Ring-billed Gull in Alberta. Wilson Bull. 48(3):169-180.

Nordstrom, W. 1978. Lake Newell: Colonial birds in jeopardy. Alta. Nat. 8(2):107-111.

Palmer, R.S. 1941. A behavior study of the Common Tern. Boston Society of Natural History 42:1-119.

Pinel, H.W. 1986. Herring Gull migration in the Calgary area, 1966-1985. Alta. Nat. 16(3):79-81.

Pittman, H.H. 1927. The Black Terns of Saskatchewan. Condor 29:140-145.

Ryder, R.A. 1966. Some movements of Ring-billed Gulls color-marked in Colorado. Blue Jay 24(2):73-75.

Salt, G.W., and D.E. Willard. 1971. The hunting behavior and success of Forster's Tern. Ecology 52:989-998.

Severinghaus, L. 1983. The case of the Common Tern. Living Bird 2(2):24-26.

Soper, J.D. 1949. Birds observed in the Grande Prairie-Peace River region of northwestern Alberta, Canada. Auk 66:233-257.

Termaat, B.M., and J.P. Ryder. 1984. Differences in skeletal characteristics between disjunct eastern and western populations of Ring-billed Gulls. Can. J. Zool. 62(6):1067-1074.

Tilgham, N.G. 1980. The Black Tern survey, 1979. Passenger Pigeon 42:1-8.

Tinbergen, N. 1953. The Herring Gull's world - a study of the social behavior of birds. Collins Clear - Type Ress, London, England. 255 pp.

Twomey, A.C. 1934. Breeding habits of Bonaparte's Gull. Auk 51(3): 291-296.

Vermeer, K. 1967. A study of two species of gulls, *Larus californicus* and *L. delawarensis*, breeding in an inland habitat. Ph.D. Thesis, University of Alberta, Edmonton. 128 pp.

Vermeer, K. 1967. Common Terns (*Sterna hirundo*) nesting at Miquelon Lake, Alberta. Can. Field-Nat. 81(4):274-275.

Vermeer, K. 1969. Egg measurements of California and Ring-billed Gulls at Miquelon Lake, Alberta, in 1965. Wilson Bull. 81(1):102-103.

Vermeer, K. 1970(a). Breeding biology of California and Ring-billed Gulls: a study of ecological adaptation to the inland habitat. Canadian Wildlife Service Report Series No. 12. 52 pp.

Vermeer, K. 1970(b). Breeding records of Herring Gulls in Alberta and California Gulls in Manitoba. Can. Field-Nat. 84(3):182.

Vermeer, K. 1970(c). Large colonies of Caspian Terns on Lakes Winnipeg and Winnipegosis, 1970. Blue Jay 28(3):117-118.

Vermeer, K. 1973. Food habits and breeding range of Herring Gulls in the Canadian prairie provinces. Condor 75(4):478-480.

Weseloh, D.V. 1971. Occurrences of the Mew Gull (*Larus canus*) in Calgary, Alberta, fall 1971. Calgary Field Nat. 8(8):202-204.

Weseloh, D.V. 1972(a). Letters to the editor. Calgary Field Nat. 3(11):162-163.

Weseloh, D.V. 1972(b). Migration of the Herring Gull in the Calgary area, 1965-1971. Calgary Field Nat. 3(9):122-123.

Weseloh, D.V. 1973. Terns in southern Alberta. Alta. Nat. 3(3):47-53.

Weseloh, D.V. 1975. Comments on a 1972 Glaucous Gull sighting in Calgary. Calgary Field Nat. 7(3):88-90.

Weseloh, D.V. 1978. Lake Newell: Colonial birds in jeopardy - further comments. Alta. Nat. 8(4):209-211.

Weseloh, D.V. 1981. A probable Franklin's x Ring-billed Gull pair nesting in Alberta. Can. Field-Nat. 95(4):474-476.

Weseloh, D.V., and H. Blockpoel. 1979. Hinterland who's who: Ring-billed Gull. Environment Canada Wildlife Service, No. 63.

Weseloh, D.V., and L. Cocks. 1979. Recent nesting of the Caspian Tern at Egg Island, Lake Athabasca, Alberta. Blue Jay 37(4):212-215.

Weseloh, D.V., and V. Lang. 1973. Glaucous-winged Gull and Thayers Gull at Calgary, Alberta. Blue Jay 31(4):230-232.

Weseloh, D.V., and L.M. Weseloh. 1975. Mew Gull at Frontier Farms landfill site. Edmonton Nat. 3(8):6.

Yuill, D., R. West, R. Leavitt, M. French and K. Henn. 1975. Ring-billed and California Gulls on Shanks Lake. Alta. Nat. 5(3):77-78.

Order Columbiformes

Family Columbidae: PIGEONS AND DOVES

Anonymous. 1959. A bed of nails. Land, Forest, Wildlife 2(3):10.

Bruggeman, P.F. 1933. Mourning Doves. Can. Field-Nat. 47(4):75.

Fimreite, N., R.W. Fyfe and J.A. Keith. 1970. Mercury contamination of Canadian prairie seed eaters and their avian predators. Can. Field-Nat. 84(3):269-276.

Goodwin, D. 1967. Pigeons and doves of the world. British Museum (Natur. Hist.) London.

Houston, C.S. 1986. Mourning Dove numbers explode on the Canadian prairies. Amer. Birds 40(1):52-54.

Kondla, N.G., and H.W. Pinel. 1973. Frequency of ground nesting by the Mourning Dove in the prairie provinces. Calgary Field Nat. 4:154.

Lang, V. 1973. Behavioral observation: Rock Dove (*Columba livia*). Calgary Field Nat. 4:148.

McGillivray, W.B. 1988. Breeding of the Rock Dove, (*Columba livia*), in January at Edmonton, Alberta. Can. Field-Nat. 102(1):76-77.

Nice, M.M. 1922-23. A study of the nesting of Mourning Doves. Auk 39:457-474; 40:37-58.

Preble, D.E., and F.H. Heppner. 1981. Breeding success in an isolated population of Rock Doves. Wilson Bull. 93:357-362.

Stokes, D, and L. Stokes. 1990. Mourning Doves. Living Bird 9(2):32-33.

Switzer, B. 1967. Breeding biology of the Mourning Dove (*Zenaida macroura*) in southern Alberta. Unpubl. report. CWS.

Switzer, B.C. 1970. Reproductive biology of the Mourning Dove, *Zenaida macroura*, in southern Alberta. M.Sc. Thesis, University of Alberta, Edmonton. 98 pp.

Weseloh, C. 1975. The co-operative breeding bird survey in Canada, 1974. Calgary Field Nat. 7(2):37-38.

Order Cuculiformes

Family Cuculidae: CUCKOOS, ROADRUNNERS, AND ANIS

Anonymous. 1973. Field trip news. Calgary Field Nat. 5(2):47.

Kondla, N. 1974. A nesting record of the Black-billed Cuckoo for the Drumheller area. Calgary Field Nat. 6(5):154.

Ross, H. 1980. Club news Lethbridge Naturalist Society July 1980. Alta. Nat. 10(3):129-130.

Wyllie, I. 1981. The Cuckoo. B.T. Batsford Ltd. London. 176 pp.

Order Strigiformes

Family Strigidae: TYPICAL OWLS

Alberts, S. 1965. Observations of nesting Short-eared Owls. Blue Jay 23(3):150.

Anonymous. 1956. Alberta controversy re: protection of birds of prey. Blue Jay 114(1):15-16.

Anonymous. 1971. Saw Whet Owl. Dinny's Digest 1(10):4.

Austing, G.R., and J.B. Holt Jr. 1966. The world of the Great Horned Owl. J.B. Lippincott Co., New York. 158 pp.

Beck, B., and J. Beck. 1988. The greater Edmonton Owl prowl. Alta. Nat. 18(3):97-100.

Beck, B., and J. Beck. 1989. 1989 Alberta owl prowl (and frog jog). Alta. Nat. 19(1):6.

Bonney, R.E. Jr. 1983. More than just a pretty face. Living Bird 2(1):10-13.

Boxall, P.C. 1986. Additional observations of the Barred Owl in Alberta. Blue Jay 44(1):41-43.

Boxall, P.C. 1988. Owls. Alberta Bird Atlas Newsletter 2(1) spring 1988.

Boxall, P.C., and M.R. Lein. 1989. Time budget and activity of wintering Snowy Owls. Jour. of Field Ornith. 60:20-29.

Boxall, P.C., and P.H.R. Stepney. 1982. The distribution and status of the Barred Owl in Alberta. Can. Field-Nat. 96(1):46-50.

Burnett, J.A., C.T. Dauphine Jr., S.H. McCrindle and T. Mosquin. 1989. On the brink: Endangered species in Canada. Western Producer Prairie Books, Saskatoon. 192 pp.

Clark, R.J. 1975. A field study of the Short-eared Owl, *Asio flammeus* (Pontoppidan) in North America. Wildl. Monog. No. 47. 67 pp.

Di Labio, B. 1990. Bird watch (Great Horned Owl). Nature Canada 19(1):48-49.

Eckert, A.W. 1974. The owls of North America (north of Mexico). Doubleday and Co., Inc., Garden City, New York.

Everett, M. 1977. A natural history of owls. Hamlyn Publ., New York. 156 pp.

Godfrey, W.E. 1947. A new Long-eared Owl. Can. Field-Nat. 61:196-197.

Godfrey, W.E. 1967. Some winter aspects of the Great Gray Owl. Can. Field-Nat. 81:99-101.

Green, G.A., and R.G. Anthony. 1989. Nesting success and habitat relationships of Burrowing Owls in the Columbia Basin, Oregon. Condor 91:347-354.

Haug, E., and L.W. Oliphant. 1990. Movements, activity patterns and habitat use of Burrowing Owls in Saskatchewan. J. Wildl. Manage. 54(1):27-35.

Haug, E.A. 1991. 1988 Manitoba Burrowing Owl conservation program: status report. **In:** Proceedings of the Second Endangered Species and Prairie Conservation Workshop. Holroyd, G.L., G. Burns and H.C. Smith, eds. Provincial Museum of Alberta Natural History Occasional Paper No. 15, Edmonton, Alberta.

Henderson, A.D. 1915. Nesting of the Great Gray Owl in central Alberta. Oologist 32(1):2-6.

Henderson, A.D. 1924. Early nesting of Arctic Horned Owl. Oologist 61(4):42.

Hoffman, W. 1990. Banding adult Great Horned Owls in the Edmonton area. Blue Jay 48(1):39-40.

Hoffmann, W. 1991. Ghost of the old timber. Wildbird 5(1):26-31.

James, P.C., T.J. Ethier, G.A. Fox and M. Todd. 1991. New aspects of Burrowing Owl biology. **In:** Proceedings of the Second Endangered Species and Prairie Conservation Workshop. Holroyd, G.L., G. Burns and H.C. Smith, eds. Provincial Museum of Alberta Natural History Occasional Paper No. 15, Edmonton, Alberta.

Johns, S. 1977. Saw Whet Owl attacks Robin. Blue Jay 35(3):172.

Johns, S., and A. Johns. 1978. Observations on the nesting behaviour of the Saw-whet Owl in Alberta. Blue Jay 36(1):36-38.

Johnsgard, P.A. 1988. North American owls. Biology and natural history. Smithsonian Institution Press, Washington.

Johnson, J. 1976. Courtship flight behavior of a Long-eared Owl at Black Diamond. Calgary Field-Nat. 7(9):241.

Johnson, J., and A. Larson. 1974. Watching Great Horned Owl nests in the Alberta foothills. Nature Canada 3(4):3-7.

Jones, E.T. 1954. The Great Gray Owl. Blue Jay 12(1):8.

Jones, E.T. 1966. Barred Owl nest record for Alberta. Blue Jay 24(3):140.

Jones, E.T. 1987. Early observations of Barred Owl in Alberta. Blue Jay 45(1):31-32.

Karpinski, D.A. 1981. Habits of the Great Horned Owl. Edmonton Nat. 9(2):9-16.

Kondla, N.G. 1973. Great Gray Owls raise two young southeast of Edmonton, Alberta. Blue Jay 31(2):98-100.

Law, C. 1960. The Great Gray Owl of the woodlands. Blue Jay 18:14-16.

Lewis, C. 1987. No home on the range. Nature Canada 16(4):41-44.

McGillivray, W.B. 1985. Size, sexual dimorphism and their measurement in Great Horned Owls in Alberta. Can. J. Zool. 63:2364-2372.

McInvaille, W.B. Jr., and L.B. Keith. 1974. Predator-prey relations and breeding biology of the Great Horned Owl and Red-tailed Hawk in central Alberta. Can. Field-Nat. 88(1):1-20.

McNicholl, M.K. 1988. Ecological and human influences on Canadian populations of grassland birds. **In:** Ecology and conservation of grassland birds. Gorivp, P.D., ed. ICBP Tech. Publ. No. 7, Cambridge, U.K.

Mikkola, H. 1983. Owls of Europe. Buteo Books, Vermilion, South Dakota.

Murray, G.A. 1976. Geographic variation in the clutch sizes of seven owl species. Auk 93(3):602-613.

Nero, R.W. 1969. The status of the Great Gray Owl in Manitoba with special reference to the 1968-69 influx. Blue Jay 27:191-209.

Nero, R.W. 1971. Spirit of the boreal forest: the Great Gray Owl. Beaver 302:25-29.

Nero, R.W. 1979. Status report on the Great Gray Owl (*Strix nebulosa*) in Canada, 1979. Committee on the Status of Endangered Wildlife in Canada, Ottawa.

Nero, R.W. 1980. The Great Gray Owl: Phantom of the northern forest. Smithsonian Institution. Press, Washington, D.C.

Nero, R.W. 1982. Leaving home Great Gray Owl. Manitoba Nature 23(2):39-48.

Nero, R.W. 1991(a). White Great Gray Owl. Blue Jay 49(1):31.

Nero, R.W. 1991(b). Focal concentration: a possible cause of mortality in the Great Gray Owl. Blue Jay 49(1):28-30.

Nero, R.W., and H.W.R. Copland. 1981. High mortality of Great Gray Owls in Manitoba - winter 1980-1981. Blue Jay 39:158-165.

Norman, E.S. 1915. Nesting of the western Horned Owl in central Alberta. Oologist 32(1):6-9.

Oeming, A.F. 1955. A preliminary study of the Great Gray Owl (*Strix nebulosa*) in Alberta. Dissertation, University of Alberta, Edmonton.

Oeming, A.F., and E.T. Jones. 1955. The Barred Owl in Alberta. Can. Field-Nat. 69:66-67.

Pinel, H.W. 1978. Brood size and food habits of Great Horned Owls near Calgary, Alberta. Blue Jay 36(3):154-156.

Plummer, D., and P.C. Christgau. 1984. Some go ... some don't. Nature Canada 13(4):22-27.

Preble, E.A. 1941. Barred Owl on Athabasca River, Alta. Auk 58(3):407-408.

Randall, T.E. 1930. Arctic Horned Owl. Oologists' Rec. 10(4):74-75.

Rees, C. 1972. Some notes on nesting Great Horned Owls. Blue Jay 30(2):98-100.

Rich, T. 1986. Habitat and nest-site selection by Burrowing Owls in the sagebrush steppe of Idaho. J. Wildl. Manage. 50:548-555.

Rusch, D.H., E.C. Meslow, P.D. Doerr and L.B. Keith. 1972. Response of Great Horned Owl populations to changing prey densities. J. Wildl. Manage. 36(2):282-296.

Salt, J.R. 1984. Hunting by Horned Owls in the "alpine zone" of the Rockies. Alta. Nat. 14(4):143-144.

Salt, W.R. 1950. The Burrowing Owl in Alberta. Can. Field-Nat. 64:221.

Schutz, F. 1978. Pygmy Owl dives on Black-capped Chickadee. Alta. Nat. 8:30-31.

Simpson, R.B. 1915. The Great Horned Owl. Oologist 32(1):10-12.

Simpson, R.B. 1915. The Barred Owl. Oologist 32(1):9-10.

Smith, H.C. 1976. Comparison of food items found in pellets of seven species of owls. Edmonton Nat. 4(2):36-38.

St. Clair, K. 1972. Report on fieldtrips: Drumheller trip. Calgary Field Nat. 3(11):148.

Taverner, P.A. 1942. Canadian races of the Great Horned Owls. Auk 59(2):234-245.

Taylor, A.L. Jr., and E.D. Forsman. 1976. Recent range extensions of the Barred Owl in western North America, including the first records for Oregon. Condor 78(4):560-561.

Trann, K. 1974. Short-eared Owls near Edmonton, 1970-1973. Blue Jay 32(3):148-153.

Trann, K. 1976. Short-eared owls in the Edmonton area 1974. Alta. Nat. 6:274-283.

Walker, L.W. 1974. The Book of Owls. Alfred A. Knopf, New York. 255 pp.

Wedgewood, J.A. 1978. The status of the Burrowing Owl in Canada. Committee on the Status of Endangered Wildlife in Canada, Ottawa.

Winn, R. 1989. Cavity nesting owls of the front range. Colorado Bird Atlas Newsletter (9):6-7.

Wood, K. 1987. Owl predation on flying squirrels. Pica 7(2):27.

Order Caprimulgiformes

Family Caprimulgidae: GOATSUCKERS

Boag, D. 1972. Poor-will from east-central Alberta. Can. Field-Nat. 86:296-297.

Bulmer, B. 1977. Common Nighthawk nesting observations. Edmonton Nat. 5(8):19.

Goulden, L.L. 1972. Notes on a captive Poor-will in Alberta. Blue Jay 30(4):222-225.

Oberholser, H.C. 1926. The migration of North American birds. Second series. XXXI. The nighthawks. Bird-Lore 28(4):255-261.

Pinel, H.W., and J.R. Riddell. 1986. First authenticated record of Common Poorwill in the Calgary area. Alta. Nat. 16(4):135.

Rand, A.L. 1948. Distributional notes on Canadian Birds. Can. Field-Nat. 62(6):175-180.

Rust, H.J. 1947. Migration and nesting of Nighthawks in northern Idaho. Condor 49:177-188.

Selander, R.K. 1954. A systematic review of the booming Nighthawks of western North America. Condor 56(2):57-82.

Order Apodiformes

Family Apodidae: SWIFTS

Holroyd, G.L. 1987. Black Swift nests at Maligne Canyon, Jasper National Park. Alta. Nat. 17(2):46-48.

Kondla, N.C. 1973. Nesting of the Black Swift at Johnston's Canyon, Alberta. Can. Field-Nat. 87(1):64-65.

Oberholser, H.C. 1926. The migration of North American birds. Second series XXIX. The Swifts. Bird-Lore 28(1):9-13.

Family Trochilidae: HUMMINGBIRDS

Brunton, D.F., S. Andrews and D.G. Paton. 1979. Nesting of the Calliope Hummingbird in Kananaskis Provincial Park, Alberta. Can. Field-Nat. 93(4):449-451.

Cassels, E. 1928. Rufous Hummingbird. Can. Field-Nat. 42(6):150.

Cristall, M. 1989. Lady rainbow, the hummingbird mom. Alta. Nat. 19(1):13-16.

Di Labio, B. 1991. Birdwatch. Nature Canada 20(2):54.

Forsyth, A. 1990. Natural liaisons: Rubythroats and Yellow-bellies. Equinox 51:37-38.

Gehlert, R.E. 1970. Rufous Hummingbird at Jasper, Alberta. Blue Jay 28(1):36-37.

Oberholser, H.C. 1924. The migration of North American Birds. Second series XXIV. Ruby-throated, Black-chinned and Calliope Hummingbirds. Bird-Lore 26(2):108-111.

Saunders, W.E. 1902. Canadian hummingbirds. Ottawa Naturalist 16(4):97-103.

Soper, J.D. 1965. Recent additions to the list of Jasper National Park birds. Can. Field-Nat. 79(3):214-215.

Stokes, D., and L. Stokes. 1989. The hummingbird book. Little, Brown and Company, Toronto.

Order Coraciiformes

Family Alcedinidae: KINGFISHERS

Fry, C.H. 1980. The evolutionary biology of kingfishers. Living Bird 18:113-160.

Oberholser, H.C. 1930. The migration of North American birds. Second series XLIV. The Kingfishers. Bird-Lore 32(6):414-417.

Prose, B.L. 1985. Habitat suitability index models: Belted Kingfisher. U.S. Dept. of the Interior, Fish and Wildlife Service, Biol. Rep. 82(10.87), Washington, D.C. 22 pp.

Rand, A.L. 1948. Distributional notes on Canadian birds. Can. Field-Nat. 62:175-180.

Order Piciformes

Family Picidae: WOODPECKERS AND WRYNECKS

Albert, M.G. 1986. Yellow-bellied Sapsucker. Edmonton Nat. 14(3):12-13.

Alexander, S.E. 1986. Unconfirmed sighting of a Three-toed Woodpecker with a white back instead of usual ladder back. Alta. Bird Record 4(2):45.

Anonymous. 1988. Wonderful woodpeckers. Alberta Bird Atlas Newsletter 2(3):6-8.

Banfield, A.W.F. 1946. Field notes on ranges and wildlife, Athabasca Valley, Jasper National Park, June 1946. Unpubl. field notes, Canadian Wildlife Service. 33 pp.

Bangs, D. 1900. A review of the three-toed woodpeckers of North America. Auk 17(2):126-142.

Bock, C.E. 1970. The ecology and behavior of the Lewis' Woodpecker (*Asyndesmus lewis*). Univ. Calif. Publ. Zool. 92:1-91.

Bock, C.E., and J.H. Bock. 1974. On the geographical ecology and evolution of the three-toed woodpeckers *Picoides tridactylus* and *P. arcticus*. Amer. Midl. Nat. 92(2):397-405.

Bull, E.L. 1987. Ecology of the Pileated Woodpecker in northeastern Oregon. J. Wildl. Manage. 51(2):472-481.

Bull, E.L., E.C. Meslow. 1977. Habitat requirements of the Pileated Woodpecker in northeastern Oregon. J. of Forestry 75(6):335-337.

Bulmer, B., and G. Bulmer. 1973. Flicker and sapsucker nesting in same tree. Edmonton Nat. 10:8.

Burns, P.L. 1900. Monograph of the flickers. Wilson Bull. 7:1-82.

Cook, G.I. 1924. Trapnesting the Sapsucker. Oologist 41(10):129.

Erskine, A.J., and W.D. McLaren. 1972. Sapsucker nest holes and their use by other species. Can. Field-Nat. 86:357-361.

Farley, F.L. 1931. Lewis's Woodpecker *Asyndesmus lewis* in central Alberta. Can. Field-Nat. 45(3):71.

Greenlee, G.M. 1972. Pileated Woodpecker. Blue Jay 30(4):258.

Hadow, H.H. 1976. Growth and development of nesting Downy Woodpeckers. N. Am. Bird Bander 1:155-164.

Hall, E. 1987. Unusual behavior of Northern Flickers. Alta. Nat. 17(4):156.

Howell, T.R. 1952. Natural history and differentiation in the Yellow-bellied Sapsucker. Condor 54(5):237-282.

Hoyt, S.F. 1957. The ecology of the Pileated Woodpecker. Ecology 38:246-256.

Keisker, D.G. 1986. Nest tree selection by primary cavity nesting birds in south-central British Columbia. M.Sc. Thesis, Simon Fraser University, Burnaby, B.C.

Kilham, L. 1962. Breeding behavior of Yellow-bellied Sapsuckers. Auk 79:31-43.

Kilham, L. 1964. The relations of breeding Yellow-bellied Sapsuckers to wounded birches and other trees. Auk 81:520-527.

Kilham, L. 1968. Reproductive behavior of Hairy Woodpeckers II. Nesting and habitat. Wilson Bull. 80:286-305.

Kilham, L. 1971. Reproductive behavior of Yellow-bellied Sapsuckers. I. Preference for nesting in *Fomes*-infected aspens, and nest hole interrelations with flying squirrels, raccoons and other animals. Wilson Bull. 83:159-171.

Koenig, W.D. 1984. Geographic variation in clutch size in the Northern Flicker (*Colaptes auratus*): support for Ashmole's hypothesis. Auk 101(4):698-706.

Lange, J. 1983. Downy Woodpecker vs. Yellow-bellied Sapsucker. Alta. Nat. 13(3):114-115.

Lawrence, L. de K. 1967. A comparative life-history of four species of woodpeckers. Ornith. Monogr., No. 5.

McGillivray, W.B., and G.C. Biermann. 1987. Expansion of the zone of hybridization of Northern Flickers in Aberta. Wilson Bull. 99(4):690-692.

McNicholl, M., and P. Taylor. 1973. Woodpecker season. Edmonton Nat. 1(4):7-8.

Oberholser, H.C. 1927. The migration of North American birds. Second series. XXXIII. The flickers. Bird-Lore 29(2):110-114.

Oberholser, H.C. 1928. The migration of North American birds. Second series. XXXVII. Yellow-bellied and Red-breasted Sapsuckers. Bird-Lore 30(4):253-257.

Patton, D. 1926. Notes on flickers in Alberta. Can. Field-Nat. 40(2):41.

Rand, A.L. 1944. A northern record of the Flicker and a note on the cline *Colaptes auratus cl. auratus - luteus*. Can. Field-Nat. 58:183-184.

Rand, A.L. 1948. Distributional notes on Canadian birds. Can. Field-Nat. 62(6):175-180.

Randall, T.E. 1925. Notes of interest from Castor, Alberta, district. Can. Field-Nat. 39(8):194-195.

Rogers, T.H. 1961. Regional reports. Fall migration August 16 to November 30, 1960. Northern Rocky Mountain-intermountain region. Audubon Field Notes 15(1):59-61.

Rogers, T.H. 1972. The nesting season. June 1, 1972 -Aug. 15, 1972. Northern Rocky Mountain-intermountain region. Amer. Birds 26(5):878-882.

Salt, J.R. 1985. A note on "condominium" nesting of the Northern Flicker *Colaptes auratus* in western Alberta. Can. Field-Nat. 99(4):534-536.

Schroeder, R.L. 1982. Habitat suitability index model: Pileated Woodpecker. U.S. Dept. Int., Fish and Wildl. Serv. 15 pp.

Scott, D.M., C.D. Ankey and C.H. Jasosch. 1976. Sapsucker hybridization in British Columbia: changes in 25 years. Condor 78(2):253-257.

Short, L.L. 1982. Woodpeckers of the world. Delaware Museum of Natural History, Delaware and Greenville.

Sieb, G. 1976. Hand feeding birds at Inglewood. Calgary Field Nat. 7(9):243.

Soper, J.D. 1965. Recent additions to the list of Jasper National Park birds. Can. Field-Nat. 79(3):214-215.

Storer, J.E., and M. Wilson. 1972. Pileated Woodpeckers near Drumheller, Alberta. Blue Jay 30:97.

Storms, R.W. 1986. The three newest "splits" that affect Alberta birders. Alta. Bird Record 4(2):46-47.

Weseloh, D.V. 1975. The co-operative breeding bird survey in Canada, 1974. Calgary Field Nat. 7(2):37-38.

Order Passeriformes

Family Tyrannidae: TYRANT FLYCATCHERS

Aldrich, J.W. 1951. A review of the races of the Traill's Flycatcher. Wilson Bull. 63(3):192-197.

Anonymous. 1970(a). *Empidonax* flycatchers. Birds of the world 6(3):1747-1750.

Anonymous. 1970(b). Olive-sided Flycatcher. Birds of the world 6(3):1744.

Anonymous. 1970(c). Eastern Phoebe. Birds of the world 6(2):1724-1725.

Anonymous. 1970(d). Eastern Kingbird. Birds of the world 6(2):1734-1755.

Barlow, J.C., and W.B. McGillivray. 1983. Foraging and habitat relationships of the sibling species Willow Flycatcher (*Empidonax traillii*) and Alder Flycatcher (*E. alnorum*) in southern Ontario. Can. J. Zool. 61:1510-1516.

Browning, M.R. 1976. The status of *Sayornis saya yukonensis* Bishop. Auk 93(4):843-846.

Butot, R. 1977. The Eastern Kingbird and the Merlin. Alta. Nat. 7(3):189-190.

Carroll, B. 1989. Elk Island birds. Alta. Nat. 19(1):10-11.

Cooke, W.W. 1908(a). The migration of flycatchers. Fourth paper. Bird-Lore 10(3):114-117.

Cooke, W.W. 1908(b). The migration of flycatchers. Fifth paper. Bird-Lore 10(4):166-170.

Cooke, W.W. 1908(c). The migration of the flycatchers. Sixth paper. Bird-Lore 10(5):210-212.

Dickson, R. 1989. Calgary area birding highlights - 1 March to 31 May 1989. Pica 9(3):35-36.

Greenlee, G.M. 1972. Great Crested Flycatcher again observed in Alberta. Blue Jay 30(2):86-87.

Greenlee, G.M. 1973. Alberta Great Crested Flycatcher sightings, 1972. Blue Jay 31(2):97.

Heath, R., and R. Lister. 1957. Western Kingbird in Alberta. Canada. Field-Nat. 71:34.

Hussell, D.J.T. 1980. The timing of fall migration and moult in Least Flycatchers. J. Field Ornith. 51(1):65-71.

Hussell, D.J.T. 1982. The timing of fall migration in Yellow-bellied Flycatchers. J. Field Ornith. 53(1):1-6.

Johnson, N.K. 1966. Morphologic stability versus adaptive variation in Hammond's Flycatcher. Auk 83(2):179-200.

Kondla, N.G. 1972. Use of Barn Swallow nests by phoebes. Can. Field-Nat. 3(11):158-159.

Lang, V. 1978. Great Crested Flycatcher near Dunstable, Alberta. Alta. Nat. 8(3):177.

Murphy, M.T. 1983. Nest success and nesting habits of Eastern Kingbirds and other flycatchers. Condor 85:208-219.

Phillips, A.R. 1948. Geographic variation in *Empidonax traillii*. Auk 65(3):507-514.

Phillips, A.R., M.A. Howe and W.E. Lanyon. 1966. Identification of the flycatchers of eastern North America, with special emphasis on the genus *Empidonax*. Bird-Banding 37:153-171.

Quinn, M.S., and G.L. Holroyd. 1989. Nestling and egg destruction by House Wrens. Condor 91(1):206-207.

Rising, J.D., and F.W. Schueler. 1980. Identification and status of wood pewees (*Contopus*) from the great plains: what are sibling species? Condor 82(3):301-308.

Rogers, T.H. 1978. The nesting season. June 1 - July 31, 1978. Northern Rocky Mountain-intermountain region. Amer. Birds 32(6):1186-1190.

Salt, J.R. 1979. Proceedings of the Alberta Ornithological Records Committee, 1977-78. Alta. Nat. 9(1):4-5.

Schreck, J.M. 1910. The Least Flycatcher. Bird-Lore 12(4):148-149.

Slater, A. 1977. Willow and Alder Flycatchers in the Highwood Valley. Calgary Field Nat. 8(7):170.

Stein, R.C. 1963. Isolating mechanisms between populations of Traill's Flycatcher. Proc. Amer. Phil. Soc. 107(1):21-50.

Thormin, T. 1977. Bird news. Edmonton Nat. 5(7):9.

Trann, K. 1976. Great Crested Flycatcher sightings. Alta. Nat. 6(4):319-320.

Ussher, R.D. 1940. The Crested Flycatcher in central Alberta. Canada. Field-Nat. 54:74-75.

Walkinshaw, L.H. 1966(a). Summer observations of the Least Flycatcher in Michigan. Jack-pine Warbler 44:151-168.

Walkinshaw, L.H. 1966(b). Summer biology of Traill's Flycatcher. Wilson Bull. 78:31-46.

Walkinshaw, L.H. 1967. The Yellow-bellied Flycatcher in Michigan. Jack-pine Warbler 45:2-9.

Walkinshaw, L.H., and C.J. Henry. 1957. Yellow-bellied Flycatcher nesting in Michigan. Auk 74:293-304.

Weber, W.C. 1974. More Great Crested Flycatchers and Short-billed Marsh Wrens in Alberta. Blue Jay 32(4):230-233.

Wershler, R. 1987. Fall 1986 - highlights. Alta. Bird Record 5(1):1.

Zink, R.M., and B.A. Fall. 1981. Breeding distribution, song and habitat of the Alder Flycatcher and Willow Flycatcher in Minnesota. Loon 53:208-214.

Family Alaudidae: LARKS

Anonymous. 1919. Horned Larks. Birds of the world 6(5):1804-1806.

Craig, G., and E. Craig. 1978. The Horned Lark. Northern Naturalist 28:5.

Lange, J. 1984. Edmonton Christmas bird count 1983. Edmonton Nat. 12(1):12.

Oberholser, H.C. 1918. The migration of North American birds. Second series. VI. Horned Larks. Bird-Lore 20(5):345-349.

Family Hirundinidae: SWALLOWS

Anonymous. 1970(a). Swallow. Birds of the World Vol. 6, Part 6, No. 66:1821-1829.

Anonymous. 1970(b). Purple Martin. Birds of the World Vol. 6, Part 5, No. 65:1811-1812.

Anonymous. 1970(c). Tree Swallow. Birds of the World Vol. 6, Part 5, No. 65:1809-1810.

Anonymous. 1970(d). Rough-winged Swallow. Birds of the World Vol. 6, Part 5, No. 65:1812-1814.

Anonymous. 1987. Announcements: Purple Martin colony registry. Alta. Nat. 17(2):66-67.

Balph, M.H. 1984(a). Significant encounters. N. Am. Bird Bander 9(1):15.

Balph, M.H. 1984(b). Significant encounters with marked birds. N. Am. Bird Bander 12(2):73.

Behle, W.H. 1976. Systematic review, intergradation and clinal variation in cliff swallows. Auk 93(1):66-77.

Beyer, L.K. 1938. Nest life of the Bank Swallow. Wilson Bull. 50:122-127.

Bilesky, A., and L. Bilesky. 1978. Tree Swallow "adopts" eggs. Edmonton Nat. 6(1):13-14.

Bird, C.D., and A. Bird. 1971. The status of the Purple Martin in the Calgary area. Calgary Field Nat. 3(4):48-49.

Brown, C.R. 1986. Cliff Swallow colonies as information centers. Science 234(4772):83-85.

Brown, C.R., and J.L. Hoogland. 1986. Risk in mobbing for solitary and colonial swallows. Anim. Beh. 34:1319-1323.

Brown, C.R., and M.B. Brown. 1987. Group-living in cliff swallows as an advantage in avoiding predators. Behav. Ecol. Sociobiol. 21:97-107.

Brown, C.R., and M.B. Brown. 1988. The costs and benefits of egg destruction by conspecifics in colonial cliff swallows. Auk 105(4):737-748.

Butler, R.W. 1988. Population dynamics and migration routes of Tree Swallows, (*Tachycineta bicolor*), in North America. J. Field Ornith. 59(4):395-402.

Butler, R.W., and C.A. Campbell. 1987. Nest appropriation and interspecific feeding between Tree Swallows, *Tachycineta bicolor* and Barn Swallows, *Hirundo rustica*. Can. Field-Nat. 101(1):433-434.

Cole, M.P. 1939. Bird notes from Red Deer, Alberta -Violet Green Swallow. Can. Field-Nat. 53(1):12.

Collister, D. 1990. Mate swapping - can you swallow that! Alta. Nat. 20(4):145.

Dannacker, E. 1988. Purple Martins, Swallows and Bluebirds. Edmonton Nat. 16(2):10-12.

Erskine, A.J. 1979. Man's influence on potential nesting sites and populations of swallows in Canada. Can. Field-Nat. 93:371-377.

Finch, D.M. 1990. Effects of predation and competitor interference on nesting success of House Wrens and Tree Swallows. Condor 92(3):674-687.

Finlay, J.C. 1971. Breeding biology of Purple Martins at the northern limit of their range. Wilson Bull. 83(3):255-269.

Finlay, J.C. 1975. Nesting of Purple Martins in natural cavities and in man-made structures in Alberta. Can. Field-Nat. 89(4):454-455.

Finlay, J.C. 1976. Some effects of weather on Purple Martin activity. Auk 93(2):231-244.

Griffiths, D. 1973. Possible northern extension of the breeding range of the Rough-winged Swallow (*Stelgiodopteryx* <sic> *ruficollis*). Edmonton Nat. 12:8.

Harper, C. 1987. Rough-winged Swallow behavior. Edmonton Nat. 6(4):93-94.

Harris, R.N. 1979. Aggression, superterritories and reproductive success in Tree Swallows. Can. J. Zool. 57:2072-2078.

Hjertaas, D.G., P. Hjertaas and W.J. Maher. 1988. Colony size and reproductive biology of the Bank Swallow, *Riparia riparia*, in Saskatchewan. Can. Field Nat. 102(3):465-470.

Hofman, D.E. 1980. Feeding of nestling Cliff Swallows by a House Sparrow. Can. Field-Nat. 94(4):462.

Holroyd, G.L. 1987. Beaverhill Bird Observatory Project Reports: Tree Swallow biology. Edmonton Nat. 15(2):9-10.

Hoogland, J.L., and P.W. Sherman. 1976. Advantages and disadvantages of Bank Swallow (*Riparia riparia*) coloniality. Ecol. Monog. 46(1):33-58.

Johnston, R.F., and J.W. Hardy. 1962. Behavior of the Purple Martin. Wilson Bull. 74:243-262.

Lang, B. 1974. Census of Cliff Swallow populations in the city of Calgary. Calgary Field Nat. 5(9):229-231.

Leffelaar, D., and R.J. Robertson. 1985. Nest usurpation and female competition for breeding opportunities by tree swallows. Wilson Bull. 97(2):221-224.

Moller, A.P. 1990. Male tail length and female mate choice in the monogamous swallow *Hirundo rustica*. Anim. Beh. 39:458-465.

Morck, K. 1986. Violet-green Swallow: a nesting record. Pica 6(3):33-34.

Oberholser, H.C. 1917. The migration of North American birds. Second series. Five. Swallows. Bird-Lore 19(6):320-330.

Pinel, H.W. 1980. Reproductive efficiency and site attachment of Tree Swallows and Mountain Bluebirds. Blue Jay 38(3):177-183.

Pletz, H. 1974. Tree Swallows accept second brood. Alta. Nat. 4(3):79-80.

Rendell, W.B., and R.J. Robertson. 1989. Nest-site characteristics, reproductive success and cavity availability for Tree Swallows breeding in natural cavities. Condor 91(4):875-885.

Rogers, T.H. 1979. The autumn migration. Aug. 1 - Nov. 30, 1978 Northern Rocky Mountain-intermountain region. Amer. Birds 33(2):196-199.

Salt, W.R. 1966. Some unusual bird records from the Peace River District. Can. Field-Nat. 80(2):114-115.

Samuel, D.E. 1971. The breeding biology of Barn and Cliff Swallows in West Virginia. Wilson Bull. 83:284-301.

Shaw, G.G. 1983. Organochlorine pesticide and PCB residues in eggs and nestlings of Tree Swallows, (*Tachycineta bicolor*), in central Alberta. Can. Field-Nat. 98(2): 258-260.

Stiles, D. 1982. Double broods of Tree Swallows in one nest box. Alta. Nat. 12(4):175.

Stiles, D., and G. Neill. 1982. Double broods of Tree Swallows in one nest box. Blue Jay 40(4):205-206.

Van Mechelen, N. 1974. An experience of hand-raising tree swallows. Alta. Nat. 4(3):76-79.

Walsh, H. 1978. Food of nestling Purple Martins. Wilson Bull. 90(2):248 260.

Wilde, G.A. 1977. Bluebirds and clearcuts. Blue Jay 35(1):42-43.

Wilde, G.A. 1979. Possible swallow response to insecticide use. Blue Jay 37(4):226.

Wilde, G.A. 1987. Unusual nesting site for Bank Swallows. Blue Jay 45(2):112.

Family Corvidae: JAYS, MAGPIES AND CROWS

Buitron, D. 1988. Female and male specialization in parental care and its consequences in Black-billed Magpies. Condor 90(1):29-39.

Dhindsa, M.S., P.E. Komers and D.A. Boag. 1988. Nest height of Black-billed Magpies: is it determined by human disturbance or habitat type? Can. J. Zool. 67(1):228-232.

Dunn, P.O., and S.J. Hannon. 1989. Evidence for obligate male parental care in Black-billed Magpies. Auk 106(4):635-644.

Fisher, R.M. 1979. 1976 eruption of Clark's Nutcracker in Cypress Hills, Alberta. Blue Jay 37(1):47.

Fisher, R.M., and M.T. Myres. 1980(a). Nutcracker in Canada. Can. Field-Nat. 94(1):43-51.

Fisher, R.M., and M.T. Myres. 1980(b). A review of factors influencing extralimital occurrences of Clark's Nutcracker in Canada. Can. Field-Nat. 94(1):43-51.

Hogan, G.G. 1975. Magpie nest site selection in the Edmonton area. Alta. Nat. 5(2):56-58.

Kennedy, D. 1988. The great transformer. Nature Canada 17(3):34-39.

Komers, P.E., and D.A. Boag. 1988. The reproductive performance of Black-billed Magpies: is it related to mate choice? Can. J. Zool. 66(7):1679-1684.

Komers, P.E., and M.S. Dhindsa. 1989. Influence of dominance and age on mate choice in Black-billed Magpies: an experimental study. Anim. Beh. 37(4):645-655.

Kondla, N., and B. Danielson. 1973. Nesting of the Common Raven in southwestern Alberta. Calgary Field Nat. 5(2):62-63.

Mewaldt, L.R. 1956. Nesting behavior of the Clark's Nutcracker. Condor 58(1):3-23.

O'Shea, M.D. 1980. Occurrence of the Blue Jay in Lethbridge: A new nesting record. Alta. Nat. 10(1):8-12.

Oeming, A.F. 1957. Steller's Jay in central Alberta. Can. Field-Nat. 71(3):153.

Reebs, S.G. 1987. Roost characteristics and roosting behavior of Black-billed Magpies, *Pica pica*, in Edmonton, Alberta. Can. Field-Nat. 101(4):519-525.

Scharf, C., and G. Clover. 1983. The importance of social display in the life cycle of the Black-billed Magpie, *Pica pica*. Amer. Birds 37(6):935-940.

Visser, S. 1986. Observations of the American Crow on the University of Calgary campus. Pica 6(3):13-23.

Waller, S. 1973. The bird that nests in winter. Manitoba Nature 14(4):26-28.

Walley, B. 1981. Riding Mountain Whiskey Jacks. Manitoba Nature 22(1):14-17,40.

Family Paridae: TITMICE

Brittingham, M.C. 1990. Should we feed birds in winter? Living Bird 9(1):19-21.

Carroll, B. 1989. Elk Island Birds. Alt. Nat. 19(1):10-11.

Hill, B.G., and M.R. Lein. 1988. Ecological relations of sympatric Black-capped and Mountain Chickadees in southwestern Alberta. Condor 90(4):875-884.

Hill, B.G., and M.R. Lein. 1989. Territory overlap and habitat use of sympatric chickadees. Auk 106(2):259-269.

McLaren, M.A. 1975. Breeding biology of the Boreal Chickadee. Wilson Bull. 87(3):344-354.

Odum, E.P. 1941(a). Annual cycle of the Black-cappedChickadee. Auk 58:314-333.

Odum, E.P. 1941(b). Annual cycle of the Black-capped Chickadee. Auk 58:518-535.

Runde, D.E., and D.E. Capen. 1986. Characteristics of northern hardwood trees used by cavity-nesting birds. J. Wildl. Manage. 51(1):217-223.

Family Sittidae: NUTHATCHES

Stokes, D., and L. Stokes. 1986. Watching: White-breasted Nuthatches. Living Bird 5(2):19.

Wetmore, A. 1964. Song and garden birds of North America. National Geographic Society.

Family Troglodytidae: WRENS

Abrahamson, H.S. 1980. Lazuli Bunting and Rock Wren at Sylvan Lake, Alberta. Blue Jay 38(1):50.

Bart, J. 1989. Male care, mate switching and future reproductive success in a double-brooded passerine. Behav. Ecol. Sociobiol. 26(5):307-314.

Belles-Isles, J.C., and J. Picman. 1986. Nesting losses and nest site preferences in House Wrens. Condor 88(4):483-486.

Drilling, N.E., and C.F. Thompson. 1988. Natal and breeding dispersal of House Wrens (*Troglodytes aedon*). Auk 105(3):480-491.

Finch, D.M. 1989. Relationships of surrounding riparian habitat to nest-box use and reproductive outcome in House Wrens. Condor 91(4):848-859.

Finch, D.M. 1990. Effects of predation and competitor interference on nesting success of House Wrens and Tree Swallows. Condor 92(3):674-687.

Jaeger, E.C. 1951. Courtship display of the Rock Wren, *Salpinctes obsoletus obsoletus*. Auk 68(4):511.

Johnson, L.S., and L.H. Kermott. 1990. Structure and context of female song in a north-temperate population of House Wrens. J. Field Ornith. 61(3):273-376.

Knowlton, G.F., and F.C. Harmston. 1942. Insect food of the Rock Wren. The Great Basin Naturalist 3(1):22.

Leonard, M.L., and J. Picman. 1987(a). The adaptive significance of multiple nest building by male Marsh Wrens. Anim. Beh. 35(1):271-277.

Leonard, M.L., and J. Picman. 1987(b). Nesting mortality and habitat selection by Marsh Wrens. Auk 104(3):491-495.

Leonard, M.L., and J. Picman. 1988. Mate choice by Marsh Wrens: the influence of male and territory quality. Anim. Beh. 36(2):517-528.

Nordstrom, W. 1978. Lake Newell: colonial birds in jeopardy. Alta. Nat. 8(2):107-111.

Picman, J. 1977(a). Destruction of eggs by the Long-billed Marsh Wren (*Telmatodytes palustris palustris*). Can. J. Zool. 55(11):1914-1920.

Picman, J. 1977(b). Intraspecific nest destruction in the Long-billed Marsh Wren, *Telmatodytes palustris palustris*. Can. J. Zool. 55(12):1997-2003.

Price, D.K., G.E. Collier and C.F. Thompson. 1989. Multiple parentage in broods of House Wrens: genetic evidence. J. Heredity 80:1-5.

Quinn, M.S., and G.L. Holroyd. 1989. Nestling and egg destruction by House Wrens. Condor 91:206-207.

Renaud, W.E. 1979. The Rock Wren in Saskatchewan: status and distribution. Blue Jay 37(3):138-148.

Rumble, M.A. 1987. Avian use of scoria rock outcrops. The Great Basin Naturalist 47(4):625-630.

Seutin, G., and B. Chartier. 1989. The Rock Wren, *Salpinctes obsoletus*, breeding at Churchill, Manitoba. Can. Field-Nat. 103:416-417.

Smith, P.W. 1904. Nesting habits of the Rock Wren. Condor 6:109-110.

Smyth, M., and G.A. Bartholomew. 1966. The water economy of the Blue-throated Sparrow and the Rock Wren. Condor 68(5):447-458.

Welter, W.A. 1935. The Long-billed Marsh Wren. Wilson Bull. 47:3-34.

Family Cinclidae: DIPPERS

Ealey, D.M. 1977. Aspects of the ecology and behavior of a breeding population of dippers (*Cinclus mexicanus*: Passeriformes) in southern Alberta. M.Sc. Thesis, Dept. of Zoology, Univ. of Alberta, Edmonton. 198 pp.

Ealey, D.M. 1978. The American Dipper - fellow of the waterfall. Alta. Nat. 8(1):35-40.

O'Keeffe, S. 1974. A summary of the 1973 Dipper census. Calgary Field Nat. 5(9):232-234.

Salt, J.R. 1977. Use of Cliff Swallow nests by Dippers, in west-central Alberta. Alta. Nat. 7(3):194.

Family Muscicapidae: OLD WORLD WARBLERS, KINGLETS, GNATCATCHERS, OLD WORLD FLYCATCHERS, THRUSHES, BABBLERS AND ALLIES

Backhouse, F. 1986. Bluebird revival. Can. Geog. 106(2):32-39.

Bittner, R.A. 1990. Effects of temperature on Bluebird and Swallow Nesting Dates. Blue Jay 48(4):222.

Butot, R. 1978. Migration patterns and occurrences of Robins, Thrushes and Mountain Bluebirds in the Calgary area. Calgary Field Nat. 9(10):272-276.

Di Labio, B. 1988. Bird watch. Nature Canada 17(4):51.

Di Labio, B. 1989. Bird watch. Nature Canada 18(2):48.

Eiserer, L. 1976. The American Robin. Nelson-Hall, Chicago.

Elphinstone, D. 1986. A bluebird hat trick in Alberta. Alta. Nat. 16(1):32.

Finlay, J.C. 1980(a). Alberta Bluebird Trails - 1978. Alta. Nat. 10(4):149-151.

Finlay, J.C. 1980(b). Alberta Bluebird Trails - 1979. Alta. Nat. 10(4):151-153.

Haddow, D.J., M.A. Skeel and K.J. Van Tighem. 1978. Gray-cheeked Thrush in the Rocky Mountains of Alberta. Alta. Nat. 8(4):205-206.

Hoffmann, W. 1991. Golden-crowned Kinglet nest record for Ft. Saskatchewan, Alberta. Alta. Nat. 21(1):29-30.

Hoffmann, W., and H. Pletz. 1988. Mountain Bluebirds show fidelity to nesting boxes and maintenance of pair bonding. Alta. Nat. 18(3):101.

Höhn, E.O., and P. Marklevitz. 1974. Noteworthy summer observations of birds in the Caribou Mountains. Can. Field Nat. 88(1)77-78.

Martin, T.E. 1988. Nest predation and nest-site selection of a western population of the Hermit Thrush. Condor 90(1):51-57.

McNicholl, M.K. 1985. Singing Gray-cheeked Thrush at Lake Louise, Alberta. Alta. Nat. 15(1):10-11.

Pinel, H.W. 1980. Reproductive efficiency and site attachment of Tree Swallows and Mountain Bluebirds. Blue Jay 38:177-183.

Sharp, M.H. 1990. America's songbird - species profile: American Robin (*Turdus migratorius*). Wildbird 4(5):22-27.

Stiles, D. 1979. History of Calgary Bluebird Trails. Pica 1(1):21-36.

Family Mimidae: MOCKINGBIRDS AND THRASHERS

Brazier, F.H. 1964(a). Status of the Mockingbird in the Northern Great Plains. Blue Jay 22(2):63-75.

Brazier, F.H. 1964(b). Additional records of the Mockingbird. Blue Jay 22(4):151-152.

Burns, H.W. 1964. Nesting of the Mockingbird at Cadogan, Alberta. Blue Jay 22(4):150-151.

Farley, F.L. 1926. Notes on the occurrence of some rare birds in central Alberta. Can. Field-Nat. 40(3):65-66 [correction in Taverner, P.A. 1928. Can. Field-Nat. 42(5):127].

Jones, E.T. 1988. A second Northern Mockingbird breeding record for Alberta. Alta. Nat. 18(4):136-137.

Lewin, V. 1965. A new northern record for the Mockingbird, *Mimus polyglottos*. Can. Field-Nat. 79(3):208.

Lister, R. 1973. Northern records of the Mockingbird in Alberta. Can. Field-Nat. 87(3):324.

O'Shea, M.D. 1991. Alberta Bird Atlas Program yields rarities: Sage Thrasher and Northern Mockingbird records for Bindloss. Alta. Nat. 21(1):26-27.

Rand, A.L., and M.A. Traylor. 1949. Variation in *Dumetella carolinensis*. Auk 66(1):25-28.

Snell, C.H. 1932. The Mockingbird (*Mimus polyglottos*) in central Alberta. Can. Field-Nat. 46(3):67.

Family Motacillidae: WAGTAILS AND PIPITS

Anonymous. 1970. Rock and Water Pipits. Birds of the World. Vol. 6, Part 7, (67):1861-1863.

King, B. 1981. The field identification of North America pipits. Amer. Birds 35(5):778-788.

Wershler, C., W.W. Smith and C. Wallis. 1991. Status of the Baird's Sparrow in Alberta - 1987/1988 - Update with notes on other grassland sparrows and Sprague's Pipit. **In:** Proceedings of the Second Endangered Species and Prairie Conservation Workshop. Holroyd, G.L., G. Burns and H.C. Smith, eds. Provincial Museum of Alberta Natural History Occasional Paper No. 15, Edmonton.

Family Bombycillidae: WAXWINGS

Anonymous. 1970. Waxwings. Birds of the world Part 9, Vol. 6, (69):1922-1926.

Crouch, J.E. 1936. Nesting habits of the Cedar Waxwing (*Bombycilla cedrorum*). Auk 54:1-8.

Di Labio, B. 1989. Bird watch. Nature Canada 18(1):50.

Höhn, E.O. 1951. Courtship behavior of the Bohemian Waxwing. Can. Field-Nat. 65:168-169.

Kerr, W.G. 1976. Bohemian Waxwing behavior. Calgary Field Nat. 7(9):240.

Krause, H. 1959. Regional reports winter season December 1, 1958 - March 31, 1959. Northern Great Plains region. Audubon Field Notes 13(3):303-304.

McNicholl, M.K. 1978. Bohemian Waxwings flycatching -in February. Alta. Nat. 8(1):27-28.

Oberholser, H.C. 1918. The migration of North American birds. Second series. IX. The waxwings and *Phainopepla*. Bird-Lore 20(3):219-222.

Putnam, L.S. 1949. The life history of the Cedar Waxwing. Wilson Bull. 61:141-182.

Rothstein, S.I. 1976. Experiments on defenses Cedar Waxwings use against Cowbird parasitism. Auk 93(4):675-691.

Family Laniidae: SHRIKES

Burnside, F.L. 1987. Long-distance movements by Loggerhead Shrikes. J. Field Ornith. 58(1):62-65.

Butot, R. 1977. Black-capped Chickadees meet Northern Shrikes - the deadly silence. Alta. Nat. 7(2):117-118.

Cadman, M.D. 1985. Status report on the Loggerhead Shrike *Lanius ludovicianus* in Canada. Committee on the Status of Endangered Wildlife in Canada, Ottawa.

DeSmet, K.D., and M.P. Conrad. 1989. The Loggerhead Shrike in Manitoba, its status and habitat needs. Dept. of Natural Resources, Winnipeg, Manitoba.

DeSmet, K.D., and M.P. Conrad. 1991. Status and habitat needs of the Loggerhead Shrike in Manitoba. **In:** Proceedings of the Second Endangered Species and Prairie Conservation Workshop. Holroyd, G.L., G. Burns and H.C. Smith, eds. Provincial Museum of Alberta Natural History Occasional Paper No. 15, Edmonton, Alberta.

Lang, V. 1973. Identification tip #3: Loggerhead and Northern Shrikes. Calgary Field Nat. 5(4):140-141.

McNicholl, M.K. 1986. Profiles on risk status of Canadian birds: 3. Loggerhead Shrike. Alta. Nat. 16(3):93-94.

Miller, A.H. 1931. Systematic revision and natural history of the American shrikes (*Lanius*). Univ. Calif. Publ. in Zool. 38(2):11-242.

Oberholser, H.C. 1918. The migration of North American birds. Second series. V. The shrikes. Bird-Lore 20(4):286-290.

Smith, W.W. 1991. The Loggerhead Shrike in Alberta. **In:** Proceedings of the Second Endangered Species and Prairie Conservation Workshop. Holroyd, G.L., G. Burns and H.C. Smith, eds. Provincial Museum of Alberta Natural History Occasional Paper No. 15, Edmonton, Alberta.

Telfer, E.S., C. Adam, K.D. DeSmet and R. Wershler. 1989. Status and distribution of the Loggerhead Shrike in western Canada. Progress Note No. 184, Canadian Wildlife Service, Environment Canada.

Family Sturnidae: STARLINGS

Cairns, A. 1975. Dispersal and roosting behavior of starlings. Calgary Field Nat. 6(7):225-230.

Dolbeer, R.A., and R.A. Stehn. 1979. Population trends of blackbirds and starlings in North America, 1966-76. U.S. Fish and Wildl. Serv., Spec. Sci. Rept., Wildl. No. 214. 99 pp.

Feare, C.J. 1984. The Starling. Oxford University Press, Oxford.

Holroyd, G.L. 1988. The Edmonton Christmas bird count, a tradition. Edmonton Nat. 16(1):14-23.

Power, H.W., E. Litowich and M.P. Lombardo. 1980. Male Starlings delay incubation to avoid being cuckolded. Auk 98:386-389.

Savard, H. 1952. Starlings rapidly spreading. Blue Jay 10(3):26.

Schneider, D. 1990. Starling wars. Nature Canada 19(4):33-39.

Wing, L. 1943. Spread of the Starling and English Sparrow. Auk 60(1):74-87.

Family Vireonidae: VIREOS

Barlow, J.C., and D.M. Power. 1970. An analysis of character variation in Red-eyed and Philadelphia Vireos (Aves:Vireonidae) in Canada. Can. J. Zool. 48(4):673-680.

Barlow, J.C., and J.C. Rice. 1977. Aspects of the comparative behaviour of Red-eyed and Philadelphia Vireos. Can. J. Zool. 55(3):528-542.

Cooke, W.W. 1909. The migration of vireos. First paper. Bird-Lore 11(2):78-82.

Cooke, W.W. 1909. The migration of vireos. Third and concluding paper. Bird-Lore 11(4):165-168.

James, R.D. 1981. Factors affecting variation in the primary song of North American Solitary Vireos (Aves: Vireonidae). Can. J. Zool. 59(10):2001-2009.

Johnson, J. 1976. Bird species and subspecies possibly or definitely breeding in 1970 in areas southwest of Calgary -further south or east than indicated in *Birds of Alberta* (Salt and Wilk). Calgary Field-Nat. 7(9):251.

Lawrence, L. deK . 1953. Nesting life and behaviour of the Red-eyed Vireo. Can. Field-Nat. 67:47-77.

Salt, W.R. 1973. Alberta vireos and wood warblers. Family Vireonidae and Parulidae distribution and breeding. Provincial Museum and Archive of Alberta Publication No. 3, Queen's Printer, Edmonton.

Family Emberizidae: WOOD-WARBLERS, TANAGERS, CARDINALS, SPARROWS, BUNTINGS, MEADOWLARKS, BLACKBIRDS AND ORIOLES

Abrahamson, H.S. 1980. Lazuli Bunting and Rock Wren at Sylvan Lake, Alberta. Blue Jay 38(1):50.

Alberts, S. 1960. Further observations of distraction display of the Western Meadowlark. Blue Jay 18(4):157.

Aldrich, J.W. 1984. Ecogeographical variation in size and proportions of Song Sparrows (*Melospiza melodia*). Ornithol. Monogr. 35:1-134.

Anonymous. 1976. Observational notes. Calgary Field Nat. 7(5):145.

Anonymous. 1984. Unusual bird sightings - winter 1983-84. Alta. Nat. 14(2):55.

Anonymous. 1991. Red-winged Blackbirds. Conservator 12(1):20.

Baugartner, A.M. 1936. Distribution of the American Tree Sparrow. Wilson Bull. 51(3):137-149.

Biermann, G.C., W.B. McGillivray and K.E. Nordin. 1987. The effect of cowbird parasitism on Brewer's Sparrow productivity in Alberta. J. Field Ornith. 58(3):350-354.

Bock, C.E., and L.W. Lepthien. 1976. Synchronous eruptions of boreal seed-eating birds. Amer. Nat. 110:559-571.

Brunton, D.F., G.L. Holroyd and H.F. Coneybeare. 1978. Cassin's Finch, Baird's Sparrow and Harris's Sparrow new to Banff National Park, Alberta. Alta. Nat. 8(2):89-92.

Carroll, B. 1989. Elk Island birds. Alta. Nat. 19(1):10-11.

Cartwright, B.W., T.M. Shortt and R.D. Harris. 1937. Baird's Sparrow. Trans. Royal Can. Inst. 21:153-197.

Clark, K.L., and R.J. Robertson. 1981. Cowbird parasitism and evolution of anti-parasite strategies in the Yellow Warbler. Wilson Bull. 93(2):249-258.

Collister, D. 1986. Summer 1986 - highlights. Alta. Bird Record 4(4):94.

Collister, D. 1988. Fall 1987 - highlights. Alta. Bird Record 6(1):1-2.

Cooke, W.W. 1910(a). The migration of North American sparrows. Second paper. Bird-Lore 12(1):12-16.

Cooke, W.W. 1910(b). The migration of North American sparrows. Third paper. Bird-Lore 12(2):67-70.

Cooke, W.W. 1910(c). The migration of North American sparrows. Fourth paper. Bird-Lore 12(3):111-112.

Cooke, W.W. 1911(a). The migration of North American sparrows. Eighth paper. Bird-Lore 13(1):15-17.

Cooke, W.W. 1911(b). The migration of North American sparrows. Ninth paper. Bird-Lore 13(2):83-88.

Cooke, W.W. 1911(c). The migration of North American sparrows. Tenth paper. Bird-Lore 13(3):144-146.

Cooke, W.W. 1912(a). The migration of North American sparrows. Fifteenth paper. Bird-Lore 14(2):98-105.

Cooke, W.W. 1912(b). The migration of North American sparrows. Sixteenth paper. Bird-Lore 14(3):158-161.

Cooke, W.W. 1912(c). The migration of North American sparrows. Eighteenth paper. Bird-Lore 14(5):287-290.

Cooke, W.W. 1913(a). The migration of North American sparrows. Twenty-first paper. Bird-Lore 15(2):104-17.

Cooke, W.W. 1913(b). The migration of North American sparrows. Twenty-third paper. Bird-Lore 15(4):236-240.

Cooke, W.W. 1913(c). The migration of North American sparrows. Twenty-fourth paper. Bird-Lore 15(5):301-303.

Cooke, W.W. 1914(a). The migration of North American sparrows. Twenty-ninth paper. Bird-Lore 16(4):267-268.

Cooke, W.W. 1914(b). The migration of North American sparrows. Thirty-first paper. Bird-Lore 16(6):438-442.

Cortopassi, A.J., and M.R. Mewaldt. 1965. The circumannual distribution of White-crowned Sparrows. Bird-Banding 36(3):141-169.

Cox, G.W. 1960. A life history of the Mourning Warbler. Wilson Bull. 72:5-28.

Cox, G.W. 1973. Hybridization between Mourning and MacGillivray's Warblers. Auk 90(1):190-191.

Cullop, J.B. 1979. Prairie Province's Region (Summer 1979). Amer. Bird 33:872-873.

DeSmet, K.D. 1991. Baird's Sparrow and miscellaneous grassland birds - summary of discussion. **In:** Proceedings of the Second Endangered Species and Prairie Conservation Workshop. Holroyd, G.L., G. Burns and H.C. Smith, eds. Provincial Museum of Alberta Natural History Occasional Paper No. 15.

DeSmet, K.D., and M.P. Conrad. 1991. Management and research needs for Baird's Sparrow and other grassland species in Manitoba. **In:** Proceedings of the Second Endangered Species and Prairie Conservation Workshop. Holroyd, G.L., G. Burns and H.C. Smith, eds. Provincial Museum of Alberta Natural History Occasional Paper No. 15.

Dennis, J.V. 1958. Some aspects of the breeding ecology of the Yellow-breasted Chat (*Icteria virens*). Bird-Banding 29:169-183.

Dewolfe, B.B., G.C. West and L.J. Peyton. 1973. The spring migration of Gambel's Sparrow through southern Yukon Territory. Condor 75:43-59.

Dolbeer, R.A., and R.A. Stehn. 1979. Population trends of blackbirds and starlings in North America, 1966-76. U.S. Fish and Wildl. Serv., Spec. Sci, Rept., Wild. No. 214. 99 pp.

DuBois, A.D. 1935. Nests of Horned Larks and Longspurs on a Montana prairie. Condor 37(2):56-72.

Duvall, A.J. 1942. Records from Lower California, Arizona, Idaho and Alberta. Auk 59(2):317-318.

Erskine, A.J. 1971. Some new perspectives on the breeding ecology of Common Grackles. Wilson Bull. 83(4):352-370.

Friedmann, H. 1971. Further information on the host relations of the parasitic cowbirds. Auk 88(2):239-255.

Gallaway, D. 1965. Lark Sparrow's nests found. Blue Jay 23(2):107.

Godfrey, W.E. 1949. Distribution of the races of the Swamp Sparrow. Auk 66(1):35-38.

Graham, D.S. 1989. Rejection, desertion, burial and the wanton layers. Living Bird 8(2):21-24.

Griscom, L., and A. Sprunt. 1979. The warblers of America. Doubleday and Co. Inc., New York.

Griscom, L., A. Sprunt, E.M. Reilly, and J.H. Dick. 1979. The warblers of America: a popular account of the wood warblers as they occur in the Western hemisphere. Doubleday, Garden City, New York. [Devin-Adair, New York]

Hann, H.W. 1937. Life history of the Oven-bird in southern Michigan. Wilson Bull. 49:145-237.

Hanners, A. 1977. Untitled. Calgary Field Nat. 9(4):106.

Harris, R.D. 1944. The Chestnut-collared Longspur in Manitoba. Wilson Bull. 56(2):105-115.

Hill, N.P., and J.M. Hagan III. 1991. Population Trends of Some Northeastern North American Landbirds: A Half Century of Data. Wilson Bull. 103(2):165-338.

Hofslund, P.B. 1957. Cowbird parasitism of the Northern Yellow-throat. Auk 74(1):42-48.

Höhn, E.O. 1959. Prolactin in the cowbird's pituitary in relation to avian brood parasitism. Nature 184:2030.

Holm, C.H. 1973. Breeding sex ratios, territoriality and reproductive success in the Red-winged Blackbird (*Agelaius phoeniceus*). Ecology 54(2):356-365.

Hussell, D.J.T. 1972. Factors affecting clutch size in arctic passerines. Ecol. Monog. 42:317-364.

Jasper, D. 1988. Cowbird breeding behavior. Colorado Bird Atlas Newsletter 6:4.

Johnson, J. 1976. Bird species and subspecies possibly or definitely breeding in 1970 in area southwest of Calgary -further south or east than indicated in *Birds of Alberta* (Salt and Wilk). Calgary Field-Nat. 7(9):251.

Johnson, J. 1988. An interesting location for a Dark-eyed Junco's nest. Alta. Nat. 18(2):64.

Jones, E.T. 1986. The passerine decline. N. Am. Bird Bander 11(3):74-75.

Kendeigh, S.C. 1945. Nesting behavior of wood warblers. Wilson Bull. 57:145-164.

Kennedy, A.J. 1980. Lark Bunting in northeastern Alberta. Blue Jay 38(1):49-50.

Knittle, C.E., G.M. Linz, B.E. Johns, J.L. Cannings, J.E. Davis Jr. and M.E. Jaeger. 1987. Dispersal of male Red-winged Blackbirds from two spring roosts in central North America. J. Field Ornith. 58:490-498.

Kondla, N.G., and H.W. Pinel. 1971. Some noteworthy records of cowbird parasitism in southern Alberta. Blue Jay 24(4):204-207.

Lawrence, L. de K. 1953. Notes on the nesting behavior of the Blackburnian Warbler. Wilson Bull. 65:135-144.

Lein, M.R., and G.M. Wagner. 1982. First breeding record for the Nashville Warbler, *Vermivora ruficapilla*, in Alberta. Can. Field-Nat. 96(1):88-89.

Lowther, J.K. 1964. A probable breeding record of the Bobolink at Vermilion, Alberta. Can. Field-Nat. 78(3):200.

Maher, W.J. 1979. Nestling diets of prairie passerine birds at Matador, Saskatchewan, Canada. Ibis 121(4):437-452.

McNicholl, M.K. 1977(a). A habitat note on Clay-colored Sparrows in Edmonton. Edmonton Nat. 5(2):30-31.

McNicholl, M.K. 1977(b). McCown's Longspur at Beaverhill Lake. Edmonton Nat. 5(6):147.

McNicholl, M.K. 1981. Flycatching by male Red-winged Blackbirds. Blue Jay 39(4):206-207.

McNicol, D.K., R.J. Robertson and P.J. Weatherhead. 1982. Seasonal, habitat and sex-specific food habits of Red-winged Blackbirds: implications for agriculture. Can. J. Zool. 60:3282-3289.

Mickey, F.W. 1943. Breeding habits of McCown's Longspur. Auk 60:181-209.

Morse, D.H. 1989. American Warblers: An Ecological and Behavioral Perspective. Harvard University Press, Cambridge, Mass. and London, Eng. 406 pp.

Murray, B.G. Jr. 1969. A comparative study of the Le Conte's and Sharp-tailed Sparrows. Auk 86(2):199-231.

Nero, R.W. 1956. A behavior study of the Red-winged Blackbird. Wilson Bull. 68:5-37, 129-150.

Nice, M.M. 1937. Studies in the life history of the Song Sparrow I. Trans. Linn. Soc. N.Y. 1-247.

Nice, M.M. 1943. Studies in the life history of the Song Sparrow II. Trans. Linn. Soc. N.Y. 6:1-328.

Nordin, K.E., W.B. McGillivray and G.C. Biermann. 1988. Courtship feeding in Brewer's Sparrow (*Spizella breweri*). J. Field Ornith. 59(1):33-36.

Oberholser, H.C. 1918. The migration of North American birds. Second series. II. The Scarlet and Louisiana Tanager. Bird-Lore(1):16-19.

Oberholser, H.C. 1920. The migration of North American birds. Second series. XIV. Cowbirds. Bird-Lore 22(6):343-345.

Oberholser, H.C. 1921(a). The migration of North American birds. Second series. XV. Yellow-headed Blackbirds and meadowlarks. Bird-Lore 23(1):78-82.

Oberholser, H.C. 1921. The migration of North American birds. Second series. XVI. Purple Grackle. Bird-Lore 23(4):192-194.

Oberholser, H.C. 1921(c). The migration of North American birds. Second series. XVII. Rusty Blackbird and Brewer's Blackbird. Bird-Lore 23(6):295-299.

Oberholser, H.C. 1922(a). The migration of North American birds. Second series. XVIII. Red-winged Blackbirds. Bird-Lore 24(1):85-88.

Oberholser, H.C. 1922(b). The migration of North American birds. Second series. XX. Baltimore Oriole. Bird-Lore 24(6):339-341.

Owens, R.A., and M.T. Myres. 1973. Effects of agriculture upon populations of native passerine birds of an Alberta fescue grassland. Can. J. Zool. 51(7):697-713.

Petrides, G.A. 1938. A life history of the Yellow-breasted Chat. Wilson Bull. 50:184-189.

Pinel, H.W., and C.J. Robinson. 1973. First report of a Golden-crowned Sparrow at Calgary, Alberta. Blue Jay 31(1):57-58.

Pitelka, F.A. 1947. British Columbian records of the Clay-colored Sparrow. Condor 49(3):128-130.

Randall, T.E. 1925. Nests of the Chestnut-collared and McCown's Longspur. Oologists' Rec. 5(2):48.

Randall, T.E. 1926. Nesting of Baird's Sparrow(*Ammodramus bairdii*). Oologists' Rec. 6(2):43-44.

Ratcliff, B. 1987. Baird's sparrow in Manitoba. **In:** Proceedings of the Workshop on Endangered Species in the Prairie Provinces. Holroyd, G.L., P.H.R. Stepney, G.C. Trottier, W.B. McGillivray, D.M. Ealey and K.E. Eberhart, eds. Provincial Museum of Alberta Natural History Occasional Paper No. 9.

Read, M. 1991. Hayfield songster. Birders' World 5(3):20-24.

Reynolds, J.O., and R.W. Knapton. 1984. Nest-site selection and breeding biology of Chipping Sparrow. Wilson Bull. 96:488-493.

Richter, W. 1984. Nestling survival and growth in the Yellow-headed Blackbird, *Xanthocephalus xanthocephalus*. Ecology 65(2):579-608.

Rising, J.D. 1973. Morphological variation and status of the Orioles, *Icterus galbula*, *I. bullockii* and *I. abeillei*, in the Northern Great Plains and Durango, Mexico. Can. J. Zool. 51(12):1267-1273.

Rotenberry, J.T., and J.A. Wiens. 1976. A method for estimating species dispersion from transect data. Amer. Midl. Nat. 95(1):64-78.

Salt, J.R. 1984. Black-and-white Warblers in Maligne Valley, Jasper National Park. Alta. Nat. 14(2):51.

Salt, W.R. 1958. Black-throated Blue Warbler in Alberta. Can. Field-Nat. 72(4):165-166.

Salt, W.R. 1966. A nesting study of *Spizella pallida*. Auk 83(2):274-281.

Salt, W.R. 1972. Western records of the Chestnut-sided Warbler. Can. Field-Nat. 86(4):390-391.

Salt, W.R. 1973. Alberta vireos and wood warblers, Families Vireonidae and Parulidae. Distribution and breeding. Provincial Museum and Archives of Alberta Publication No. 3, Queen's Printer, Edmonton.

Schmidt, C.L. 1990. The Baird's Sparrow in southwestern Alberta. Alta. Nat. 20(4):129-130.

Sedgwick, J.A. 1987. Avian habitat relationships in Pinyon-Juniper woodland. Wilson Bull. 99(3):413-431.

Smith, A. 1977. The Sharp-tailed Sparrow at BeaverhillLake. Edmonton Nat. 5(6):146.

Smith, W.W. 1967. Alberta Indigo Bunting record. Blue Jay 25(4):187-188.

Smith, W.W. 1987. The Baird's sparrow in Alberta. **In:** Proceedings of the Workshop on Endangered Species in the Prairie Provinces. Holroyd, G.L., P.H.R. Stepney, G.C. Trottier, W.B. McGillivray, D.M. Ealey and K.E. Eberhart, eds. Provincial Museum of Alberta Natural History Occasional Paper No. 9.

Spaw, C.D., and S. Rohwer. 1987. A comparative study of eggshell thickness in Cowbirds and other passerines. Condor 89(2):307-318.

Stepney, P.H.R. 1975. Wintering distribution of Brewer's Blackbird: historical aspect, recent changes, and fluctuations. Bird Banding 46(2):106-125.

Stepney, P.H.R. 1979. Brewer's Blackbird breeding in the Northwest Territories. Can. Field Nat. 93(1):76-77.

Stepney, P.H.R., and D.M. Power. 1973. Analysis of the eastward breeding expansion of Brewer's Blackbird plus general aspects of avian expansions. Wilson Bull. 85(4):452-464.

Stewart, R.E. 1953. A life history of the Yellow-throat. Wilson Bull. 65:99-115.

Stone, W. 1929. Nesting of a Connecticut Warbler in Alberta. Auk 46:552-553.

Walley, W.J. 1985. Breeding range extension of the Lark Sparrow into west-central Manitoba. Blue Jay 43:18-24.

Watkins, P.B. 1982. Fearless cowbird. Blue Jay 40:231-232.

Weeden, J.S. 1965. Territorial behaviour of the Tree Sparrow. Condor 67:193-209.

Wershler, C. 1983. American Tree Sparrow nesting record and other bird notes from the Cameron Hills, northwestern Alberta. Alta. Nat. 13(4):144-145.

Wershler, C. 1987. Baird's sparrow. **In:** Proceedings of the Workshop on Endangered Species in the Prairie Provinces. Holroyd, G.L., P.H.R. Stepney, G.C. Trottier, W.b. McGillivray, D.M. Ealey and K.E. Eberhart, eds. Provincial Museum of Alberta Natural History Occasional Paper No. 9.

Wershler, C. 1990. Status of the Baird's Sparrow in Alberta - 1989 (draft). Report for World Wildlife Fund.

Wershler, C., W.W. Smith and C. Wallis. 1991. Status of the Baird's Sparrow in Alberta - 1987/1988. Update with notes on other grassland sparrows and Sprague's Pipit. **In:** Proceedings of the Second Endangered Species and Prairie Conservation Workshop. Holroyd, G.L., G. Burns and H.C. Smith, eds. Provincial Museum of Alberta Natural History Occasional Paper No. 15, Edmonton, Alberta.

Weston, H.G. 1947. Breeding behavior of the Black-headed Grosbeak. Condor 49:54-73.

Williams, L. 1952. Breeding behavior of the Brewer's Blackbird. Condor 54:3-47.

Willson, M.F. 1965. Breeding ecology of the Yellow-headed Blackbird. Ecol. Monog. 36(1):51-77.

Zammuto, J.G., R.M. Zammuto and J.D. Schieck. 1982. Grasshopper Sparrows in Rocky Mountain pasture. Alta. Nat. 12(1):5-6.

Family Fringillidae: FINCHES

Adkisson, C.S. 1981. Geographic variation in vocalizations and evolution of North America. Pine Grosbeaks. Condor 83(4):277-288.

Anonymous. 1984. Unusual bird sightings - winter 1983-84. Alta. Nat. 14(2):55.

Baillie, J.L. Jr. 1940. The summer distribution of the eastern Evening Grosbeak. Can. Field-Nat. 54:15-25.

Bancroft, J., and R.J. Parsons. 1991. Range expansion of the House Finch into the Prairie Provinces. Blue Jay 49(3):128-136.

Brunton, D.F., G.L. Holroyd and H.F. Coneybeare. 1978. Cassin's Finch, Baird's Sparrow and Harris's Sparrow new to Banff National Park, Alberta. Alta. Nat. 8(2):89-92.

Collister, D. 1989. Migration, wintering range, longevity and mortality of Purple Finches as evidenced by banding data. Alta. Bird Record 7:3-8.

Cooke, W.W. 1910. The migration of North American sparrows. Fifth paper. Bird-Lore 12(4):139-141.

Cooke, W.W. 1912. The migration of North American sparrows. Nineteenth paper. Bird-Lore 14(6):345-346.

Cooke, W.W. 1913. The migration of North American sparrows. Twenty-fifth paper. Bird-Lore 15(6):364-365.

Cooke, W.W. 1914. The migration of North American sparrows. Twenty-sixth paper. Bird-Lore 16(1):19-23.

Edwards, R.Y., and Stirling, D. 1961. Range expansion of the House Finch into British Columbia. Murrelet 42:38-42.

Farley, F.L. 1921. A Pine Siskin invasion. Can. Field-Nat. 35(7):141.

Fenna, L. 1984. Nesting of Evening Grosbeaks near Edmonton. Alta. Nat. 14(4):145.

Godfrey, W.E. 1976. Breeding status of the Common Redpoll in Alberta and Saskatchewan. Can. Field-Nat. 90(2):199-200.

Hoffmann, W., and H. Pletz. 1988. 1988 nest record for an Evening Grosbeak near Elk Island Park. Alta. Nat. 18(3):101-103.

James, P.C., T.W. Barry, A.R. Smith and S.J. Barry. 1987. Bill crossover ratios in Canadian crossbills (*Loxia* spp.) Ornis. Scand. 18(4):310-312.

Johnson, H. 1964. Gray-crowned Rosy Finch record. Blue Jay 22(3):111.

Johnson, R.E. 1977. Seasonal variation in the genus *Leucosticte* in North America. Condor 79(1):76-86.

Lister, R. 1975. Common Redpolls nesting at Edmonton, Alberta. Can. Field-Nat. 89(1):64-65.

McNicholl, M.K. 1976. Evening Grosbeaks on Alberta Christmas bird count. Alta. Nat. 6(4):253-264.

McNicholl, M.K. 1977. Alberta wintering populations of Evening Grosbeaks. N. Am. Bird Bander 2:164-165.

McNicholl, M.K. 1978. Early record of redpoll nesting in southern Alberta. Blue Jay 36(3):166.

Middleton, A.L.A. 1979. Influence of age and habitat on reproduction by the American Goldfinch. Ecology 60(2):418-432.

Nordstrom, W.R., and J.R. Butler. 1976. Notes on roosting habits of Gray-crowned Rosy Finches. Alta. Nat. 6(2):105-107.

Pittman, H.H. 1917. The American Crossbill. Bird-Lore 19(6):332-333.

Rand, A.L. 1946. A new race of the Purple Finch *Carpodacus purpureus* (Gmelin). Can. Field-Nat. 60:95-96.

Salt, G.W. 1952. The relation of metabolism to climate and distribution in three finches of genus *Carpodacus*. Ecol. Monog. 22(2):121-152.

Salt, J.R. 1984. Feeding of a Pine Siskin by White-winged Crossbill. Alta. Nat. 14(3):104.

Stelter, D. 1987. The elusive White-winged Crossbill. Alta. Nat. 17(1):16.

Stokes, D., and L. Stokes. 1986. American Goldfinches. Living Bird 5(3):22.

Taverner, P.A. 1921. The Evening Grosbeak in Canada. Can. Field-Nat. 35:41-45.

Family Passeridae: OLD WORLD SPARROWS

Hamilton, S., and R.F. Johnston. 1978. Evolution in the House Sparrow - VI. Variability and niche width. Auk 93(2):313-323.

Hofmann, D.E. 1980. Feeding of nestling Cliff Swallows by a House Sparrow. Can. Field-Nat. 94(4):462.

Houston, C.S. 1978. Arrival of the House Sparrow of the prairies. Blue Jay 36(2):99-102.

Johnston, R.F., and R.K. Selander. 1971. Evolution in the House Sparrow. II. Adaptive differentiation in North American populations. Evolution 25(1):1-28.

Lohr, L. 1967. Communal nesting of House Sparrows with birds of prey. Blue Jay 25(2):100-101.

McGillivray, W.B. 1978(a). The effects of nest position on reproductive performance in the House Sparrow. M.A. Thesis, University of Kansas, Lawrence.

McGillivray, W.B. 1978(b). House Sparrows nesting near a Swainson's Hawk nest. Can. Field-Nat. 92(2):201-202.

McGillivray, W.B. 1980. Communal nesting in the House Sparrow. J. Field Ornith. 51(4):371-372.

McGillivray, W.B. 1981. The breeding ecology of House Sparrows. Ph.D. Thesis, University of Kansas. 83 pp.

McGillivray, W.B. 1983. Intraseasonal reproductive costs for the House Sparrow (*Passer domesticus*). Auk 100(1):25-32.

McGillivray, W.B. 1988. The birds of Bistcho Lake. Pp. 93-99. **In:** The Natural History of the Bistcho Lake Region, Northwest, Alberta. McGillivray, W.B., and R.I. Hastings, eds. Provincial Museum of Alberta Natural History Occasional Paper No. 10, Edmonton, Alberta.

Murphy, E.C. 1977. Breeding ecology of House Sparrows. Ph.D. Thesis, University of Kansas, Lawrence.

Robbins, C.S. 1973. Introduction, spread and present abundance of the House Sparrow in North America. pp. 3-9. **In:** Kendeigh, S.C., ed. A Symposium of the House Sparrow (*Passer domesticus*) and European Tree Sparrow (*P. montanus*) in North America. Ornithol. Monog. No. 14. 121 pp.

Summers-Smith, J.D. 1988. The sparrow: a study of the genus *Passer*. Poyser, Calton.

Underhill, J.P. 1981. A solution for House Sparrow problems. Alta. Nat. 11(2):80-81.

INDEX OF BIRD NAMES - ENGLISH

INDEX OF BIRD NAMES - FRENCH

INDEX OF BIRD NAMES - LATIN

INDEX OF MIGRANT SPECIES

English

Latin

French

Notes

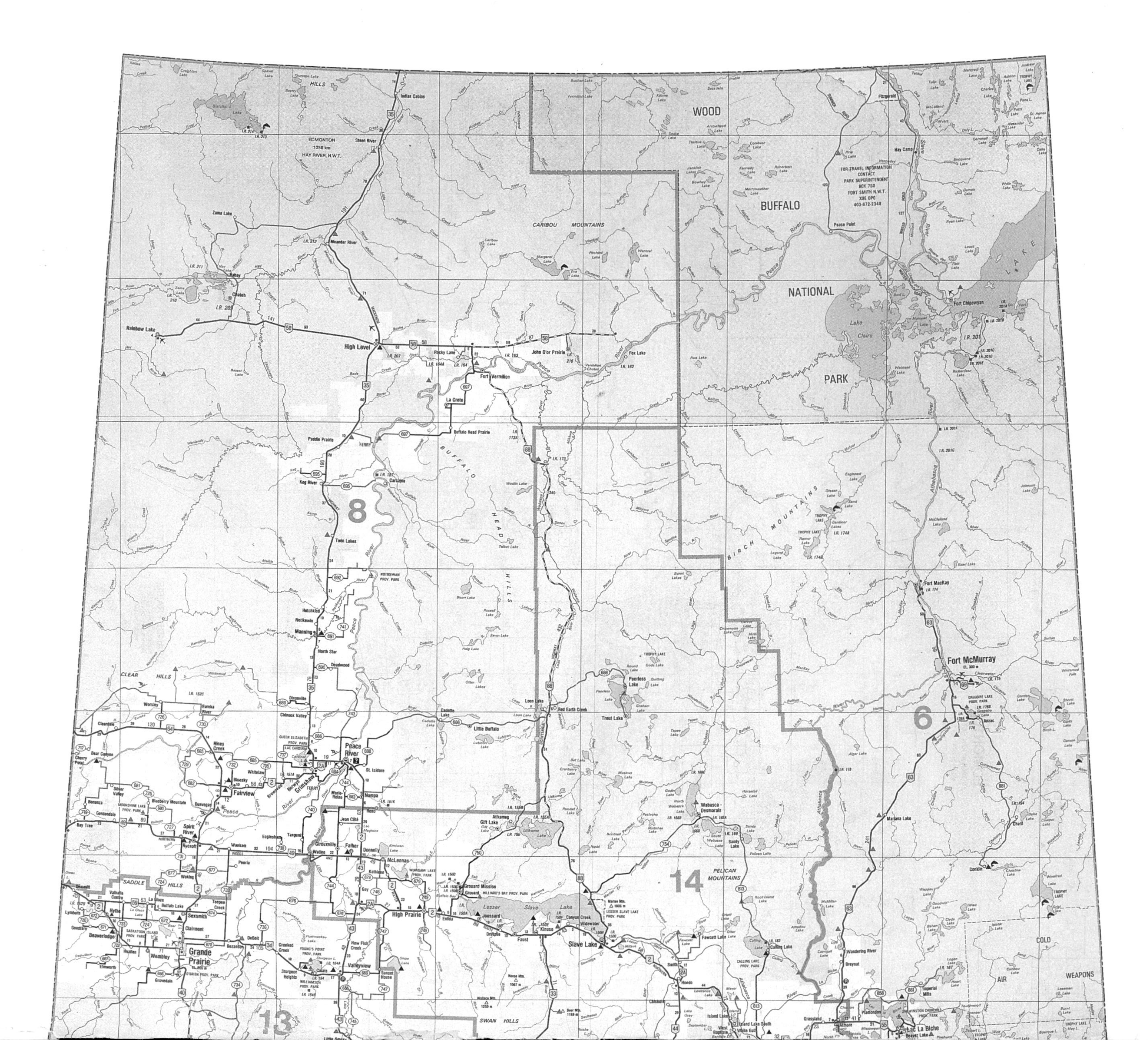

WOOD BUFFALO NATIONAL PARK
CARIBOU MOUNTAINS
BIRCH MOUNTAINS
BUFFALO HEAD HILLS
PELICAN MOUNTAINS
CLEAR HILLS
SADDLE HILLS
SWAN HILLS
COLD LAKE AIR WEAPONS
8
6
14
13
EDMONTON
1050 km
HAY RIVER, N.W.T.
FOR TRAVEL INFORMATION
CONTACT
PARK SUPERINTENDENT
BOX 750
FORT SMITH N.W.T.
X0E 0P0
403-872-2349
Fort Chipewyan
Fort McMurray
Fort MacKay
High Level
Fort Vermilion
La Crete
Rainbow Lake
Zama Lake
Meander River
Indian Cabins
Steen River
Fitzgerald
Hay Camp
Peace Point
Lake Claire
Paddle Prairie
Keg River
Manning
Peace River
Grimshaw
Fairview
Spirit River
Grande Prairie
Beaverlodge
Valleyview
High Prairie
McLennan
Falher
Slave Lake
Lesser Slave Lake
Red Earth Creek
Wabasca - Desmarais
Mariana Lake
Lac La Biche